Economic Effects of Natural Disasters

Theoretical Foundations, Methods, and Tools

Economic Effects of Natural Disasters

Theoretical Foundations, Methods, and Tools

Edited by

Taha Chaiechi

College of Business, Law and Governance,
James Cook University, Cairns, QLD, Australia

Academic Press is an imprint of Elsevier
125 London Wall, London EC2Y 5AS, United Kingdom
525 B Street, Suite 1650, San Diego, CA 92101, United States
50 Hampshire Street, 5th Floor, Cambridge, MA 02139, United States
The Boulevard, Langford Lane, Kidlington, Oxford OX5 1GB, United Kingdom

Notices
Knowledge and best practice in this field are constantly changing. As new research and experience broaden our
understanding, changes in research methods, professional practices, or medical treatment may become necessary.

Practitioners and researchers must always rely on their own experience and knowledge in evaluating and using any
information, methods, compounds, or experiments described herein. In using such information or methods they
should be mindful of their own safety and the safety of others, including parties for whom they have a professional
responsibility.

To the fullest extent of the law, neither the Publisher nor the authors, contributors, or editors, assume any liability
for any injury and/or damage to persons or property as a matter of products liability, negligence or otherwise, or
from any use or operation of any methods, products, instructions, or ideas contained in the material herein.

British Library Cataloguing-in-Publication Data
A catalogue record for this book is available from the British Library

Library of Congress Cataloging-in-Publication Data
A catalog record for this book is available from the Library of Congress

ISBN: 978-0-12-817465-4

For Information on all Academic Press publications
visit our website at https://www.elsevier.com/books-and-journals

Publisher: Brian Romer
Editorial Project Manager: Lindsay Lawrence
Production Project Manager: Punithavathy Govindaradjane
Cover Designer: Miles Hitchen

Typeset by MPS Limited, Chennai, India

Contents

List of Contributors

Achiransu Acharyya
Department of Economics, Visva-Bharati (A Central University), Santiniketan, India

Alex O. Acheampong
School of Business, The University of Newcastle, Newcastle, NSW, Australia; Faculty of Business and Law, The University of Newcastle, Newcastle, NSW, Australia

Ogechi Adeola
Lagos Business School, Pan-Atlantic University, Lagos, Nigeria

Prince C. Agwu
Department of Social Work, University of Nigeria, Nsukka, Nigeria

Prakash Andugula
Centre of Studies in Resource Engineering, Indian Institute of Technology Bombay, Mumbai, India

Simona Azzali
School of Science and Technology, James Cook University, Singapore, Singapore

Hurriyet Babacan
Rural Economies Centre of Excellence and The Cairns Institute, James Cook University, Cairns, QLD, Australia

Lawal Billa
Environment and Geographical Sciences, University Nottingham, Selangor, Malaysia

Mwansa Chabala
School of Business, The Copperbelt University, Kitwe, Zambia

Taha Chaiechi
College of Business, Law and Governance, James Cook University, Cairns, QLD, Australia

Tanuj Chawla
Tata-Cornell Institute for Agriculture and Nutrition, New Delhi, India

Progress Choongo
School of Business, The Copperbelt University, Kitwe, Zambia

Susan Ciccotosto
College of Business, Law and Governance, James Cook University, Cairns, QLD, Australia

Winn Costantini
Williams College, Williamstown, MA, United States

Graeme Cotter
College of Business, Law and Governance, James Cook University, Cairns, QLD, Australia

Alison Cottrell
School of Earth & Environmental Sciences, Centre for Disaster Studies, James Cook University, Townsville, QLD, Australia

W.S. (Bill) Cummings
Cummings Economics, Cairns, QLD, Australia

Allan P. Dale
The Cairns Institute, James Cook University, Cairns, QLD, Australia

Michael O. Dioha
Department of Energy and Environment, TERI School of Advanced Studies, New Delhi, India

Janet Dzator
School of Business, The University of Newcastle, Newcastle, NSW, Australia; Australia Africa Universities Network (AAUN) Partner, Newcastle, NSW, Australia

Michael Dzator
Australia Africa Universities Network (AAUN) Partner, Newcastle, NSW, Australia; School of Access Education, Central Queensland University, Mackay, QLD, Australia

Emiel L. Eijdenberg
Business, IT and Science Department, James Cook University, Singapore, Singapore

Emeh Ikechukwu Eke
Department of Public Administration, University of Nigeria, Nsukka, Nigeria

Nnaemeka V. Emodi
Future Energy Research Group, Tasmanian School of Business and Economics, University of Tasmania, Hobart, TAS, Australia

Daniel R.E. Ewim
Department of Mechanical Engineering, Mechatronics and Industrial Design, Tshwane University of Technology, Pretoria, South Africa

Simon Feeny
Centre for International Development, School of Economics, Finance and Marketing, RMIT University, Melbourne, VIC, Australia

Oludele Folarin
Department of Economics, University of Ibadan, Ibadan, Nigeria

Haripriya Gundimeda
Department of Humanities and Social Sciences, Indian Institute of Technology Bombay, Mumbai, India

Yetta Gurtner
Centre for Disaster Studies, College of Science and Engineering, James Cook University, Townsville, QLD, Australia

Meegan Hardacker
Science and Engineering Faculty, Queensland University of Technology, Brisbane, QLD, Australia

T.P. Harshan
School of Habitat Studies, Tata Institute of Social Sciences, Mumbai, India

S. Jagadeesh
Karnataka State Natural Disaster Monitoring Centre, Bengaluru, India

Maneka Jayasinghe
Asia Pacific College of Business and Law, Charles Darwin University, Darwin, NT, Australia

Olivia Jensen
LRF Institute for the Public Understanding of Risk, National University of Singapore, Singapore

Rupak Kumar Jha
Department of Humanities and Social Sciences, Indian Institute of Technology Bombay, Mumbai, India

Archibald James Juniper
Conjoint Academic, University of Newcastle, Callaghan, NSW, Australia

Zilmiyah Kamble
School of Business, James Cook University, Singapore, Singapore

Lenin Babu Kamepalli
Karnataka State Natural Disaster Monitoring Centre, Bengaluru, India

Shrutidhara Kashyap
Department of Economics, Arya Vidyapeeth College, Guwahati, India

Ben Katoka
Hankuk University of Foreign Studies, Yongin-si, Republic of Korea

David King
Centre for Disaster Studies, College of Science and Engineering, James Cook University, Townsville, QLD, Australia

Florine M. Kuijpers
Independent Researcher, Amsterdam, The Netherlands

Atul Kumar
Department of Energy and Environment, TERI School of Advanced Studies, New Delhi, India

John Lungu
School of Graduate Studies, The Copperbelt University, Kitwe, Zambia

Ratul Mahanta
Department of Economics, Gauhati University, Guwahati, India

Enno Masurel
School of Business and Economics, Vrije Universiteit Amsterdam, Amsterdam, The Netherlands

Trang Nguyen
College of Business, Law and Governance, James Cook University, Cairns, QLD, Australia

Kalu T.U. Ogba
Department of Psychology, University of Nigeria, Nsukka, Nigeria

Luke Emeka Okafor
School of Economics, University of Nottingham Malaysia, Jalan Broga, Semenyih, Malaysia

Uzoma O. Okoye
Department of Social Work, University of Nigeria, Nsukka, Nigeria

Charles Nnamdi Olise
Department of Public Administration & Local Government, University of Nigeria, Nsukka, Nigeria

Francisca N. Onah
Social Science Unit, School of General Studies, University of Nigeria, Nsukka, Nigeria

Vikrant Panwar
Department of Humanities and Social Sciences, Indian Institute of Technology, Roorkee, India

Petina L. Pert
CSIRO Land and Water Flagship and Division of Tropical Environments and Societies, James Cook University, Cairns, QLD, Australia

Alberto Posso
Centre for International Development, School of Economics, Finance and Marketing, RMIT University, Melbourne, VIC, Australia

Ruth Potts
School of Geography and Planning, Cardiff University, Cardiff, NSW, Australia

C.N. Prabhu
Karnataka State Natural Disaster Monitoring Centre, Bengaluru, India

Josephine Pryce
College of Business, Law, and Governance, James Cook University, Cairns, QLD, Australia

G.S. Srinivasa Reddy
Karnataka State Natural Disaster Monitoring Centre, Bengaluru, India

David A. Savage
Newcastle Business School, University of Newcastle, Newcastle, NSW, Australia

E.A. Selvanathan
Griffith Business School, Griffith University, Nathan Campus, QLD, Australia

Saroja Selvanathan
Griffith Business School, Griffith University, Nathan Campus, QLD, Australia

Subir Sen
Department of Humanities and Social Sciences, Indian Institute of Technology, Roorkee, India

Ashish Sharma
Department of Humanities and Social Sciences, Indian Institute of Technology, Roorkee, India

Thomas K. Taylor
School of Graduate Studies, The Copperbelt University, Kitwe, Zambia

K. Thirumaran
School of Business, James Cook University, Singapore, Singapore

Chitranjali Tiwari
LRF Institute for the Public Understanding of Risk, National University of Singapore, Singapore

Benno Torgler
School of Economics and Finance, Queensland University of Technology, Brisbane, QLD, Australia

Trong-Anh Trinh
Centre for International Development, School of Economics, Finance and Marketing, RMIT University, Melbourne, VIC, Australia

Chioma S. Ugwu
Department of Political and Administrative Studies, University of Port Harcourt, Port Harcourt, Nigeria

Chukwuma Felix Ugwu
Department of Social Work, Faculty of the Social Sciences, University of Nigeria, Nsukka, Nigeria

Christopher Onyemaechi Ugwuibe
Department of Public Administration & Local Government, University of Nigeria, Nsukka, Nigeria

Karen J. Vella
Science and Engineering Faculty, Queensland University of Technology, Brisbane, QLD, Australia

Pengji Wang
School of Business, James Cook University, Singapore, Singapore

Caroline Wong
School of Business, James Cook University, Singapore, Singapore

Jacob Wood
School of Business, James Cook University, Singapore, Singapore; Visiting Professor of International Trade, Chungnam National University, Daejeon, South Korea

Huiping Zhang
School of Business, James Cook University, Singapore, Singapore

Editor Biographies

Editor-in-Chief:

Dr. Taha Chaiechi is Australia Director, Centre for International Trade and Business in Asia, at JCU where she is also an Associate Professor of Economics. In the past several years, Taha has contributed to the governance and the Teaching and Learning profile of the College in different capacities. Taha served JCU as the Head, Economics and Marketing Academic Group from October 2014 to March 2019, she is also the Program Convener for Master of Economics. Furthermore, she served the University in the acting position of Associate Dean, Learning and Teaching from July 2018 to January 2019.

Taha is an expert in systematic modeling of dynamic relationships between economic, environmental, and social variables. Taha's research attitude is holistic and inspired by issues in climate change and natural disasters, and their impact on different economic sectors such as public health, tourism, environmental, energy, and urban economics, which makes it especially suitable for sustainability analysis. Since 2011, she has been collaborating in several research projects exceeding $1million in value.

Associate editors:

Nnaemeka Vincent Emodi is a Research Fellow in the Future Energy Research Group within the Tasmanian School of Business and Economics at the University of Tasmania, Hobart, Australia. He completed his PhD degree with a focus on energy economics and climate policy at the Cairns campus of James Cook University, Australia. He has a BSc degree in Chemistry from Michael Okpara University of Agriculture, Umudike, Nigeria and a MSc degree in Engineering from Seoul National University, Republic of Korea, focusing on energy policy and management. Prior to his current role, Nnaemeka was a Graduate Policy Analyst at the Australian Academy of Science. He has authored and coauthored articles, reviews, a monograph, and some book chapters. His current research interest includes energy economics with a focus on energy system decarbonization, energy consumer research, and climate policy.

Diana Castorina is a sessional academic in the College of Business, Law and Governance at James Cook University. Her teaching experience in the higher education sector spans 10 years across both undergraduate and postgraduate offerings.

Diana is a member of the Centre for International Trade and Business in Asia and a Postgraduate Fellow at The Cairns Institute at JCU. Her research interests are in the area of understanding individual decision-making, sustainability of regions, and spatial econometrics. She is currently completing her PhD in economics in the area of interregional migration in Australia where her overall research objective seeks to understand what makes people want to stay, move away from, or move into a region. Diana also draws on her experience within the private sector where she worked as a research economist undertaking social and economic impact assessments.

Preface

One can apply Frederic Bastiat's parable of the broken window to any amount of destruction in the form of intended or unintended consequences, with the former extensively explored in the works of mainstream economists. The Austrian–American economist, Joseph Schumpeter (1942), for instance, derived the term "creative destruction" to refer to outdated production units that are replaced by new production mechanisms through the innovation process. Schumpeter introduced this term in his book *Capitalism, Socialism and Democracy* in 1942 and used it to refer to the disruptive practice of industrial transformation that accompanies revolutionary modernization and innovations. While natural disasters are different from other economic events; research about the effects of disasters on macroeconomic performance is growing. Some consider disasters similar to economic frustration (Okuyama, 2003) such as a recession phase in a business cycle, while others argue that natural disasters can bring about some long-term economic "benefits" that might lead to Schumpeterian "creative destruction."

In recent studies, the economic impact of climate change and natural disasters has been broadly discussed, and climate change has been ascribed an increasing influence over economic development. However, most of these discussions fail to adequately investigate these effects within a general/multisectoral macroeconomic model. In the absence of such evidence, this book aims to draw on principles of different theories of growth and distribution to propose a framework for capturing economic sectors' response to devastating natural disasters. Accordingly, the fundamental objective of this book is to explore the mechanism through which natural disasters affect sources of economic growth and development using theoretical econometrics and real-world data.

Economic Effects of Natural Disasters: Theoretical Foundations, Methods, and Tools will show scholars and researchers how to use different research methods and techniques to investigate a natural disaster. To teach readers "how to do economics," the contributors present evidence about the economic effects of natural disasters. The aim is to discuss the economic impacts of natural disasters on the sources of sustainable economic growth, covering different areas of environmental economics, development economics, society (including issues such as employment), tourism, gender economics, stock markets, socio-economic resilience, disaster management, and FDI. No other book presents empirical frameworks for the evaluation of the quality of macroeconomic research practice with a focus on climate change and natural disasters.

Readers are provided with an invaluable collection of theoretical and empirical frameworks using different econometrics and statistics methods and estimation techniques to tackle this real-world problem. Furthermore, readers can access highly effective information about sources of research data shared by the contributors. This makes research more controllable and increases the credibility of future research within this context. Finally, this book offers the audience a variety of skills such as research methodology, appropriate estimation techniques, and a deeper understanding of the application of a selection of theoretical frameworks.

Many of these subjects are so large that different regions of the world use significantly different approaches to them. To attain a global approach, a selection of chapters is allocated to evidence from developing countries, and a selection is allocated to lessons from developed nations.

Taha Chaiechi

REFERENCE

Okuyama, Y. (2003). Economics of natural disasters: A critical review. Regional Research Institute Publications and Working Papers. 131. https://researchrepository.wvu.edu/rri_pubs/131.

THE ECONOMIC IMPACT OF NATIONAL DISASTER RELIEF AND RECOVERY FUNDING FOR LOCAL GOVERNMENT INFRASTRUCTURE IN TROPICAL NORTH QUEENSLAND

1

W.S. (Bill) Cummings
Cummings Economics, Cairns, QLD, Australia

1.1 INTRODUCTION

As a first-world country, Australia has a well-developed framework to achieve relief and recovery after natural disasters—the National Disaster Relief and Recovery Arrangements (NDRRA). However, there is always discussion and debate about appropriate levels of assistance and administrative detail.

The Tropical North Queensland region (which essentially covers the Peninsula Australia geographic region in Australia's northeast) is deep into the tropics and subject to occurrence of intense tropical cyclones and heavy rainfall events caused by intense tropical lows that can cause substantial wind and flooding damage.

In 2017 Cummings Economics was asked by the Far North Queensland Regional Organisation of Councils (FNQROC) to carry out a study into the economic impact of national disaster funding to local government in the region to help with discussion about funding levels and administration. The following presents some of the key findings from this study.

1.2 BACKGROUND OF THE REGION

The 13 councils in the FNQROC cover 92% of the population and 83% of the area of the Peninsula Australia geographic region (see Map 1.1).

The Peninsula Australia geographic region (also referred to as the Tropical North or Far North Queensland region) has an area about the size of the British Isles, is 1.5 times the area of Victoria, is as deep from north to south as the rest of Queensland and as New South Wales (see Map 1.2).

Economic Effects of Natural Disasters. DOI: https://doi.org/10.1016/B978-0-12-817465-4.00001-7

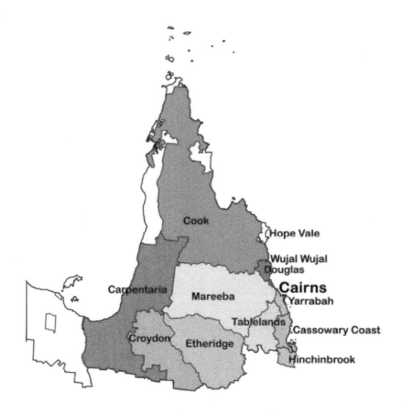

MAP 1.1

Cairns/Tropical North Queensland—Local Governments Areas (LGAs) covered by Far North Queensland Regional Organisation of Councils.

Cummings Economics.

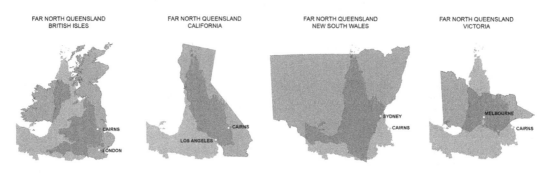

MAP 1.2

Comparative areas and distances.

Cummings Economics.

Being deep into the tropics, the region faced early challenges for a young Australian nation with most of its population and technology derived from Northwestern Europe. However, over the past four decades, the region has been outpacing in growth most regional areas of Australia. It is now the largest in population and fastest growing in northern Australia. However, population is still low and approaching only 300,000.

1.3 TWO ZONES OF ANALYSIS

Analysis of impacts of NDRRA funding to local government was broken into two zones (see Map 1.3).

A core zone includes the regional capital of Cairns and most of the six (6) surrounding local government areas in the southeast of the region. This area has relatively well-developed infrastructure accounting for only 5% of the FNQROC area but 92% of the population.

A remote zone comprises local government areas covering the rest of the region comprising 95% of the area but only 7% of the population. This area has poorly developed infrastructure, including large lengths of unsealed roads and poorly developed stream crossings.

MAP 1.3

Zonal regions.

Cummings Economics.

1.4 NATURAL DISASTER WEATHER EVENTS

Over the period analyzed, weather events identified as natural disasters were recorded every year as follows:

2010	Tropical cyclones, Charlotte, Tasha, Olga, Neville
2011	Severe tropical cyclone, Yasi
2012	Intense monsoonal rains
2013	Tropical cyclone, Oswald
2014	Tropical cyclones, Ita and Fletcher
2015	Tropical cyclone, Nathan
2016	Intense monsoonal rains

Events varied in their location in the region. In 2010 effects of various events were widespread. Severe tropical cyclone Yasi in 2011 especially affected the Cassowary Coast and Hinchinbrook areas. Cyclones Ita and Fletcher in 2014 especially affected Cook and Douglas areas.

1.5 DISASTER EXPENDITURE FUNDED

Expenditure by local governments funded by the NDRRA scheme was analyzed over the 7 years 2010−16 inclusive. Local authorities applied for NDRRA funding through the Queensland Reconstruction Authority (QRA). Subsequent outlays by QRA were analyzed over the 6 years 2011−12 to 2016−17. They indicate the same total level of expenditure but with the QRA expenditure lagging behind the Local Government Area (LGA) expenditure.

LGA expenditure identified over the 7 years totaled $1058 million with $408 million in the core zone and $651 million in the remote zone, an average of $58 million a year in the core zone and $93 million a year in the remote zone (see Table 1.1).

Expenditure per square km was much higher in the core zone. However, on a per capita of population basis, expenditure in the remote zone was very high. Chart 1.1 shows distribution of the LGA expenditure funded by years.

Table 1.1 LGA disaster expenditure funded Far North Queensland Regional Organisation of Councils area 2010−16.

	Amount total (million)	Per km²	Per capita
Core zone	$408	$24,420	$1627
Remote zone	$651	$1853	$31,942
Total	**$1058**	**$3348**	**$3907**
Average per annum	**$151**	**$478**	**$558**

LGA, Local Government Area.
Cummings Economics.

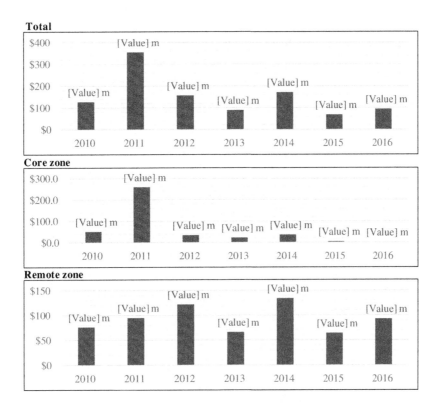

CHART 1.1

NDRRA—LGA expenditure funded by years—Tropical North Queensland: *total, core zone, and remote zone.* *LGA*, Local Government Area, *NDRRA*, National Disaster Relief and Recovery Arrangement.

Expenditure in the core zone was heavily concentrated in 2011 due to the severe tropical cyclone Yasi. Expenditure in the large area of the remote zone was more evenly distributed over the years.

1.6 INFRASTRUCTURE AFFECTED

Almost all NDRRA supports were for local authorities in the region related to local government roads and associated infrastructure of bridges, culverts, floodways, and, including the Daintree River Ferry.

Data available indicates that in the FNQROC region, there are over 15,000 km of local government roads (87% rural) with 60% in the remote zone. Cook Shire had the largest mileage at 2900 km, equivalent to the road distance between Cairns and Melbourne. However, reflecting the more developed standard of the road system in the core zone, there were more bridges and major culverts in that zone, but many more minor culverts and floodways in the remote zone. Total traffic kilometers over the local government roads in the FNQROC region was estimated at 1.6 billion vehicle km in 2014—15.

1.7 ANNUAL DISASTER SPENDING COMPARED WITH CAPITAL VALUE

The disaster funding played an important role in keeping the roads open and efficient in their role. Capital value of local government roads in the two zones as measured by replacement cost was estimated by the Queensland Department of Local Government in 2014—15 as follows:

Core zone	$2.7 billion ($446,000 per km)
Remote zone	$0.9 billion ($93,000 per km)

With average annual disaster funding at $150 million a year, over this period, the investment in keeping roads open averaged about 4% per annum of their capital value as measured by replacement cost. This of course fluctuated strongly over the years depending on events and ranged from less than 2% to about 10%.

1.8 RECONSTRUCTION WORKS IMPACTS

Estimated direct expenditure of an average of $150 million per annum was estimated to result in:

- impact on gross regional product (GRP) of the order of $130 million a year, including "flow-on" effects, that is, about 0.9% of total GRP of the FNQROC region;
- creation of about 260 direct jobs, and with "flow-on" effects, of the order of 800 jobs per annum.

The reconstruction work is particularly important to the Remote Zone regional economy. The NDRRA reconstruction work averaging $93 million per annum in the remote zone compared with:

- average building approvals of $17 million per annum over the period 2009—10 to 2015—16;
- an estimated GRP of the order of $800 million in 2016—17.

1.9 SCALE OF ECONOMIC IMPACTS IF THE RESTORATION WORK DID NOT TAKE PLACE

1.9.1 DIRECT IMPACTS

The aim of the reconstruction works was to restore activity to levels as if the event had not occurred.

The extent and scale of the works was such that it was not possible to estimate the economic impacts in detail. The following aims to give some appreciation however, of how impacts would have occurred.

Economic analysis of impacts of road funding normally involves two levels of analysis:

- economic efficiency impacts
- impacts on aggregate levels of economic activity

Economic efficiency measures the costs to users mainly in terms of "travel time," "vehicle operating costs," and "accidents."

Clearly, failure to restore efficiency in the transport system will not just affect activity currently, but indefinitely over time. The normal approach to this is to quantify the savings in user costs due to restoration and discount future effects over a project period at a discount rate to produce a present value (PV) of the savings. For road upgrading benefit cost analysis, it is common to use a project period of 30 years and a "real" discount rate of 4%. On the basis of these parameters, a $1 saving per annum will have a PV of $17.29.

As an example, if an event caused deterioration of a road surface that caused delay of 30 seconds in travel time with mixed private and business cars and heavy vehicles, along with an increase in vehicle operating cost due to a rougher road surface over 500 m, and the traffic on that section had an average annual daily traffic of 300 with traffic growing at an average of 2% per annum, PV of savings over a 30-year project period, no further deterioration of the surface, would calculate at about $1.0 million using standard national road assessment parameters. (With further deterioration taken into account, it would be much higher.) If the cost of rebuilding that sector was $0.5 million, the benefit–cost ratio would be 2.0 or higher if a further deterioration factor was taken into account.

This is a relatively minor case of damage. However, it is likely to add a cost burden to any activity in the area serviced by the road in question. At this level it would be unlikely to have effects of raising costs so much as to reduce viability of industries dependent on the road to a point where it resulted in a curtailment of activities.

However, a more serious road closure, with an alternative route resulting in say 10 minutes time loss and extra mileage adding to a rise in operating costs, could start having an effect on the level of economic activity in the area.

A road closure with no alternative, such as occurred in 2014 with the closure of the Daintree Ferry in Douglas Shire, results in a "sky rocketing" of impacts on activity in the area serviced. In the case of the ferry, impact was estimated to be of the order of $1 million every 9 days in the low tourism activity season, making most activity and residences north of the Daintree River unviable.

1.9.2 FLOW-ON EFFECTS

Apart from direct negative impact on economic entities, there will be "flow-on" effects. Thus the cessation of operations in a rural area will have "flow-on" effects on those activities that:

- supply inputs (materials and labor);
- subcontract services;
- supply various consumer goods;
- provide services, including education, medical, and the like; and
- support further processing.

These "flow-on" effects are usually estimated with the use of input/output multiplier tables. These effects are often on activities in district towns and the regional city.

1.9.3 **DEVELOPMENTAL/CATALYTIC EFFECTS**

It also needs to be recognized that impacts on transport efficiency can have "developmental" "catalytic" effects. A relatively small rise in costs can make types of businesses uncompetitive and unprofitable and lead to cessation in an area. For instance, reduction in supply of cane to a sugar mill can result in mill closure affecting all growers in that district. Reduction of population in an area due to economic activity decline can result in closure of a local school adding further to costs of living and operating in an area.

Thus in calculating economic impacts, there are four levels involved:

Level 1—Direct effects on road user costs. These will be absorbed by the users and diminish their profitability/disposable income. (It can be assumed that these costs will usually at least equal the level of expenditure on the restoration work and generally exceed it.)

Level 2—Direct impacts on the level of economic activity in the area affected by a road. This will range from negligible when only minor costs are imposed by the road damage to complete cessation of activity where an activity is vulnerable and resulting costs are very high.

Level 3—"Flow-on" impacts to Level 2 effects. These will especially come from rural activity affected, reducing demand for goods and services delivered from urban centers.

Level 4—Catalytic effects where reduction in activity causes some activities in the area to become unviable with wider impacts (the mill closure and country school closure effect if restoration does not take place).

Chart 1.2 illustrates the above.

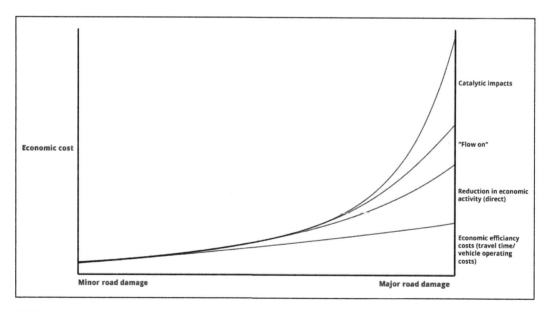

CHART 1.2

Diagram of scale of effects of road damage.

1.10 COMPARISON OF DISASTER SPENDING WITH OUTPUT OF VULNERABLE INDUSTRIES

While the situation was too complex to enable a calculation of actual impact if the restoration work did not take place, an appreciation of the likely scale of impact can be derived from looking at the scale of industries in the area likely to be vulnerable to local government roads becoming inoperable if the restoration did not take place.

Particularly vulnerable to the efficiency of local government roads are pastoral and farming operations, part tourism, and recreation and mining. Contraction of these activities will then have "flow-on" effects to suppliers of inputs to industry and consumers in the area. Decline of activity can then have "catalytic" effects.

In the core zone, value of agriculture and part tourism and recreation was estimated to total of the order of $2.1 billion a year with impact on GRP of about $1.76 billion (including "flow-on" effects). This is some 28 times the average amount of NDRRA funding per annum. In other words, failure to repair affected road infrastructure indefinitely into the future would only need to lead to a contraction of 3.7% in these industries to equal the investment in restoring the roads.

In the remote zone, vulnerable industries in grazing, farming, mining, tourism, and recreation were identified with an estimated value of approximately $1 billion per annum and addition to GRP of about $770 million. This is eight times the average annual cost of NDRRA funding of $93 million per annum. In other words, failure to restore the road infrastructure indefinitely into the future would only need to result in a 12.5% contraction in these activities to result in a cost to the economy equals to the cost of the restoration works.

1.11 IMPORTANCE OF OUTSIDE FUNDING

Examination of local government rate and utility income indicates that local governments in the remote zone are in a particularly poor position to meet natural disaster restoration costs.

A large part of local government areas can be taken up by state government national parks and reserves that are not rateable. In Cook Shire, total national parks and government reserves are estimated at about 26,000 km^2 or 25% of the Cook Shire area.

There is a good economic case for "betterment" works to make infrastructure more resilient, especially in the remote zone where lower levels of development of infrastructure make it vulnerable to repeat event damage.

THE EFFECTS OF NATURAL DISASTERS ON STOCK MARKET RETURN AND VOLATILITY IN HONG KONG

2

Trang Nguyen and Taha Chaiechi

College of Business, Law and Governance, James Cook University, Cairns, QLD, Australia

2.1 INTRODUCTION

After a formal transfer of sovereignty from the United Kingdom to China in 1997, Hong Kong has been a Special Administrative Region with a high level of autonomy. For decades, Hong Kong has been reputed to be a highly developed economy and a world-leading financial center. Hong Kong's stock exchange and foreign currency exchange have successfully developed into the world's sixth- and fourth-largest markets, respectively [Bank for International Settlements (BIS), 2018]. This vibrant megacity is also ranked seventh in global competitiveness and merchandise exports, making it the world's most unfettered economy and a primary economic powerhouse of the global economy [World Trade Organisation (WTO), 2018; World Economic Forum, 2019].

Hong Kong is located at the Pearl River Delta, which is among the urban areas that are most vulnerable to natural disasters in the world. Consequently, it confronts different types of natural disaster threats such as typhoons, tropical cyclones, earthquakes, and tsunamis (Murphy, 2015). These incidents engender floods, landslides, casualties, and severe destruction of transportation and infrastructures. According to Arcadis Sustainable Cities Index 2018, Hong Kong poses the highest risk to natural catastrophes in Asia. It is fortunate that this coastal metropolis has yet faced a severe weather event that caused substantial damages or casualties since the 1970s (Sim, Wang, & Han, 2018). The last severe one was Typhoon Wanda in 1962 which was responsible for 434 casualties and ranked third in the list of natural disasters by death toll in Hong Kong since 1884 (Time Out Hong Kong, 2018).

Being a great challenge for sustainable development, climate change has made the global weather more unpredictable. Cities across the globe, therefore, have been preparing adaptation and mitigation plans for climate change and Hong Kong has not stayed out of this movement. In its first Climate Change Report in 2015, Hong Kong government placed emphasis on the challenges and efforts to deal with this regard. In 2016 a Steering Committee on Climate Change was constituted to be responsible for designing long-term policies and procedures responding to the consequences of climate change such as an increase in temperature, sea-level rise, and cyclones.

On the other hand, stock market return and volatility can be heavily affected by environmental factors such as extreme weather events. In Australia the research suggests that extreme weather events such as bushfires and cyclones have much stronger effects on the equity market returns, than severe storms and floods (Nguyen & Chaiechi, 2019). In the United States, the stock markets volatilities increase by twofold when hurricanes, floods, storms, and severe heat waves happen. However, this scenario is not always the case, for example, other research shows that the capital markets of the United States, the United Kingdom, Canada, Germany, Hong Kong, and Australia seem to be unaffected by earthquakes. As such, inconclusive research indicates that further work is required in this area. In addition, postdisaster impact analysis, which focuses on only a very short period of time after the event, may not reveal the true effects of disasters on the stock market due to lag effects (Nguyen & Chaiechi, 2019).

Therefore this study is intended to explore the impacts of natural catastrophes on the return and volatility of a stock market, investigating a wide range of extreme events using postdisaster windows from 1 day to 2 months. Hong Kong was selected as a case study because this coastal megacity bears the highest risk of being affected by natural catastrophes in Asia, as noted previously. For the analysis an autoregressive moving average with exogenous regressor (ARMAX) model and the extended exponential generalized autoregressive conditional heteroskedastic (EGARCH) model were adopted to conduct an event study. Accordingly, a dummy variable or an intervention variable representing disaster events was integrated into the ARMA–EGARCH model structure to investigate the impacts of those events on stock market return and volatility.

2.2 LITERATURE REVIEW

While the body of literature on the effects of natural catastrophes on stock markets is growing, the empirical findings are found to be contradictory. On the one hand, for instance, Worthington and Valadkhani (2004) indicated that Australian equity market returns were significantly affected by bushfires, cyclones, and earthquakes rather than severe storms and floods. Taimur and Khan (2015) found the mean return of the Karachi Stock Exchange influenced by floods and earthquakes during an event window of 5 days. Bourdeau-Brien and Kryzanowski (2017) reported a surge in the volatility of the US stock market returns in the occurrence of hurricanes, floods, winter storms, and severe heat waves. Tavor and Teitler-Regev (2019) collected data on the world's 88 significant natural disasters and observed that the corresponding stock market indices dropped 3 consecutive days including the day of the disaster events. Moreover, Siddikee and Rahman (2017) showed the effects of natural catastrophes in Australia on the transmission of stock market volatility from Australia to India, New Zealand, Hong Kong, China, Taiwan, and Japan.

On the other hand, Worthington (2008) showed no significant evidence that the Australian stock market return is affected by storms, floods, cyclones, earthquakes, and bushfires. Luo (2012) reported insignificant effects of the Japanese earthquake 2011 on the capital markets of the United States, the United Kingdom, Canada, Germany, and Hong Kong. Wang and Kutan (2013) also found immaterial fluctuation in the American and Japanese capital markets in the aftermath of earthquakes, tsunamis, and volcano eruptions. The divergent findings, thus, call for further investigation in the economic impacts of natural catastrophes on capital markets.

Furthermore, most of the previous studies on the impacts of natural catastrophes on capital markets limit the event window to just a couple of days. Using a short period to assess the effects of disasters may not expose the actual effects which are often delayed due to the following reasons. First, it usually takes longer than a couple of days to gather accurate information on catastrophe-related damages. The precision of damages estimation in the early stage is especially low but appears to be reasonably reliable over long periods of time (Downton & Pielke, 2005). Second, several extreme weather events such as floods and droughts are relatively long-lasting, probably for months. Therefore restraining the event window to just a couple of days would likely underestimate the consequences of such extreme events. And third, the suspension of manufacturing in the short run can be derived from supply-chain disruptions rather than from catastrophic destructions. It may take some time for the effects of supply-chain disruptions to become material due to the implementation of the inventory risk management process (Norrman & Jansson, 2004).

This study, thus, is intended to fill the knowledge gaps in the literature by investigating the impacts of natural catastrophes on a stock market's return and volatility over different postdisaster periods from 1 day to 2 months.

2.3 ARMAX—EGARCHX MODEL

As noted earlier, this study aims to examine the dynamic impacts of severe weather events in Hong Kong on the local stock market return and volatility. For this intention, the authors extended the classic intervention analysis of Worthington and Valadkhani (2004), which only involves an ARMA process, with an EGARCH process of Nelson (1991) and further incorporated an intervention variable into both ARMA and EGARCH processes. The intervention variable in this study represents the occurrence of catastrophic weather events. Accordingly, our augmented model becomes ARMAX—EGARCHX model. Of which, the ARMAX model or conditional mean model captures the effects of catastrophic events on stock market returns while the EGARCH model or conditional variance model quantifies the effects of catastrophic events on the underlying return volatility. The augmented model can be written in the following equations:

$$r_t = \mu + \sum_{i=1}^{p} \varphi_i r_{t-1} + \sum_{i=1}^{p} \theta_j \varepsilon_{t-i} + \alpha Dis_t + \varepsilon_t \tag{2.1}$$

$$\varepsilon_t = z_t \sigma_t; z_t \sim D(0, 1) \tag{2.2}$$

$$\ln\left(\sigma_t^2\right) = \omega + \eta \ln\left(\sigma_{t-1}^2\right) + \psi \frac{\varepsilon_{t-1}}{\sigma_{t-1}} + \delta\left(\frac{|\varepsilon_{t-1}|}{\sigma_{t-1}} - E\left[\frac{|\varepsilon_{t-1}|}{\sigma_{t-1}}\right]\right) + \beta Dis_t \tag{2.3}$$

Eq. (2.1) describes the conditional mean model in which p and q represent AR terms and MA terms, respectively. Coefficient α quantifies the effect of natural disasters on the market returns. Dis_t is a dummy variable that is assigned a value of one if a natural disaster occurs and a value of zero otherwise. ε_t is the residuals of the mean model.

Eq. (2.2) defines the standardized residuals $z_t = \varepsilon_t / \sigma_t$. The standardized residuals follow a probability density function (D) of zero mean and unit variance. This function is restricted to Gaussian distribution or Student-t distribution for simplicity.

Eq. (2.3) models the natural logarithm of the conditional variance of the residuals ε_t using the EGARCH process. Superior to a standard GARCH, EGARCH captures the asymmetric conditional variance−covariance of market returns, which also known as asymmetric volatility or the leverage effect observed in market returns. Asymmetric volatility refers to the situation that a decrease in market price induces higher volatility of market returns than an equivalent increase in market price. η and ψ represent the coefficients for GARCH and ARCH terms, respectively. δ denotes the coefficient that quantifies the asymmetric effect which refers to a situation in which "bad" news tends to have a larger effect on return volatility than "good" news of an equivalent level of impact. Presuming the parametric form of errors follow a Gaussian distribution, the EGARCHX model is estimated by a maximum log-likelihood function (*LLF*) as follows:

$$LLF = -\frac{N}{2}\ln(2\pi) - \frac{1}{2}\sum_{t=1}^{N}\ln\sigma_t^2 - \frac{1}{2}\sum_{t=1}^{N}\frac{\varepsilon_t^2}{\sigma_t^2} \qquad (2.4)$$

where N indicates the number of observations.

In addition, Hansen and Lunde (2005) compared the performance of more than 300 conditional heteroscedasticity models and asserted that the models with more than one order of the ARCH and GARCH terms perform no better than the parsimonious model of GARCH(1,1). Moreover, EGARCH(1,1) effectively outperform the simple GARCH(1,1). Therefore this study rationally used an EGARCH(1,1) model for subsequent analysis.

2.4 DATA

This study used daily data of the Hang Seng Composite Index (HSI) and natural disasters in Hong Kong. The sample period started from January 2, 2008 to September 30, 2019, yielding almost 2900 observations. HSI daily closing prices were retrieved from Bloomberg Database and the information on natural disasters were obtained from Hong Kong Observatory. The natural disasters in this study refer to the events of major storm surge and floods, tropical cyclones, and earthquakes and tsunamis. Other kinds of natural disasters, such as bushfires and episodes of extreme temperature were not included as Hong Kong is not inclined to such disasters.

The daily HSI price series was converted into daily logarithmic return series defined as $r_t = \log(IP_t) - \log(IP_{t-1})$, where IP_t and IP_{t-1} are index closing prices at day t and $t-1$, respectively. Table 2.1 reports the descriptive statistics for HSI log returns. Hong Kong stock market experienced negative mean returns and low standard deviation, suggesting risk and return trade-off in this market. The positive skewness indicated that the return series has fatter tails and longer right tail compared to the normal distribution. This nonnormality was also confirmed by the significant Jarque−Bera statistic. The return series is leptokurtic and has a sharp peak given a large kurtosis. The Ljung−Box Q2 statistics and Engle ARCH statistics were significant up to lag 10 and 20, implying the presence of serial correlation and conditional heteroscedasticity in the variance of the return series. Therefore a model that contains ARCH or GARCH terms may be well-suited for the data. In addition, the HSI log return series was also tested for stationarity using common unit root tests: Augmented Dickey and Fuller (1981) test, Phillips and Perron (1988) test, and Ng and Perron

Table 2.1 Descriptive Statistics of Log Return Series.

Obs.	2894	Jarque−Bera	2876.6[a]
Min	−0.059	Q(10)	7.94
Mean	−9.6E-06	Q(20)	13.19
Max	0.058	Q^2(10)	2266.52[a]
Std. dev.	0.007	Q^2(20)	3435.01[a]
Skewness	0.026	ARCH(10)	111.41[a]
Kurtosis	9.82	ARCH(20)	68.29[a]

[a]*The test statistic is significant at 1%; Q(q) and Q^2(q) are the Ljung and Box (1979) test statistics for serial correlation up to lag q in logarithmic returns and squared logarithmic returns, respectively; ARCH(q) is the Engle (1982) ARCH test statistic for unconditional heteroscedasticity up to lag q in logarithmic returns.*

Table 2.2 Stationarity Tests of Log Return Series.

ADF	Statistic	NP-C	Statistic	NP—C&T	Statistic
C	−54.86[a]	MZ_a^d	−459.03[a]	MZ_a^d	−596.54[a]
C&T	−54.86[a]	MZ_t^d	−15.14[a]	MZ_t^d	−17.27[a]
PP		MSB^d	0.03[a]	MSB^d	0.03[a]
C	−54.93[a]	MP_T^d	0.06[a]	MP_T^d	0.15[a]
C&T	−54.93[a]				

ADF represents Augmented Dickey and Fuller (1981); PP represents Phillips and Perron (1988); NP represents Ng and Perron (2001); C is constant; C&T is constant and trend; $MZ_\alpha^d, MZ_t^d, MSB^d,$ and MP_T^d are the test statistics of the NP test.
[a]*The test statistic is significant at 1%.*

(2001) test. The results consistently show that the HSI log return series is stationary because the unit root test statistics were all insignificant (see Table 2.2), thus it is valid for modeling.

2.5 FINDINGS AND DISCUSSION

Initially, it is essential to test for the asymmetry in Hong Kong stock market return volatility to avoid model misspecification. Accordingly, Engle and Ng (1993) size and sign bias tests were performed. Table 2.3 reports the asymmetric test statistics for the log return series. The results show that the Hong Kong Stock Market exhibits both size bias and sign bias in return volatility given all test statistics were significant. This exposes the presence of asymmetry in the market return volatility. Thus a return volatility model that captures asymmetric volatility should be an ideal fit for the log return series.

Consequently, an ARMA−EGARCH model was utilized for further analysis. To explore the dynamic effects of natural disasters on Hong Kong stock market return and volatility, a dummy variable representing the events of natural disasters was inserted in the conditional mean equation and conditional variance equation of the ARMA−EGARCH model. To assess the persistent effect

Table 2.3 Asymmetric Tests for Log Return Series.

	Size Bias (*t*-Test)	Negative Sign Bias (*t*-Test)	Positive Sign Bias (*t*-Test)	Joint Effects (*F*-Test)
r_t	2.13**	4.74***	1.85*	12.17***

Note: *,**, *** *show the test statistic is significant at 10%, 5%, and 1%, respectively.*

Table 2.4 ARMAX−EGARCHX Model Estimation.

	Disaster Event Window						
Coefficient	1 Day	5 Days	10 Days	12 Days	15 Days	30 Days	60 Days
Conditional Mean Equation							
Constant (μ)	0.0002	0.0002	0.0002	0.0002	0.0002	0.0002	0.0001
	1.87*	2.18**	1.93*	1.82*	1.76*	1.73*	1.63
Dis (α)	−0.0008	−0.0005	−0.0002	−0.0002	−0.0001	−0.0001	−0.0001
	−1.83*	−1.82*	−1.93*	−1.38	−0.27	−0.67	−0.78
AR(1) (φ)	0.2089	0.2321	0.1657	0.1874	0.2375	0.2431	0.2523
	0.39	0.50	0.34	0.40	0.54	0.56	0.62
MA(1) (θ)	−0.1883	−0.2122	−0.2044	−0.2097	−0.2168	−0.2226	−0.2367
	−0.35	−0.45	−0.42	−0.44	−0.49	−0.50	−0.55
Conditional Variance Equation							
Constant (ω)	0.0006	0.0006	0.0006	0.0006	0.0006	0.0006	0.0006
	4.18***	4.25***	4.34***	4.37***	4.38***	4.91***	4.98***
GARCH(1) (η)	0.9611	0.9605	0.9602	0.9600	0.9599	0.9599	0.9581
	125.5***	122.9***	122.0***	122.0***	122.2***	124.8***	125.3***
ARCH(1) (ψ)	0.2516	0.2538	0.2554	0.2559	0.2563	0.2568	0.2571
	10.74***	10.71***	10.76***	10.75***	10.74***	10.91***	11.01***
Asymmetry (δ)	−0.0025	−0.0024	−0.0024	−0.0023	−0.0023	−0.0022	−0.0021
	8.74***	8.71***	8.76***	8.75***	8.74***	8.91***	9.31***
Dis (β)	−0.0015	−0.0007	−0.0005	−0.0005	0.0001	0.0000	0.0000
	−3.06***	−2.56**	1.98**	1.98**	1.25	0.59	0.69

Note: *,**, *** *indicate the* t-*statistic is significant at 10%, 5%, and 1%, respectively;* Dis *denotes natural disaster.*

of extreme weather events on the stock market return and volatility, the corresponding ARMAX−EGARCHX model was estimated for different disaster event windows starting from 1 day to 2 months. The one-day event window is the peak date of the catastrophe that is most broadcasted on Hong Kong news coverage. The 2-month-long event window starts from the peak date until 2 months following the peak date.

Table 2.4 reports the model estimation for seven event windows of natural disasters: 1, 5, 10, 12, 15, 30, and 60 days. The results determine that the coefficients of natural disasters (α) in the conditional mean equation were significant at 10% and increased in magnitude as the model was estimated for the event window of 1 (-0.0008), 5 (-0.0005), and 10 days (-0.0002). This indicates that natural disasters in Hong Kong negatively impact the local stock market return with the growing magnitude up to 10 days following the event peak date. At the same time, in the conditional variance equation, the coefficients of natural disasters (β) also remained significant and increased in their degree as the model was estimated for the event window of 1 (-0.0015), 5 (-0.0007), 10 (-0.0005), and 12 days (-0.0005). The results indicate that natural disasters in Hong Kong have a negative effect on the local stock market volatility with a rising degree up to 12 days following the event peak date. In addition, the asymmetry coefficients (δ) were negative and statistically material up to 60-day-long event window, supporting the fact that negative events have larger impacts on the stock market return volatility than positive events. As such, the presence of asymmetry in return volatility indicates that return volatility is higher during the catastrophic weather events. Higher return volatility induces a higher probability of a bear market while lower return volatility induces a higher probability of a bull market.

To ensure the efficiency and consistency of the model estimation, postestimation diagnostic tests were performed and are reported in Table 2.5. The test outcomes show that all statistics of Ljung–Box Q test and Engle ARCH test up to lag 5, 10, and 20 were immaterial for the seven event windows of natural disasters. This confirmed no serial correlation and heteroscedasticity in the innovation terms when a dummy variable was included in the conditional mean and conditional variance equations of the models. The models are also covariance stationary since the summation of ARCH(1)^2 and GARCH(1)^2 terms were less than unity. Therefore the ARMAX–EGARCHX models have no sign of misspecification because it satisfied the conditions of no serial correlation and homoscedasticity in the residuals, as well as the stationarity in covariance.

As discussed earlier, natural disasters apparently have negative impacts on the Hong Kong Stock Market return and volatility. However, the effects survived only for a short period of 12 days following the peak date and quickly died out afterward. This may have been due to the

Table 2.5 Postestimation Diagnostics.

Test Statistics	Disaster Event Window						
	1 Day	**5 Days**	**10 Days**	**12 Days**	**15 Days**	**30 Days**	**60 Days**
Q(5)	1.85	1.90	1.90	1.83	1.75	1.72	1.70
Q(10)	6.21	6.30	6.18	6.10	6.04	6.04	5.94
Q(20)	20.13	20.28	20.21	20.17	20.16	20.02	19.72
Q^2(5)	9.88	8.69	8.77	8.75	8.72	8.54	8.24
Q^2(10)	11.93	10.81	10.93	10.83	10.75	10.59	10.12
Q^2(20)	13.57	12.26	12.31	12.11	12.05	11.85	11.06
ARCH(5)	1.86	1.91	1.91	1.83	1.75	1.73	1.69
ARCH(10)	6.22	6.31	6.18	6.08	6.05	6.04	5.92
ARCH(20)	20.12	20.27	20.21	20.16	20.15	20.01	19.88
ARCH(1)^2 + GARCH(1)^2	0.99	0.99	0.99	0.99	0.99	0.99	0.98

Note: $Q(q)$ and $Q^2(q)$ are statistics of the Ljung and Box (1979) test for serial correlation up to lag q in the residuals and in the squared residuals, respectively; ARCH(q) is statistic of the Engle (1982) ARCH test for conditional heteroscedasticity up to lag q.

effectiveness of the emergency response management system in Hong Kong. The system, which was established in 1996, consists of three-tier emergency response operations according to the level of severity of given crises including but not limited to natural disasters. The Tier-One response requires Police Force and Fire Services Department functioning under their own commands and control facilities. The Tier-Two response is activated in an event that poses risks to life, property, and security and which could be in need of Government Secretariat involvement. The Tier-Three response is triggered in an event that poses pervasive risks to life, property, and security and which requires extensive government responses to emergencies. In addition, Sim et al. (2018) recently evaluated the disaster resilience of Hong Kong using the Sendai Framework Local Urban Indicators Scorecards. They asserted that Hong Kong effectively recorded 4.2 out of 5 points, implying a satisfactory level of disaster resilience of the country.

2.6 CONCLUSION

This study investigated the impacts of natural disasters on stock market return and volatility in Hong Kong over various event windows ranging from 1 day to 2 months following the disaster event. For empirical analysis, a standard ARMA process was extended with an EGARCH process and a dummy variable X indicating the incident of natural disasters. Consequently, the extended model called ARMAX−EGARCHX was estimated for seven event windows to assess the persistence of natural disasters on the Hong Kong Stock Market return and volatility.

The results determined that natural disasters have negative impacts on Hong Kong's stock market return and volatility with increasing magnitude. Nonetheless, the disaster impacts on return and volatility only persist up to 10 and 12 days, respectively, after the event. The increasing magnitude of the impact appears to align with the fact that natural disasters can have enduring consequences and it often takes more than a couple of days to accurately estimate the economic losses. The relatively short impact period is perhaps owing to the effectiveness of the emergency response management system in Hong Kong. The outcomes of this study are likely to assist Hong Kong policymakers in scheduling the postdisaster reconstruction programs. The findings may also be relevant to Hong Kong stock market investors in considering appropriate insurance coverage during times of severe weather to minimize their investment losses.

While this study concluded that the disaster impacts on the Hong Kong Stock Market return and volatility survive in a relatively short period, it did not target to investigate the link between the short impact period and the effectiveness of the emergency response system in Hong Kong. This area is, therefore, recommended for future research.

REFERENCES

Arcadis. (2018). *Citizen centric cities, the Sustainable Cities Index*. Retrieved from <https://www.arcadis.com/media/1/D/5/%7B1D5AE7E2-A348-4B6E-B1D7-6D94FA7D7567%7DSustainable_Cities_Index_2018_Arcadis.pdf>.
Bank for International Settlements (BIS). (2018). *Annual report: Promoting global monetary and financial stability*. Retrieved from <https://www.bis.org/about/areport/areport2018.pdf>.

Bourdeau-Brien, M., & Kryzanowski, L. (2017). The impact of natural disasters on the stock returns and volatilities of local firms. *The Quarterly Review of Economics and Finance, 63*, 259–270.

Dickey, D., & Fuller, W. (1981). Likelihood ratio statistics for autoregressive time series with a unit root. *Econometrica: Journal of the Econometric Society, 49*(4), 1057–1072.

Downton, M., & Pielke, R. (2005). How accurate are disaster loss data? The case of US flood damage. *Natural Hazards, 35*(2), 211–228.

Engle, R. (1982). Autoregressive conditional heteroscedasticity with estimates of the variance of United Kingdom inflation. *Econometrica: Journal of the Econometric Society, 50*(4), 987–1007.

Engle, R., & Ng, V. (1993). Measuring and testing the impact of news on volatility. *The Journal of Finance, 48*(5), 1749–1778.

Hansen, P., & Lunde, A. (2005). A forecast comparison of volatility models: does anything beat a GARCH (1, 1)? *Journal of Applied Econometrics, 20*(7), 873–889.

Ljung, G., & Box, G. (1979). The likelihood function of stationary autoregressive-moving average models. *Biometrika, 66*(2), 265–270.

Luo, N. (2012). *The impact of natural disasters on global stock market: The case of the Japanese 2011 earthquake* (Master Research Project). Halifax: Saint Mary's University.

Murphy, P. (2015). *Four Asian cities amongst world's most prone to natural disasters.* Arcadis, Design and Consultancy for Natural and Built Assets. Retrieved from <https://www.arcadis.com/en/asia/news/latest-news/2015/5/four-asian-cities-amongst-world-s-most-prone-to-natural-disasters/>.

Nelson, D. (1991). Conditional heteroskedasticity in asset returns: A new approach. *Econometrica: Journal of the Econometric Society, 59*(2), 347–370.

Ng, S., & Perron, P. (2001). Lag length selection and the construction of unit root tests with good size and power. *Econometrica, 69*(6), 1519–1554.

Nguyen, T., & Chaiechi, T. (2019). The impacts of natural hazards on stock market return and volatility. *Research Newsletter, James Cook University, CBLG Bulletin.* Retrieved from <https://www.jcu.edu.au/__data/assets/pdf_file/0004/947830/Aug-Sep-CBLG-Research-Bulletin.pdf>.

Norrman, A., & Jansson, U. (2004). Ericsson's proactive supply chain risk management approach after a serious sub-supplier accident. *International Journal of Physical Distribution & Logistics Management, 34*(5), 434–456.

Phillips, P. C., & Perron, P. (1988). Testing for a unit root in time series regression. *Biometrika, 75*(2), 335–346.

Siddikee, N., & Rahman, M. (2017). Effect of catastrophic disaster in financial market contagion. *Cogent Economics & Finance, 5*(1), 1288772. Available from https://doi.org/10.1080/23322039.2017.1288772.

Sim, T., Wang, D., & Han, Z. (2018). Assessing the disaster resilience of megacities: The case of Hong Kong. *Sustainability, 10*(4), 1137. Available from https://doi.org/10.3390/su10041137.

Taimur, M., & Khan, S. (2015). Impact of political and catastrophic events on stock returns. *VFAST Transactions on Education and Social Sciences, 6*(1), 21–32.

Tavor, T., & Teitler-Regev, S. (2019). The impact of disasters and terrorism on the stock market. *Jàmbá: Journal of Disaster Risk Studies, 11*(1), 1–8.

Time Out Hong Kong. (2018). *Hong Kong's worst typhoons.* Retrieved from <https://www.timeout.com/hong-kong/things-to-do/hong-kongs-worst-typhoons>.

Wang, L., & Kutan, A. (2013). The impact of natural disasters on stock markets: Evidence from Japan and the US. *Comparative Economic Studies, 55*(4), 672–686.

World Economic Forum. (2019). *The global competitiveness report 2019.* Cologny, Geneva: World Economic Forum. Retrieved from <http://www3.weforum.org/docs/WEF_TheGlobalCompetitivenessReport2019.pdf>.

World Trade Organisation (WTO). (2018). *World Trade statistical review*. Retrieved from <https://www.wto. org/english/res_e/statis_e/wts2018_e/wts2018_e.pdf>.

Worthington, A. (2008). The impact of natural events and disasters on the Australian stock market: A GARCH-M analysis of storms, floods, cyclones, earthquakes and bushfires. *Global Business and Economics Review, 10*, 1−10.

Worthington, A., & Valadkhani, A. (2004). Measuring the impact of natural disasters on capital markets: An empirical application using intervention analysis. *Applied Economics, 36*(19), 2177−2186.

CLIMATE CHANGE AND EFFECTS: A QUALITATIVE EXPERIENCE OF SELECTED OLDER ADULTS

3

Prince C. Agwu[1], Nnaemeka V. Emodi[2] and Uzoma O. Okoye[1]

[1]*Department of Social Work, University of Nigeria, Nsukka, Nigeria* [2]*Future Energy Research Group, Tasmanian School of Business and Economics, University of Tasmania, Hobart, TAS, Australia*

3.1 INTRODUCTION

Climate change poses the greatest socioeconomic and environmental challenges in human history and will have implications for the future human society (Wade & Jennings, 2015). This has necessitated global partnerships in dealing with this change capable of leading to the extinction of our world (Babagana, 2009; IPCC (International Panel on Climate Change), 2014; Kovats & Ebi, 2006). With vivid unusual changes in the air, water, and land contents of the earth, it is no exaggeration to state that consequences of economic, social, political, and cultural concerns are implied for societies (Ogbo, Ndubuisi, & Ukpere, 2013; Okunola & Ikuomola, 2010; Sayne, 2011). Against the backdrop, varying population distribution is disproportionately affected by these changes, given their adaptive capacities. Thus while the vibrant and active youth populations tend to have features that could enable them resiliently adapt to changes in the world's climate, older adults seem to lack such resilience due to their frail nature (HelpAge International, 2015; Wells, Calvi-Parisetti, & Skinner, 2013). As a result of such disproportionate effect, there is a need to pay special attention to the older adults (Dominelli, 2011; Oven et al., 2011).

Definitions of climate change have succeeded in arriving at a consensus that the global atmosphere is altering with deleterious effects on water, land, and air, which exercise threatening consequences for human life (Achstatter, 2014; Bennet, 2010; Environment Canada EC, 2008). Human activities and natural factors buoy these alterations and deleterious consequences. Although human activities have been acknowledged to be at the fore of the issue, but if controlled to be eco-friendly, will form a panacea to this inimical change and effects (Odemerho, 2015; Okunola & Ikuomola, 2010; Oven et al., 2011). The African continent is highly vulnerable to climate change with temperatures increase by about 0.7°C across its nations (United Nation, 2006). Temperature predictions show an increase in warming across African nations that will increase climatic events such as drought and flooding resulting in an increase in food scarcity, inundation in coastal areas, the spread of water-borne diseases, and changes in the natural ecosystem. While the changes in future weather events will greatly have an impact on the African continent, the design and implementation of policies and strategic measures to assist adaptation and mitigation to the vulnerability of climate change will be essential (African Climate Policy Centre (ACPC), 2013).

Economic Effects of Natural Disasters. DOI: https://doi.org/10.1016/B978-0-12-817465-4.00003-0

As mentioned previously, remedial actions against climate change should be led by citizens at both the grassroots- and government levels. Nigeria as a country has a good number of its citizens involved in climatic-dependent jobs such as agriculture, and essential infrastructures such as its power generation are tied to climatic supplies. The country is among those nations marked as having poor adaptive capacity and resilience to combat effects of climate change, owing to its level of socioeconomic development and civility (Enete, Officha, Ezezue, & Agbonome, 2012; Jackson, 2009; United Nations Development Program (UNDP), 2010; UNEP, 2011). Amidst these climatic challenges the older adults are bound to face heightened environmental risks such as extreme weather, compromised agricultural livelihood, reduced availability of unpolluted air and water, and decreased habitability of human population centers (Filiberto et al., 2009). Unfortunately, as people age, their susceptibility to diseases and stress increases.

The exposure of the elderly to climate threat can be reduced with the provision of reliable and clean energy access (HelpAge International, 2015; Ketlhoilwe & Kanene, 2018). On the one hand, this can reduce the level of heat stress experienced by the elderly and, on the other hand, ensures the operation of medical facilities that can address their health needs (Tawatsupa et al., 2012). This study investigates the health implications of climate change and its impact on the elderly in Nigeria using the University of Nigeria, Nsukka as a study area. The results of the interviews suggest the need for vanguards of ecological rights, promotion of eco-friendly culture for Nigeria, and promulgation of policies addressing climate change, in order to cushion the impacts of climate change on older adults. Therefore there is a need to enhance Nigeria's current climate change mitigation policies and increase public awareness of climate change. The rest of this paper is organized as follows. Section 3.2 presents a theoretical insight into the energy and climate crisis in Nigeria from a social perspective. Section 3.3 describes the data and methods applied in this study. The results are presented in Section 3.4 followed by the discussion and conclusion in Section 3.5.

3.2 ENERGY AND CLIMATE CRISIS IN NIGERIA FROM A SOCIAL PERSPECTIVE

It is clear that older adults aged 65 years and above tend to grapple with Nigerian issues of epileptic social protection in policies and aids (Haq, Brown, & Hards, 2010; Okoye & Asa, 2011). This has forced a good number of them to retire to their villages usually in rural areas, while some exclusively become dependent on filial care as urban dwellers (Okoye, 2012). More so, they are known for their huge involvement in domestic agro-activities that are climate dependent (HelpAge International, 2015). Given their health conditions, the older adults quest for reliable power supply that tends to be epileptic in Nigeria, partly owing to its overreliance on gas-fired and hydropower plants. Issues of inadequate gas supply and low rainfall in hydrodams are challenges faced by the current set of electricity generators in Nigeria (Chala, Ma'Arof, & Sharma, 2019; Emodi, 2016; Nwanya, Mgbemene, Ezeoke, & Iloeje, 2018; Oyerinde et al., 2016). This has led to many citizens resorting to carbon-emitting generators, carbon lanterns, and wood fuels as alternatives. Although these alternatives are acknowledged to be eco-unfriendly, they are considered domestic additions to depleting the climate (Okoye & Ijiebor, 2012; Sharma, Thakur, & Kaur, 2012; UNEP, 2011).

The obtainable culture in Nigeria encourages parents to see to the welfare of their children even at their own cost. Thus in terms of disasters occasioned by climate change, older adults might first prefer having the younger ones safe before they even think of themselves. Where migration becomes an option, they might choose to stay back because of the stress that comes with migration, as well as their agelong attachment to their lands (Doherty & Clayton, 2011; Ogbo et al., 2013; Wells et al., 2013). This makes them become victims of poor care, attention, and isolation. To this end, it is obvious that older adults in Nigeria need adequate socioeconomic protection and care in events of problems arising from climate change. Such should be targeted at improving their adaptive capacities on preventive and curative grounds, with a fundamental objective of developing resilience (Moth & Morton, 2009; Negi & Furman, 2010). It is in this vein that social workers are the professionals readily coming to mind (Dominelli, 2012).

The involvement of social workers in climate change response is to manage vulnerable conditions while strengthening resilience and protection (Alston, 2015; Bobby, 2014; Dominelli, 2011). IPCC (International Panel on Climate Change) (2014) defines vulnerability as a case of being incapable of grappling with the adverse challenges of climate change. The vulnerable populations are those classified as having the above-discussed shortcomings and increasingly susceptible to consequences of the changing conditions. This implies that the vulnerable populations face some level of eco-marginalization since they are disproportionately affected by the ecological crisis (HelpAge International, 2015; Oven et al., 2011). With the ideals of social work founded on principles of democracy, social justice, and humanitarianism, those who are vulnerable as a result of inimical climatic experience attract the attention of social workers. Involvements of social workers are encouraged to enable mitigation of climate change occurrences while enabling adaptive competencies for older adults (Achstatter, 2014; Peters, 2012). Social workers in developed countries have commenced and are doing so well in an aspect of practice called "green social work" (Dominelli, 2011). They are involved in educating persons on eco-friendly behaviors, curriculum development of green social work courses, and responding to ecological crisis situations. Unfortunately, this is lacking in Nigeria.

Furthermore, social workers can advocate for policies that will protect vulnerable groups from the possibility of harms while mobilizing social actions and charting dialectic paths with industries that constitute destructive occupations to our ecology. Clinically, social workers meet the individual needs of older adults who are affected by climate change. This they do by using resource building and referral skills, listening skills, counseling, and behavior modification techniques (Charles Sturt University, 2016; Negi & Furman, 2010). They could help older adults with adaptive skills and ideas to climate change situations, as well as initiate rapport between older adults and their communities, where necessary should community response be required. Generally, social workers target social care and protection for older adults in climate change scenarios and do so following clinical, structural, and curriculum approaches (International Federation of Social Work (IFSW), 2014).

In view of the foregoing, studies abound on climate change in Nigeria (Enete et al., 2012; Odemerho, 2015; Okunola & Ikuomola, 2010; Oyero, Oyesomi, Abioye, Ajiboye, & Kayode-Adedeji, 2018; Sayne, 2011). However, there are scarcely published literature on climate change and the involvement of the social work profession within Nigeria. Albeit, there are a good number of foreign studies (Achstatter, 2014; Alston, 2015; Bennet, 2010; Bobby, 2014; Cumby, 2016; Dominelli, 2012). Therefore given the novel direction of this study in Nigeria, it brings to the knowledge of stakeholders in policy formulation and strategic management of Nigeria, social dimensions of climate change, and the extent to which older adults are vulnerable in such situations. This further inspires fulfillment toward the need for adaptation and resilience of Africa to climate change, particularly vulnerable populations, as mentioned in the "Aspiration 1 (16) and Call to Action 72f" of the AU 2063 Agenda.

From a theoretical perspective, crisis intervention theory explains the degree of challenges and threats faced by older adults in climate change situations and the exigency of response that should be made by social workers. The theory asserts that crisis is best used to describe experiences where people are met with disequilibrium, severely reduced functioning, and ineffectiveness of traditional coping methods (Rapoport, 1970; Roberts, 2000). This pictures the situation of older adults in Nigeria who appraise challenges and threats of climate change as severely hazardous. Fueling such appraisal is their frailty and disengagement from socioeconomic profiting activities, which implies insufficient coping capacity (Doherty & Clayton, 2011; Oven et al., 2011). Poor responses from government and lack of eco-friendly culture among Nigerians leave older adults more helpless and deeply crisis situated.

From the positions of this theory, social workers are expected to apply calculated attempts in improving adaptive capacities of older adults (Teater, 2010). They follow a two-way approach. One of which will be engaging older adults through education and social support as measures to contain the crisis and, on the other hand, utilizing policy approaches, curriculum development, and social workers acting in the capacity of eco-vanguards. The second is relevant in preserving the eco-structure, through regulating and kicking against ecological destructive behaviors. By this theory's assertion, social workers in Nigeria must take into consideration the urgent demand for action following the climate crisis found to be experienced by older adults.

3.3 DATA AND METHODS

3.3.1 STUDY AREA

The study area is the University of Nigeria, Nsukka. It involved 11 older adults of 65 years and above who live within the campus. There is no sourced statistics as regards the number of older adults who dwell within the campus. The climate profile of Nsukka LGA as obtained from the Department of Geography, University of Nigeria, Nsukka, reads that the area is tropically wet and dry, with a latitudinal location of $6 - 7$ degrees north of the equator. Its temperature is high, although it varies with altitude and seasons. The area usually experiences its rainy season between March and October, while its dry season occurs between November and February.

3.3.2 SAMPLING PROCEDURE

The participants were sampled using purposive and snowball techniques. This applied as the researchers specifically targeted just older adults of the needed age level and also on referral. Those who consented to participate in the study were interviewed. Information was elicited from 11 older adults of the specified age made up of 5 women and 6 men. The interviews were conducted over a period of 1 month between December 2017 and January 2018. Participants accepted the interviews based on a scheduled time fixed by them, either at their homes or offices. Timing for each interview never exceeded 1 hour. Participants were free in narrating their experiences in English language and sometimes having a blend of Igbo language. With participants' permission, interviews were tape-recorded while a notetaker equally took notes.

3.3.3 DATA ANALYSIS

The interviews were all transcribed in English language. Attention was paid to their emotional expressions (Creswell, 2007). The researchers further made use of thematic analysis in building responses into themes.

3.4 RESULTS

3.4.1 DEMOGRAPHIC CHARACTERISTICS OF PARTICIPANTS

No respondent was less than 65 years, and none was above 81 years. Eight of the respondents were in active service, three of them were retirees of which two were dependents who live with university staff. See Table 3.1 for full demographic description.

3.4.2 OLDER ADULTS AND THEIR UNDERSTANDING OF CLIMATE CHANGE

Respondents expressed good knowledge over the subject. This cannot be farfetched from their level of education or association with family members who are for the most educated. Some went ahead to call the subject matter in their indigenous language. A consensus was reached that climate

Table 3.1 Demographic Characteristics of Respondents.

Code	Age	Sex	Occupation	Educational Level
001	71	Female	Retiree	BA
002	66	Male	Professor	PhD
003	69	Male	Retiree	MSc
004	67	Female	Trader	NCE
005	79	Female	Retiree	SSCE
006	66	Female	Self Employed	CHEW
007	65	Male	Retiree	BSc
008	65	Male	Professor	PhD
009	81	Male	Retired farmer	No formal education
010	65	Female	Retiree	Diploma
011	68	Male	Retiree	BEng

Field Survey (2017).

change implies the way the weather has not been the same over a long period of time. This they discussed from their experiences within the Nsukka area. Few attributed the change to religion while others maintained human activities as causes. A respondent said:

> I believe you mean 'mgbenwe uboch'. Of course, if you notice seasons are no longer static. We can't say for sure that rainy season starts from a particular time and ends at a particular time again and so for dry season. Is this not a change? Also, if you notice nowadays, the weather is usually very hot even in the supposed cool hours of the night and it was never like this while I was growing. The climate or weather is indeed changing [001].

Another responded:

> Nothing in this world will be same forever including the climate. Though, most things that change in this world move from good to worse and an example is the climate. It is very bad now because of the way we do things in this country. We burn bushes anyhow, we use generators constantly, our industries do not follow environmental protection laws if we even have any that is functional. All these things have spoilt the way the weather is [...] Nigeria's climate is in a worse condition... [003]

In addition, is another narrative: "why won't our climate change when we no longer plant trees but rather choose to use them as firewood (charcoal) or something else... in fact, it will be changing more [010]."

From a differing perspective, a male respondent narrated:

> I know science introduced this climate change of a thing. I don't disagree with what science said especially the area of global warming but I must point out that the Bible predicted these changes even before science started thinking about it. The changes in our climate realities are not just scientific, they are equally spiritual and a sign that the world will soon come to an end and we must acknowledge it [009].

Lastly, another respondent when asked on her take regarding religion and climate change further commented:

> Do you think God will want to destroy the world he created? Same way we are destroying our bodies by not eating natural foods again, same way we have destroyed the climate by not doing what we are supposed to do to protect it. When I was a child, our parents were accurate when predicting the weather but now it is not possible because industries have scattered our climate with their harmful gases, smoked cars are everywhere... In fact, if you go to "Ogige market" (the town's market), you just cannot breathe fresh air because of all these nonsenses in the name of development. We might all die one day because of the nonsenses we inhale [010].

These narratives confirm that climate change is indeed evident in Nsukka area, and the respondents are abreast of the fundamental meanings of the subject. Very common across responses is the fact that human activities contribute in no small amount to climate change conditions. We recorded religious interpretations to climate change that indicates the belief of Christians is that the world is at the verge of coming to an end. This was, however, disregarded by other respondents.

3.4.3 **EFFECTS OF CLIMATE CHANGE ON OLDER ADULTS**

The respondents narrated their individual experiences. Their experiences were more negative than positive. They cited experiences of the effects of the change and maintained that older adults are the worst hit of the situation. This they blamed on their frailty. Their narratives confirm the health, economic, and social implications of climate change. A respondent said:

> Several times the doctor diagnosed me of respiratory problems. Even the doctor agrees with me that the air is bad and ventilation is not as it used to be during our younger days [...] I easily get thirsty too because of how the weather is these days. My dear, one just has to manage because it can't be better [003].

In corroboration, another respondent added:

> This weather is not good for us older people at all. Do you know that even at night, that one is supposed to be feeling cool breeze, I still feel hot? You can see the number of rashes on my skin [004].

Commenting further by another:

> If there is steady power supply, we will not be feeling this bad weather as much. I can't use the money I will use in buying food to buy petrol for my generator. So, my dear, I am just managing the weather. I hope for improvement [005].

Another respondent said:

> The kind of discomfort I witness these days is very saddening. Heat is everywhere and when cold finally comes, it will be so extreme that it can even cause pneumonia if you don't cover your body very well. This was not so about twenty years ago. Nsukka usually has one of the best climates but now it seems the worst in Nigeria [002].

A female respondent added:

> You know that as I am old, I should be eating enough fruits and vegetables. If it was some years back, I easily get enough of them at very cheap prices. But now, I don't know if it is because of this climate change, their prices are now high. I just manage to get them in little quantities [004].

From a social perspective, one said:

> During my days as a child, I respected the intelligence of my parents because they could tell you when rain will fall and when it will not fall. Because of the way my parents accurately predicted the weather, we were having good plans for our farming business [011].

Another supported the above view by narrating his own experience thus:

> we knew when our crops will start yielding and when to harvest. But this time, our crops suffer because even during the rainy season, you will see very heavy sun and you don't expect me to be constantly watering because after watering, the heavy sun will still dry them. Our children do not care about farming if not they would have been helping out in watering the crops. Again, we

don't enjoy that kind of respect our parents enjoyed since they could accurately predict the weather [009].

One of the respondents was encouraging a little:

I know the climate in Nsukka is not too good but who knows about other places. At least we do not experience flooding in residential areas here because of the hilly nature of the area unlike other areas in Nigeria. I visit places outside Nsukka and Nigeria even. But sincerely, Nsukka is a very livable place and I am trying to adapt to its new weather condition [008].

From the responses earlier, it is clear that climate change impacts very vulnerable populations. As narrated by respondents, areas of impact include health, economic sustenance through agriculture, cost of food materials, domestic comfort, and even the respect they are supposed to enjoy as older adults owing to longtime familiarity with the weather and the ability to predict it. The majority of respondents agreed that the situation was better some years ago. However, few participants believe that the climate situation of Nsukka is better when compared to their experiences and reports from other places within and outside Nigeria. They generally believe that the best they can do is to adapt to the situation, which they are of the view might not change.

3.4.4 SUGGESTIONS TO OVERCOME CLIMATE CHANGE

The study participants offered their suggestions as regards what could be done to preserve the climate. Their suggestions were, however, drawn from their agelong experiences as they wish to have a better country with policies that protect the ecology. They encouraged human rights activists to shift their focus to also cover the environmental rights of people. The participants mentioned the roles of media and education in encouraging individuals to desist from eco-unfriendly activities. Though some of them had given up on the wherewithal of Nigeria to deal with the issue, few resorted to solace in religion. They also used the opportunity to canvass for quality care for older adults. One of the respondents said, "These smoked cars that move everywhere should be impounded. They should not be allowed to move freely [001]."

Corroborating her is another who said:

You must be aware of the time Volkswagen made a pronouncement that they had issues with some cars they produced which have issues capable of spoiling the climate. Now, how many of such cars were returned to the company from Nigeria? In fact, when the company collects back those cars from other developed countries, they will ship it into Nigeria. For me, Nigeria does not have what it takes to deal with climate change. When they do, I will make my suggestions [003].

Another narrative was put forward:

It is not a lie when we say Nigeria has failed to protect its environment. There are no implemented environmental policies. Industries still emit bad gases and individuals still burn bushes and cut trees down for firewood and they dispose of their refuse in drainage systems. I think nongovernmental organizations and other social-related professions should make efforts using the media to educate people including the government on ways to protect the environment from climate change disasters [008].

One of the respondents in the same vein said:

> Environmental policies made by intelligent people and well implemented by strong people are what we need. People who default should be punished severely. In addition, there should be emergency numbers to call so that we can report these defaulters [007].

A female respondent said:

> Since the climate is already bad and the government is not ready to do anything, they should just try and improve power supply and also assist farmers so that we can access foods at cheap price. I wonder what these government people go to do at climate change programmes outside this country [004].

Another respondent commented further:

> If the government have older adults in mind, they will try and help reduce these changes in our weather. If they cannot, they should improve health care delivery for us. At our age, we are supposed to be having 100% free medical care. You don't expect us to take care of the health of our children who also suffer some of these problems and also take care of our own in these difficult times. It is not fair [005].

Lastly, a male respondent said, "people should just learn how to cope by themselves. There is nothing that can be done by anybody to change what the Bible has already predicted as I said earlier [009]."

From responses, we found out that some respondents felt efforts can be made to curb the situation but argued that the Nigerian government seems not able to take up the challenge. They believe that policies and programs can be initiated in order to create awareness on climate change as well as ways to contain it. They mentioned efforts social service professionals could make to help the situation, as well as the need to have emergency numbers to contact whenever they find culprits of ecological destruction. Severe punishment for culprits was advocated, as they believe that would serve as deterrence. While they expressed dissatisfaction at the ways governmental authorities are going about climate change, they advocated that healthcare, power, and agriculture should be improved. That way, the effects of climate change could be cushioned.

3.5 DISCUSSION AND CONCLUSION

Our findings came from 11 older adults of 65 years and above, who reside within the campus of the University of Nigeria, Nsukka. The respondents expressed satisfactory knowledge about what climate change is. This is expected given that the majority of them were educated. They accepted that climate change experiences are real and further blamed it on human activities. However, we recorded religious interpretations of climate change that is equally available in the literature (Pepper & Leonard, 2016; Wardekker, Petersen, & Van der Sluijs, 2008). Respondents agreed that they are disproportionately affected by climate change in terms of their health, economic, and social life (HelpAge International, 2015; Oven et al., 2011). This confirms the position of the AU 2063 Agenda that vulnerable populations should be the exigency target in alleviating climate change effects (African Union Commission AUC, 2015).

Among outlined eco-unfriendly behaviors by respondents include bush burning, carbon-emitting vehicles and power supply generators, industrial activities, and cutting down of trees for wood fuels. There is a consensus between the positions of the respondents and what similar studies have equally concluded in detail of climate unfriendly attitudes in Nigeria (Babagana, 2009; Odemerho, 2015; Okunola & Ikuomola, 2010; Sayne, 2011).

In another development, findings show the effects of climate change on older adults' health and economic well-being. The respondents lamented failing health conditions with reference to respiratory diseases, skin rashes, insufficient ventilation, and feeding deficiencies resulting from lack of access or inflation of certain food materials. They buttressed the fact that the excessive heat in their homes and epileptic nature of power supply have made their homes unconducive. To accommodate these health challenges owing to climate change, respondents noted that they are forced to spend from the little they have to offer themselves and their children quality healthcare and spend money to power their generators.

They also were of the position that the climate has not been supportive of their crops, especially due to unpredictability of the weather. Participants also noted that their inability to predict the weather like their parents tends to have reduced their respect in the face of the younger generation. Nevertheless, some of the respondents sounded encouraging, as they believe that the climate of Nsukka, though worse than what they had in earlier years, is better to be compared with the climatic conditions and reports of other places within and outside Africa.

Participants recommended ideas that could go a long way in preserving the climate. Topping the list of said recommendations is the issue of Nigerian government enforcing strict ecological protective legislation and severely punishing defaulters. They also highlighted the importance of having emergency contacts of designated offices to report issues arising from climate change, including reporting culprits of ecological destruction. Some participants suggested impounding vehicles that gravely emit carbons. They as well wished for quality infrastructural development in terms of power supply, healthcare, and improved food production/supplies to help contain the effects of the change on them.

Some participants felt resorting to self-generated coping mechanisms should be the best as they perceive that the government seems not ready. They believed that the Nigerian government lacks the wherewithal to combat climate change and its effects. Therefore they called on professionals such as social workers and organizations who can champion ecological rights culture (Cumby, 2016; Peters, 2012). This supports our theoretical framework that argues that social workers and significant entities should see this situation as a crisis and should be approached as such. This is because, a good number of older adults interviewed described the situation as a helpless one and more or less a crisis while explaining how it is of exigency to be attended to. Participants also recommended the involvement of the media to educate individuals and even the government on eco-friendly approaches as a panacea to climate change and its effects (Dominelli, 2011; Dominelli, 2012).

We conclude by quoting "Aspiration 1 (16)" of the AU 2063 Agenda:

> [...] Africa shall address the global challenge of climate change by prioritizing adaptation in all our actions, drawing upon **skills of diverse disciplines** with adequate support [...] to ensure implementation of actions for survival of the **most vulnerable populations** [...] (African Union Commission AUC, 2015: 4)

However, the lack of political will by African leaders to address the above has been noted by participants to be the core of the problem. This lack of political will is also manifest in Nigeria where the implementation of agreed-upon strategies is not forthcoming (Nzeadibe, Uchem, & Nzeadibe, 2018). In Nigeria today, there is little or no government funds available for research into various facets of climate change and its social implications. Rather, what is obtainable are individual efforts that amount to very little. Also, there is a general lack of awareness on some government policies targeting adaptation to climate change (Oyero et al., 2018). Therefore there is a need for government to step up by not only introducing achievable policies on climate change in Nigeria but also creating awareness among the populace.

Social workers are well placed to create such awareness if well mobilized. Nigerians, especially older adults, need to know about the effects of climate change and global warming and also the human activities causing these changes. Currently, social workers are branching into specialties of "green" and "environmental" social work and so can sensitize Nigerians on environment-friendly/unfriendly behaviors. It is important to note that "Aspiration 1 (16)" of the AU 2063 Agenda emphasized the implementation of actions to ensure the survival of the most vulnerable populations. Social workers by the nature of their training are in a position to help the government implement this action. They could advocate for policies that are environment-friendly and support the fight against inequities and inequalities that lead to the destruction of the environment (Okoye & Agwu, 2019).

Finally, the study was not void of certain limitations. The focus of the study within the Nsukka Campus of the University of Nigeria and with only 11 older adults might not be entirely representative of the views of older adults that are within Nsukka and neighboring areas. Therefore, there is the need to replicate similar studies outside the university as well as other areas in Enugu and neighboring states with a larger number of respondents. The study can equally be done with younger population groups to enable a comparative study of climate change perceptions and experiences.

REFERENCES

Achstatter, L. C. (2014). Climate change: Threats to social welfare and social justice requiring social work intervention. *21st Century Social Justice, 1*, 1−22.

Alston, M. (2015). Social work, climate change and global cooperation. *International Social Work, 58*, 355−363.

African Climate Policy Centre (ACPC). (2013). Vulnerability to climate change in Africa: Challenges and recommendations for Africa. In: *ClemDev-Africa policy brief*. Available from: <https://www.uneca.org/sites/default/files/PublicationFiles/policy_brief_2_vulnerability_to_climate_change_in_africa_challenges_and_recommendations_for_africa.pdf>.

African Union Commission (AUC). (2015). *Agenda 2063: The Africa we want*. Addis Ababa: AUC.

Babagana, A. (2009). The impacts of global climate change in Africa: The Lake Chad, adaptation and vulnerability. *Journal of South-South Studies, 5*, 109−123.

Bennet, A. (2010). *Climate change response strategy*. Western Australia: Department of Agriculture and Food.

Bobby, J. (2014). *Climate change and vulnerable people: Time for eco-social work practice*. Available from: <https://prezi.com/0daelbgyy0d4/climate-change-and-vulnerable-people-time-for-eco-social-work-practice/>.

Chala, G. T., Ma'Arof, M. I. N., & Sharma, R. (2019). Trends in an increased dependence towards hydropower energy utilization—A short review. *Cogent Engineering*, 1631541. (just-accepted).

Charles Sturt University. (2016). *Ecological social work*. Available from: <https://www.csu.edu.au/faculty/arts/humss/research/ecological--social--work>.

Creswell, J. W. (2007). *Qualitative inquiry and research design: Choosing among five approaches* (2nd ed.). Thousand Oaks, CA: Sage.

Cumby, T. (2016). *Climate change and social work: Our roles and barriers to action*. Available from: <http://scholars.wlu.ca/cgi/viewcontent.cgi?article = 2935andcontext = etd>.

Doherty, T. J., & Clayton, S. (2011). The psychological impacts of global climate change. *American Psychologist*, *66*, 265−276.

Dominelli, L. (2011). Climate change: A social work perspective. *International Journal of Social Welfare*, *20*, 430−438.

Dominelli, L. (2012). *Green social work*. Cambridge: Polity Press.

Emodi, N. V. (2016). *Energy policies for sustainable development strategies* (pp. 9−67). *Springer*.

Enete, I. C., Officha, M. C., Ezezue, A. M., & Agbonome, P. C. (2012). Adapting Nigeria cities to climate change using design options: A review. *British Journal of Applied Science and Technology*, *2*, 367−378.

Environment Canada (EC). (2008). *What is climate change?* Available from: <http://www.ec.gc.ca/climate/overview-trends-e.html>.

Filiberto, D., Wethington, E., Pillemer, K., Wells, N., Wysocki, M., & Parise, J. T. (2009). Older people and climate change: vulnerability and health effects. *Generations*, *33*(4), 19−25.

Haq, G., Brown, D., & Hards, S. (2010). *Older people and climate change: The case for better engagement*. New York: Stockholm Environment Institute.

HelpAge International (2015). *Climate change in an ageing world*. Available from: <http://www.prevention-web.net/files/47086_cop21helpagepositionpaperfinal.pdf>.

International Federation of Social Work (IFSW). (2014). *Global definition of social work*. Available from: <http://ifsw.org/get-involved/global-definition-of-social-work/>.

IPCC (International Panel on Climate Change). (2014). *Climate change: Impacts, adaptation and vulnerability*. Cambridge: Cambridge University Press.

Jackson, T. (2009). *Prosperity without growth: Economics for a finite planet*. London: Earthscan.

Ketlhoilwe, M. J., & Kanene, K. M. (2018). Access to energy sources in the face of climate change: Challenges faced by women in rural communities. *Jàmbá: Journal of Disaster Risk Studies*, *10*(1), 1−8.

Kovats, R. S., & Ebi, K. L. (2006). Heatwaves and public health in Europe. *European Journal of Public Health*, *16*, 592−599.

Moth, R., & Morton, D. (2009). *Social work and climate change: A call to action*. Available from: <http://www.socialworkfuture.org/articles-resources/uk-articles/101-social-work-and-climate-change-a-call-to-action-rich-moth-a-dan-morton>.

Negi, N. J., & Furman, R. (2010). *Transnational social work practice*. New York: Columbia University Press.

Nwanya, S. C., Mgbemene, C. A., Ezeoke, C. C., & Iloeje, O. C. (2018). Total cost of risk for privatized electric power generation under pipeline vandalism. *Heliyon*, *4*(7), e00702.

Nzeadibe, A. C., Uchem, R. N., & Nzeadibe, T. C. (2018). Beyond "traditional geographies": Integrating urban political ecology and cultural sustainability into undergraduate geographical education in Nigeria. *The Journal of Environmental Education*, *49*(3), 228−241.

Odemerho, F. (2015). Building climate change resilience through bottom-up adaptation to flood risk in Warri, Nigeria. *Environment and Urbanization*, *27*, 139−160.

Ogbo, A., Ndubuisi, E. L., & Ukpere, W. (2013). Risk management and challenges of climate change in Nigeria. *Journal of Human Ecology*, *41*, 221−235.

Okoye, U. O., & Agwu, P. C. (2019). Sustainable and healthy communities: The medical social work connection. *Journal of Social Work in Developing Societies, 1*, 30–45.

Okoye, U. O., & Asa, S. S. (2011). Caregiving and stress: Experience of people taking care of older adults relations in South-Eastern Nigeria. *Arts and Social Sciences Journal, 2*, 1–9.

Okoye, U. O. (2012). Family care-giving for ageing parents in Nigeria: Gender differences, cultural imperatives and the role of education. *International Journal of Education and aging, 2*, 139–154.

Okoye, U. O., & Ijiebor, E. E. (2012). Problems associated with climate change and implications for social work practice in Nigeria. In: *Paper presented at the Nigerian Meteorological Society (NMetS) 2012 annual conference with the theme climate change and variability: Saving our tomorrow today*, University of Benin, Benin-city.

Okunola, R. A., & Ikuomola, A. D. (2010). The socioeconomic implication of climate change, desert encroachment and communal conflicts in Northern Nigeria. *American Journal of Social and Management Sciences, 1*, 88–101.

Oyerinde, G., Wisser, D., Hountondji, F., Odofin, A., Lawin, A., Afouda, A., & Diekkrüger, B. (2016). Quantifying uncertainties in modeling climate change impacts on hydropower production. *Climate, 4*(3), 34.

Oyero, O., Oyesomi, K., Abioye, T., Ajiboye, E., & Kayode-Adedeji, T. (2018). Strategic communication for climate change awareness and behavioural change in Ado-Odo/Ota Local Government of Ogun State. *African Population Studies, 32*(1), 4057–4067. (Suppl.).

Oven, K., Curtis, S., Reaney, S., Riva, M., Ohlemüller, R., Dunn, C. E., … Holden, R. (2011). Climate change and health and social care: Defining future hazard, vulnerability and risk for infrastructure systems supporting older people's health care in England. *Journal of Applied Geography, 33*, 16–21.

Pepper, M., & Leonard, R. (2016). Climate change, politics and religion: Australian churchgoers' beliefs about climate change. *Religions, 7*, 1–18.

Peters, J. (2012). The place of social work in sustainable development: Towards ecosocial practice. *International Journal of Social Welfare, 21*, 287–298.

Rapoport, L. (1970). Crisis intervention as a mode of brief treatment. In R. W. Roberts, & R. H. Nee (Eds.), *Theories of social casework*. Chicago, IL: University of Chicago Press.

Roberts, A. R. (2000). *An overview of crisis theory and crisis intervention*. New York: Oxford University Press.

Sayne, A. (2011). *Climate change adaptation and conflict in Nigeria*. Available from: <https://www.usip.org/sites/default/files/Climate_Change_Nigeria.pdf>.

Sharma, S., Thakur, M., & Kaur, S. (2012). Health problems and treatment seeking behaviour among older adults. *Help Age India-Research and Development Journal, 18*, 21–27.

Tawatsupa, B., Yiengprugsawan, V., Kjellstrom, T., Seubsman, S. A., Sleigh, A., & Thai Cohort Study Team. (2012). Heat stress, health and well-being: findings from a large national cohort of Thai adults. *BMJ open, 2*(6), e001396.

Teater, B. (2010). *Applying social work theories and methods*. Berkshire: Open University Press.

United Nations Development Program (UNDP). (2010). *Human development report 2010: The real wealth of nations: Pathways to human development*. 20th anniversary edition. Available from: <http://hdr.undp.org/en/reports/global/hdr2010/chapters/en/>.

UNEP. (2011). *Environmental assessment of Ogoni land*. Nairobi: United Nations Environment Programme.

United Nation. (2006). *Africa is particularly vulnerable to the expected impacts of global warming*. United Nations Fact Sheet on Climate Change. Available from: <https://unfccc.int/files/press/backgrounders/application/pdf/factsheet_africa.pdf>.

Wardekker, A., Petersen, A. C., & Van der Sluijs, J. P. (2008). *Religious positions on climate change and climate policy in the United States*. Available from: <http://citeseerx.ist.psu.edu/viewdoc/download?doi = 10.1.1.824.8313andrep = rep1andtype = pdf>.

Wade, K., & Jennings, M. (2015). The impact of climate change on the global economy. *Schroders TalkingPoint.*

Wells, J., Calvi-Parisetti, P. & Skinner, M. (2013). *The neglected generation: The impact of displacement on older people.* Available from: <https://scribd.hulkproxy.online/document/97534627/The-neglected-generation-the-impact-of-displacement-on-older-people>.

FURTHER READING

Golaz, V., & Rutaremwa, G. (2011). The vulnerability of older adults: what do census data say? An application to Uganda. *African Population Studies, 25*(2), 605−622.

Haq, G., Brown, D., & Hards, S. (2008). Growing old in a changing climate: Meeting the challenges of an ageing population and climate change. New York: Stockholm Environment Institute.

Seeley, J., & Ekoru, K. (2010). Mitigating the impact of the epidemic on the households and families of older people in rural Uganda: Lessons for social protection. *African Population Studies, 24*(1&2), 113−129.

NATURAL DISASTERS AND LABOR MARKETS: IMPACTS OF CYCLONES ON EMPLOYMENT IN NORTHEAST AUSTRALIA

Josephine Pryce and Graeme Cotter

College of Business, Law and Governance, James Cook University, Cairns, QLD, Australia

4.1 INTRODUCTION

Cyclones, or "tropical revolving storms" as they may be known, are part of the fabric of life in the Tropics. Not surprisingly, the region of Northern Australia lies in an active cyclone-prone area with cyclone season officially stretching from November to April during what is typically known as the "wet season." The Bureau of Meteorology (BoM) (2019) has maintained a database of tropical cyclones (TCs) since 1970 and has estimated that there are on average 11 cyclones that form across the top of Australia (90−160°E). Of these, an average of 25% will make landfall. In 2018 BoM partnered with the National Energy Resources Australia (NERA) and the oil and gas industry to undertake a "reanalysis" of TCs to better understand the historical data and risk posed by cyclone events. This ongoing partnership has produced its first report (BoM, 2018) and database (BoM, 2019). The latter has collated information on cyclones from 1981 to 2016.

With the increasing focus today on climate change, research examining natural disasters is advancing. It is thought that climate change will increase the frequency and intensity of extreme weather events, such as cyclones. The United Nations Intergovernmental Panel on Climate Change (IPCC) highlighted in their Fifth Assessment Report (IPCC, 2014, p. 3) the "risks for human and natural systems" to support decision making in light of vulnerability, exposure, and impacts of hazards and so, to inform socioeconomic pathways, governance and actions related to adaptation and mitigation. The report presents the possible impacts of phenomena purportedly associated with climate change (e.g., extreme weather events) on various sectors such as terrestrial and marine ecosystems, food security and food production systems, urban and rural areas, human health and security, and livelihoods and poverty.

A search in Google Scholar shows that academic articles with "natural disasters" in the title have risen from 294 on average per year in the previous decade (1999−2008) to 591 per year in the past decade (2009−18), see Table 4.1. A similar trend is evident when searching using the term "cyclones" in the title. With the latter search the average for the period 1999−2008 is 260 articles per year and 440 per year for the decade of 2009−18. Despite this increasing interest in the literature on both natural disasters and cyclones, Table 4.1 shows that research associated with these phenomena in Australia is limited.

Economic Effects of Natural Disasters. DOI: https://doi.org/10.1016/B978-0-12-817465-4.00004-2

Table 4.1 Google Scholar Search for Papers With "Natural Disasters" and "Cyclones" in Title.

	Year	"Natural Disasters"	"Natural Disasters" + Australia	"Cyclones"	"Cyclones" + Australia
Overall		**11,800**	**50**	**12,000**	**93**
1	2018	639	5	547	2
2	2017	615	1	430	2
3	2016	603	4	487	7
4	2015	575	2	423	8
5	2014	561	1	468	5
6	2013	628	1	413	4
7	2012	631	9	424	2
8	2011	640	6	373	3
9	2010	553	1	464	3
10	2009	469	2	375	3
Total		**5840**	**32**	**4540**	**38**
Average		**591**		**440**	
11	2008	461	1	421	7
12	2007	446	2	363	4
13	2006	439	1	312	1
14	2005	367	0	242	2
15	2004	252	0	241	2
16	2003	237	0	228	0
17	2002	170	2	213	1
18	2001	197	3	180	0
19	2000	190	0	202	1
20	1999	180	0	195	0
Total		**2710**	**9**	**2620**	**18**
Average		**294**		**260**	

When the search is narrowed to "Cairns," it is evident that there is further academic literature that examines cyclones specific to this area. In reference to labor markets and associated terms (such as livelihoods and employment), there is some literature that explores these topics in relation to natural disasters in general, but very few that pursue associated research specific to cyclones (see Table 4.2).

This chapter seeks to address this gap in the literature and add to the understanding of the impacts of cyclones in Northeast Australia with a focus on employment, workers, and businesses.

4.2 BACKGROUND TO CYCLONES IN NORTHEAST AUSTRALIA

Cyclones, hurricanes, and typhoons! In the Northwest Pacific, they are known as "typhoons"; in the North Atlantic, central North Pacific, and eastern North Pacific, they are called "hurricanes"; and,

Table 4.2 Google Scholar Search for Research Relating to Cyclones/Natural Disasters and Livelihoods/Employment/Labor Markets.

	"Natural Disasters"	"Natural Disasters" + Australia	"Natural Disasters" + Cairns	"Natural Disasters" + Livelihoods/Labor Markets/Employment
Total	11,800	50	0	36
2009−2018	5840			
1999−2008	2710			
	"Cyclones"	"Cyclones" + Australia	"Cyclones" + Cairns	"Cyclones" + Livelihoods/ Labor Markets/Employment
Total	12,000	93	4	4
2009−2018	4540			
1999−2008	2620			

in the South Pacific and Indian Ocean, they are referred to as "cyclones." These powerful storms develop in oceans across the world and encroach on coastal areas, wreaking damage through various mechanisms, such as strong gusty winds, torrential rain, storm surges, flooding, and landslides.

It is anticipated that as climate change accelerates, the frequency and intensity of natural disasters will increase. Rising sea temperatures are said to impact on the direction, strength, development and impacts of cyclones. In 2018 the Northern Hemisphere experienced one of the most active seasons recorded, with 22 major hurricanes noted within 3 months (Nunez, 2019). In 2017 Cyclone Debbie was dubbed "Lazy Cyclone Debbie" as it progressed slowly across Queensland and NSW (SBS News, 2017). The slow movement of this system meant that exposure to high winds and heavy rain was extended and the potential for destruction intensified. Officially, Cyclone Debbie is recognized as Australia's second most expensive cyclone (Insurance Council of Australia, 2018). In this category, four cyclones crossed the coast at Airlie Beach on March 28 and in the week that followed caused extensive damage across the two states. In its wake, it left a path of physical destruction to homes, roads, businesses, community assets, and motor vehicles. As it traversed populated regions, its social impacts were complex and ongoing in the lives of affected people (Deloitte, 2017).

It was noted that the Queensland east coast has been hit by 207 cyclones since 1858, with the majority making landfall in North Queensland (Hind, 2011). Of these, 53 cyclones have hit Cairns, most with reported destruction to the city. The damaging impacts of a TC are caused by accompanying strong winds, coastal flooding brought on by heavy rain, and tidal surges. Undoubtedly, the greatest impact being loss of human life.

Cairns, founded in 1876, is affected by TCs almost every 2 years. The first recorded cyclone to hit was on March 8, 1878. McLeish (2017) discusses five of the most devastating cyclones to hit Queensland: TC Mahina (March 4, 1899), TC Mackay (January 21, 1918), TC Innisfail (March 10, 1918), TC Larry (March 20, 2006), and TC Yasi (February 3, 2011). Subsequent to Yasi, there have been smaller cyclones to hit the Northeast Australia region: TC Low (January 26, 2012), TC Marcia (February 20, 2015), Cyclone Debbie (March 28, 2017). The BoM (2019) contends that the incidence and severity of cyclones are reduced during El Niño years, especially those crossing the coast of Queensland. Currently, the El Niño—Southern Oscillation Index (ENSOI) for the tropical Pacific Ocean is "neutral" and that the number of TCs expected for the Coral Sea Region is below average, but the BoM (2019) warns that there may be other complicating global climate influences. So, for residents and communities of Northeast Australia, the potential for cyclones to develop and strike the region is an ever-present reality; and, being prepared for extreme weather conditions has become a part of life. Locals have grown to understand that such events will impact on all aspects of their lives, including work, income, and well-being.

4.3 IMPACT OF CYCLONES ON EMPLOYMENT AND LABOR MARKETS

Existing literature supports the relationship between labor markets and natural disasters, with some studies presenting research specific to cyclones (e.g., Belasen & Polachek, 2008; Chang-Richards, Seville, Wilkinson, & Walker, 2019; Wu, Xu, Liu, Guo, & Zhou, 2019). It suggests that there are direct and indirect effects that can have both temporary and long-term impacts on the workforce, businesses, and infrastructure, with effects varying between sectors/industries depending on the cyclone's strength and ensuant water damage. It is the interplay of environmental factors with socioeconomic ones that determines the impacts of cyclones. With disruption and damage to people's natural and physical environments the cyclone's influence over time on employment and the associated labor markets, earnings, and livelihoods is realized.

4.3.1 PRECYCLONE: CYCLONE WARNING ISSUED

Less commonly discussed is that the effects of cyclones can be experienced from the time that a cyclone warning has been issued and the cyclone begins to show signs of approaching the coastline. The unpredictability of the cyclone's landfall, path, and strength can make for a stressful and unsettled time for individuals. The safety and well-being of people are of immediate concern, beginning as soon as the threat of a cyclone is perceived. From the outset, individuals are encouraged to prepare for the impending cyclone. This activity includes getting oneself and family members home safely, packing an evacuation kit, and securing of properties. The latter can entail ensuring that there are no potential flying objects in one's property. People may also be asked to assist with preparing their workplaces. These activities can bring disruption to workplaces and people's ability to work and earn income, even before the cyclone makes landfall and irrespective of whether it eventuates (see Box 4.1).

It is not only individuals and workplaces but communities too that need to be prepared for natural disasters. Mallick, Rahaman, and Vogt (2011) maintain, "Communities that are well trained locally, culturally, socially and psychologically are better prepared and are more effective in responding to the aftermath effect of disasters" (p. 23). They argue that awareness of risks and strategies to alleviate those risks can empower individuals and communities to prepare and manage in the event of an impending natural disaster. Individual and collective disaster risk management plans can be of great value but are tested when cyclones arrive at peak intensity or take their time to make landfall or even perversely approach the coast and then retreat. Fitchett (2019) talked of how TC Idai sat for 6 days in the Mozambique Channel before making landfall, when flooding, gusty winds, and storm surges displaced thousands of people.

BOX 4.1 SCENARIO HIGHLIGHTING PREPARATION FOR CYCLONE AND IMPACT ON EMPLOYMENT

Dilemmas in the Face of Cyclone Debbie

"Hey Jo, we have just received notice that a cyclone is heading our way and we need to prepare the premises."

I looked at Gayle, the Restaurant Manager, in bewilderment and concern. My thoughts were racing. As the Restaurant Supervisor, I had responsibilities and actions that I now had to follow, despite that I worked only on weekends. I immediately realized that if the cyclone hit, it would mean that I would need to stay in the hotel so that the guests who were visiting the region were stranded here could be looked after. But there was my family to think of as well. With four young children in the home, there were preparations that needed attending. Fortunately, we were stocked up on essential supplies—dried and canned food, torches, batteries, radio, and water containers. The emergency kit was also well stocked—I always dreaded the panic run on supermarket shelves when a cyclone is on its way—but it was the yard and house that needed securing and locking down.

"Jo, are you listening!" Gayle continued, noticing that I had slid into my own reverie. "It's a Category 3 and it's moving fast. So, we need to make sure that all is secured." She began to walk away and motioned to grab the attention of the staff nearby who were clearing up the aftermath of a busy breakfast service. As an afterthought, she turned around and added, "And, it will mean that we will probably have to stay at the hotel until it is over." With that, she moved toward the waiting staff.

I followed her so that we could coordinate our activities. Gayle was an expert in these situations, and we valued her diligence and ability to coordinate and motivate us with fairness and firmness. Jojee and I were assigned to ensuring that the furniture in the al-fresco area was secured. Jojee was a casual who worked the breakfast shift three to four mornings per week. She was a long-term resident in Cairns, although originally from Thailand. We had connected because we were both university students. I knew that she would be worried about her dog, Sami. I was reminded of my own concerns, now having completed preparations at the restaurant, I began to think that if the cyclone hit, the river at the Freshwater junction would flood and it would mean that I wouldn't be able to get home for a couple of days. My thoughts jumped ahead. That would mean that I wouldn't see my 4-month-old little boy for days. I cursed the need to work!

The need for money and enhancement of career prospects had meant an early return to work. As a casual on permanent weekend work, I had no eligibility for maternity leave. That aside, I had a great opportunity for securing permanent full-time work and a career pathway with the hotel once I finished my degree. As it was part of an international chain, I was forever mindful of the career that could be carved, if I played my cards right. All that aside, I loved my work and I loved the people that I worked with.

"You worried about your family, Jo?" Jojee knew me too well.

"Yes," I said. "You must be worried about Sami?" I added.

Jojee nodded. We both fell silent as, lost in our own thoughts; we continued to busy ourselves with bringing in the tables and chairs.

It has been agreed that disaster preparedness and risk mitigation build more resilient communities (e.g., Mallick et al., 2011). Dora and Padhee (2019) reflected on Cyclone Fani that hit eastern India. They argue that there are four key areas where lessons need to be learnt: building of relief cyclone shelters, accuracy of early warning systems, clear communication plan, and effective coordination of groups. The impacts of TC Marcus when it hit Darwin in March 2018 highlight the consequences of not being adequately prepared and the lessons that should have been learnt from previous cyclone events (Surjan, Mathur, Lassa, & Mathew, 2018). For example, the extent and cost of the clean-up in Darwin was amplified by large trees with shallow roots that had been brought down and fallen overhead powerlines. Surjan et al. (2018) argue that such lessons should have been learnt from Cyclone Tracey—the trees were planted after the latter event and the powerlines should have been evaluated post-Cyclone Tracy.

4.3.2 **DURING THE CYCLONE**

As the cyclone advances on the coastline and people wait in anticipation of its landfall, most businesses are generally closed, and residents are confined to their homes or shelters. During this time, the degree of impact on employment and income is variable on different individuals. Some workers will be unable to access work as their workplaces are closed. For others (e.g., emergency workers, health workers, service workers), their work may begin to escalate in demand, intensity, and urgency, and those who volunteer their help may need to be ready to respond.

In Australia Government Fair Work Ombudsman (2019) recognizes, "Australians can experience a range of natural disasters, such as floods, bushfires, tropical cyclones, severe storms and even earthquakes. These events can cause devastation to communities and financial hardship for individuals and businesses." Subsequent to this statement is detailed for employers and employees, respectively, entitlements and reasonable actions in the event of natural disasters.

The Australia Government Fair Work Ombudsman (2019) also acknowledges that some employees may be members of recognized emergency organizations and that they are entitled to "unpaid community service leave" for activities associated with natural disasters. In some cases, awards and workplace agreements accommodate entitlements to income for employees engaged in community service. In other cases, employees forgo any income during the period of community service and still volunteer.

4.3.3 **POSTCYCLONE**

The postcyclone recovery period can bring both negative and positive impacts on labor markets and employment. As the effects of the cyclone are felt by individuals, businesses, and industries within affected regions, there can be an influx of workers and businesses as repairs and reconstruction gets underway. Equally, as homes and businesses are destroyed, there may be mass evacuations causing severe disruptions to individuals' and businesses' capacity for "having a job" (employment) and ability to "make a living" (earnings/income).

The recovery period is driven by several factors that influence the demand and supply of labor and determine the capacity, profile, and characteristics of the labor market. Factors include the individual's ability and willingness to work, the workplace and surrounds (e.g., damage sustained, and repairs needed), market demand, temporary employment availability, and economic recovery schemes. This supply and demand in the labor market fluctuate during the recovery period. Wu et al. (2019, p. 273) found that impacts of TCs on labor markets have an immediate negative impact, which is "gradually restored through postdisaster reconstruction." Similarly, Belasen and Polachek (2008) found that the impacts of hurricanes on labor markets are variable with time. Focusing on hurricanes in Florida, they noted that repairs to damages are generally completed within 2 years, but some effects can be enduring and widespread. The largest effects were consistent with the most severe hurricanes. Belasen and Polachek (2008) point out that impacts on labor markets are observed in the demand shocks experienced postdisaster. Specifically, they noted that while employment increased immediately after the hurricane and oscillated between decreases and increases from the first year, it eventually settled at a lower level than the precyclone point, by up to 4.76%.

The consequences to businesses evolve from the initial physical destruction to later productivity and labor availability (Leiter, Oberhofer, & Raschky, 2009). There is agreement that impacts on productivity and labor are negative, at least in the short-term. Business performance is compromised differently, dependent on the industry. For example, primary industries are particularly vulnerable to cyclone effects. They may suffer great losses due to physical damage to crops. The associated availability of work and flow on for employee remuneration may be severely affected. By comparison, secondary and tertiary industries (such as construction, trade, public service, facilities, transportation, and retail) can experience positive economic benefits. These may be short-lived for some businesses and industries, and in some cases, they can provide opportunities for change, update of material capital, implementation of new technologies, and innovation (Leiter et al., 2009; Wu et al., 2019). Leiter et al. (2009, p. 346) point out that after a natural disaster "physical capital accumulation increases for [some] companies ... and [the] natural catastrophe may induce investment activities in production factors that go beyond the sole replacement of disaster losses." This suggests that productivity and employment could increase in the long-term, subsequent to the "duration of the adjustment process" (Leiter et al., 2009, p. 347). This situation presents dilemmas and challenges in the face of forecasts predicting increases in frequency and intensity of natural disasters induced by climate change. There needs to be an expedient implementation of appropriate strategies that will boost the resilience and existence of businesses and industries. To not to do so will perpetuate negative impacts on labor markets and ultimately, individuals' employment and income, and the well-being of the community.

These fluctuations in labor markets can result in gain or loss of income and have implications for the purchasing power of individuals. Indeed, the literature indicates that disasters may weigh negatively on wages, with people being faced with potential unemployment, reemployment, and relocation (Groen, Kutzbach, & Polivka, 2015; Mueller & Quisumbing, 2011). Each of these, in turn, affects individuals' capacity to purchase and may impose budgetary constraints (Wu et al., 2019). Groen et al. (2015, 2) explored the long-term impact of hurricanes Katrina and Rita on individuals' employment and earnings. In a study that spanned 9 years, they found that individuals' earnings were reduced during the first-year posthurricane by 2.2% of average prehurricane income. They explained that this decline was reflective of disruptions that prevented individuals from working, such as "job separations, migration to other areas, and business contractions." As is generally, observed in storm-affected areas, physical damages can cause individuals to migrate out of the area and find accommodation in other areas. Such action may result in individuals being separated from their prestorm employment. Chang-Richards et al. (2019) also confirmed that relocation can result in changes to individuals' employment and options. Relocation requires settling into new neighborhoods and accessing opportunities there. It may be facilitated by a level of government support and aid, and by the broader economic climate, in the relocation site.

Storm damage can also cause disruptions to businesses and force them to close or reduce their operations because of direct and/or indirect causes, for example, reduction in demand for businesses' products/services. The extent of disruption may be variable between individuals and businesses and is dependent on the extent of the damage. Those individuals or businesses suffering major damage will experience the most severe short-term earnings losses and possible loss of employment for workers.

Belasen and Polachek (2008) also found that the workforce composition changed over time, largely through migration of people into and out of affected areas. They reported that this migration was dependent on factors such as individuals' income, skills, and educational levels. It was found that individuals with higher incomes had the means to move to areas where the damage was minimal. By contrast, individuals with lower incomes were confined to remaining in the disaster area. This migration of people impacted local labor markets. Examination of earnings relating to industry showed that the greatest short-term loss of earnings was associated with individuals employed in tourism-related organizations and those in population-related industries, such as healthcare and education (Belasen & Polachek, 2008). These losses were eventually followed by noticeable increases in earnings for these individuals. Groen et al. (2015) reported that for affected individuals the increase in earnings was 3.2% by the third year and 7.2% by the seventh year. This increase was consistent with wage growth rather than a rise in employment. Belasen and Polachek (2008) presented consistent findings for individuals and industry/sector in their study that examined the 19 hurricanes, which had hid Florida between 1988 and 2005. They found that employment decreased (up to 4.76%) and earnings increased (up to 4.35%), with more severe storms having the greatest consequences. Assessment of five industrial sectors showed that each sector behaved differently from the collective responses. Construction and service industries experienced positive outcomes. Negative shocks were found in the following: manufacturing; trade, transport, and utility; and, finance, investment, and real estate. The results consistently show that the effects of storms on the workforce and associated labor markets can be short-term as well as enduring and widespread. Hence, the response and coordination of initiatives relating to employment and livelihood is important and should be implemented in a timely manner.

Chang-Richards et al. (2019) emphasized the importance of appropriate and timely postdisaster recovery priorities. Their findings showed, "early recovery income support, the ability to transfer to other jobs/skills, physical and mental health, availability and timeliness of livelihood support, together with its cultural sensitivity and governance structure are amongst the most important factors" (Chang-Richards et al., 2019, p. 181). They talked of "sustained livelihoods" and how access to infrastructure and ensuring the safety of neighborhoods can impact positively on the resilience of the people and of livelihoods. The authors emphasized the importance of coordinated efforts in provision of external assistance and that these should be effectively managed as the needs of affected people changes. Such assistance should include checks on the health and well-being of individuals. The poststorm recovery period can result in loss of income and additional participation barriers, such as individuals' well-being and ability to cope. Leiter et al. (2009) highlight, "adverse consequences of natural disasters due to climate change will intensify human suffering" and that future research needs to examine "their impact on social welfare" (p. 347). Such research is necessary for the development of sustainable communities, which Mallick et al. (2011) describe as, "one that balances: social equality and equity, economic vitality, environmental responsibility and infrastructural effectiveness ... as well as community resilience" (p. 234). This research contributes to these discussions by focusing on the impacts of cyclones on employment and associated aspects, such as work and earnings.

4.4 METHODOLOGY

This chapter presents findings from a study that is part of a larger mixed-method research project. In this study, qualitative document analysis was used to find, collate, and analyze media articles about "cyclone damage" in Cairns. The data was collected by searching the State Library of

Queensland for relevant media articles in the local newspaper: *The Cairns Post*. This 132-year-old newspaper services the Cairns region exclusively and publishes an edition every weekday and *The Weekend Post* on Saturday.

Sourcing articles from *The Cairns Post* provided a unique opportunity to gather data on cyclones through the lenses of journalists and afforded a perspective that is often overlooked. In particular, in the event of cyclones, journalists themselves are subject to the forces of nature and face logistically demanding challenges in their attempt to report the news. In a 2007, news article, journalist Kylie Reghenzani, reported interviews with four journalists and their associated crew to present insights into the challenges faced by them when out in the field in the lead up to, during and after a cyclone. This authenticity of experiences can bring value to portrayal of events.

The validity of using newspaper articles as data is confirmed by the literature. For example, Albrecht (2017) highlighted the key role the media play as vital sources of information for the public during natural disasters. Often, they are the only providers of "in-depth analysis of critical issues" (Albrecht, 2017, p. 43). Bohensky and Leitch (2014) also supported media analysis in their examination of resilience in the Brisbane floods of 2011. They made similar points to Albrecht (2017) and added that newspaper articles capture a range of perspectives because while written by journalists, articles include "multiple actors as sources of information" (p. 478).

Separate searches were conducted in the newspaper using the key terms of "cyclone," "cyclone damage," "disaster management," and "disaster." Of these searches the one using the term "cyclone damage" turned out to be the most relevant and useful. It returned 208 results for the period 2004−19. This period was chosen because it was the time span available on the State Library of Queensland site.

The 208 articles were saved in NVivo, where the documents were further assessed in terms of relevance and analyzed to create nodes of the themes that were emerging from the articles. Both authors worked individually and then, collaboratively to collect and collate the articles and to verify the relevance of the articles in NVivo and, the veracity of the nodes. In the end, 161 articles were considered valid (Table 4.3).

The greater number of articles were published in years when the most severe and destructive cyclones hit. For example, Table 4.3 shows that nearly a quarter of the articles analyzed were written in 2006, the year that TC Larry hit. The media coverage continued into 2007, when close to another 10% of articles relating to the cyclone were written. Similarly, 21.74% of the articles were written in 2011, the year that TC Yasi hit. Cyclones noted in Table 4.3 were highlighted in respective articles. The next section discusses the results in more detail.

4.5 FINDINGS AND DISCUSSION

4.5.1 OVERALL RESULTS

In an initial analysis of the articles a word cloud was formed to capture the most common words that were presented in all collated articles (Fig. 4.1). These results show that the words appearing most frequently were: insurance, place, Innisfail, conditions, 2006, Larry, people, 2017, Yasi, million, coast, council, industry, region, and homes. Each of these had a count of greater than 100 mentions.

Table 4.3 Media Articles Analyzed per Year, Highlighting the Most Severe and Destructive Cyclones.

Year	Cyclone	Count	Percentage
2004		2	1.24
2005		1	0.62
2006	Larry	38	23.6
2007		16	9.94
2008		2	1.24
2009	Charlotte	3	1.86
2010	Olga	5	3.11
2011	Yasi	35	21.74
2012		7	4.35
2013		6	3.73
2014	Ita	12	7.45
2015	Marcia	8	4.97
2016		13	8.07
2017	Debbie	5	3.11
2018		3	1.86
2019		5	3.11
Total		**161**	**100**

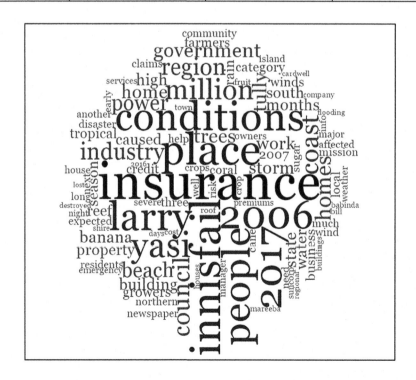

FIGURE 4.1

Most common words appearing in all collated articles.

Further analysis showed that there were 32 main themes that emerged from the data. These included: dwellings and construction, businesses, ecosystem damage, employment, food supplies, feelings aroused by cyclones, historical occurrence, insurance, primary production, and public infrastructure (Fig. 4.2). Fig. 4.2 presents a cluster analysis of the themes, capturing the similarity between the themes. The most populated theme was "damage" with 69 references; followed by "employment & workers" with 44 references; and "recovery" with 40 references (see Fig. 4.3). The NVivo diagram reported that the strongest Pearson correlation coefficient was for the similarity between "dwellings & construction" and "businesses" at $r = 0.575$. All other coefficients were less than 0.476 ("recovery" with "damage").

From this overview of the results the chapter continues with a focus on the theme of "employment & workers," drawing on the areas highlighted in collated articles for this theme.

4.6 EMPLOYMENT—WORK, WORKERS, AND BUSINESSES

The newspaper articles showcase the extent of physical damage, portraying destruction to people's homes, trees, buildings, business' premises, crops, fishing vessels, and public buildings (e.g., schools and hospitals). While the articles don't articulate specifically impacts of physical destruction, they paint a picture of the extent and magnitude of that destruction. From this portrayal, it makes sense that the everyday and working lives of people involved in these affected communities were disrupted, as the literature above had indicated. As an example, a 2011 article notes, "Yasi didn't just damage houses. It crippled entire industries and ruined livelihoods." Yet, despite the destruction, there was optimism. The article added, "With the arrival of the dry season the rebuilding effort was well underway, and local builders had taken on extra staff to complete the enormous workload." And so, the commentary painted a stark picture of damage and destruction in the aftermath of cyclones but equally, their narratives spoke of the "phoenix arising from the ashes" as the clean-up and rebuilding ensued. The fortitude of the people is best captured in the 2007 article "One tinnie replaces another . . ." in which the publican is noted as saying: "We've got a few boats tied up here and some others have phoned in to say they're on their way . . . It's a bit of a ritual here in flood times, while we've still got power on, we'll be pouring beers." This example of the positive outlook of the people in Tully when flooding prevented them from accessing their workplaces highlights the spirit and resilience of the people. It was no surprise then that the highest Pearson correlation coefficient (Table 4.4) was observed between "recovery" and "employment & workers" ($r = 0.380$). Of note is that there were no negative correlations between "employment & workers" and any of the other themes.

To better understand the impacts of cyclones on employment, articles coded for "employment & workers" were analyzed for occurrence of key words. The results (Fig. 4.4) show that reporting of the following was most frequent: industry, people, staff, banana, work, Yasi, builders, council, emergency, Innisfail, Larry, and local. Following close behind was business, government, and growers. The reference to industry was strong in many articles. Statistics for the region show that the industry with the most employment in 2018/19 is Healthcare and Social Assistance (National Institute of Economic and Industry Research, 2019) with 14.9% of the workforce. Following that industry, there is construction (11.5%), retail (10.8%), tourism (accommodation and food services, 9.9%), and education and training (8.6%) as the largest industries in the region.

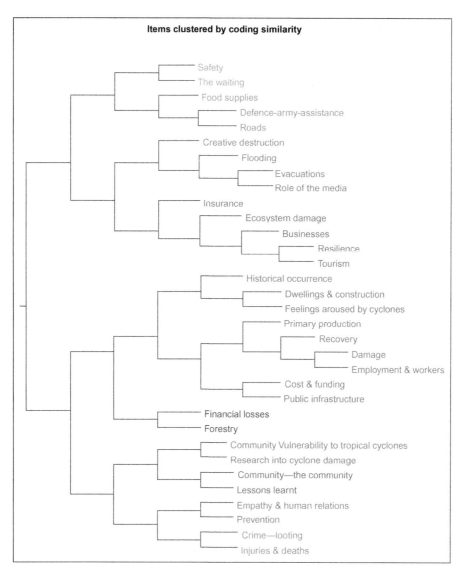

FIGURE 4.2

Themes clustered by coding similarity.

In the aftermath of cyclones, "people" want to get back to "work" as soon as possible. With potentially substantial losses to property and possessions, people are keen to rebuild and have the income to do so. They are also concerned about the potential loss of their jobs. Equally, employees and employers want to return to ongoing employment and business,

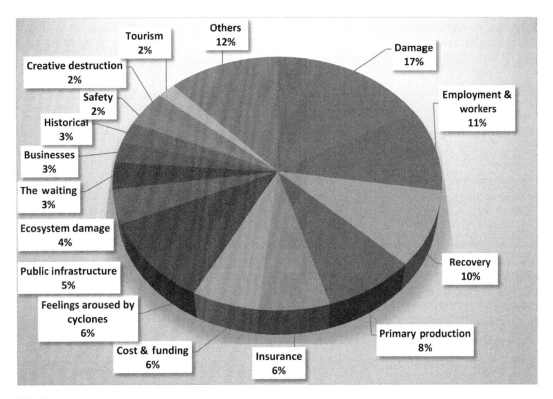

FIGURE 4.3

Themes presented as percentage of overall importance.

respectively, as soon as possible. Government support is often sought and forthcoming as it can markedly assist and expedite businesses and people during the recovery period. A 2011 article presented a statement from a tourism business owner, evidencing the plight of businesses. It stated:

> At the moment, we have about 30 staff we have to let go unless the Government steps in with wage assistance like they did for the banana industry after cyclone Larry … [and] owners are seeking government help to keep the gates open and their staff in a job after the tourism icon suffered significant damage when cyclone Yasi struck.

This plea echoes the far-reaching impacts of cyclones and the similarities and differences in severity of damage and recovery of natural disasters between different industries and communities, a point emphasized by Ranke (2016).

This latter sequence of key findings led to further exploration of the data for the themes of "businesses" (Fig. 4.4) and "primary production" (Fig. 4.5). Fig. 4.5 shows that the most prominent

Table 4.4 Correlations Between "Employment & Workers" and Various Themes.

Code A	Code B	Pearson Correlation Coefficient
Recovery	Employment & workers	0.380
Primary production	Employment & workers	0.360
Cost & funding	Employment & workers	0.339
Creative destruction	Employment & workers	0.306
Damage	Employment & workers	0.255
Businesses	Employment & workers	0.224
Public infrastructure	Employment & workers	0.218
Feelings aroused by cyclones	Employment & workers	0.217
Tourism	Employment & workers	0.187
Community	Employment & workers	0.176
Ecosystem damage	Employment & workers	0.174
Insurance	Employment & workers	0.169
Forestry	Employment & workers	0.149
Dwellings & construction	Employment & workers	0.139
The waiting	Employment & workers	0.136
Crime—looting	Employment & workers	0.099
Injuries & deaths	Employment & workers	0.094
Roads	Employment & workers	0.092
Safety	Employment & workers	0.080
Role of media	Employment & workers	0.079
Prevention	Employment & workers	0.075
Food supplies	Employment & workers	0.049
Historical occurrence	Employment & workers	0.044
Empathy & human relations	Employment & workers	0.041
Lessons learnt	Employment & workers	0.036

aspects associated with businesses are industry (again), private, bookings, business, open, and buildings. Sometimes, businesses experienced greater effects as one roadhouse owner commented, "I've now got to tell my workers—one who's been here for 17 years—that there's going to be no wages for some months."

Similarly, impacts were severe for primary industries, with one article reporting:

> The massive loss means up to 4000 banana industry workers could find themselves without a job ... [and quoting a farmer] ... "There's going to be a flow-on through to the towns and the supermarkets, people are going to move away, your transport operators - they're going to be out of production for six, seven months, the same as us."

FIGURE 4.4

Word cloud for theme of "employment & workers."

Articles also indicated that once businesses were up and running again, they were keen to draw customers back and let people know that they were "open." For example, a 2011 article highlighted the urgency of the tourism industry to regain visitation. It reported:

> ... serious damage was done to the [tourism] industry when images of destruction were beamed across the world. Interstate advertising campaigns urging Australians to holiday in the region to support locals have gone some way to helping keep businesses afloat. "We've got to understand that basically we're not going to break any records, we're basically trying to maintain the level of visitation we have," Tourism Tropical North Queensland Chief said.

The reality is that businesses do not always come back to the scale of existence prior to the cyclone or that they come back at all, as pointed out in a 2011 article, "Over half of the businesses affected by Cyclone Yasi are now open or plan to be by the end of the month—some on a smaller scale than before as they rebuild." Another article, from 2013, added to this aspect and highlighted impacts on employment when people leave the region because of cyclones:

> In Tully and Cardwell, two of the worst hit communities, there are concerns that people have left and are not coming back. There is a sense that this drift in population is continuing and the economies of these centres will suffer further unless something is done to reinvigorate employment.—January. 28, 2013

FIGURE 4.5

Word cloud for theme of "businesses."

Such comments support points that were raised in the literature earlier.

Fig. 4.6 shows that the news was focused on aspects such as: cane, farmers, industry, crops, Larry, growers, sugar, banana, Yasi, season, lost, and region. Much was written about Cyclone Larry (2006), which was a large system that brought severe destruction, with articles highlighting impacts on farmers, farmworkers, and income. One reporter noted, "Cyclone Larry obliterated crops from the coast to the Tableland, costing millions in lost income, and forcing farmers to the brink." The plight of farmers was a repeated story, with one standout being a 2011 newspaper article—"Farms of wrath and heartache" (anonymous)—which presented comprehensive coverage of all the primary industries affected, for example, timber, pineapples, cattle, sugar, and coffee. It also draws attention to "growers," "industry," "region," "worth," and "production."

Mallick et al. (2011) discuss the increasing focus on society in research that examines natural hazards and disasters, often referred to as "social disaster." The impacts of natural disasters are also determined by the social situation and capability of people and communities. Ranke (2016) warns, "When a natural disaster strikes, the most vulnerable groups are the poor, disabled, elderly, and the young" (p. 38). He compares developing countries to those with high-income and adds that coping capacities, adaptation, and resilience will vary with different societies. Similarly, Mallick et al. (2011) contend, "Though the resilience of an individual depends on socioeconomic and political conditions in the respective community, it explores also the degree of vulnerability of the individual

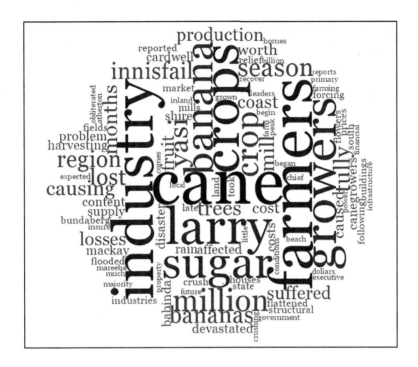

FIGURE 4.6

Word cloud for theme of "primary production."

or the family to a disastrous situation" (p. 223). The communities of Northeast Australia have experienced varying degrees of destruction over the years, with support often presented by the national government and sometimes, opportunities to rebuild arising from unexpected areas. The concept of "creative destruction" affords an explanation for such chances as the latter.

Chaiechi (2014) draws on the 1940s derivation by Joseph Schumpeter of "creative destruction" to apply the term to recovery after a disastrous event. She argues that while the immediate impact of natural disasters can be destructive, they can provide opportunities for long-term economic benefits. In terms of "employment & workers," the articles revealed that while some businesses and workers were unduly impacted, the business of other organizations (e.g., insurance companies) and the lives of other workers were enriched. A 2006 article noted:

Although there will be a population outflow due to job losses in the farm industries, there will also be a large influx of construction workers for the rebuilding process.

With thousands of homes requiring repairs and possibly another 500 needing to be demolished and rebuilt as a result of irreparable damage, the size of the reconstruction task will be enormous.

The construction worker influx will maintain the pressure on an already depleted supply of housing while the rebuilding takes place.

Similar comments were made in 2011 article:

> A bit of it is the revival in the building industry, particularly in cyclone-related work ... one good sign was that people who had received insurance payouts for cyclone damage were now employing local people to do the rebuilding work ... [and] jobs in the banana industry were rebounding as well.

Another 2011 article notes, "hundreds of emergency service workers clearing debris and Ergon Energy crews restoring power" and that even the Army arrived to assist with clean-up and repairs to roads. One article reported, "Twenty-one extra paramedics have been flown to the region − 17 for Innisfail, and four in Cairns and the Tableland."

Chaiechi's (2014) work showed that natural disasters promote innovation postdisaster and represent progress. This result is evident in a 2011 article that featured a story on an initiative to "export cyclone-damaged forest products" because commercial use had been found for "Thousands of plantation trees, knocked over in the cyclone ... to be exported as logs or woodchip." The author of the article added, "Cyclone Yasi's devastation has delivered a $20 million export trade and a jobs boost for the port of Mourilyan [and] the latest development would create at least 40 fulltime and casual jobs." Another article summed up the postcyclone phase by emphasizing:

> ... the new money in the district - from an influx of tradesmen, government employees and government funding - has benefitted everyone from cafe owners to accommodation houses and building suppliers... We lost some businesses to cyclone damage and only a low percentage of businesses were adequately insured for business interruption, so it's been tough for people ... But we've also gained new businesses, and while we've lost people who never returned after Cyclone Larry took homes and jobs, we've also gained people.

Articles also indicated that cyclones can provide ongoing work for some people. For example, a 2013 article reported that researchers at the cyclone testing station had developed removable units, "designed to record and store data" relating to cyclonic weather "anchored to concrete housings in strategically placed locations along the northern coastline," in an attempt to better understand cyclones. Such stories of resilience were evident throughout the articles. A 2016 article recognized the importance of "mitigation and resilience as the only long-term solution for protecting north Queensland communities from cyclone risk." Ultimately, it is the spirit of the people that will sustain communities, as depicted in separate 2007 and 2011 articles showcasing Innisfail's Feast of the Senses, with the former noting, "Twelve months after category 5 Cyclone Larry shut the event down early and wrought havoc throughout the region, a bigger Feast of the Senses will be staged to celebrate Innisfail's resilience." These stories showcase the tenacity of the people and communities of Northeast Australia.

4.7 CONCLUSION

This chapter has presented a snapshot of the impacts of cyclones on employment and associated issues for workers, businesses, and communities in Northeast Australia. By capturing data from newspaper articles, it has been able to portray an understanding of the extent and complexity of the issues facing people in the path and aftermath of cyclones. The tapestry of accounts weaves a picture of the uncertain nature of cyclones. Their wrath and power are unpredictable, and survival can be a precarious balance between life, home, and work. For the people of Northeast Australia, cyclones are a part of life and, therefore, they experience interruptions to employment, livelihoods, and income. The disruption is sometimes short-lived, and the recovery is quick; at other times, the recovery is long, painful, and costly. For some, for example, emergency personnel and construction workers, the recovery period is a time of heightened activity when their expertise is called upon to assist. Throughout it all, the spirit and resolve of the people are evident.

Newspaper articles afforded a grassroots view of the challenges, emotions, and resilience of those impacted. This lens of the journalist is often neglected in research but here it shows that their positionality lends advantage for in-depth capturing of critical events with coverage that the public trust and seek. The role of traditional media is crucial at times of cyclone events and other natural disasters. They are an invaluable source of information and sometimes, the only source for the public and researchers, alike.

This document analysis derived from the newspaper articles showed that cyclones frequent Northeast Australia and highlighted the effects on businesses and workers. It disclosed the vulnerability of people in the face of the most destructive cyclones, such as TC Larry and TC Yasi. It also emphasized that the lessons to be learnt are ongoing, especially as climate change increasingly affects the occurrence and severity of cyclones. Furthermore, the results underscore that a collective approach from all levels—local, regional, state, and national—is required to rebuild and expedite recovery so that wages and livelihoods can be resurrected expediently. From here, sustained holistic and shared involvement should guide preparative and innovative practices for dealing with future cyclones.

REFERENCES

Albrecht, F. (2017). *The social and political impact of natural disasters. Investigating attitudes and media coverage in the wake of disasters. Digital comprehensive summaries of Uppsala dissertations from the Faculty of Social Sciences.* Uppsala: Acta Universitatis Upsaliensis.

Australia Government Fair Work Ombudsman. (2019). *Employment conditions during natural disasters and emergencies.* Retrieved from <https://www.fairwork.gov.au/how-we-will-help/templates-and-guides/fact-sheets/rights-and-obligations/employment-conditions-during-natural-disasters-and-emergencies>.

Belasen, A., & Polachek, S. (2008). *How hurricanes affect employment and wages in local labour markets. Discussion paper no. 347.* Bonn: Institute of Labour Economics (IZA).

Bohensky, E., & Leitch, A. (2014). Framing the flood: A media analysis of themes of resilience in the 2011 Brisbane flood. *Regional Environmental Change, 14,* 475−488.

Bureau of Meteorology (BoM). (2018). *Joint industry project for objective tropical cyclone reanalysis: Final report.* Retrieved from <http://www.bom.gov.au/cyclone/history/database/OTCR-JIP_FinalReport_V1.3_public.pdf>.

Bureau of Meteorology (BoM). (2019). *Past tropical cyclones.* Retrieved from <http://www.bom.gov.au/cyclone/tropical-cyclone-knowledge-centre/history/past-tropical-cyclones/>.

Chaiechi, T. (2014). The broken window: Fallacy or fact—A Kaleckian—Post Keynesian approach. *Economic Modelling, 39,* 195−203.

Chang-Richards, A., Seville, E., Wilkinson, S., & Walker, B. (2019). Effects of disasters on displaced workers. In A. Asgary (Ed.), *Resettlement challenges for displaced populations and refugees* (pp. 185−195). Sustainable Development Goals Series. Available from https://doi.org/10.1007/978-3-319-92498-4_14.

Deloitte. (2017). *A silver lining: Queensland Business Outlook*. Retrieved from <https://www2.deloitte.com/content/dam/Deloitte/au/Documents/Economics/deloitte-au-economics-queensland-business-outlook-silver-lining-030517.pdf>.

Dora, M., & Padhee, A. (May 13, 2019). India's Cyclone Fani recovery offers the world lessons in disaster preparedness. *The Conversation*. Retrieved from <https://theconversation.com/indias-cyclone-fani-recovery-offers-the-world-lessons-in-disaster-preparedness-116870>.

Fitchett, J. (March 21, 2019). Tropical Cyclone Idai: The storm that knew no boundaries. *The Conversation*. Retrieved from <https://theconversation.com/tropical-cyclone-idai-the-storm-that-knew-no-boundaries-113931>.

Groen, M., Kutzbach, J., & Polivka, A. (2015). Storms and jobs: The effect of hurricanes on individuals' employment and earnings over the long term. In *U.S. Census Bureau Center for economic studies paper CES-WP-15-21R*. https://doi.org/10.2139/ssrn.2782038.

Hind, K. (2011). *Far North Queensland cyclones*. Retrieved from <http://blogs.slq.qld.gov.au/jol/2011/11/14/far-north-queensland-cyclones/>.

Insurance Council of Australia. (2018). *ICA response to ACCC issues paper   Northern Australia Insurance Inquiry*. Retrieved from <https://www.insurancecouncil.com.au/assets/submission/2018/ICA_ACCC_ SUB_FINAL.pdf>.

IPCC. (2014). Summary for policymakers. In C. B. Field, V. R. Barros, D. J. Dokken, K. J. Mach, M. D. Mastrandrea, T. E. Bilir, & ... L. L. White (Eds.), Climate Change 2014: Impacts, adaptation, and vulnerability. Part A: Global and sectoral aspects. Contribution of working group II to the fifth assessment report of the Intergovernmental Panel on Climate Change (pp. 1−32). Cambridge and New York: Cambridge University Press.

Leiter, A., Oberhofer, H., & Raschky, P. (2009). Creative disasters? Flooding effects on capital, labour and productivity within European firms. *Environmental and Resource Economics*, *43*(3), 333−350. Available from https://doi.org/10.1007/s10640-009-9273-9.

Mallick, B., Rahaman, K. R., & Vogt, J. (2011). Social vulnerability analysis for sustainable disaster mitigation planning in coastal Bangladesh. *Disaster Prevention and Management*, *20*(3), 220−237.

McLeish, K. (2017). *Cyclones in Queensland: Learning from five of the worst ever recorded*. Retrieved from <https://www.abc.net.au/news/2017-03-27/learning-from-history-five-of-the-worst-recorded-cyclones/8389558>.

Mueller, V., & Quisumbing, A. (2011). How resilient are labour markets to natural disasters? The case of the 1998 Bangladesh flood. *Journal of Development Studies*, *47*(12), 1954−1971.

National Institute of Economic and Industry Research. (2019). *Employment by industry*. Retrieved from <https://economy.id.com.au/cairns/employment-by-industry>.

Nunez, C. (2019). *Hurricanes, cyclones, and typhoons, explained*. Retrieved from <https://www.nationalgeographic.com/environment/natural-disasters/hurricanes/>.

Ranke, U. (2016). *Natural disaster risk management: Geosciences and social responsibility*. London: Springer.

Reghenzani, K. (March 19, 2007). The media. *The Cairns Post*. Retrieved from <http://ezproxy.slq.qld.gov.au/login?url = https://search.proquest.com/docview/376890444>.

SBS News. (2017). *Lazy Cyclone Debbie to pack punch: Expert*. Retrieved from <https://www.sbs.com.au/news/lazy-cyclone-debbie-to-pack-punch-expert>.

Surjan, A., Mathur, D., Lassa, J., & Mathew, S. (March 27, 2018). Lessons not learned: Darwin's paying the price after Cyclone Marcus. *The Conversation*. Retrieved from <https://theconversation.com/lessons-not-learned-darwins-paying-the-price-after-cyclone-marcus-93862>.

Wu, X., Xu, Z., Liu, H., Guo, J., & Zhou, L. (2019). What are the impacts of tropical cyclones on employment? An analysis based on meta-regression. *Weather, Climate, and Society*, *11*(2), 259−275.

NATURAL DISASTERS, GEOGRAPHY, AND INTERNATIONAL TOURISM

Luke Emeka Okafor[1], Ogechi Adeola[2] and Oludele Folarin[3]

[1]*School of Economics, University of Nottingham Malaysia, Jalan Broga, Semenyih, Malaysia* [2]*Lagos Business School, Pan-Atlantic University, Lagos, Nigeria* [3]*Department of Economics, University of Ibadan, Ibadan, Nigeria*

5.1 INTRODUCTION

Tourism sector plays a vital role as a driver of economic growth and development in most countries (Khalid, Okafor, & Aziz, 2019; Khalid, Okafor, & Shafiullah, 2019; Okafor & Teo, 2019; Shafiullah, Okafor, & Khalid, 2018), as it helps to create employment, increase income as well as generate demand and growth in many other sectors. Globally, this assertion is consistent with the report by WTTC (2019) showing that the sector contributed 10.4% of global GDP and 10% of total employment in 2018 using data from 185 countries and 25 regions. The tourism sector is, however, heavily dependent on climate, seawater, sunlight, and natural landscape (Wijaya & Furqan, 2018). While there exists empirical evidence that natural disasters have a negative impact on economic growth (Bergholt & Lujala, 2012; Noy, 2009), little is known about the link between natural disasters weighted by population affected and international tourist flows as well as the role of geography in the underlying relationship. The tourism sector is potentially vulnerable to natural disasters, as a significant fraction of tourists would likely cancel their trips and postpone or change their destinations to safe alternative destinations in the event of natural disasters. In general, natural disasters dampen international tourism. This is consistent with the evidence that the occurrence of natural disaster reduces tourist arrivals (e.g., Chiou, Huang, Tsai, Lin, & Yu, 2013; Park & Reisinger, 2010). Previous studies, however, fail to account for the time-weighted population affected by natural disasters. This suggests that the effects of natural disasters on tourism may not necessarily be negative. If a smaller fraction of the time-weighted population is affected by a natural disaster and the effect of the natural disaster was temporary, then, tourism flow is likely to increase and/or be unaffected.

In view of the scant literature on the effects of different natural disasters on tourist flows, the purpose of this chapter is to explore the impacts of natural disasters on international tourist flows. This includes examining the moderating role of geography on the natural disaster–tourism nexus. This is especially important as the effects of natural disasters on tourist flows are likely to be region specific. This is consistent with the notion that time-weighted populations affected by different natural disasters in different regions are likely to be different.

The tourism sector is regarded as the largest nongovernmental economic activity in the world, and it is important for local economies globally (Maditinos & Vassiliadis, 2008). The natural characteristics of local environments are a key tourist attraction, and the occurrence of natural disasters in such geographical locations might hinder the inward movement of people into such territories. The impact of tourism on economic growth and development might be adversely affected if the impacts of natural disasters on tourism flows and their implications on the attraction of tourists are not addressed.

Tourist areas are vulnerable to disasters and crises due to high rate of movement of people inward and outward, and with the interdependent nature of the world as a result of globalization, the impacts of natural disasters are going to be more severe in the coming years (Maditinos & Vassiliadis, 2008; Scott, Gössling, & Hall, 2012). In the presence of increasing population, environmental degradation, and climate change, natural disasters occur more frequently (Hastley, 2011). Understanding the impacts of different natural disasters on international tourism, especially when adjusted by population affected, has significant policy implications. It is likely that the effects of different natural disasters on international tourism flows vary by regions. This suggests that a one-size-fits-all approach in terms of policy formulation is potentially undesirable. Hence, understanding natural disaster life cycle, vulnerable regions, and how it can be managed to prevent economic distress is important. The multifaceted scope of tourism enhances its socio-structural impact as people travel around the world not only for pleasure but also for religious, health, educational, and business purposes. The multifaceted scope of modern travel has contributed to the growth of international tourism (Scott et al., 2012). Over the years, international tourism has experienced tremendous growth, with the United Nation World Tourism Organization (UNWTO) (2011) estimating that tourist arrivals would exceed 1.8 billion by 2030. This represents around four times the number of tourists traveling presently. The significant growth in international tourism has benefited local tourist sites with infrastructural amenities and expansion in local businesses, though the increased presence of environmental hazards cannot be ignored. Consequently, as the tourism sector grows at a geometric rate, the perceived risk of natural disasters by travelers is also on the increase (Park & Reisinger, 2010). Furthermore, the perception of risk differs by gender, age, culture, and nationality (Park & Reisinger, 2010). Studies such as Huan, Beaman, and Shelby (2004) and Seraphin (2019) reveal that fear and uncertainty reduce tourist attraction to countries with issues of an earthquake, hurricane, and other climate-related natural disasters, especially in situations where low rate of survival had been recorded.

In the past few years, natural disasters have been on the increase in Africa, Asia, and Europe, according to findings and predictions from literature (e.g., Chiou et al., 2013; Huan et al., 2004; Park & Reisinger, 2010; Seraphin, 2019). The global rise in the frequency of natural disasters and its impact on tourist flows and by extension economic growth presents new challenges in terms of effective international tourism management.

This chapter contributes to the literature that relates to natural disasters and tourism in three ways. First, the study shows that the effects of natural disasters on international tourism are region dependent. For instance, volcanoes have a negative impact on tourist flows in America, Latin America and Caribbean (LATCA), Asia, Middle East, North Africa, and Oceania, whereas its impact in sub-Saharan Africa (SSA) is positive. This finding can be explained by a larger fraction of the time-weighted population affected by volcanoes in America, LATCA, and other regions compared to SSA. This suggests that volcanic eruptions are less devastating to property and human life in SSA compared to other regions. This finding is consistent with the evidence that over the period 1900−2008 the Americas and Southeast Asia accounted for over 20% of volcanic eruptions, whereas Europe and Africa's regions accounted for less than 10%, respectively. In terms of loss of life, 73% of the deaths over the period occurred in the Americas. This represents a significant share of the number of deaths (Doocy, Daniels, Dooling, & Gorokhovich, 2013). Second, the study shows that the effects of some natural disasters on tourist flows disappear after a year in some regions. For instance, population affected by storm has a negative impact on tourist flows in Europe contemporaneously, but its impact is statistically insignificant after 1 year. Third, a larger number of proxies of natural disasters are employed in this study compared to extant studies in this strand of literature. In addition, time-weighted population affected by the natural disasters is used in the study to identify a more robust effect of natural disasters on tourist flows.

The next section presents the review of related literature, Section 5.3 discusses the data and measures, Section 5.4 presents the empirical model, and Section 5.5 discusses the findings of the chapter, while the final section concludes.

5.2 A REVIEW OF THE LITERATURE

It has been established in the literature that climate change and/or natural disaster has detrimental effects on the environment and to some extent tourism (Breiling, 2016; Gossling, Scott, & Hall, 2015). Some of these detrimental effects include the risk of flood and forest fires, rise in beach erosion, and scarcity of drinkable water, among others (Isik, Dogru, & Turk, 2018). In addition, the Intergovernmental Panel on Climate Change (2007) report that there is practical evidence from all continents and oceans, indicating the effect of regional climate changes on natural systems. Earlier studies (Gössling, Bredberg, Randow, Sandström, & Svensson, 2006; Lise & Tol, 2002) have found that climate-related variables such as humidity, temperature, and rain determine tourist travel flows. Likewise, Becken and Wilson (2013) investigated the impact of weather on international tourist visits to New Zealand and found that there were high levels of changes made on trips.

Interestingly, geographical size is among the several factors highlighted by the World Bank that affect a country's vulnerability to natural disasters (Jones, 2014). For example, studies such as March, Saurí, and Llurdés (2014) and Wijaya and Furqan (2018) posit that coastal areas, in particular, are destinations that are often adversely affected by natural disasters such as flood, rise in sea level, extreme weather, storm, and coastal erosion. However, empirical study of the effects of natural disasters on international tourism as well as the underlying role of geography has remained far underresearch. This study thus fills the gap.

Tourist arrivals in a destination may not be limited to the availability of recreation sites alone as people travel in and out of nations for different purposes. Nevertheless, the importance of recreation sites in attracting travelers cannot be underemphasized (Chiou et al., 2013). The source of tourist attraction of a nation can be built around monumental sites, economic activities, or religious institutions; change in weather, transportation accessibility, economic environment, travel duration. Natural disasters, however, tend to discourage international tourism (Albalate & Bel, 2010; Chiou et al., 2013; Cho, 2003; Taylor & Ortiz, 2009). In recent times, studies are beginning to give attention to natural disasters as they increase in rate, causalities, and perceived effects on the larger society. This is imperative because of the role tourism plays in the economic growth and development of nations around the world (Adnan Hye & Ali Khan, 2013; Fayissa, Nsiah, & Tadasse, 2008). Unexpected natural disaster in a nation that depends on tourism for its economic growth and development would have multiple socioeconomic impacts on their developmental agenda and would require conscious effort to alleviate fear and communicate safety to intending tourists (Park & Reisinger, 2010).

Consequently, a natural disaster can cause severe damage to human and economic resources, and for this reason, tourists are conscious of the environment with the antecedent of natural disaster (Park & Reisinger, 2010; Tembata & Takeuchi, 2019). In the case of major devastation resulting from a natural disaster, adequate destination disaster management strategies are needed if tourist trust in destination would be restored (Murphy & Bayley, 1989; Seraphin, 2019). A natural disaster is an outcome of nature such as weather, which causes huge damage to the physical environment and threatens the safety of people (De Almeida & Machado, 2019). Natural disasters occur suddenly, and as a result, it is difficult to put in place adequate plans to deal with different types of natural disasters. For instance, a total of 6457 weather-related disasters have been recorded worldwide between 1995 and 2015 (Tembata & Takeuchi, 2019). Natural disasters have resulted in economic damages worth $1891 billion and affected over 250 million people annually (Tembata & Takeuchi, 2019), some of which are floods and wildfires (Tembata & Takeuchi, 2019); hurricane (Seraphin, 2019); wildfires (Maditinos & Vassiliadis, 2008); and storms, earthquakes, and extreme temperature (Botzen, Deschenes, & Sanders, 2019; Felbermayr & Gröschl, 2014). According to Tembata and Takeuchi (2019), flood is the most frequent type of natural disaster with a 43% rate of occurrence. They reported that 2.3 million people have been affected by floods with economic loss of over $662 billion and with a projection of an increase in flooding in developing nations and some other parts of the world. This trend calls for urgent action to limit the occurrence of natural disasters, especially those that are caused by human actions. As shown in Fig. 5.1, there is a link between climate change, natural disaster, geography, and tourism. In the modern world, natural

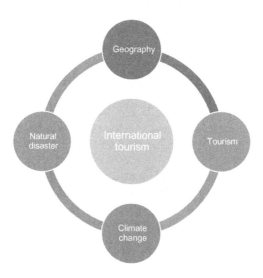

FIGURE 5.1

Showing interaction between geography, tourism, climate change, and natural disaster. Proposed by the authors. Authors in this context refer to the authors of this manuscript.

disasters occur frequently due to climate change. As noted earlier, the frequency and intensity of natural disasters are likely to vary by regions. As a result, the population affected by natural disasters and concomitant impacts on tourist flows potentially varies by regions.

In terms of geography, one of the few regions that are tourism dependent and are a victim of natural disaster is the Madeira (De Almeida & Machado, 2019). The tourism industry has been responsible for generating employment and adding value to the community, which has made the region vulnerable to climatic and socioeconomic changes. The natural endowment of an environment might make it a tourist hub such as having natural habitats, nondisruptive weather, a natural landscape, and a host of other comfort appealing factors. Hitherto, tourism and natural disasters are like two mutually exclusive outcomes (Beattie, 1992; Santana, 2004; De Almeida & Machado, 2019). The images of tourist environment in the mind of tourists would always be about fun, relaxation, beautiful surroundings, and sightseeing (De Almeida & Machado, 2019), while natural disaster would bring the image of death, horror, and tragedy (De Almeida & Machado, 2019; Santana, 2004). A recent event has shown that the Islands that attract tourists are now the most vulnerable to natural disasters such as hurricane and flash floods (De Almeida & Machado, 2019; Fragoso et al., 2012). The story of Madeira Island in 2010 is a typical example (De Almeida & Machado, 2019).

Places that use to be a haven for international tourists are gradually becoming a scary environment to visit because of the perceived risk to life resulting from an increase in the incidence of natural disasters. Specifically, the frequency and nature of natural disasters have been on the increase in recent years, and the implication is becoming greater with nations and communities relying on tourism activities for survival. Natural disasters such as flooding, wildfires, earthquakes, and hurricane that have occurred in recent times have shown how much nature that unites people from different walks of life can also be harmful to mankind. The geographical characteristic of an environment is supposed to attract people and serve as an economic resource at the micro- and macro-level. The tourism sector might cease to drive the development it has provided for nations over the years if adequate strategies are not put in place to mitigate the negative effects of natural disasters on geographical locations where tourism thrives. In light of the review, a gap exists in the strand of literature that relates to tourism and natural disasters. The issue of whether the effects of natural disasters on tourism flows vary by the time-weighted affected population or by regions has not been explored. The present study aims to answer this question.

5.3 DATA AND MEASURES

The dataset consists of a panel of 145 origin countries, 200 destination countries, and 11,499 country pairs over the period of 1995−2007. Multiple datasets, such as UNWTO data, gravity data, were merged for the empirical analysis yielding an unbalanced panel of 107,089 observations. Some data were lost during the empirical analysis due to missing relevant variables.

5.3.1 DEPENDENT VARIABLE

The dependent variable for the analysis is bilateral tourist arrivals. The data are obtained from UNWTO database. The database consists of flows of visitors distinguished by place of origin and destination for 222 countries over the period 1995–2015 (World Tourism Organisation, 2017). Tourist flow is one of the commonly used measures of international tourism demand. Tourist flow is a robust proxy of international tourism as it is highly correlated with tourism receipts (Neumayer, 2004). It is, however, a reliable measure of international tourism flows compared with tourism receipts. This is because it is relatively easier to ascertain the number of individuals arriving in a country compared with the expenditure of tourists that can be inferred from estimated econometric models. Besides, published tourism receipts from the balance of payments are less reliable due to inaccuracies resulting from measurement error and under coverage (Sinclair, 1998).

5.3.2 EXPLANATORY VARIABLES

In this study, natural disaster data from Bergholt and Lujala (2012) is used for the empirical analysis. The data are obtained from the Emergency Events Database (EM-DATA, https://www.cred.be/) of the Centre for Research on the Epidemiology of Disasters. It is a worldwide database containing information on disasters since 1900. The criteria for an event to be included in the database include at least one of the following: 10 or more casualties, declaration of a state of emergency, 100 or more people affected, or call for international aid (for more detail, see Bergholt & Lujala, 2012).

The explanatory variables used in the empirical analysis include population share affected by floods (POPFLOOD), population share affected by storms (POPSTORM), population share affected by earthquakes (POPQUAKE), population share affected by volcanoes (POPVOL), and drought, which is set to 1 if drought year and 0 otherwise. The explanatory variables are event time corrected. Correcting for event timing is important as tourist flow is measured on an annual basis. For instance, an event that happened at the beginning of the year potentially has a greater impact on current year tourist flows compared with an event that happened toward the end of the year. An event that happened toward the end of the current year is likely to affect next year's tourist flows. To adjust for event timing, the time that has passed since the occurrence of an event is weighted by the devaluation rate, $[12 - event\ month_{ijt}]/12$. The subscript, ijt refers to country, natural disaster, and year. For instance, for a natural disaster that happened in January, the event month is 1, and thus the normalized population affected is multiplied by 11/12. Similarly, for an event that happened in June, the normalized population affected is multiplied by 6/12. The correction for event timing assigns more weight to disasters occurring at the beginning of the year compared to those that occur toward the end of the year, and similarly, their relative impact on tourist flows (Bergholt & Lujala, 2012).

Similar to Bergholt and Lujala (2012), the normalized annually time-corrected size of the population affected over the year, Ω, is measured as follows:

$$\sum_{j=1}^{n} \Omega_{ijt} = \sum_{j=1}^{n} \left[\frac{PA_{ijt}}{PS_{it-1}} \times \frac{12 - Event\ month_{ijt}}{12} \right] \tag{5.1}$$

where PA_{ijt} denotes the number of individuals in country i affected by disaster j in year t, and PS is population size, $(12 - Event\ month)/12$, is the time weight. The left-hand side of Eq. (5.1) captures the aggregated population affected by a particular disaster during a given year t.

5.3.3 **CONTROL VARIABLES**

The relative size of the source and destination countries is captured using the population of the source country, *POPORI*, and destination country, *POPDES*, respectively. Similarly, real GDP per capita at PPP for the source country, *RGDPORI*, and destination country, *RGDPDES*, is controlled for to account for the levels of economic development. The population and real GDP data are obtained from the World Development Indicators (World Bank, 2018).

The gravity variables such as distance, colonial ties, island, and language are also controlled for in the empirical analysis where appropriate. The distance between the source country and the destination country is captured with the use of the distance variable. The distance is used as a proxy for transportation costs between a country pair. High transportation costs are likely to lower tourist flows, whereas low transportation costs are likely to promote tourist flows. The geographical characteristics of the source country and the destination country are captured with the use of a landlocked dummy and an island dummy. Landlocked is set to 1 if the source and destination countries are landlocked and 0 otherwise. Island is set to 1 if the source and destination countries are an island and 0 otherwise. In general, island countries tend to be more attractive for international tourists compared to landlocked countries (Khalid et al., 2019; Okafor, Khalid, & Then, 2018).

Furthermore, additional gravity variables such as colonial ties, contiguity, and official language are controlled in the empirical analysis. Colonial ties take the value 1 if the source country is a former colony of the destination or vice versa. Contiguity is set to 1 if the source and the destination countries share a border and 0 otherwise. The official language is a dummy that is set to 1 if the source and destination countries share a common official language and 0 otherwise. These dummies are included in the estimation where appropriate, taking into account that colonial linkage, sharing of a border or official language, tends to promote international tourism. Costs of tourism services tend to be lower if country pair shares a colonial relationship, border or official language (Okafor et al., 2018). The gravity data from Head, Mayer, and Ries (2010) are sourced from CEPII. The dataset consists of 224 countries for variables such as distance, colonial ties, island, and language distance.

The influence of relative prices on tourist decision is captured by using a real effective exchange rate (REERA), as a proxy for relative prices. Real exchange has been used in a previous study as a proxy for destination prices (Adeola & Evans, 2019). REERA is measured using the nominal effective exchange rate and the measure of relative price between a country and its trading partners (Darvas, 2012; Okafor, Bhattacharya, & Apergis, 2019). REERA data are obtained from Bruegel. REERADES refers to the growth rate of real effective exchange of the destination country, while REERAORI refers to the growth rate of real effective exchange of the source country.

5.4 **EMPIRICAL MODEL**

In line with extant studies that relate to international tourism flows (Khalid et al., 2019; Okafor et al., 2018; Page, Song, & Wu, 2012; Shafiullah et al., 2018; Wang, 2009), a gravity equation, which is estimated with the use of fixed-effects estimator, is specified to capture the contemporaneous effects of different types of natural disasters on international tourism flows as follows:

$$LnTOU_{odt} = \alpha_0 + \alpha_1 POPFLOOD_{dt} + \alpha_2 POPSTORM_{dt} + \alpha_3 POPQUAKE_{dt} + \alpha_4 POPQVOL_{dt}$$
$$+ \alpha_5 DROUGHT_{dt} + \alpha_6 LnPOPORI_{ot-1} + \alpha_7 LnPOPDES_{dt-1} + \alpha_8 LnRGDPORI_{ot-1}$$
$$+ \alpha_9 LnRGDDES_{dt-1} + \alpha_{10} LnREERORI_{ot-1} + \alpha_{11} LnREERDES_{dt-1} + \lambda_t + \theta_{od} + \varepsilon_{odt} \quad (5.2)$$

where Ln denotes natural logarithm; TOU is bilateral tourist flows; odt indexes country of origin, destination country, and year; α_0 is a constant; $POPFLOOD$ is time-weighted population share affected by floods; $POPSTORM$ is time-weighted population share affected by storms; $POPQUAKE$ is time-weighted population share affected by earthquakes; $POPQVOL$ is time-weighted population share affected by volcanoes; and $DROUGHT$ is a dummy variable for drought. $POPORI$ refers to population of the origin country, $POPDES$ is population of the destination country, $RGDPORI$ is real GDP per capita of the country of origin at PPP, and $RGDDES$ is real GDP per capita of the destination country at PPP. $REERORI$ denotes the growth rate of real effective exchange rate of the country of origin, $REERDES$ is the growth rate of real effective exchange rate of the destination country, λ is year-specific effects, θ is country pair−specific effects, ε is an error term, and $\alpha's$ are parameters to be estimated.

Similarly, a gravity equation, which is estimated with the use of fixed-effects estimator, is specified to capture the dynamic effects as follows:

$$LnTOU_{odt} = \alpha_0 + \alpha_1 POPFLOOD_{dt-1} + \alpha_2 POPSTORM_{dt-1} + \alpha_3 POPQUAKE_{dt-1}$$
$$+ \alpha_4 POPQVOL_{dt-1} + \alpha_5 DROUGHT_{dt-1} + \alpha_6 LnPOPORI_{ot-1} + \alpha_7 LnPOPDES_{dt-1}$$
$$+ \alpha_8 LnRGDPORI_{ot-1} + \alpha_9 LnRGDDES_{dt-1} + \alpha_{10} LnREERORI_{ot-1} + \alpha_{11} LnREERDES_{dt-1}$$
$$+ \lambda_t + \theta_{od} + \varepsilon_{odt} \quad (5.3)$$

To check if the estimates are susceptible to the inclusion of standard gravity variables, an alternative gravity equation, which is estimated with the use of OLS, is specified as follows:

$$LnTOU_{odt} = \beta_0 + \beta_1 POPFLOOD_{dt} + \beta_2 POPSTORM_{dt} + \beta_3 POPQUAKE_{dt} + \beta_4 POPQVOL_{dt}$$
$$+ \beta_5 DROUGHT_{dt} + \beta_6 LnPOPORI_{ot-1} + \beta_7 LnPOPDES_{dt-1} + \beta_8 LnRGDPORI_{ot-1}$$
$$+ \beta_9 LnRGDDES_{dt-1} + \beta_{10} LnRGDDES_{dt-1} + \beta_{11} LnRGDDES_{dt-1} + \beta_{12} LnREERORI_{ot-1}$$
$$+ \beta_{13} LnREERDES_{dt-1} + \beta_{14} DIS_{od} + \beta_{15} LAND_{od} + \beta_{16} ISLA_{od} + \beta_{17} COLO_{od} + \beta_{18} CONT_{od}$$
$$+ \beta_{19} OFLA_{od} + \lambda_t + \gamma_o + \Gamma_d + \epsilon_{odt} \quad (5.4)$$

where DIS denotes distance, $LAND$ is landlocked dummy, $ISLA$ is island dummy, $COLO$ is colonial linkage dummy, $CONT$ is contiguity dummy, γ is country of origin specific effects, Γ is country of destination specific effects, and $OFLA$ is common official language dummy.

In the OLS dynamic specification, not shown here, all the explanatory variables are lagged, similar to FE specification. The models will be estimated for all countries covered in the sample. The models will be also estimated by regions, in order to ascertain if the effects of natural disasters differ by regions.

5.5 SUMMARY STATISTICS

Table 5.1 presents the summary statistics by regions. Similar to Okafor et al. (2018), we classify countries covered in the sample into four geographical locations: Europe; America and LATCA; Asia, Middle East, North Africa, and Oceania; and SSA. This classification will allow us to test if the impacts of different measures of natural disasters on tourist flows differ by geographical

Table 5.1 Summary Statistics.

	1	2	3	4	5
Variable	Europe	America and LATCA	Asia, Middle East, North Africa, and Oceania	SSA	Whole Sample
Ln Tourist Flows	7.7697	6.6907	6.6483	6.3143	6.8594
	(3.5622)	(3.2363)	(3.1065)	(2.8089)	(3.2383)
POPFLOOD	0.2662	0.5130	0.9356	0.8191	0.6930
	(2.3884)	(2.5146)	(4.6248)	(7.3963)	(4.4314)
POPSTORM	0.1228	1.5062	0.7398	0.1204	0.6675
	(2.0506)	(8.7397)	(4.6946)	(1.6120)	(5.1102)
POPQUAKE	0.0103	0.2584	0.1583	0.0007	0.1238
	(0.1286)	(7.0321)	(1.0268)	(0.0043)	(3.1481)
POPVOL	–	0.0655	0.0064	0.0006	0.0155
	–	(0.6779)	(0.4812)	(0.0327)	(0.4404)
Drought	0.0314	0.1169	0.0646	0.0659	0.0673
	(0.1744)	(0.3214)	(0.2459)	(0.2480)	(0.2505)
LN POPORI	16.1641	15.9447	16.1135	16.4652	16.1404
	(1.8904)	(2.1166)	(1.9199)	(1.7365)	(1.9347)
LN POPDES	16.2674	15.9969	16.5871	16.1807	16.3514
	(1.5272)	(1.9996)	(2.0810)	(1.5187)	(1.9007)
LN RGDPORI	9.5306	9.3716	9.2238	9.1319	9.3077
	(1.1530)	(1.1021)	(1.2411)	(1.3739)	(1.2242)
LN RGDDES	9.8599	9.4557	9.2996	7.9462	9.2725
	(0.6947)	(0.8107)	(1.0732)	(0.8762)	(1.0848)
Ln REERORI$_t$	0.0086	0.0052	0.0057	0.0053	0.0062
	(0.0995)	(0.0978)	(0.1047)	(0.1034)	(0.1021)
Ln REERDES	0.0227	0.0057	0.0037	0.0043	0.0084
	(0.0963)	(0.0753)	(0.0905)	(0.1802)	(0.1061)
Number of observations	23,632	20,147	49,024	14,213	107,089

Notes: "–" POPVOL variable for indicates that the population share affected by volcanoes in Europe over the sample period is zero. LATCA, Latin America and Caribbean; SSA, sub-Saharan Africa.

locations. As shown in Table 5.1, Europe tends to attract more tourists followed by countries in America, LATCA, countries in Asia, Middle East, North Africa, and Oceania, while countries in SSA are likely to attract a less number of tourists.

Furthermore, countries in Asia, Middle East, North Africa, and Oceania are more likely to have a higher incidence of flooding as captured by weighted population share affected by floods followed by countries in SSA, countries in America and LATCA, whereas countries in Europe are likely to be the least affected. In general, countries in America and LATCA are more likely to be affected by storms, earthquakes, and volcanoes all weighted by population, followed by countries in Asia, Middle East, North Africa, and Oceania compared with countries in SSA. Countries in Europe tend to be affected by storms and earthquake weighted by population to a lesser extent compared to countries in America and LATCA. Drought tends to be more prevalent in countries located in America and LATCA, followed by countries in SSA, countries in Asia, Middle East, North Africa, and Oceania, while Countries in Europe are least likely to be affected by drought.

As reported in Table 5.1, origin countries in SSA as a whole tend to be heavily populated followed by countries in Europe, Asia, Middle East, North Africa, and Oceania, while America and LATCA are likely to be the least populated. It is important to control for the population of the origin country in the gravity framework as it helps to account for the source of demand for tourism services. Similarly, the population in the destination country is essential as the potential source of the supply of tourism services. Countries in Asia, Middle East, North Africa, and Oceania as a whole tend to be more populous followed by countries in Europe, SSA, whereas countries in America and LATCA tend to be least populated.

Real GDP per capita is one of the key determinants of tourism demand. For instance, individuals in wealthier countries tend to demand more tourism services compared with those in poorer countries. Similarly, wealthier countries tend to have a higher level of tourism infrastructure development compared with poorer countries. Preliminary evidence from Table 5.1 suggests that a typical individual in Europe either in origin or the destination country tends to be wealthier, followed by a typical individual in America and LATCA, before a typical individual in Asia, Middle East, North Africa and Oceania, while a typical individual in SSA tends to less wealthy. This preliminary evidence is consistent with the notion that SSA has the largest proportion of people that are poor (Folarin & Adeniyi, 2020).

Growth in real effective exchange rate growth in the host country is relatively similar across the regions. In contrast, the growth in the real effective exchange rate in destination country varies slightly across the regions with noticeable growth in Europe followed by America and LATCA, while slow growth was observed in Asia, Middle East, North Africa, and Oceania. Since the exchange rate mirrors the price of tourism product, a significant growth in exchange rate suggests a potential increase in the price of tourism product. Thus changes in exchange rate are expected to influence international tourist flows.

Table 5.2 reports the tests for multicollinearity. The estimates reported in columns 1 and 2 were obtained using the contemporaneous natural disaster measures and other relevant variables, whereas those reported in columns 3 and 4 use lagged natural disaster measures in lieu of current natural disaster measures. Following the rule of thumb of less than 0.1 for tolerance level or a variance inflation factor (VIF) of 10 based on econometric literature, our estimates show no evidence of severe multicollinearity.

Table 5.2 Tests for Multicollinearity: Variance Inflation Factors (VIFs) and Tolerance-Dependent Variable: Ln Tourist Flows.

	1	2		3	4
Variable	**VIF**	**1/VIF**	**Variable**	**VIF**	**1/VIF**
POPFLOOD	1.11	0.897	POPFLOOD$_{t-1}$	1.12	0.89
POPSTORM	1.11	0.903	POPSTORM$_{t-1}$	1.10	0.91
LN POPDES$_{t-1}$	1.10	0.912	LN POPDES$_{t-1}$	1.10	0.91
LN RGDPDES$_{t-1}$	1.07	0.934	LN RGDPDES$_{t-1}$	1.08	0.93
Drought	1.03	0.975	Drought$_{t-1}$	1.03	0.97
LN POPORI$_{t-1}$	1.01	0.987	LN POPORI$_{t-1}$	1.01	0.99
LN RGDPORI$_{t-1}$	1.01	0.990	LN RGDORI$_{t-1}$	1.01	0.99
Ln REERDES$_{t-1}$	1.01	0.994	Ln REERDES$_{t-1}$	1.01	0.99
Ln REERORI$_{t-1}$	1.00	0.995	Ln REERORI$_{t-1}$	1.01	0.99
POPQUAKE	1.00	0.997	POPQUAKE$_{t-1}$	1.00	1.00
POPVOL	1.00	0.999	POPVOL$_{t-1}$	1.00	1.00
Mean VIF	1.04		Mean VIF	1.04	

5.6 DISCUSSION OF EMPIRICAL RESULTS

5.6.1 THE CONTEMPORANEOUS LINK BETWEEN NATURAL DISASTERS AND INTERNATIONAL TOURIST FLOWS

The parameter estimates of the contemporaneous effects of natural disasters on international tourist flows are presented in Table 5.3. In columns 1−5 the effects of different measures of natural disasters are estimated separately while controlling for relevant variables. In column 6 the effects of different measures of natural disasters are estimated jointly. As reported in Table 5.3, natural disasters have ambiguous effects on tourist flows. For instance, based on the pooled sample, the effect is positive when floods and drought are used to capture a natural disaster, while it is negative when storms and volcanoes are used. These results give an indication that the effects of natural disasters on tourist flows depend on the severity of the impacts, especially in terms of the weighted population affected as well as the negative global image associated with a country affected by a natural disaster. Besides, these estimates suggest that the effects of natural disaster on tourist flows are likely to be region dependent.

More specifically, as shown in column 1 of Table 5.3, a 1 standard deviation change in the weighted population affected by floods results in a 0.0029 standard deviation increase in the number of tourist arrivals. Similarly, as reported in column 4 of Table 5.3, on average, tourist arrival is higher by 1.643% for countries affected by drought compared with those that were not affected. The positive effects of the weighted population affected by flood and drought suggest that tourist flow is not especially vulnerable to flooding and drought on average. This is likely to hold if the share of population affected by flooding is negligible, and reconstruction of affected areas is completed in the shortest possible time.

Table 5.3 Contemporaneous Link Between Natural Disasters and International Tourist Flows. Dependent Variable: Ln Tourist Flows.

Variable	1	2	3	4	5	6
POPFLOOD	0.0021***					0.0026***
	(0.0004)					(0.0005)
POPSTORM		−0.0009*				−0.0016***
		(0.0005)				(0.0005)
POPQUAKE			−0.0009			−0.0010
			(0.0007)			(0.0007)
POPVOL				−0.0304***		−0.0304***
				(0.0102)		(0.0102)
Drought					0.0163**	0.0173**
					(0.0082)	(0.0083)
LN POPORI$_{t-1}$	0.1601	0.1604	0.1600	0.1589	0.1766	0.1601
	(0.1260)	(0.1260)	(0.1260)	(0.1259)	(0.1252)	(0.1259)
LN POPDES$_{t-1}$	0.9850***	0.9828***	0.9839***	0.9867***	0.9855***	0.9873***
	(0.1278)	(0.1279)	(0.1278)	(0.1278)	(0.1272)	(0.1278)
LN RGDPORI$_{t-1}$	0.6699***	0.6698***	0.6700***	0.6688***	0.6646***	0.6689***
	(0.0720)	(0.0720)	(0.0720)	(0.0719)	(0.0714)	(0.0719)
LN RGDPDES$_{t-1}$	1.0696***	1.0716***	1.0704***	1.0699***	1.0794***	1.0696***
	(0.0695)	(0.0696)	(0.0696)	(0.0696)	(0.0692)	(0.0695)
Ln REERORI$_{t-1}$	0.0493*	0.0491*	0.0492*	0.0490*	0.0468*	0.0493*
	(0.0269)	(0.0269)	(0.0269)	(0.0269)	(0.0267)	(0.0269)
Ln REERDES$_{t-1}$	−0.0323	−0.0314	−0.0312	−0.0317	−0.0410	−0.0340
	(0.0267)	(0.0267)	(0.0267)	(0.0267)	(0.0271)	(0.0268)
Number of observations	82,800	82,800	82,800	82,800	83,900	82,800
R-squared (within)	0.1717	0.1716	0.1716	0.1720	0.1724	0.1724

Notes: Robust standard errors, clustered on countries, are reported in parentheses. Constant and year effects included in all the regressions but not reported. *P < .10, **P < .05, and ***P < .001. Country-pair fixed effects are included all the estimations.

Moreover, minor incidence of drought is unlikely to discourage tourists from visiting a country. In general, effective management of flood and drought events can promote tourist flows. For instance, the Cornish village of Boscastle was hit by flash floods in 2004 that dropped a record amount of 200 mm of rain in just 24 hours leaving behind a vast trail of destruction. The central government undertook massive reconstruction that helped to create more jobs and attracted more businesses to the area. The images and videos about Boscastle were disseminated through the media, helping to create much-needed publicity, which, in turn, led to increase in the number of tourists (Tourismembassy, 2014). These findings also suggest that the impacts of weighted population affected by flood and drought are potentially region dependent, especially taken into account that countries in different regions have different capacity to effectively mitigate the negative impacts of these events.

As shown in columns 2, 3, 4, and 6, the weighted population affected by storms, earthquakes, or volcanoes has a negative impact on tourist flows, though the impact of earthquake is not statistically significant. For instance, as reported in column 2, a 1 standard deviation change in the weighted population affected by storms results in a 0.0014 standard deviation decrease in the number of tourist arrivals. Similarly, as reported in column 4, a 1 standard deviation change in the weighted population affected by volcanoes results in a 0.0041 standard deviation drop in tourist flows. These findings suggest that the adverse effect of volcanoes on tourist flows is higher when compared to storms. In general, volcanoes tend to create more tension and uncertainty over a relatively long period, compared with storms, thus, instilling a higher level of negative international publicity that dissuades tourists from traveling to affected countries.

The parameter estimates obtained when all the proxies of natural disasters are included in the model are reported in column 6. Overall, the estimates are relatively similar to the ones reported in columns 1−5 in terms of magnitudes and signs. This suggests that the model is stable as it is not susceptible to inclusion or omission of at least one of the proxies of natural disaster. The estimates reported in Table 5.3 suggest that lagged population in the destination is a key determinant of tourist flows. The larger population in the destination countries would be associated with larger supply of tourism services, which in turn promote international tourism. In contrast, the impact of lagged population in the origin countries is positive as expected but statistically insignificant. Lagged GDP per capita in the origin or destination countries has a positive and statistically significant impact on tourist flows. In general, these findings are consistent with the notion that richer countries demand more tourism services as well as supply greater tourism services.

The parameter estimates presented in Table 5.3 show that the growth in the real effective exchange rate in origin and destination country affects tourist flows in opposite direction as expected. The effect of growth in real effective exchange rate in origin country is statistically significant at 10% whereas its effect in destination country is insignificant. Hence, an increase in growth in real effective exchange rate by 10% contributes to an increase in tourism flows by 0.49%. This suggests that as the exchange rate in origin country appreciates, the price of tourism services reduces, which in turn leads to a reduction in the cost of tourism and an increase in demand for international tourism.

5.6.2 THE DYNAMIC LINK BETWEEN NATURAL DISASTERS AND INTERNATIONAL TOURIST FLOWS

Table 5.4 reports the parameter estimates of the dynamic impact of different measures of natural disasters on tourist flows. As shown in Table 5.4, the impact of lagged weighted population affected by floods or storms is insignificant, whereas the impact of lagged weighted population affected by volcanoes and drought still persists after a year. These findings indicate that the impact of floods or storms adjusted by the weighted population is short-lived on average, while the effect of volcanoes adjusted by weighted population or drought persists after a year.

As shown in columns 3 and 6 of Table 5.4, lagged weighted population affected by earthquake has a negative and statistically significant impact on tourist flows. This suggests that the effect of earthquakes on dampening tourist flows is more noticeable a year after the disaster has occurred. More specifically, a 1 standard deviation change in the lagged weighted

Table 5.4 Dynamic Link Between Natural Disasters and International Tourist Flows. Dependent Variable: Ln Tourist Flows.

	1	2	3	4	5	6
POPFLOOD$_{t-1}$	0.0001					0.00004
	(0.0004)					(0.0004)
POPSTORM$_{t-1}$		0.0005				0.0006
		(0.0004)				(0.0004)
POPQUAKE$_{t-1}$			−0.0010***			−0.0011***
			(0.0004)			(0.0004)
POPVOL$_{t-1}$				−0.0148***		−0.0147***
				(0.0034)		(0.0034)
Drought$_{t-1}$					0.0302***	0.0306***
					(0.0081)	(0.0081)
Other control variables	Yes	Yes	Yes	Yes	Yes	Yes
Number of observations	82,800	82,800	82,800	82,800	83,900	82,800
R-Squared (within)	0.1715	0.1715	0.1716	0.1716	0.1725	0.1718

Notes: *Robust standard errors, clustered on countries, are reported in parentheses. Constant and year effects included in all the regressions but not reported. *P<.10, **P<.05, and ***P<.001. Country-pair fixed effects are included all the estimations.*

population affected by earthquakes results in a 0.00115 standard deviation decrease in the number of tourist arrivals.

5.6.3 THE CONTEMPORANEOUS LINK BETWEEN NATURAL DISASTERS AND INTERNATIONAL TOURIST FLOWS BY REGIONS

Table 5.5 presents the parameter estimates of the effect of natural disasters on tourist flows by regions. Given that the number of people affected by the natural disasters varies across regions as well as the capacity of countries in different regions to effectively manage the aftermaths of natural disasters, it is expected that the effects of different measures of natural disasters on tourist flows potentially differ by regions.

The results show that the effect of the weighted population share affected by floods on international tourism flows is only negative and statistically significant in Europe, whereas its impact in other regions is positive. These findings suggest that flooding is more likely to affect a larger number of people as well as cause more reputational damage in Europe compared with other regions.

As reported in columns 1–4 of Table 5.5, the weighted population share affected by storms has a negative impact on tourist flows across regions. These findings reinforce the notion that storms have the potential to affect a larger share of population as well as inflicting reputational damage across different regions. The negative image resulting from damages caused by storms lowers international tourism attractiveness of affected countries, regardless of geographical location compared with those not affected.

Furthermore, the weighted population affected by earthquakes has a positive and significant effect in Europe and SSA, whereas its effect is negative and significant in Asia, Middle East, North

Table 5.5 Contemporaneous Link Between Natural Disasters and International Tourist Flows by Regions.

Variable	1	2	3	4
	Europe	America and LATCA	Asia, Middle East, North Africa, and Oceania	SSA
POPFLOOD	− 0.0026***	0.0082***	0.0070***	0.0013**
	(0.0010)	(0.0020)	(0.0009)	(0.0006)
POPSTORM	− 0.0094***	− 0.0015**	− 0.0022***	− 0.0050***
	(0.0028)	(0.0007)	(0.0007)	(0.0015)
POPQUAKE	0.0473***	−0.0009	− 0.0127***	15.0896***
	(0.0158)	(0.0007)	(0.0037)	(2.4248)
POPVOL	−	− 0.0326***	− 0.0306*	0.2099***
	−	(0.0061)	(0.0156)	(0.0696)
Drought	0.0473*	0.0973***	0.0102	− 0.1135***
	(0.0243)	(0.0140)	(0.0118)	(0.0296)
Other Control Variables	Yes	Yes	Yes	Yes
Number of observations	19,300	14,800	37,700	10,900
R-squared (within)	0.1617	0.1450	0.2263	0.1652

Notes: "-" POPVOL variable for indicates that the population share affected by volcanoes in Europe over the sample period is zero. Robust standard errors, clustered on countries, are reported in parentheses. Constant and year effects included in all the regressions but not reported. *P < .10, **P < .05, and ***P < .001. Country-pair fixed effects are included all the estimations. LATCA, Latin America and Caribbean; SSA, sub-Saharan Africa.

Africa, and Oceania. The impact is also negative in America and LATCA but statistically insignificant. These findings indicate that earthquakes are more likely to affect a larger fraction of the population in Asia, Middle East, North Africa, and Oceania in addition to the reputational damage resulting from their occurrence compared with other regions. Besides, earthquakes are less likely to affect a larger number of individuals in SSA and to extent Europe compared to other regions as suggested by the summary statistics. As a result, earthquakes adjusted by weighted population do not depress tourist flows in Europe and SSA.

The effect of the weighted population affected by volcanoes is also region dependent. For instance, the weighted population affected by volcanoes has a negative effect on tourist flows in America and LATCA as well as Asia, Middle East, North Africa, and Oceania, while its impact is positive in SSA. The positive impact of volcanoes in SSA is potentially associated with the smaller fraction of the population that is affected compared with other regions. The higher the number of people affected by volcanoes, the larger the damage caused in terms of loss of property and human life as well as the reputation damage resulting from it, which would in turn depress tourist flows, such as in America and LATCA. The quest for safety explains why tourists will avoid traveling to countries and/or regions that are prone to danger such as, disasters, terrorism attacks (Ajogbeje, Adeniyi, & Folarin, 2017).

The effect of drought on tourist flows is also region dependent. For example, there is strong evidence that drought depresses tourist flows in SSA, whereas its impact in Europe and America and LATCA is positive. This can be explained to some extent by the observation that countries in SSA are less integrated with the global economy compared with countries in other regions. They also have limited foreign exchange earnings and managerial know-how to curtail the negative effect of drought on tourist flows compared with countries in other regions.

5.6.4 THE DYNAMIC LINK BETWEEN NATURAL DISASTERS AND INTERNATIONAL TOURIST FLOWS BY REGIONS

The estimates of the dynamic effects of natural disasters on tourist flows by regions are reported in Table 5.6. The effect of the lagged weighted population affected by floods on tourist flows is relatively similar to the contemporaneous impacts. For instance, the negative effect of floods on tourist flows in Europe persists a year after the disaster. This suggests that the adverse effect of flooding in Europe is not short-lived. This could be explained in terms of sensitivity of international tourists to climatic conditions in Europe.

The lagged effect of the weighted population affected by storms on tourist flows differs by regions. For example, a year after storm-related disaster, its impact on tourist flows in Europe and

Table 5.6 Dynamic Link Between Natural Disasters and International Tourist Flows by Regions. Dependent Variable: Ln Tourist Flows.

	1	2	3	4
	Europe	America and LATCA	Asia, Middle East, North Africa, and Oceania	SSA
POPFLOOD$_{t-1}$	−0.0144***	0.0055**	0.0043***	0.0007*
	(0.0026)	(0.0028)	(0.0008)	(0.0004)
POPSTORM$_{t-1}$	0.0019	−0.0018***	−0.0011*	0.0026
	(0.0013)	(0.0006)	(0.0007)	(0.0023)
POPQUAKE$_{t-1}$	−0.0207	0.0001	−0.0308***	7.2725***
	(0.0192)	(0.0002)	(0.0029)	(2.5519)
POPVOL$_{t-1}$	−	−0.0246***	−0.0106***	−0.0997
	−	(0.0067)	(0.0034)	(0.0676)
Drought$_{t-1}$	0.0194	0.1030***	0.0301**	−0.0388
	(0.0247)	(0.0135)	(0.0119)	(0.0313)
Other Control variables	Yes	Yes	Yes	Yes
Number of observations	19,300	14,800	37,700	10,900
R-Squared (within)	0.1635	0.1441	0.2275	0.1570

Notes: "−" POPVOL variable for indicates that the population share affected by volcanoes in Europe over the sample period is zero. Robust standard errors, clustered on countries, are reported in parentheses. Constant and year effects included in all the regressions but not reported. *P < .10, **P < .05, and ***P < .001. Country-pair fixed effects are included all the estimations. LATCA, Latin America and Caribbean; SSA, sub-Saharan Africa.

Table 5.7 Contemporaneous Link Between Natural Disasters and International Tourist Flows by Regions (Robustness Check).

Variable	1	2	3	4
POPFLOOD	− 0.0030***	0.0046	0.0042***	0.0016**
	(0.0010)	(0.0040)	(0.0012)	(0.0007)
POPSTORM	−0.0092***	−0.0009	0.0006	0.0038
	(0.0031)	(0.0009)	(0.0009)	(0.0061)
POPQUAKE	0.0470***	−0.0001	−0.0135**	15.3834***
	(0.0168)	(0.0008)	(0.0054)	(2.4637)
POPVOL	−	−0.0479***	−0.0288*	0.2466**
	−	(0.0081)	(0.0160)	(0.1014)
Drought	0.0611**	0.0874***	−0.0019	− 0.1105***
	(0.0261)	(0.0151)	(0.0134)	(0.0324)
LN POPORI$_{t-1}$	−0.0520	−0.2705	−0.1252	0.3750
	(0.3041)	(0.3856)	(0.2094)	(0.4538)
LN POPDES$_{t-1}$	− 3.6388***	2.4862***	0.3623	1.1767**
	(0.6033)	(0.6256)	(0.2261)	(0.4793)
LN RGDPORI$_{t-1}$	0.3998**	0.5719***	0.9240***	0.7899***
	(0.1763)	(0.1671)	(0.1272)	(0.2036)
LN RGDPDES$_{t-1}$	− 0.7859***	1.0796***	1.2579***	0.8769***
	(0.1623)	(0.2214)	(0.1092)	(0.2009)
Ln REERORI$_{t-1}$	0.2635***	0.0034	0.1100**	−0.0916
	(0.0688)	(0.0713)	(0.0532)	(0.0976)
Ln REERDES$_{t-1}$	0.3920***	−0.1204	−0.6089***	0.0584
	(0.0878)	(0.0764)	(0.0539)	(0.0440)
Ln distance	− 1.4039***	−1.7762***	−1.5334***	−1.8526***
	(0.0871)	(0.0648)	(0.0443)	(0.0869)
Landlocked	− 0.6516***	−0.1917	0.5885***	0.2114
	(0.1786)	(0.1590)	(0.1670)	(0.2151)
Island	−0.0689	0.4960***	0.3563***	0.2926
	(0.2080)	(0.1469)	(0.1079)	(0.2433)
Colonial ties	1.2869***	0.1917	0.4291**	1.0527***
	(0.1965)	(0.3243)	(0.1742)	(0.2387)
Contiguity	0.8927***	0.9529***	1.6487***	0.9282***
	(0.1640)	(0.1754)	(0.2015)	(0.1676)
Official language	0.3957***	1.3208***	1.2105***	1.0201***
	(0.1318)	(0.0715)	(0.0781)	(0.1071)
Number of observations	19,300	14,800	37,700	10,900
R-Squared	0.9004	0.9165	0.8220	0.8030

Notes: "−" POPVOL variable for indicates that the population share affected by volcanoes in Europe over the sample period is zero. Robust standard errors, clustered on countries, are reported in parentheses. Constant and year effects included in all the regressions but not reported. *P < .10, **P < .05, and ***P < .001. Country of origin and country of destination fixed effects are included all the estimations.

Table 5.8 Dynamic Link Between Natural Disasters and International Tourist Flows by Regions (Robustness Check).

Variable	1	2	3	4
POPFLOOD$_{t-1}$	− 0.0159***	0.0031	0.0019**	0.0009*
	(0.0028)	(0.0039)	(0.0009)	(0.0005)
POPSTORM$_{t-1}$	0.0016	−0.0010	0.0017**	0.0050
	(0.0015)	(0.0008)	(0.0008)	(0.0037)
POPQUAKE$_{t-1}$	0.0236	0.0008	−0.0304***	5.8026**
	(0.0208)	(0.0005)	(0.0049)	(2.6565)
POPVOL$_{t-1}$	−	−0.0343***	−0.0090***	−0.0635
	−	(0.0088)	(0.0034)	(0.1010)
Drought$_{t-1}$	0.0421	0.0962***	0.0182	− 0.0866**
	(0.0267)	(0.0153)	(0.0136)	(0.0339)
LN POPORI$_{t-1}$	−0.0738	−0.2751	−0.1190	0.3922
	(0.3036)	(0.3856)	(0.2092)	(0.4548)
LN POPDES$_{t-1}$	− 3.1200***	2.3795***	0.3942*	1.1067**
	(0.5842)	(0.6296)	(0.2257)	(0.4748)
LN RGDPORI$_{t-1}$	0.3970**	0.5710***	0.9244***	0.7904***
	(0.1754)	(0.1672)	(0.1272)	(0.2042)
LN RGDPDES$_{t-1}$	− 0.6543***	0.9587***	1.2412***	0.7906***
	(0.1582)	(0.2172)	(0.1103)	(0.2021)
Ln REERORI$_{t-1}$	0.2706***	0.0049	0.1131**	−0.0779
	(0.0691)	(0.0722)	(0.0531)	(0.0981)
Ln REERDES$_{t-1}$	0.3749***	−0.0590	−0.5857***	0.0911**
	(0.0905)	(0.0770)	(0.0544)	(0.0442)
Ln distance	− 1.4040***	−1.7761***	−1.5334***	− 1.8515***
	(0.0870)	(0.0647)	(0.0443)	(0.0869)
Landlock	− 0.6519***	−0.1914	0.5890***	0.2146
	(0.1786)	(0.1578)	(0.1670)	(0.2151)
Island	−0.0700	0.4960***	0.3560***	0.2919
	(0.2076)	(0.1469)	(0.1080)	(0.2433)
Colonial ties	1.2867***	0.1905	0.4281**	1.0535***
	(0.1964)	(0.3243)	(0.1742)	(0.2388)
Contiguity	0.8928***	0.9524***	1.6489***	0.9291***
	(0.1640)	(0.1755)	(0.2016)	(0.1675)
Official language	0.3960***	1.3215***	1.2105***	1.0195***
	(0.1317)	(0.0714)	(0.0781)	(0.1071)
Number of observations	19,300	14,800	37,700	10,900
R-Squared	0.9005	0.9165	0.8220	0.8025

Notes: "−" POPVOL variable for indicates that the population share affected by volcanoes in Europe over the sample period is zero. Robust standard errors, clustered on countries, are reported in parentheses. Constant and year effects included in all the regressions but not reported. *P < .10, **P < .05, and ***P < .001. Country of origin and country of destination fixed effects are included all the estimations.

SSA is statistically insignificant, whereas its impact is negative in America and LATCA, Asia, Middle East, North Africa, and Oceania. The higher incidence of storm-related disasters in America and LATCA as well as Asia, Middle East, and Oceania as suggested by the summary statistics can help to explain in part why the adverse effect lingers after a year. The effects of lagged weighted population affected by earthquake on tourist flows in Asia, Middle East, North Africa, Oceania, and SSA are relatively similar to the contemporaneous effects. It still dampens tourist flows in Asia, Middle East, North Africa, and Oceania while promoting tourism in SSA. Similarly, lagged weighted population affected by volcanoes continues to depress tourist flows in America and LATCA as well as Asia, Middle East, North Africa, and Oceania, whereas its impact in SSA is statistically insignificant.

Finally, the adverse effect of drought on tourist flows in SSA is short-lived as the negative effect of occurrence of drought was found to be insignificant after a year. This suggests that tourist's decision whether to visit countries in SSA is not influenced by incidence of drought after a year. Past drought is found to promote tourism in America and LATCA as well as Asia, the Middle East, North Africa, and Oceania. International tourism attractiveness of countries in these regions is not highly susceptible to incidence of drought. Access to goods and services through international trade can to some extent help some countries cope with the adverse effect of drought.

5.7 ROBUSTNESS CHECKS

5.7.1 THE CONTEMPORANEOUS AND DYNAMIC LINK BETWEEN NATURAL DISASTERS AND INTERNATIONAL TOURIST FLOWS REGIONS

Following Okafor et al. (2018), additional standard gravity variables were controlled for to check the sensitivity of the parameter estimates. The standard gravity variables include distance, island, landlocked, colonial ties, contiguity, and official language. In general, the parameter estimates of the variables of interest are qualitatively similar based on the contemporaneous and dynamic specifications by regions as presented in Tables 5.7 and 5.8.

5.8 CONCLUSION

Natural disasters have increased in frequency due to climate change and man-made causes. This may hinder the expected growth of international tourism, which in turn would hinder economic growth and development. This chapter demonstrates that the impacts of natural disasters on international tourist flows depend on the weighted population affected. The higher the weighted population affected by a natural disaster, the higher the losses in terms of human lives and property and the higher the negative impact on tourist flows. This suggests that the effects of natural disasters on tourism flows can be positive or negative. In addition, the impacts of natural disasters are also time and region dependent. For instance, lagged weighted population affected by storms has negative impact on tourist flows in America and LATCA as well as Asia, Middle East, North Africa, and

Oceania, whereas its impact in Europe and SSA is insignificant. Policies aimed at eliminating or reducing the number of weighted population affected by natural disasters, such as inclusion of disaster mitigation in the development process would help to promote international tourism over time.

REFERENCES

Adeola, O., & Evans, O. (2019). ICT, infrastructure, and tourism development in Africa. *Tourism Economics, OnlineFirst*. Available from https://doi.org/10.1177/1354816619827712.

Adnan Hye, Q. M., & Ali Khan, R. E. (2013). Tourism-led growth hypothesis: A case study of Pakistan. *Asia Pacific Journal of Tourism Research*, *18*(4), 303−313.

Ajogbeje, K., Adeniyi, O., & Folarin, O. (2017). The effect of terrorism on tourism development in Nigeria: A note. *Tourism Economics*, *23*(6), 1673−1678.

Albalate, D., & Bel, G. (2010). Tourism and urban public transport: Holding demand pressure under supply constraints. *Tourism Management*, *31*(3), 425−433. Available from https://doi.org/10.1016/j.tourman.2009.04.011.

Beattie, M. A. (1992). *The effect of natural disasters on tourism a study of Mount Saint Helens and Yellowstone National Park*. Rochester Institute of Technology, Rochester Institute of Technology. Retrieved from <http://scholarworks.rit.edu/cgi/viewcontent.cgi?article = 8433&context = theses>.

Becken, S., & Wilson, J. (2013). The impacts of weather on tourist travel. *Tourism Geographies*, *15*(4), 620−639.

Bergholt, D., & Lujala, P. (2012). Climate-related natural disasters, economic growth, and armed civil conflict. *Journal of Peace Research*, *49*(1), 147−162. Available from https://doi.org/10.1177/0022343311426167.

Botzen, W. J., Deschenes, O., & Sanders, M. (2019). The economic impacts of natural disasters: A review of models and empirical studies. In: *Review of environmental economics and policy*.

Breiling, M. (2016). *Tourism supply chains and natural disasters: The vulnerability challenge and business continuity models for ASEAN countries*. Economic Research Institute for ASEAN and East Asia.

Chiou, C. R., Huang, M. Y., Tsai, W. L., Lin, L. C., & Yu, C. P. (2013). Assessing the impact of natural disasters on tourist arrivals: The case of Xitou Nature Education Area (XNEA), Taiwan. *International Journal of Tourism Sciences*, *13*(1), 47−64.

Cho, V. (2003). A comparison of three different approaches to tourist arrival forecasting. *Tourism Management*, *24*(3), 323−330. Available from https://doi.org/10.1016/S0261-5177(02)00068-7.

Darvas, Z. (2012). Real effective exchange rates for 178 countries: a new database. In: *Bruegel working paper, 2012/06, Brussels, Belgium*. Retrieved from <http://www.bruegel.org/publications/publication-detail/publication/716-real-effective-exchange-rates-for-178-countries-a-new-database/>.

De Almeida, A. M. M., & Machado, L. P. (2019). Madeira Island: Tourism, natural disasters and destination image. In: *Climate change and global development* (pp. 285−301). <https://doi.org/10.1007/978-3-030-02662-2_14>.

Doocy, S., Daniels, A., Dooling, S., & Gorokhovich, Y. (2013). The human impact of volcanoes: A historical review of events 1900−2009 and systematic literature review. In: *PLOS currents disasters*. <https://doi.org/10.1371/currents.dis.841859091a706efebf8a30f4ed7a190>.

Fayissa, B., Nsiah, C., & Tadasse, B. (2008). Impact of tourism on economic growth and development in Africa. *Tourism Economics*, *14*(4), 807−818.

Felbermayr, G., & Gröschl, J. (2014). Naturally negative: The growth effects of natural disasters. *Journal of Development Economics*, *111*, 92−106. Available from https://doi.org/10.1016/j.jdeveco.2014.07.004.

Folarin, O., & Adeniyi, O. (2020). Does tourism reduce poverty in sub-Saharan African countries? *Journal of Travel Research*, *59*(1), 140−155.

Fragoso, M., Trigo, R., Pinto, J., Lopes, S., Lopes, A., Ulbrich, S., et al. (2012). The 20 February 2010 Madeira flash-floods: Synoptic analysis and extreme rainfall assessment. *Natural Hazards and Earth System Sciences, 12*, 715−730.

Gössling, S., Bredberg, M., Randow, A., Sandström, E., & Svensson, P. (2006). Tourist perceptions of climate change: a study of international tourists in Zanzibar. *Current Issues in Tourism, 9*(4−5), 419−435.

Gossling, S., Scott, D., & Hall, C. M. (2015). Inter-market variability in CO_2 emission-intensities in tourism: Implications for destination marketing and carbon management. *Tourism Management, 46*, 203−212.

Hastley, R. (Producer). (2011). *The economic impact of natural disasters*. Business Pundit. <http://www.businesspundit.com/> Accessed on 09.06.19.

Head, K., Mayer, T., & Ries, J. (2010). The erosion of colonial trade linkages after independence. *Journal of International Economics, 81*(1), 1−14.

Huan, T.-C., Beaman, J., & Shelby, L. (2004). No-escape natural disaster: Mitigating impacts on tourism. *Annals of Tourism Research, 31*(2), 255−273. Available from https://doi.org/10.1016/j.annals.2003.10.003.

Intergovernmental Panel on Climate Change. (2007). Fourth assessment report (AR4). In: *Summary for policymakers. Synthesis report. Contribution of working groups I, II and III to the fourth assessment report of the intergovernmental panel on climate change*. Geneva, Switzerland. Available from: <https://www.ipcc.ch/site/assets/uploads/2018/02/ar4_syr_full_report.pdf>.

Isik, C., Dogru, T., & Turk, E. S. (2018). A nexus of linear and non-linear relationships between tourism demand, renewable energy consumption, and economic growth: Theory and evidence. *International Journal of Tourism Research, 20*(1), 38−49.

Jones, A. (2014). Natural disaster and the economic impacts on tourism. In: *The Centre for Labour Studies and the Institute of Earth Systems University of Malta in collaboration with the University of Prince Edward Island*, Canada Smithsonian Conservation Biology Institute, Washington, DC. Retrieved from <https://www.um.edu.mt/library/oar/bitstream/123456789/17193/3/2014%20ENVIRONMENTAL%20CHANGE%20DEC%20%202014%20natural%20disaster%20impacts%20tourism%20the%20med%20and%20malta%20-%20%20%20%20%281%29.pdf> Journal of Travel Research. Doi: 10.1177/0047287518821736.

Khalid, U., Okafor, L. E., & Aziz, N. (2019). Armed conflict, military expenditure and international tourism. *Tourism Economics, OnlineFirst*. Available from https://doi.org/10.1177/1354816619851404.

Khalid, U., Okafor, L. E., & Shafiullah, M. (2019). The effects of economic and financial crises on international tourist flows: A cross-country analysis. *Journal of Travel Research, OnlineFirst*. Available from https://doi.org/10.1177/0047287519834360.

Lise, W., & Tol, R. S. (2002). Impact of climate on tourist demand. *Climatic Change, 55*(4), 429−449.

Maditinos, Z., & Vassiliadis, C. (2008, July). Crises and disasters in tourism industry: Happen locally, affect globally. In Management of international business and economics systems, MIBES conference (67-76).

March, H., Saurí, D., & Llurdés, J. C. (2014). *Perception of the effects of climate change in winter and summer tourist areas: The Pyrenees and the Catalan and Balearic coasts*.

Murphy, P. E., & Bayley, R. (1989). Tourism and disaster planning. *Geographical Review, 79*(1), 36−46. Available from https://doi.org/10.2307/215681.

Neumayer, E. (2004). The impact of political violence on tourism: Dynamic cross-national estimation. *Journal of Conflict Resolution, 48*(2), 259−281.

Noy, I. (2009). The macroeconomic consequences of disasters. *Journal of Development Economics, 88*(2), 221−231. Available from https://doi.org/10.1016/j.jdeveco.2008.02.005.

Okafor, L. E., & Teo, W. L. (2019). 2018 WTO Trade Policy Review of Malaysia. *The World Economy, OnlineFirst*. https://doi.org/10.1111/twec.12846.

Okafor, L. E., Bhattacharya, M., & Apergis, N. (2019). Bank credit, public financial incentives, tax financial incentives and export performance during the global financial crisis. *The World Economy, OnlineFirst*. https://doi.org/10.1111/twec.12848.

Okafor, L. E., Khalid, U., & Then, T. (2018). Common unofficial language, development and international tourism. *Tourism Management, 67,* 127−138. Available from https://doi.org/10.1016/j.tourman.2018.01.008.

Page, S., Song, H., & Wu, D. C. (2012). Assessing the impacts of the global economic crisis and Swine Flu on inbound tourism demand in the United Kingdom. *Journal of Travel Research, 51*(2), 142−153. Available from https://doi.org/10.1177/0047287511400754.

Park, K., & Reisinger, Y. (2010). Differences in the perceived influence of natural disasters and travel risk on international travel. *Tourism Geographies, 12*(1), 1−24.

Santana, G. (2004). Crisis management and tourism: Beyond the rhetoric. *Journal of Travel & Tourism Marketing, 15*(4), 299−321.

Scott, D., Gössling, S., & Hall, C. M. (2012). International tourism and climate change. *Wiley Interdisciplinary Reviews: Climate Change, 3*(3), 213−232.

Seraphin, H. (2019). Natural disaster and destination management: the case of the Caribbean and hurricane Irma. *Current Issues in Tourism, 22*(1), 21−28.

Shafiullah, M., Okafor, L. E., & Khalid, U. (2018). Determinants of international tourism demand: Evidence from Australian states and territories. *Tourism Economics.*. Available from https://doi.org/10.1177/1354816618800642, OnlineFirst.

Sinclair, M. T. (1998). Tourism and economic development: A survey. *The Journal of Development Studies, 34*(5), 1−51.

Taylor, T., & Ortiz, R. A. (2009). Impacts of climate change on domestic tourism in the UK: A panel data estimation. *Tourism Economics, 15*(4), 803−812.

Tembata, K., & Takeuchi, K. (2019). Floods and exports: An empirical study on natural disaster shocks in Southeast Asia. *Economics of Disasters and Climate Change, 3*(1), 39−60.

Tourismembassy. (2014). *Impact of natural disasters on the tourism industry.* London: World Tourism Fields Venture Ltd. <https://tourismembassy.com/en/news/tourismology-by-tourismembassy/impact-of-natural-disasters-on-the-tourism-industry> Accessed 29.11.19.

United Nation World Tourism Organization (UNWTO). (2011). *Tourism towards 2030: Global overview.* United Nation World Tourism Organization (UNWTO). Accessed from <https://www.globalwellnesssummit.com/wp-content/uploads/Industry-Research/Global/2011_UNWTO_Tourism_Towards_2030.pdf>.

Wang, Y.-S. (2009). The impact of crisis events and macroeconomic activity on Taiwan's international inbound tourism demand. *Tourism Management, 30*(1), 75−82. Available from https://doi.org/10.1016/j.tourman.2008.04.010.

Wijaya, N., & Furqan, A. (2018). Coastal tourism and climate-related disasters in an archipelago country of Indonesia: Tourists' perspective. *Procedia Engineering, 212,* 535−542. Available from https://doi.org/10.1016/j.proeng.2018.01.069.

World Bank. (2018). *World development indicators.* Washington, DC: World Bank.

World Tourism Organisation. (2017). *UNWTO Tourism Statistics, CD-ROM.* Madrid, Spain: World Tourism Organisation.

WTTC. (2019). *Travel and Tourism: Economic Impact 2019.* Retrieved from London, United Kingdom. <https://www.wttc.org/-/media/files/reports/economic-impact-research/regions- 2019/world2019.pdf>.

WOMEN AND ECONOMIC DIMENSIONS OF CLIMATE CHANGE

6

Hurriyet Babacan

Rural Economies Centre of Excellence and The Cairns Institute, James Cook University, Cairns, QLD, Australia

6.1 INTRODUCTION

Climate change is one of the greatest ecological events of our time. Scientists continue to give dire warnings about climate change amidst major global debate about the nature and extent of climate change. Terry (2009:6) reminds us climate change should not be viewed in a vacuum and that it takes place "in the context of other risks, including economic liberalization, globalization, conflict, unpredictable government policies, and risks to health." Global inequality and poverty persists across the world. Gender disparities persist across countries in the world with varying degrees of inequality for women in life chances and opportunities. While appearing as disconnected, these phenomena are intricately linked and impact on the way we can respond, adapt, and be more resilient in the future.

The Intergovernmental Panel on Climate Change (IPCC) (2015:2) argues that "human influence on the climate system is clear, and recent anthropogenic emissions of green-house gases are the highest in history." Climate change includes both incremental changes such as droughts and sea level rises and catastrophic events such as bushfires, floods, and other disasters. Frequency and intensity of extreme weather and climate events caused by the anthropogenic impacts of climate change include extreme temperatures (heat waves and cold weather), droughts, cyclones, coastal erosion, acidity of oceans, wildfires, heavy precipitation and flooding, and high sea levels which lead to significant vulnerability and exposure for natural and human ecosystems (IPCC, 2015; Mignaquy, 2013).

Climate change has influenced food and water security, incidence of disease, and livelihoods. While climate change is negatively affecting the whole of humanity, the impacts are not equally distributed. As pointed out by IPCC (2015:13), "climate change will amplify existing risks and create new risks for natural and human systems. Risks are unevenly distributed and are generally greater for disadvantaged people and communities in countries at all levels of development." It is accepted that climate change disproportionately affects the world's poor, the majority of whom are women and children (Alam, Bhatia, & Mawby, 2015; Terry, 2009).

The link between gender and climate change has taken decades of work in the international arena. In 2018 the Committee on the Elimination of Discrimination Against Women (CEDAW) (2018:7) recommended that "States parties should ensure that all policies, legislation, plans, programs, budgets and other activities related to disaster risk reduction and climate change are gender responsive and grounded in human-rights based principles." In 2017 at the annual Conference of the Parties (COP23) to the Paris Agreement, the first ever Gender Action Plan was adopted. These agreements demonstrate the international recognition of the differentiated impacts of climate change on men and women.

This chapter focuses attention on the gendered economic impacts of climate change. Drawing on key themes from past research projects across six countries in rural contexts, this chapter identifies the economic impacts of climate change on women. Gender also plays a critical role in individual, family, and community vulnerability, with women typically more likely to be negatively affected by the impacts of climate change and extreme weather events. Women are affected by climate change in major ways, including livelihoods and wellbeing in areas such as food security, poverty, health, violence access to resources, and power.

6.2 METHODOLOGICAL APPROACH

This chapter draws upon findings from climate change and gender research and development projects conducted between 2014 and 2019 in six countries: Australia, Papua New Guinea (PNG), Kenya, Cambodia, Vanuatu, and Fiji. The studies were in rural contexts and the findings reflected here do not cover urban environments. The projects were focused on exploring the gendered dimensions of climate change, adaptation, and disaster risk reduction. While these projects were distinct and unique in each country, two overall approaches guided the methodology of these projects:

1. *Intersectionality*: Intersectionality is a concept developed by Crenshaw (1991) to locating and analyzing multiple constructs of oppression and marginalization. It stems from theories of social location theory which excavates the notion that the intersection of different markers of identity and difference map one's "social location." Intersectionality contends that the distinguishing categories within a society, such as race/ethnicity, gender, religion, sexual orientation, class, and other markers of identity and difference, do not function independently but, rather, act in tandem (Manuel, 2006:175).

    As Hill-Collins (1995:491) writes:

    We clearly need new models that will assist us in seeing how structures of power organized around intersecting relations of race, class, and gender frame the social positions occupied by individuals and work in explaining how interlocking systems of oppression produce social locations for us all.

    Denis (2008:677) emphasizes that the focus is on a matrix of power relations and the need to have concurrent analyses of multiple, intersecting sources of subordination/oppression. Structural forms of intersectionality refer to the ways in which individuals with intersecting identities find themselves marginalized because of structural barriers (language, gender, governance, institutions, poverty, and citizenship). Structural intersectionality is contextual and dynamic.

    Intersectionality is also important in that it enables a noncategorical analysis of individual positions and community locations as women experience them (Makkonen, 2002). It prevents seeing women who experience intersectional discrimination as victims and allows us to see them as actors with agency, a position which is critical when we consider climate change vulnerability for women.

2. *Inclusive Systemic Evaluation for Gender equality, Environments and Marginalized voices* (ISE4GEMs). The ISE4GEMs approach is grounded in both systems thinking and complexity. Systems thinking does not separate individual parts of what is being studied to gain understanding, rather focusing the analysis on the interaction between the individual parts giving a more expansive understanding and offering different conclusions. Complexity refers to "situations of change and uncertainty, in which many forces interact simultaneously, so that not only is each place and situation completely different from the next one, the same place is completely different from how it was before" (Stephens, Lewis, & Reddy, 2018:11). This considers the interrelationships between gender equality, environments, and marginalized voices (human and nonhuman) using systemic thinking. The ISE4GEMs methodology enables a boundary story to be developed in which intersectional analysis of the gender, environments, and marginalized voices could be included through a range of lenses, including interrelationships, boundaries, and perspectives (Stephens et al., 2018). Data collection is undertaken within specific ethical research values and frameworks. Data triangulation occurs through an analysis of values, boundaries, and findings (Stephens et al., 2018).

It is not the intention of this chapter to report on each country. Individual research reports, written elsewhere, provide the specific findings. Rather this chapter draws upon the metafindings from the research projects to draw a broader conclusion. The research is also supported by an extensive literature review, in academic and gray literature, using key words such as gender, women, climate change, and economic development. The results provided here analyses the findings to identify themes, common issues, cross-linkages, and emerging patterns, as part of data analysis process (Sarantakos, 2013).

6.3 GENDER BARRIERS, VULNERABILITY, AND ADAPTATION

Gender: Gender is socially constructed and is one of the most complex and important concepts in social theory (Babacan, 2013). Gender assigns men and women characteristics such as roles, tasks, functions, and roles in public and private life (SDC, 2003) and determines what is expected, permitted, and valued in men and women in specific contexts (UNDP, 2009).

Women are not a homogeneous group and their lives vary depending on the place in which they live as well as their age, social class, ethnic origin, and religion. In 1948 the Universal Declaration of Human Rights reaffirmed the belief in the equal rights of men and women. Despite this, women experience considerable disadvantage and discrimination in society perpetuated by gender-differentiated structures. Kimmel (2004:1) points out that gender also determines structures of social power:

Gender is not simply a system of classification by which biological males and biological females are sorted, separated and socialized into equivalent sex roles. Gender also expresses the universal inequality between women and men. When we speak about gender, we also speak about hierarchy, power, and inequality, not simply difference.

In all societies, female subordination is a common denominator of the female gender, although the relations of power between men and women may be experienced and expressed differently in different societies and at different times (Babacan, 2014, 2009; Lengermann & Niebrugge, 2013; Rowbottom & Linkogle, 2001). Gender disparities remain among the most persistent forms of inequality across all countries (UNDP, 2019:147). The global gender gap parity index measures gender-based gaps for 153 countries in four key areas: economic participation and opportunity, educational attainment, health and survival, and political empowerment. The 2019 results show that the average distance is 68.6% with a 31.4% gender gap parity that remains to be closed at the global level [World Economic Forum (WEF), 2019:5]. However, the parity is not evenly distributed across the four areas. Educational attainment as well as health and survival have improved over the years and enjoy much closer to parity between men and women (96.1% and 95.7%, respectively) (WEF, 2019). However, women's economic participation and opportunity has regressed with gender parity being 57.8% impacting on labor force participation and economic inclusion. In political participation of women, there is little progress. At the global level, only 24.3% of all national parliamentarians were women in 2019, with only 11 women are serving as Head of State (UN Women, 2019b). The Human Development Index shows that political participation of women in parliament in 2018 is 32.7% for Australia; 23.2% for Kenya; 19.3% for Cambodia; 16% for Fiji, and Vanuatu and PNG do not have any female political representatives. The highest percentage of women in parliament is in Norway with 41.4% (UNDP, 2019:316−318). Lack of participation of women in decision-making forums continues into the corporate sector. For example, only 5% of Fortune 500 Chief Executive Officers (CEOs) are women (UN Women, 2019a) and representation of women in corporate boards is very low.

The World Economic Forum estimates that it will take 247 years before gender parity can be reached. There is significant evidence from around the world to demonstrate women's inequality. The Human Development Index (UNDP, 2019:148) argues that "there are troubling signs of difficulties and reversals on the path towards gender equality" in areas such as female heads of state and government, for women's participation in the labor market, even where there is a buoyant economy, and gender parity in access to education.

6.4 GENDER AND ECONOMIC EXCLUSION

Economic development, inclusion, and participation for women are explicitly linked with gender-based equalities outlined previously. The World Bank (2014:2) argues that "in virtually every global measure, women are more economically excluded than men." The key areas of economic exclusion of women include the following:

Economic exclusion and access to productive resources: UN Women (2019a) identifies key factors in women's economic empowerment: women's ability to participate equally in existing markets; their access to and control over productive resources, access to decent work, control over their own time, lives and bodies; and increased voice, agency, and meaningful participation in economic decision-making at all levels from the household to international institutions. Men and women have differentiated access to the means of production. There is gender inequality globally in the distribution of assets, services, information, for example, secure and adequate land, credit, employment, mobility, climate and market information, access to markets, education and training, information and communication technologies, economic support services, and other resources (Simelton & Ostwald, 2019; World Bank, 2014). In agriculture, where women are dependent on for livelihoods, the World Bank identified lack of land ownership, or long-term user rights, access to agricultural credit, access to productive farm inputs, access to timely labor, support from extension and advisory services, access to markets and market information, and access to weather and climate information (World Bank, 2015:8).

Land ownership: Globally, women own less than 10% of land, and in agriculture, receive only 5% of all agricultural extension services, and have access to only 10% of total agricultural aid, even though they contribute significantly to agricultural and food production [Food and Agriculture Organization (FAO), 2019]. There is considerable variation across women in the world. The share of female agricultural land holders ranges from 0.8% in Saudi Arabia to 51% in Cape Verde, with an overall global share of 12.8%. In many countries, land is owned predominantly by men and transferred intergenerationally to men or there are customary gendered land-owning practices (Alston, Clarke, & Whittenbury, 2018; Deering, 2019; FAO, 2018).

Labor force participation: There is a steady decline in the labor force participation of women globally. In 2018 the labor force participation of women was 48.5% which was 26.5 percentage points below that of men (ILO, 2018:6). ILO identifies that the gender gap in labor force participation is not uniform and is narrowing in developed countries and increasing in emerging and developing countries. The levels of unemployment for women are globally higher than men, with 6% unemployment across the world [compared to men at 5% (ILO, 2018:7)]. Restrictions to female employment, norms limiting women's employment in general or in particular sectors, poor work conditions, high fertility, and low female wages are major aspects of women's participation in the labor market (Klasen & Gaddis, 2013).

Vulnerable employment: Greater numbers of women are in informal, casual, and vulnerable employment. For example, women constitute approximately 17% of family workers compared to 6% of men (ILO, 2018:9). UN Women (2016:71) identifies that women make up disproportionate numbers of informal employment with 80% in South Asia, 74% in sub-Saharan Africa, and 54% in Latin America and the Caribbean, in jobs such as street vendors and domestic workers to subsistence farmers and seasonal agriculture workers (UN Women, 2016:71). In developed nations, women are in casual, part-time, and insecure jobs. Vulnerable employment leaves women often without any protection of labor laws, social benefits such as pension, health insurance or paid sick leave, unsafe work conditions, and at risk of harassment and violence. Even in developed economies, such as in France, Germany, Australia, women's average pension is more than 30% lower than men's (UN Women, 2016:147).

Income and wage gaps: No country in the world has yet achieved gender parity in wages (UN Women, 2019a; WEF, 2019). The global gender wage gap is estimated to be 23%. This means that women earn 77% of what men earn (UN Women, 2019a), although there are significant differences across countries. The World Economic Forum (2019) outlines that purchasing power parity is a useful way to look at average incomes. In 2018 the global average of woman's income was about $11,000 (in purchasing power parity), while the average income of a man was $21,000 (WEF, 2019:17). The Gender Development Index 2019 identified the major income differentials between men and women, including in developed countries. For example, in 2018, estimated gross national income per capita varied for men and women: in Australia, for men it was $52,359, while for women it was $ 35,900; in Fiji, for men $12,292 and for women it was $5,839; in Kenya, for men $3490 and women's income was $2619; in PNG, men's income was $4106, while the income for women was $3248; in Cambodia, $4089 was for men and for women it was $3129; and in Vanuatu, it was $3413 for men and $2185 for women (UNDP, 2019:312−314). These figures demonstrate that there are major gaps in per capita income between men and women around the world. They also show the global inequalities and disparities between men and men and women and women in different parts of the world.

Financial exclusion: Financial inclusion is a key factor in women's economic participation. Globally, 65% of women have an account compared with 72% of men (World Bank, 2017:vi). Of the 1.7 billion unbanked adults in the world, 56% are women (World Bank, 2017:4). Savings patterns also vary by gender and income. In developing economies, men are 6 percentage points more likely than women to save at a financial institution (World Bank, 2017:9). Saving in financial institutions was much higher for developed nations (77%) compared to developing nations (43%) (World Bank, 2017:8). Alternative forms of saving were more common in developing nations, including investing in other assets, cash at home, and semiformal savings clubs (community-based). Women had less access to credit and borrowing. Worldwide, women as borrowers receive just 5% of available credit (USAID, 2013:8).

Gender and the law: There are numerous countries across the world where women are not protected or obstructed by the law. These include freedom to travel and mobility, right to work, property ownership and inheritance, labor laws, for example, wage rates and maternity leave, domestic violence protections, social protections, and freedom from gender discrimination at work or in business transactions. Reviewing progress in women, business, and the law, the World Bank 2019 (2019:3) identifies a score of 74.71 indicating that the global economy gives women only 3/4s of legal rights as men. While progress and reforms are made, there is considerable variability across the world. Of 189 economies assessed in 2018, 104 economies still have laws preventing women from working in specific jobs (UN Women, 2019b). In 72 countries, women are barred from opening bank accounts or obtaining credit or opening a business without male permission (WEF, 2019).

Digital connectivity: Women have less access to digital connectivity. A study conducted by the McKinsey Global Health Institute (2015:12) found 4.4 billion are offline, with 52% are women. The World Bank identifies that digital connectivity is critical for economic and financial inclusion of women. For example, to mobile money services delivers major benefits for women, for example, increased savings, options for leaving farming, developing businesses, and reducing poverty among women-headed households by 22% (World Bank, 2017:1).

6.5 DIFFERENTIATED CLIMATE CHANGE IMPACTS: EMERGING THEMES FROM RESEARCH

6.5.1 CLIMATE CHANGE AND GENDER

Climate change impacts are disproportionate and influence lives and livelihoods. One crucial determinant of these disproportionate impacts is gender (Bhadwal, Sharma, Gorti, & Sen, 2019). It is argued that climate change will hit women disproportionately hard (Bjornberg & Hansson, 2013;

Eastin, 2018). Women and children are 14 times more likely than men to die during natural disasters. 70% of fatalities in the 2004 Asian tsunami, and 96% in the 2014 Solomon Island floods were women and children (FAO, 2018). Different subgroups of people have varying interests, priorities, levels of power, and capacities to access critical resources for climate change adaptation. Climate change and gender inequalities are intertwined. As noted by UNDP (2019:179) "Part of the reason climate change and disasters are disequalizing is that inequality exists in the first place; they run along, exploit and deepen existing social and economic fault lines." The International scholarship confirms that women are disproportionately vulnerable to the effects of climate change than men (Alam et al., 2015; Alston & Whittenbury, 2013; Eastin, 2018; UNDP, 2009; UN Women, 2016). This, in turn, can exacerbate existing gender disparities. However, it is important to remember that vulnerability and exposure of women and girls to disaster risk and climate change are economically, socially, and culturally constructed and can be reduced (CEDAW, 2018).

Vulnerability comprises exposure and sensitivity to environmental threats, and capacity to cope with environmental crises and climate change (Eastin, 2018: 289). *Adaptation* is defined by the IPCC (2014) as "initiatives and measures to reduce the vulnerability of natural and human systems against actual or expected climate change effects." Gender issues interface with different facets of climate change. The vulnerability, exposure, and adaptation to climate change by women and men differ due to differences in resources, adaptive capacities, decision-making participation, levels of knowledge and information, and power relations. Ampaire et al. (2019:1) points out that "sociocultural and gender norms, the gender division of labor and differing levels of access to productive resources and cash, not only make women more vulnerable but also affect women's ability to respond and adapt to climate impacts and shocks." The IPCC (2007) recognizes that gender roles and relations shape vulnerability and people's capacity to adapt to climate change. It acknowledges the vulnerability of rural women, particularly in developing countries, who are often dependent on natural resources for their livelihoods, do most of the agricultural work, and bear responsibility for collecting water and fuel (IPCC, 2007). Gender disparities in vulnerability have been found to magnify inequalities in intrahousehold bargaining power, with implications for gender equality and women's rights at both the micro- and macro-levels (Eastin, 2018:291). Bhadwal et al. (2019:69) identify four key areas of gender differences in vulnerability and adaptation: gendered perception of risk and exposure, gender roles and responsibilities, gender entitlements and capabilities, and gender interactions with multiple actors. The discourses on vulnerability have focused on women as being victims, for example, poor or higher mortality rates in disasters (Arora-Jonsson, 2011:745). Women are more vulnerable to the effects of climate change than men due to socially constructed roles and face unequal power relations tend to make them more disadvantaged (Mignaquy, 2013). In the discourses of vulnerability, it is essential not lose sight of the demonstrated capacity of women to be proactive and be change agents in adapting to climate change (Alston et al., 2018; Tanyag & True, 2019; Terry, 2009).

There is general consensus that greater inclusion of women will improve environmental decisions and social and economic outcomes (Alston et al., 2018; Babacan & Lewis, 2019; Grillos, 2018). There are considerable barriers to women's leadership and decision-making, at different scales from global to the household level. The United Nations Framework Convention on Climate Change (UNFCCC) identified that only 40% of nations explicitly mention "gender" or "women" in the context of their national priorities and ambitions for reducing emissions (UN Women, 2016:16). The Paris Agreement mandates gender-responsive adaptation actions. In article 7.5, "Parties acknowledge that adaptation action should follow a country-driven, gender-responsive, participatory and fully transparent approach ... with a view to integrating adaptation into relevant socioeconomic and environmental policies and actions." In addition, Article 11.2 recognizes the need to build capacity for participation and states that "... Capacity-building should be guided by lessons learned ... and should be an effective, iterative process that is participatory, cross-cutting and gender-responsive." Despite recognition of the importance of women in climate change, gender gaps are evident in climate governance in practice. The [International Union for Conservation of Nature (IUCN), 2015:2] observes "that despite national and international agreements on gender equality, leadership positions continue to be heavily dominated by men at all levels." In an analysis of the percentage of women's representation across the major international environmental forums such COPs of the three Rio Conventions COPs (UNCBD, UNCCD, and UNFCCC) revealed that in six out of nine decision-making processes analyzed, women represent less than one-third (IUCN, 2015:2).

In recent years, there has been a focus on climate change and economic impacts. There is a broadening of approach toward economic development and growth. *Sustainable livelihoods approach* emerges as a framework in which to view economic development. The sustainable livelihoods framework is based on understanding that people operate within contexts of vulnerability and have access to assets that typically include natural, human, social, physical, and financial capital to support their means of living. A livelihood is sustainable "when it can cope with and recover from stresses and shocks and maintain or enhance its capabilities and assets both now and in the future, while not undermining the natural resource base" (Chambers & Conway, 1992). This approach is beneficial to traditional economic frameworks as it can provide basis for understanding how livelihood strategies can build adaptive capacity to enable people to better cope with change, and diversify their activities to increase resilience in responding to climate change (Reed et al., 2013).

6.5.2 EMERGING THEMES FROM RESEARCH ON CLIMATE CHANGE AND GENDER

The findings from the research projects in six different countries revealed that women take a wholistic approach to climate change and disaster risk reduction. Women were clearly able to articulate an understanding of the impacts of climate change in their local situations, regardless of age, education, and location. They identified climate change through an economic lens of their livelihoods and took a critical bridging approach between crisis (disaster management) and development work (livelihoods). Key issues identified included impact of climate change of food production, lack of income, lack of opportunity and jobs, food security, increased workloads, household and domestic conflict, and social issues such as illness and health.

Women demonstrated rich and localized knowledge about environmental management, which was gender-differentiated knowledge due to division of labor and role segmentation. Their traditional practices focused at the household or micro-level which were often not recognized by decision-makers.

A number of key themes emerge from the research:

Food security: Women are responsible for the production of food and food security at the household level. Women, on average, comprise 43% of the agricultural labor force in developing countries and account for an estimated two-thirds of the World's 600 million poor livestock keepers (FAO, 2019). In some countries this figure is higher; for example, women account for 60%−80% of smallholder farmers and produce 90% of food in Africa and about half of all food worldwide (FAO, 2019). Climate change is significantly changing agricultural production and food security across the world (IPCC, 2014), with negative impact on agriculture, affecting major crops, livestock production, and fisheries. Climate change will cause greater uncertainty and risk, changing workforce and productivity, and agricultural production patterns (Alston et al., 2018; IPCC, 2014; Torquebiau, Tissier, & Grosclaude, 2016). There is a risk of food insecurity and breakdown of food supply systems in the face of climate change (Torquebiau et al., 2016:8).

A few issues emerged in food security for women, including unequal access to land, information, water rights, livestock, and inputs such as improved seeds and fertilizer. Traditional food sources were more unpredictable and scarce as the climate changed. Risks to water resources, marine systems, and food supply systems mean threats to food security. Women bear the burden of poverty, often referred to as feminization of poverty (Klasen, Lechtenfeld, & Povel, 2015). When poverty prevails, the climate change impact exacerbates other stress factors and depreciates the wellbeing of vulnerable people. While economic growth is important in addressing food security, it is not enough (Deering, 2019; FAO, 2017). Approaches to food security have become nuanced with recognition of the importance of food supply and distribution chains at the macro-level (Oxfam, 2014) as well as the availability, acquirement, and utilization of food at micro- and intra-household levels (Deering, 2019; FAO, 2014) with both scales being critical in defining food security. Women, especially in development contexts, tend to be more dependent on the products of their local production systems for their food security, energy, water and other products, and services and are likely to be more vulnerable to the local-scale effects of climate change (FAO, 2017).

While women's work in the private sphere is generally recognized, their contributions to overall food security are often not recognized (Alston et al., 2018). Women are treated as home producers and not recognized as economic agents, limiting their access to support for food production. Studies show that women are more likely than men to prioritize the food security needs of their families. Women are referred to as "gate keepers" (Brown et al., 1995) as they ensure that household members receive adequate share of food that is available and contribute their time, income and reduced own consumption of food in times of food insecurity (Alston & Akhter, 2016; Brown et al., 1995).

Analysis from our research findings show that climate change will disproportionally affect those who depend on natural resources for their livelihoods, men and women. Many of the women in the study were identified as food producers and responsible for cocontribution to food security in their households. They identified challenges, including prolonged periods of drought, floods, fire, and storms, diminishing their ability to grow food. In PNG, Fiji, and Vanuatu, women identified that their vegetable gardens, major sources of their food, have been impacted, limiting the quality and amount of food available. The key impacts on livelihoods included reduction in crop yields, changing times of harvest due to climatic conditions, working longer to get similar yields, not having knowledge to change cropping patterns, livestock poor health, and washing away of salts by rains (salt harvesting). An overwhelming majority of the women in this study (both in the developed and developing nation context) identified that they were responsible for food preparation, domestic duties, and securing food for their families. Access to water had been affected in many instances, including drying up of water sources, less allocation of water, water pollution, and lack of infrastructure for water capture. For many women, there has been an increase in their workloads in finding water, wood for fire, and in trying to grow crops.

6.5.3 LIVELIHOODS AND ECONOMIC DEVELOPMENT

It is important to recognize that risk and vulnerability have a fundamental social dimension: men's and women's livelihoods differ in relation to specific natural hazards, climate change, and disaster risk reduction (FAO, 2016). The research identified that a number of strategies are used by women and men for livelihoods and economic development, including smallholder entrepreneurship, participation in agricultural and nonfarm labor market, and outmigration, detailed next.

Division of labor: In all countries of the research, there was a strong division of labor and role segmentation in economic activity. In developing countries, women were more reliant on agriculture and livestock for their food security and income. Gendered patterns over income security varied. In some instances the division of labor resulted in women as producers and men market the crops and had control over the income. In other situations, women control the markets (e.g., Vanuatu). The division of labor showed differences across countries but one important factor was that women were overwhelming responsible for domestic duties and household chores. Majority of women were primary carers and identified as having less mobility than men. Women identified spending more time in collecting water, firewood, and preparing meals for the family. There is no country where men spend the same amount of time on unpaid work as women. In countries where the ratio is lowest, it is still 2:1. On average, women spend about 2.5 times as much time on unpaid care and domestic work as men do (ILO, 2018). 75% of global unpaid work is undertaken by women (McKinsey Global Health Institute, 2015:2).

Despite this the global gender income gap is 44% (UNDP, 2019:161). This impacts on women's labor force participation, lowers economic productivity, and limits their opportunities (Alonso, Brussevich, Dabla-Norris, Kinoshita, & Kochhar, 2019). Women in developing nations identified increased workloads of managing climate change impacts, work which men did not undertake such as food security, fetching water, and cleaning after disasters. In Australia the size of farm ownership is changing form small-scale family farms to larger farms, women informants identified as increasingly working longer hours. Women who participated in the agriculture and family work identified that their contributions were not explicitly recognized. Women also identified the need to diversify incomes but there was a lack of work and unemployment. They identified lack of job opportunities in their area, lack of relevant qualifications (education and training), and lack of childcare options.

Access to land: Women identified access to land as a major issue as they needed to expand planting of crops or diversify crops. In many countries, there were strict customary rules about land ownership and use, limiting women's access to means of economic and food production. In traditional societies, such as PNG and Vanuatu, land ownership is through complex customary arrangements, often marginalizing women as decision-making is by men. Women's strategies to adapt to climate change, for example, plant diverse crops, in different locations, have multiple vegetable gardens and developed enterprises were linked with access to land. In the Australian context, land is often jointly owned by husband and wife, but decision-making processes on the land showed variability with factors such generational practices ("we have done it like that for three generations"; male making the decisions; and joint decision making about livelihoods). Overall, women's contribution to agriculture and other livelihoods did not translate into ownership of land or participation in key decisions about livelihoods.

Diversification of income: The negative effects of climate change on their livelihoods have meant that many women have had to look for other sources of income. Women identified having to work on other farms, nonfarm laboring jobs, cleaning, retail, making and selling handcrafts, sewing, other agricultural work to supplement their income. For some, it has meant leaving family to work in factories in towns. Climate change and the need to diversify income generation exhibited pressure on the social norms. For example, women who did not work outside their homes were forced to seek work in towns (e.g., Cambodia, Vanuatu). Women in developed nation contexts, such as Australia, identified that their work on farms has increased the workload of men who worked harder on the farms in the face of drought, floods, and bushfires. The women's workloads were concentrated around farm management, domestic chores, and finding other sources of income as farms struggled. Women identified finding part time or casual work to supplement incomes or started small-scale enterprises as to generate income. Overall, it was identified that climate change increased food insecurity, income insecurity and generated harder work for men and women (in differentiated ways). Women identified the need for technical support in agricultural management, diversifying crops, access to microcredit, and developing new skills and learning enterprise development in order to continue their livelihoods.

Outmigration: The outmigration was cited for men and women. Women have also left to work in big cities, for example, garment factories, or overseas in domestic work. Female outmigration often means women leaving families to go to cities and other areas for work, creating other challenges. Issues of safety, accommodation, low wages, and poor work conditions were identified for women who migrated out. Male outmigration created significant work for women. Examples were provided in fishing villages where fish sources have been impacted or unproductive agriculture forced men to leave for cities or overseas to work, leaving women with greater household and farm management. In Australia, it was identified that young women left rural areas for economic and educational opportunities in bigger cities. Men often took fly-in-fly-out roles where they are away to work in mines or other farms, leaving the women as defactor sole household heads to manage households, businesses, and farms.

Market access and entrepreneurship: Women in the study had great insights and knowledge of what was needed to adapt to changing environmental circumstances and had practical solutions to impact on their economic situations. Women demonstrated economic adaptation strategies. Some examples included modifying crop varieties in agriculture, water management strategies, business management strategies, financial adjustments, changing products, and changes in marketing strategies where relevant for the sale of their products/crops. They identified significant challenges, including social norms and male biases about women's economic leadership, restricted land and water rights, lack of access to training, lack of transport and mobility, poor access to technology and digital skills and financial resources, particularly credit. Women identified the lack of critical infrastructure, roads, ICT, and storage to be able to capitalize on markets within and outside their region. Others identified the need for further training in key aspects of business development, communication, and improving productivity. Some identified lack of knowledge and confidence to take up entrepreneurial roles and market-based roles. The research showed that asymmetries in access to land, monetary resources, skills and capabilities, and unequal power relations determined how women participate in the market, including choice of work, access to finance, choice of investment and risk, and product development. Some of the women identified the need for value adding to their produce in order to have improved price, product diversity, and market options. Market access provided opportunities and challenges. Access to markets was identified as an importatnt consideration to provide certainty to climate-related risks for women by reducing the risk of perishable product storage losses and providing new market opportunities but also increased risk exposure to market volatility. Numerous barriers were identified for accessing markets, including lack of decision-making power, lack of finances, lack of technical knowledge, and lack of confidence. Where women were empowered, through a range of initiatives targeting women, men, and social structures, key transformations were identified, including increased savings, access to credit, increased knowledge about production, increased household autonomy, personal autonomy-increased independent income, increased education and training opportunities, and greater say in the control of strategic household and community decision-making and assets and leadership in the community. Women identified the need for training in book-keeping and financial management, business development, leadership, savings and credit schemes, and marketing.

Participation in decision-making: Numerous accounts of societal expectations of women's roles were identified in households, communities, local authorities, workplaces, and overall society. Women identified numerous examples of exclusion from decision-making that limit their opportunities and choices. At the household level decision-making about personal autonomy, access to money, strategic decision-making about household assets, investment decisions, education of children, reproductive choice and right to work are some key areas of exclusion. At local or regional level, women identified issues of lack of representation of local committees and decision-making forum, investment decisions of public authorities, lack of knowledge about formal processes and expectations that women should not be in the public sphere (e.g., expectations of more feminine and submissive roles, women not speaking before males, women not seen as educated, poor status as widows). For example, in Vanuatu, the "nakamals" are where local decisions are made by Chiefs and women have no representation. Prioritization about decisions for infrastructure was made by men at local or commune level in Cambodia and women's priorities for public spending was different (e.g., water storage, health clinic by women compared to men who prioritize large infrastructure and road projects). Examples were given by women in the different countries of women being undermined for taking economic initiatives, some by family and other times by authorities. Women leaders identified that they could gain respect and decision-making status via achieving outcomes. Women were seen to have a choice in the "women's issues," but these areas had very limited "soft" areas with small budgets. At the national level, women identified lack of representation, lack of advocacy, lack of gender mainstreaming, and lack of knowledge about policy processes to influence budgetary and policy outcomes.

Conflict and gender-based violence: Climate change has the potential to exacerbate conflict over limited natural resources, threatening lives and security of individuals and communities

(Lewis & Babacan, 2019). In some instances, it can also lead to political instability. In situations of increased conflict, women are most vulnerable and excluded from decision-making processes (Alam et al., 2015). It is also known that domestic and family violence increases in times of socioeconomic pressure and disaster events (Babacan & Lewis, 2019; Lewis & Babacan, 2019a). Examples of gender-based violence were provided in the research. These included lack of safety when women have to travel further distances to gather water or firewood or to go to the markets to sell produce, domestic violence at home from husbands in conflicts over money, food and work, vulnerability due to access to transport and roads, and risks for women during disasters and displacement (e.g., sexual violence). Women in the research identified the devastating impacts of violence, including loss of confidence, fear, negative health consequences, and barriers to full social, economic, and political participation.

Livelihood approaches: The women in the study took a livelihoods approach to economic adaptation. At the household level, men were more engaged with formal financial transactions, whereas women had greater informal exchanges that were critical adaptation strategies for food security, savings schemes, and building capacity. Women who had participated in development programs identified that they had more bargaining power and equal household decision-making. Women used kin and social networks to build social protection and market mechanisms (e.g., savings and enterprise development). Women who are part of collective networks identified multiple benefits of building social capital and safety nets. Although each of the relationships was context specific, there was the capacity to develop a portfolio of strategies for coping with the effects of climate change. In the context of power imbalance, these processes enabled building of confidence and provided empowerment. Examples were given of where women collectively established a women's market, set up women's enterprises to diversify livelihoods, set up savings scheme, influenced infrastructure development, and influenced investment decisions (at household and village/regional level). They advocated for physical capital and for the prioritization of infrastructure that they deemed essential, for example, schools, health clinic, roads, and market. They developed ways to raise finance and credit, although this was identified as a major area of challenge. There was an acute awareness of the need to build human capital and capacities. Some women had attended training on climate change and disaster risk reduction and vulnerability mapping. Through women's safe spaces to organize on livelihoods and climate change, women could organize in a culturally appropriate manner to build skills but more importantly, develop empowering strategies for economic independence and participation in decision-making at multiscalar levels.

6.6 ANALYSIS AND DISCUSSION

Climate change has differential impacts on men and women due to differing roles, resources, rights, knowledge, and time with which to cope with climate change (Babacan & Lewis, 2019; Resurrección et al., 2019). Women are disproportionately represented in bearing the burden of climate change and are relatively more reliant on climate-sensitive livelihoods (FAO, 2018). Complex factors make women more vulnerable to the adverse effects of climate change. Women tend to have limited adaptive capacities, and their livelihoods and economic prosperity are sensitive to climatic change (Dankelman, 2010; FAO, 2011; UNDP 2019). In general, women have lower levels of access to resources (e.g., information, land, assets, education, and development services) to capture opportunities to diversify their livelihood options and to lessen dependencies on stressed natural resources (Dankelman, 2010). Gender inequalities undermine climate change economic and livelihoods adaptation (Babacan & Lewis, 2019; Resurrección et al., 2019).

Women leaders understand climate change adaptation and resilience as a continuum and interlinked with ongoing development issues such as access to reliable water, health, and livelihoods (Lewis & Babacan, 2019). Key issues that are identified by women in the research countries have been resources for recovery from climate-induced disasters and securing livelihoods. Therefore their priorities and needs are vastly different from that of men and often from that of policymakers.

For effective climate change adaptation, it is critical to harness the knowledge and capability of women (WEF, 2019). The Food and Agriculture Organization found that giving women the same access as men to agricultural resources could increase production on women's farms in developing countries by 20%−30% (FAO, 2014, 2018). Women's contribution can enhance economic growth if gender parity could be narrowed. For example, if could women play an identical role in labor markets to men's, 26% could be added to global annual GDP in 2025 (McKinsey Global Health Institute, 2015:1). Climate change will impact negatively on economic growth. An increase in average global temperature by 0.04°C per year, in the absence of mitigation policies, reduces world real GDP per capita by more than 7% by 2100 (Kahn et al., 2019:47). Maximizing women's knowledge and productivity is critical to broader macroeconomic climate change strategies.

Studies on gender and livelihoods often focus at micro-level (household level). Research confirms that responses to climate change must be integrated with other development and protection mechanisms and address constraints on livelihood options and gendered social practices. Micro-level home-based inequalities exacerbate market-based gender inequality (UNDP, 2019) and have flow on macroeconomic impacts. Women face different levels of constraint in achieving economic outcomes: (1) those rooted in the informal and intrinsically gendered institutions of family, kinship, and community and (2) those embodied in the formal domains of states, markets, and civil society (Kabeer, 2018:i). These gendered structures of constraint enforced within the family and community are also reproduced and reinforced within the public domains. Women experiences of these constraints show variation in geographical ways intersectional across different life stages. This means that economic empowerment of women will require complex interventions at different levels. While focusing effort at the individuals at the micro-level is beneficial, it is also critical to have laws and policies which address the larger constraints within public domains of states, markets, and civil society. The broader interventions increase the likelihood that micro-level interventions will succeed (Kabeer, 2018). There are significant gaps in policy domain in relation to gender and economic development. The focus on public policy has been selective, and more gendered analyses are needed in macroeconomics, supply chains and trade, and capital flows, as well as consideration of land reform, gender, and macroeconomic impacts (Alston et al., 2018; Kabeer, 2018; UN Women, 2016).

Supporting and empowering women's capabilities are critical for economic resilience. There is a need to identify ways to create livelihood that strategically reinforces women's leadership and voice in household, local and national economies. Governments have a critical role to play. The policy-making processes can reinforce or mitigate the biases at market, household, and societal levels. Creating gender-sensitive governance processes, supporting women's participation in decision-making formal processes, and providing safe spaces for women's leadership are critical (Goetz & Jenkins, 2005). Initiatives that enable women to organize collectively to follow economic opportunity and develop economic independence are fundamental. Collective spaces also enable women to have access to information, develop confidence, develop capacities and skills for leadership, and independence (Agarwal, 2009). Women's voice, participation in decision-making, and agency are critical in the management of natural resources and responding to climate change. Evidence from around the world shows that women are not well represented in decision-making processes. Women are often excluded from decision-making at many levels, including in relation to land use planning, prioritization of investment, and management/risk decisions. Evidence suggests that when women are involved in decision-making processes, the involvement results in positive outcomes, including better environmental management, improved productivity, and less conflict (Agarwal, 2009; Coleman & Mwangi, 2013). Therefore the centrality of women-led approaches, women articulating and expressing their needs and priorities, capacity building (skills and knowledge) and influencing and advocacy capacity cannot be underestimated as fundamental requisites for economic resilience.

The economic empowerment of women is defined as a process whereby women and girls experience transformation in power and agency, as well as economic advancement (Pereznieto & Taylor, 2014:234). Ultimately it has been identified as addressing power disparities: *power within* (knowledge and individual capabilities, self-esteem); *power to* make economic decisions in household, local and other levels (including outside areas what is regarded as women's business); *power over* (access and control over resources, e.g., financial, physical, and knowledge); and *power with* [to organize with others to enhance economic activity (Pereznieto & Taylor, 2014:236)]. Addressing power inequities requires an intersectional approach for gender justice and gender rights (Chant & Sweetman, 2012).

6.7 CONCLUSION

It is projected that by 2100, the compounded threats faced due to climate change will be in multitudes across five main human systems: human health, water, food, economy, and infrastructure and security (Mora et al., 2018). The consequences of climate change will require complex approaches that address the interaction effects of different risks and hazards, develop multiscalar mitigation, adaptation, and resilience strategies. It is critical to recognize that climate change is intertwined with existing and emerging inequalities and crises such as poverty, armed conflicts, health pandemics, and economic recessions (Tanyag & True, 2019). Women face multiple and compounding barriers in responding to climate change impacts. We need more nuanced gendered analysis of the impacts of climate change, especially the economic impacts. We need to take an intersectional lens to understand climate change as a pervasive economic issue. Evidence identifies that the climate change should not be viewed in a vacuum but within existing structural inequalities of class, race, gender, and ability. Climate change can exacerbate further entrench economic inequality as climate change burdens are not equally distributed. Gender also plays a critical role in individual, family and community vulnerability, with women typically more likely to be negatively affected by the impacts of climate change and extreme weather events. Evidence shows that women are disproportionately impacted by climate change and have fewer resources for adaption.

According to the Human Development Index, based on current trends, it would take 202 years to close the gender gap in economic opportunity (UNDP, 2019:147). The key economic parities are in areas such as labor force participation, control and ownership of means of production, participation in decision-making processes at different levels, land ownership or tenure, access education and training, personal autonomy, digital connectivity, and access to capital and markets. Our research indicates that women take holistic approaches to economic development, particularly taking a livelihoods approach. The economic empowerment of women will need to focus not on how women can access market-based systems, but how can women be empowered in climate change adaptation and resilience as a continuum, interlinked with ongoing development issues such as access to reliable water, health, and livelihoods. There is a strong need together economic agendas and development agendas in the response to climate change risk reduction, resilience, and adaptation.

REFERENCES

Agarwal, B. (2009). Gender and forest conservation: The impact of women's participation in community forest governance. *Ecological Economics, 68*(11), 2785−2799.

Alam, M., Bhatia, R., & Mawby, B. (2015). *Women and climate change: Impact and agency in human rights, security and economic development*. Washington, DC: Georgetown Institute for Women, Peace and Security.

Alonso, C., Brussevich, M., Dabla-Norris, M. E., Kinoshita, Y., & Kochhar, M. K. (2019). *Reducing and redistributing unpaid work: Stronger policies to support gender equality*. Washington, DC: International Monetary Fund, IMF Working Paper 19/225.

Alston, M., & Akhter, B. (2016). Gender and food security in Bangladesh: The impact of climate change. *Gender, Place & Culture, 23*(10), 1450−1464.

Alston, M., Clarke, J., & Whittenbury, K. (2018). Contemporary feminist analysis of Australian farm women in the context of climate changes. *Social Sciences, 7*(16), 2−15.

Alston, M., & Whittenbury, K. (2013). *Research, action and policy: Addressing the gendered impacts of climate change*. Springer.

Ampaire, E. L., Acosta, M., Huyer, S., Kigonya, R., Muchunguzi, R. M., & Jassogne, L. (2019). Gender in climate change, agriculture, and natural resource policies: Insights from East Africa. *Climatic Change*. Available from https://doi.org/10.1007/s10584-019-02447-0.

Arora-Jonsson, S. (2011). Virtue and vulnerability: Discourses on women, gender and climate change. *Global Environmental Change, 21*, 744−751.

Babacan, H. (2009). Women and religion. In J. Jupp (Ed.), *The encyclopaedia of religion in Australia* (pp. 695−711). Melbourne, VIC: Cambridge University Press.

Babacan, H. (2013). Gender, diversity and public policy. In A. Jakubowicz, & C. Ho (Eds.), *Australian multiculturalism: Theory, policy and practice* (pp. 179−186). Sydney, NSW: Australian Scholarly Press.

Babacan, H. (2014). Gender inequalities in mental health: Towards resilience and empowerment. In A. Francis (Ed.), *Social work in mental health: Contexts and theories for practice* (pp. 229−247). New Delhi: Sage Publications.

Babacan, H., & Lewis, E. (2019). *Gender responsive alternatives to climate change, Vanuatu project evaluation report*. Sydney, NSW: Action Aid.

Bhadwal, S., Sharma, G., Gorti, G., & Sen, S. M. (2019). Livelihoods, gender and climate change in the Himalayas. *Environmental Development*, *31*, 68–77.

Bjornberg, K. E., & Hansson, E. (2013). Gendering local climate adaptation. *Local Environment*, *18*(20), 217–232.

Brown, L.R., Feldstein, H., Haddad, L., Pe, C., Quisumbing, A., (1995). Generating food security in the year 2020: Women as producers, gatekeepers and shock absorbers, 2020 Brief, 17 International Food Policy Research Institute.

Committee on the Elimination of Discrimination Against Women (CEDAW). (2018). *General recommendation no. 37 on gender-related dimensions of disaster risk reduction in the context of climate change*. New York: United Nations, CEDAW/C/GC/37.

Chambers, R., & Conway, G. (1992). *Sustainable rural livelihoods: Practical concepts for the 21st century*. Institute of Development Studies Brighton.

Chant, S., & Sweetman, C. (2012). Fixing women or fixing the world? 'Smart economics', efficiency approaches, and gender equality in development. *Gender & Development*, *20*(3), 517–529.

Coleman, E. A., & Mwangi, E. (2013). Women's participation in forest management: A cross-country analysis. *Global Environmental Change*, *23*(1), 193–205.

Crenshaw, K. W. (1991). Mapping the margins: Lntersectionality, identity politics, and violence against women of colour. *Stanford Law Review*, *43*(6), 1241–1299.

Dankelman, I. (2010). *Gender and climate change: An introduction*. London: Earthscan.

Deering, K. (2019). *Gender-transformative adaptation: From good practice to better policy*. Geneva: CARE.

Denis, A. (2008). Intersectional analysis. A contribution of feminism to sociology. *International Sociology*, *23*(5), 677–694.

Eastin, J. (2018). Climate change and gender equality in developing states. *World Development Journal*, *107*, 289–305.

Food and Agriculture Organization (FAO). (2011). *Gender and climate change research in Agriculture and Food Security for development*. Rome: Food and Agriculture Organization of the United Nations.

Food and Agriculture Organization (FAO). (2014). *State of food security in the world*. Rome: Food and Agriculture Organization of the United Nations.

Food and Agriculture Organization (FAO). (2016). *Gender-responsive disaster risk reduction in the agricultural sector*. Rome: Food and Agriculture Organization of the United Nations.

Food and Agriculture Organization (FAO). (2017). *Big roles, little powers: The reality of women in agriculture in ECOWAS region*. Rome: Food and Agriculture Organization of the United Nations.

Food and Agriculture Organization (FAO). (2018). *Tackling climate change through rural women's empowerment*. Rome: Food and Agriculture Organization of the United Nations.

Food and Agriculture Organization (FAO). (2019). *The female face of farming*. Rome: Food and Agriculture Organization of the United Nations.

Goetz, A. M., & Jenkins, R. (2005). *Reinventing accountability: Making democracy work for human development*. New York: Palgrave.

Grillos, T. (2018). Women's participation in environmental decision-making: Quasi-experimental evidence from northern Kenya. *World Development*, *108*, 115–130.

Hill-Collins, P. (1995). Symposium on west and fensternmakers, "doing difference". *Gender & Society*, *9*(4), 491–494.

ILO. (2018). *World employment social outlook: Trends for women 2018*. Geneva: International Labour Office.

Intergovernmental Panel on Climate Change (IPCC). (2007). *Fourth assessment report: Synthesis report*. Geneva: Intergovernmental Panel on Climate Change.

Intergovernmental Panel on Climate Change (IPCC). (2014). *Climate change: Impacts, adaptation and vulnerability*. Geneva: Intergovernmental Panel on Climate Change.

Intergovernmental Panel on Climate Change (IPCC). (2015). *Climate change 2014: Synthesis report.* Geneva: Intergovernmental Panel on Climate Change.

International Union for Conservation of Nature (IUCN). (2015). *Women's participation in global environmental decision making.* Washington, DC: International Union for Conservation of Nature.

Kabeer, N. (2018). *Gender, livelihood capabilities and women's economic empowerment, gender and adolescence global evidence programme.* London: Overseas Development Institute.

Kahn, M. E., Mohaddes, K., Ng, R. N., Pesaran, M. H., Raissi, M., & Yang, J. C. (2019). *Long term macroeconomci effects of climate change: A cross-country analysis.* Washington, DC: International Monetary Fund, IMF Working Paper no WP/19/215.

Kimmel, M. S. (2004). *The gendered society reader.* New York: Oxford University Press.

Klasen, S., Lechtenfeld, T., & Povel, F. (2015). A feminization of vulnerability? Female headship, poverty, and vulnerability in Thailand and Vietnam. *World Development, 77*(1), pp. 36-3.

Klasen S., Gaddisv, I., (2013). Structural change, economic development, and women's labor force participation. Journal of Population Economics, 27(3), 639−681.

Lengermann, P. M., & Niebrugge, G. (2013). Contemporary feminist theory. In G. Ritzer (Ed.), *Contemporary sociological theory and its classical roots: The basics* (4th ed.). St Louis, MO: McGraw-Hill.

Lewis, E., & Babacan, H. (2019). *Gender responsive alternatives to climate change, Kenya project evaluation report.* Sydney, NSW: Action Aid.

Lewis, E., & Babacan, H. (2019a). *Gender responsive alternatives to climate change, Cambodia project evaluation report.* Sydney, NSW: Action Aid.

Makkonen, L. T. (2002). *Multiple, compound and Intersectional discrimination: Bringing the experience of the most marginalised to the fore.* Turku, Finland: Abo Akademi University, Institute for Human Rights, Report 1102.

Manuel, T. (2006). Envisioning the possibilities for a good life: Exploring the public policy implications of intersectionality theory. *Journal of Women, Politics & Policy, 28*(3−4), 173−203.

McKinsey Global Health Institute. (2015). *The power of parity: How advancing women's equality can add $14 trillion to the global economy.* Washington, DC: McKinsey and Company.

Mignaquy, J. (2013). *Gender perspectives on climate change.* Sydney, NSW: University of NSW.

Mora, C., Pridandelli, D., Franklin, E. C., Franklin, J. L., Kantar, M. B., Miles, W., . . . Hunter, C. L. (2018). Broad threat to humanity from cumulative climate hazards intensified by greenhouse gas emissions. *Nature Climate Change, 8,* 1062−1071.

Oxfam. (2014). *Monitoring for better impact: Why the Asian Development Bank should monitor food security at household and individual level.* Carlton, VIC: Oxfam Australia.

Pereznieto, P., & Taylor, G. (2014). A review of approaches and methods to measure economic empowerment of women and girls. *Gender & Development, 22*(2), 233−251.

Reed, M. S., Podesta, G., Fazey, I., Geeson, N., Hessel, R., Hubacek, K., . . . Thomas, A. D. (2013). Combining analytical frameworks to assess livelihood vulnerability to climate change and analyse adaptation options. *Ecological Economics, 94,* 66−77.

Resurrección, B. P., Bee, B. A., Dankelman, I., Park, C. M. Y., Halder, M., & McMullen, C. P. (2019). Gender-transformative climate change adaptation: Advancing social equity. In: *Background paper to the 2019 Report of the Global Commission on Adaptation, Rotterdam and Washington, DC.*

Rowbottom, S., & Linkogle, S. (2001). *Women resist globalization: Mobilizing for livelihood and rights.* London: Zed Books.

Sarantakos, S. (2013). *Social research* (4th ed.). London: Springer.

SDC. (2003). *Gender in practice.* Benn, Switzerland: Swiss Agency for Development and Cooperation.

Simelton, E., & Ostwald, M. (2019). *Multifunctional land uses in Africa: Sustainable food security solutions.* London: Routledge.

Stephens, A., Lewis, E., & Reddy, S. (2018). *Inclusive systemic evaluation (ISE4GEMs): A new approach for the SDG era*. New York: UN Women.

Tanyag, M., & True, M. (2019). *Gender responsive alternatives to climate change*. Melbourne, VIC: Monash University and ActionAid Australia.

Terry, G. (2009). Climate justice without gender justice: An overview of the issues. *Gender and Development*, *17*(1), 5−18.

Torquebiau, E., Tissier, J., & Grosclaude, J. Y. (2016). How climate change reshuffles the cards for agriculture. In E. Torquebiau (Ed.), *Climate change and agriculture worldwide*. Dordrecht: Springer.

UNDP. (2009). *Resource guide on gender and climate change*. New York: United Nations Development Programme.

UNDP. (2019). *Human development report 2019: Beyond income, beyond averages, beyond today*. New York: United Nations Development Programme.

UN Women. (2016b). *Leveraging co-benefits between gender equality and climate action for sustainable development*. New York: UN Women.

UN Women. (2019a). *Economic empowerment: Facts and figures*. New York: UN Women. <https://www.unwomen.org/en/what-we-do/economic-empowerment/facts-and-figures> Accessed 20.12.19.

UN Women. (2019b). *Women's leadership and participation*. New York: UN Women.

USAID. (2013). *Gender: Regional agricultural trade environment summary*. Washington, DC: USAID. <https://www.usaid.gov/sites/default/files/documents/1861/Gender.pdf> Accessed 29.12.19.

World Economic Forum (WEF). (2019). *Global gender gap report 2020*. Geneva: World Economic Forum.

World Bank (2014). *Gender at work: A companion to the world development report on jobs*. Washington, DC.

World Bank. (2015). *Gender in climate smart agriculture*. Washington, DC: World Bank, Food and Agriculture Organization of the United Nations and IFAD.

World Bank (2017). *Global Findex database: Measuring financial inclusion*. Washington, DC.

World Bank (2019). *Women, business and the law 2019: A decade of reform*, Washington, DC.

FOREIGN DIRECT INVESTMENT, NATURAL DISASTERS, AND ECONOMIC GROWTH OF HOST COUNTRIES

7

Pengji Wang[1], Huiping Zhang[1] and Jacob Wood[1,2]

[1]*School of Business, James Cook University, Singapore, Singapore* [2]*Visiting Professor of International Trade, Chungnam National University, Daejeon, South Korea*

7.1 INTRODUCTION

This chapter examines the impact of inward foreign direct investment (FDI) on a host country's economic recovery following the occurrence of a natural disaster. An assessment of the extant literature shows that little has been done to assess the nexus between inward FDI and natural disasters, with most FDI-based studies examining the determinants of FDI inflows (Dunning, 1993; Resmini, 2000; Singh & Jun, 1995) and the relationships between FDI and economic growth (Chowdhury & Mavrotas, 2006; Nair-Reichert & Weinhold, 2001). Of those with a specific natural disaster focus, it remains inconclusive whether and how FDI interacts with the host country's economy following the occurrence of a natural disaster.

On the one hand, FDI is usually characterized by its "footlooseness" and is extremely sensitive to the negative shocks in host countries. That is, multinational enterprises (MNEs) are more likely to enter and exit a host country than are local firms (Bernard & Jensen, 2007; Görg & Strobl, 2003). The negative shocks could destroy local upstream suppliers and, thus, MNEs might find it profitable to leave the host country. While it is unclear whether FDI has lower entry barriers to an industry than the local capital of the host country, it is definitely easier for FDI to exit a host country than for its domestic-based equivalents when facing a negative shock such as a natural disaster. In support of such an argument, White and Fan (2006) note that disaster events could initiate or trigger changes in attitudes or perception of disaster risks by raising the level of risk aversion and therefore the likelihood of an avoidance response, sentiments that are shared by Escaleras and Register (2011). Moreover, from an empirical perspective, Anuchitworawong and Thampanishvong (2015) find in their Thai-based study that major natural disasters reduce FDI inflows.

On the other hand, FDI, presumably from developed to emerging countries, can contribute to the state of the local economy by boosting average productivity and employment opportunities. In the context of a natural disaster, after a negative shock occurs, MNEs may sustain their physical capital capabilities through the support of parent companies located in their respective home

Economic Effects of Natural Disasters. DOI: https://doi.org/10.1016/B978-0-12-817465-4.00007-8

countries. In this instance, parent firms can immediately provide much-needed capital to their foreign affiliates so as to ensure a quick recovery, thereby also contributing to the recovery of the host country. In-line with such logic, Oh and Oetzel (2011) find that there is no significant impact on the number of MNEs in a host country following a natural disaster.

In addition, a natural disaster may also act as a disruptive creation by readjusting the industry configuration of MNEs and local firms in a host country. There is a concern that high levels of FDI may crowd out local firms and deter the entry of local entrepreneurs (Caves, 1996). Following a natural disaster, more physical capital replacement is needed, thereby adjusting the state of equilibrium within the economy. In this readjustment process, FDI may play an important role in strengthening or weakening local value chains. Kato and Okubo (2018) show that when natural disasters seriously damage capital, the industrial configurations in the host country switch from one that is dominated by MNEs to one that is dominated by local firms.

With these different strands of literature in mind, we conclude that whether and how FDI helps to revive the host country's economy after natural disasters remains an imperative empirical question. We therefore plan to provide evidence of the impact of FDI on a local economy following a natural disaster using a multinational dataset of natural disasters, FDI inflow, and economic growth. More specifically, this chapter addresses the following research questions:

1. Does a natural disaster lead to a withdrawal of FDI from the local economy?
2. Does inward FDI stock moderate against the potential negative impact a natural disaster has on the economic growth of the local economy?

The remainder of the chapter is outlined as follows. Section 7.2 assesses the literature that examines the influence that natural disasters play on FDI. Section 7.3 details the econometric model adopted for our study, while Section 7.4 documents the key results from our empirical tests. Finally, Section 7.5 provides some concluding remarks and policy implications.

7.2 LITERATURE REVIEW

The literature on FDI determinants is rich, with a significant number of studies examining the relationship between FDI and various macroeconomic and institutional variables (Jabri, 2015; Sabir, Rafique, & Abbas, 2019). However, little has been done to address the influence that natural disasters may have on FDI decisions, the exceptions being Escaleras and Register (2011), Anuchitworawong and Thampanishvong (2015), and Kato and Okubo (2018). These subsequent studies form an important part of the basis of this study and are explained in greater detail later in the review.

The first significant area of literature involving FDI and natural disasters examines the "immediate" economic effects of natural disasters and the determinants of FDI. Within this, a broad area of research lies in three distinct areas, the earliest of which focuses specifically on the impact that natural disasters have on the gross domestic product (GDP). Work by Albala-Bertrand (1993) is pivotal, when an examination of the macroeconomic impacts of disasters during the 1960s/70s shows a positive impact on GDP. However, subsequent studies by Auffret (2003) and Rasmussen (2004) draw different conclusions, finding that natural disasters led to reductions in GDP performance in the Caribbean and Latin American regions. The second area of study focuses on the "longer term" macroeconomic implications of natural disasters. Similar to the first strain, the findings here are also mixed. Skidmore and Toya (2002) show that disasters are linked with positive GDP growth, the likes of which are caused by greater reductions in physical to human capital accumulation. Subsequent investments in human capital lead to increases in total factor productivity, which in turn grows GDP performance. However, other research by Tavares (2004) and Noy (2009) find no relationship between natural disasters and GDP growth. The third area of study examines the role that institutions play in mitigating the effects of natural disasters. Toya and Skidmore (2007) indicate that higher levels of educational attainment, sophisticated financial marketplaces, and a higher degree of trade openness provide a significant amount of insulation from the economic impacts of disasters, while Escaleras, Anbarci, and Register (2007) show that higher levels of corruption in the public sector lead to higher death tolls in the aftermath of a natural disaster. Meanwhile, Noy (2009) notes that death rates from disasters are negatively correlated with literacy rates, bigger government institutions, and greater support for international trade openness.

Secondly, an assessment of the empirical literature reveals another important thematic area of research, the determinants of FDI. In this regard, Escaleras and Register (2011) note that the primary determinants of FDI inflows are the size of the national market and a country's openness to international trade. From a national market perspective, a large national market helps with the efficient use of resources and the development of well-structured transportation infrastructure such as ports, rail, and road networks that allow the country to take advantage of any economies of scale that may exist (Resmini, 2000). In terms of trade openness, Singh and Jun (1995) show a complementary relationship between FDI flows and trade export orientation. Meanwhile, Walsh and Yu (2010) demonstrate that the income level, the quality of the institution, and the level of development are also important determinants. While Froot and Stein (1991) note the importance of foreign exchange rate; in particular, a weaker currency in the host country can lead to larger inward FDI as the purchase of production facilities and acquisition of assets become more affordable in the host country. Other studies by Root and Ahmed (1979) and Campos and Kinoshita (2003) find the existence of "clustering effects" whereby foreign firms tend to cluster together due to close business relationships that they share, such effects can also influence FDI decisions. Quality of institutions was also found to be an important factor with Wei (2000) concluding that corrupt institutional bodies add significant costs to firms and in doing so impede FDI inflows. Finally, a series of other determinants of FDI has also been assessed, including local wages (Pistoresi, 2000), trade barriers (Blonigen & Feenstra, 1996), and taxes (De Mooij & Ederveen, 2003).

Finally, as noted earlier, the previous literature does little to take into account the influence of natural disasters on FDI decisions. As Anuchitworawong and Thampanishvong (2015) note, natural disasters are an important consideration as they can influence the behavior of investors in a way that influences how they perceive risk and the extent to which they can tolerate such events. Moreover, the occurrence of such an event could affect businesses that affiliate their production in the host countries in several ways. Natural disasters can impose additional costs on businesses that reduce revenues and the production capabilities of the firm. Moreover, as White and Fan (2006) state, these events can also trigger changes in attitudes surrounding the perception of disaster risks by increasing the level of risk aversion and a desire to avoid further problems. Empirical evidence on their impact is limited. Escaleras and Register (2011) consider the relationship between FDI in 94 countries between 1984 and 2004 and the number of disasters occurring in these countries. Using empirical tests that control for two-way fixed effects, Escaleras and Register show that natural disasters are negatively and statistically significantly related to a country's FDI. Anuchitworawong and Thampanishvong (2015), using FDI flows in Thailand from 1971 to 2012, find that natural disasters lower FDI flows, in particular, a 1% increase in the level of severity of the disaster leads to a 0.54% reduction in the FDI flow. More recently, Kato and Okubo (2018) examine how MNEs affect the host country through their vertical industrial linkages when large natural disasters occur. Using a simple theoretical framework, they show that as trade costs decline, the host country is first dominated by MNEs and then later by local businesses. Therefore when natural disasters have seriously impacted on capital flows, the industrial configurations in the host country change from one that is dominated by the FDI of MNEs to one that is dominated by investments made by local firms.

With the literature highlighting the "footlooseness" of FDI and its extreme sensitivity to negative shocks in host countries, our study builds on previous empirical works by examining both the impact of natural disasters on FDI and also its ability to moderate against the potential impact the disaster might have on economic growth in the local marketplace moving forward.

7.3 METHODOLOGY

Given the limited scope of previous research efforts identified within the literature, our study seeks to provide evidence on the impact of FDI on a local economy following a natural disaster using a dataset of natural disasters, inward and outward FDI stock, and economic growth across countries. In an attempt to do so, we develop the following methodological framework.

Econometric model

$$\text{GDPg}_{c,y} = \text{Disaster}_{c,y} + \text{FDI}_{c,y} + \text{Disaster}_{c,y} \times \text{FDI}_{c,y} + \text{control variables} \qquad (7.1)$$

$\text{GDPg}_{c,y}$ is the per capita GDP growth rate of country c in year y. $\text{FDI}_{c,y}$ is the stock of inward FDI into a focal country c in year y. Disaster is measured using either total or average deaths in thousands incurred in all types of disasters.

To assess the link between international capital flows and economic growth and its sources, we control for other growth determinants. *The inflation rate* is a measure of macroeconomic stability. *The fertility rate* is controlled to consider the induced effect of fertility rate on the economy. It has a negative effect on the steady-state ratio of capital on effective workers using the neoclassical growth model. Higher fertility also reflects greater resources devoted to child-rearing, as shown in models of endogenous fertility (Barro & Sala-i-Martin, 1995), raising discount rates of future consumption and reduce the degree of altruism of each child for parents, which will not be conducive to human capital investment on their children (Becker & Barro, 1989). *Government consumption* is controlled with the prediction that a higher government consumption ratio leads to a lower steady-state level of output per effective worker and, hence, to a lower growth rate for given values of the state variables (Barro, 1991; Levine & Zervos, 1993; Sachs & Warner, 1997). The assumption is that the government consumption variable measures expenditures that do not directly affect productivity but that entail distortions of private decisions. These distortions can reflect the governmental activities themselves and also involve the adverse effects of the associated public finance. *Private credit* is calculated as being the level of credit provided to the private sector by financial intermediaries as a share of GDP (Beck, Levine, & Loayza, 2000). *Savings to GDP* is controlled because Houthakker, Hendrik, and Robinson (1961), Houthakker, Hendrik, and Robinson (1965), Modigliani (1970), and Carroll and Weil (1994) have shown that there is a large and highly significant positive correlation between saving and growth in the cross section of countries. *Labor force* is controlled as labor input is an important influencer for GDP growth. *Initial GDP per capita*, measured at the end of year $y - 1$ and in current USD, is also controlled. In terms of the given values of the other explanatory variables, the neoclassical model predicts a negative coefficient on initial GDP (Barro, 1996). A list of the variables and measures are provided in Table 7.A1 in the Appendix.

The model utilizes natural disaster data from 1980 to 2019, obtained from the International Disaster Database (IDD) that is available at https://www.emdat.be/emdat_db/. Table 7.A2 shows the categories of the natural disaster. Moreover, data for FDI flows, GDP, and other control variables from 1980 to 2019 are obtained from the World Bank.

7.4 RESULTS

7.4.1 OVERVIEW OF NATURAL DISASTER AROUND THE WORLD

Natural disasters, also called "acts of God," include, but are not limited to, droughts, earthquakes, epidemics, extreme temperatures, floods, insect infestations, mudslides, volcanic eruptions, tsunamis, and wildfires. Examples include Cyclone Nargis, which devastated Myanmar in May of 2008 and leftover 130,000 people dead (Lin, Chen, Pun, Liu, & Wu, 2009), and the December 26, 2004 tsunami that ravaged the Southeast Asian region causing more than 220,000 deaths (Stone, 2006).

In this study, we follow IDD and categorize the disasters into six big categories, that is, biological, climatological, extraterrestrial, geophysical, hydrological, and meteorological disasters with 13 types of disasters. A detailed categorization of the disasters is in Table 7.A2. As there was only one extraterrestrial disaster being reported in the IDD database in the observation period, we do not include this category into our analysis.

Based on the IDD data from 1980 to 2018, the number of natural disasters has been generally increasing as in Fig. 7.1, with a sharp increase between the mid-1990s and 2011. During the same period, hydrological disasters are the most prominent of the five categories of natural disasters measured, which are followed by meteorological disasters, biological disasters, geophysical disasters, and climatological disasters, as shown in Fig. 7.2.

In Fig. 7.3, among the five continents, Asia has the largest number of natural disasters, while Oceania has the smallest number. It is worth noting that in all continents, hydrological disasters are the most prevalent, while in Africa, biological disasters are the second most frequent form of disaster. This is consistent with the poorer surveillance and health intervention programs that exist in Africa due to its lower economic development level and unstable political system (Batuo, Mlambo, & Asongu, 2018).

Fig. 7.4 shows the average total population affected and the total size of economic damage caused from 1980 to 2018. In general, the impact of natural disasters is becoming larger over time, in that they cause more economic loss and have the propensity to affect more people. This is due to the urbanization that leads to a higher concentration of population and economic activities. Such a

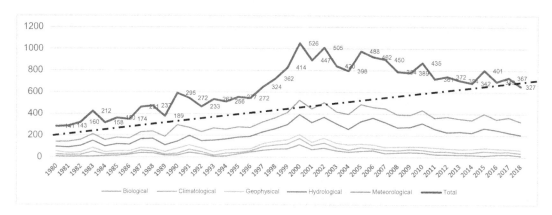

FIGURE 7.1

Types of natural disasters (1980−2018).

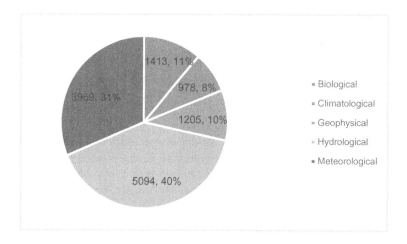

FIGURE 7.2

The occurrence of five categories of natural disasters.

result highlights the role that population growth and rapid urbanization play in increasing the risk of disasters occurring (López-Peláez and Pigeon, 2011).

7.4.2 NATURAL DISASTER AND INWARD FOREIGN DIRECT INVESTMENT STOCK IN COUNTRIES WITH DIFFERENT ECONOMIC DEVELOPMENT LEVELS

An assessment by the country economic development level is then conducted using four World Bank Analytical Classification levels, including low-income countries (LC), low- and middle-income countries (LMC), upper middle−income countries (UMC), and high-income countries

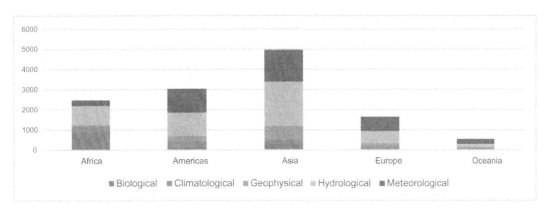

FIGURE 7.3

Natural disaster groups in five continents.

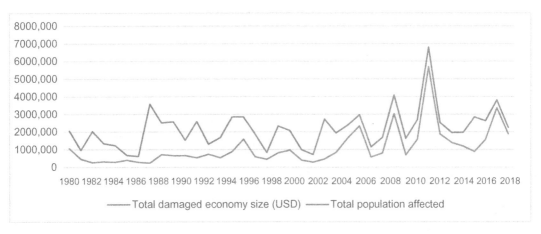

FIGURE 7.4

Total affected population and the total size of economic damage.

(HC). Table 7.1 shows that the number of disasters occurring per year is similar across the four categories of countries. However, the population affected and death tolls are much more in LC and LMC than in UMC and HC. Such results are consistent with the prior findings that natural hazards have a greater impact on property and loss of life in developing countries than they do in developed nations (Oh and Oetzel, 2011). This again reflects the lack of proper health intervention programs and resilient infrastructure needed to shelter and support people during and after disasters. Disasters hurt the poor and vulnerable the most (World Bank, 2018).

In terms of the total size of economic damage, the data show that UMC and HC suffer more than LC and LMC, which is probably owing to the urbanization and clustering of economic activities in HC.

Table 7.1 Disaster and Inward Foreign Direct Investment (FDI) Stock of Countries in Different Income Brackets.

	Low Income	Lower Middle Income	Upper Middle Income	High Income
Inward FDI stock/GDP (%)	25.22	28.87	41.22	86.69
Disaster occurrence	3.06	3.40	2.89	3.42
Population affected	744,526.8	499,521.4	307,154	86,776.27
Death tolls	425.91	219.19	96.28	96.51
Economic value damaged (000 USD)	311,273.2	423,331.7	611,449.6	1,552,957

GDP, Gross domestic product.

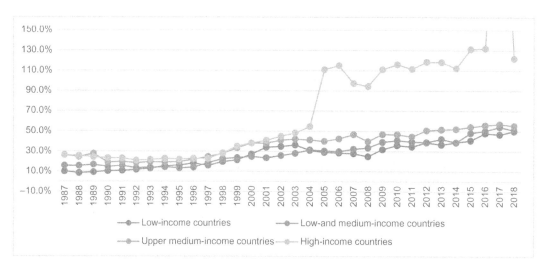

FIGURE 7.5

Ratio of inward FDI stock to GDP. *FDI*, Foreign direct investment; *GDP*, gross domestic product.

Fig. 7.5 shows that the inward FDI stock has generally increased in four categories of countries from 1987 onward. The average inward FDI stock to GDP ratio is at least 50% in countries of all income categories as of 2018, suggesting that MNEs are playing an increasingly important role in the host country economy.

7.4.3 NATURAL DISASTER: DESTRUCTION TO THE ECONOMY OR A DESTRUCTIVE GROWTH OPPORTUNITY?

We examine the level of GDP growth in the current and year immediately following the disaster, benchmarking it with the nondisaster years (Table 7.2). In the year of the disaster, the GDP growth

Table 7.2 Gross Domestic Product (GDP) Growth Rate After Disaster and Nondisaster Years. (p-value that is no more than 0.05 indicates a significant difference.)

	GDP Growth Rate in Year t			GDP Growth Rate in Year $t+1$		
	Disaster	No Disaster	P-Value	Disaster	No Disaster	P-Value
All sample	2.09	1.93	.169	2.17	1.61	.000
N	3374	2625		3800	5069	
Low-income countries	1.82	2.31	.120	2.29	2.54	.291
N	1070	398		1053	394	
Low- and middle-income countries	2.28	2.27	.490	2.51	1.95	.038
N	985	701		957	693	
Upper middle–income countries	2.57	2.34	.250	2.55	1.97	.061
N	642	593		610	572	
High-income countries	1.77	1.27	.005	1.68	1.14	.003
N	677	933		641	899	

rate does not show any significant difference, compared with when there is no disaster, except for HC. In the year following the disaster, the GDP growth rate becomes higher. This is the case for countries of all income categories except for LC where the GDP growth rate after disasters seems to be lower than normal, although the difference is not significant. Such results suggest that the disaster constitutes a destructive growth opportunity for countries of low- and middle-income levels and above. In the case of HC, the economy picks up faster than in UMC and LMC, due to their greater domestic economic capacity.

7.4.4 DOES FOREIGN DIRECT INVESTMENT CONTRIBUTE TO ECONOMIC RECOVERY AFTER DISASTERS?—A REGRESSION ANALYSIS

7.4.4.1 Overall Results

We estimate Model (1) with country and year fixed effects and report the regression results in Table 7.3. The dependent variable *GDPg* is shown as a percentage. The main explanatory variable of *disaster* is measured by either total or average deaths in thousands incurred in all types of disasters. *FDI* is measured by the inward FDI stock divided by the GDP of the country in the same year, all of which are in US dollars. To remove data errors and outliers, we truncate *GDPg* and *FDI* at the top and bottom 1%. To mitigate the multicollinearity issue, we deflate both *disaster* and *FDI* using their average values. The first two columns in Table 7.3 report the impact of disasters on the GDP growth in the concurrent year *y*, while the last two columns report the impact of disasters on the GDP growth in the year following the disaster.

Consistent with our prior findings, concurrent disasters affect a country's economy negatively, evidenced by the negative sign of *disaster* across the two measures and by a small *P*-value. More importantly, the significant and negative sign of the interaction term between *disaster* and *FDI* indicates that a country experiencing disasters in a year tends to have even worse GDP growth if its FDI is higher. This finding is in-line with the crowding-out effect of foreign investments to the

Table 7.3 The Economic Effect of Foreign Direct Investment (FDI) After Disasters.

	Disaster Is Measured by			
	Total Deaths (y)	**Average Deaths** (y)	**Total Deaths** (y − 1)	**Average Deaths** (y − 1)
Constant	4.701	4.708	4.132	4.121
	(.00)	(.00)	(.00)	(.00)
Disaster	− 0.137	− 0.219	− 0.034	− 0.035
	(.05)	(.03)	(.51)	(.71)
FDI	0.225	0.246	0.296	0.291
	(.20)	(.16)	(.09)	(.09)
Disaster × FDI	− 0.356	− 0.475	− 0.122	− 0.253
	(.06)	(.06)	(.43)	(.30)
LagGDPPCg	0.277	0.277	0.264	0.264
	(.00)	(.00)	(.00)	(.00)
Fertility	− 0.354	− 0.351	− 0.307	− 0.305
	(.00)	(.00)	(.00)	(.00)
Inflation	− 0.001	− 0.001	− 0.001	− 0.001
	(.00)	(.00)	(.03)	(.03)
Government consumption	− 0.094	− 0.094	− 0.076	− 0.076
	(.00)	(.00)	(.00)	(.00)
Labor	0.009	0.009	0.010	0.009
	(.01)	(.02)	(.01)	(.01)
Private credit	− 0.013	− 0.013	− 0.012	− 0.012
	(.00)	(.00)	(.00)	(.00)
Savings to GDP	0.026	0.026	0.038	0.038
	(.00)	(.00)	(.00)	(.00)
LagGDPPC	− 0.030	− 0.030	− 0.035	− 0.035
	(.00)	(.00)	(.00)	(.00)
Year fixed effect	Yes	Yes	Yes	Yes
Country fixed effect	Yes	Yes	Yes	Yes
Adj R-square (%)	37.3	37.3	40.3	36.6
N	2672	2672	2681	2681

Note: P-Values are in brackets. *GDP,* Gross domestic product.

host country's economy. On the other hand, focusing on GDP growth after disasters as shown in the last two columns of Table 7.3, the negative effect of disasters on GDP growth has become insignificant. FDI generally contributes to the host country's GDP growth positively, but its impact is weakened as the severity of natural disasters increases.

In addition, the signs of the control variables are mostly consistent with those indicated in the literature, namely, fertility rate, inflation rate, government consumption, and private credit are negatively related to GDP growth while GDP growth in the previous year, labor and savings positively relate to economic growth.

7.4.4.2 Different economic development levels and different categories of disasters

We also estimate Model (1) for countries with different levels of economic development to examine the differential effect of FDI on GDP growth. For the sake of simplicity, we report the results using the total deaths measure of *disaster* hereafter. The results using the average deaths measure of *disaster* are qualitatively similar. Table 7.4 shows that the conditional effect of FDI on GDP growth is significantly negative only in LMC of the same disaster year, as shown in column 1 of Table 7.4. This finding suggests that the crowding-out effect of FDI is more of an issue for developing countries when they face negative shocks such as disasters. The economic growth of such countries might be too dependent on foreign investments, which are typically sensitive to the host country's unexpected shocks. They need to implant domestic capital and enterprises to support stable long-term growth in their economy.

Model (1) is also reestimated for an individual type of disasters in the sample. As shown in Figs. 7.1 and 7.2, the IDD lists five categories of natural disasters: biological, climatological, geophysical, hydrological, or meteorological. Among all disasters, biological disasters account for 0.5% of country-years in our final dataset with valid data points for all the variables. Hence we focus on the rest four categories of natural disasters for our analysis in this subsection. The unreported regression results suggest that the overall negative effect of FDI on GDP growth of countries in the disaster year is not particularly driven by one single type of natural disaster.

7.4.5 "FOOTLOOSENESS" OF FOREIGN DIRECT INVESTMENT

Our findings suggest that with more inward FDI stock in a host country, the slower the host country's economic recovery following a natural disaster. To explore the reasons, we conduct some robustness tests.

First, we investigate whether MNEs withdraw or slow down their investments into the host countries experiencing disasters. Table 7.5 shows the increment of inward FDI in the year of the disaster and 1 year after disasters and compares it to the increment of inward FDI in nondisaster years. In the disaster year itself, the inward FDI increment is significantly larger than those nondisaster years, except in LC, where the difference is not significant. But in the following year after disasters, the inward FDI increment is not significantly different from nondisaster years; in HC the inward FDI increment after the disaster years is even less than the nondisaster years at a marginally significant level.

Table 7.6 examines the inward FDI increment as a ratio of GDP. The results suggest that the inward FDI increment as a ratio of GDP is lower in the year of disaster and the next year of disaster, as compared to nondisaster years, in countries with different income levels.

Combining the results in Tables 7.5 and 7.6, we conclude that although inward FDI stock increases in and following the disaster year, its relative size to the host country economy (the ratio to GDP) has decreased. Such results suggest that following natural disaster, MNEs still mobilize resources to the host countries; however, due to their concern that disasters may worsen business environments, they tend to slow down the investment increment, at least temporarily. The

Table 7.4 The Economic Effect of Foreign Direct Investment (FDI) After Disasters in Countries With Different Economic Development Status.

	Different Economic Development Status			
	LC or LMC (Disaster in y)	HC or UMC (Disaster in y)	LC or LMC (Disaster in y-1)	HC or UMC (Disaster in y − 1)
Constant	7.327	5.474	6.686	4.925
	(.00)	(.00)	(.00)	(.00)
Disaster	− 0.152	− 0.050	− 0.037	− 0.169
	(.07)	(.71)	(.53)	(.19)
FDI	0.306	0.249	0.523	0.075
	(.54)	(.17)	(.29)	(.67)
Disaster × FDI	− 0.401	− 0.010	− 0.121	− 0.616
	(.04)	(.98)	(.49)	(.16)
LagGDPg	0.254	0.262	0.237	0.275
	(.00)	(.00)	(.00)	(.00)
Fertility	− 0.554	− 0.839	− 0.474	− 1.031
	(.00)	(.04)	(.00)	(.01)
Inflation	− 0.001	− 0.057	− 0.001	− 0.033
	(.00)	(.23)	(.02)	(.48)
Government consumption	− 0.139	− 0.062	− 0.109	− 0.045
	(.00)	(.09)	(.00)	(.21)
Labor	0.004	0.008	0.003	0.010
	(.61)	(.05)	(.72)	(.01)
Private credit	− 0.010	− 0.007	− 0.008	− 0.008
	(.13)	(.03)	(.22)	(.02)
Savings to GDP	0.033	0.029	0.044	0.041
	(.00)	(.07)	(.00)	(.01)
LagGDP	− 0.786	− 0.037	− 0.813	− 0.032
	(.00)	(.00)	(.00)	(.00)
Year fixed effect	Yes	Yes	Yes	Yes
Country fixed effect	Yes	Yes	Yes	Yes
Adj R-square (%)	35.8	53.3	35.0	52.4
N	1444	696	1447	705

Note: P-values are in brackets. GDP, Gross domestic product; HC, high-income countries; LC, low-income countries; LMC, low- and middle-income countries; UMC, upper middle–income countries.

investment might be just enough to replace the facilities being damaged, but not large enough to expand production. In that sense, MNEs show some "footlooseness" that could slow down the recovery in the host countries considering their rising importance in host countries' economy, as shown in the results in Section 7.4.6.

Table 7.5 Increment of Inward Foreign Direct Investment (USD).

	Current Year of Disaster			Following 1 Year After the Disaster		
	Disaster	No Disaster	P-Value	Disaster	No Disaster	P-Value
Full sample	4679.82	836.01	.006	2918.98	3029.14	.472
N	3603	2753		2975	3188	
LC	488.92	126.91	.6535	351.99	485.20	.104
N	1086	417		1059	417	
LMC	2243.33	412.75	.000	1354.07	1677.33	.129
N	900	578		827	610	
UMC	7131.10	1202.73	.000	4443.33	4266.60	.454
N	568	448		474	491	
HC	25858.42	9713.00	.0005	12,294.52	20,951.44	.0507
N	639	734		482	830	

HC, High-income countries; LC, low-income countries; LMC, low- and middle-income countries; UMC, upper middle−income countries.

Table 7.6 Inward Foreign Direct Investment Increment to Gross Domestic Product.

	Current Year of Disaster			Following 1 Year After the Disaster		
	Disaster	No Disaster	P-Value	Disaster	No Disaster	P-Value
All sample	29.72%	56.76%	.000	30.70%	56.22%	.036
N	3535	2549		3442	2527	
Low-income countries	23.93%	28.69%	.0443	25.09%	28.61%	.1075
N	1029	383		1017	378	
Low- and middle-income countries	25.57%	34.10%	.000	25.89%	34.48%	.000
N	901	568		879	567	
Upper middle−income countries	36.15%	47.65%	.000	36.78%	48.15%	.000
N	567	446		535	434	
High-income countries	46.22%	124.93%	.0003	50.29%	124.46%	.0008
N	633	670		601	650	

7.4.6 "CROWDING-OUT" OR "CROWDING-IN" EFFECT OF FOREIGN DIRECT INVESTMENT FOLLOWING DISASTER: EFFECT OF FOREIGN DIRECT INVESTMENT ON THE HOST COUNTRY'S DOMESTIC CAPITAL FORMATION

We further examine the impact of inward FDI on domestic capital formation. We regress the domestic capital formation as a ratio of GDP on the increment of inward FDI and disaster, controlling several variables as shown in Table 7.7. The positive relationship between the increment of

Table 7.7 Impact of Inward Foreign Direct Investment (FDI) Increment on Domestic Capital Formation During Disasters.

Gross Fixed Capital Formation, Private Sector (% of GDP)	Model Full Sample	Model Low Income	Model Lower Middle Income	Model Upper Middle Income	Model High Income
Inward FDI increment/ GDP $\times$ disaster	1.49*	9.42***	1.45	− 0.04	− 0.23
	(0.74)	(1.94)	(2.65)	(2.09)	(0.24)
Disaster	− 0.1 +	0.06***	− 0.22*	0.03	0.02
	(0.06)	(0.12)	(0.1)	(0.14)	(0.03)
Inward FDI increment/GDP	− 0.08	− 9.29***	0.25	9.03	1.11
	(1.7)	(2.79)	(9.33)	(7.3)	(0.81)
Inflation	0	0.02***	0	0	0.16
	(0)	(0.02)	(0)	(0)	(0.1)
Fertility rate	− 0.81*	− 1.27***	− 0.7	3.95***	2.01
	(0.33)	(0.61)	(0.51)	(1.09)	(1.32)
Private credit	0.01	0***	0.06**	0.06**	0.03***
	(0.01)	(0.04)	(0.02)	(0.02)	(0.01)
Save to GDP	0.18***	0.26***	0.05	0.1	0.28***
	(0.03)	(0.04)	(0.04)	(0.08)	(0.07)
Trade to GDP	0.03**	0.11***	0	− 0.05**	− 0.02**
	(0.01)	(0.02)	(0.01)	(0.02)	(0.01)
Government consumption	− 0.02	0.2***	− 0.24*	− 0.62**	− 0.21
	(0.07)	(0.11)	(0.1)	(0.18)	(0.13)
Labor force	0 +	0***	0**	0***	0 +
	(0)	(0)	(0)	(0)	(0)
GDP per cap growth$_{t−1}$	0.13***	0.01***	0.11**	0.23***	0.07
	(0.03)	(0.07)	(0.04)	(0.06)	(0.05)
Constant	11.85***	6***	23.76***	6.91	8.48 +
	(2.12)	(4.13)	(3.55)	(4.87)	(4.54)
R-sq: within	.14	.38	.14	.23	.64
Between	.12	.11	.00	.26	.02
Overall	.13	.25	.02	.02	.14
F	15.53***	20.88***	4.81***	6.07***	11.04***
	(11,1055)	(11,369)	(11,327)	(11,220)	(11,68)
N	1155	422	386	259	88

Note: ***$P < .001$; **$P < .01$; *$P < .05$; *standard errors in brackets*. GDP, Gross domestic product.

inward FDI and fixed capital formation during a disaster in an LC ($\beta = 9.42$; $P < .001$) suggests the high dependence of domestic capital recovery and accumulation on foreign investment. This could be explained by the crowding-out effect of FDIs in the host country economy.

Foreign investment can stimulate domestic investment activity through downstream or upstream linkages. For instance, an MNE may source raw materials from domestic suppliers or it may outsource particular activities to firms in the host country. By engaging local firms into their supply chain, MNEs not only create demands (Cardoso and Dornbusch, 1989) and loosen financial constraints (Harrison, Love, & McMillan, 2004) for local firms but also transfer advanced technologies, production processes, management skills, and human capital (through on-the-job training and labor flow) to local firms to increase their productivity (Markusen & Venables, 1999). All of these encourage and enable domestic firms to build up their own capital accumulation and production capacity. On the other hand, foreign investments can also crowd out domestic investments because MNEs, with better access to financial resources, have higher efficiency levels which in turn allow them to produce higher quality products/services, thus posing additional competitive disadvantages on local enterprises (Spencer, 2008). Increased competition may eventually lead local enterprises to abandon investment projects or even reduce existing production capacities (Jude, 2016). Foreign investors may also raise interest rates when they borrow from domestic financial markets, which results in the crowding out of domestic firms (Polat, 2017). This is particularly the case if MNEs do not engage local enterprises into their supply chain.

In many LC such as many of those in Africa, the bulk of FDI has flowed into the natural resources sector, which has few linkages with local firms, and therefore the indirect effect of FDI on domestic investment accumulation is likely to be marginal. During a disaster in which domestic facilities and credit system in the host country are damaged, MNEs' investment will become particularly important for the host country economy to recover. However, as revealed in Section 7.4.4, MNEs slow down their investment in the host country; this leads to slower economic recovery of host LC. This is supported by a significantly lower increase in domestic fixed asset formation rate in LC during disaster years than nondisaster years, as shown in Table 7.8.

However, in LMC, UMC, and HC, the change in the domestic fixed assets formation is not significantly different across disaster and nondisaster years. This could be attributed to the fact that the domestic firms do not depend on too much on MNEs' injection of resources. When MNEs slow

Table 7.8 Change in the Domestic Fixed Asset to Gross Domestic Product Ratio (Year *t*).

	Disaster	No Disaster	*P*-Value
Full sample	0.08%	0.14%	.3226
N	1561	791	
LC	0.13%	0.59%	.083
N	635	243	
LMC	0.22%	0.10%	.3213
N	503	254	
UMC	− 0.01%	− 0.05%	.452
N	316	194	
HC	− 0.03%	0.00%	.4699
N	107	100	

HC, High-income countries; *LC*, low-income countries; *LMC*, low- and middle-income countries; *UMC*, upper middle−income countries.

down their investment, domestic firms are forced to obtain more resources locally to build up capital accumulation; this is the time for them to reconfigure the domestic value chain (Kato and Okubo, 2018).

7.5 CONCLUSION

This study provides evidence of the impact of FDI on a local economy following a natural disaster using a multinational dataset of natural disasters, FDI, and economic growth. Key findings show natural disaster seems to have a larger negative impact on the growth opportunity for low- or middle low–income countries when they have more inward FDI stock. While in the case of HC, the effects are less pronounced with the economy able to rebound more quickly than less developed economies due to their greater domestic economic capacity. An assessment of disaster types shows that the overall negative effect of inflow FDI on GDP growth of countries in the disaster year is not particularly driven by one single type of natural disaster. From an MNE perspective, in the period following a natural disaster, international businesses mobilize resources in the host country operation. However, as disasters may temporarily worsen business environments, MNEs often choose, at least in the short term, to slow down the investment increment. This example highlights the "footlooseness" of MNEs, which could slow down the economic recovery of the host country. Finally, we examine the impact of inward FDI on domestic capital formation. Our regression results demonstrate a positive relationship between the increment of inward FDI and fixed capital formation during a disaster, particularly in a low-income economy, a finding that could be explained by the crowding-out effect of FDIs in host country economy.

Overall, these findings highlight the important role MNEs can potentially play in helping host countries' recovery from the impacts of a natural disaster. On the one hand, noting the fact that MNEs seem to have crowded out domestic investment, and the fact that the local economy is highly dependent on MNEs' investment, particularly in low-income and less developed countries, it is recommended that the host government introduce favorable and supportive packages and policies that incentivize MNEs' contributions during and after a natural disaster. Such a package and policies could be, but not limited to, tax rebates, low-interest loans, and so on. On the other hand, it is also important for the host government to involve MNEs into their economic development plan in a more strategic way, so as to convert the so-called crowding-out effect to one that provides a crowding-in effect. Policies enticing MNEs to engage domestic players into their value chain, be it upstream or downstream, would be helpful in building up domestic production capacity.

REFERENCES

Albala-Bertrand, J. M. (1993). *Political economy of large natural disasters*. New York: Oxford University Press.

Anuchitworawong, C., & Thampanishvong, K. (2015). Determinants of foreign direct investment in Thailand: Does natural disaster matter? *International Journal of Disaster Risk Reduction, 14*, 312–321.

Auffret, P. (2003). *High consumption volatility: The impact of natural disasters*. Policy research working paper *2962*. Washington, DC: Word Bank.

Barro, R. J. (1991). Economic growth in a cross-section of countries. *Quarterly Journal of Economics*, *106*(2), 407−443.

Barro, R. J. (1996). Democracy and growth. *Journal of Economic Growth*, *1*(1), 1−27.

Barro, R. J., & Sala-i-Martin, X. (1995). *Economic growth*. New York: McGraw Hill.

Batuo, M., Mlambo, K., & Asongu, S. (2018). Linkages between financial development, financial instability, financial liberalisation, and economic growth in Africa. *Research in International Business and Finance*, *45*, 168−179.

Beck, T., Levine, R., & Loayza, N. (2000). Financial intermediation and growth: Causality and causes. *Journal of Financial Economics*, *46*(1), 31−77.

Becker, R. J., & Barro, G. S. (1989). Fertility choice in a model of economic growth. *Econometrica*, *57*(2), 481−501.

Bernard, A., & Jensen, B. (2007). Firm structure, multinationals and manufacturing plant deaths. *The Review of Economics and Statistics*, *89*(2), 193−204.

Blonigen, B. A., & Feenstra, R. C. (1996). Effects of U.S. trade protection and promotion policies. In: *NBER working paper* 5285. Cambridge.

Cardoso, E., & Dornbusch, R. (1989). Foreign private capital flows. *Handbook of Development Economics*, *2*, 1387−1439.

Campos, N., & Kinoshita, Y. (2003). Foreign direct investment and structural reforms: Panel evidence from east Europe and Latin America. In: *Proceedings of the IMF research conference on the causes and consequences of structural reforms*.

Carroll, C. D., & Weil, D. N. (1994). Saving and growth: A reinterpretation. In: *Carnegie−Rochester conference series on public policy, 40, 133−192*.

Caves, R. E. (1996). *Multinational enterprise and economic analysis* (2nd ed.). Cambridge: Cambridge University Press.

Chowdhury, A., & Mavrotas, G. (2006). Growth FDI: What causes what? *The World Economy*, *29*(1), 9−19.

De Mooij, R., & Ederveen, S. (2003). Taxation and foreign direct investment: A synthesis of empirical research. *International Tax and Public Finance*, *10*(6), 673−693.

Dunning, J. H. (1993). *Multinational enterprises and the global economy*. Harlow: Addison-Wesley.

Escaleras, M., Anbarci, N., & Register, C. A. (2007). Public sector corruption and major earthquakes: A potentially deadly interaction. *Public Choice*, *132*(1), 209−230.

Escaleras, M., & Register, C. A. (2011). Natural disasters and foreign direct investment. *Land Economics*, *87*(2), 346−363.

Froot, K. A., & Stein, J. (1991). Exchange rates and foreign direct investment: An imperfect capital markets approach. *Quarterly Journal of Economics*, *106*(4), 1191−1217.

Görg, G., & Strobl, E. (2003). Footloose multinationals. *The Manchester School*, *71*(1), 1−19.

Harrison, A. E., Love, I., & Mc Millan, M. S. (2004). Global capital flows and financing constraints. *Journal of Development Economics*, *75*, 269−301.

Houthakker, H. S. (1961). An international comparison of personal saving. *Bulletin of the International Statistical Institute*, *38*, 55−60.

Houthakker, H. S., & Hendrik, S. (1965). On some determinants of saving in developed and underdeveloped countries. In E. A. G. Robinson (Ed.), *Problems in economic development*. London: Macmillan.

Jabri, A. (2015). Institutional determinants of foreign direct investment in MENA region: Panel co-integration analysis. *Journal of Applied Business Research*, *31*(5), 1−12.

Jude, C. (2016). Employment effects of foreign direct investment: New evidence from Central and Eastern European countries. *International Economics*, *145*(C), 32−49.

Kato, H., & Okubo, T. (2018). The impact of a natural disaster on foreign direct investment and vertical linkages. In: *Keio-IES discussion paper series*. Institute for Economics Studies, Keio University.

Levine, R., & Zervos, S. (1993). Looking at facts: What we know about policy and growth from cross-country analysis. In: *World bank policy research papers, WPS 1115.*

Lin, I.-I., Chen, C.-H., Pun, I.-F., Liu, T., & Wu, C.-C. (2009). Warm ocean anomaly, air sea fluxes, and the rapid intensification of tropical cyclone Nargis (2008). *Geophysical Research Letters, 36*(3), LO3817.

López-Peláez, J., & Pigeon, P. (2011). Co-evolution between structural mitigation measures and urbanization in France and Colombia: A comparative analysis of disaster risk management policies based on disaster databases. *Habitat International, 35*(4), 573−581.

Markusen, J., & Venables, A. (1999). Foreign direct investment as a catalyst for industrial development. *European Economic Review, 432,* 335−356.

Modigliani, F. (1970). The life cycle hypothesis of saving and inter-country differences in the saving ratio. In W. A. Eltis (Ed.), *Induction, Growth, and trade:* Essays in honor of Sir Roy Harrod. London: Clarendon Press.

Nair-Reichert, U., & Weinhold, D. (2001). Causality tests for cross-country panels: a new look at FDI and economic growth in developing countries. *Oxford Bulletin of Economics and Statistics, 63*(2), 153−171.

Noy, I. (2009). The macroeconomic consequences of natural disasters. *Journal of Development Economics, 88* (2), 221−231.

Oh, C.-H., & Oetzel, J. (2011). Multinationals' response to major disasters: How does subsidiary investment vary in response to the type of disaster and the quality of country governance? *Strategic Management Journal, 32*(6), 658−681.

Pistoresi, B. (2000). Investimenti Diretti Esteri e Fattori di Localizzazione: L' America Latina e il Sud Est Asiatico. *Rivista di Politica Economica, 90*(2), 27−44.

Polat, B. (2017). Rate of return on foreign investment income and employment labour protection: A panel analysis of thirty OECD countries. *Cogent Economics and Finance, 5*(1). Available from https://doi.org/10.1080/23322039.2016.1273588.

Rasmussen, T. (2004). Macroeconomic implications of natural disasters in the Caribbean. In: *IMF working paper* WP/04/224. Washington, DC: International Monetary Fund.

Resmini, L. (2000). The determinants of foreign direct investment in the CEECs. *Economics of Transition, 8,* 665−689.

Root, F., & Ahmed, A. (1979). Empirical determinants of manufacturing direct foreign investment in developing countries. *Economic Development and Cultural Change, 27*(4), 751−767.

Sabir, S., Rafique, A., & Abbas, K. (2019). Institutions and FDI: Evidence from developed and developing countries. *Financial Innovation, 5*(8), 1−20.

Sachs, J. D., & Warner, A. M. (1997). Sources of slow growth in African economies. *Journal of African Economies, 6,* 335−376.

Singh, H., & Jun, K. W. (1995). Some new evidence on determinants of foreign direct investment in developing countries. In: *World Bank policy research working paper, No. 1531.*

Skidmore, M., & Toya, H. (2002). Do natural disasters promote long-run growth? *Economic Inquiry, 40*(4), 664−687.

Stone, R. (2006). Facing a tsunami with no place to run. *Science, 314*(5798), 408−409.

Tavares, J. (2004). The open society assesses its enemies: Shocks, disasters and terrorist attacks. *Journal of Monetary Economics, 51*(5), 1039−1070.

Spencer, J. W. (2008). The impact of multinational enterprise strategy on indigenous enterprises: Horizontal spillovers and crowding out in developing countries. *Academy of Management Review, 33*(2), 341−361.

Toya, H., & Skidmore, M. (2007). Economic development and the impacts of natural disasters. *Economic Letters, 94*(1), 20−25.

Walsh, J. P., & Yu, J. (2010). Determinants of foreign direct investment: A sectoral and institutional approach. In: *IMF working paper, WP/10/187.*

Wei, S.-J. (2000). Local corruption and global capital flows. Brookings Paper. *Economic Activities*, *0*(2), 303–346.

White, C., & Fan, M. (2006). *Risk and foreign direct investment*. Hampshire: Palgrave MacMillan.

World Bank. (2018). Building resilience into development: pioneering earthquake bonds reinforce World Bank leadership in providing financial protection against natural disasters. In: *World Bank Press Release*. World Bank. Retrieved online December 15, 2019 from <https://www.worldbank.org/en/news/press-release/2018/02/08/building-resilience-intodevelopment-pioneering-earthquake-bonds-reinforce-world-bank-leadership-in-providing-financial-protection-against-natural-disasters>.

APPENDIX

See Tables 7.A1 and 7.A2.

Table 7.A1 Data Source.

Variables	Measure	Source
GDP per capita growth	GDP per capita growth (annual %). The annual percentage growth rate of GDP per capita based on constant local currency. Aggregates are based on constant 2010 US dollars. GDP per capita is GDP divided by midyear population. GDP at purchaser's prices is the sum of gross value added by all resident producers in the economy plus any product taxes and minus any subsidies not included in the value of the products. It is calculated without making deductions for depreciation of fabricated assets or depletion and degradation of natural resources.	World Bank
Inward FDI stock	Inward FDI stock in US dollars at current prices in millions.	UNCTAD (United Nations Conference on Trade and Development)
Outward FDI stock	Outward FDI stock in US dollars at current prices in millions.	UNCTAD
GDP (current USD)	GDP at purchaser's prices is the sum of gross value added by all resident producers in the economy plus any product taxes and minus any subsidies not included in the value of the products. It is calculated without making deductions for depreciation of fabricated assets or depletion and degradation of natural resources. Data are in current US dollars. Dollar figures for GDP are converted from domestic currencies using single year official exchange rates. For a few countries where the official exchange rate does not reflect the rate effectively applied to actual foreign exchange transactions, an alternative conversion factor is used.	World Bank

(Continued)

Table 7.A1 Data Source. *Continued*

Variables	Measure	Source
Inflation	Inflation as measured by the consumer price index reflects the annual percentage change in the cost to the average consumer of acquiring a basket of goods and services that may be fixed or changed at specified intervals, such as yearly. The Laspeyres formula is generally used.	World Bank
Labor force	Labor force comprises people who meet the ILO definition of the economically active population.	World Bank
	Labor force comprises people ages 15 and older who supply labor for the production of goods and services during a specified period. It includes people who are currently employed and people who are unemployed but seeking work as well as first-time job seekers. Not everyone who works is included, however. Unpaid workers, family workers, and students are often omitted, and some countries do not count members of the armed forces. Labor force size tends to vary during the year as seasonal workers enter and leave.	
GDS	Gross savings (% of GDP).	World Bank
	Gross savings are calculated as gross national income less total consumption, plus net transfers.	
Private sector credit	Domestic credit to the private sector (% of GDP).	World Bank
	Domestic credit to private sector refers to financial resources provided to the private sector by financial corporations, such as through loans, purchases of nonequity securities, and trade credits and other accounts receivable, that establish a claim for repayment. For some countries, these claims include credit to public enterprises. The financial corporations include monetary authorities and deposit money banks, as well as other financial corporations where data are available (including corporations that do not accept transferable deposits but do incur such liabilities as time and savings deposits). Examples of other financial corporations are finance and leasing companies, money lenders, insurance corporations, pension funds, and foreign exchange companies.	
Fertility rate	Log [fertility rate, total (births per woman)]: total fertility rate represents the number of children that would be born to a woman if she were to live to the end of her childbearing years and bear children in accordance with age-specific fertility rates of the specified year.	World Bank
Government consumption ratio	General government final consumption expenditure (formerly general government consumption) includes all government current expenditures for purchases of goods and services (including compensation of employees). It also includes most expenditures on national defense and security but excludes government military expenditures that are part of government capital formation.	World Bank

Table 7.A1 Data Source. *Continued*

Variables	Measure	Source
High- or middle-income countries	Dummy, 1 indicating high- or middle-income countries and 0 indicating low- and middle low−income countries.	World Bank
Natural disaster	1980−2019.	https://www.emdat.be/emdat_db/
Population	Total population is based on the de facto definition of population, which counts all residents regardless of legal status or citizenship. The values shown are midyear estimates.	World Bank

FDI, Foreign direct investment; *GDP*, gross domestic product; *GDS*, gross domestic savings.

Table 7.A2 Categories of the Natural Disaster.

Disaster Category	Disaster Type	Disaster Subtype
Biological: "scenarios involving disease, disability or death on a large scale among humans, animals and plants due to toxins or disease caused by live organisms or their products. Such disasters may be natural in the form of epidemics or pandemics of existing, emerging or re-emerging diseases and pestilences or man-made by the intentional use of disease-causing agents in Biological Warfare (BW) operations or incidents of Bioterrorism (BT)" (National Disaster Management Guidelines, 2008).	Animal accident	
	Epidemic	Bacterial disease, parasitic disease, viral disease
	Insect infestation	Grasshopper, locust
Climatological	Drought	Drought
	Wildfire	Forest fire, land fire (brush, bush, pasture)
Geophysical	Earthquake	Ground movement, tsunami
	Mass movement (dry)	Avalanche, landslide, rockfall, subsidence
	Volcanic activity	Ashfall, lava flow, pyroclastic flow
Hydrological	Flood	Coastal flood, flash flood, riverine flood
	Landslide	Avalanche, landslide, mudslide, rockfall, subsidence
Meteorological	Extreme temperature	Cold wave, heat wave, severe winter conditions
	Storm	Convective storm, extratropical storm, tropical cyclone

MITIGATING IMPACT FROM NATURAL DISASTERS, BUILDING RESILIENCE IN TOURISM: THE CASE OF KERALA

Simona Azzali[1], Zilmiyah Kamble[2], K. Thirumaran[2], Caroline Wong[2] and Jacob Wood[2,3]

[1]*School of Science and Technology, James Cook University, Singapore, Singapore* [2]*School of Business, James Cook University, Singapore, Singapore* [3]*Visiting Professor of International Trade, Chungnam National University, Daejeon, South Korea*

8.1 INTRODUCTION

The World Travel and Tourism Council (2019) reported that the travel and tourism sector accounted for 10.4% of global GDP and created 319 million jobs or 10% of total employment in 2018. The upward trend in international tourist arrivals will reach more than 1.8 billion by 2030, with tourist arrivals in emerging economy destinations projected to grow at double the rate of that in advanced tourism economies (UNWTO, 2015). While such economic projections signal optimism, conversely, this Asia-Pacific region is also one of the hardest hit by natural catastrophes. Natural disasters in recent years have resulted in loss of lives, mass evacuations, and a tremendous impact on the tourism industry for countries caught in these natural calamities. An understanding of how governments and disaster management agencies respond to these natural disasters within the context of mitigating harm to people with a focus on tourists, would inform practitioners and scholars alike of ways future policies can be developed to assist in prevention and to lessen the impact on loss of lives and assets.

The "Weather, Climate & Catastrophe Insight: 2018 Annual Report" revealed that there were 394 natural catastrophic events globally in 2018, generating economic losses of US$225 billion, of which 144 were in Asia-Pacific, the impact of which inflicted more than US$89 billion in economic losses, or around 40% of the global total (Olano, 2019). Some of the deadliest natural disasters occurred in India, Indonesia, and Japan. Globally, Indonesia recorded nearly half the total deaths from disasters in 2018, while India recorded nearly half the total number of individuals affected [Centre for Public Policy Research (CPPR), 2018a].

Overall, floods have affected more people than any other type of natural hazard in the 21st century, including 2018 (CPPR, 2018a). In southwestern Japan the floods in July 2018 set off landslides killing at least 176 people, forcing millions to evacuate and leaving dozens missing in the country's worst weather disaster in 36 years (Sieg, 2018). In a similar catastrophe, the flash flooding in India's Kerala State in August 2018 was the worst humanitarian disaster the region had experienced since the 1920s, with two-thirds of the state's residents affected (over 23 million people), at a total economic cost of US$2.85 billion (CRED Crunch 54, 2019). These floods also destroyed a thriving tourism industry, a key source of growth and prosperity for the region [KTDC (Kerala Tourism Development Corporation), 2006]. In response to the disaster the Kerala Tourism Board (KTB) devised a plan to revive the industry through rebuilding accessibility to popular tourist attractions using aggressive marketing campaigns. Despite the initiatives, the tourism inflow post floods has been sluggish.

While tourism is considered a way to revitalize a sluggish economy (Shondell, 2008), the fact that the tourists are in the eye of a major disaster, calls for a different kind of safety approach. Faulkner (2001) in the very early days on studies for the need for preempting natural disasters to prevent severe impact suggested that an assessment on the community's resources be done ahead, so that necessary measures can be implemented. This chapter records an understanding of the impact of the 2018 cyclone and flooding on tourism in Kerala. The strong inverse relationship between tourism and natural disasters leads to a greater focus on disaster mitigation in Kerala pertaining to impact, risk management, and preventive measures. The study unravels the ability of this popular tourist destination to manage the impact of the disaster on tourists, efforts in loss recovery and post–disaster marketing, thereby providing an understanding of future policy and risk management projections. The findings have applications for policymakers and other stakeholders to consider impact-limiting measures and construct built-in resilience in the tourism industry.

8.2 METHODOLOGY, DATA COLLECTION, AND ANALYSIS

This study uses secondary data to comprehend the impact and management of the August 2018 flood on the tourist industry in Kerala. It reviews theoretical insights drawn from tourism and disaster management literature as well as relevant media coverage and government documentations to examine post–disaster policy changes and outcomes. To this end, it offers some policy and planning considerations in conclusion. It addresses three specific research questions (RQs):

RQ1: How did the 2018 Kerala flood affect various aspects of the tourism industry in Kerala?
RQ2: What are the challenges faced by the state government and organizations in managing the 2018 Kerala flood?
RQ3: What are the policy lessons and changes that government and organizations can take away from this disaster to mitigate the impact of future flooding?

8.3 KERALA AND THE TOURIST INDUSTRY

Kerala is a southwestern Indian state located in the tropical Malabar Coast with 580 km of its shoreline facing the Arabian Sea. With an area of 38,863 km^2, it lies between the Western Ghats in the east and the Arabian Sea in the west and has many connected lagoons and backwaters (Government of Kerala, 2018). It experiences tropical climatic conditions with a summer season and two different monsoon seasons (June to September of southwest monsoon and October to December of northeastern monsoon). Approximately 90% of rainfall occurs during these 6 monsoon months. Nearly 14.8% of the state is prone to flooding, and the proportion is as high as 50%

for certain districts. It has a population of 3.34 million and Trivandrum is the capital (Government of Kerala, 2018). Effectively positioned and promoted as "God's own country," Kerala boasts of lush green tropical rainforests, grasslands, coconut groves, tea gardens, and diverse wildlife. Kerala's economy depends on agriculture (paddy, pulses, sugarcane, rubber, tea, coconut, ground-nut, pepper, cardamom, and ginger) and its allied sectors (livestock, forest, fishing, and aquaculture), manufacturing industries, and the tourism industry (Kerala State Planning Board, 2019).

Tourism constitutes 10% of Kerala's GDP and reportedly contributes around 23.5% to the total employment in the state between 2009 and 2012 (CPPR, 2018b). Kerala is the first state in India to declare tourism as an industry. Foreign tourists' arrival to the state has been increasing in the last 10 years and statistics indicate 1.09 million tourists visiting the country in 2017, with an increase of 5.15% over 2016. The increase in domestic tourist arrivals is even more significant. Kerala Tourism Statistics states the number of visitors to Kerala was 14.67 million in 2017, showing an increase of 11.39% over the previous year (CPPR, 2018b).

Tourism is emphasized in the development agenda of the state due to its implication in terms of regional development, employment, and foreign exchange revenue. The diverse tourism offered in Kerala includes heritage/cultural/religious sites and events, beaches, hill stations, house boating or backwater tourism, pilgrimage, ecotourism sites such as wildlife sanctuaries, and Ayurveda/wellness-related activities (Ministry of Tourism, 2003). It is a popular tourist destination for both domestic and international tourists. Kerala's Tourism Vision 2025 envisages a growth rate of 7% per annum in foreign tourist arrivals and 9% annual growth in domestic tourist arrivals (Ministry of Tourism, 2003). This projected growth is most likely to result in a difficult situation mainly due to imbalances in general infrastructure. At the proposed growth rates, it is anticipated that the number of tourists would become comparable to the entire state's population by the year 2021−22 (Ministry of Tourism, 2003). The growth of Kerala's tourism in the last 20 years has resulted in the mushrooming of hotels, resorts, and other buildings in key tourist spots with implications for ecologically sensitive zones [Rebuild Kerala Development Programme (RKDP), 2019].

However, growth in international tourist arrivals indicated a slowdown in recent years dropping to about 5% in 2017 from 11% between the years 2007 and 2012. While 2018 witnessed a healthy growth in the first few months, the trend was broken due to the outbreak of the Nipah virus in May−June, followed by the floods in July−August 2018 (RKDP, 2019).

8.4 KERALA 2018 FLOOD AND ITS IMPACT

With many internal water bodies and a large coastline, Kerala is prone to disasters like flooding. It experienced recent floods in August 2019 and one of the most severe floodings took place in 2018.

In about 30 days, during August 2018, the flood caused the death of 483 people with 140 people missing and evacuation of about a million people (New Indian Express, 2018). The disaster was catastrophic in terms of loss of lives, property, and infrastructure due to massive destruction on 110,000 houses, 60,000 ha of farmland, 130 bridges, 83,000 km of roads, and death of countless livestock thereby threatening the livelihoods of millions. The major airport, Cochin International, one of India's busiest airports, was shut down for almost 9 days between August 15 and 29, 2018.

Table 8.1 outlines the extent of the damages for the 2018 and 2019 floods in Kerala.

Table 8.1 Comparison Between August 2018 and August 2019 Flood.

	Flooding in August 2018	**Flooding in August 2019**
Rainfall	August 2018: 821.9 mm was recorded in August 2018 (New Indian Express, 2018)	August 2019: The month received a record-breaking rainfall of 951.4 mm (New Indian Express, 2018)
Casualties	483 People died, and 140 are missing, about a million people were evacuated (Times of India, 2019)	121 People (New Indian Express, 2018)
Alert	All 14 districts of the state were placed on red alert (Times of India, 2018)	Red alert in the 9 districts in northern and central Kerala, orange alert in 3 districts of central Kerala, and yellow alert in the 2 districts of southern Kerala (BusinessToday, 2019)
House damage	A total of 10,319 urban houses in urban areas have been affected by floods [State Relief Commissioner, Disaster Management (Additional Chief Secretary) Government of Kerala, 2018]	1789 Houses had been damaged fully in between 8 and 19 August, while the number of partially damaged houses is 14,542 (NewsClick, 2019)

The damage caused by the deluge has affected most industries and sectors supporting Kerala's economy especially the private sector has borne most of the impact of the disasters. The share of private sector damages and losses is equivalent to 90% of the total, while that of the public sector constitutes the remaining 10% (RKDP, 2019, p. 61). The PDNA (Post Disaster Needs Assessment) indicates that nearly 90% of the damages/losses were sustained by the state's productive and social sectors (primarily housing) and employment and livelihoods. From an economic perspective, the impact was felt most by the agriculture, tourism sector, and micro-, small-, and medium-sized enterprises, who normally have limited or no access to formal sector financing (RKDP, 2019, p. 61).

The tourism hot spot Kochi and the surrounding areas and the Alappuzha, known for its famed boathouses and backwaters, bore the brunt of the floods bringing the industry to a complete standstill (Hindustan Times, 2019). The floods of 2018 led to heavy damages to the roads sector. Many plantations and ecotourism centers were also badly affected by the floods. There are 60 such centers under the control of the Forest Department. Among these, 23 centers suffered the most. Many tribal sectors were cut off from the main land due to landslides, flood, mudslides, and heavy rains. In different territorial circles the deluge adversely affected 131 settlements. Having a good road transport infrastructure and services is key to promote tourism and seamless passenger and freight movement. As a result, the tourism industry suffered greatly due to cancellations of tourist booking in response to damages to hotels, monuments, roads, and airports.

To get tourists back, the KTB begun targeting domestic tourists first, using aggressive marketing campaigns and partnership meetings with stakeholders in 10 different cities in India. These partnership meetings showcased village life, folklore, and traditional dance and were successful in increasing the domestic arrivals in Kerala from 3.87 million in 2018 to 4.19 million in the first quarter of 2019 (Hindustan Times, 2019). Other initiatives adopted by KTB to attract tourists and investors include Champions Boat League in the backwaters and other tailor-made experiences for domestic tourists. To attract international tourists, a 4-day Kerala Travel Mart was organized, which featured 1600 buyers from 66 countries along with the participation of local hotels, houseboats, homestays, Ayurveda centers, and tour operators (Hospitality Biz, 2018). Apart from these marketing efforts, restoration efforts for roads and transportation were also initiated.

However, the international arrivals showed negative trends till the end of the first quarter (January–March) of 2019 even though tourist footfalls registered a growth of 14.81% in the second quarter of 2019 (The Economic Times, 2019). Competitions by safer destinations are partly to explain for the negative numbers. Tourists are mindful about disaster-prone destinations and will always consider those when planning their next destination. Considering the state's overdependence on tourism, it is significant to have a public and private collaborated effort to reduce the period of recovery, mitigate the impact from natural disasters, and build resilience in the tourist industry.

8.5 POST–DISASTER MANAGEMENT

The Government of Kerala responded to the devastating August 2018 Floods with immediate relief operations through the National Disaster Management Authority that managed rescue efforts with the help of their National Disaster Response Force (RKDP, 2019). The Kerala State fire and rescue services, army, police, navy, air force, coast guard, and boarder security forces worked alongside them and tremendous spontaneous help from artisanal fishers supported the rescue efforts in saving 65,000 lives (PDNA, 2018). Immediate relief was provided to the displaced and affected at the relief camps. Local Indian civilians and organizations made voluntary donations, along with central government and state government funds supported the rescue efforts, compensation, and relief. Humanitarian aid also came in the form of foreign help especially from the Middle East.

In the post disaster of mid-September, the Kerala state government steered a PDNA with the help of international development partners such as the World Bank, European Union, Asian Development Bank, and the UN. The PDNA (2018) report highlighted the recovery needs and suggested policy choices and official measures for recovery in Kerala in the subsequent 5 years. In May 2019 a visionary road map proposed in the RKDP (2019) encompasses a resilient recovery policy framework and action plan for shaping Kerala's resilient, risk-informed development and recovery from the 2018 floods. The question remains: Is Kerala prepared for another disaster after 2018?

8.6 POST-2018 KERALA FLOODS

While Kerala is recovering from the setback of the 2018 floods, disaster struck Kerala again in August 2019 in the form of floods, landslides, and debris fallout. In the 2019 flood, 121 people died with 1789 houses damaged fully while the number of partially damaged houses stood at 14,542 (NewsClick, 2019). The 2019 flood disaster was labeled a "man-made disaster" (India Today, 2019) and the criticism revolved around the following reasons:

- Laws are being ignored or diluted. The unchecked mining and soil-piping phenomenon in Kerala has resulted in landslides and floods in the state. The Chief Minister of Kerala had only recently announced that the state would implement restrictions on constructing buildings or houses in landslide-prone areas and undertake surveys to identify vulnerable and hazardous spots to keep people informed about safe zones where construction activities can be undertaken. This effort was criticized as taken too late because had such a survey been undertaken in 2018, it would have helped reduce the impact of the damages that occurred in 2019.

- Extreme rainfall because of climate change, unscientific quarrying, irresponsible tourism in the hills, improper constructions at the slopes, agricultural malpractices, and many more such reasons contribute to Kerala's present vulnerability, according to experts.
- The lack of efficient warning and alert systems took the people by surprise. The most disastrous incidents were in Kavalappara in Malappuram district and Puthumala in Wayanad district, where several families were buried under the mud in massive landslips. People in Kavalappara recalled that last year authorities had warned them of a possible landslide; however, this time they had no clue. In Malappuram district, alone 60 people died, and 11 are still missing.
- Despite the existence of government-funded scientific institutions, the state does not have proper flood maps that could have helped in sending alerts that are more accurate to the people. It is a matter of concern that very little scientific data are being collected on the various phenomena that had occurred during and after the flood.
- The lack of a long-term dam management policy. Frequent floods and debris flowing into the dams decrease their capacity and reduce the longevity of the dams. There is a need for a policy on long-term dam management.
- The concentration on post–disaster activities rather than preparedness meant that people are not equipped to deal with an unexpected situation such as the 2019 floods. Though the state's immediate post–disaster intervention was to be appreciated, its preparedness to face another disaster was questionable.
- The centralized disaster management plans of the state and the lack of long-term plans are areas of concern. There is a need for a decentralized disaster management system, which includes risk and vulnerability mapping at the district level. An expert advisory scientific committee comprising scientists and other experts ought to be set up to deal with the disaster. Task forces have to be constituted at the community level rather than depending on individuals to step in to help.
- The lack of coordination between the departments concerned. Ideally, there should be liaison committee or coordination among irrigation department, electricity board, disaster management organizations, and other scientific communities, something that is not there now.
- It is yet to be analyzed whether the discussions post-2018 floods, on climate change, extreme rainfall, and irresponsible human interventions had helped the state to formulate better strategies to deal with natural disasters. Even if strategies had been drawn up such as the recommendations in the RKDP (2019), their application at the ground level was not clearly visible when floods struck the state again in August 2019.

8.7 CHALLENGES FACED

The 2018 and 2019 Kerala floods highlighted challenges in key institutional coordination, policy guidelines, contingency planning, disaster risk management programs, public infrastructure services, financing programs, and data collection.

The floods highlighted a number of structural constraints that left Kerala unprepared for major natural disasters. This included inadequate policies and institutional frameworks to manage and monitor critical natural resources such as water and land. There is also the absence of risk-informed spatial and sectoral planning policies and frameworks that led to extensive urban sprawl, unmanaged construction in hazard-prone areas. Another challenge faced at the time of the disaster was the lack of disaster risk preparedness in key socioeconomic sectors and weak capacity of institutions and individuals to anticipate and respond to extreme events. More importantly, there is a lack of availability and sharing of reliable data for disaster risk planning and management due to inadequate hydromet system, and limited fiscal resources and financing modalities for risk pooling and sharing (RKDP, 2019).

8.8 RECOMMENDATIONS

There is not only a need to address the fundamental drivers of floods but also be better prepared for future disasters. A comprehensive well-coordinated multisectoral program is required to address those challenges in order to ensure a resilient recovery and development pathway for Kerala. Despite having an inclusive and comprehensive road map for a green and resilient Kerala as featured in the RKDP Report 2019, the floods highlighted the need for a comprehensive approach to address various aspects of disaster management. More needs to be done to improve the capabilities of disaster risk management in Kerala by considering models of disaster management that have been used by other countries.

For instance, the systems thinking approach has been used to achieve more sustainable disaster management outcomes (Rehman, Sohaib, Asif, & Pradhan, 2019). This integrative approach helps to identify all active stakeholders and then examines various organizational, infrastructural, technical, social, and environmental factors (Ha 2016, 2019). In doing so, the systems thinking approach is able to provide a complete picture of the disaster management process.

From a practical perspective, flood disasters triggered by geomorphological conditions often overwhelm the institutional capacity of governments and local communities to respond as exemplified by the flood situations in Kerala. The nature of this risk is nonstationary, as are the flood risk management approaches adopted by countries as they seek to overcome changing circumstances (Bubeck et al., 2017). Flood disaster risk represents a probabilistic occurrence of floods of varying intensity and losses associated with them. Hence, the need for risk identification and assessment process taking on the following four stages: (1) hazard identification and assessment, (2) exposure assessment, (3) vulnerability assessment, and (4) risk assessment. From a disaster management perspective the prediction, prevention, or mitigation of the consequences of the disaster can be deconstructed into four phases: prevention, preparedness, response, and recovery (Halldin & Bynander, 2012), although the boundary between these phases, is by no means distinct with each being very much interlinked (Djimesah, Okine, & Mireku, 2018). It is evident from the 2018 and 2019 floods that the first two phases are highly inadequate to cope with the unexpected natural disaster.

Montz (2000) examined the issue of reducing flood-related vulnerabilities by emphasizing the need to improve hydrological systems, while also addressing relevant political, economic, and sociocultural factors that add to these vulnerabilities. Studies have been conducted to predict and manage disasters with strategies that will reduce the impact on tourists and the economy (Bernard & Cook, 2015; Fitchett, Hoogendoorn, & Swemmer, 2016; Yang, 2016). Yang (2016) for example explains that through the 3D GIS digital visualization of the coastal areas prone to flooding, there can be better management and communication channels opened for rescue efforts. In another case the Egyptian authorities in the tourism city of Hurghada continue to use control measures by calibrating the floodgates and walls (Abd-Elhamid, Fathy, & Zeleňáková, 2018). Apart from strategies to prevent and manage disasters, strengthening rules and regulations can be a solution on a long-term basis.

Despite such a focus, changes in climatic conditions, ecological disturbances, and rapid urban development have further exacerbated the problem. To that end, Norén, Hedelin, Nyberg, and Bishop (2016) note that the effects of global socioeconomic and environmental changes, when combined with inadequate institutional support, can lead to increased risks and therefore threatens the quality and way of life for many people.

Despite having an inclusive and comprehensive road map for a green and resilient Kerala as featured in the RKDP Report 2019, dealing with disaster is also a social issue that needs a partnership between different stakeholders in the society. Social acceptance of a particular strategy is the major challenge a government faces when dealing with environmental issues. People generally consider environmental protection strategies a hindrance to their livelihood that it may affect their agricultural practices or business. However, when a state does not take up the responsibility to implement a proper policy, social acceptance becomes just an excuse. Besides educating the population of the need for environmental protection and sustainability, there is a need to involve the community in framing some of those initiatives and strategies. K.V. Thomas, a retired scientist at the National Centre for Earth Science Studies, Thiruvananthapuram, pointed out that climate-resilient action within the local communities could be a sustainable solution in the present situation in Kerala (John, 2019). The local communities, entrepreneurs, and businesses need to join hands to safeguard the ecology of prime destinations and promote responsible tourism (The Economic Times, 2018). Kerala's tourism sector is showing resilience by getting itself ready to reinvent itself as a destination for meetings, incentives, conferences, and exhibitions after the floods (Matthew, 2019). Shondell (2008) observes that in the category of disaster tourism, encouraging a return of tourists to a previously natural disaster destination takes more than policymakers' efforts in marketing. Tourists themselves have to reflect on the place as a site of disaster and how their presence can help the lives of survivors and ways in which they can protect themselves should they be caught in a major disaster.

REFERENCES

Abd-Elhamid, H. F., Fathy, I., & Zeleňáková, M. (2018). Flood prediction and mitigation in coastal tourism areas, a case study: Hurghada, Egypt. *Natural Hazards*, *93*(2), 559–576. Available from https://doi.org/10.1007/s11069-018-3316-x.

Bernard, K., & Cook, S. (2015). Luxury tourism investment and flood risk: Case study on unsustainable development in Denarau Island resort in Fiji. *International Journal of Disaster Risk Reduction*, *14*, 302–311. Available from https://doi.org/10.1016/j.ijdrr.2014.09.002.

Bubeck, P., Kreibich, H., Penning-Rowsell, E. C., Botzen, W. J. W., de Moel, H., & Klijn, F. (2017). Explaining differences in flood management approaches in Europe and in the USA − A comparative analysis. *Journal of Flood Risk Management*, *10*(4), 436−445.

BusinessToday. *Monsoon rains: Red alert in 9 Kerala districts; floods wreak havoc in Maharashtra, Karnataka*. (Aug 9, 2019). From <https://www.businesstoday.in/current/economy-politics/monsoon-rains-live-updates-kerala-floods-maharashtra-karnataka-india/story/371336.html> Retrieved 03.11.19.

Centre for Public Policy Research (CPPR). *Review of disaster events − Supplementary information*. (2018a). From <https://www.google.com/url?sa = t&rct = j&q = &esrc = s&source = web&cd = 2&ved = 2ahUKEwjYgPq 72M_lAhWS8XMBHau3ALkQFjABegQIABAC&url = https%3A%2F%2Fcred.be%2Fsites%2Fdefault% 2Ffiles%2FReview2018.pdf&usg = AOvVaw0xed1h_reSwtJ722Cmb_dz> Retrieved 03.11.19.

Centre for Public Policy Research (CPPR). *Kerala tourism—The role of the government and economic impacts*. (2018b). From <https://www.keralatourism.org/touriststatistics/> Retrieved 02.09.19.

CRED (Centre for Research on the Epidemiology of Disaster) Crunch. (2019). *Disasters 2018: Year in review*. Issue No. 54.

Djimesah, I. E., Okine, A. N. O., & Mireku, K. K. (2018). Influential factors in creating warning systems towards flood disaster management in Ghana: An analysis of 2007 Northern flood. *International Journal of Disaster Risk Reduction*, *28*, 318−326.

Faulkner, B. (2001). Towards a framework for tourism disaster management. *Tourism Management*, *22*(2), 135−147. Available from https://doi.org/10.1016/S0261-5177(00)00048-0.

Fitchett, J. M., Hoogendoorn, G., & Swemmer, A. M. (2016). Economic costs of the 2012 floods on tourism in the Mopani district municipality, South Africa. *Transactions of the Royal Society of South Africa*, *71*(2), 187−194. Available from https://doi.org/10.1080/0035919X.2016.1167788.

Government of Kerala. *About Kerala*. (2018). From <https://kerala.gov.in/about-kerala> Retrieved 03.09.19.

Ha, K.-M. (2016). Facilitating redundancy-oriented management with gene-therapy-oriented management against disaster. *Risk Analysis*, *36*(6), 1262−1276.

Ha, K.-M. (2019). A mechanism of disaster management in Korea: Typhoons accompanied by flooding. *Heliyon*, *5*, 1−9.

Halldin, S., & Bynander, F. (2012). A strategic Swedish initiative for disaster risk reduction. In: *Proceedings of the 4th international disaster and risk conference: Integrative risk management in a changing world − Pathways to a resilient society, IDRC Davos* (pp. 290−292).

Hindustan Times. *Kerala tourism eyes rise in footfalls, earnings in 2020*. (Oct 7, 2019). From <https://www. hindustantimes.com/travel/kerala-tourism-eyes-rise-in-footfalls-earnings-in-2020/story-tjIfGFGLoFQE2kwggqSZwJ.html> Retrieved 03.11.19.

Hospitality Biz. *Kerala's tourism industry showed strong character post floods*. (2018). From <https://search. proquest.com/docview/2113728739?accountid = 16285> Retrieved 03.09.19.

India Today. *Death toll in flood-hit Kerala rises to 121, 40 injured*. (2019). From <https://www.indiatoday. in/india/story/death-toll-in-flood-hit-kerala-rises-to-121-40-injured-1582258-2019-08-19> Retrieved 03.11.19.

John, H. (Oct 14, 2019). Preparedness remains inadequate even as floods become an annual affair in Kerala by Haritha John. In: *Mongabay series: Flood and drought*. From <https://india.mongabay.com/ 2019/10/preparedness-remains-inadequate-even-as-floods-become-an-annual-affair-in-kerala/> Retrieved 03.11.19.

Kerala State Planning Board. *Economic review 2018*. (2019). From <http://spb.kerala.gov.in/ER2018/index. php> Retrieved 03.11.19.

KTDC (Kerala Tourism Development Corporation). *Tourist statistics for Kerala*. (2006). From <https://web. archive.org/web/20060701140613/http://www.keralatourism.org/php/media/data/tourismstatistics/ TOURISTSTATISTICS2005.pdf> Retrieved 03.11.19.

Matthew, A. (Aug 31, 2019). *In the wake of floods, Kerala tourism seeks revival through MICE*. World Asia. From <https://gulfnews.com/world/asia/india/in-wake-of-floods-kerala-tourism-seeks-revival-through-mice-1.6612398> Retrieved 12.11.19.

Ministry of Tourism. *20 Year perspective plan for Kerala tourism (2002-03 to 2021-22)*. (2003). From <http://tourism.gov.in/perspective-plans-states-uts-1> Retrieved 03.11.19.

Montz, B. E. (2000). The generation of flood hazards and disasters by urban development of floodplains. *Routledge Hazards Disasters Series, 1,* 116−127.

New Indian Express. *Kerala floods: Death toll rises to 483, says CM Pinarayi Vijayan*. (Aug 30, 2018). From <https://indianexpress.com/article/india/483-dead-in-kerala-floods-and-landslides-losses-more-than-annual-plan-outlay-pinarayi-vijayan-5332306/> Retrieved 03.11.19.

NewsClick. *Kerala floods 2019: 121 dead, 1,789 houses collapsed*. (Aug 20, 2019). From <https://www.newsclick.in/kerala-floods-2019-121-dead-1789-houses-collapsed> Retrieved 03.11.19.

Norén, V., Hedelin, B., Nyberg, L., & Bishop, K. (2016). Flood risk assessment − Practices in flood prone Swedish municipalities. *International Journal of Disaster Risk Reduction, 18,* 206−217.

Olano, G. *Asia-Pacific bears brunt of 2018 economic costs*. (Jan 23, 2019). From <https://www.insurancebusinessmag.com/asia/news/regional-news/asiapacific-bears-brunt-of-2018-nat-cat-economic-costs--aon-122933.aspx> Retrieved 10.06.19.

PDNA. *Kerala: Post disaster needs assessment: Floods and landslides − August 2018*. (2018). From <https://www.undp.org/content/undp/en/home/librarypage/crisis-prevention-and-recovery/post-disaster-needs-assessment---kerala.html> Retrieved 10.06.19.

Rehman, J., Sohaib, O., Asif, M., & Pradhan, B. (2019). Applying systems thinking to flood disaster management for a sustainable development. *International Journal of Disaster Risk Management, 36,* 1−10.

Rebuild Kerala Development Programme (RKDP). *Rebuild Kerala Development Programme: A resilient recovery policy framework and action plan for shaping Kerala's resilient, risk-informed development and recovery from 2018 floods*. (May 2019). From <https://kerala.gov.in/documents/10180/5f7872f9-48ab-40c8-bf3f-283d8dceacfe> Retrieved 3 Nov.

Shondell, M. D. (2008). Disaster tourism and disaster landscape attractions after hurricane Katrina: An autoethnographic journey. *International Journal of Culture, Tourism and Hospitality Research, 2*(2), 115−131. Available from https://doi.org/10.1108/17506180810880692.

Sieg, L. (Jul 12, 2018). This is why Japan's floods have been so deadly. In: *World economic forum agenda*. From <https://www.weforum.org/agenda/2018/07/japan-hit-by-worst-weather-disaster-in-decades-why-did-so-many-die> Retrieved 03.11.19.

State Relief Commissioner, Disaster Management (Additional Chief Secretary) Government of Kerala. (2018). *Kerala floods − 2018 1 August to 30 August 2018*. Thiruvananthapuram: Govt. Secretariat.

The Economic Times. *Kerala chief minister stresses on responsible tourism after the floods*. (Sep 28, 2018). From <https://economictimes.indiatimes.com/news/politics-and-nation/kerala-chief-minister-stresses-on-responsible-tourism-after-the-floods/articleshow/65991638.cms> Retrieved 12.11.19.

The Economic Times. *Kerala records 6% rise in tourist arrivals despite floods and Nipah virus scare*. (Feb 14, 2019). From <https://economictimes.indiatimes.com/industry/services/travel/kerala-records-6-rise-in-tourist-arrivals-despite-floods-and-nipah-virus-scare/articleshow/67995390.cms?from = mdr> Retrieved 3 Nov.

Times of India. *Kerala floods live updates: Death toll rises to 79; Kochi airport to remain closed till August 26*. (2018). From <https://timesofindia.indiatimes.com/city/kochi/kerala-floods-live-updates-more-ndrf-teams-rushed-to-kerala-as-flood-situation-worsens/liveblog/65403405.cms> Retrieved 03.11.19.

Times of India. *August rain in Kerala highest in IMD data*. (2019). From <http://timesofindia.indiatimes.com/articleshow/70941896.cms?utm_source = contentofinterest&utm_medium = text&utm_campaign = cppst> Retrieved 03.11.19.

UNWTO. *UNWTO tourism highlights 2015 edition, World Tourism Organization, Madrid.* (2015). From <www.e-unwto.org/doi/pdf/10.18111/978928441689> Retrieved 02.06.19.

World Travel and Tourism Council. *Travel and tourism economic impact 2019.* (2019). From <https://www.wttc.org/-/media/files/reports/economic-impact-research/regions-2019/world2019.pdf> Retrieved 10.06.19.

Yang, B. (2016). GIS based 3-D landscape visualization for promoting citizen's awareness of coastal hazard scenarios in flood prone tourism towns. *Applied Geography*, *76*, 85−97. Available from https://doi.org/10.1016/j.apgeog.2016.09.006.

SHOWCASING ENTREPRENEURS' RESPONSES TO SEVERE DROUGHT: QUALITATIVE FINDINGS FROM CAPE TOWN, SOUTH AFRICA

Florine M. Kuijpers[1] and Emiel L. Eijdenberg[2]

[1]*Independent Researcher, Amsterdam, The Netherlands*
[2]*Business, IT and Science Department, James Cook University, Singapore, Singapore*

9.1 CONTEXTUALIZATION: DROUGHT AND ENTREPRENEURSHIP

That South Africa science pushes to improve daily life does not come as a surprise. Apart from the emerging country's social and economic struggles, climate change leaves its traces as exemplified by severe drought (Wild, 2018). "Drought is a natural hazard and recurrent feature of southern Africa's climate" (Baudoin, Vogel, Nortje, & Naik, 2017, p. 129). Without doubt, South Africa—especially the Western Cape region, in and around Cape Town—has been suffering from drought for years. Some blame solely climate change (e.g., Baudoin et al., 2017; Botai, Botai, Wit, de, Ncongwane, & Adeola, 2017; Chersich et al., 2018; Favretto, Dougill, Stringer, Afionis, & Quinn, 2018; Schiermeier, 2018); others blame failure of people in the government and market sector acting upon ecological challenges (Muller, 2018). The fact is that the environment is changing and scholars call for more research on how society should respond to those changes (Flatø, Muttarak, & Pelser, 2017; Sershen et al., 2016; Vogel & Olivier, 2018; Zwane, 2019).

Different research perspectives (e.g., development economics, see Mare, Bahta, & van Niekerk, 2018; Mdungela, Bahta, & Jordaan, 2017; ecological, see Dos Santos, 2016; Vogel & Olivier, 2018; societal, see Flatø et al., 2017; sustainability, see Favretto et al., 2018) have shed light on the effects of drought and society's response to it. However, the entrepreneurship perspective has largely been overlooked. Entrepreneurship, seen as the establishment of organizations and those who establish them are the *entrepreneurs* (Gartner, 1988), is key for social and economic development (Audretsch, Keilbach, & Lehmann, 2006; Baumol & Strom, 2007; Powell, 2008). Admittedly, entrepreneurship "impacts economic performance at the individual, firm and societal

Economic Effects of Natural Disasters. DOI: https://doi.org/10.1016/B978-0-12-817465-4.00009-1

levels, affecting personal wealth, firm profitability, and economic growth" (Wennekers, Uhlaner, & Thurik, 2002, p. 27). This impact is especially important for the world's emerging contexts such as South Africa.

Understandings of entrepreneurship are not a one size fit all (Annink, Den Dulk, & Amorós, 2016; Eijdenberg, 2016). In recent years, contextualization has received an increased attention from scholars in entrepreneurship research (Welter, 2011; Welter, Baker, Audretsch, & Gartner, 2017; Welter, Kautonen, Chepurenko, Malieva, & Venesaar, 2004; Zahra, Wright, & Abdelgawad, 2014). One way of contextualizing entrepreneurship is examining the role of institutional environments, especially in ecological terms. This type of research has facilitated perspectives on entrepreneurship in different ways, for example, in terms of its challenges and how entrepreneurs respond to these challenges (Eijdenberg, Thompson, Verduijn, & Essers, 2019). However, most of this research is conducted on macro- (i.e., society) and meso- (i.e., sector, community) levels, often based on large sample surveys and cross-country comparative studies (Manolova, Eunni, & Gyoshev, 2008; Stenholm, Acs, & Wuebker, 2013).

Despite the valuable insights that the macro- and meso-level studies have brought forth (e.g., cultural and economic effects, see Aparicio, Urbano, & Audretsch, 2016; Naudé, Gries, Wood, & Meintjies, 2008; Urbano & Alvarez, 2014), little is known about entrepreneurial activities on the micro-level, that is, the personal stories and experiences of entrepreneurs who act in, for example, emerging countries. Institutional theory helps to understand these activities in terms of untangling a context into different components (e.g., cultural, economic, and political) (Shane, 2003; Welter & Smallbone, 2011).

Institutional theory is instrumental indeed to link the macro—micro-level dimensions, and, therefore, shed more light on how macro-level factors affect micro-level activities (Oliver, 1991; Welter & Smallbone, 2011; Wicks, 2001). This way, entrepreneurial activities are made more mundane and this is vital to know from the world's under-researched areas: emerging countries. The aim of this chapter is to reveal entrepreneurial activities by focusing on the responses of 12 entrepreneurs to the institutional challenge of a natural disaster: severe drought. This is done in the emerging context of Cape Town, South Africa, during and just after three years of severe drought. The following research question will be answered: "How do entrepreneurs in an emerging context deal with the ecological challenge of severe drought?"

Following calls for more qualitative insights (Bruton, Ahlstrom, & Li, 2010), this chapter makes two contributions to theory. First, the main contribution is following up to calls for more contextualization of entrepreneurship (Weerawardena & Mort, 2006; Welter & Smallbone, 2011; Welter et al., 2017; Welter, 2011). By doing so, the mundane activities of entrepreneurs will overcome the disconnection of macro- and micro-level research (Eijdenberg et al., 2019; Greenman, 2013; Manolova & Yan, 2002; Stenholm et al., 2013; Urbano & Alvarez, 2014; Wicks, 2001). Second, this study expands the "conventional" institutional challenges (i.e., cultural, economic, political) (cf. Eijdenberg et al., 2019; Griffiths, Gundry, & Kickul, 2013) by adding ecological challenges into the equation. Entrepreneurs respond in three ways: (1) personal solutions; (2) improvised solutions of water management; and (3) professional solutions of water management. This chapter unfolds these types of responses by highlighting many different personal stories and experiences. Hereafter, the relevant literature will be discussed, followed by the methodology. Subsequently, the study's findings are discussed. The chapter closes with a discussion and conclusion.

9.2 BACKGROUND LITERATURE

9.2.1 INSTITUTIONAL THEORY IN ENTREPRENEURSHIP RESEARCH

Institutional theory introduces a unique approach regarding the study of social, economic, and political dynamics (DiMaggio & Powell, 2000). Institutions form the rules of the game within society (North, 1991). Formal as well as informal institutions, for example, the extended family, the private sector and the government, are part of those rules. New institutional insights are building on sociological traditional theories (DiMaggio & Powell, 1983; Scott, 1995). These new insights emphasize the role of the operating institutional context, which are considered to shape enterprises and the behavior of entrepreneurs (Scott, 1995).

When the institutional structure is operating appropriately, it can reduce transaction costs, uncertainty, and risk for entrepreneurs. Legal structures also determine the ease of entering markets and influence bankruptcy laws. Contrarily, a weak or unsatisfactory legal structure may constrain development. Particularly, where institutional gaps leave room for arbitrary behavior, this may lead to corruption and objective behavior of entrepreneurs (Puffer, McCarthy, & Boisot, 2010; Smallbone & Welter, 2001). In addition, informal relationships, such as local networks, arise as a result of institutional constraints (Khanna & Palepu, 1997). As a result, entrepreneurship can appear while not being legally recognized (Klapper, Laeven, & Rajan, 2006).

Institutional theory has received increased attention in entrepreneurship research; however, most of them taking place in developed countries. Conversely, emerging contexts are gaining stage regarding institutional theory-led research. Most of this type of research has shed light on the aspects of the market and government failure of countries, therefore, providing room for different types of entrepreneurs to operate (Littlewood & Holt, 2018; Rivera-Santos, Holt, Littlewood, & Kolk, 2015). Examples of aspects are the aging infrastructure; static-centric systems; retained government control in the private sector and land ownership; limited access to information; and bureaucratic procedures to start new enterprises (Central Intelligence Agency, 2019; Saini & Bhatia, 1996; World Bank, 2019a). Although interviewing every entrepreneur in a country is unrealistic, attempts have been made to hear personal stories of them about operating enterprises in emerging contexts (cf. Abdallah & Eijdenberg, 2019; Eijdenberg et al., 2019; Khavul, Bruton, & Wood, 2009; Littlewood & Holt, 2018). By hearing these stories the link between macro and micro is made, making entrepreneurial activities in the world's underrepresented research areas come alive.

9.2.2 ECOLOGICAL CHALLENGES OF ENTREPRENEURS IN SOUTH AFRICA

From an institutional theory perspective, effective drought management can be hampered by complicated bureaucratic procedures, static-centric institutions, slow top-down management, corruption, lack of transparency and accountability of the institutions, and poor infrastructure (Baudoin et al., 2017). The government, as guardian of the country's citizens, plays a key role in this. South Africa has been suffering from many of the previously mentioned factors that all led to dramatic societal consequences such as destroyed farmlands and malnutrition of people. In the past, drought management has only been implemented *after* times of drought, not during or as a precautionary measure (Baudoin et al., 2017; Favretto et al., 2018). Recently, an increasing number of official, government-led arrangements have been made such as new legislation on water management, policies, warning systems, support programs, trainings, and water management equipment for farmers, private sector and citizens—all with moderate success (Bahta, Jordaan, & Muyambo, 2016; Baudoin et al., 2017; Mare et al., 2018).

Nevertheless, from an entrepreneurship perspective, commercial farmers have often been subjected to broad-sample surveys on the effects of drought and their responses thereof (e.g., Bahta et al., 2016; Mdungela et al., 2017; Mutero, Munapo, & Seaketso, 2016). Indeed, as being highly dependent on climate, farmers—if perceived as *entrepreneurs* anyways—are probably the ones hit first by drought. Multiple consequences of drought to farmers are reported, for example, lack of resources, unemployment, price sensitivity, and the lack of financial safety nets (Mdungela et al., 2017); and inadequate government support, psychological stress, and high levels of stock theft (Bahta et al., 2016). The provision of basic tools is often the key to acquire some sort of irrigation and improved management, and, hence, business viability and continuation (Mutero et al., 2016). Similar patterns (i.e., business viability and continuation based on basic entrepreneurial tools and skills) are found by different types of entrepreneurs in other African emerging countries (cf. Choongo, 2017; Eijdenberg et al., 2019; Khavul et al., 2009). However, the entrepreneurship perspective—especially connecting macro-level events to micro-level activities, based on qualitative data—is hardly applied in the context of ecological challenges such as drought.

9.3 METHODOLOGY

9.3.1 RESEARCH CONTEXT: CAPE TOWN, SOUTH AFRICA

South Africa has one of the oldest notions of human civilization and has a long history of colonization before it received full independence from the United Kingdom in 1931. Through history, the country has had different ethnic groups (e.g., Asians, Europeans, indigenous Bantu peoples) living together on 1.2 million square kilometers (Central Intelligence Agency, 2019).

With respect to all aspects of South Africa's history in the 20th century, a huge leap brings us to 2010: the year that South Africa entered the league of so-called BRICS (Brazil, Russia, India, China, and South Africa), that is, the five major emerging economies of the world (Bremmer, 2017). To illustrate the economic situation, South Africa's estimated population of 55 million makes up for a gross domestic product of USD 767 billion (i.e., USD 13,600 per capita—in comparison with, e.g., the United States: USD 59,800). The country is rich in natural resources, and well-developed manufacturing and services sectors. However, in recent years, economic growth has decelerated and unemployment (i.e., 27.5%), poverty (i.e., 16.6% lives below the poverty line) and inequality in the post-apartheid era have hindered the country's social and economic progression (Carter & May, 2001; Central Intelligence Agency, 2019; Durrheim, Mtose, & Brown, 2011).

South Africa is one of the world's most unequal countries, having a 63.0 score on the Gini Index in 2014 (World Bank, 2019b). Typical social issues in South Africa are significant numbers of an unskilled and low-educated workforce; high crime rates; and the highest HIV/AIDS numbers of the world (Central Intelligence Agency, 2019). Moreover, the country has been showing concerning numbers in comparison with the entire sub-Saharan region in terms of perceived opportunities, entrepreneurial intentions, entrepreneurial activity, and established business ownership

(Global Entrepreneurship Monitor, 2019b). For example, the percentage of the 18−64 population (individuals involved in any stage of entrepreneurial activity excluded) who are latent entrepreneurs and who intend to start a business within three years is 12.76% (2013)—in comparison, Botswana is 20.85%, Malawi is 28.11%, Namibia 33.34%, and Zambia is 39.91%, in the same year (Global Entrepreneurship Monitor, 2019a).

Conversely, entrepreneurs in emerging contexts such as South Africa are important because they move economic development forward (Naudé et al., 2008). Like elsewhere, entrepreneurs in South Africa address global trends (e.g., innovation, mobility, sustainability) and provide improvised solutions to them (e.g., frugal innovations[1]).

9.3.2 RESEARCH DESIGN

As the aim of this explorative study is to showcase the personal responses of entrepreneurs to drought in an emerging context, a qualitative research approach was chosen. "The qualitative researchers study things in their natural settings, attempting to make sense of, or interpret, phenomena in terms of the meanings people bring to them" (Denzin & Lincoln, 2011, p. 3). This approach fits with the aim of the study to get an understanding of entrepreneurs' experiences, feelings, and perceptions. For the same reason, this study follows a phenomenological approach (Groenewald, 2004; Yin, 2009).

The data were collected from individual interviews with entrepreneurs, active in different industries. Taking Gephart (2004) as the point of departure, the aim of the individual interviews was to collect inductive data on environmental challenges and entrepreneurs' responses to those challenges. Semi-structured interviews with open-ended questions were conducted, which led to an in-depth interview, which was recorded and transcribed. The advantage of the individual interview is the in-depth, detailed, and personal aspect. In addition, the interviewer is able to elaborate extensively on personal stories. The use of probing questions and follow-up questions results in a deeper understanding of the studied phenomena (Bryman, 2016; Saunders, Lewis, & Thornhill, 2016).

9.3.3 SAMPLE SELECTION

The data stems from entrepreneurs in Cape Town. Amongst other major urban areas such as Johannesburg, Durban, and Pretoria, Cape Town is the country's legislative capital of 4.4 million inhabitants. In addition, Cape Town is one of the major tourist destinations of South Africa. Much entrepreneurial activity has grown in this city because of its branches of different multinational enterprises and its ideal central location with a large harbor and airport (Cape Town Tourism, 2018; Central Intelligence Agency, 2019; World Bank, 2017). Cape Town is leading in ease of doing business compared to other cities in South Africa (World Bank, 2019a). Apart from its strategic location, since 2015 Cape Town has been suffering from severe drought that has peaked in mid-2017 and mid-2018 (City of Cape Town, 2019; Climate System Analysis Group, 2019) impacting daily life significantly.

---

[1]Frugal innovations are typically not perfect, rather acceptable. They represent affordable products or services that sufficiently meet the needs of resource-constrained consumers who have no access to more expensive alternatives, see Hossain (2018) and Khan (2016).

A generic purposive sampling technique was used to select respondents for the interviews (Saunders et al., 2016). The "Social Enterprise Academy South Africa" facilitated a list of enterprises based in Cape Town (Social Enterprise Academy South Africa, 2019). This initiative monitors social entrepreneurial activity in Cape Town with the objective of learning for enterprises to increase their social impact and financial sustainability. This initial sample of 80 enterprises was the point of departure for further selection: 50 enterprises were contacted randomly of which 20% responded to cooperate. In addition, two other interviews were conducted based on snowball sampling (Mey & Smit, 2013): "snowballing involves recruiting individuals to collect data from other individuals whom they think meet certain inclusion criterion defined by the researcher" (McGee, Peterson, Mueller, & Sequeira, 2009, p. 974). Thus, a total of 12 interviews was conducted, a sufficient number to achieve saturation level in qualitative research (Saunders et al., 2016). The respondents' selection criteria were (1) the total number of respondents had to represent an equal gender-balance of female and male entrepreneurs; (2) each respondent had to be the founder of the enterprise, but in case the founder was not anymore involved, the successor or board member was selected; (3) the respondents had to be from different industries to achieve heterogeneity of the sample; and (4) both the enterprises and its entrepreneurs had to be based in Cape Town. Table 9.1 presents the most relevant pertinent information of the respondents.

9.3.4 DATA COLLECTION

After the sample selection, the first author of this chapter (hereafter referred to as "the researcher") collected the qualitative data by online, face-to-face video interviews using Facetime, Google Hangouts, and Skype. The data were collected in December 2018 and January 2019. The researcher used one semi-structured interview guide. The interviews revolved around the following main question: "How have you and your business dealt with the effects of drought in Cape Town, South Africa?". Naturally, deviations from this question occurred (e.g., follow up questions and probes) to structure the interview.

The 12 interviews were conducted in English. The interviews took on average 45 minutes; however, significant deviations in time occurred because certain entrepreneurs conducted the interview during operation hours of their business. While all entrepreneurs were located in the wider Cape Town region, the majority of them were located in the business district of Cape Town (i.e., "Cape Town CBD" or "City Bowl" area). Exceptions were townships and other parts of the city, usually with poorer Internet connectivity. All interviews were recorded and afterward transcribed intelligent verbatim.

9.3.5 DATA ANALYSIS

After the data were collected and transcribed the analyses were conducted. An inductive analysis approach was used, following the "Gioia methodology" (Gehman et al., 2018; Gioia, Corley, & Hamilton, 2013). In this way, credible interpretations of data were established through systematic conceptual and analytical discipline. This methodology brings qualitative rigor and transparency to inductive research and it has previously been applied in other African contexts (cf. Eijdenberg et al., 2019). First, the first-order concepts were formulated (a reduced number of words of a quotation or observation) by coding the 12 interviews. Second, these first-order concepts were

Table 9.1 Pertinent Information of the Respondents.[a]

Respondent Number	Gender	Age	Position Within Business	Founding Year of Business	Number of Employees at Time of Interview	Industry
1[b]	Male	26	Responsible for social enterprises	1981	15	Education
2	Female	49	Founder	1981	15	Education
3	Male	24	Founder	2018	2–10	Agriculture
4	Male	24	Founder	2014	2–10	Education
5	Male	45	Founder	2004	15	Entertainment
6	Female	39	Founder	2012	5	Entertainment
7	Male	44	Founder	2014	2–10	Education
8	Male	43	Director	1997	25	Education
9	Female	60	Founder	1992	51–200	Disability service
10	Female	34	Cofounder	2017	2–10	Education
11	Female	49	Cofounder	2010	75	Retail
12	Female	28	Founder	2016	9	Education

[a]The information in the table is as indicated by the respondents. Therefore certain information (e.g., "Number of employees at time of interview: 51–200") is factual, yet estimated.
[b]Respondent 1 and respondent 2 are both from the same business. As opposed to respondent 2 who is the founder of the business, respondent 1 is responsible for the initiative "Social Enterprises". Hence, respondent 1 was included as one of the respondents.

categorized into 15 second-order codes. Third, based on the second-order codes three aggregate dimensions were generated: "personal solutions," "improvised solutions of water management," and "professional solutions of water management."

9.4 FINDINGS

The three themes (i.e., "personal solutions," "improvised solutions of water management," and "professional solutions of water management") are presented in Fig. 9.1. This figure is supported by exemplary quotes. These quotes are presented in Table 9.2.

Cape Town's drought has influenced respondents in effectively running their enterprises. To illustrate, all inhabitants of Cape Town were allowed to use only 20 L of water per person per day and businesses had water restrictions imposed. Respondents reported that they were not able to manage their businesses as before the drought. For example, donated clothes could not be washed properly since washing machines were using too much water (respondent 8). In addition, Cape Town's environment changed: respondents were not able to enter certain parts of townships anymore because of safety reasons due to drought. This has limited business operations, causing declining sales.

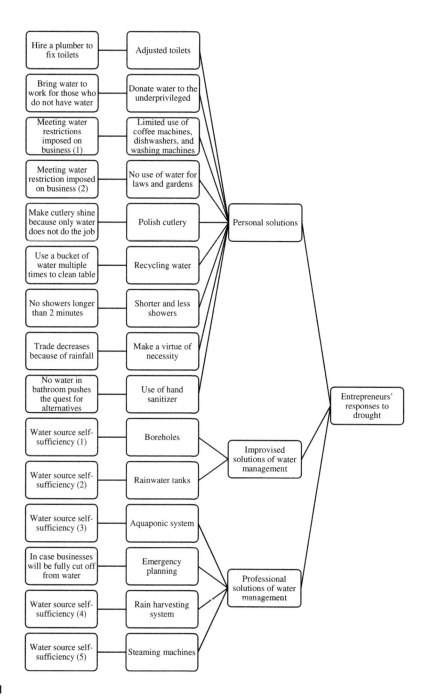

FIGURE 9.1

Entrepreneurs' responses to drought.

Table 9.2 Exemplary Quotes of Entrepreneurs' Responses.

Exemplary Quote of First-Order Quote	Second-Order Code
"We had the plumber coming in so that the toilets wouldn't flush as strongly, because we had really old toilets in this building, it would still flush properly, but not as fast, like those type of things."	Adjusted toilets
"Yeah! Saving the water, bringing water to work to help people who didn't have water; it's just part of work."	Donate water to the underprivileged
"So, we were going from doing 20 loads of washing a day to doing six or seven loads of washing a day, because we couldn't fit into the water restrictions that were imposed on us. We had to change all of our machines because the previous machines were using too much water."	Limited use of coffee machines, dishwashers and washing machines
"For the business what we did was, so normally young adults would polish the cutlery and stuff like that, and they do that with water. So, the cutlery is washed in the dishwasher, but to make it shiny, we still polish it."	Polish cutlery
"Normally they would fill a whole bucket and then just clean the table once and then throw that water away. So, same story, I would tell them: 'No, you must use that at least a couple of times until it's dirty, otherwise it's a waste of water.'"	Recycling water
"I mean, up until a couple of weeks ago and don't think I've had a shower longer than two minutes for more than a year."	Shorter and less showers
"What did happen was with the sales booming in our program, they had a really good sales winter, because in Cape Town it rains in winter. So, most traders are informal. So, in winter their sales often drops because of the raining, and so they can't get our things traded. So, they actually did quite well with their trading in the winter that we didn't had [have] much rain. So, it's sort of a benefit of being in a kind of funny sense."	Make a virtue of necessity
"With regards to the droughts, obviously there was the general things of not having water in the bathrooms and having to use hand sanitizer for example, and that sort of thing."	Use of hand sanitizer
"We're far more water source self-sufficient and we've got a borehole so and we're still trying to come close to Table Mountain so we got beautiful mountain water into the borehole."	Boreholes
"The strength of the drought was obviously a concern for us which we responded to immediately so well, our properties have proper rainwater tanks and so that should safe some."	Rainwater tanks
"An aquaponic system is a method to grow food, in which aquaculture, the growth of aquatic animals, is combined in a symbiotic way with hydroponics, the cultivation of plants in water. Fish provide the necessary nutrients for plant growth and the plant roots filter the water for the fish. Therefore, aquaponic creates its own ecosystem, without the need of adding non-ecological plant nutrients. Water with the nutrients from the fish is pumped into a grow bed. Plants filter the water, which is being pumped into the sump tank and back into the fish tank. And one of the major benefits of the system is that it saves up to 90% of water in comparison to growing vegetables in regular fertile soil."	Aquaponic system

(Continued)

Table 9.2 Exemplary Quotes of Entrepreneurs' Responses. *Continued*

Exemplary Quote of First-Order Quote	Second-Order Code
"Because it would be a nightmare if we were cut off, imagine all the bathrooms, and with like 100 people in and out here all day, it would be a disaster. So, we made a plan in case 'Day Zero' came."	Emergency planning
"We had to put in a rain harvesting system, so that when it does rain, we can catch that in a tank and use that."	Rain harvesting system
"We had to buy steaming machines so that we could steam stuff that weren't […] And it costed us 60.000 Rand to do that. A few months later, we did two more water tanks at our cellar centre that was another 10.000 Rand. Let's call it 70.000 Rand. It's not insignificant for us, but we just had to do that, we couldn't avoid it."	Steaming machines

Personal solutions from entrepreneurs were exemplified by the many things they could do with their bare hands or required little time and resources. For example, hiring a plumber to adjust the toilets not to flush as strongly (respondent 12); donating water on work to those who do not have access to water (respondent 9); and using hand sanitizer as an alternative for traditional washing hands in the bathroom (respondent 10).

At the business level, two types of solutions of water management were discussed. First, improvised solutions were made to contribute with as limited resources as possible such as making boreholes (respondent 2) and installing "Jojo" rainwater tanks (respondent 11) (see for a description: Jojo, 2019). Second, more professional solutions were made such as an aquaponic system (see Table 9.2 for a description from the data and see, e.g., Mchunu, Lagerwall, & Senzanje, 2018 for a literary reference) (respondent 3), rain harvesting systems and steaming machines (respondent 8). However, the professional solutions often come at a financial premium that means they are not feasible for all entrepreneurs.

9.5 DISCUSSION

9.5.1 CONTRIBUTION TO THE LITERATURE

The main contribution of this chapter is following up to calls for more contextualization of entrepreneurship (Weerawardena & Mort, 2006; Welter & Smallbone, 2011; Welter et al., 2017; Welter, 2011). By doing so, the mundane activities of entrepreneurs as showcased in the Findings section overcome the disconnection of macro- and micro-level research (Eijdenberg et al., 2019; Greenman, 2013; Manolova & Yan, 2002; Stenholm et al., 2013; Urbano & Alvarez, 2014; Wicks, 2001). This is of great value because it provides more lively and tangible examples, catering for better theoretical and practical solutions. Within contextualization, this chapter adds onto the calls for more research of emerging countries in Africa (George, 2015; George, Corbishley, Khayesi,

Haas, & Tihanyi, 2016; Khavul et al., 2009; Khayesi, George, & Antonakis, 2014; Littlewood & Holt, 2015, 2018; Rivera-Santos et al., 2015).

On a more granular level the second contribution is that this chapter expands the "conventional" institutional challenges (i.e., cultural, economic, political) (cf. Eijdenberg et al., 2019; Griffiths et al., 2013) by adding ecological challenges into the equation. Obviously, ecological challenges themselves are not new, nor the sustainable solutions to them in emerging countries (e.g., Askham, & Van der Poll, 2017; Choongo, 2017; Choongo, van Burg, Masurel, Paas, & Lungu, 2017; Mabhaudhi et al., 2019). However, the entrepreneurs' responses to those challenges, from an institutional theory perspective, are novel—especially in the case of the recent drought in the Cape Town region. These responses contain many detailed, contextual examples, providing a closer look at real activities of entrepreneurs during and after times of a natural disaster.

9.5.2 LIMITATIONS AND RECOMMENDATIONS FOR FUTURE RESEARCH

Taking the findings of the second-order codes and the three aggregate themes as such, they are limited by the used methodology; sample selection, size, and location; and time of the data collection. Obviously, different findings are produced when these limitations are overcome. Therefore, future researchers are encouraged to fill out more qualitative findings by, for example, conducting triangulation and researching other cases in different settings and times (Saunders et al., 2016). Especially, the aggregated theme of ecological challenges and the entrepreneurs' responses thereof should receive more scholarly attention because this is clearly found as an addition to previously mentioned "conventional" institutional challenges (i.e., cultural, economic, political) (cf. Abdallah & Eijdenberg, 2019; Eijdenberg et al., 2019).

Moreover, as this study follows up on institutional theory and entrepreneurship in emerging countries (e.g., Eijdenberg et al., 2019; Littlewood & Holt, 2015, 2018; Rivera-Santos et al., 2015), future researchers are encouraged to move this type of research from a qualitative to a quantitative methodological fashion (Edmondson & McManus, 2007). Indeed, qualitative research such as the current study yields rich, proto-typical, and exemplary findings from South Africa's emerging context; however, future research should also contain events and patterns that are generalizable to a certain extent, and, therefore, surveying entrepreneurs would be the next methodological step.

9.5.3 IMPLICATIONS FOR PRACTICE

The practical relevance of this chapter reflects in multiple ways. First, on a general level, this chapter stimulates the debate about what it means to entrepreneurs to overcome the effects of drought in South Africa. This can be done, for example, already on primary and secondary school: developing children's entrepreneurial skills to think of and create improvised solutions (e.g., boreholes, rainwater tanks) of water management. Moreover, also the current existing cohort of entrepreneurs in South Africa, or elsewhere in the world, can take lessons from their peers to overcome the effects of natural disasters. For example, educators can teach entrepreneurs to become more self-sufficient and efficient by using water (e.g., less use of coffee machines, dishwashers, and washing machines; using hand sanitizer; polish cutlery as a way to use less water). In that way, children and entrepreneurs will be more prepared and resilient to natural crises which is something, unfortunately, has to be taken into account for the future because of climate change.

9.6 CONCLUSION

The aim of this chapter was to give an answer to the research question "How do entrepreneurs in an emerging context deal with the ecological challenge of severe drought?" by studying the interplay between the institutional context and entrepreneurship from a micro-level perspective within the context of Cape Town, South Africa. The ecological challenges due to recent drought play a key role in everyday entrepreneurial life. Entrepreneurs respond to the challenges in three ways: (1) personal solutions; (2) improvised solutions of water management; and (3) professional solutions of water management. In addition to the literature, this chapter showcases the mundane activities of entrepreneurs in an emerging context during and after times of a natural disaster, that is, severe drought. By doing so, more contextualization of entrepreneurship is given (Welter & Smallbone, 2011; Welter et al., 2017; Welter, 2011), bridging the gap between macro- and micro-level research (Eijdenberg et al., 2019; Manolova et al., 2008; Stenholm et al., 2013).

REFERENCES

Abdallah, G. K., & Eijdenberg, E. L. (2019). Entry and stay in the informal economy: Qualitative findings from a least developed country. *Journal of Enterprising Culture*, 27(2), 115−145.

Annink, A., Den Dulk, L., & Amorós, J. E. (2016). Different strokes for different folks? The impact of heterogeneity in work characteristics and country contexts on work-life balance among the self-employed. *International Journal of Entrepreneurial Behavior & Research*, 22(6), 880−902.

Aparicio, S., Urbano, D., & Audretsch, D. (2016). Institutional factors, opportunity entrepreneurship and economic growth: Panel data evidence. *Technological Forecasting and Social Change*, 102, 45−61.

Askham, T. M., & Van der Poll, Huibrecht, M. (2017). Water sustainability of selected mining companies in South Africa. *Sustainability*, 9(6), 957.

Audretsch, D. B., Keilbach, M. C., & Lehmann, E. E. (2006). *Entrepreneurship and economic growth*. Oxford: Oxford University Press.

Bahta, Y. T., Jordaan, A., & Muyambo, F. (2016). Communal farmers' perception of drought in South Africa: Policy implication for drought risk reduction. *International Journal of Disaster Risk Reduction*, 20, 39−50.

Baudoin, M. A., Vogel, C., Nortje, K., & Naik, M. (2017). Living with drought in South Africa: Lessons learnt from the recent El Niño drought period. *International Journal of Disaster Risk Reduction*, 23, 128−137.

Baumol, W. J., & Strom, R. J. (2007). Entrepreneurship and economic growth. *Strategic Entrepreneurship Journal*, 1(3-4), 233−237.

Botai, C. M., Botai, J. O., De Wit, J. P., Ncongwane, K. P., & Adeola, A. M. (2017). Drought characteristics over the Western Cape Province, South Africa. *Water*, 9(11), 876.

Bremmer, I. (2017). *The mixed fortunes of the BRICS countries, in 5 facts*. Retrieved from <http://time.com/4923837/brics-summit-xiamen-mixed-fortunes/>.

Bruton, G. D., Ahlstrom, D., & Li, H.-L. (2010). Institutional theory and entrepreneurship: Where are we now and where do we need to move in the future? *Entrepreneurship Theory and Practice*, 34(3), 421−440.

Bryman, A. (2016). *Social research methods*. Oxford: Oxford University Press.

Cape Town Tourism. (2018). *Annual report 2017/2018*. Retrieved from <https://www.capetown.travel/wp-content/uploads/2018/10/Annual-Report-20172018.pdf>.

Carter, M. R., & May, J. (2001). One kind of freedom: Poverty dynamics in post-apartheid South Africa. *World Development*, 29(12), 1987−2006.

Central Intelligence Agency. (2019). *The world factbook: South Africa.* Retrieved from <https://www.cia.gov/library/publications/the-world-factbook/geos/sf.html>.

Chersich, M., Wright, C., Venter, F., Rees, H., Scorgie, F., & Erasmus, B. (2018). Impacts of climate change on health and wellbeing in South Africa. *International Journal of Environmental Research and Public Health, 15*(9), 1884–1898.

Choongo, P. (2017). A longitudinal study of the impact of corporate social responsibility on firm performance in SMEs in Zambia. *Sustainability, 9*(8), 1300.

Choongo, P., van Burg, E., Masurel, E., Paas, L. J., & Lungu, J. (2017). Corporate Social Responsibility motivations in Zambian SMEs. *International Review of Entrepreneurship, 15*(1), 29–62.

City of Cape Town. (2019). *Water outlook 2018 report.* Retrieved from <https://resource.capetown.gov.za/documentcentre/Documents/City%20research%20reports%20and%20review/Water%20Outlook%202018%20-%20Summary.pdf>.

Climate System Analysis Group. (2019). *Big six monitor.* Retrieved from <http://cip.csag.uct.ac.za/monitoring/bigsix.html>.

Denzin, N. K., & Lincoln, Y. S. (2011). *The Sage handbook of qualitative research.* Thousand Oaks, CA: Sage.

DiMaggio, P. J., & Powell, W. W. (1983). The iron cage revisited: Collective rationality and institutional isomorphism in organizational fields. *American Sociological Review, 48*(2), 147–160.

DiMaggio, P. J., & Powell, W. W. (2000). *The iron cage revisited institutional isomorphism and collective rationality in organizational fields. Economics meets sociology in strategic management* (pp. 143–166). Bingley: Emerald Group Publishing Limited.

Dos Santos, M. J. P. L. (2016). Smart cities and urban areas—Aquaponics as innovative urban agriculture. *Urban Forestry and Urban Greening, 20,* 402–406.

Durrheim, K., Mtose, X., & Brown, L. (2011). *Race trouble: Race, identity and inequality in post-apartheid South Africa.* Lanham, MD: Lexington Books.

Edmondson, A. C., & McManus, S. E. (2007). Methodological fit in management field research. *Academy of Management Review, 32*(4), 1246–1264.

Eijdenberg, E. L. (2016). Does one size fit all? A look at entrepreneurial motivation and entrepreneurial orientation in the informal economy of Tanzania. *International Journal of Entrepreneurial Behavior and Research, 22*(6), 804–834.

Eijdenberg, E. L., Thompson, N. A., Verduijn, K., & Essers, C. (2019). Entrepreneurial activities in a developing country: An institutional theory perspective. *International Journal of Entrepreneurial Behavior and Research, 25*(3), 414–432.

Favretto, N., Dougill, A., Stringer, L., Afionis, S., & Quinn, C. (2018). Links between climate change mitigation, adaptation and development in land policy and ecosystem restoration projects: Lessons from South Africa. *Sustainability, 10*(3), 779.

Flatø, M., Muttarak, R., & Pelser, A. (2017). Women, weather, and woes: The triangular dynamics of female-headed households, economic vulnerability, and climate variability in South Africa. *World Development, 90,* 41–62.

Gartner, W. B. (1988). "Who is an entrepreneur?" is the wrong question. *American Journal of Small Business, 12*(4), 11–32.

Gehman, J., Glaser, V. L., Eisenhardt, K. M., Gioia, D., Langley, A., & Corley, K. G. (2018). Finding theory—method fit: A comparison of three qualitative approaches to theory building. *Journal of Management Inquiry, 27*(3), 284–300.

George, G. (2015). Expanding context to redefine theories: Africa in management research. *Management and Organization Review, 11*(1), 5–10.

George, G., Corbishley, C., Khayesi, J. N. O., Haas, M. R., & Tihanyi, L. (2016). Bringing Africa in: Promising directions for management research. *Academy of Management Journal, 59*(2), 377–393.

Gephart, R. P., Jr (2004). Qualitative research and the Academy of Management Journal. *Academy of Management Journal, 47*(4), 454–462.

Gioia, D. A., Corley, K. G., & Hamilton, A. L. (2013). Seeking qualitative rigor in inductive research: Notes on the Gioia methodology. *Organizational Research Methods, 16*(1), 15–31.

Global Entrepreneurship Monitor. (2019a). *Data: South Africa*. Retrieved from <https://www.gemconsortium.org/data/key-aps>.

Global Entrepreneurship Monitor. (2019b). *South Africa 2016-2017 national report*. Retrieved from <https://www.gemconsortium.org/report/gem-south-africa-2016-2017-report>.

Greenman, A. (2013). Everyday entrepreneurial action and cultural embeddedness: An institutional logics perspective. *Entrepreneurship & Regional Development, 25*(7-8), 631–653.

Griffiths, M. D., Gundry, L. K., & Kickul, J. R. (2013). The socio-political, economic, and cultural determinants of social entrepreneurship activity: An empirical examination. *Journal of Small Business and Enterprise Development, 20*(2), 341–357.

Groenewald, T. (2004). A phenomenological research design illustrated. *International Journal of Qualitative Methods, 3*(1), 42–55.

Hossain, M. (2018). Frugal innovation: A review and research agenda. *Journal of Cleaner Production, 182*, 926–936.

Jojo. (2019). *Jojo: For water, for life*. Retrieved from <https://www.jojo.co.za/>.

Khan, R. (2016). How frugal innovation promotes social sustainability. *Sustainability, 8*(10), 1034.

Khanna, T., & Palepu, K. (1997). Why focused strategies may be wrong for emerging markets. *Harvard Business Review, 75*(4), 41–43.

Khavul, S., Bruton, G. D., & Wood, E. (2009). Informal family business in Africa. *Entrepreneurship Theory and Practice, 33*(6), 1219–1238.

Khayesi, J. N. O., George, G., & Antonakis, J. (2014). Kinship in entrepreneur networks: Performance effects of resource assembly in Africa. *Entrepreneurship Theory and Practice, 38*(6), 1323–1342.

Klapper, L., Laeven, L., & Rajan, R. (2006). Entry regulation as a barrier to entrepreneurship. *Journal of Financial Economics, 82*(3), 591–629.

Littlewood, D., & Holt, D. (2015). *Social and environmental enterprises in Africa: Context, convergence and characteristics. The business of social and environmental innovation* (pp. 27–47). Cham: Springer.

Littlewood, D., & Holt, D. (2018). Social entrepreneurship in South Africa: Exploring the influence of environment. *Business and Society, 57*(3), 525–561.

Mabhaudhi, T., Chibarabada, T., Chimonyo, V., Murugani, V., Pereira, L., Sobratee, N., ... Modi, A. (2019). Mainstreaming underutilized indigenous and traditional crops into food systems: A South African perspective. *Sustainability, 11*(1), 172.

Manolova, T. S., Eunni, R. V., & Gyoshev, B. S. (2008). Institutional environments for entrepreneurship: Evidence from emerging economies in Eastern Europe. *Entrepreneurship Theory and Practice, 32*(1), 203–218.

Manolova, T. S., & Yan, A. (2002). Institutional constraints and entrepreneurial responses in a transforming economy: The case of Bulgaria. *International Small Business Journal, 20*(2), 163–184.

Mare, F., Bahta, Y. T., & van Niekerk, W. (2018). The impact of drought on commercial livestock farmers in South Africa. *Development in Practice, 28*(7), 884–898.

McGee, J. E., Peterson, M., Mueller, S. L., & Sequeira, J. M. (2009). Entrepreneurial self-efficacy: Refining the measure. *Entrepreneurship Theory and Practice, 33*(4), 965–988.

Mchunu, N., Lagerwall, G., & Senzanje, A. (2018). Aquaponics in South Africa: Results of a national survey. *Aquaculture Reports, 12*, 12–19.

Mdungela, N. M., Bahta, Y. T., & Jordaan, A. J. (2017). Indicators for economic vulnerability to drought in South Africa. *Development in Practice, 27*(8), 1050–1063.

Mey, Ld, & Smit, D. (2013). *Advanced research methods*. London: Sage Publications.

Muller, M. (2018). Lessons from Cape Town's drought. *Nature, 559*(7713), 174–176.

Mutero, J., Munapo, E., & Seaketso, P. (2016). Operational challenges faced by smallholder farmers: A case of Ethekwini Metropolitan in South Africa. *Environmental Economics*, *7*(2), 40−52.

Naudé, W. A., Gries, T., Wood, E., & Meintjies, A. (2008). Regional determinants of entrepreneurial start-ups in a developing country. *Entrepreneurship and Regional Development*, *20*(2), 111−124.

North, D. C. (1991). Institutions. *Journal of Economic Perspectives*, *5*(1), 97−112.

Oliver, C. (1991). Strategic responses to institutional processes. *Academy of Management Review*, *16*(1), 145−179.

Powell, B. (2008). *Making poor nations rich: Entrepreneurship and the process of economic development*. Palo Alto, CA: Stanford University Press.

Puffer, S. M., McCarthy, D. J., & Boisot, M. (2010). Entrepreneurship in Russia and China: The impact of formal institutional voids. *Entrepreneurship Theory and Practice*, *34*(3), 441−467.

Rivera-Santos, M., Holt, D., Littlewood, D., & Kolk, A. (2015). Social entrepreneurship in sub-Saharan Africa. *Academy of Management Perspectives*, *29*(1), 72−91.

Saini, J. S., & Bhatia, B. S. (1996). Impact of entrepreneurship development programmes. *Journal of Entrepreneurship*, *5*(1), 65−80.

Saunders, M., Lewis, P., & Thornhill, A. (2016). *Research methods for business students*. Essex: Pearson Education Limited.

Schiermeier, Q. (2018). Climate as culprit. *Nature*, *560*, 20−22.

Scott, W. R. (1995). *Organizations and institutions*. Thousand Oaks, CA: Sage.

Sershen, S., Rodda, N., Stenström, T. A., Schmidt, S., Dent, M., Bux, F., ... Fennemore, C. (2016). Water security in South Africa: Perceptions on public expectations and municipal obligations, governance and water re-use. *Water SA*, *42*(3), 456−465.

Shane, S. A. (2003). *A general theory of entrepreneurship: The individual-opportunity nexus*. Cheltenham: Edward Elgar Publishing.

Smallbone, D., & Welter, F. (2001). The distinctiveness of entrepreneurship in transition economies. *Small Business Economics*, *16*(4), 249−262.

Social Enterprise Academy South Africa. (2019). *Learning and development for social chdngemakers in South Africa*. Retrieved from <https://www.socialenterprise.academy/za/>.

Stenholm, P., Acs, Z. J., & Wuebker, R. (2013). Exploring country-level institutional arrangements on the rate and type of entrepreneurial activity. *Journal of Business Venturing*, *28*(1), 176−193.

Urbano, D., & Alvarez, C. (2014). Institutional dimensions and entrepreneurial activity: An international study. *Small Business Economics*, *42*(4), 703−716.

Vogel, C., & Olivier, D. (2018). Re-imagining the potential of effective drought responses in South Africa. *Regional Environmental Change*, *19*(6), 1561−1570.

Weerawardena, J., & Mort, G. S. (2006). Investigating social entrepreneurship: A multidimensional model. *Journal of World Business*, *41*(1), 21−35.

Welter, F. (2011). Contextualizing entrepreneurship—Conceptual challenges and ways forward. *Entrepreneurship Theory and Practice*, *35*(1), 165−184.

Welter, F., Baker, T., Audretsch, D. B., & Gartner, W. B. (2017). Everyday entrepreneurship—A call for entrepreneurship research to embrace entrepreneurial diversity. *Entrepreneurship Theory and Practice*, *41*(3), 311−321.

Welter, F., Kautonen, T., Chepurenko, A., Malieva, E., & Venesaar, U. (2004). Trust environments and entrepreneurial behavior—Exploratory evidence from Estonia, Germany and Russia. *Journal of Enterprising Culture*, *12*(4), 327−349.

Welter, F., & Smallbone, D. (2011). Institutional perspectives on entrepreneurial behavior in challenging environments. *Journal of Small Business Management*, *49*(1), 107−125.

Wennekers, S., Uhlaner, L., & Thurik, R. (2002). Entrepreneurship and its conditions: A macro perspective. *International Journal of Entrepreneurship Education*, *1*(1), 25−64.

Wicks, D. (2001). Institutionalized mindsets of invulnerability: Differentiated institutional fields and the antecedents of organizational crisis. *Organization Studies*, 22(4), 659–692.

Wild, S. (2018). South Africa pushes science to improve daily life. *Nature*, 561(7722), 157–159.

World Bank. (2017). *Tourism for development: 20 reasons sustainable tourism counts for development.* Retrieved from <http://documents.worldbank.org/curated/en/558121506324624240/pdf/119954-WP-PUBLIC-SustainableTourismDevelopment.pdf>.

World Bank. (2019a). *Doing business: South Africa*. Retrieved from <http://www.doingbusiness.org/en/data/exploreeconomies/south-africa#DB_sb>.

World Bank. (2019b). *Gini Index: South Africa*. Retrieved from <https://data.worldbank.org/indicator/SI.POV.GINI?locations = ZA>.

Yin, R. K. (2009). *Basics of qualitative research: Techniques and procedures for developing grounded theory.* Thousand Oaks, CA: Sage Publications.

Zahra, S. A., Wright, M., & Abdelgawad, S. G. (2014). Contextualization and the advancement of entrepreneurship research. *International Small Business Journal*, 32(5), 479–500.

Zwane, E. M. (2019). Impact of climate change on primary agriculture, water sources and food security in Western Cape, South Africa. *Jàmbá: Journal of Disaster Risk Studies*, 11(1), 1–7.

THE INFLUENCE OF ENVIRONMENTAL POLLUTION AND DROUGHT ON THE SATISFACTION WITH LIFE OF ENTREPRENEURS IN ZAMBIA'S MINING SECTOR

10

Progress Choongo[1], Emiel L. Eijdenberg[2], John Lungu[3], Mwansa Chabala[1], Thomas K. Taylor[3] and Enno Masurel[4]

[1]*School of Business, The Copperbelt University, Kitwe, Zambia* [2]*Business, IT and Science Department, James Cook University, Singapore, Singapore* [3]*School of Graduate Studies, The Copperbelt University, Kitwe, Zambia* [4]*School of Business and Economics, Vrije Universiteit Amsterdam, Amsterdam, The Netherlands*

10.1 POINT OF DEPARTURE: ENTREPRENEURSHIP IN SUB-SAHARAN AFRICA

The creation of organizations (in this chapter also referred to as "businesses") is known as entrepreneurship and those who create them are the entrepreneurs (Gartner, 1988). Generally, entrepreneurs are considered as economic actors and their actions as the factors of economic development (Kirzner, 1974, 1997; Schumpeter, 1934). "Entrepreneurial activities" involve all activities that occur in and around the creation of businesses, for example, starting, managing, and exiting a business.

Entrepreneurship and societal development go hand in hand. Entrepreneurship has taken place since the existence of humankind in Africa, from hunter-gatherers in the Early Stone Age who sought commercial exchange to Greeks, Romans and Vikings who linked cultures through trade. In modern times, entrepreneurship is embodied by successful examples such as Steve Jobs, Mark Zuckerberg, and Larry Page and Sergey Brin. Hence, today, most people understand entrepreneurship as something from the Western, developed world.

Conversely, much of the world's entrepreneurial activities take place in sub-Saharan Africa (SSA): an overlooked but developing region of 46 countries in which 41% of the people live in extreme poverty (United Nations, 2019). These countries are typically factor-driven economies, characterized by a lack of infrastructure, poverty, low life expectancies, and government and market failure (African Economic Outlook, 2019; Rivera-Santos, Holt, Littlewood, & Kolk, 2015). Although constrained resources—especially skilled human capital rather than natural resources (e.g., minerals)—factor-driven economies contain more often than not heavy polluting industries

Economic Effects of Natural Disasters. DOI: https://doi.org/10.1016/B978-0-12-817465-4.00010-8

such as mining and petroleum factories. Apart from the heavy polluting industries, such resource-constraints areas create low market entry and exit barriers (Khavul, Bruton, & Wood, 2009) that pave the way for countless innovative entrepreneurs who find opportunities to start and manage businesses (Eijdenberg, Thompson, Verduijn, & Essers, 2019; Rivera-Santos et al., 2015).

These businesses are often so-called micro and small enterprises (MSEs) or small- and medium-sized enterprises (SMEs). The difference between MSEs and SMEs is that the former are "one-person operations, poorly managed, sometimes temporary, less productive, and undercapitalized" (Kiggundu, 2002, p. 248), while the latter are better managed, sustainable, and generally more productive. Both MSEs and SMEs are small businesses and the entrepreneurs who start and manage them are called "small business owners" (i.e., the type of entrepreneurs in this chapter). A small business owner is a "(a person or group of people) who creates a new business (for profit) and employs at least one other paid employee" (see also Kirkwood, 2009, p. 350). In SSA, high numbers of small businesses exist due to small large-scale sectors and large small-scale sectors—a typical characteristic of factor-driven economies (McDade & Spring, 2005).

Entrepreneurs start businesses from different motivations: opportunity motivations (i.e., taking advantage of an observed opportunity), necessity motivations (i.e., starting a business because one has to, it is forced), or a mix of these two (Eijdenberg & Masurel, 2013; Eijdenberg, 2016; Eijdenberg, Paas, & Masurel, 2015). Although there is not one conclusive answer, developing countries such as many in SSA have more necessity-motivated entrepreneurs than opportunity-motivated entrepreneurs (Eijdenberg, 2016; Wennekers, van Stel, Thurik, & Reynolds, 2005). Conversely, entrepreneurs in SSA's developing countries seize contextual opportunities that are contributory for decision-making and the implementation of innovative and sustainable products or services (Rooks, Sserwanga, & Frese, 2016). These types of innovations are "design innovation process in which the needs and context of citizens in the emerging world are put first in order to develop appropriate, adaptable, affordable, and accessible services and products for emerging markets" (Basu, Banerjee, & Sweeny, 2013, p. 64). Such frugal innovations are inexpensive, easy to use and purchase, portable, and both economically and socially sustainable (Basu et al., 2013; Rao, 2013). Examples of frugal innovations in SSA are homebuilt water purifiers, especially applicable in slums; no-frills cell phone chargers; and many basic self-made improvised applications (e.g., utensils and cutlery).

10.2 ARRIVING AT ENVIRONMENTAL POLLUTION, DROUGHT, AND SATISFACTION WITH LIFE OF ENTREPRENEURS

The development of humankind has come at the expense of the world's natural conditions. In the quest to achieve economic development, nations have been increasing exploiting natural resources. This has resulted in environmental harm such as natural resources overuse, air and water pollution, shortage of rainfall, climate change, and global warming (Arndt, Asante, & Thurlow, 2015; Ogalleh, Vogl, Eitzinger, & Hauser, 2012; Thompson, Berrang-Ford, & Ford, 2010). As drivers of economic and societal development, entrepreneurs have their stake in environmental harm as well.

Social consequences of these climate-related environmental problems include flooding, drought, which lead to the displacement of population, famine, hunger, and loss of economic livelihood. These effects occur worldwide; however, vulnerable regions in SSA such as Zambia are hit the most because of their limited economic resilience (Arndt et al., 2019; Fayiga, Ipinmoroti, & Chirenje, 2018; Gannon et al., 2018). In these regions, resource scarcity from water and land leads to food insecurity and forced migration (Nkomoki, Bavorová, & Banout, 2019). This suggests that the natural environment is threatened, and human society is at greater risk because its survival is threatened (Bose-O'Reilly et al., 2018; Mihaljevič et al., 2018). A human being's survival is dependent on the natural environment capacity to sustain well-being, including food, shelter, and clothing (Brundtland, Khalid, Agnelli, & Al-Athel, 1987; Krefis, Augustin, Schlünzen, Oßenbrügge, & Augustin, 2018). Scholars and international regulatory bodies have called for action to reduce the negative environmental and social impacts (Babiak & Trendafilova, 2011; Libanda, Zheng, & Ngonga, 2019; Mulenga, 2019). Because, for example, it is known that natural disasters cause a decrease in entrepreneurial activity in the short run, especially in developing countries (Boudreaux, Escaleras, & Skidmore, 2019). As such, entrepreneurs should take measures of how environmental disasters, such as environmental pollution and drought caused by industry, impact the well-being (i.e., measured in this chapter as *satisfaction with life*: Diener, Emmons, Larsen, & Griffin, 1985) of entrepreneurs. But before taking measures, first, the impact of natural disasters on entrepreneurs' satisfaction with life has to be determined.

Following urgent global trends such as climate change, and consequently natural disasters, the relationship between entrepreneurs, on the one hand, and well-being, on the other hand, is not new. This relationship has been researched from different global perspectives (Lepeley, Kuschel, Beutell, Pouw, & Eijdenberg, 2019), including Zambia (see Eijdenberg & Ehmann, 2019), showcasing that entrepreneurship is a defining factor for societal well-being. From the entrepreneurs' perspective, taking care of the environment is one way of taking care of one's own and others' well-being. Giving back to the community, engaging underprivileged people in the business, recycling products, offering sustainable solutions to customers, and using resources efficiently are examples of corporate social responsibility of entrepreneurs in Zambia (Choongo, Paas, Masurel, van Burg, & Lungu, 2018; Choongo, van Burg, Masurel, Paas, & Lungu, 2017; Choongo, van Burg, Paas, & Masurel, 2016). More often than not, these studies showed how corporate social responsibility determined business performance, and, hence, the well-being (i.e., welfare, expressed in monetary terms) of entrepreneurs. However, how environmental disasters, such as environmental pollution and drought caused by industry, affect the well-being of entrepreneurs has hardly been researched.

10.2.1 THE RESEARCH QUESTION

Although the number of studies on the economic effects of natural disasters has increased in recent years (Ishizawa & Miranda, 2019; Karbownik & Wray, 2019; Mohan, Ouattara, & Strobl, 2018; Oliva & Lazzeretti, 2018), most often these studies are conducted in (1) developed countries and (2) often by using large quantitative, macro-level surveys. However, a showcase of environmental disasters in an economically important sector of a developing country—the mining sector in Zambia—has been missing hitherto. Especially the one propelling economic and societal development, *the entrepreneurs*, have largely been overlooked, including their satisfaction with life. The satisfaction with life of these acting individuals is important because, eventually, it contributes to the well-being of individuals, in particular, and society in general. Hence, the research question (*RQ*) of this study is "What is the influence of environmental pollution and drought on the satisfaction with life of entrepreneurs in Zambia's mining sector?"

Based on primary data, the influence of environmental disasters (i.e., environmental pollution and drought) on the satisfaction with life of entrepreneurs in Zambia's major mining cities Chambeshi, Chingola, Kitwe, Luanshya, Mufulira, Ndola, and Solwezi, is examined. A sample of 132 entrepreneurs is surveyed. Besides descriptive statistics and correlations, regression analyses (Hair, Black, Babin, Anderson, & Tatham, 2006) are conducted on the entrepreneur's satisfaction with life, and their perception of how this has been affected by environmental pollution and drought in Zambia.

10.3 METHODOLOGY

10.3.1 RESEARCH CONTEXT: ZAMBIA

The Republic of Zambia (or short Zambia) is the context of this study. Zambia is a country in the southern regions of SSA with Lusaka as its capital city. Zambia has an estimated population of 17 million. It is typically a "developing country" with increasing gross domestic product (GDP) growth rates per annum and an average GDP per capita of USD 4000, while high inflation rates on consumer prices were recorded in 2017, at around 6.6% (17.9% in 2016). More than half of Zambia's population live below the poverty line and is younger than 17 years old. Urbanization is high: approximately 43.5% of the population live in the larger urban areas, a comparable percentage with other SSA countries (Central Intelligence Agency, 2020; Choongo, Eijdenberg, Chabala, Lungu, & Taylor, 2020).

Like other SSA countries, Zambia is still in premature stages of economic development. Countries such as Zambia depend on natural resources to achieve development. They, for instance, introduce extractive industries such as mining to generate resources to fulfill their basic needs, improve the standard of living, and reduce poverty. In the case of Zambia, minerals extracted such as copper and cobalt are melted in furnaces in the smelter to transform concentrates in copper cathodes ready for sale. In the past century, Zambia's mining industry has been the economy's main driver, and it has considerably contributed to economic and societal development (Choongo et al., 2020). As a heavy industry, mining is among the most polluting (e.g., air, water, and soil) (Mulenga, 2019). The mining activities are mainly in the country's northern regions: the Copperbelt Province and the North-Western Province. Often related sectors to mining, such as construction, oil processing, and manufacturing, are also great contributors to the country's economy.

Among the other heavy polluting industrial activities, the melting of copper leads to the emission of large amounts of carbon dioxide into the atmosphere causing air and water pollution. When sulfur dioxide combines with rainwater, it forms acidic rain that corrodes metal roofs, destroys plantation, and affects rivers. Furthermore, air pollution contributes to global warming and climate change that consequently results in natural disasters such as floods, droughts, and cyclones. Flood also results in crop failure. Thus, the changes induced to the natural environment make people living in those communities prone to natural disasters. Therefore this study is significant because the findings will inform scholars and practitioners on the importance of taking measures to reduce negative environmental effects, thereby improving society's satisfaction with life and the natural environment.

10.3.2 DATA COLLECTION

Following the *RQ*, we collected data on the influence of environmental pollution and drought on the satisfaction with life of entrepreneurs within Zambia's mining sector. All items in the questionnaire were in English. After the process of items' development, the questionnaire was pilot tested on a group of 10 entrepreneurs to check on comprehensibility, consistency and validity, and English language. The final version of the questionnaire was administered on a sample of $n = 132$ entrepreneurs in the cities and outskirts of Chambeshi, Chingola, Kitwe, Luanshya, Mufulira and Ndola; in the Copperbelt Province; and Solwezi in the North-Western Province. As mentioned earlier, mining is the dominant industry in this region, and many entrepreneurs are active either within the mining industry or in related areas such as distributors, suppliers, or service providers. Table 10.1 presents the items of the final questionnaire.

The procedures for generating suitable items for the questionnaire evolved as follows:

1. Sociodemographic items were selected to control for pertinent information of the respondents (cf. Choongo, 2017; Eijdenberg et al., 2015; Eijdenberg, 2016);
2. Diener et al. (1985)'s scale[1] was adopted to measure satisfaction with life, the proxy of well-being. Principal Components Factor analysis (PCA) was used to check the suitability of the scales in the Zambian context. All the items from the original scale loaded on a single-item Eigenvalue greater than one, explaining the variance of 57.2%. The factor loadings were above 0.6, and the scale showed reliability with a Cronbach's alpha value of 0.81.
3. The items measuring environmental pollution and drought were developed on the basis of, first, a literature review of these two topics and, second, an iterative process of the Zambian authors of this chapter to shape the items in such a way that they would capture the intended meaning and be applicable to the Zambian context. From the PCA results the Kaiser—Meyer—Olkin score, a measure of sampling adequacy, for environmental pollution was 0.795 and for drought was 0.728. The Bartlett's Test of Sphericity values were significant at $P < .001$ for both variables. All commonalities for the variables were above the critical value of 0.3 (Hair et al., 2006). The items on the scale loaded as expected, explaining a total variance of 69.4% for environmental pollution and 79.4% for drought. Both variables showed reliability with the Cronbach's alpha of environmental pollution ($\alpha = 0.849$) and drought ($\alpha = 0.864$) and were above the recommended threshold of 0.7 (Hair et al., 2006). We calculated the composite scores for the constructs of environmental pollution and drought, respectively, by calculating the average score on all scale items.

The sample consisted of entrepreneurs who were managing and owning SMEs, strongly related to the mining industry. These types of businesses are manufacturers of finished and semifinished products but especially contractors and suppliers. From the sample, 34% of the businesses were female-owned, while 66% were owned by men. The majority of the respondents were below the age of 40 years (i.e., 77%), while all respondents had completed secondary school education and 74% had some form of tertiary education (i.e., educational qualifications higher than secondary school).

---

[1]To illustrate its impact, according to Scholar Google the paper that contains this scale has been cited 27,544 times on September 7, 2020.

Table 10.1 Final Questionnaire.

Number	Item	Scale
Sociodemographics		
1	Gender	1 = male; 2 = female
2	Age	1 = <20, 2 = 20–30, 3 = 31–40, 4 = 41–50, 5 = 51–60, 6 = >60
3	Education	1 = no education, 2 = primary school, 3 = secondary school, 4 = craft certificate, 5 = diploma, 6 = bachelor's degree, 7 = master's degree, 8 = doctorate, 9 = other
4	Number of current employees, including yourself	Ratio
Satisfaction With Life		
5	In most ways, my life is close to my ideal	1 = strongly disagree, 2 = disagree, 3 = slightly disagree, 4 = neither agree nor disagree, 5 = slightly agree, 6 = agree, 7 = strongly agree
6	The conditions of my life are excellent	
7	I am satisfied with my life	
8	So far I have gotten the important things I want in life	
9	If I could live my life over, I would change almost nothing	
Environmental Pollution		
10	In Zambia's mining region, I observe much environmental pollution such as air pollution	1 = strongly disagree, 2 = disagree, 3 = slightly disagree, 4 = neither agree nor disagree, 5 = slightly agree, 6 = agree, 7 = strongly agree
11	In Zambia's mining region, I observe much environmental pollution such as noise pollution	
12	In Zambia's mining region, I observe much environmental pollution such as water pollution	
13	In Zambia's mining region, I observe much environmental pollution such as garbage	
Drought		
14	In Zambia's mining region, I observe that environment suffers from drought	1 = strongly disagree, 2 = disagree, 3 = slightly disagree, 4 = neither agree nor disagree, 5 = slightly agree, 6 = agree, 7 = strongly agree
15	In Zambia's mining region, I observe that temperature is rising	
16	In Zambia's mining region, I observe that water levels in rivers are dropping	
Effects From Environmental Pollution on Profitability		
17	To what extent has environmental pollution in Zambia's mining region, such as air pollution, affected the profitability of your business?	1 = very negative, 2 = negative, 3 = somewhat negative, 4 = neither negative nor positive, 5 = somewhat positive, 6 = positive, 7 = very positive
18	To what extent has environmental pollution in Zambia's mining region, such as noise pollution, affected the profitability of your business?	

(Continued)

Table 10.1 Final Questionnaire. *Continued*

Number	Item	Scale
19	To what extent has environmental pollution in Zambia's mining region, such as water pollution, affected the profitability of your business?	
20	To what extent has environmental pollution in Zambia's mining region, such as garbage, affected the profitability of your business?	

Effects From Drought on Profitability

Number	Item	Scale
21	To what extent has drought in Zambia's mining region had a profound effect on the profitability of your business?	1 = very negative, 2 = negative, 3 = somewhat negative, 4 = neither negative nor positive, 5 = somewhat positive, 6 = positive, 7 = very positive
22	To what extent has rising temperatures in Zambia's mining region had a profound effect on the profitability of your business?	
23	To what extent has dropping water levels in rivers in Zambia's mining region had a profound effect on the profitability of your business?	

Effects From Environmental Pollution on Decision-Making

Number	Item	Scale
24	To what extent has environmental pollution in Zambia's mining region, such as air pollution, influenced the decision-making practices of your business?	1 = very negative, 2 = negative, 3 = somewhat negative, 4 = neither negative nor positive, 5 = somewhat positive, 6 = positive, 7 = very positive
25	To what extent has environmental pollution in Zambia's mining region, such as noise pollution, influenced the decision-making practices of your business?	
26	To what extent has environmental pollution in Zambia's mining region, such as water pollution, influenced the decision-making practices of your business?	
27	To what extent has environmental pollution in Zambia's mining region, such as garbage, influenced the decision-making practices of your business?	

Effects From Drought on Decision-Making

Number	Item	Scale
28	To what extent has drought in Zambia's mining region influenced the decision-making practices of your business?	1 = very negative, 2 = negative, 3 = somewhat negative, 4 = neither negative nor positive, 5 = somewhat positive, 6 = positive, 7 = very positive
29	To what extent has rising temperatures in Zambia's mining region influenced the decision-making practices of your business?	

(Continued)

Table 10.1 Final Questionnaire. *Continued*

Number	Item	Scale
30	To what extent has dropping water levels in rivers in Zambia's mining region influenced the decision-making practices of your business?	

Table 10.2 Descriptive Statistics.

Composite Constructs	Effects on Profitability: Item Numbers 17–23		Effects on Decision-Making: Item Numbers 24–30	
	Mean	**Standard Deviation**	**Mean**	**Standard Deviation**
Environmental pollution	2.40	1.12	1.94	1.04
Drought	1.97	1.08	2.10	1.06

10.3.3 DATA ANALYSIS

To answer the *RQ*, the collected data were analyzed following the conventional steps of quantitative methodologies (Hair et al., 2006) as previously applied in other SSA countries (cf. Eijdenberg & van Montfort, 2017; Eijdenberg et al., 2015) and in Zambia (cf. Choongo et al., 2016, 2020). After the data collection, descriptive statistics and correlation analyses were conducted. To test the effects of environmental pollution and drought on the satisfaction with life of entrepreneurs, regression analyses were performed. The next section presents the results.

10.4 RESULTS

10.4.1 DESCRIPTIVE STATISTICS

We computed the Means and Standard Deviations of the items numbered 17–23 and 24–30—as per Table 10.1. These items reflect the effects of environmental pollution and drought on profitability and the effects of environmental pollution and drought on decision-making, respectively. The results are presented in Table 10.2.

From Table 10.2 can be drawn that the effects of environmental pollution and drought on both business' profitability and decision-making are negative: all means are far below the midpoint of 4.0.

10.4.2 CORRELATION ANALYSES

A partial correlation was conducted to establish the relationship between satisfaction with life and environmental pollution and drought while controlling for gender, age, education, and business size (i.e., measured by "number of current employees, including yourself"). There was a high, negative partial correlation between satisfaction with life and environmental pollution ($t = -0.79$, $P < .001$)

and drought ($t = -.66$, $P < .001$). Zero-order correlations also showed that there was a statistically significant, negative correlation between satisfaction with life and environmental pollution ($t = -.73$, $P < .001$) and drought ($t = -0.66$, $P < .001$), indicating that the control variables did not have much influence in controlling for the relationship satisfaction with life and environmental pollution and drought.

10.4.3 REGRESSION ANALYSES

Multiple linear regression analyses were used to establish the effects of environmental pollution and drought on satisfaction with life of entrepreneurs in Zambia's mining sector. Hence, in addition to the sociodemographics (i.e., items numbered 1−4) (i.e., "controls"), the composite constructs of environmental pollution (i.e., items numbered 10−13) and drought (i.e., items numbered 14−16) are defined as the independent variables (i.e., "main effects"), and the composite construct of satisfaction with life (i.e., items numbered 4−9) is the dependent variable. The results are shown in Table 10.3. In "Model 1," only the controls were considered, while "Model 2" included all independent variables. To ensure that multicollinearity was not a problem, we tested for possible collinearity among all variables by using the variance inflation factors (VIF). All VIF values were below the critical value of 10, indicating that multicollinearity was excluded (Hair et al., 2006).

The results show that the "Model 2" that included both the controls and the main effects was statistically significant with an adjusted R^2 value of .64 ($F = 39.90$; $P < .001$). Both environmental

Table 10.3 Regression Analyses.

	β Coefficients	
Independent Variables	**Model 1**	**Model 2**
Controls		
Gender	0.07	0.04
Age	0.09	0.02
Education	0.07	0.05
Number of current employees, including yourself	−0.08	0.01
Main Effects		
Environmental pollution		−0.49[a]
Drought		−0.14[b]
Model Performance Statistics		
R^2	0.03	0.66
R^2		0.63
Adjusted R^2		0.64
F	1.02	39.90

[a]*Effect is significant at the 0.01 level (two-tailed).*
[b]*Effect is significant at the 0.05 level (two-tailed).*

pollution ($\beta = -0.49$; $P < .001$) and drought ($\beta = -0.14$; $P < .05$) had a statistically significant negative effect on the entrepreneurs' satisfaction with life.

10.5 CONCLUDING DISCUSSION

The aim of this chapter was to answer the *RQ*, therefore, examining the effects of environmental pollution and drought on satisfaction with life (as a proxy of well-being) of entrepreneurs in Zambia's mining sector. The analyses of the primary data of 132 respondents showed that environmental pollution and drought have a negative effect on satisfaction with life. Additional analyses show that the effects of environmental pollution and drought on both business' profitability and decision-making are negative as well.

The results involve contributions to the existing literature. On a general level, this chapter may serve among the first to overcome a scholarly disconnection between research fields, which is entrepreneurship (i.e., a social science) and natural sciences. Although in the past, knowledge spillovers within grand science streams such as social sciences (including entrepreneurship and development economics) have little but increasingly occurred (Naudé, 2010) (especially journals, among others, *Small Business Economics* and *Journal of African Economies* have met this trend); however, entrepreneurship—as part of social sciences—meeting disaster science—as part of natural sciences—is rather new.

Furthermore, on a granular level, more often than not, entrepreneurs' measures to the environment are studied (Choongo et al., 2016, 2017; Choongo, 2017). In addition, creative responses to natural disasters, often showing that entrepreneurs are willing, able, and capable of starting up and running (new) businesses, are abundantly researched as well (Dutta, 2017; Linnenluecke & McKnight, 2017; Monllor & Murphy, 2017). However, the effects of a changing natural environment, viz., effects from climate change such as natural disasters to the entrepreneur—*as a person*—are largely overlooked. The focus on satisfaction with life is important because it defines well-being, and, hence, the performance of the entrepreneurs' businesses (Eijdenberg et al., 2019; Lepeley et al., 2019) that, consequently, contributes to overall regional social and economic development (Gries & Naudé, 2010, 2011; Naudé, 2010; Wennekers et al., 2005). The results of this study provide the detailed insights of what types of environmental pollution (e.g., water pollution, garbage) and drought (e.g., rising temperatures, dropping water levels of rivers) affect the entrepreneurs, in terms of their satisfaction with life and, additionally, their businesses' profitability and decision-making.

What do the results imply for practitioners? Based on the results of environmental pollution and drought, entrepreneurs should know which factors (and to what extent) affect their satisfaction with life, business' profitability, and decision-making. Entrepreneurs could take measures to combat these specific factors of environmental pollution and drought, for example, by not polluting water and arranging garbage collection. At the same time, other practitioners, such as policymakers and educators, should focus on those determining factors of satisfaction with life, business' profitability, and decision-making, as well. By creating policies and education/training programs that mitigate the environmental pollution and drought (e.g., reducing emissions, recycling, and considerate waste disposal), entrepreneurs' satisfaction with life will be improved—which leads as a positive spiral to social and economic development.

10.5.1 LIMITATIONS AND RECOMMENDATIONS FOR FUTURE RESEARCH

This study is constrained by its limitations such as sample selection, size, and location, and the time of the data collection. Without a doubt, different results are found when these constraints are overcome. Therefore, we encourage researchers to defeat these limitations by studying a similar topic with different entrepreneurs, a larger sample and on different locations. Naturally, longitudinal data collections bear different results, allowing observing the effects over time. This would especially be interesting in a scenario of a (sudden) significant change of environmental pollution and drought (e.g., in the midst of and shortly after the peak of the global COVID-19 crisis in 2020), for example, in the case of zero emissions, no water pollution, and no waste disposal. Assessing the scores between entrepreneurs' satisfaction with life of *pre*environmental pollution and drought with *post*environmental pollution and drought would allow for making invaluable comparisons.

Moreover, future researchers are encouraged to study the topic of this chapter in different methodological fashions, for example, qualitatively. Focus groups, interviews, and observations would bear deeper and more personal insights into the effects of environmental pollution and drought on satisfaction with life. Ideally, these qualitative insights would provide more contextual information about the topic under study (Saunders, Lewis, & Thornhill, 2009). This could possibly lead to a reconceptualization of, for instance, satisfaction with life, especially in the context of Zambia where certain traditional African values (i.e., "Ubuntu") might be still relevant for everyday life and business practices (West, 2014).

REFERENCES

African Economic Outlook. (2019). *Zambia*. Retrieved from <https://www.africaneconomicoutlook.org/zambia/>.

Arndt, C., Asante, F., & Thurlow, J. (2015). Implications of climate change for Ghana's economy. *Sustainability*, 7(6), 7214−7231.

Arndt, C., Chinowsky, P., Fant, C., Paltsev, S., Schlosser, C. A., Strzepek, K., ... Thurlow, J. (2019). Climate change and developing country growth: The cases of Malawi, Mozambique, and Zambia. *Climatic Change*, 154(3−4), 335−349.

Babiak, K., & Trendafilova, S. (2011). CSR and environmental responsibility: Motives and pressures to adopt green management practices. *Corporate Social Responsibility and Environmental Management*, 18(1), 11−24.

Basu, R. R., Banerjee, P. M., & Sweeny, E. G. (2013). Frugal innovation. *Journal of Management for Global Sustainability*, 1(2), 63−82.

Bose-O'Reilly, S., Yabe, J., Makumba, J., Schutzmeier, P., Ericson, B., & Caravanos, J. (2018). Lead intoxicated children in Kabwe, Zambia. *Environmental Research*, 165, 420−424.

Boudreaux, C. J., Escaleras, M. P., & Skidmore, M. (2019). Natural disasters and entrepreneurship activity. *Economics Letters*, 182, 82−85.

Brundtland, G. H., Khalid, M., Agnelli, S., & Al-Athel, S. (1987). *Our common future*. New York City: United Nations.

Central Intelligence Agency (2020). *The World Factbook: Zambia*. Retrieved from <https://www.cia.gov/library/publications/the-world-factbook/geos/za.html>.

Choongo, P. (2017). A longitudinal study of the impact of corporate social responsibility on firm performance in SMEs in Zambia. *Sustainability*, 9(8), 1300.

Choongo, P., Eijdenberg, E. L., Chabala, M., Lungu, J., & Taylor, T. K. (2020). The evolution of urban entrepreneurship in Zambia. In M. Iftikhar, J. Justice, & D. Audretsch (Eds.), *The urban book series. Urban studies and entrepreneurship* (pp. 249−269). Cham, Switzerland: Springer.

Choongo, P., Paas, L. J., Masurel, E., van Burg, E., & Lungu, J. (2018). Entrepreneurs' personal values and CSR orientations: Evidence from SMEs in Zambia. *Journal of Small Business and Enterprise Development*, *26*(4), 545−570.

Choongo, P., van Burg, E., Masurel, E., Paas, L. J., & Lungu, J. (2017). Corporate social responsibility motivations in Zambian SMEs. *International Review of Entrepreneurship*, *15*(1), 29−62.

Choongo, P., van Burg, E., Paas, L., & Masurel, E. (2016). Factors influencing the identification of sustainable opportunities by SMEs: Empirical evidence from Zambia. *Sustainability*, *8*(1), 81.

Diener, E. D., Emmons, R. A., Larsen, R. J., & Griffin, S. (1985). The satisfaction with life scale. *Journal of Personality Assessment*, *49*(1), 71−75.

Dutta, S. (2017). Creating in the crucibles of nature's fury: Associational diversity and local social entrepreneurship after natural disasters in California, 1991−2010. *Administrative Science Quarterly*, *62*(3), 443−483.

Eijdenberg, E. L. (2016). Does one size fit all? A look at entrepreneurial motivation and entrepreneurial orientation in the informal economy of Tanzania. *International Journal of Entrepreneurial Behavior & Research*, *22*(6), 804−834.

Eijdenberg, E. L., & Ehmann, L. (2019). Exploring wellbeing indicators of women micro entrepreneurs in Zambia. In M.-T. Lepeley, K. Kuschel, N. Beutell, N. Pouw, & E. L. Eijdenberg (Eds.), *The wellbeing of women in entrepreneurship: A global perspective* (pp. 359−373). New York City: Routledge.

Eijdenberg, E. L., & Masurel, E. (2013). Entrepreneurial motivation in a least developed country: Push factors and pull factors among MSEs in Uganda. *Journal of Enterprising Culture*, *21*(1), 19−43.

Eijdenberg, E. L., Paas, L. J., & Masurel, E. (2015). Entrepreneurial motivation and small business growth in Rwanda. *Journal of Entrepreneurship in Emerging Economies*, *7*(3), 212−240.

Eijdenberg, E. L., Thompson, N. A., Verduijn, K., & Essers, C. (2019). Entrepreneurial activities in a developing country: An institutional theory perspective. *International Journal of Entrepreneurial Behavior & Research*, *25*(3), 414−432.

Eijdenberg, E. L., & van Montfort, K. (2017). Explaining firm performance in African Least Developed Countries: Evidence from Burundi and Tanzania. *International Review of Entrepreneurship*, *15*(3), 375−394.

Fayiga, A. O., Ipinmoroti, M. O., & Chirenje, T. (2018). Environmental pollution in Africa. *Environment, Development and Sustainability*, *20*(1), 41−73.

Gannon, K. E., Conway, D., Pardoe, J., Ndiyoi, M., Batisani, N., Odada, E., . . . Nyambe, M. (2018). Business experience of floods and drought-related water and electricity supply disruption in three cities in sub-Saharan Africa during the 2015/2016 El Niño. *Global Sustainability*, *1*, 1−15.

Gartner, W. B. (1988). "Who is an entrepreneur?" is the wrong question. *American Journal of Small Business*, *12*(4), 11−32.

Gries, T., & Naudé, W. A. (2010). Entrepreneurship and structural economic transformation. *Small Business Economics*, *34*(1), 13−29.

Gries, T., & Naudé, W. A. (2011). Entrepreneurship and human development: A capability approach. *Journal of Public Economics*, *95*(3−4), 216−224.

Hair, J. F., Black, W. C., Babin, B. J., Anderson, R. E., & Tatham, R. L. (2006). *Multivariate data analysis*. Upper Saddle River, NJ: Pearson Prentice Hall.

Ishizawa, O. A., & Miranda, J. J. (2019). Weathering storms: Understanding the impact of natural disasters in Central America. *Environmental and Resource Economics*, *73*(1), 181−211.

Karbownik, K., & Wray, A. (2019). Long-run consequences of exposure to natural disasters. *Journal of Labor Economics*, *37*(3), 949−1007.

Khavul, S., Bruton, G. D., & Wood, E. (2009). Informal family business in Africa. *Entrepreneurship Theory and Practice*, *33*(6), 1219−1238.

Kiggundu, M. N. (2002). Entrepreneurs and entrepreneurship in Africa: What is known and what needs to be done. *Journal of Developmental Entrepreneurship*, *7*(3), 239−258.

Kirkwood, J. (2009). Motivational factors in a push-pull theory of entrepreneurship. *Gender in Management: An International Journal*, *24*(5), 346−364.

Kirzner, I. M. (1974). *Competition and entrepreneurship*. Chicago, IL: University of Chicago Press.

Kirzner, I. M. (1997). Entrepreneurial discovery and the competitive market process: An Austrian approach. *Journal of Economic Literature*, *35*(1), 60−85.

Krefis, A., Augustin, M., Schlünzen, K., Oßenbrügge, J., & Augustin, J. (2018). How does the urban environment affect health and well-being? A systematic review. *Urban Science*, *2*(1), 21.

Lepeley, M. T., Kuschel, K., Beutell, N., Pouw, N., & Eijdenberg, E. L. (Eds.), (2019). *The wellbeing of women in entrepreneurship: A global perspective*. New York City: Routledge.

Libanda, B., Zheng, M., & Ngonga, C. (2019). Spatial and temporal patterns of drought in Zambia. *Journal of Arid Land*, *11*(2), 180−191.

Linnenluecke, M. K., & McKnight, B. (2017). Community resilience to natural disasters: The role of disaster entrepreneurship. *Journal of Enterprising Communities: People and Places in the Global Economy*, *11*(1), 166−185.

McDade, B. E., & Spring, A. (2005). The 'new generation of African entrepreneurs': Networking to change the climate for business and private sector-led development. *Entrepreneurship & Regional Development*, *17*(1), 17−42.

Mihaljevič, M., Jarošíková, A., Ettler, V., Vaněk, A., Penížek, V., Kříbek, B., ... Svoboda, M. (2018). Copper isotopic record in soils and tree rings near a copper smelter, Copperbelt, Zambia. *Science of the Total Environment*, *621*, 9−17.

Mohan, P. S., Ouattara, B., & Strobl, E. (2018). Decomposing the macroeconomic effects of natural disasters: A national income accounting perspective. *Ecological Economics*, *146*, 1−9.

Monllor, J., & Murphy, P. J. (2017). Natural disasters, entrepreneurship, and creation after destruction: A conceptual approach. *International Journal of Entrepreneurial Behavior & Research*, *23*(4), 618−637.

Mulenga, C. (2019). Judicial mandate in safeguarding environmental rights from the adverse effects of mining activities in Zambia. *PER: Potchefstroomse Elektroniese Regsblad*, *22*(1), 1−33.

Naudé, W. A. (2010). Entrepreneurship, developing countries, and development economics: New approaches and insights. *Small Business Economics*, *34*(1), 1−12.

Nkomoki, W., Bavorová, M., & Banout, J. (2019). Factors associated with household food security in Zambia. *Sustainability*, *11*(9), 2715.

Ogalleh, S., Vogl, C., Eitzinger, J., & Hauser, M. (2012). Local perceptions and responses to climate change and variability: The case of Laikipia District, Kenya. *Sustainability*, *4*(12), 3302−3325.

Oliva, S., & Lazzeretti, L. (2018). Measuring the economic resilience of natural disasters: An analysis of major earthquakes in Japan. *City, Culture and Society*, *15*, 53−59.

Rao, B. C. (2013). How disruptive is frugal? *Technology in Society*, *35*(1), 65−73.

Rivera-Santos, M., Holt, D., Littlewood, D., & Kolk, A. (2015). Social entrepreneurship in sub-Saharan Africa. *Academy of Management Perspectives*, *29*(1), 72−91.

Rooks, G., Sserwanga, A., & Frese, M. (2016). Unpacking the personal initiative−performance relationship: A multi-group analysis of innovation by Ugandan rural and urban entrepreneurs. *Applied Psychology*, *65*(1), 99−131.

Saunders, S., Lewis, P., & Thornhill, A. (2009). *Research methods for business students* (5th ed.). Essex: Pearson Education Limited.

Schumpeter, J. A. (1934). *The theory of economic development*. Cambirdge, MA: Harvard University Press.

Thompson, H. E., Berrang-Ford, L., & Ford, J. D. (2010). Climate change and food security in sub-Saharan Africa: A systematic literature review. *Sustainability*, *2*(8), 2719−2733.

United Nations. (2019). *About sub-Saharan Africa*. Retrieved from <http://www.africa.undp.org/content/rba/en/home/regioninfo.html>.

Wennekers, S., van Stel, A., Thurik, R., & Reynolds, P. (2005). Nascent entrepreneurship and the level of economic development. *Small Business Economics*, *24*(3), 293−309.

West, A. (2014). Ubuntu and business ethics: Problems, perspectives and prospects. *Journal of Business Ethics*, *121*(1), 47−61.

ENVIRONMENTAL RESILIENCE OF BOTTOM OF THE PYRAMID STRATEGIES TOWARD SINGLE-USE PLASTICS: A RECIPE FROM AN EMERGING ECONOMY

11

Tanuj Chawla[1], Emiel L. Eijdenberg[2] and Jacob Wood[3,4]

[1]*Tata-Cornell Institute for Agriculture and Nutrition, New Delhi, India* [2]*Business, IT and Science Department, James Cook University, Singapore, Singapore* [3]*School of Business, James Cook University, Singapore, Singapore* [4]*Visiting Professor of International Trade, Chungnam National University, Daejeon, South Korea*

11.1 SETTING THE SCENE: SINGLE-USE PLASTICS, BOTTOM OF THE PYRAMID, AND FRUGAL INNOVATION

The industrialized world has seen an increase in the production, consumption, and disposal of plastic materials in recent years. The use of plastic has become commonplace in our everyday lives, and as such, they can be found just about everywhere including vast areas of our oceans (Lebreton et al., 2018). Less spoken about are the countless number of landfills, drains, and choked rivers that affect millions of people in the socioeconomic constrained Bottom of the Pyramid (BoP) areas such as in India. The toxic debris released from these plastics has the propensity to enter food chains and microbodies in the form of microplastics (Rajmohan, Ramya, & Varjanic, 2019). The dioxins released from the plastic polymers contained within the plastic material has been found to cause a range of serious illnesses such as cancer, neurological damage, and impairs the development of reproductive systems (Kavlock et al., 2002). If unmanaged, the social, environmental, and political consequences would be widely felt and catalyst for a long-term environmental disaster.

Despite the potential health and environmental risks associated with its production, annual global plastic consumption has reached over 320 million tonnes with the last decade of use is being equal to the total amount of plastic production across human history (European Plastics, 2017). Of this, India generates 25,940 tonnes of plastic waste every day (Government of India, 2018). Another estimate put forward by the Federation of Indian Chambers of Commerce and Industry (2017) puts these numbers at 16.5 million tonnes of annual plastic use, 43% of it being single-use plastic. A substantial portion of this single-use plastic figure refers to the class of plastic known as polypropylene and polystyrene, which are either difficult to recycle or cannot be recycled at all because of the relevant material properties and/or the cost associated with the recovery of the material (Miller, Soulliere, Sawyer-Beaulieu, Tseng, & Tam, 2014). These two forms of plastic are most commonly used to make takeaway cutlery and are not only the least recyclable but also the most difficult to trace and recover from a pile of garbage, creating a situation that may form the basis for potential environmental disasters in the future.

The lack of India's waste management infrastructure and awareness puts the country at a higher risk of environmental contamination and the need to have an increasing number of landfill sites. Given the environmental challenge associated with single-use plastics and the unorganized nature of the production, this chapter presents an empirical assessment of India's single-use disposables economy and how it can be transformed into a biomass-based plate[1] ecosystem. Such developments would undoubtedly ease the environmental concerns that are associated with traditional forms of production.

Entrepreneurs—seen as the creators of organizations (Gartner, 1988)—are examined in their role of key actors in the value chain of plant-based plates as an alternative of single-use plastic cutlery. The BoP includes not only consumers and recipients of aid but also "talented" entrepreneurs—striving to generate new ideas, routines, products and services with minimal resources (George, McGahan, & Prabhu, 2012; Hall, Matos, Sheehan, & Silvestre, 2012; Karnani, 2007). In light of these characteristics, entrepreneurs in India have played a crucial role in mapping the "frugal innovation" phenomenon (Lim, Han, & Ito, 2013; Tiwari & Herstatt, 2012). Frugal innovations are cheap, convenient, portable, and both economically and socially sustainable products that provide improvized solutions to local resource-constrained problems (Basu, Banerjee, & Sweeny, 2013; Rao, 2013). Examples of frugal innovations are self-made water purifiers, to be used in slums; cell phone chargers of dynamos on bicycles; and many improvized applications (e.g., utensils, board games, plant-based cutlery).

Frugal innovations like leaf plates or biomass-based single-use plates have been a part of Indian culture since Vedic times (Ahuja & Ahuja, 2011). They have been used in a variety of areas including serving food at home to community food services at weddings, temples, and during other rituals almost across the entire Indian continent. The raw leaf is plucked from forests as nontimber forest produce and, hence, are not grown for it. They are a by-product of the forest growth and, as an example, we take the Siali leaf plate, which is found in the Sal jungles in certain parts of India (Professional Assistance for Development Action, 2017). The process is highly labor intensive, from plucking the leaves to sewing them into shape and processing them into different shapes for a variety of usage depending upon the local requirement and the innovation activities around the leaf plate supply chain. Despite their labor-intensive nature, their biodegradable qualities (Haider, Völker, Kramm, Landfester, & Wurm, 2019) help reduce the environmental risks associated with traditional forms of plastic materials. It is important to understand that the raw material for these leaf plates is a nontimber forest produce and the regeneration depends completely on the health of the forests. In recent years, forests have been rapidly degrading due to human activities and as an indirect action of human-caused climate change (ENVIS Centre of Odisha's State of Environment, 2017; Reddy, Rao, Pattanaik, & Joshi, 2009).

Scholars, practitioners, and policymakers have understood the concept of innovation in various ways with studies focusing predominantly on the electronics, energy, finance, healthcare, information and communication technology, and transportation sectors (Rao, 2013; Tiwari & Herstatt, 2012). However, despite the high importance of specific sectors such as agriculture, the grassroots

---

[1]In this chapter, we use different terms of "biomass-based plate(s)," such as "plant-based plates" or "leaf plates". Although used interchangeably, the terms refer to the same.

and social enterprises related to biomass-based businesses have received limited attention (Hossain, 2017, 2018).

The BoP market population being served through cheap and cost-effective products may or may not be directly involved in the process of innovation (Nari Kahle, Dubiel, Ernst, & Prabhu, 2013), while in most cases, the products designed for these markets are improvized from local usage and demand patterns. Similar is the case of the single-use plastic cutlery evolution in the North Indian market, which started through the introduction of biomass-based plates and is now dominated by the use of disposable plastic cutlery (Gautam & Caetano, 2017). This market evolution through the use of frugal innovation activities has caused a drift in the demand of products supplied from and to the BoP markets, which has affected the whole supply chain of sustainable options across the single-use disposable market. Hence, this chapter explores the following research question: "How are plant-based plates a potential solution to single-use plastics?".

The next section provides the methodology of this chapter. The section thereafter discusses relevant findings. The chapter ends with a concluding discussion.

11.2 **METHODOLOGY**

11.2.1 **RESEARCH CONTEXT: INDIA'S BOTTOM OF THE PYRAMID**

India (or as it is formally referred to, the Republic of India) is one of the world's most populated countries per square kilometer. The country's capital is New Delhi with more than 28 million people, followed by megacities Mumbai (almost 20 million) and Kolkata (more than 14 million). Haryana, where the data collection took place, is one of the 29 states in India and is located in the northern part of the country surrounding New Delhi on three sides. India has a diverse economy with about 8.5% unemployment, growing manufacturing, and tech industries as well as a large agricultural sector. Socioeconomic differences are significant, separating higher and lower social, and richer and poorer economic, classes apart. The gross domestic product per capita is USD 7200 (Central Intelligence Agency, 2019).

The BoP provides an economic classification of the world's population according to its per capita income. The world economic pyramid uses the analogy of a three-tiered pyramid, dividing the world's population into the top, middle, and the BoP (Prahalad, 2006). The top tier consists of the highest income population with earnings higher than USD 20,000 annually and has a population range of 75−100 million. The second tier houses a theoretical second and third-tier population of 1500−1750 million people with an income range from USD 2000 to USD 20,000 annually. The last tier of the economic pyramid is the most populous. This tier houses the actual bottom of the economic pyramid (BoP, as referred to in this chapter) and has a population that earns less than USD 2000 annually and approximately four billion people falling within this income category. In India, 96.7% of its 1221 million population fall into the bottom of the economic pyramid (Prahalad, 2006, 2012).

11.2.2 **DATA COLLECTION**

A total of 16 entrepreneurs associated with the disposable cutlery business were interviewed using the methodology set by Burnard (1994). The sample size chosen was 30% of the number of entrepreneurs in the district of Hisar, the state of Haryana (a minimum of 15 interviews was set out of 55 total). All entrepreneurs were male, with an average business age of 13 years, the most recent one started two years ago while the oldest was a second-generation owner of a 40-year-old business operation. The primary information was gathered through a conversation with a local market committee representative, who provided specific address data for 55 entrepreneurs. Out of these, only 43 could be traced through online searches and local visits. Only eight agreed to have a conversation and provided references to other entrepreneurs dealing with similar products. With 20 additional references, eight more agreed to speak for this research. Table 11.1 provides the most important pertinent information of the entrepreneurs.

Short structured interviews were used to collect empirical data. Structured interviews, in contrast to semi-structured and unstructured interviews, provide a rigid empirical data collection method. These require not only the use of various skills, such as intensive listening and note-taking but also careful planning and sufficient preparation, a structured interview allows for better organization and quantification of the findings (Saunders, Lewis, & Thornhill, 2009).

The interviews were based on an interview template that was developed out of the theoretical foundation gained during a literature analysis. The emerging topics and themes were used to concentrate on the vital thematic elements missing in the literature. The interviews were conducted in February and March 2019. The interviews took 15−20 minutes due to their standardized nature. The voice-recorded interviews were conducted over telephone calls and a mix of Hindi, Haryanvi, and Punjabi languages were spoken as these are widely used languages in Hisar (Mann & Mann, 2015). The recordings were simultaneously translated and transcribed into English. Keeping in

Table 11.1 Entrepreneurs' Interview Number, Age of Business and Interview Date.

Interview Number	Age of Business	Date of Interview
1	8	February 13, 2019
2	22	February 16, 2019
3	20	February 16, 2019
4	2	February 16, 2019
5	22	February 16, 2019
6	9	February 17, 2019
7	2	February 17, 2019
8	15	March 7, 2019
9	20	March 7, 2019
10	4	March 7 and 8, 2019
11	10	March 7, 2019
12	5	March 7, 2019
13	7	March 10, 2019
14	20	March 10, 2019
15	40	March 10, 2019
16	4	March 10, 2019

mind the possibility of additional information being supplied by the entrepreneurs, additional data were recorded as remarks to the closest matching question section.

11.2.3 DATA ANALYSES

The method chosen for this research was a contextual content analysis (McTavish & Pirro, 1990). So as to dive deeper into the responses, this was followed by coding the interview segments using an inductive approach where the coding and theme generation are directed by the content of the data (Braun & Clarke, 2006; Saunders et al., 2009). This approach was especially useful in establishing the generalizability of inferences and for developing more vibrant and nuanced interpretations of a phenomenon (Eisenhardt & Graebner, 2007).

The interview analyses were done using a qualitative analysis tool called MAXQDA Analytics Pro. The tool helped in organizing relevant code arrangements and provided visualizations of critical elements using various built-in tools (MAXQDA, 2019).

The phases of thematic analysis as described by Braun and Clarke (2006) allow for a systematic way of seeing, as well as processing qualitative information using "coding". The several phases of the thematic analysis as used in the current research are described in the following (Braun & Clarke, 2006):

1. Data familiarization: this step involves transcribing the data, reading and re-reading the data, and noting down the initial ideas. Major ideas were highlighted and written down for each transcript.
2. Generating initial code: "Coding interesting features of the data in a systematic fashion across the entire data set, collecting data relevant to each code" (Braun & Clarke, 2006, p. 87). While translating and transcribing, features were coded as a small phrase or keyword representing a specific idea. Memos were written down to keep track of the condensed information.
3. Searching for themes across the data: "Collating codes into potential themes, gathering all data relevant to each potential theme" (Braun & Clarke, 2006, p. 87). The data were read and re-read, and the cycle was repeated several times to narrow down the number of codes and categorized them into identifiable themes. The codes were then analyzed and grouped into four central themes as stated in the next section.
4. Reviewing themes: "Checking if the themes work in relation to the coded extracts at the first level and then the entire data set at the second level, generating a thematic map of the analysis" (Braun & Clarke, 2006, p. 87). The complete interview data were re-read to validate the codes. In-built tools in MAXQDA were used to see patterns within the data, and these were used to draw thematic maps.
5. Producing the report: "The final analysis; selection of vivid, compelling extract examples, the final analysis of selected extracts, relating to the analysis to their search question and literature, producing a scholarly report of the analysis" (Braun & Clarke, 2006, p. 87). Several vital statements/features representing the data were extracted to showcase the resulting outcomes both as statements in the form of ideas and feelings, and visual representations are drawn using interconnections between codes.

11.3 FINDINGS

The coding process as explained in the Methodology was deployed using an inductive approach for code development (Gioia, Corley, & Hamilton, 2013; Maguire & Delahunt, 2017; McTavish & Pirro, 1990). These codes were used to develop thematic elements from codes falling into a similar logical category of actions (Maguire & Delahunt, 2017). A total of 158 segments from the 16 interviews were coded and grouped into three major themes. One standard set of themes entitled *Socio-Political Factors and Themes* was developed as these could be considered exclusive themes. An effort was made to reduce the thematic complexity and improve the general visualization of the code frequencies and their respective themes.

The next subsections provide a portrayal of codes versus their respective frequency among the interview responses. This portrayal is followed by an explanation of the most significant elements and specific examples from the theme, thereby providing a clear illustration of it. Due to the overlapping nature of the coded elements and a complete document overview for the statistics generation, the outcomes are not presented as a score out of 100%.

11.3.1 CONSUMER BEHAVIOR

The theme derives its foundation from interview responses that were analyzed as consumer actions leading to a change in the understanding of consumer demand (Ali, Kapoor, & Moorthy, 2010; Dikgang, Leiman, & Visser, 2012). The highest stated code under this theme is "dislike toward biomass-based plates," which is featured in over 66% of the text. One respondent attributed the reason for declining demand for biomass-based plates to the decreasing quality of the product; this, in turn, leads to changing consumer perception about the product. Such a finding may also provide evidence of the belief that bioplastics may not be as eco-friendly as many believe (Cho, 2019). It is followed by a 50% presence of both "status symbol/finding the plastic/paper better in stature" and "lack of awareness." This status of consumption has been positively correlated with ownership of products reputed to be higher in status and social appeal (Eastman, Goldsmith, & Flynn, 1999).

Forty-one percent of the replies found the plastic disposables superior to the leaf plates and claimed it the reason for the change of consumer perceptions. "No environmental concern" is featured 33% of the time, while "increased wealth of consumer" and "awareness" is found at 16% and 8%, respectively. Islam (2015) demonstrates a negative correlation between income and environmentally responsible outcomes. One respondent acknowledges the changing consumer wealth and choices using the following statement: "People have become rich, and they hence wish to buy quality material." Table 11.2 provides the contextual explanation of each code, extenuating the necessity of the code.

11.3.2 MARKET FORCES

Market forces are expressed in the context of this research as demands mediated through the marketplace (Mowery & Rosenberg, 1979). According to Mowery and Rosenberg (1979, p. 140), demand "is a precise concept, denoting a systematic relationship between prices and quantities, one devolving from the constellation of consumer preferences and incomes." Hence, the second

Table 11.2 "Consumer Behavior" Codes and Their Explanations.

Consumer Behavior	Number of Interviews Containing This Code	Code Explanation
Dislike toward biomass-based plates (quality/performance concerns with biomass disposables)	11	The code signifies all statements or responses mentioning a decline in quality of the available product as perceived from consumer feedback
Lack of awareness	6	Responses on environmental awareness were consolidated under this code
Status symbol/finding the plastic/paper better in stature (more liking toward plastic/paper plates)	6	Social stature related responses that impacted the consumer side demand were accounted for under this code
Finding the plastic disposable superior performing	5	Product-specific superiority claims
No environmental concern	4	Rational decision making without having an environmental concern
Increased wealth of consumers	2	Change in consumer income
Awareness	1	Environmental or related impact consideration examples

essential theme developed during the analysis refers to the market forces. The theme analyses market factors leading to a change in the product availability landscape from both the demand and production sides. As explained by Adner and Levinthal (2001), changing consumer behavior leads to a demand heterogeneity, allowing explanations for an alternative to supply-side changes. The "local and easy availability of plastic disposables" is the prime code in this thematic category with 73% presence, with one respondent stating the reduction in the availability of biomass-based disposables since the year 2002. At 60%, "local and easy availability of paper disposables" follows the plastic plate availability; several respondents maintained the presence of local suppliers for paper disposables.

"Reduced demand of biomass-based disposables" holds a 53% relative presence. "Reduced supply of biomass-based disposables" and "inexpensive plastic/paper disposables" each show a relative presence of 27%. "Lack of raw material for biomass-based disposables" and "increasing costs," both stand at 13% each, while "expensive biomass-based disposables" is mentioned 7% of the time. The respondents recognized new demand is met through newly established local supply chains, which is driven by changing consumer behavior. Table 11.3 provides the contextual explanation of each code.

11.3.3 ENTREPRENEURIAL/INNOVATION ACTIVITIES

The theme focuses on the entrepreneurial and innovation activities within the BoP markets. Market demands are stimulants for innovation and are even qualified to concede new markets through new technologies (Gilpin, 1975; Mowery & Rosenberg, 1979). As has been previously reported in the

Table 11.3 "Market Forces" Codes and Their Explanations.

Market Forces	Number of Interviews Containing This Code	Code Explanation
Local and easy availability of plastic disposables (increased demand for plastic disposables)	11	Easy market availability characterized by local producers or suppliers of plastic disposables
Local and easy availability of paper disposables (increased demand for paper plates/local and easy availability of paper plates)	9	Easy market availability characterized by local producers or suppliers of paper disposables
Reduced demand for biomass-based disposables	8	Lack of consumer demand defined with reasons mentioned in the theme "consumer demand"
Reduced supply of biomass-based disposables	4	Lack of availability of biomass-based disposable especially leaf plates
Inexpensive plastic/paper disposables	4	Positive change in price for sellers to sell more and at higher margins
Lack of raw material for biomass-based disposables	2	Reduction in raw material resource for production
Increasing costs	2	A negative change in price for sellers leading to decreased sales
Expensive biomass-based disposables	1	A negative change in price for sellers leading to decreased sales specifically for biomass-based disposables

literature, entrepreneurial activities thrive through increased product performances at stable prices (Adner & Levinthal, 2001), the previously mentioned consumer behavior choices and market forces have successfully lead to several entrepreneurial outcomes. Many respondents claimed to have set-up their equipment to correspond with changing market trends.

Hence, the theme features three additional codes. "Small-scale manufacturing for paper and plastic disposables" features strongest with 56% significance. "Change in the business model" is represented at 44%. At 33% presence, "household industry catering to the demand of paper disposables" is represented, followed by 22% for "product/process innovation" code. See Table 11.4 for the contextual explanation of the codes.

11.3.4 SOCIO-POLITICAL FACTORS AND THEMES

A total of 31 segments were categorized across four additional themes listed under *Socio-Political Factors and Themes*. For coding and reducing the code complexity, these themes were consolidated under one header. The importance of each of these themes as constraints and enablers is equally represented further, with important segments from each theme quoted in this section.

Table 11.4 "Entrepreneurial/Innovation Activities" Codes and Their Explanations.

Entrepreneurial and Innovation Activities	Number of Interviews Containing This Code	Code Explanation
Small-scale manufacturing for paper or plastic disposables	5	Mention of a new industrial manufacturing facility for paper or plastic disposable production
Change in business model	4	A transition from an earlier business model to a new one because of any of the market forces or consumer demands
Household industry catering to the demand of paper disposables	3	Mention of a new household industry facility for paper disposables production
Product/process innovation	2	Any new process or product improvement featured in the disposable product range

Table 11.5 Themes Under Socio-Political Factors, Segment Frequencies, and Their Explanations.

Socio-Political Factors and Themes	Number of Interviews Containing This Code	Theme Explanation
Case examples and comparisons	17	Examples from interviews representing a situation or explanation of the situation
Religious demand for biomass-based disposables	6	Special mentions of a religious demand of biomass-based disposables
Government initiative to ban plastic	5	Segments featuring government initiatives to control plastic pollution and regulate single-use disposables
Negative impact on biomass-based disposable manufacturers	3	Specific mentions for a (negative) impact of a change in the market or consumer behavior on the producers of biomass-based disposable manufacturers

Several respondents provided one or more examples of situations explaining the current disposable scenario, the changing markets, and consumer demands. These statements are used as contexts for activities at the BoP, leading to noticeable changes in entrepreneurial strategies. The theme "case examples and comparisons" highlights these examples. Six statements documented the current level of demand for biomass-based disposables, whose demand was much driven by religious activities well documented in the Indian ethnocentric literature (Mohan, 2009; Sharma & Pegu, 2011). Institutional factors were also mentioned five times that highlighted the role new policy infrastructure plays in shaping the producer side landscape in the market (Campbell, 2007; George et al., 2012), although the impact was not seen due to failed policy measures and poor implementation (Goyal, Esposito, Kapoor, Jaiswal, & Sergi, 2014). The limited response to the question on livelihood impact at BoP is summarized in the theme entitled *Negative Impact on Biomass-Based Disposable Manufacturers*. Table 11.5 provides the contextual explanation of the codes.

11.3.5 **TRENDS AND INTERCONNECTIONS**

One common agreeable statement on a specific type of innovation is that of inclusive innovation "as primarily dealing with business model breakthroughs that enable participation in high-growth, high-profit ventures by previously disenfranchised poor people, [including forms of] ownership, managerial control, employment, consumption, and supply-chain involvement" (George et al., 2012, p. 663). Breaking down the code relations with each thematic fragment versus the rest of the themes reveals several other exciting elements that are driving the market trends. The code relations analysis, as shown in Fig. 11.1, highlights the impact of government policies on changes in business models on the ground, while the quality/performance concerns with biomass-based disposables and reduced demand for the same have led to a negative impact on the biomass-based disposable manufacturers. The numeral in the bracket shows the frequency of the respective coded segment. The frequency is equivalent to the number of coded segments assigned to this specific code or sub-code, across the 16 interviews. One code or sub-code can/has been assigned multiple times within the same interview—and has been accounted for. We used the same approach for the proposed theoretical model as shown in Fig. 11.2.

The factors that have shaped the BoP markets are equally interconnected as shown in the figure. These examples highlight local manufacturing infrastructure development that supports the changing market demand. In addition to the local manufacturing set-ups, there was inductively derived evidence that provides information toward a household industry being present to cater to the local demands.

The consumer preferences that have led to increases in plastic disposable demand are attributed to three different elements that provide an insight into the educational status about single-use plastic outcomes. Lack of awareness and not having an environmental concern illustrate this phenomenon across consumers. The increased demand is additionally supplemented by consumer preference for plastic goods as they find them superior to other alternatives such as paper and biomass-based disposables as shown in Fig. 11.1.

A collective analysis of leaf plate mentions across all interviews reveals stark facts about the disappearance of the sustainable resource, as well as the impact of climate change on the product supply chain, which allows for a potential means of reviving the sustainable disposable option. The following are the key extracts from the interview explaining the aforementioned: only 13% of the respondents believe that price is a reason for the trends away from sustainable leaf plates to plastic single-use disposables. This is supported by 38% of respondents agreeing to quality concerns with biomass-based disposables being the major consumer complaint. Several of these entrepreneurs agreed to have sold biomass-based at one point in time; 44% of overall responses state the lack of raw materials, increasing labor costs and an overall reduction in supply of primary raw materials for leaf plates as the reasons toward a decline in the leaf plate disposable sales and an eventual transformation toward the plastic-based supply chain. This is corroborated further by the following interview statements:

- Interview 2: "The quality of leaf plates was very low, they used to easily break and leak and hence people chose the paper (with aluminum) option and now there are plastic-based and some Chinese alternatives. Additionally, I believe that the raw material for the leaf plates might have reduced and hence they are not coming in the market."
- Interview 3: "The raw materials (i.e., the leaves) have all been cut. The resource has been exploited heavily. The raw material used to come from South India, but as of now the resources have become extinct, so we do not sell them anymore."

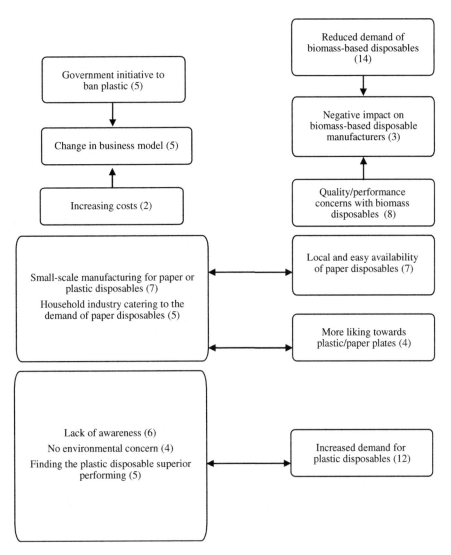

FIGURE 11.1

Code relationship analysis. The number in the bracket reflects the frequency of the respective coded segment.

- Interview 4: "We now receive even worse quality of leaf plates because of declining resource and producer interests."

Currently, only one respondent confirmed to have sold biomass-based disposable in the last year with reference to a local religious event that required the leaf plates.

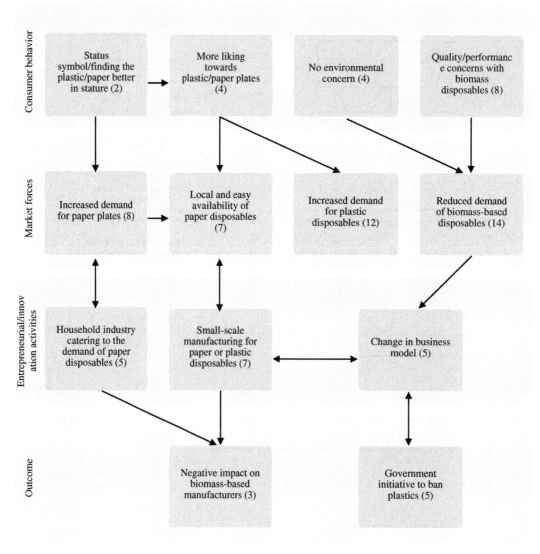

FIGURE 11.2

Proposed theoretical model. Similar as Fig. 11.1, the number of the bracket reflects the frequency of the respective coded segment.

11.4 **CONCLUDING DISCUSSION**

This chapter aimed to answer the research question based on the analyses of 16 qualitative interviews with entrepreneurs in India. It is evident that a demand exists for sustainable biomass-based cutlery, although currently driven through other, religious sentiments and not a sustainability mindset across consumers. While government actions have been able to instill a change, a true sustainable bio-based economy is yet to be seen.

It is to be recognized that climate change and other anthropogenic activities have caused a decline in the forest resource for leaf plates arising out of the state of Orissa. The entrepreneurs' responses validate the climate impact and prove that customer mindset has played an equivocal role in the change in demands, although the possibilities of transformation exist.

The possible alternatives to plastic single-use cutlery shall be found in the form of new innovative products made out of plants or easy to decompose biodegradable plastics. The key results provide a good illustration of the need for governments and businesses to work together to achieve profitability commercial outcomes for entrepreneurs that are wanting to adopt more environmentally friendly manufacturing techniques. Therefore more research is needed to better understand all aspects of the bioplastic production process, with some studies finding that bioplastics production resulted in greater amounts of pollutant residue. Due mainly to the fertilizers and pesticides used in growing the crops, the chemical processing is used to transform the organic material into plastic (Tabone, Cregg, Beckman, & Landis, 2010). Nonetheless, a failure to make the necessary changes and leave things as the status quo could undoubtedly lead to significant degradation of important natural resources in India.

11.4.1 THEORETICAL IMPLICATIONS

The findings confirm that consumer behavior actions drive market forces toward specific entrepreneurial activities. These activities are enforced by environmental constraints at both the consumer and producer level. The outcome reflects the research of several scholars approving of frugal innovation products and services that meet the needs of resource-constrained consumers while creating new demands in BoP markets through local entrepreneurs (Altmann & Engberg, 2016; George et al., 2012; Hall et al., 2012; Karnani, 2007; Leliveld & Knorringa, 2018; Zeschky, Widenmayer, & Gassmann, 2011).

The thematic literature analysis (Anderson, Lees, & Avery, 2015) proves that a large share of the previous research endeavors has focused on utilizing the BoP literature to multinational corporation (MNC) advantage. BoP has been considered as a source of breakthrough innovation for MNCs to explore new markets and build new products (Prahalad, 2012; Tasavori, Zaefarian, & Ghauri, 2015), allowing access to the opportunities presented at the BoP (Prahalad, 2006, 2012). On the contrary, only a small number of researchers have considered BoP knowledge as being potentially utilized to benefit the poor and develop cost-effective innovations that provide better socioeconomic outcomes for the deprived (Goyal, Sergi, & Jaiswal, 2016; Goyal, Sergi, & Kapoor, 2014; Singh, Gupta, & Mondal, 2012).

From a market demand perspective, the chapter finds a noticeable increase in demand and the availability of different supply chains serving the BoP, revealing novel processes being built to deliver to the changing marketplace demands. Hence, this chapter responds to the call of researchers seeking to measure the impact of frugal actions on the people working at the BoP (Carmel, Lacity, & Doty, 2016; Hart, Sharma, & Halme, 2016; Hossain, 2017, 2018; Hossain, Simula, & Halme, 2016), to create deeper insights by way of a narrative analysis of BoP markets through the eyes of local entrepreneurs.

The responses directly correlate to the impact of people living at the BoP, which points toward a negative impact on the biomass-based disposable manufacturers for reasons including the availability of alternative disposable sources such as plastics, a lack of biomass-resources and reduction in demand for their products due to quality issues. A few other respondents were unsure about the state of the original biomass-based disposable producers. However, additional responses provide insights into the current state of innovative approaches deployed by either himself or herself or other related entrepreneurs in delivering to the changing demands. One of the respondents claimed to have employed over 500 people through a home-based model involving women workers. This and several other contexts mentioned supplement the call made to conduct more research by scholars focusing on frugal innovations (George et al., 2012; Hall et al., 2012; Knorringa, Peša, Leliveld, & van Beers, 2016; Nari Kahle et al., 2013), which is conducive to inclusive growth, reduced inequality, and increased participation and empowerment.

Entrepreneurial activities thrive through increased product performances at stable prices. This chapter confirms the practice for cost-effective innovations arising out of consumer demands and not price challenges. The interview responses highlight the absence of price as a factor for consumer choice and instead press the increasing need for better products irrespective of changing prices (Ramani & Mukherjee, 2014). In addition, 50% of the respondents mentioned user preference based on finding the paper and plastic products being higher in stature when compared to a biomass-based alternative. This causation can be attributed to social factors that primarily include social reputation where consumers find new products that look clean, over a relatively unstable and unpleasant looking product (Luchs & Kumar, 2017). Such a finding also highlights the risks that a continuation of such practices could pose a challenge on important environmental resources in India, potentially increasing the probability of a significant natural or humanitarian disaster for the country.

Furthermore, BoP entrepreneurs must deal with several sociopolitical factors that also affect the consumers in the supply chain. The role of state policies therefore becomes imperative in shaping the landscapes of BoP markets (Campbell, 2007; George et al., 2012). Although several respondents mentioned a government ban on plastic as a reason to change their business model and adapt to the updating regulations, it was claimed ineffective due to lack of policy implementation (Goyal, Esposito et al., 2014).

The chapter also proves that there is a general lack of environmental awareness. Fifty percent of the respondents claimed that their consumers are not aware of the environmental implications of plastic plates and the nonexistence of a recycling system at the municipality (Ayub & Khan, 2011). Nulkar (2016) argues that the proliferation of innovative products for the BoP benefits businesses by opening new markets and provides more extensive choices for the poor, this process of consumption and disposal has an equally high environmental impact. It is further supported by 33% of responses referring to a lack of environmental concern among consumers; this can be attributed to social features in the society as consumers generally find better-looking products as a status symbol and hence are inclined toward choosing a product irrespective of its environmental costs (Luchs & Kumar, 2017). These results suggest that more needs to be done to educate both the entrepreneurial community in particular, and society in general. With so much research illustrating the potentially damaging health and environmental effects of plastic use (Kavlock et al., 2002), businesses in India need to be more aware of the repercussions if the status quo is not changed.

The consolidated findings are presented as a visual outcome in Fig. 11.2. The visual outcome provides factors that lead to innovation capabilities and improve the competitiveness of local entrepreneurs, away from the research focused on MNCs (Lim et al., 2013). The outcome reflects the research approving of frugal products that meet the needs of resource-constrained consumers while creating new demand opportunities in emerging markets through local entrepreneurs (Altmann & Engberg, 2016; Leliveld & Knorringa, 2018; Zeschky et al., 2011; Zeschky, Winterhalter, & Gassmann, 2014). Using inputs from the findings and specifically the interconnected themes and codes, the visual outcome presents a composite analysis of the strength of various codes under their respective thematic segments. This output serves as a proposed theoretical model to be validated for other BoP supply chains for their entrepreneurial and frugal rigor.

The use of resources in a more sustainable manner can reduce the threat of environmental disasters occurring through excessive plastic production. Moreover, the use of biodegradable materials can address concerns over plastic waste disasters (Narancic & O'Connor, 2017). Another important factor that plays an influential role within Indian markets is the price Pareto, and hence, a revival of sustainable biomass-based disposables requires value chain interventions that either reduce the leaf plate price down to generic plastic price levels, or by bringing in additional taxes to set both products at similar price bars. These additional taxes could then be utilized by the state's central government agencies to improve the recycling infrastructure as a whole.

11.5 FUTURE OUTLOOK

This chapter reveals unique features of the market, cost-effective innovation factors, and their relationship with consumer perception. The findings validate the frameworks put across by George et al. (2012), by providing contextual examples of businesses that enfranchise individuals and communities. However, the findings cannot be universally applied as product cases that differ in several variables, and hence, other cases of local products that have made a global impact through reverse innovation or other streams need to be analyzed.

The current findings examine the impact on the BoP population qualitatively, which allows for findings that corroborate with existing theoretical innovation models. This chapter can provide some interesting points of reference. The first would be a quantitative measurement of income in the specific case of disposable cutlery in both the plastic and biomass-based supply chains. Second, one could specifically look at improving the biomass-based production stream for a sustainable product to be made available in the Indian market and even serving the global bioeconomy. This study also referred to the environmental problems that are associated with using both traditional and bio-based forms of production. It also detailed the implications that a failure to educate and better understand the risks that are associated with the production process may have on businesses and the Indian society as a whole.

REFERENCES

Adner, R., & Levinthal, D. (2001). Demand heterogeneity and technology evolution: Implications for product and process innovation. *Management Science*, 47(5), 611–628.

Ahuja, S. C., & Ahuja, U. (2011). Betel leaf and betel nut in India: History and uses. *Asian Agri-History*, 15(1), 13–35.

Ali, J., Kapoor, S., & Moorthy, J. (2010). Buying behaviour of consumers for food products in an emerging economy. *British Food Journal, 112*(2), 109−124.

Altmann, P., & Engberg, R. (2016). Frugal innovation and knowledge transferability: Innovation for emerging markets using home-based R&D Western firms aiming to develop products for emerging markets may face knowledge transfer barriers that favor a home-based approach to frugal innovation. *Research-Technology Management, 59*(1), 48−55.

Anderson, D., Lees, B., & Avery, B. (2015). *Reviewing the literature using the Thematic Analysis Grid.* Retrieved from <https://search.proquest.com/docview/1721001703/fulltextPDF/EB1164C8F9943AFPQ/1?accountid = 14116>.

Ayub, S., & Khan, A. H. (2011). Landfill practice in India: A review. *Journal of Chemical and Pharmaceutical Research, 3*(4), 270−279.

Basu, R. R., Banerjee, P. M., & Sweeny, E. G. (2013). Frugal innovation. *Journal of Management for Global Sustainability, 1*(2), 63−82.

Braun, V., & Clarke, V. (2006). Using thematic analysis in psychology. *Qualitative Research in Psychology, 3*(2), 77−101.

Burnard, P. (1994). The telephone interview as a data collection method. *Nurse Education Today, 14*(1), 67−72.

Campbell, J. L. (2007). Why would corporations behave in socially responsible ways? An institutional theory of corporate social responsibility. *Academy of Management Review, 32*(3), 946−967.

Carmel, E., Lacity, M. C., & Doty, A. (2016). The impact of impact sourcing: Framing a research agenda. In B. Nicholson, R. Babin, & M. C. Lacity (Eds.), *Socially responsible outsourcing* (pp. 16−47). London: Palgrave Macmillan.

Central Intelligence Agency. (2019). *The World Factbook: India.* Retrieved from <https://www.cia.gov/library/publications/the-world-factbook/geos/in.html>.

Cho, R. (2019). *Sustainability: The truth about bioplastics.* Retrieved from <https://blogs.ei.columbia.edu/2017/12/13/the-truth-about-bioplastics/>.

Dikgang, J., Leiman, A., & Visser, M. (2012). Elasticity of demand, price and time: Lessons from South Africa's plastic-bag levy. *Applied Economics, 44*(26), 3339−3342.

Eastman, J. K., Goldsmith, R. E., & Flynn, L. R. (1999). Status consumption in consumer behavior: Scale development and validation. *Journal of Marketing Theory and Practice, 7*(3), 41−52.

Eisenhardt, K. M., & Graebner, M. E. (2007). Theory building from cases: Opportunities and challenges. *Academy of Management Journal, 50*(1), 25−32.

ENVIS Centre of Odisha's State of Environment. (2017). *Climate.* Retrieved from <http://orienvis.nic.in/index1.aspx?lid = 24&mid = 1&langid = 1&linkid = 22>.

European Plastics. (2017). *Plastics—The facts 2017. An analysis of European plastics, demand and waste data.* Retrieved from <https://www.plasticseurope.org/application/files/5715/1717/4180/Plastics_the_facts_2017_FINAL_for_website_one_page.pdf>.

Federation of Indian Chambers of Commerce and Industry. (2017). *Sustainable infrastructure with plastics: Knowledge and strategic partner plastic industry for infrastructure.* New Delhi: Federation of Indian Chambers of Commerce and Industry. Retrieved from <http://ficci.in/spdocument/20872/report-Plastic-infrastructure-2017-ficci.pdt>.

Gartner, W. B. (1988). "Who is an entrepreneur?" is the wrong question. *American Journal of Small Business, 12*(4), 11−32.

Gautam, A. M., & Caetano, N. (2017). Study, design and analysis of sustainable alternatives to plastic take-away cutlery and crockery. *Energy Procedia, 136*, 507−512.

George, G., McGahan, A. M., & Prabhu, J. (2012). Innovation for inclusive growth: Towards a theoretical framework and a research agenda. *Journal of Management Studies, 49*(4), 661−683.

Gilpin, R. (1975). *Technology, economic growth, and international competitiveness.* Washington, DC: United States Government Print Office.

Gioia, D. A., Corley, K. G., & Hamilton, A. L. (2013). Seeking qualitative rigor in inductive research: Notes on the Gioia methodology. *Organizational Research Methods*, *16*(1), 15−31.

Government of India. (2018). *Annual report 2017−2018*. Retrieved from <http://cpcb.nic.in/uploads/plastic-waste/Annual_Report_2017-18_PWM.pdf>.

Goyal, S., Esposito, M., Kapoor, A., Jaiswal, M. P., & Sergi, B. S. (2014). Linking up: Inclusive business models for access to energy solutions at base of the pyramid in India. *International Journal of Business and Globalisation*, *12*(4), 413−438.

Goyal, S., Sergi, B. S., & Jaiswal, M. P. (2016). Understanding the challenges and strategic actions of social entrepreneurship at base of the pyramid. *Management Decision*, *54*(2), 418−440.

Goyal, S., Sergi, B. S., & Kapoor, A. (2014). Understanding the key characteristics of an embedded business model for the base of the pyramid markets. *Economics and Sociology*, *7*(4), 26−40.

Haider, T. P., Völker, C., Kramm, J., Landfester, K., & Wurm, F. R. (2019). Plastics of the future? The impact of biodegradable polymers on the environment and on society. *Angewandte Chemie International Edition*, *58*(1), 50−62.

Hall, J., Matos, S., Sheehan, L., & Silvestre, B. (2012). Entrepreneurship and innovation at the base of the pyramid: A recipe for inclusive growth or social exclusion? *Journal of Management Studies*, *49*(4), 785−812.

Hart, S., Sharma, S., & Halme, M. (2016). Poverty, business strategy, and sustainable development. *Organization & Environment*, *29*(4), 401−415.

Hossain, M. (2017). Mapping the frugal innovation phenomenon. *Technology in Society*, *51*, 199−208.

Hossain, M. (2018). Grassroots innovation: The state of the art and future perspectives. *Technology in Society*, *55*, 63−69.

Hossain, M., Simula, H., & Halme, M. (2016). Can frugal go global? Diffusion patterns of frugal innovations. *Technology in Society*, *46*, 132−139.

Islam, S.N. (2015). Inequality and environmental sustainability. In: *DESA working papers, 145* (pp. 1−30).

Karnani, A. (2007). The mirage of marketing to the bottom of the pyramid: How the private sector can help alleviate poverty. *California Management Review*, *49*(4), 90−111.

Kavlock, R., Boekelheide, K., Chapin, R., Cunningham, M., Faustman, E., Foster, P., ... Little, R. (2002). NTP Center for the Evaluation of Risks to Human Reproduction: Phthalates expert panel report on the reproductive and developmental toxicity of di (2-ethylhexyl) phthalate. *Reproductive Toxicology*, *16*(5), 529−653.

Knorringa, P., Peša, I., Leliveld, A., & van Beers, C. (2016). Frugal innovation and development: Aides or adversaries? *European Journal of Development Research*, *28*(2), 143−153.

Lebreton, L., Slat, B., Ferrari, F., Sainte-Rose, B., Aitken, J., Marthouse, R., ... Levivier, A. (2018). Evidence that the Great Pacific Garbage Patch is rapidly accumulating plastic. *Scientific Reports*, *8*(1), 4666.

Leliveld, A., & Knorringa, P. (2018). Frugal innovation and development research. *European Journal of Development Research*, *30*(1), 1−16.

Lim, C., Han, S., & Ito, H. (2013). Capability building through innovation for unserved lower end mega markets. *Technovation*, *33*(12), 391−404.

Luchs, M. G., & Kumar, M. (2017). "Yes, but this other one looks better/works better": How do consumers respond to trade-offs between sustainability and other valued attributes? *Journal of Business Ethics*, *140*(3), 567−584.

Maguire, M., & Delahunt, B. (2017). Doing a thematic analysis: A practical, step-by-step guide for learning and teaching scholars. *AISHE-J: The All Ireland Journal of Teaching and Learning in Higher Education*, *9*(3), 1−14.

Mann, J., & Mann, P. (2015). A review of language and literature of Haryana. *International Journal of Reviews and Research*, *3*(3), 1−6.

MAXQDA. (2019). *Powerful and easy-to-use qualitative analysis software*. Retrieved from <https://www.maxqda.com/>.

McTavish, D. G., & Pirro, E. B. (1990). Contextual content analysis. *Quality and Quantity*, *24*(3), 245−265.

Miller, L., Soulliere, K., Sawyer-Beaulieu, S., Tseng, S., & Tam, E. (2014). Challenges and alternatives to plastics recycling in the automotive sector. *Materials, 7*(8), 5883−5902.

Mohan, S. (2009). *Siali: A document on the Siali leaf-plate making activity of Kandhamal district, Orissa.* New Delhi: PRADAN.

Mowery, D., & Rosenberg, N. (1979). The influence of market demand upon innovation: A critical review of some recent empirical studies. *Research Policy, 8*(2), 102−153.

Narancic, T., & O'Connor, K. E. (2017). Microbial biotechnology addressing the plastic waste disaster. *Microbial Biotechnology, 10*(5), 1232−1235.

Nari Kahle, H., Dubiel, A., Ernst, H., & Prabhu, J. (2013). The democratizing effects of frugal innovation: Implications for inclusive growth and state-building. *Journal of Indian Business Research, 5*(4), 220−234.

Nulkar, G. (2016). The environmental costs of serving the bottom of the pyramid. *Sustainability: The Journal of Record, 9*(1), 31−38.

Prahalad, C. K. (2006). *The fortune at the bottom of the pyramid.* Upper Saddle River, NJ: Wharton School Publishing.

Prahalad, C. K. (2012). Bottom of the pyramid as a source of breakthrough innovations. *Journal of Product Innovation Management, 29*(1), 6−12.

Professional Assistance for Development Action. (2017). *A document on the Siali leaf-plate making activity.* Retrieved from <http://www.pradan.net/portfolio-item/a-document-on-the-siali-leaf-plate-making-activity/>.

Rajmohan, K. S., Ramya, C., & Varjanic, S. (2019). Plastic pollutants: Waste management for pollution control and abatement. *Current Opinion in Environmental Science & Health, 12*, 72−84.

Ramani, S. V., & Mukherjee, V. (2014). Can breakthrough innovations serve the poor (BoP) and create reputational (CSR) value? Indian case studies. *Technovation, 34*(5-6), 295−305.

Rao, B. C. (2013). How disruptive is frugal? *Technology in Society, 35*(1), 65−73.

Reddy, C. S., Rao, K. R. M., Pattanaik, C., & Joshi, P. K. (2009). Assessment of large-scale deforestation of Nawarangpur district, Orissa, India: A remote sensing based study. *Environmental Monitoring and Assessment, 154*(1-4), 325.

Saunders, M., Lewis, P., & Thornhill, A. (2009). *Research methods for business students* (5th ed.). Essex: Pearson Education Limited.

Sharma, U. K., & Pegu, S. (2011). Ethnobotany of religious and supernatural beliefs of the Mising tribes of Assam with special reference to the 'Dobur Uie'. *Journal of Ethnobiology and Ethnomedicine, 7*(16), 1−13. Available from https://doi.org/10.1186/1746-4269-7-16.

Singh, R., Gupta, V., & Mondal, A. (2012). Jugaad—From "making do" and "quick fix" to an innovative, sustainable and low-cost survival strategy at the bottom of the pyramid. *International Journal of Rural Management, 8*(1−2), 87−105.

Tabone, M. D., Cregg, J. J., Beckman, E. J., & Landis, A. E. (2010). Sustainability metrics: Life cycle assessment and green design in polymers. *Environmental Science & Technology, 44*(21), 8264−8269.

Tasavori, M., Zaefarian, R., & Ghauri, P. N. (2015). The creation view of opportunities at the base of the pyramid. *Entrepreneurship & Regional Development, 27*(1-2), 106−126.

Tiwari, R., & Herstatt, C. (2012). Assessing India's lead market potential for cost-effective innovations. *Journal of Indian Business Research, 4*(2), 97−115.

Zeschky, M., Widenmayer, B., & Gassmann, O. (2011). Frugal innovation in emerging markets. *Research-Technology Management, 54*(4), 38−45.

Zeschky, M. B., Winterhalter, S., & Gassmann, O. (2014). From cost to frugal and reverse innovation: Mapping the field and implications for global competitiveness. *Research-Technology Management, 57*(4), 20−27.

SEMANTIC TECHNOLOGIES FOR DISASTER MANAGEMENT: NETWORK MODELS AND METHODS OF DIAGRAMMATIC REASONING

12

Archibald James Juniper

Conjoint Academic, University of Newcastle, Callaghan, NSW, Australia

12.1 INTRODUCTION

This chapter is concerned with developments in applied mathematics and theoretical computing that can provide a formal and technical support for practices of disaster management. To this end it will draw on recent developments in applied category theory (ACT), which inform semantic technologies. In the interests of brevity, it will be obliged to eschew formal exposition of these techniques, but to this end, comprehensive references will be provided. The justification for what might at first seem to be an unduly narrow focus is that ACT facilitates translation between different mathematical, computational, and scientific domains.

For its part, semantic technology (ST) can be loosely conceived as an approach treating the World Wide Web as a "giant global graph," so that valuable and timely information can be extracted from it using rich structured-query languages and extended description logics. These query languages must be congruent with pertinent (organizational, application, and database) ontologies so that the extracted information can be converted into intelligence. Significantly, database instances can extend beyond relational or graph databases, to include Boolean matrices, relational data embedded within the category of linear relations, and data pertaining to the initial conditions and parameter values for systems of differential equations within a finite vector space, or even for quantum tensor networks within a finite Hilbert space.

More specifically, this chapter will introduce the formalism of string diagrams, which were initially derived from the work of the mathematical physicists, Penrose (1969) and Feynman (1948). However, this diagrammatic approach has since been extended and reinterpreted by category theorists such as Joyal and Street (1988, 1991). For example, Feynman diagrams can be viewed as morphisms in the category *Hilb* of Hilbert spaces and bounded linear operators (Westrich, 2006, fn. 3: 8), while Baez and Lauda (2009) interpret them as "a notation for intertwining operators between positive-energy representations of the Poincaré group." Penrose diagrams can be viewed as a representation of operations within a tensor category.

Economic Effects of Natural Disasters. DOI: https://doi.org/10.1016/B978-0-12-817465-4.00012-1

Joyal and Street have demonstrated that when these string diagrams are manipulated in accordance with certain axioms—the latter taking the form of a set of equivalence relations established between related pairs of diagrams—the movements from one diagram to another can be shown to reproduce the algebraic steps of a nondiagrammatic proof. Furthermore, they can be shown to possess a greater degree of abbreviative power. This renders an approach using string diagrams extremely useful for teaching, experimentation, and exposition.

In addition to these conceptual and pedagogical advantages, however, there are additional *implementation* advantages associated with string diagrams including (1) those of compositionality and layering (e.g., in Willems's, 2007 behavioral approach to systems theory, complex systems can be construed as the composites of smaller and simpler building blocks, which are then linked together in accordance with certain coherence conditions); (2) a capacity for direct translation into functional programming (and, thus, into propositions within a linear or resource-using logic); and (3) the potential for the subsequent application of software design and verification tools. It should be appreciated that these formal attributes will become increasingly important as the correlative features of what some have described as the *digital* economy.

This chapter will consider the specific role of string diagrams in the development and deployment of semantic technologies, which in turn have been developed for applications of relevance to disaster management practices. Techniques based on string diagrams have been developed to encompass a wide variety of dynamic systems and application domains, such as Petri nets, the π-calculus, and bigraphs (Milner, 2009), Bayesian networks (Kissinger & Uijlen, 2017), thermodynamic networks (Baez & Pollard, 2017), quantum tensor networks (Biamonte & Bergholm, 2017), as well as reaction−diffusion systems (Baez & Biamonte, 2012). Furthermore, they have the capacity to encompass graphical forms of linear algebra (Sobociński), universal algebras (Baez, 2006), and signal flow graphs (Bonchi, Sobociński, & Zanasi, 2014, 2015), along with computational logics based on linear logic and graph rewriting (on his see Melliès, 2018, and Fong and Spivak for additional references).

12.2 APPLIED CATEGORY THEORY

Category theory and topos theory have taken over large swathes in the field of formal or theoretical computation, because categories serve to link together the structures found in algebraic topology, and with the logical connectives and inferences to be found in formal logic, as well as with recursive processes and other operations in computation.

Bell (1988: 236) succinctly explains why it is that category theory also possesses enormous powers of generalization:

> A category may be said to bear the same relation to abstract algebra as does the latter to elementary algebra. Elementary algebra results from the replacement of constant quantities (i.e. numbers) by variables, keeping the operations on these quantities fixed. Abstract algebra, in its turn, carries this a stage further by allowing the operations to vary while ensuring that the resulting mathematical structures (groups, rings, etc) remain of a prescribed kind. Finally, category theory allows even the kind of structure to vary: it is concerned with structure in general.

Baez and Stay (2008) demonstrate that category theory can also be interpreted as a universal approach to the analysis of process, across various domains including (1) mathematic practice (theorem proving), (2) physical systems (their evolution and measurement), (3) computing (data types and programs), (4) chemistry (chemicals and reactions), (5) finance (currencies and various transactions), and (6) engineering (flows of materials and production).

This way of thinking about processes now serves as a unifying interdisciplinary framework that researchers within business and the social sciences have also taken up. Alternative approaches to those predicated on optimizing behavior on the part of individual economic agents include the work evolutionary economists and those in the business world who are obliged to work with computational systems designed for the operational management of commercial systems. However, these techniques are also grounded in conceptions of process.

Another way of thinking about dynamic processes is in terms of circuit diagrams, which can represent displacement, flow, momentum, and effort—phenomenon modeled by the Hamiltonians and Lagrangians of Classical Mechanics. It can be appreciated that key features of economic systems are also amenable to diagrammatic representations of this kind, including asset pricing based on notion of arbitrage, a concept initially formalized by Augustin Cournot in 1838. Cournot's analysis of arbitrage conditions is grounded in Kirchhoff voltage law (Ellerman, 2000). The analogs of displacement, flow, momentum, and effort are depicted later for a wide range of disciplines (Fig. 12.1).

In the United States and Europe, contemporary developments in ACT have been spurred along and supported by a raft of EU, DARPA, and ONR Grants. A key resource on ACT is Fong and Spivak's (2018) downloadable text on compositionality. This introductory publication explores the

	Displacement q	Flow q'	Momentum p	Effort p'
Electronics	Charge	Current	Fluxlinkage	Voltage
Translation	Position	Velocity	Momentum	Force
Rotation	Angle	Angular velocity	Angular momentum	Torque
Hydraulics	Volume	Flow	Pressure momentum	Pressure
Thermodynamics	Entropy	Entropy flow	Temperature momentum	Temperature
Chemistry	moles	Molar flow	Chemical momentum	Chemical potential

FIGURE 12.1

Analogues for circuit diagrams.

Modified from Baez, J. C., & Fong, B. (2018). A compositional framework for passive linear networks (Baez & Fong, 2018), Table, p. 2.

relationship between wiring diagrams or string diagrams and a wide variety of mathematical and categorical constructs, including as a means for representing symmetric monoidal preorders and signal flow graphs, graphical linear algebra, hypergraph categories, and operads—the latter applied to electric circuits and network compositionality—as well as a detailed description of functorial translations between signal flow graphs and matrices, along with other aspects of functorial semantics. Topos theory is introduced to characterize the logic of system behavior on the basis of indexed sets, glueings, and sheaf conditions for every open cover.

12.3 **DIAGRAMMATIC REASONING**

Authors such as Sáenz-Ludlow and Kadunz (2015), Shin (1994), Sowa (2000), and Stjernfelt (2007), who have published research on knowledge representation (KR) and diagrammatic approaches to reasoning, tend to work within a philosophical trajectory that stretches from F.W. Schelling and C.S. Peirce to E. Husserl and A. N. Whitehead, then on to M. Merleau-Ponty and T. Adorno. Where Kant and Hegel privileged symbolic reasoning over the iconic or diagrammatic, Peirce, Whitehead, and Merleau-Ponty followed the lead of Schelling for whom "aesthetics trumps epistemology"! It is, in fact, this shared philosophical allegiance that not only links diagrammatic research to the semantic (or embodied) cognition movement (Stjernfeld himself refers to the embodied cognition theorists Eleanor Rosch, George Lakoff, Mark Johnson, Leonard Talmy, Mark Turner, and Gilles Fauconnier) but also to those researchers who have focused on issues of educational equity in the teaching of mathematics and computer science, including ethnomathematics and critical work on "Orientalism" specialized to emphasize a purported division between the "West and the Rest" in regard to mathematical and computational thought and practice.

As such, insights from this research carry over to questions of ethnic "marginalization" or "positioning" in the mathematical sciences (see the papers reproduced in Forgasz and Rivera, eds., 2012 and Herbel-Eisenmann, Choppin, Wagner, & Pimm, 2012). In a nutshell, diagrammatic reasoning (DR) is sensitive to both context and positioning and, thus, is closely allied to this critical axis of mathematics education.

To explain the various elements and flows associated with diagrammatic forms of reasoning, I have drawn on Hoffman's (2011) explication of the concept first outlined by the American philosopher and logician Charles Sanders Peirce in his largely unpublished research on diagrammatic logic.

Hoffman describes three stages of DR: (1) constructing a diagram conceived as a formal and consistent representation of key relations, (2) analyzing a problem on the basis of this representation, and (3) experimenting with the diagram and then observing the results. Consistency is ensured in two ways. First, the researcher or research team develop an ontology specifying elements of the problem and the relations holding between these elements, along with pertinent rules of operation. Second, a diagrammatic language is specified in terms of both syntactical and semantic properties. Furthermore, in association with this language, a rigorous axiomatic system is specified, which both constrains and enables any pertinent diagrammatic transformations. In Chapter 5 of their online introduction to applied category theory, Fong and Spivak (2018) provide a user-friendly introduction to the axioms of relevance to string diagrams (also see the section on Frobenius structure outlined in Definition 6.5.2 of the text).

Hoffman (2011) contends that this diagrammatic and collaborative approach can clarify and coordinate confused ideas about a problem, clarify implicit assumptions and any background knowledge that might be insufficient or inadequate, structure a problem space, enable a change in perspectives, support any playing with interpretations, thus permitting researchers to discover contradictions and distinguish the essential from the peripheral.

12.3.1 CASE STUDY ONE

A 2010 paper by SAP Prof., Paulheim and Probst reviews an application of STs to the management and coordination of emergency services in the Darmstadt region of Germany. The aim of the following diagram, reproduced from their work, is to highlight the fact that, from a computational perspective, the integrative effort of STs can apply to different organizational levels: that of the common user interface, shared business logics, and that of data sources (Fig. 12.2).

In their software engineering application the upper level ontology DOLCE is deployed to link a core domain ontology together with a user-interface interaction ontology. In turn, each of these ontologies draws on inputs from an ontology on deployment regulations and various application ontologies. Improved search capabilities, across this hierarchy of computational ontologies, are achieved through the adoption of the ONTOBROKER and F-Logic systems.

12.3.2 CASE STUDY TWO

An important contribution to the field of network modeling (NM) has come from the DARPA-funded CASCADE Project (Complex Adaptive System Composition and Design Environment),

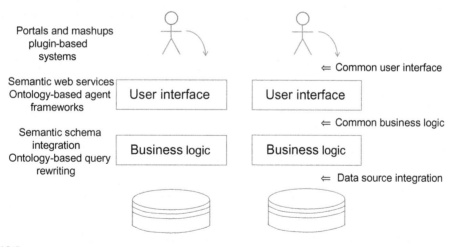

FIGURE 12.2

Levels of system integration.

Modified from Paulheim, H., & Probst, F. (2010). Application integration on the user interface level: An ontology-based approach.
Data and Knowledge Engineering, 69, *1103–1116 (Paulheim & Probst, 2010), Fig. 12.1, p. 2.*

which has invested in long-term research into the "system-of-systems" perspective (see John Baez's extended discussion of this project on his Azimuth blog). This research has been influenced by Willems's (2007) behavioral approach to systems, which, in turn, is based on the notion that large and complex systems can be built up from simple building blocks.

Baez et al. (2017) introduce "network models" to encode different ways of combining networks both through overlaying one model on top of another and by setting each model side by side. In this way, complex networks can be constructed using simple networks as components. Vertices in the network represent fixed or moving agents, while edges represent communication channels.

The components of their networks are constructed using colored operads, which include vertices representing entities of various types and edges representing the relationships between these entities. Each network model gives rise to a typed operad with an associated canonical algebra, whose operations represent ways of assembling a more complex network from smaller parts. The various different ways to compose these operations characterize a more general notion of an operation, which must be complemented by ways of permuting the arguments of an operation a process yielding a permutation group of inputs and outputs.

In research conducted under the auspices of the CASCADE Project, Baez, Foley, Moeller (2019) have worked out how to combine two formalisms. First, there are Petri nets, commonly used as an alternative to process algebras as a formalism for business process management. The vertices in a Petri net represent collections of different types of entities (species) with morphisms between them used to describe processes (transitions) that can be carried out by combining various sets of entities (conceived as resources or inputs into a transition node or process of production) together to make new sets of entities (conceived as outputs or vertices are positioned after the relevant transition node). The stocks of each type of entity that is available are enumerated as a "marking" specific to each type or color together with the set of outputs that can be produced by activated the said transition.

Second, there are network models that describe processes that a given collection of agents (say, cars, boats, people, and planes in a search-and-rescue operation) can carry out. However, in this kind of network, while each type of object or vertex can move around within a delineated space, they are not allowed to turn into other types of agent or object.

In these networks, morphisms are functors (generalized functions) which describe everything that can be done with a specific collection of agents. Fig. 12.3 depicts this kind of operational network in an informal manner, where icons represent helicopters, boats, victims floating in the sea, and transmission towers with communication thresholds.

By combining Petri nets with an underlying network model resource-using operations can be defined. For example, a helicopter may be able to drop supplies gathered from different depots and packaged into pallets, onto the deck of a sinking ship or to a remote village cut off by an earthquake or flood.

The formal mechanism for combining a network model with a Petri net relies on treating different type of entities as catalysts, in the sense that the relevant species are neither increased nor decreased in number by any given transition. The derived category is symmetric monoidal and possesses a tensor product (representing processes for each catalyst that occur side by side), a coproduct (or disjoint union of amounts of each catalyst present), and within each subcategory of a particular catalyst, an internal tensor product describes how one process can follow another while reusing the pertinent catalysts.

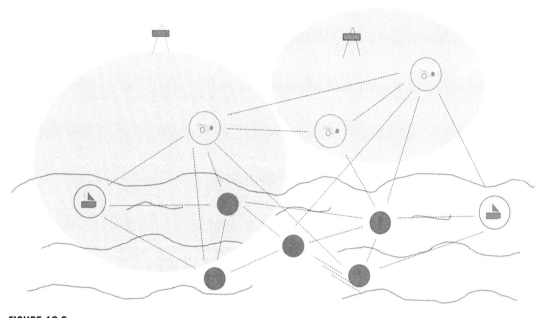

FIGURE 12.3

A search-and-rescue network (flight control, container port, etc.). Baez, J., Foley, J., Moeller, J., & Pollard, B. (2020). *Network models. arXiv:1711.00037v1 [math.CT]*. March 27, 2020 (Baez et al., 2020).

The following diagram taken from Baez et al. (2020) illustrates the overlaying process which is one way of assembling more complex networks from their simpler components. Relevant constraints can also be imposed over both the vertices and edges of any given network. The use of the Grothendieck construction in this research ensures that when two or more diagrams are overlayed, there will be no "double-counting" of edges and vertices (Fig. 12.4).

Each network model is characterized by a "plug-and-play" feature based on an algebraic component called an operad. Each operad then serves as a kind of "building block" for a canonical algebra, whose operations are ways of assembling a network of the given kind from smaller parts. This canonical algebra, in turn, accommodates a set of types, a set of operations, ways to compose these operations to arrive at more general operations, and ways to permute an operation's arguments (i.e., via a permutation group), along with a set of relevant distance constraints (e.g., pertinent communication thresholds for each type of entity). See Section 6.5. of Fong and Spivak (2018) for an introduction to the structure, meaning, and practical deployment of operads.

One of Baez's coauthors, John Foley, works for Metron, Inc., Virginia, a company specializing in applications of advanced mathematics to "search-and-rescue" operations, the detection of network incursions, and sports analytics. Their 2017 paper mentions a number of formalisms that have relevance to "search-and-rescue" applications, especially the ability to distinguish between different communication channels (different radio frequencies and capacities) and vertices (e.g., planes, boats, walkers, individuals in need of rescue etc.) and the capacity to impose distance constraints over those agents who may fall outside the reach of communication networks.

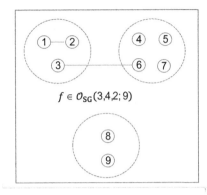

$f \in \mathcal{O}_{SG}(3,4,2; 9)$

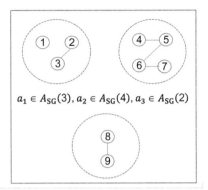

$a_1 \in A_{SG}(3), a_2 \in A_{SG}(4), a_3 \in A_{SG}(2)$

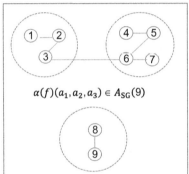

$\alpha(f)(a_1, a_2, a_3) \in A_{SG}(9)$

FIGURE 12.4

Compositional overlays.

Modified from Baez, J., Foley, J., Moeller, J., & Pollard, B. (2020). Network models. arXiv:1711.00037v1 [math.CT]. March 27, 2017 (Baez, Foley, Moeller, & Pollard, 2020).

In related research paper Schultz, Spivak, Vasilakopoulou, and Wisnesky (2016) argue that dynamical systems can be gainfully thought of as "machines" with inputs and outputs, carrying some sort of signal that occurs through some notion of time. Special cases of this general approach include discrete, continuous, and hybrid dynamical systems. The authors deploy lax functors out of monoidal categories, which provide them with a language of compositionality. As with Baez and his coauthors, Spivak et al. (2016) exploit an operadic construct so as to understand systems that result from an "arbitrary interconnection of component subsystems." They also draw on the mathematics of sheaf theory, to flexibly capture the crucial notion of time. The resulting sheaf-theoretic perspective relates continuous- and discrete-time systems together via functors. Their approach can also account for synchronized continuous time, in which each moment is assigned a specific phase within the unit interval.

12.4 RELATED DEVELOPMENTS IN SOFTWARE ENGINEERING

Contemporary developments in software engineering that have implications for system-of-systems approaches in ST. This section will examine on-going work on Statebox software by researchers who are mainly located at Oxford University along with the work of researchers in the MIT company, Categorical Informatics.

The Statebox team based at Oxford University have developed a language for software engineering that uses diagrammatic representations of generalized Petri nets. In this context, transitions in the net are morphisms between data-flow objects, which, in turn, represent terminating functional programming algorithms. In Statebox (integer and semiinteger) Petri nets are constructed with both positive and negative tokens to account for contracting. Negative tokens represent borrowing, while positive tokens represent lending and, likewise, the taking of short and long positions in asset markets. This allows for the representation of smart contracts, conceived as separable nets. Nets are also endowed with interfaces that allow for channeled communications through user-defined addresses. Furthermore, guarded and timed nets, with side effects (which are mapped to standard nets using the Grothendieck construction), offer greater expressive power in regard to the conditional behavior affecting transitions (The Statebox Team, 2018).

Researchers in Categorical Informatics have developed rich structured query languages for KR. Patterson (2017) provides an introductory discussion of description logics (e.g., OWL, WC3), which function as the calculi and substrates responsible for the World Wide Web (WWW). He notes that these logics lie somewhere between propositional logic and first-order predicate logic, insofar as they possess the capability for expressing the $(\exists, \wedge, T, =)$ fragment of first-order logic. Patterson highlights the trade-off that must be made between computational tractability and expressivity before introducing a third KR formalism that interpolates between description logic and ontology logs or ologs. Ologs express key constructs from category theory, such as products and coproducts, pullbacks, and pushforwards, along with the representations of recursive operations using diagrams that have been labeled with concepts drawn from everyday conversation (see Spivak & Kent, 2012).

Patterson (2017) calls his extended construct the "relational ontology log" or relational olog, because it is based on *Rel*, the category of sets and relations and, as such, draws on relational algebra, which is the $(\exists, \wedge, \vee, T, \perp, =)$ fragment of first-order logic. He calls Spivak and Kent's (2012) version a "functional olog" to avoid any confusion, because the latter are solely based on *Set*, the category of sets and functions. Relational ologs achieve their expressivity through a variety of categorical constructs, including limits and colimits, products, pullbacks, and pushforwards.

The advantages of Patterson's framework are that functors allow instance data to be associated with a computational ontology in a mathematically precise way, by interpreting it as a relational or graph database, Boolean matrix, or category of linear relations. Moreover, relational ologs are, by default, typed, which he suggests can mitigate the "maintainability challenges" posed by the open world semantics of description logic.

While string diagrams have been deployed to represent a variety of network structures and processes, including models of backpropagation in machine-learning, they also have the capacity to represent more strictly computational aspects. For example, influenced by the program idioms of machine-learning (as embodied in Microsoft's TensorFlow software), Muroya and Ghica (2017) have developed what they choose to call a "Dynamic Geometry of Interaction Machine," which can be defined as a state transition system operating with transitions for token passing as well as those for graph rewriting (loosely conceived as graph-based approach to diagrammatic proving of hypotheses) that fulfills this purpose and can be supported by diagrammatic implementation based on the proof structures of the multiplicative and exponential fragment of linear logic. In Muroya, Cheung, and Ghica (2017), this approach is complemented by a sound call-by-value lambda calculus, which has been inspired by Peircean abductive inference. The resulting bimodal programming model operates in both: (1) direct mode, with new inputs provided, new outputs obtained and (2) learning mode, with special inputs applied for which outputs are known; to achieve optimal tuning of parameters to ensure desired outputs approach actual outputs. They contend that their holistic approach is superior to that of TensorFlow, which they assert to be a shallow embedding of a domain specific language into "PYTHON" rather than a stand-alone programming language.

Adopting a somewhat different approach, Cruttwell and MacAdam (2019) extend Plotkin's differential programming framework, which is itself a generalization of differential neural computers, where arbitrary programs with control structures encode smooth functions also represented as programs. Within this generalized domain the derivative can be directly applied to programs or to algorithmic steps and, furthermore, can be rendered entirely congruent with categorical approaches to Riemannian and Differential geometry such as Lawvere's Synthetic Differential Geometry. Crucially, within this programming context, they develop a rigorous diagrammatic language for both derivative and integral operators.

The authors go on to observe that, when working in a simple neural network, back-propagation takes the derivative of the error function and then uses the chain rule to push errors backward. They insist that, for convolution neural networks, the necessary procedure is less straightforward due to the presence of looping constructs.

In this context the authors further note that attempts to work with the usual "if-then-else" and "while" commands can also be problematic. To overcome all of these problems associated with recursion, they deploy what have been called "join restriction tangent categories," which express the requisite domain of definition and detect and achieve disjointness of domains, while expressing iteration using the join of disjoint domains (i.e., in technical terms, this is the trace of a coproduct in the idempotent splitting). The final mathematical construct they arrive at is that of a differential join restriction category along with the associated join restriction functor, which admits a coherent interpretation of differential programming.

It should be stressed that each of these category-theoretic initiatives to formalize the differential of an algorithmic step will become important in future efforts to develop improvements to diagrammatically based forms of software that achieve greater capacity and efficiency than existing software suites. The fact that both differential and integral categories have a coherent string diagram formalism (Lemay, 2017) provides a link back to the earlier discussion about DR in semantic technologies.

It is clear that techniques of this kind would also carry over to a variety of NM domains (e.g., for the centralized and decentralized control of hybrid cyber-physical systems), where optimization routines are required.

12.5 CONCLUSION

In conclusion the innovations in software engineering described previously have obvious implications for those attempting to develop new semantic technologies for the effective management of emergency services and search-and-rescue operations in the aftermath of a major disaster. Hopefully, the material surveyed in this chapter should serve to highlight the advantages of a category-theoretic approach to the issue at hand, along with the specific benefits of adopting an approach that is grounded in the pedagogical and formal representational power of string diagrams, especially within a networked computational environment that will increasingly be characterized by Big Data, parallel processing, hybridity, and some degree of decentralized control.

REFERENCES

Atterson, Evan (2017). Knowledge Representation in Bicategories of Relations. ArXiv. 1706.00526v1 [cs.AI] 2 Jun 2017.

Baez, J. (2006). *Course notes on universal algebra and diagrammatic reasoning.* Available from <http://math.ucr.edu/home/baez/universal/>.

Baez, J. C., & Biamonte, J. D. (2012). *A course on quantum techniques for stochastic mechanics. arXiv:1209.3632v1 [quant-ph].* September 17, 2012.

Baez, J., Foley, J., Moeller, J., & Pollard, B. (2020). *Network models. arXiv:1711.00037v1 [math.CT].* March 27, 2020.

Baez, J. C., & Fong, B. (2014). *Quantum techniques for studying equilibrium in reaction networks.*

Baez, J. C., & Lauda, A. (2009). *A prehistory of n-categorical physics.* <https://arxiv.org/abs/0908.2469> Accessed 05.02.18.

Baez, J. C., & Stay, M. (2008). *Physics, Topology, Logic and Computation: A Rosetta Stone. New Structures for Physics, ed. Bob Coecke, Lecture Notes in Physics vol. 813, Springer, Berlin, 95-174. Table 4.* John Baez's website. <http://math.ucr.edu/home/baez/rosetta.pdf> Accessed 22.02.09.

Bell, J. T. (1998). *A primer of infinitesimal analysis.* Cambridge: Cambridge University Press.

Biamonte, J., & Bergholm, V. (2017). *Quantum tensor networks in a nutshell. arXiv:1708.00006v1 [quant-ph].* July 31, 2017.

Bonchi, F., Sobociński, P., & Zanasi, F. (2014). A categorical semantics of signal flow graphs. *CONCUR 2014, Ens de Lyon.*

Bonchi, F., Sobociński, P., & Zanasi, F. (2015). Full abstraction for signal flow graphs. In: *Principles of programming languages, POPL'15, 2015.*

Cruttwell, G., & MacAdam, B. (2019). Towards formulating and extending differential programming using tangent categories. In: *Extended abstract, ACT 2019.* <http://www.cs.ox.ac.uk/ACT2019/preproceedings/Jonathan%20Gallagher,%20Geoff%20Cruttwell%20and%20Ben%20MacAdam.pdf>.

Ellerman, David (2000). *Towards an arbitrage interpretation of optimization theory* (accessed 1/7/20), http://www.ellerman.org/Davids-Stuff/Maths/Math.htm.

Feynman, R. P. (1948). Space-time approach to nonrelativistic quantum mechanics. *Review of Modern Physics, 20,* 367.

Fong, B., & Spivak, D. I. (2018). *Seven sketches in compositionality: An invitation to applied category theory.* Available from <http://math.mit.edu/~dspivak/teaching/sp18/7Sketches.pdf>.

Forgasz, Helen, & Ferdinand Rivera (Eds.) (2012). *Towards equity in mathematics education: Gender, culture, and diversity.* Advances in Mathematics Education Series. Dordrecht, Heidelburg: Springer.

Herbel-Eisenmann, B., Choppin, J., Wagner, D., & Pimm, D. (Eds.), (2012). *Equity in discourse for mathematics education theories, practices, and policies. mathematics education library* (Vol. 55). Dordrecht, Heidelberg: Springer.

Hoffman, M. H. G. (2011). Cognitive conditions of diagrammatic reasoning. *Semiotica, 186*(1/4), 189–212.

Joyal, A., & Street, R. (1988). *Planar diagrams and tensor algebra.* Unpublished manuscript, available from Ross Street's website: <http://maths.mq.edu.au/~street/>.

Joyal, A., & Street, R. (1991). The geometry of tensor calculus, I. *Advances in Mathematics, 88,* 55–112.

Kissinger, A., & Uijlen, S. (2017). *A categorical semantics for causal structure.* <https://arxiv.org/abs/1701.04732v3>.

Lemay, J.-S. P. (2017). *Integral categories and calculus categories.* Alberta: University of Calgary, PhD thesis.

Melliès, P.-A. (2018). *Categorical semantics of linear logic.* Accessible from <https://www.irif.fr/~mellies/mpri/mpri-ens/biblio/categorical-semantics-of-linear-logic.pdf>.

Milner, R. (2009). *The space and motion of communicating agents.* Cambridge University Press.

Muroya, K., & Ghica, D. (2017). *The dynamic geometry of interaction machine: A call-by-need graph rewriter. arXiv:1703.10027v1 [cs.PL].* March 29, 2017.

Muroya, Koko, Cheung, Steven, & Ghica, D. R. (2017). *Abductive functional programming, a semantic approach.* arXiv:1710.03984v1 [cs.PL] 11 Oct 2017.

Paulheim, H., & Probst, F. (2010). Application integration on the user interface level: An ontology-based approach. *Data and Knowledge Engineering, 69,* 1103–1116.

Patterson, Evan (2017). *Knowledge Representation in Bicategories of Relations.* ArXiv. 1706.00526v1 [cs.AI] 2 Jun 2017.

Penrose, R. (1969). Applications of negative dimensional tensors. In: Combinatorial mathematics and its applications: Proceedings of a Conference Held at *the Mathematical* Institute, Oxford: Academic Press, pp. 221–244.

Sáenz-Ludlow, A., & Kadunz, G. (2015). *Semiotics as a tool for learning mathematics.* Berlin: Springer.

Schultz, P., Spivak, D., Vasilakopoulou, C., & Wisnesky R. (2016). *Algebraic Databases.* arXiv:1602.03501v2 [math.CT] 15 Nov 2016.

Shin, S.-J. (1994). *The logical status of diagrams.* Cambridge: Cambridge University Press.

Sobociński, P. *Blog on graphical linear algebra blog.* <http://graphicallinearalgebra.net/>.

Sowa, John F. (2000). *Knowledge Representation: Logical, Philosophical, and Computational Foundations.* Pacific Grove, CA: Brooks Cole Publishing.

Spivak, David I., Christina Vasilakopoulou, & Patrick Schultz (2019). *Dynamical Systems and Sheaves.* arXiv:1609.08086v4 [math.CT] 15 Mar 2019.

Statebox Team, (2018) University of Oxford. Statebox. <https://statebox.org/>.

Stjernfelt, F. (2007). *Diagrammatology: An investigation on the borderlines of phenomenology, ontology, and semiotics, synthese library* (Vol. 336). Dordrecht, The Netherlands: Springer.

Westrich, Q. (2006). Lie Algebras in Braided Monoidal Categories. Thesis, Karlstads Universitet, Karlstad, Sweden. http://www.diva-portal.org/smash/get/diva2:6050/FULLTEXT01.pdf

Willems, J. C. (2007). The behavioral approach to open and interconnected systems: Modeling by tearing, zooming, and linking. *Control Systems Magazine, 27*(46), 99.

METHODS AND INSIGHTS ON HOW TO EXPLORE HUMAN BEHAVIOR IN THE DISASTER ENVIRONMENT

13

David A. Savage[1] and Benno Torgler[2]

[1]*Newcastle Business School, University of Newcastle, Newcastle, NSW, Australia* [2]*School of Economics and Finance, Queensland University of Technology, Brisbane, QLD, Australia*

> Disaster planning is only as good as the assumptions it is based upon. Unfortunately this planning is often based upon a set of conventional beliefs that have been shown to be inaccurate or untrue when subjected to empirical analysis [. . .] It is more efficient to learn what people tend to do naturally in disasters and plan around that, rather than design your plan and then expect people to conform to it.
>
> **van der Heide (2004, p. 340)**

> This little volume on Halifax is offered as a beginning; don't let it be the end. Knowledge will grow scientific only after the most faithful examination of many catastrophes
>
> **Prince (1920, p. 23)**

13.1 INTRODUCTION

Disasters, by their very definition, are "... a sudden, calamitous event that seriously disrupts the functioning of a community or society and causes human, material, and economic or environmental losses that exceed the community's or society's ability to cope using its own resources" (International Federation of Red Cross, 2019).[1] Whether man-made or natural, disasters represent a danger to life and are a real and ever-present threat to individuals, the destruction of property, cultures, and society in general. There are two fundamental reasons why we should research disasters from a behavioral perspective. First, decisions made in this extreme environment can significantly affect the probability of survival, not only for an individual but also for family, friends, or even impact on other members of that society. Second, due to scarcity constraints, resource allocation is often a problem, which unfortunately extends to the provision of humanitarian aid and disaster

---

[1]Accessed from https://www.ifrc.org/en/what-we-do/disaster-management/about-disasters/what-is-a-disaster/

Economic Effects of Natural Disasters. DOI: https://doi.org/10.1016/B978-0-12-817465-4.00013-3

support, as there are too many in need with too few resources to go around. In order to maximize our limited resources, we require a better understanding of behavior and extreme environments so as to better predict the resource needs and better facilitate their deployment. This requires us to comprehend the actions, behavior, and decisions of individuals within extreme environments, when choices are neither minor nor simple. In fact, these decisions are likely to be some of the most difficult, costly choices an individual could be forced to make in their lifetime. Understanding the reasons/rationale behind these decisions requires innovative analysis from data that is difficult to obtain, as individual-level data from these events are not commonly available (or may in fact not exist). We need to understand why people choose to stay or leave, what factors influence this decision process, and when they choose to do so. If people choose to flee a conflict or disaster zone (Savage, 2016), do they have a specific demographic identity; are they somehow different from those who chose to stay? Where do these people go, who do they take or leave behind, in what direction do they flee and is their behavior (ir)rational or in some way predictable?

An economist's traditional interest in disasters has been to calculate the economic effects and costs associated with the event, from the short to long run (see, e.g., Albala-Bertrand, 1993; Dacy & Kunreuther, 1969; De Alessi, 1975; Gross, 1996; Hirshleifer, 1963; Kunreuther & Pauly, 2005; Sorkin, 1982), with particular attention being paid to insurance (Kunreuther, 1996; Kunreuther & Roth, 1998). On those occasions where economists have discussed human behavior, it was in reference to the homo economicus model which assumes that individuals are rational, utility maximizing, and self-interested. However, it has become increasingly clear that this does not explain the observed behavior in many disasters or life-and-death events, forcing a review of behavioral assumptions and models. Other fields such as psychology and sociology have focused on the behavior of individuals involved in the disaster itself, such as collective action problems or as a mass panic (see, e.g., Aguirre, Wenger, & Vigo, 1998; Elster, 1985; Fehr, Fischbacher, & Gachter, 2002; Johnson, 1988; Kelley, Condry, Dahlke, & Hill, 1965; Mawson, 1980; Quarantelli, 2001; Smelser, 1963). However, both these fields have generally relied on analytic tools that suffer from well-known problems such as biased data or small sample sizes, which lack predictive power.

Attempts to analyze human behavior in extreme life-or-death or disaster environments are inherently problematic, as generally the analyses are done ex post (as opposed to ex ante or during the event). This means the researcher does not directly observe the event and cannot formulate questions or model the event beforehand, nor can they reduce perception or recall biases with data and information happening during the event. However, a full and faithful examination of catastrophes (disasters) needs to begin with an understanding not only of *how* people behave, but also *why* they behave in a particular way. Exploring the "why" has long presented a significant challenge in economics and other social science disciplines (Wilson, 2017). Trying to understand *why* helps to provide contextual insights into both the ultimate and proximate causes of behavior, but also requires a new kind of methodological knowledge. A promising way is to apply counterfactuals: "My headache is gone now, but why? Was it the aspirin I took? The food I ate? The good news I heard? These queries take us to . . . the level of counterfactuals, because to answer them we must go back in time, change history, and ask, 'What would have happened if I had not taken the aspirin?'" (Pearl & Mackenzie, 2018, p. 33). The inability to directly observe the event often results in a loss of information which then requires rigorous analysis to estimate what has been lost or is missing. Postdisaster surveys (and interviews) are one of the most common tools for understanding behavior of those caught up in an event. However, this methodology has a significant bias, that is, interviewing individuals' postevent (by its very nature) creates an overwhelming survivor bias, and therefore individuals who perished during the event cannot be represented in the sample. This means that the results are internally valid for that specific group of survivors but have no external validity (i.e., the results cannot be generalized to the broader population or to any other group other than the sample). The lack of rigor and the absence of control groups have fatally flawed much of this research, weakening the otherwise excellent theoretical contributions made by the broader noneconomic social sciences.

Therefore we require additional research methodologies if we want to gain insights into the way in which humans behave in disasters—one that does not rely on the questionable recollections of a subgroup of the population who are under extreme stress, but one built on the behavior or revealed preferences of the whole. This chapter explores and presents a number of methods that can and have been used to understand individuals' behavior and preferences within a disaster. This includes comparative and narrative analysis, experiments (field, laboratory and natural), individual panel analysis, simulations, and the use of technology such as apps, sensing systems, and virtual reality (VR).

13.2 EXPERIMENTATION

Studying the behavior of individuals during such calamitous events is extremely difficult, but experimentation offers one way of exploring and testing models of human behavior, as it is possible to create simulations and test hypothesis under controllable conditions. We know that humans are capable of a wide range of behavior in response to these events, from fairness, altruism, helping behavior, or social and moral norms (see, e.g., Amato, 1990; Andreoni & Miller, 2002; Batson et al., 1979; Becker, 1974; Bolton & Ockenfels, 2000; Dufwenberg & Kirchsteiger, 2004; Eagly & Crowley, 1986; Fehr & Schmidt, 1999; Frey, 1997; Harrell, 1994; Rabin, 1993; Sobel, 2005). While the emerging literature suggests that altruistic and prosocial responses predominate in disaster environments (Frey, Savage, & Torgler, 2010a, 2010b, 2011; Quarantelli, 2001), research has demonstrated that it is also important to understand strategic incentives and whether self-regarding or other-regarding preferences dominate (Camerer & Fehr, 2006). However, in order to understand the motivations for action and behavior in extreme or life-and-death environments, we need to be able cleanly elicit and measure if or when individuals deviate from the traditional homo economicus model of rational self-interest. This has led to economists using three types of experiments to explore economic behavior and preferences, they are *laboratory*, *natural*, and *field* experiments.

13.2.1 LABORATORY EXPERIMENTS

Laboratory experiments, as the name implies, are those which occur within the confines of a designated experimental or laboratory space—quite often a bespoke computer room. These laboratory experiments were initially adopted to test economic theory for the same reasons' experiments are used in the physical sciences: the ability to control confounding factors and manipulate single variables to determine their effect. Because these environments are highly controllable and allowed the researcher the ability to isolate every variable and manipulate the one of interest, the laboratory became the workhorse for testing behavioral economic theories (List & Cherry, 2008), as its design construct limits that allows the scientist to engage with a small piece of the world (Friedman &

Cassar, 2004, p. 12). It allows to closely monitor the environment and the institutional conditions through initial endowments, monetary rewards, language used, and defined rules (Smith, 1994). Negative shocks in a lab experiment in the area of disaster research can be modeled via increasing the show-up fees (initial endowment)[2] so that individuals do not end with a financial loss which would otherwise be problem from an ethics perspective. However, lab experiments have problems in exploring emotional and psychological aspects related to disasters. In addition, using experimental methods in this way has allowed researchers to test theoretical models or generate (substitute) data that is normally unavailable from the field, which is often the case for disasters (Abbink, 2012). The experimental method particularly allows exploration of cognitive aspects and discrimination between potential theories or understands the causes of a theory failure. This allows to compare between environments, as well as between formal and information institutions, and therefore use the laboratory as a testing ground for institutional, environmental, and policy design strategies (Smith, 1994). Laboratory experiments can help to understand the implications of extreme parameters which are important for disaster research (e.g., extreme environmental conditions). Furthermore, the dynamics in an experiment can identify whether some properties (e.g., social norms) may begin to break down.

However, unlike the physical sciences, our particles (aka humans) are heterogeneous in nature and can alter their choices or behaviors dependent upon many (sometimes unobservable) factors. This is very different from the physical sciences where the aggregate laws of nature are more absolute, such that an identical force applied to the same object will always react in the same manner, that is, they will always behave in the same way when exposed to the same stimuli. Therefore it is important to replicate experiments to check the robustness of the results, although it should be noted that few replications in social science are able to exactly replicate former conditions due to many contextual aspects that cannot be controlled for (e.g., different subject pool). However, external validity can be enhanced if the experiment simulates cognitive processes that are relevant in the real world. Laboratory experiments focused on understanding individual behavior in disaster research would also benefit from enlisting a broader population set; that is, recruiting beyond students.[3] Researchers who apply experiments in the area of disaster research will also need to be more inventive in the way they design their experiments. They cannot rely on common and widely used experiments. This induces the Duhem−Quine problem that each test of a theory is a joint test of the theory tested and the way the experiment is conducted. As the design needs to be subjected to a test, it may attract scholars who are less risk averse. Smith (1994) points out that a "theory always swims in the rough water of anomaly. You don't abandon a theory because of a (or many) falsifying observation(s)" (p. 129). He refers to Einstein who emphasized that "Only after a diverse body of observations becomes available will it be possible to decide with confidence whether systematic deviations are due to a not yet recognized source of error or to the circumstances that the foundations of the theory [of relativity] do not correspond to the facts" (Einstein in Smith, 1994, p.

---

[2]However, this can also create a "house money" effect (Thaler & Johnson, 1990) where the prior gains mitigate the individual normal sense of loss aversion and facilitate risk-seeking, the term comes from gambling at casino's where if people gamble when they are ahead are said to be "gambling with the houses money."

[3]We must carefully consider the impact of research being done involving university students, as the vast majority of papers published in top behavioral journals are based upon samples from Western, Educated, Industrialized, Rich, and Democratic (WEIRD) societies (Henrich, Heine, & Norenzayan, 2010) and are likely not to be representative of the majority of the world populations.

129). Thus Smith concludes that the procedures under which a theory is tested are also required to be part of theory itself, a procedural aspect that researchers have strongly neglected so far.

Another issue with conducting experiments on humans is that they are self-aware and will be cognizant that in laboratory settings their actions are being observed, which may result in them deviating from their normal behavior or preferences. This is known as the Hawthorne[4] effect and it difficult to generalize results because the actions or choices made in the lab may not be an accurate or representative reflection of that individual's normal behavior or true preferences (Levitt & List, 2009). However, this instability of action/choice makes the testing of causal and treatment effects more complicated; therefore, economists have addressed this through the implementation of a range of instrumental variables and other econometric techniques. While one could argue that laboratory experiments are of limited use due to the conditions under which the experiments occur, that is, nonnatural and sterile environments which may evoke nonnatural responses from participants, it must be noted that the absolute inability of experimenters to access real disasters has made the laboratory environment very attractive. Furthermore, while it may not be possible to replicate or even approximate the levels of stress and danger to provide sufficient threat of death in test subjects it does allow for a highly controllable environment in which to explore low cost decisions. As such, it remains somewhat unclear how much of the insights gained in the laboratory can be extrapolated to the world beyond it (i.e., a lack of generalizability). When looking at average effects one also needs to be aware of further selection problems (e.g., those who select themselves into experiments may have a higher willingness to seek social approval or a higher risk tolerance to accept environmental or institutional changes). The relative differences in a between-treatment design between the control and treatment group may still hold, but care must be exercised in interpreting the overall average effects.

13.2.2 NATURAL EXPERIMENTS

Due to the limitations of laboratory experiments, economists search for naturally occurring events to exploit, such as policy changes, unforeseen market shocks or even the shifting of arbitrary boundaries that by chance, luck or serendipity, have provided a control/treatment effect with almost perfect randomness (Rosenzweig & Wolpin, 2000, p. 828). In these natural situations, individuals react to the changes in a way that reveals their true preferences, as such changes occur within the normal environment and "participants" are unaware that their actions (choices) may be observed or analyzed. These events are "real world" meaning that individuals will incur real costs and are motivated by real incentives, and individuals in disaster situations find themselves competing in a high stakes contest (life and death) in a specific controlled environment. This naturalistic reaction and revelation of true preferences offer "a unique inside view that will not only provide a glimpse into the decision-making black box but permit a deeper empirical exploration into problems that excite economists, practitioners, and policymakers" (Levitt & List, 2009: p. 2). It is this realism that provides researchers with an advantage over laboratory, self-reporting, and other forms of experiments while maintaining the randomness of natural data (Reiley & List, 2007). While it may appear to be counter intuitive, disasters can be thought of as a near perfect natural experiment because the

---

[4]The so-named Hawthorn effect comes from the location of the research a plant of the Western Electric Company, where the findings suggested that workers increased output in response to the knowledge of being observed.

events are relatively bounded (controlled) and all the participants encounter the same environmental variables and shocks. In this way, studying disasters is very much like utilizing sports data, as they "take place in a controlled environment and generate outcomes that come very close to holding 'other things equal' ... supply real-world laboratories for testing economic theories. The data supplied in these labs have some advantages over the data normally used in economic research ... the economist can perform controlled experiments similar to those performed by the physical and life scientists" (Goff & Tollison, 1990, pp. 6–7). This approach can be observed in the studies by Frey et al. (2010a, 2010b), who used the sinking of the Titanic and Lusitania as naturally occurring field experiments to investigate the decision process and determinants of survival. Individuals were not able to abstain from or remove themselves from the event (experiment), meaning that everyone must engage (unable to opt out) and those not willing to participate in the event will generally receive the worst possible outcome, in many cases this would result in death.

Of course, all economists know there is no such thing as a free lunch, as everything comes at a cost, and natural experiments are no different. The price that must be paid for randomness and a real environment is an absolute lack of control by the experimenter! This can result in aspects of the experiment being less than desirable: missing data, a number of unobservables, and the simultaneous shifts of multiple variables which invariably muddy the results. Natural experiments, such as disaster analysis, are done postmortem, as opposed to perimortem and unlike laboratory experimentation, disasters do not afford the researcher the ability to formulate questions or model the event beforehand. Given that it was not possible to observe the event directly, any analysis of this event must in a sense work backward, here the outcome is known and the analysis attempts to determine the factors that are most likely to be responsible for it. The resulting data is less than perfect, as it often contains missing observations or uncaptured information that is vital to a pertinent question, as it was not thought of at the time, or not considered relevant. As a result, economists have turned to complex or multivariate analysis to formulate a clearer picture about what occurred during the event with the postevent analysis.

13.2.3 FIELD EXPERIMENTS

There is a middle ground between laboratory and natural experiments, that is, field experiments, which has some of the randomnesses of a natural experiment but retains a much greater level of control by the experimenter (Levitt & List, 2009; Reiley & List, 2007). One way of using this approach is to enter the field with a prepared experiment, which maintains the reality of the environment but has the advantage of having additional control. This method can be observed in Page, Savage, and Torgler (2014), where the experimenters ran experiments in the weeks following a devastating widespread flooding that inundated the city of Brisbane (Australia) in 2011. This field experiment utilized the naturally occurring disaster to explore the changes in risk behavior of individual along the peak of the flood line, as the flood peak of the event was completely unknown prior to the event affected and unaffected households were randomized—combining the randomness of a natural event and the design of a controlled laboratory experiment. However, field experiments also suffer from their own issues: First, it is likely to be difficult if not dangerous for the researcher to be in the field during or immediately following a disaster event, but it will likely generate the most complete data, especially if it could be collected intertemporally. Second, it can be very expensive to get into the field to run experiments and collect data, especially considering that normal facilities and operations are likely to be disrupted and access would be limited. One possible solution is to undertake fieldwork after some period of time has elapsed, and while this is likely a much safer prospect, there is a risk that it will suffer from bias due to large numbers of missing individuals (deaths). A potential solution for the missing observations could be to generate a synthetic control group, that is, creating an artificial control group out of the population at hand (Abadie, Diamond, & Hainmueller, 2010; Chan, Frey, Gallus, & Torgler, 2014). However, if one wants to obtain true behavioral insights from individuals in an extreme environment, there is often no other choice but to go to the field.[5] While the survey approach can be fruitful, one must be aware of the limitations of using such an approach in the estimation of casualties in conflict zones (Johnson, Spagat, Gourley, Onnela, & Reinert, 2008; Spagat, 2012).[6]

There has been criticism of the use of field data in a multiple regression context in the place of nonrandom experimental data, where multiple regressions are not fully able to estimate (without noise) the effect on choices as it is impossible to measure all the variables that might conceivably affect it. This often leaves us with a dilemma, as on the one hand, we want the cleanest and best estimations possible to ensure our results are statistically significant, but on the other, we desire to capture the real world and how it influences our behavior. This contest of wills is captured by Amato (1990) and Allison (1999) in stating that "researchers who value the rigor of the laboratory have been reluctant to extend the study of prosocial behaviour to everyday life, where the possibility of control is minimal…" (Amato, 1990, p. 31) and "… No matter how many variables we include in a regression equation, someone can always come along and say, Yes, but you neglected to control for variable X and I feel certain that your results would have been different if you had done so" (Allison, 1999, p. 20). Unfortunately, the real world is difficult and messy, resisting all efforts to fit into simple "as−if" models, but and running experiments out there in the real world doubly so—as the propensity toward missing data or not perfectly capturing everything has created a long line of economists uninterested in exploring reality. However, recent years have witnessed an increase in the use of internet-based data collection for research, most notably in the experimental domain (see, e.g., Rubinstein, 2013). This approach is blurring the distinctions between laboratory and field experiments, as individuals can participate in the experiments from remote locations (e.g., home, university, work). Although this approach does cause concern with some of the experimental community on the grounds that participant conditions are not standardized, and they are not cash incentivized (something that has long been a standard of economic experiments). Yet, the positive side is that it may be possible to leverage the online delivery to facilitate the running of experiments in these dangerous regions.

The strengths and weaknesses of each of the experimental types are to some degree complementary and should be pursued together as a way of verifying the findings of the others. The noise and lack of control generated in natural experiments are compensated by their realism and randomization, but it remains virtually impossible to isolate single factors for causal impact. Thus an

---

[5]In order for a researcher to study or explore an obscure or elusive phenomenon they must first obtain data on it, as such one should go to where the data is (and not expect it to come to them). Armchair economists are able to marvelous things with models and econometrics, but secondary data is by its nature designed for something else entirely. Primary data is messy and often very difficult to collect, but the rewards are also great.
[6]For an extended discussion see Asher (2009, 2008), Biemer and Lyberg (2003), Brunborg, Tabeau, and Urdal (2006), Rosenblum and van der Laan (2009), and Spagat (2009).

experimental approach should not be adopted as a one size fits all or one-shot game, where we conduct only a single field, laboratory, or natural experiment. Furthermore, the findings of one should be verified by the results from another type, that is, a laboratory experiment should not be verified by another laboratory experiment, but with a field (or natural) experiment.[7] However, for the study of any extreme or high-stress situations, laboratory experiments lack the realism and the ability to generate the required factors (believable stress or life-threatening danger) but these shortcomings are outweighed by the ability to control just about every other variable and isolate the factor of interest. Field experiments fall in between these extreme situations and may become the master of none as a consequence; the field experiment is able to engage in a more natural environment with more rigorous control, but it does so in a less real way than natural with less control than laboratory experiments. However, when it comes to testing behavior in extreme environments or disasters, technology has provided a solution—specifically that of VR.

13.3 SIMULATIONS

Simulations are useful for understanding and predicting the behaviors of systems (Simon, 1996) and therefore are useful for analysis of disasters. They are also particularly interesting from a behavioral perspective, as the behavior of the large system is modeled in the small. They provide new knowledge to understand the implications of premises and when it is hard to understand what governs the behavior of the inner system which in particular a challenge in disasters due to its complexity. A simulation is an intuitive and explorative tool that can achieve an algorithmization of counterfactuals (via answering what—ifs, which is also a very human way of thinking about the world) (Pearl & Mackenzie, 2018).

A disaster is a shock event that is highly likely to impact individuals and negatively affect their decision-making processes across all aspects, possibly simultaneously. Disasters seriously disrupt the functioning of a community or society and cause human, material, and economic or environmental losses that exceed the community's or society's ability to cope using its own resources. Social scientists can benefit from the use of agent-based models (ABM) (Epstein, 2006). The strength of such ABM is their ability to visualize system dynamics that are otherwise difficult to solve mathematically (Reilsback & Grimm, 2012). In particular, one can use a microspecification to explore the evolution of norms and conflict dynamics in a disaster based on predefined (local) communication channels or rules (bottom-up approach) that control for the contextual factors while still observing the macrostructure (system implication of a disaster) and the time required for a disaster environment to cope and recover. Such a system can be best modeled with an out-of-equilibrium dynamics that requires transient behaviors (Epstein, 2006). It allows realistic assumptions by assuming that decision-making is achieved from an internal state of knowledge, while recognizing the agent's limitations in gathering complete knowledge and as an individual. In making decisions, individuals may act as satisficers or boundedly rational agents; that is, they sometimes fail to calculate the best course of action when cost function complexity or other environmental

---

[7]This is not to suggest that laboratory experiments should not be validated by numerous replication studies, but the results should be compared to those gathered outside the lab to ensure that what is being captured is not an artifact of the laboratory itself.

constraints are large (Simon, 1962), and instead satisfice with "good enough" rather than "optimal" decisions (Simon, 1956). For example, a higher level of complexity can lead to increased reliance on simple rules of thumb (Stadelmann & Torgler, 2013), allowing to model information selection in an adaptive environment. One can also emphasize communication transmission as an underlying structure among individuals and groups that act as virtual entities with forms of networked intelligence comparable to electronic brains (Morgan, 2006). The disaster event and its recovery have an organic structure that proceeds as an irreversible force in line with growth principles; it may develop in a manner similar to complex organizations and organisms (Boulding, 1961). ABM have successfully explored challenging topics such as the rise and fall of nations. For example, Axtell et al. (2002) explored the growth and collapse of the Kayenta Anasazi in Long House Valley.

In addition, experimental data can help calibrate the ABM designed to identify interaction dynamics. The pure laboratory settings can be expanded adding exploration of physiological data [e.g., heart rate variability (HRV), see Dulleck, Ristl, Schaffner, & Torgler, 2011; Dulleck, Schaffner, & Torgler, 2014; Dulleck et al., 2016] to identify emotional and cognitive factors, as well as moving to field experiments that observe participants in their natural environments in order to improve assessment not only of how individuals spend their time but also of how they experience various activities (Kahneman, Krueger, Schkade, & Stone, 2004). Such information will provide important insights into agents' emotional states and foci of attention in extreme situation, aspects that are crucial to understanding human problem solving (Simon, 1983).

13.4 **VIRTUAL REALITY**

In order to elicit a realistic response from experimental participants, the laboratory environment would need to (re)create a similar and believable extreme high-stress environment. Not only is this extremely difficult, but to intentionally put participants in simulated or realistic experimental conditions that could result in physical or psychological harm raises the specter of infamous research from the past. For example, Zimbardo and Cross (1971) recreated a prison environment into which they assigned experimental subjects (students) as either prisoners or guards, which caused extreme distress for the prisoners when brutalized by the guards. In his PhD thesis, Howard (1966) threatened to electrocute his female student participants if they failed to successfully navigate his experiment. The ground-breaking prison and electric shock experiments run by Milgram (1963) and Milgram and van Gasteren (1995) motivated the formation of ethical guidelines in research. Attempts to run experiments in disaster zones or in the aftermath of such events raise questions about the ethics or morality of how we approach the participants in this field of study. While the immediacy of events undoubtedly provides the most authentic data, we must question the stress it could cause to individuals who have already undergone such extremely traumatic upheavals. However, this is as they say, where the rubber meets the road; science can only move forward if we push and ask the difficult questions, but we have a responsibility to our fellow man not to exacerbate their distress. Here VR will also be able to assist—while it may not be possible or ethical to actually subject participants to sufficient threat in order to provoke a sense of the necessary stress and danger in a traditional laboratory, VR (as its name implies) can create a virtual world or environment in which individuals are able to interact and explore. The most recent iterations of the technology have become highly immersive and interactive, such that it can replicate the setting and the sense of danger without the physical risk. This is where the VR environment is similar to the controlled laboratory—where researchers are able to strictly control the settings and level of interaction in order to manipulate the variable of interest—but like a natural experiment, it can have a high degree of realism and context of the natural world. As Harrison, Haruvy, and Rutström (2011) stress, "[t]he potential benefit to experimental and behavioral research of utilizing virtual reality is the cues provided are naturalistic, allowing respondents to get immersed in the task in ways that may not be possible using standard text and picture interactions" (p. 88). Naturalistic field cues are particularly important when exploring disasters. Studies have already started, for example, to apply virtual experiments involving forest fires, involving an important visualization and attendant element via illumination, charring, denuding, and smoke (Fiore, Harrison, Hughes, & Rutström, 2009). In addition, the VR environment is able to generate a more natural interaction with other participants. This makes it a much better environment in which to explore the complex behavioral interactions between individuals, groups, or the environment, which until now has been all but impossible in any controlled manner. Thus VR has the ability to generate the internal validity of lab experiments and the external validity of field or natural experiments (Fiore et al., 2009).

Another advantage of adopting VR to explore disaster behavior is the ability to run repeated experiments with large subject pools, which is all but impossible in normal disaster environments. By adopting VR, experimenters would be able to implement the scientific rigors of the laboratory and run an unlimited number of identical experiments with a large number of participants, providing the realistic but repeatable results that have thus far eluded disaster research. Just like the traditional laboratory experiments, the researchers will be able to subtly manipulate single factors or interactions in order to explore the questions at hand, and unlike real disaster events, participants could also be included in a number of these types of experiments, which will allow researchers to explore learning behavior and thus not be limited to one-shot event analysis. As Fiore et al. (2009) point out, "participants can experience long-run scenarios in a very short amount of time, and many counterfactual scenarios can be generated" (p. 84). Furthermore, unlike real-world disaster events, VR experiments would not suffer from participant attrition (dropouts), which is a major problem for natural research studies and has the potential to create a significant bias in the results. Where the laboratory was central in driving behavioral economics forward (see Camerer, Loewenstein, & Rabin, 2004; List & Levitt, 2007), VR will be vital in driving forward inclusive multidisciplinary behavioral research in more realistic environments.

13.5 COMPARATIVE AND NARRATIVE ANALYSIS

As we have discussed, earlier data is one of the major issues when it comes to analyzing disaster behavior, which can come from biased samples, missing or incomplete data, uncaptured variables, or loss of individual-level data through aggregation. In addition, when we are exploring the analysis of historical events, we can only create estimates based upon the available data, which are the outcomes of events and not the event themselves. Narrative analysis (Bates, Greif, Levi, Rosenthal, & Weingast, 1998) is a technique created as a method to work around these limitations by considering the impact of strategic interactions, beliefs, cultural features and social structures on behavior using, for example, a classical game theoretic (or a behavioral) approach (see, e.g., Greif, 1998; Greif & Laitin, 2004). It is an attempt to understand the mechanisms that generate behavior by piecing together the story where data limitations and holes are numerous, through the marriage of empirical and narrative data (such as historical documents, eye-witness accounts). Essentially, it "seek[s] to understand the actors' preferences, their perceptions, their evaluation of alternatives, the information they possess, the expectations they form, the strategies they adopt, and the constraints that limit their actions" (Bates et al., 1998), by "cut[ing] deeply into the specifics of a time and place, and to locate and trace the processes that generate the outcome of interest." The approach is not theory driven as most normal economics papers may be but rather is very context specific as it is a study of that particular event in time and place. The empirical analysis extends the work well beyond the simple narrative account of the event, which increases the validity of the study and thus complementing the narrative aspects. A narrative analysis follows Aristotle's dictum "the whole is greater than the sum of its parts," where together they contribute to a much fuller (richer) understanding of the event that either the data or the narrative could have alone. However, the shortcomings of a narrative analysis stem from its advantage as an analytical tool, the lack of observational data means that its analysis does not provide solid causal evidence that would be easier to achieve with a stronger empirical data set. The major benefit of an analytical narratives approach is that it allows the researcher to piece together a story that would have been otherwise impossible.

13.6 HIGH-QUALITY INDIVIDUAL-LEVEL PANEL DATA ANALYSIS

Much research in the social sciences has focused on highly aggregated levels to understand the after-effects of disasters. For example, several studies have evaluated the impact of natural disasters on macroeconomic indicators, primarily GDP or on regional-level data such as employment, crime, and mortality (Cavallo, Galiani, Noy, & Pantano, 2013; Fomby, Ikeda, & Loayza, 2013). Comparing outcomes before and after a disaster is also problematic due to selection effects. For example, people may move after a disaster, leaving a nonrandom selection of individuals (e.g., more disadvantaged or less flexible people). Thus there are substantial endogeneity concerns in the area of disaster research (Kousky, 2014). Empirical research on disasters in the area of economics has relied heavily on US disasters, studying factors such as employment, health, or education (Currie & Rossin-Slater, 2013; Groen & Polivka, 2008; Sacerdote, 2012; Sastry & Gregory, 2013; Zissimopoulos & Karoly, 2010). More evidence is required with data that go beyond the use of cross-sectional designs and convenience samples. As pointed out by Bonanno, Brewin, Kaniasty, and Greca (2010), "many of the studies are cross-sectional or based on retrospective assessments garnered from survivors well after the disaster had taken place... this sort of evidence is problematic because it confounds predictors with outcomes" (p. 14). Econometric techniques such as panel data modeling (including random effects, fixed effects, and duration analysis) and propensity score matching should be frequently used in disaster research. As discussed beforehand, propensity score matching (or a synthetic control group approach) allows formation of a comparison group that is as similar as possible to those from the disaster environment with respect to predisaster characteristics. For example, Abadie and Gardeazabal (2003) construct a synthetic control region that allowed estimating the economic costs of the terrorist conflict in Spain's Basque Country. In their model the counterfactuality is a weighted combination of other Spanish regions whose relevant economic variables are closest to those of the Basque Country before the onset of terrorist activity. Dynamic approaches offer a better opportunity to control for endogeneity.

To conduct a proper panel data analysis when exploring disasters, it is important to have large-scale representative samples that may even go beyond the currently available panel datasets such as the Household, Income and Labour Dynamics in Australia, the German Socio-Economic Panel, or the British Household Panel Survey. This would allow researchers to have enough observations of individuals who have been directly affected by a disaster event and a comparable group that has not been affected and/or has been indirectly affected (residing in a disaster without experiencing damage). Weather-related disaster such as bushfires, floods, extreme heat, dust storms, droughts, or cyclones is well-suited for such studies due to their frequency and regional distribution. Large-scale surveys also offer the opportunity to collect data on a large number of different factors related to health, well-being, stress, socioeconomic changes, or in general attitudes, emotions, and perceptions. In addition, various sociodemographic groups can be explored independently (e.g., kids and elderly people).

Panel data also allows us to understand whether the effects on various aspects are temporary and how quickly individuals return to homeostasis or adjust back to normality, and how heterogeneous the adjustments are (e.g., what drives resilience). Collecting data directly after a disaster, and also over time can also provide valuable insights. One could closely monitor which individuals suffer devastating losses and which individuals prosper after the disaster, thereby allowing an analysis of the determinants of postdisaster coping, adjusting, and adapting. It is important to gain insights into the dynamics using a wide range of outcomes. Collecting information on migration, volunteering and community engagement, job absenteeism, expectations, optimism, and vulnerability can provide insights into how cooperation and coordination evolve over time, as well as their duration and strength, which are not yet fully understood. Panel data allows for the monitoring of temporal development (timeline) of postdisaster stresses on individuals via the decision-making process, as ongoing stressors hamper recovery and exacerbate the likelihood of mental distress slowing recovery and resiliency. This contributes to a better understanding of the human costs of disasters not only for the coping and adaption of resilient or nonresilient individuals, but for community coordination and cooperation. For example, in large-scale or widespread events, there is a significant impact on social capital and support systems, which have been shown to act as buffers against both chronic and acute forms of stress (Aldrich, 2012).

Collaborations with government organizations may also help to deal with omitted variable or endogeneity biases. Scandinavian countries have shown us how to achieve a productive collaboration between government agencies and scholars to explore challenging questions via opening up large-scale administrative data including data on income, age, gender, marital status, occupation, or home-ownership status.

13.7 APPS AND PORTABLE SENSING SYSTEMS

Temporal analysis of individuals affected by a disaster over a period of several months can be facilitated with the development and utilization of a mobile technology application (app) tracking individuals who participated in the initial survey, that will communicate with and record ongoing data in real time after the event. Implementation of this technology can enable real time access to participants in the field and provide a steady stream of new knowledge and data, by being nonintrusive and nondisruptive to participants over the duration of the project. The greater frequency of data will enable a longer time frame than normally offered by a field experiment and will reduce any systematic differences in household characteristics between both groups of participants in the study. New technologies such as smartphone applications (app) allow the flexible and cost-effective administration of weekly and monthly questionnaires. In general, surveys provide the opportunity to collect several behavioral indicators for mental stress and well-being to gauge their impact on decision-making, including The Impact of Events Scale (IES) or the revised version (IES-R) (Weiss & Marmar, 1997); The Oxford Happiness Questionnaire (Hills & Argyle, 2002), The Patient Health Questionnaire incorporating the APA's DSM-IV depression criteria; and The Life Orientation Test-Revised (LOT-R) measure of optimism or pessimism (Carver, 2013).

In addition to this primary function, the app can serve as a feedback mechanism through which participants could pose questions and obtain answers from other users, thereby revealing their priorities and/or states of mind, as well as their interests and priorities in areas outside the direct line of research enquiry. Access to both of these data sources is vital to understanding the temporal nature of postdisaster stressors and their impact on emotional, mental, and physical well-being of individuals. The combination of these two data sources and the setting of a disaster circumvent many of the limitations experienced either in a laboratory or a nonexperimental field study. Less time-sensitive or structural information can be collected less frequently, which would include, for example, labor market participation and changes in migration attitudes. An advantage of the periodic data collection is the capture of coping mechanisms as they are being enacted, for example, community engagement or support roles may cause a significant decrease in levels of stress, which would be captured. For example, Savage and Torgler (2013) explored text messages to understand how individuals communicated throughout the day during the 9/11 attacks and found that positive and prosocial communications emerge first, followed by slower increasingly negative communication after the collapse of the second tower and shortly thereafter a strong uptake of religious sentiments began to emerge as individuals try to make sense of the event.

Technological advances in neuroscience, in particular—such as wearable, nonintrusive, and noninvasive instruments—have opened new possibilities for conducting research on disasters. Monitoring physiological processes through nonintrusive means via pocket-sized reality mining badges which are also termed "social fMRI" (Aharony, Pan, Ip, Khayal, & Pentland, 2011) is attractive for its potential to identify psychological or mental processes (Torgler, 2019) that are essential in extreme situations such as disasters, and that are otherwise hard to measure. The dense continuous data produces a "second by second picture" (Pentland, Lazer, Brewer, & Heibeck, 2009, p. 4) and therefore offers new ways of understanding human dynamics (Eagle & Pentland, 2006) and better equips researchers to deal with the messiness and challenges of exploring disasters. Smartphones or wearable electronic badges with integrated sensors allow (upon consent) tracking of the digital footprint of a large number of disaster-affected individuals over days, and even months (Torgler, 2019). These wearables can collect information on individual's habits, movements, conversation patterns, health status, and social network, as well as contextual and environmental factors (Eagle & Greene, 2014) and bio-data such as HRV, blood pressure, skin conductivity, and sleep patterns. Such data complements the use of surveys, helping to compensate for their inherent problems of reporting biases, memory errors, and sparsity of continuous data (Eagle & Pentland, 2006; Pentland et al., 2009) which can be quite dominant in disaster situations where subjects are affected by information and stimulation overload.

13.8 CONCLUSION

Traditional disaster research has relied on ex post surveys and interviews, but this methodology has an inherent survivor bias. Obviously only survivors can be interviewed, which excludes those who perished during the event and the resulting sample is not representative of the population. This means that any findings are not generalizable and are only valid for that particular group of interviewed survivors. However, the survey method is highly compatible with the use of experiments, as it can be simultaneously collected and applied as an extension to the experimental data set. For example, surveys can often include standard psychology questions to identify an individual's emotional or psychological state of mind.[8] However, like any data source, it is only as good as the methodology that was used to elicit or collect it, and poor methodology can result in issues like unrealistic findings, sampling biases, extrapolations from small populations, or interviewer incentive bias (Spagat, 2012). Hence, a greater range of methods are needed to understand human behavior in disaster situations that can handle the complex interdependencies, and in this book chapter, we have provided an overview of empirical methodological avenues for social scientists. As Buss (2012) emphasizes we need to work like a carpenter who uses a large set of instruments out of the toolbox: The carpenter gains flexibility not by having one "highly general tool" that can be used to cut, poke, saw, screw, twist, wrench, plane, balance, and hammer. Instead, the carpenter gains flexibility by having a large number of highly specific tools in the toolbox. These highly specific tools can then be used in many combinations that would not be possible with one highly "flexible" tool. Indeed, it is difficult to imagine what a "general" tool would even look like, since there is no such thing as a "general carpenter's problem" (p. 53). Unfortunately, many social scientists specialize on one single tool. As we know, for man with a hammer all problems may look like nails. Even if the used hammer is a brilliant tool, it is just one tool and still a hammer (Gintis, 2014). Thus scientists exploring disasters need to avoid using only a hammer as there are, as we have shown in this chapter, many great occupants in a social scientist's toolbox.

REFERENCES

Abadie, A., Diamond, A., & Hainmueller, J. (2010). Synthetic control methods for comparative case studies: Estimating the effect of California's tobacco control program. *Journal of the American Statistical Association, 105*(490), 493−505.

Abadie, A., & Gardeazabal, J. (2003). The economic costs of conflict: A case study of the Basque Country. *American Economic Review, 93*(1), 113−132.

[8]This can include the Big 5 personality (Barrick & Mount, 1991; Eysenck, 1991; Hough, 1992), locus of control (Lefcourt, 1991), or anxiety and depression (Bjelland, Dahl, Haug, & Neckelmann, 2002; Clark & Watson, 1991).

Abbink, K. (2012). Laboratory experiments on conflict. In M. R. Garfinkel, & S. Skaperdas (Eds.), *The Oxford handbook of the economics of peace and conflict* (pp. 532–555). Oxford University Press.

Aguirre, B. E., Wenger, D., & Vigo, G. (1998). A test of the emergent norm theory of collective behavior. *Sociological Forum, 13*(2), 301–319.

Aharony, N., Pan, W., Ip, C., Khayal, I., & Pentland, A. (2011). Social fMRI: Investigating and shaping social mechanisms in the real world. *Pervasive and Mobile Computing, 7*(6), 643–659.

Albala-Bertrand, J. (1993). *The political economy of large natural disasters: With special reference to developing countries.* Oxford: Clarandon Press.

Aldrich, D. P. (2012). *Building resilience: Social capital in post-disaster recovery.* Chicago, IL: University of Chicago Press.

Allison, P. D. (1999). *Multiple regression: a primer.* Thousand Oaks, CA: Pine Forge Press.

Amato, P. R. (1990). Personality and social network involvement as predictors of helping behavior in everyday life. *Social Psychology Quarterly, 53*(1), 31–43.

Andreoni, J., & Miller, J. H. (2002). Giving according to GARP: An experimental test of the consistency of preferences for altruism. *Econometrica, 70*(2), 737–753.

Asher, J. (2009). Developing and using surveys to estimate casualties' post-conflict: Developments for the developing world. In *International conference on recording and estimation of casualties.* Carnegie Mellon University and University of Pittsburgh.

Asher, J., Scheuren, F., & Banks, D. (2008). *Statistical methods for human rights.* Dordrecht: Springer.

Axtell, R. L., Epstein, J. M., Dean, J. S., Gumerman, G. J., Swedlund, A. C., Harburger, J., … Parker, M. (2002). Population growth and collapse in a multiagent model of the Kayenta Anasazi in Long House Valley. *Proceedings of the National Academy of Sciences of the United States of America, 99*(Suppl. 3), 7275–7279.

Barrick, M., & Mount, M. (1991). The big five personality dimensions and job performance: A meta-analysis. *Personnel Psychology, 44*, 1–26.

Bates, R. H., Greif, F. A., Levi, M., Rosenthal, J.-L., & Weingast, B. R. (1998). *Analytic narratives.* Princeton, NJ: Princeton University Press.

Batson, D., Pate, S., Lawless, H., Sparkman, P., Lambers, S., & Worman, B. (1979). Helping under conditions of common threat: Increased we-feeling or ensuring reciprocity. *Social Psychology Quarterly, 42*(4), 410–414.

Becker, G. S. (1974). A theory of social interaction. *Journal of the Political Economy, 41*(1), 54–73.

Biemer, P., & Lyberg, L. (2003). *Introduction to survey quality.* Hoboken, NJ: Wiley.

Bjelland, I., Dahl, A. A., Haug, T. T., & Neckelmann, D. (2002). The validity of the hospital anxiety and depression scale: An updated literature review. *Journal of Psychosomatic Research, 52*(2), 69–77.

Bolton, G. E., & Ockenfels, A. (2000). ERC: Theory of equity, reciprocity, and competition. *American Economic Review, 90*(1), 166–193.

Bonanno, G. A., Brewin, C. R., Kaniasty, K., & Greca, A. M. L. (2010). Weighing the costs of disaster: Consequences, risks, and resilience in individuals, families, and communities. *Psychological Science in the Public Interest, 11*(1), 1–49.

Boulding, K. E. (1961). *The image: Knowledge in life and society.* Ann Arbor, MI: University of Michigan Press.

Brunborg, H., Tabeau, E., & Urdal, H. (Eds.), (2006). *The demography of armed conflict.* Dordrecht: Springer.

Buss, D. (2012). *Evolutionary psychology: The new science of the mind.* Boston, MA: Pearson.

Camerer, C. F., & Fehr, E. (2006). When does economic man dominate social behavior? *Science, 311*(5757), 47–52.

Camerer, C. F., Loewenstein, G., & Rabin, M. (2004). *Advances in behavioral economics.* Princeton, NJ: Princeton University Press.

Cavallo, E., Galiani, S., Noy, I., & Pantano, J. (2013). Catastrophic natural disasters and economic growth. *Review of Economics and Statistics, 95*(5), 1549−1561.

Chan, H. F., Frey, B. S., Gallus, J., & Torgler, B. (2014). Academic honors and performance. *Labour Economics, 31*, 188−204.

Clark, L. A., & Watson, D. (1991). Tripartite model of anxiety and depression: Psychometric evidence and taxonomic implications. *Journal of Abnormal Psychology, 100*(3), 316−336.

Currie, J., & Rossin-Slater, M. (2013). Weathering the storm: Hurricanes and birth outcomes. *Journal of Health Economics, 32*(3), 487−503.

Dacy, D., & Kunreuther, H. (1969). *The economics of natural disasters: Implications for federal policy.* New York: Free Press.

De Alessi, L. (1975). Towards an analysis of post-disaster cooperation. *American Economic Review, 65*(1), 127−138.

Dulleck, U., Ristl, A., Schaffner, M., & Torgler, B. (2011). Heart rate variability, the autonomic nervous system, and neuroeconomic experiments. *Journal of Neuroscience, Psychology, and Economics, 4*(2), 117−124.

Dulleck, U., Schaffner, M., & Torgler, B. (2014). Heartbeat and economic decisions: Observing mental stress among proposers and responders in the ultimatum bargaining game. *PLoS One, 9*(9), e108218.

Dulleck, U., Fooken, J., Newton, C., Ristl, A., Schaffner, M., & Torgler, B. (2016). Tax compliance and psychic costs: Behavioral experimental evidence using a physiological marker. *Journal of Public Economics, 134*, 9−18.

Dufwenberg, M., & Kirchsteiger, G. (2004). A theory of sequential reciprocity. *Games and Economic Behavior, 47*(2), 268−298.

Eagly, A. H., & Crowley, M. (1986). Gender and helping behavior: A meta-analytic review of the social psychological literature. *Psychological Bulletin, 100*(3), 283−308.

Eagle, N., & Greene, K. (2014). *Reality mining: Using big data to engineer a better world.* Cambridge, MA: MIT Press.

Eagle, N., & Pentland, A. S. (2006). Reality mining: Sensing complex social systems. *Personal and Ubiquitous Computing, 10*(4), 255−268.

Elster, J. (1985). Rationality, morality, and collective action. *Ethics, 96*(1), 136−155.

Epstein, J. M. (2006). *Generative social science: Studies in agent-based computational models.* Princeton, NJ: Princeton University Press.

Eysenck, H. J. (1991). Dimensions of personality: 16, 5 or 3?—Criteria for a taxonomic paradigm. *Personality and Individual Differences, 12*(8), 773−790.

Fehr, E., & Schmidt, K. M. (1999). A theory of fairness, competition, and cooperation. *Quarterly Journal of Economics, 114*(3), 817−868.

Fehr, E., Fischbacher, U., & Gachter, S. (2002). Strong reciprocity, human cooperation and the enforcement of social norms. *Human Nature, 13*(1), 1−25.

Fiore, S. M., Harrison, G. W., Hughes, C. E., & Rutström, E. E. (2009). Virtual experiments and environmental policy. *Journal of Environmental Economics and Management, 57*(1), 65−86.

Fomby, T., Ikeda, Y., & Loayza, N. V. (2013). The growth aftermath of natural disasters. *Journal of Applied Econometrics, 28*, 412−434.

Frey, B. S. (1997). *Not just for the money: An economic theory of personal motivation.* London: Cheltenham.

Frey, B. S., Savage, D. A., & Torgler, B. (2010a). Interaction of natural survival instincts and internalized social norms exploring the Titanic and Lusitania disasters. *Proceedings of the National Academy of Sciences of the United States of America, 107*(11), 4862−4865.

Frey, B. S., Savage, D. A., & Torgler, B. (2010b). Noblesse oblige? Determinants of survival in a life and death situation. *Journal of Economic Behavior and Organization, 74*(1), 1−11.

Frey, B. S., Savage, D. A., & Torgler, B. (2011). Behavior under extreme conditions: The Titanic disaster. *Journal of Economic Perspectives*, *25*(1), 209–222.

Friedman, D., & Cassar, A. (2004). *Economics lab: An intensive course in experimental economics*. Routledge.

Gintis, H. (2014). *The bounds of reason: Game theory and the unification of the behavioral sciences*. Princeton, NJ: Princeton University Press.

Goff, B., & Tollison, R. (1990). *Sportometrics*. College Station, TX: Texas A&M University Press.

Groen, J. A., & Polivka, A. E. (2008). The effect of Hurricane Katrina on the labor market outcomes of evacuees. *American Economic Review*, *98*(2), 43–48.

Greif, A. (1998). Historical and comparative institutional analysis. *The American Economic Review*, *88*(2), 80–84.

Greif, A., & Laitin, D. D. (2004). A theory of endogenous institutional change. *American Political Science Review*, *98*(4), 633–652.

Gross, R. (1996). *Psychology: The science of mind and behavior* (3rd ed.). London: Hodder and Stoughton.

Harrell, W. A. (1994). Effects of blind pedestrians on motorists. *Journal of Social Psychology*, *134*(4), 529–539.

Harrison, G. W., Haruvy, E., & Rutström, E. E. (2011). Remarks on virtual world and virtual reality experiments. *Southern Economic Journal*, *78*(1), 87–94.

Henrich, J., Heine, S. J., & Norenzayan, A. (2010). The weirdest people in the world? *Behavioral and Brain Sciences*, *33*(2/3), 1–75.

Hills, P., & Argyle, M. (2002). The oxford happiness questionnaire: A compact scale for the measurement of psychological well-being. *Personality and Individual Differences*, *33*(7), 1073–1082.

Hirshleifer, J. (1963). Disaster and recovery: A historical survey. In: *Technical report, Rand Corporation Memorandum*.

Howard, R. (1966). *Life and death decision analysis* (Ph.D. thesis). Stanford, CA: Stanford University.

Hough, L. (1992). The 'big five' personality variables — Construct confusion: Description versus prediction. *Human Performance*, *5*, 139–155.

International Red Cross (2020). What is a disaster? Retrieved from https://www.ifrc.org/en/what-we-do/disaster-management/about-disasters/what-is-a-disaster/.

Johnson, N. R. (1988). Fire in a crowded theater: A descriptive analysis of the emergence of panic. *International Journal of Mass Emergencies and Disaster*, *6*(1), 7–26.

Johnson, N., Spagat, M., Gourley, S., Onnela, J., & Reinert, G. (2008). Bias in epidemiological studies of conflict mortality. *Journal of Peace Research*, *45*(5), 653–663.

Kahneman, D., Krueger, A. B., Schkade, D. A., Schwarz, N., & Stone, A. A. (2004). A survey method for characterizing daily life experience: The day reconstruction method. *Science*, *306*, 1776–1780.

Kelley, H. H., Condry, J. C., Dahlke, A. E., & Hill, A. H. (1965). Collective behavior in a simulated and panic situation. *Journal of Experimental Social Psychology*, *1*(1), 20–56.

Kousky, C. (2014). Informing climate adaptation: A review of the economic costs of natural disasters. *Energy Economics*, *46*, 576–592.

Kunreuther, H. (1996). Mitigating disaster losses through insurance. *Journal of Risk and Uncertainty*, *12*(2), 171–187.

Kunreuther, H., & Pauly, M. (2005). Insurance decision making and market behavior. *Microeconomics*, *1*(2), 63–127.

Kunreuther, H., & Roth, R. (1998). *Paying the price: The status and the role of insurance against natural disaster in the United States*. Washington, DC: John Henry Press.

List, J. A., & Levitt, S. D. (2007). What do laboratory experiments measuring social preferences reveal about the real world? *The Journal of Economic Perspectives*, *21*(2), 153–174.

Lefcourt, H. M. (1991). Locus of control. In J. P. Robinson, P. R. Shaver, & L. S. Wrightsman (Eds.), *Measures of personality and social psychological attitudes* (Vol. 1, pp. 413−499). Academic Press, Chapter 9.

Levitt, S. D., & List, J. A. (2009). Field experiments in economics: The past, the present and the future. *European Economic Review, 53*(1), 1−18.

List, J. A., & Cherry, T. L. (2008). Examining the role of fairness in high stakes allocation decisions. *Journal of Economic Behavior & Organization, 65*(1), 1−8.

Mawson, A. R. (1980). Is the concept of panic useful for study purposes? In B. Levin (Ed.), *Behavior in fires*. Boston, MA: National Fire Protection Agency.

Milgram, S. (1963). Behavioral study of obedience. *The Journal of Abnormal and Social Psychology, 67*(4), 371.

Milgram, S., & van Gasteren, L. (1995). *Das Milgram—Experiment*. Hamburg: Rowohlt.

Morgan, G. (2006). *Imagines of organization*. Thousand Oaks, CA: Sage.

Page, L., Savage, D. A., & Torgler, B. (2014). Variation in risk seeking behaviour following large losses: A natural experiment. *European Economic Review, 71*, 121−131.

Pearl, J., & Mackenzie, D. (2018). *The book of why: The new science of cause and effect*. New York: Basic Books.

Pentland, A., Lazer, D., Brewer, D., & Heibeck, T. (2009). Using reality mining to improve public health and medicine. *Studies in Health Technology and Informatics, 149*, 93−102.

Prince, S.H. (1920). Catastrophe and social change. In *Studies in history, economics, and public law* (Vol. 94). London: P.S. King & Son Ltd.

Quarantelli, E. L. (2001). The sociology of panic. In S. Baltes (Ed.), *International encyclopedia of the social and behavioral sciences* (30th ed.). London: Pergamon Press.

Rabin, M. (1993). Incorporating fairness into game theory and economics. *The American Economic Review, 83* (5), 1281−1302.

Reiley, D. H., & List, J. A. (2007). Field experiments in economics. In S. N. Durlaf, & L. E. Blume (Eds.), *The new Palgrave dictionary of economics* (2nd ed.). London: Palgrave Macmillan.

Reilsback, S. F., & Grimm, V. (2012). *Agent-based and individual-based modeling*. Princeton, NJ: Princeton University Press.

Rosenblum, M., & van der Laan, M. (2009). Confidence intervals for the population mean tailored to small sample sizes, with applications to survey sampling. *International Journal of Biostatistics, 5*(1), 1−46.

Rosenzweig, M. R., & Wolpin, K. I. (2000). Natural "natural experiments" in economics. *Journal of Economic Literature, 38*(4), 827−874.

Rubinstein, A. (2013). Response time and decision making: An experimental study. *Judgment and Decision Making, 8*(5), 540−551.

Sacerdote, B. (2012). When the saints go marching out: Long-term outcomes for student evacuees from Hurricanes Katrina and Rita. *American Economic Journal: Applied Economics, 4*(1), 109−135.

Sastry, N., & Gregory, J. (2013). The effect of Hurricane Katrina on the prevalence of health impairments and disability among adults in New Orleans: Differences by age, race, and sex. *Social Science & Medicine, 80*, 121−129.

Savage, D. A., & Torgler, B. (2013). The emergence of emotions and religious sentiments during the September 11 disaster. *Motivation and Emotion, 37*(3), 586−599.

Savage, D. A. (2016). Surviving the storm: Behavioural economics in the conflict environment. *Peace Economics, Peace Science and Public Policy, 22*(2), 105−129.

Simon, H. A. (1956). Rational choice and the structure of the environment. *Psychological Review, 63*, 129−138.

Simon, H. A. (1962). The architecture of complexity. *Proceedings of the American Philosophical Society*, *106*, 467−482.

Simon, H. A. (1983). *Reason in human affairs*. Stanford, CA: Stanford University Press.

Simon, H. A. (1996). *The sciences of the artificial*. Cambridge, MA: MIT Press.

Smelser, N. (1963). *Theory of collective behavior*. New York: Free Press.

Smith, V. L. (1994). Economics in the laboratory. *Journal of Economic Perspectives*, *8*(1), 113−131.

Sobel, J. (2005). Interdependent preferences and reciprocity. *Journal of Economic Literature*, *43*(2), 392−436.

Sorkin, A. L. (1982). *Economic aspects of natural disasters*. Lexington, KY: Lexington Books.

Spagat, M. (2009). The reliability of cluster surveys of conflict mortality: Violent deaths and non-violent deaths. In *International conference on recording and estimation of casualties*. Carnegie Mellon University and University of Pittsburgh.

Spagat, M. (2012). Estimating the human costs of war: The sample survey approach. In M. R. Garfinkel, & S. Skaperdas (Eds.), *The Oxford handbook of the economics of peace and conflict* (pp. 318−340). Oxford: Oxford University Press.

Stadelmann, D., & Torgler, B. (2013). Bounded rationality and voting decisions over 160 tears: Voter behavior and increasing complexity in decision-making. *PLoS One*, *8*(12), e84078.

Torgler, B. (2019). Opportunities and challenges of portable biological, social, and behavioral sensing systems for the social sciences. In G. Foster (Ed.), *Biophysical measurement in experimental social science research* (pp. 197−224). Academic Press.

Thaler, R. H., & Johnson, E. J. (1990). Gambling with the house money and trying to break even: The effects of prior outcomes on risky choice. *Management Science*, *36*(6), 643−660.

van der Heide, E. A. (2004). Common misconceptions about disasters: Panic, the "disaster syndrome", and looting. In M. O'Leary (Ed.), *The first 72 hours: A community approach to disaster preparedness* (pp. 340−380). Lincoln, NE: iUniverse Publishing.

Wilson, E. O. (2017). *The origins of creativity*. New York: W. W. Norton & Company.

Weiss, D. S., & Marmar, C. R. (1997). The Impact of Event Scale-Revised. In J. P. Wilson, & T. M. Keane (Eds.), *Assessing psychological trauma and PTSD* (pp. 399−411). Guilford Press.

Zimbardo, P., & Cross, A. (1971). *Stanford prison experiment*. Stanford, CA: Stanford University.

Zissimopoulos, J., & Karoly, L. A. (2010). Employment and self-employment in the wake of Hurricane Katrina. *Demography*, *47*(2), 345−367.

STATE-WIDE EFFECTS OF NATURAL DISASTERS ON THE LABOR MARKET

14

Taha Chaiechi[1], Josephine Pryce[1], Susan Ciccotosto[1] and Lawal Billa[2]

[1]*College of Business, Law and Governance, James Cook University, Cairns, QLD, Australia*
[2]*Environment and Geographical Sciences, University Nottingham, Selangor, Malaysia*

14.1 INTRODUCTION

In an event of a natural disaster that truncates the availability of specific resources (such as raw materials and physical capital), infrastructure may become scarce, and many communication networks are likely to be disrupted. This could lead to a large segment of the population experiencing business disruption. However, it is expected that a well-functioning and -resilient economy improves individuals' (e.g., workers' and business owners') abilities to adapt somewhat successfully when facing significant changes subsequent to natural disasters. Nevertheless, there is a limit to which adaptation can be undertaken primarily in the presence of large-scale disasters, and the exogenous shocks are probable to have a substantial impact on different sectors of the economy, including labor market outcomes (such as the unemployment rate, average earnings, and participation rates). Furthermore, when one specific local labor market is affected by these exogenous shocks, it is likely that nearby labor markets are affected too, due to probable regional population and resource shifts in response to unexpected changes.

In some studies, it is a maintained argument that natural disasters could potentially have an adverse impact on regional labor supply and demand due to such factors as:

- the outflow of workers (Belasen & Polachek, 2007; RAI, 2013);
- population displacement (RAI, 2013);
- destruction of business properties, technology, and physical capital (Chaiechi, 2012); and
- degradation and loss of productive capacity (Adam, 2013).

Some researchers, however, argue that the labor market effects of natural disasters could potentially be positive since the reduced expected return on physical capital will have a substitution effect toward human capital as a replacement (Skidmore & Toya, 2002). In essence, it is argued that natural disasters can increase economic growth rates because as physical capital is damaged or destroyed, there may be a higher demand for labor after a natural disaster, causing relative human capital to increase (Skidmore & Toya, 2002). These conflicting views call for an in-depth assessment to answer key questions regarding links between labor market outcomes and natural disasters.

Economic Effects of Natural Disasters. DOI: https://doi.org/10.1016/B978-0-12-817465-4.00014-5

However, in examining this link there is a need to be mindful that communities, businesses, and economies do not experience disasters similarly.

People are employed by either private or public organizations. In first-world countries, such as Australia, government organizations (national, state, and local) are better able to withstand the financial impact of natural disasters; however, private organizations employ 83.7% of the working population. Because of this high employment rate, it is essential to consider how different private organizations react to the impact of natural disasters. Wilson, Brankicki, Sullivan-Taylor, and Wilson (2010) noted that different businesses in various sectors were affected in different ways by extreme events. Factors that were found to influence the reaction of the business included not only the sector (for instance, primary production, manufacturing, retail) and their location. For instance, Su, Saimy, and Bulgiba (2013) in their investigation of Malayan households found that, after the 2004 tsunami, those from households with higher incomes were better off after 2.2 years, while households, where the head of the family was unemployed, had a reduced likelihood of recovery. They go on to observe that households, which lost their means of support (in Su et al.'s case a fishing vessel), had less chance of recovering their previous status. In cases where a business is completely destroyed and there are little resources available for reconstruction efforts, there is likely to be an increase in unemployment as both the owner and the employees will be out of work.

Wilson et al. (2010) noted the different ways that businesses reacted to the possibility of a natural disaster in English organizations. Many organizations were reluctant to commit resources that could be used elsewhere in the business to make a profit. Managers were unable to gain sufficient information regarding the likelihood of different scenarios and could not then be analytically rational in making such decisions. The researchers went on to note that one of the reactions of many small-to-medium businesses was to "muddle through," as the businesses considered that it was not financially possible to take preemptive action for all possible scenarios. These organizations considered it appropriate to decide a course of action after the extreme event had taken place. The researchers went on to note an increased reliance on other agencies such as emergency services and local government organizations to guarantee a return to "business as usual." Others noted that disasters dealt with real people (Matilal & Höpfl, 2009). The full cost of cyclones is represented not only in financial damage to buildings and livelihoods, but there are also losses to natural, social, and human capital (Brown Gaddis, Miles, Morse, & Lewis, 2007).

In relation to human capital, research suggests that natural disasters affect labor markets in both the short run and long run, and in a variety of related ways, such as the impact on wages and migration (Adam, 2013; Skidmore & Toya, 2002). Adam (2013, p. 99) notes that there is a consensus that the economic damage inflicted by natural disasters is pronounced in relation to "forced migration" and "loss of output." He also comments on the widespread impacts on the global economy of natural disasters and provides the example of how the outfall of the Japanese earthquake and tsunami of March 2011 demolished manufacturing and industrial units, which then affected inventories in the production of electronics and vehicles in Europe and other parts of the world. This invariably affects local labor markets and those of regions, the industries of which rely on inventories from affected areas. Aside from this effect, one of "the most visible" aspects (as Adams, 2013, p. 100 says) of extreme natural events is the dislocation of people through "forced or involuntary migration."

Involuntary migration has short-run and long-run effects on productivity growth. In the short run there is a loss of income, but in the long run skills, knowledge, and experience accrued in the labor markets are eroded and possibly permanently lost (Adam, 2013). This is where macroeconomic policy, especially short-run policy, can mitigate against such shocks, which can markedly affect labor markets and subsequently, the long-run growth path of economies. Many authors argue that public policy dramatically determines the extent to which individuals and groups are affected by extreme natural disasters (e.g., Adam, 2013; Rodríguez-Oreggia, 2013). Macroeconomic policies tend to target the aggregate economy and subsequently the lives of more impoverished people as their livelihoods are most vulnerable and at greatest risk. Cutter (2011) discusses this point about Hurricane Katrina in New Orleans and points out that unfortunately, social inequalities relating to opportunities and income are aggravated by extreme environmental events, where involuntary immobility exacerbate the impacts of natural disasters.

It is inevitable that the short-run impacts of natural disasters cause negative disruptions to economic activity, mainly as a result of falls in aggregate output and employment (Adams, 2013). The extent of impacts on labor markets is evidenced by the extent and effects of the disruptions and the state of employment before disasters. These factors determine the degree of voluntary and forced migration out of regions and the level of in-migration to assist with recovery processes. For example, labor supply can be drastically reduced if the number of mortalities caused by a natural disaster is high; and combined with shrinking of labor demand because of destruction to businesses caused by the disaster, the displacement of labor within the local economy can be potentially significant.

Adam (2013) notes that labor markets shrink as individuals disengage from their normal jobs to migrate or as they focus on securing their own families and properties. He argues that these "(re-) construction booms" can affect economies by forcing wages and related costs to rise, especially where there is a demand for skilled labor. A study conducted in Indonesian households by Kirchberger (2017) showed that reconstruction demands cause local wages to increase and drain the tradable goods sector of skilled labor. The implications of this labor drain are that local labor market conditions are affected as the tradable sectors are undermined, and subsequently, the region's competitiveness in relation to output growth and exports is unduly affected. Black, Neil Adger, Arnell, Dercon, Geddes, & Thomas (2011) discuss the complexities of these "mass migrations" and highlight the "barriers" and "facilitators" of these movements of people. They observe that not all groups of people have the potential to migrate and some people are "trapped" by their capabilities in responding to environmental shocks.

In contrast to these studies, Skidmore and Toya (2002) contend that the use of an endogenous growth framework can show how local natural disasters may increase economic growth. They argue that as physical capital is damaged or destroyed, there is a greater demand for labor in the aftermath of natural disasters, principally to support recovery. In addition, the authors' reason that there is the possibility of increased investment is as a result of rebuilding processes that are linked explicitly to monies available through disaster management and associated with targeted requirements for the rebuilding of infrastructure to withstand future environmental shocks. Furthermore, the adoption of new technologies as capital is replaced can contribute to increased labor demands and enhance economic growth. In this way, while natural disasters could create a negative return on the physical capital, in the long run there is an increase in human capital, which can result in an increase in returns and lead to higher economic growth rates.

Three major natural disasters are prevalent in Australia that includes droughts, forest fires, and tropical cyclones. The extent and frequencies of events are secluded and vary from one region to another and do not generally affect the country as a whole owing to the vast extent of Australia.

Costs of damages to humans and animal lives and materials post disaster are generally difficult to quantify, though estimates are usually made to assess the economic impacts. Although droughts are without a doubt the most complicated natural disaster problem in Australia, they are gradual and prolonged because of the vast areas of deserts, and very low annual precipitation in the country. Impacts on materials and human population are minimal as droughts occur in regions with a limited population. Tropical cyclones on the other hand are quick to occur and may have double jeopardy to their costs on material damage as wind impacts are sometimes followed by floods. Cyclones in Australia are frequent, mostly affecting coastal areas in the northeast and northwest area of the Gulf of Carpentaria (Figs. 14.1 and 14.3). Fortunately, they do not occur too often in very populated areas (Australia Government, Bureau of Meteorology); however, the cost to material damage can be high (Table 14.1). North Queensland is particularly vulnerable to tropical cyclones due to the November to April summer warming of the Coral Sea and parts of the South Pacific Ocean.

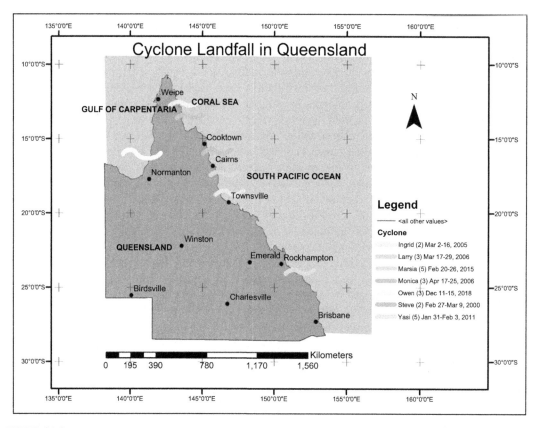

FIGURE 14.1

Selected tropical cyclone landfalls in Queensland, Australia.

Authors own using data from Australia Government, Bureau of Meteorology.

Table 14.1 Damage Costs of Severe Tropical Cyclones in Queensland 2005–17.

Season	Strongest Storms (Name)	Cost of Damages
2004–05	Ingrid	> $14.4 million (2005 USD)
2005–06	Larry	$1.1 billion (2006 USD)
2010–11	Yasi	$3.56 billion (2011 USD) (costliest cyclone in Australian region)
2014–15	Marcia	$587 million (2015 USD)
2018–19	Owen	> $25 million

Data obtained from Global Tropical Cyclones Summaries and Operational Track Data. <http://www.australiasevereweather. com/cyclones/tropical_cyclone_summaries_track_data.htm>.

Fig. 14.1 reports major cyclone events that caused significant damage between years 2000 and 2018 recorded by the Bureau of Meteorology have been mapped to visually present the main areas of cyclone landfalls in Australia.

14.2 RESEARCH GAP

Research into socioeconomic impacts of natural disasters has emerged back in the 1980s, exploring how the predisaster socioeconomic conditions of a country influences the response to the disaster when it occurs (see Hallegatte & Ghil, 2008; Tang, Wu, & Ye, 2019 for full reviews). However, the consequence of natural disasters on the labor market is still largely unknown. Rodríguez-Oreggia (2013) acknowledges that there is limited evidence on the impact of natural disasters on labor markets. Labor market analysis in this chapter aims to extend this knowledge and does so by complementing the productivity analysis and shock scenarios discussed in Chaiechi (2014), as it helps to provide a more holistic view of potentially vulnerable economic sectors. Moreover, this chapter examines data from Queensland and so is believed to be the first one to directly examine the influences of natural disasters on the labor market outside the United States (i.e., the most commonly studied region in disaster studies).

To achieve the research objective outlined previously, we use time series economic and disaster data for the state of Queensland and adopt intervention analysis [i.e., impulse response functions (IRFs)] in a vector autoregressive (VAR) framework to explore the links through which natural disasters affect the regional labor market.

14.3 NATURAL DISASTER PROFILE OF THE STUDY AREA

Queensland is the second largest state positioned in the northeast of Australia, and because of its size, its climate considerably varies across the state (Fig. 14.2).

Queensland is traversed by the Tropic of Capricorn that divides the states into tropical and subtropical regions. This particular geographic location exposes the state to tropical climatic weather events. As shown in Fig. 14.3, Queensland is prone to tropical cyclones. Major and sever tropical

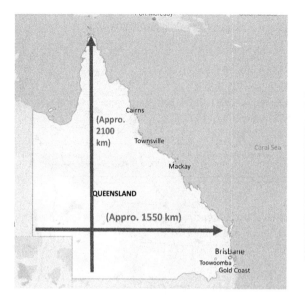

FIGURE 14.2

Geography and demographics of Queensland.

Map data © 2020 Google, Facts data sourced from https://www.qld.gov.au/.

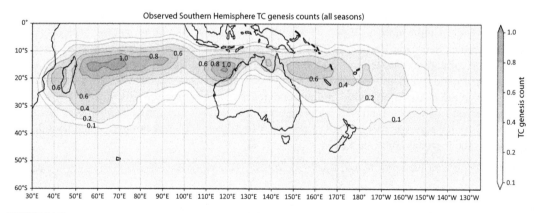

FIGURE 14.3

Average annual number of tropical cyclones through the Australian region in El Niño, La Niña and neutral years (48-year period from the 1969/70 to 2017/18 tropical cyclone season).

Retrieved on December 18, 2019, from Bureau of Meteorology (BoM, 2008)—Attribution 3.0 Australia (CC BY 3.0 AU).

cyclones typically are harmful to the state economy, businesses, and communities. The cyclone season runs between November and April each year. However, the exact timing and path of cyclones cannot be determined in advance. According to Climate Statistics for Australian Sites by Bureau of Meteorology (BoM) (2011), there have been 207 recorded impacts (including erosion, flooding, property damage, and loss of life) from cyclones rampaging the east coast of Queensland between 1858 and 2010. This indicates that over the course of an "average" year the state witnessed at least one tropical cyclone during monsoon season, many of which have led to emergency scenarios in the region; however, there are years when the state has not been hit by any major tropical cyclones.

After tropical cyclones, flood is the second largest disaster to affect the state of Queensland seriously. Historically, riverine flooding (including floods caused by decaying cyclones) had swamped large areas of the state, isolated towns, and disrupted business activities and transportation networks.

Particularly, the past decade has brought more frequent and intensified natural disasters, many of which have resulted in the loss of lives, productivity, and properties and have caused severe social and economic devastation. In the aftermath of every disaster, affected communities and local governments took actions to deal with the shock and to improve responsiveness to future natural disasters. However, with increased frequency and intensity (a trend that is expected to continue) governments, businesses, and people are faced with the challenge of adaptation and the cost of reconstruction.

14.4 THEORETICAL MODEL

The labor market, in this study, is modeled by a version of Okun's Law (Okun, 1962) that empirically demonstrates a trade-off between the country's output and rate of unemployment. In this chapter, following the works of Stockhammer (2000a, 2000b), in modeling unemployment, we also allow for Keynesian effects.

Accordingly, a change in unemployment rate is assumed to be a positive function of growth of capital stock (a measure that is used in the literature to represent the trade increase in productivity that defines Okun's law); capacity utilization (Bhaduri & Marglin, 1990; Bowles & Boyer, 1995); accumulation and productivity growth (Stockhammer & Onaran, 2004); and past rates of unemployment.

Therefore the model can be written as follows:

$$\frac{\Delta E}{N} = f\{I, z, pr, un\} \tag{14.1}$$

where $\Delta E/N$ is the change in the employment rate calculated as employment/working-age population, I is the accumulation, z is the indicator of capacity utilization, pr is a measure of productivity growth rate, and un is the past unemployment rate.

Keynesian and post-Keynesian literature pays significant attention to capacity utilization, investment, and unemployment variables, as their coefficients in the estimated models measure the impacts of market variables and unemployment perseverance. Capacity utilization variable is an important variable from neoclassical points of view as well since labor demand predominantly is a function of wages. Furthermore, capacity utilization assists in capturing the effect of real wage per worker after controlling for labor productivity. In Kaleckian theory the productivity of technological progress is another imperative variable since, if it does not align with an increase in effective demand, it will lead to increased unemployment.

14.5 ECONOMETRIC MODEL

For analyzing the dynamic effect of natural disasters on the labor market, and preceding discussions, we utilize an unemployment function.

$$\text{Unemployment} \qquad un_t = n - e_1 I_t^i - e_2 z_t + e_3 un_{t-1} + e_4 pr_t + e_5 Dis_t \qquad (14.2)$$

where I_t^i is the accumulation (i.e., investment normalized by gross state product (GSP)), z_t is the capacity utilization (capital productivity), un is the rate of unemployment, pr is the rate of labor productivity growth, and Dis is the indicator of natural disasters

14.6 DATA

14.6.1 DISASTER DATA

Although flood and cyclones are not entirely unpredicted shocks to the state, each of these extreme events is exogenous in that the exact timing of their occurrence and path they will follow inland cannot be determined a priori nor can the extent of the destruction. Data on disaster events in this chapter are collected from the Bureau of Meteorology (BOM) of Australian databases. Following Chaiechi (2014), the natural disaster variable enters the model as a dummy exogenous variable. The dummy variable (i.e., a numerical variable) demonstrates if, in each period, natural disasters (i.e., major floods and tropical cyclones) have occurred in Queensland during the period of 2002: Q1−2016: Q4. Floods in Queensland, in the past decades, impacted around 70% of the landmass and almost 60% of the state's population (SCOR, 2011). Consequently, the application of dummy variable inclines to exclude seasons that are scientifically different from other periods, where periods with no occurrence of natural disaster are given 0, and quarters during which a natural disaster is recorded are given 1. The numerical values of 0 or 1 for natural disasters' dummy variable merely capture nonnumerical attributes that includes the incidence of disaster events that are categorized as "major" or "sever" by Bureau of Meteorology of Australia.

14.6.2 OTHER VARIABLES AND SOURCES

The socioeconomic labor-market-related data are obtained from various sources such as Reserve Bank of Australia, the Office of Economic and Statistical Research—QLD government, and Australia Bureau of Statistics. All data are seasonally adjusted. Table 14.2 provides ways in which variables are generated and quantified

Table 14.2 Variable Definitions.

Variables	Indicators
Accumulation	[Physical capital stock + (1 − depreciation rate of capital) × grossed fixed capital formation]/GSP
Unemployment rate	State unemployment rates understood as % per annum
Productivity growth	The growth rate of capital stock/labor
Capacity Utilization	Capital stock/GSP
Profit rate	(1-wage rate × labor/GSP) × capacity utilization
Natural disaster	1 if a disaster occurred in a quarter, 0 otherwise

14.7 METHODS

In this chapter, we use an econometric time series technique as a simulation tool to measure the effects of natural disasters on labor markets that are within historical magnitudes. The ultimate goal is to estimate the labor market's temporal reaction from the time of an extreme event through to its full recovery. This is done by impulse response analysis that is preceded by vector autoregression (VAR) modeling.

14.7.1 PRELIMINARY ANALYSIS

14.7.1.1 Trend analysis and cointegration

One of the primary assumptions in time series analysis is that the statistical properties of variables do not change over time; therefore some form of stationarity must be assumed when dealing with time series data. Accordingly, the chapter primarily focuses on testing for this assumption with a focus on mean values and variance of data series over time.

If a trend is observed in the data, then the first differencing is required. Therefore this chapter applies unit root tests to determine if data are trending over time and if they should be differenced. This is an important step because if a variable is found to be not stationary then any further regression analysis can misleadingly propose existence of meaningful relationships.

Table 14.3 shows the result of augmented Dickey–Fuller unit root test (stationary test) conducted both at the level and first differenced, once with the inclusion of time trend and once without a time trend behavior for the periods of 2002: Q1−2016: Q3. The results show that all of the variables are integrated at order one [$I(1)$], implying that stationarity condition is achieved by differencing the data once.

Given nonstationarity of our series (since series are integrated at order 1), our next step is to determine whether the series are cointegrated and if they are, to identify the cointegrating relationships and ranks. This is important, since if the variables are found to be cointegrated, then vector error correction (VEC) model should be applied, and if there are no cointegrating relations, standard unrestricted VAR may be used to the first differences of the data. We, therefore, apply the Johansen's methodology following Johansen and Juselius (1990) to test for cointegration. Table 14.4 reports the cointegration test results and indicates that the null hypothesis of "no

Table 14.3 Stationarity Test Result.

Augmented Dickey–Fuller	Level		First D	
	No Trend	Trend	No Trend	Trend
Investment	0.62	0.78	0.002	0.001
Productivity growth	0.000	0.000	0.000	0.000
Unemployment	0.13	0.88	0.000	0.000
Capacity utilization	0.56	0.85	0.000	0.001
Profit rate	0.94	0.63	0.000	0.001

Table 14.4 Johansen Cointegration Test.

Sample: 2002: Q1–2016: Q4
Included Observations: 58
Lags Interval: 1 to 1

Selected (0.001 Level[a]) Number of Cointegrating Relations by Model

Data trend	None	None	Linear	Linear	Quadratic
Test type	No intercept	Intercept	Intercept	Intercept	Intercept
	No trend	No trend	No trend	Trend	Trend
Trace	1	0	0	1	1
Max-Eig	1	0	1	0	0

[a]*Critical values based on MacKinnon–Haug–Michelis (1999).*

cointegration" cannot be rejected in most case scenarios, especially when the VAR model includes intercepts and no trend.

14.7.2 VAR AND STRUCTURAL VAR MODELS IDENTIFICATION

For a set of n time series variables $y_t = (y_{1t}, y_{2t}, \ldots, y_{nt})$, a VAR model of order p [i.e., VAR(p)] can be written as:

$$A_0 y_t = a + b_t + A_1 y_{t-1} + A_2 y_{t-2} + \ldots + A_p y_{t-p} + A_j x_t + u_t \tag{14.3}$$

where y_t is a k vector of endogenous variables, x_t is a d vector of exogenous variables, the A_i are $(n \times n)$ coefficient matrices to be estimated, and u_t is a vector of structural shocks.

After the VAR model is set up, a structural VAR model is estimated. SVAR models can be viewed as a bridge between economic theory and multiple time series analysis to determine the dynamic response of variables to various disruptions, or shocks, and therefore, they are sometimes are referred to as the analysis of disturbances (McCoy, 1997).

Multiplication with A_0^{-1} leads to the VAR representation:

$$y_t = \alpha + \beta_t + B_1 y_{t-1} + B_2 x_t + A_0^{-1} u_t \tag{14.4}$$

The unobserved structural shocks are related to the observed-reduced form residuals by the following relation:

$$e_t = A_0^{-1} u_t, \tag{14.5}$$

$$y_t = \alpha + \beta_t + B_1 y_{t-1} + B_2 x_t + e_t \tag{14.6}$$

14.7.3 IMPULSE RESPONSE ANALYSIS

Impulse IRF depicts reactions of a dynamic system to externally influenced changes (i.e., shocks). The IRF looks at departure points of moving average representation in a VAR model to capture the effects of exogenous shocks on the time series variable.

IRF analysis is used in this chapter to show the reaction of current and future values of investment and labor market variables to a one-time one-standard-deviation shock (when the scale matters) in the natural disaster variable. Plotting IRF (as shown in Fig. 14.4) is used as a practical way to visually represent the behavior and interactions of labor market variables in response to natural disaster shocks. The vertical axis shows the magnitude of the response, while the horizontal axis shows future time periods after the shock is imposed.

Fig. 14.4 shows that the effect of an unexpected extreme weather event would impact investment rate and unemployment rate immediately. These effects would last about a year before

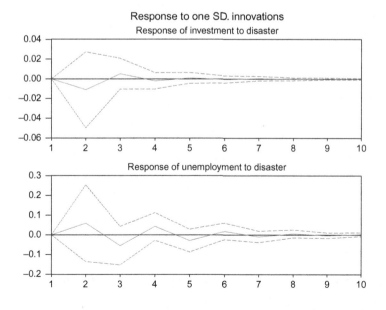

FIGURE 14.4

Results of impulse response analysis.

investment rate returns to its preshock level; however, the life of shock of a disastrous event in the labor market is longer; it takes about 2 years before the response to initial response becomes zero. Notably, as is evident by the scales of measures on the vertical axes in Fig. 14.4; the magnitude of impact on the unemployment rate is found too much higher than the impact on investment.

14.8 CONCLUSION

As the rising incidence of natural disasters is still debatable, understanding of the impact of natural disasters is growing increasingly firmer to indicate that as the world's population grows, increasing numbers of people are affected; and, as economies become more interconnected, the disastrous effects of environmental shocks can be devastating on other people, communities, and economies (local and global). Equally, using an endogenous growth framework, it can be argued that natural disasters may positively influence economic growth.

The fundamental objective of this chapter is to explore the links through which natural disasters affect the regional labor market. A VAR model is populated by the collected data that allowed us to simultaneously model changes among variables. Impulse response analysis conducted in this chapter reveals notable sensitivity in both investment and unemployment rate to disasters. It is found that natural disasters had a strong and long-lived impact on the unemployment rate, indicating the labor market adjustment process is significantly slowed down, and workers are out of a job in affected areas immediately after the occurrence of natural disasters. However, this negative effect is predicted to disappear in a couple of years post disaster; this can be due to ongoing improvements in technology, which is directly linked to new investments as suggested by Chaiechi (2014).

Words of caution: The demarcation between natural and man-made disasters is not distinct and the impact of disasters can vary greatly, depending on the location and societies affected. The challenges and options will be different, and these will subsequently afford policy recommendations specific to the situation and region. The total impact of a natural disaster is tempered by how business and governments react to the challenges presented by the breakdown of normal systems. Businesses are, generally, not well prepared for natural disasters because of the need to be able to measure the risk and uncertainty, which organizations find challenging. At the same time, natural disasters affect complex systems composed not only of governments, infrastructure, and businesses, but also people and relationships, which can have an impact on employment prospects. We cannot fully investigate this system by examining numbers that basically have been subject to measurement by accountants, economists, and others. Other forms of capitals require our attention, social, human, and natural capital if we wish to fully understand how societies react to the stresses of external shocks. In light of the complexities of the effects of natural disasters, it makes sense that a multidisciplinary research team approach brings together a number of different lenses that will provide appropriate responses of societies to environmental shocks and enable better and more meaningful decision-making by policymakers and business people alike.

REFERENCES

Adam, C. (2013). Coping with adversity: The macroeconomic management of natural disasters. *Environmental Science & Policy, 27*, 99–111.

Belasen, A. R., & Polachek, S. W. (2007). *IZA DP No. 2976: How disasters affect local labor markets: The effects of hurricanes in Florida.* Institute for the Study of Labor.

Bhaduri, A., & Marglin, S. (1990). Unemployment and the real wage: The economic basis for contesting political ideologies. *Cambridge Journal of Economics, 14*, 375–393.

Black, R., Neil Adger, W., Arnell, N. W., Dercon, S., Geddes, A., & Thomas, D. (2011). The effect of environmental change on human migration. *Global Environmental Change, 21*, S3–S11. Available at: https://www.sciencedirect.com/science/article/pii/S0959378011001531.

Bureau of Meteorology (BoM). *Maps.* (2008). From <http://www.bom.gov.au/jsp/ncc/climate_averages/tropical-cyclones/index.jsp> Retrieved 01.10.13.

Bureau of Meteorology (BoM). *Climatic statistics for Australian sites.* (2011). Available from <http://www.bom.gov.au/climate/averages/tables/ca_qld_names.shtml>.

Bowles, S., & Boyer, R. (1995). Wages, aggregate demand, and employment in an open economy: An empirical investigation. In E. Gerald & G. Herbert (Eds.), *Macroeconomic policy after the conservative era.* Cambridge.

Brown Gaddis, E., Miles, B., Morse, S., & Lewis, D. (2007). Full-cost accounting of coastal disasters in the United States: Implications for planning and preparedness. *Ecological Economics, 63*(2–3), 307–318.

Chaiechi, T. (2012). Macroeconomic impact assessment of natural disasters shocks: A Queensland case study. In: *Third Asian conference on the social sciences and the second Asian conference on sustainability, energy and the environment (proceeding)* (pp. 72–84).

Chaiechi, T. (2014). The broken window: Fallacy or fact—A Kaleckian–Post Keynesian approach. *Economic Modelling, 39.*

Cutter, S., (2011). The Katrina exodus: internal displacements and unequal outcomes. In: *Foresight review on migration and global environmental change, case study CS1.* London: Government Office for Science.

Kirchberger, M. (2017). Natural disasters and labor markets. *Journal of Development Economics, 125*, 40–58.

Hallegatte, S., & Ghil, M. (2008). Natural disasters impacting a macroeconomic model with endogenous dynamics. *Ecological Economics, 68*(1–2), 582–592.

Johansen, S., & Juselius, S. (1990). Maximum likelihood estimation and inference on cointegration—With application to the demand for money. *Oxford Bulletin of Economics and Statistics, 52*, 169–210.

Matilal, S., & Höpfl, H. (2009). Accounting for the Bhopal disaster: Footnotes and photographs. *Accounting, Auditing and Accountability Journal, 22*(6), 953–972.

McCoy, D. (1997). How useful is structural VAR analysis for Irish economics?. In: *Research technical papers 2/RT/97.* Central Bank of Ireland.

Okun, A. (1962). Potential GNP: Its measurement and significance. In: *Proceedings of the business and economic statistics section* (pp. 98–103). American Statistical Association.

RAI. (2013). The costs and benefits of geographic labour mobility: A regional perspective. In: R. A. Institute (Ed.), *Submitted to Productivity commission inquiry into geographic labour mobility.*

Rodríguez-Oreggia, E. (2013). Hurricanes and labor market outcomes: Evidence for Mexico. *Global Environmental Change, 23*, 351–359. Available from: https://doi.org/10.1016/j.gloenvcha.2012.08.001.

SCOR. (2011). *Economic impact of Queensland's natural disasters.* Global Risk Center.

Skidmore, M., & Toya, H. (2002). Do natural disasters promote long-run growth? *Economic Inquiry, 40*(4), 664–687.

Stockhammer, E. (2000a). Explaining European unemployment: Testing the NAIRU theory and a Keynesian approach. In: *Vienna University of Economics and Business Administration working papers (no.68).*

Stockhammer, E. (2000b). Is there an equilibrium rate of unemployment in the long run?. In: *Vienna University of Economics & B.A. working papers in growth and employment in Europe: Sustainability and competitiveness (no.10).*

Stockhammer, E., & Onaran, O. (2004). Accumulation, distribution and employment: A structural VAR approach to a Kaleckian macro-model. *Structural Change and Economic Dynamics, 15,* 421—447.

Su, T. T., Saimy, B. I., & Bulgiba, A. M. (2013). Socioeconomic consequences of the 2004 Tsunami: Policy implications for natural disaster management. *Preventive Medicine, 57*(Suppl.), S74—S76.

Tang, R., Wu, J., Ye, M., et al. (2019). Impact of economic development levels and disaster types on the short-term macroeconomic consequences of natural hazard-induced disasters in China. *International Journal of Disaster Risk Science, 10,* 371—385. Available from https://doi.org/10.1007/s13753-019-00234-0.

Wilson, D. C., Brankicki, L., Sullivan-Taylor, B., & Wilson, A. D. (2010). Extreme events, organizations and the politics of strategic decision making. *Accounting, Auditing & Accountability Journal, 23*(5), 699—721.

COSTING A NATURAL DISASTER: AN ACCOUNTING PERSPECTIVE

15

Susan Ciccotosto

College of Business, Law and Governance, James Cook University, Cairns, QLD, Australia

> Damage assessments of natural hazards supply crucial information to decision support and policy development in the fields of natural hazard management and adaptation planning to climate change.
>
> **Merz, Kreibich, Schwarze, & Thieken (2010, p. 1697)**

Climate change and its effect on the environment is very much in the political space, with many arguments for and against anthropomorphic involvement despite climatologists' claims that the science is settled. As the climate changes, it is anticipated that there will be more, and greater, natural disasters occurring. There is now some discussion around, not only increasing hurricanes, droughts, and storms but also the stress that changes in climate (including the anthropomorphic use of water) may be linked to earthquakes (Buis, 2019). Natural disasters use up resources to respond to the emergency they create and resources to rebuild after the emergency is over. Both public authorities, businesses and individuals, who may be affected by a natural disaster, need to understand what kinds of resources are needed and how to plan to respond to a disaster that may affect them. This is not the only use for this knowledge. Part of any plan is to assign resources now, so that the future crisis is minimized. Levees are built. Fire engines purchased. People are trained. The costs of preparedness are weighed against the cost of the disaster. Costing matters. This chapter will examine the role of the calculative practice called accounting in the way the values are placed in the context of a natural disaster.

Decision-makers need to know how a natural disaster affects national and local economies in order to plan and, when necessary, execute disaster relief and recovery. Information with greater accuracy and reliability improves the information base on which these decisions are made (Downton & Pielke, 2005). Decision-makers use information produced by the Economics discipline to inform policy decisions so that the response is seen to be efficient and effective. The economic numbers produced for this activity utilize expenditure data produced by organizations in times of crisis. The expenditure data is collated from a number of sources within society, both private and public. Accounting plays the role of recording these expenditures at the micro-level. This chapter explores how public and private entities record their expenditures as they undertake disaster rescue and recovery activities. Although this area is underresearched, we consider current literature from a critical perspective as academics have brought into question whether accounting is limited to the passive recording of events (Vosslamber, 2015) and suggest future areas of research.

Economic Effects of Natural Disasters. DOI: https://doi.org/10.1016/B978-0-12-817465-4.00015-7

It may be a flood, or an earthquake, or a hurricane (cyclone). Some natural disasters give authorities, businesses, and people some warning, others do not. As the immediate crisis passes, authorities make decisions regarding the appropriate level of response. Questions are raised by governmental bodies, not only about available resources, but also about how much of, and where, those resources are to be used. In the meantime, businesses and individuals assess the damage to property at a corporate and personal level. Families have lost homes and personal items; businesses have lost buildings, property plant and equipment and inventories; and local governments (municipalities) have damaged infrastructure such as roads, water, and sewage systems (Lai, Leoni, & Stacchezzini, 2014). Different losses are measured and calculated using accounting systems and valuations: present value, discounted value, or replaceable value, and then assigned a dollar amount.

The role of accounting is to render calculable and, therefore, visible (Miller, 1990) activities undertaken when decisions are made. Accounting enables the various costs associated with a natural disaster to be measured and then used in a number of ways. During the crisis response the accounting systems are used by governments to control the amount spent, the zone of the response, and to validate the use of resources. Businesses and citizens use accounting information to make business decisions and actions such as completing insurance claims or finding alternative funding and, eventually, decision-making about their future. After the natural disaster has taken place, these numbers are combined and utilized to calculate the impact of the disaster on the larger socioeconomic environment. Authorities and members of the community then use these costings, included with those calculated from previous disasters, to estimate the risk of loss of income and assets from a similar catastrophe within a business itself, in the insurance industry and within governmental bodies.

In the very beginning, accounting data is produced at the micro-level. Whether it is as an individual, in a business, or in a governmental authority, a person enters a number into the accounting system, indicating in some way the use or loss of a resource brought about by the natural disaster. These values are then collated moving upward from the individual, through local organizations and into government departments. To understand the final product of the collation of accounting data that is used by the Economic discipline, the first step is to discuss how these numbers are produced and collated and what is, and what is not, included. This next section considers how individual transactions are produced.

15.1 COUNTING THE COST

The importance of accurate accounting for the cost of natural disasters cannot be understated, simply because of the reliance placed upon the number which is produced by the systems used. There is tension, however, between the concept of accuracy and the ability, in practical terms, to measure, collect, and calculate relevant data. A natural disaster may take only seconds to occur or may take place over a period of several days. The impact, however, and the subsequent recovery period may take months or years (consider, for instance, that rebuilding is still taking place for the 2011 earthquake in Christchurch, New Zealand, at the time of writing 8 years later). Resources are often scarce and projects large and this will have an impact on completion dates. From an accounting perspective, this often requires costs to be estimated, with the possibility of an over, or under, valuation. This can challenge the system if accuracy is important as a cost is only truly accurately known and recorded when the expenditure is completed.

The tension between accuracy and efficiency is best shown in how authorities choose to include as a cost and what not to include. From a rational perspective, one would expect that any expenditure which takes place and is due to the natural disaster should be included in the overall cost. Authorities are concerned that only costs actually derived from the disaster are to be included, and there should be a system in place to ensure that this is so. Accounting systems are used to control the expenditure and have rules built into them to prevent nondisaster spending taking place with the funds and resources assigned to the recovery and rebuild. The authorities have decided that there are three basic questions to answer: what was the money spent on, where was it spent and when? This would seem, to begin with, to be unproblematic.

15.1.1 WHAT IS CONSIDERED RELEVANT

What was the money spent on? Authorities, such as governments and insurance bodies, have decided that only costs that are directly attributable to the event should be taken into account. These costs (referred to as "direct costs") are favored because they can be traced to an immediate use of a resource, such as medical supplies or a new bridge. There could be a loss of assets such as destroyed buildings. There could be increased costs as repairs are required. All these are direct costs. They can be directly associated with the impact of the natural disaster.

Where was the money spent? Authorities also define the area that is considered to be affected by the natural disaster, from surrounding areas which are not. This activity clarifies which data are relevant and which are not (Sargiacomo, 2015). Defining the area of concern enables the authorities to decide who is affected from who is not. The relevant cost allocations confined to those assets within the area of concern are included. Any assets damaged or destroyed outside the area are not. Resources spent on persons who live, work, or visited at the time of the disaster are included. Any resources spent or used for persons outside the area of concern are not.

When was the money spent? When the cost takes place is also relevant for decision-making purposes. Not all costs take place at the time of the disaster. Whether it is rebuilding infrastructure or repairing homes, the demand on resources may be such that there is a long time frame between the event itself and the resulting rebuilds and repairs. For practical purposes, it is necessary to amalgamate the costs at some point. Where the activity has not yet taken place, this will necessitate that the relevant authorities need to utilize the estimations of costs, rather than actual costs in the overall calculations as they attempt to arrive at a final figure (Downton & Pielke, 2005).

15.1.2 WHAT IS NOT INCLUDED

It is important to ensure that costs not considered relevant are not included in the calculations. Obviously, if the transaction does not meet the above criteria, it should not be included; however, not all monies spent during this period are due to the disaster itself. In many ways, life goes on and daily activities take place that would have occurred in any case. Normal costs are classified as those which would have occurred if the crises had not developed. They are not considered as part of the cost of the natural disaster itself, because they would have been incurred regardless.

Extraordinary costs are either a new cost or an increase of a normal cost that can be attributed to the disaster. Expenditures, as they are recorded, are then classified as either extraordinary or normal. Normal costs are then excluded from the calculations and the eventual collation of data.

As previously mentioned, authorities are also concerned that funds are not expended, regardless if they are from government funds or donations, if any claim is fraudulent or people are attempting to benefit in some way from the crisis. Claims that are for damage or loss to nonexistent assets, for instance, should not be paid. People who were not in the area and were claiming injury would be another example of an invalid claim. To ensure this does not happen and only valid claims are paid, authorities put into place accounting procedures and controls as checks and balances into the accounting system.

15.1.3 AD HOC ACCOUNTING SYSTEMS

Now that the authorities have created boundaries around the costs that will be attributed to the natural disaster, transactions are recorded in an ad hoc manner and collated into reports that have been designed to reflect the cost of the disaster itself. Individual people record individual transactions into a database. The data may relate to the payment of someone to search for the injured, or the supply of drugs. It could be for the loss of an asset, such as a building or a vehicle. It may be entered into the database by a business owner, an insurance assessor, or a government employee. This individual datum is then amalgamated into totals that are deemed to represent resources used by others and enter the world in the context of information.

The collection of data is widespread, with many different individuals involved in the work. Although the eventual information is collated usually by the national government, the groundwork is often undertaken by people who are members of authorities in the local area. This is to take advantage of their local knowledge and connections and their ability to organize face-to-face meetings. These people may be members of the local government areas (municipalities) (see for instance Lai et al., 2014, or members of a not-for-profit organization such as the Red Cross, Perkiss & Moerman, 2020). Because the data is in many different databases, this obviously makes the collation of data more difficult (Miley & Read, 2013).

Numbers are produces and collated, moving from the individual transaction to the national government database. Authorities have ensured that only those resources that have been applied to the rescue, recovery, and rebuild from the natural disaster have been included into the final cost of the natural disaster. The next section explores the areas of previous accounting research into natural disasters.

15.2 PREVIOUS RESEARCH

Accounting is presented to the world as a neutral and reliable procedure to measure and understand how resources are utilized by various organizations in society. According to the accounting profession, financial accounts, properly prepared, are to present to external agents *a true and fair view* of the organization of matter at hand [International Accounting Standard (IAS) 1]. A true and fair view implies that the financial report is free from material misstatements and faithfully represents the financial performance which it represents. A material misstatement is one, which, if known, would change the decision that relies on the numbers in the report. Accounting deals with financial numbers, in the same way that Economics does. It represents that the numbers are neutral.

In the scenario following, there were numerous steps taken to ensure that the questions of what, where, and when have been checked, authorized, and answered. They have been presented to the appropriate authorities as providing accounting numbers that can be relied upon for decision-making purposes.

Lai et al. (2014), in their investigation of the accounting communications after a disastrous flood in Italy, describe in great detail the system of accounting used to collect damage claims, check the information for accuracy, and then collate it. The purpose of the research was to understand how accounting promotes communication between different actors using the disaster scenario. They noted that there were clear communication channels, not only between the victims and authorities but also between the victims themselves so that they could gain an understanding of the process required to claim reimbursement for losses due to the disaster. They also noted that clear and regular communication between the authorities and the victims was developed to ensure that the procedures were transparent and efficient. The authorities had designed the accounting system so that, as the claims were submitted, it would be possible to check details such as asset valuations and costs against an extensive database which used data from a number of different sources. The authors of the paper emphasized how the governmental agency required that the accounting system was designed to ensure that fraudulent claims were not successful.

There is little accounting research in the role that accounting plays as a community responds to a natural disaster. Lai et al.'s (2014) research has confirmed that the authorities have used the accounting system to ensure that those questions (what, where, and when) have been given appropriate attention. It should be noted that this research concentrated on communication channels and did not take a critical view of costing procedures. Other researchers have concentrated on answering a two different questions. How is accounting used as a tool in politico/power relationships and how does the classification of valuations or costs affect the final cost of the natural disaster?

15.2.1 POWER AND ACCOUNTING

When authorities place constraints upon the permitted costs that will be considered as a "cost" of a natural disaster, they define what, and what is not considered a cost. This then filters into the system, ensuring that individuals and businesses respond to the requirements to gain relief and benefit from a financial perspective. In Lai et al.'s (2014) study, those who wished to make a claim regarding specific losses communicated with each other to ensure that the form, produced from the accounting system, was correctly completed. The system assured that the claims made were valid. There was, apparently, control. This control can lead to trials of strength between different actors involved in responsibilities for the disaster responses.

Miley and Read (2013) examined the relationships between national and local government areas (municipalities) following three earthquakes in New Zealand: Murchison (1929), Napier/Hawke's Bay (1931), and Christchurch (2011). The differing authorities, national, local, and CERA (Canterbury Earthquake Recovery Authority) required a complex, negotiated relationship to manage the recovery of the region from the 2011 earthquake. At the time of the earthquake the political climate expected financial efficiency, which led to the requirement for national and local authorities to balance their budgets. This increased the tension between the New Zealand national government and the Christchurch City Council (the local municipality). The New Zealand national government had the legal ability to direct local Council actions while, at the same time, giving the appearance that the council was able to make its own decisions. As the costs of the reconstruction mounted, this placed a financial burden on the Council because the costs were included in the Council's financial reports, making it more difficult for them to present a balanced report. The Council was faced with a difficult decision to either increase rates or sell infrastructure, which would have been politically unpopular in either case. Alternatively, if the national government had borrowed to cover costs, this also would have affected their financial report, also politically unacceptable. As the argument grew over the accounting numbers for the disaster and who was to be responsible, the rebuild was affected.

Accounting took an unexpected role in the resettlement of farm families into more productive areas (Walker, 2014). The US government offered families, financially crippled by drought, new land that was more sustainable. The offer came with a number of stipulations, one of which was to keep a set of accounting records showing income and expenditure. These records were then submitted to Resettlement Administration Authority and the Farm Security Administration as an indicator of the farms financial viability which would lead to the eventual ability of the farmer to purchase the land which he was working. The accounting records were used in a number of ways to support the resettlement strategy. The farmers were able to see progress and understand better farm management and, most, eventually came to see that keeping a set of accounts was good management practice. Interestingly, the accounts were also used by the authority to justify its actions to a skeptical national government and underpin its strategy moving forward. This is a noted example of accounting numbers being used in post disaster government intervention.

In his interesting case study of the earthquake in Abruzzo, Italy, Sargiacomo (2015) investigated how the classification of costs into normal and extraordinary was open to strategy and gaming. He also observed that a State Commissioner presented an estimate of the costs associated with the earthquake to the Federal Ministries of Health and of Economics and Finance. His estimates of costs included in the list not only the funds spent on specific actions undertaken by his section but also the costs associated with the payment of employees who were unable to work and those associated with revenue not received because the hospitals were not operational. The cost of the employees should not have been included in the estimate as it was a normal cost ("an unproductive fixed cost" p. 76) and was rejected. Sargiacomo noted that the ministries did not accept the loss of income as a cost of the disaster, which should have been included, because it meets the definition of a direct cost. The above example highlights the difficulty of defining a direct cost. The purchase of tangible items is more easily traced to the natural disaster. Other costs and loss of income must be delineated into what is normal income and expenses and what is not. The discussion between the authorities regarding what should or should not be included is not only an example of power relationships but also highlights that costs can be incorrectly included or excluded from the overall calculations.

15.2.2 CLASSIFICATION AND COSTS

Cost classification, limits what is counted, simply because each cost in each transaction must meet the definition of the classification, or it cannot be entered as a cost for the natural disaster. This is the very purpose for which the classification was initiated. It is part of how the accounting system ensures that only those costs that can be attributed to the natural disaster will be recorded, and collated. Some researchers noted that the method of classification caused problems when they attempted to claim for specific crisis-related costs.

It was important to the Italian authorities that those who were injured in the Abruzzi earthquake had access to suitable medical care. The cost of such care was assigned a special code in the accounting system, so that the injured could receive treatment without the normal costs. General medical care was not, normally, free and required some outlay to receive medical and pharmaceutical benefits. The outcome of this decision was that the inhabitants of the earthquake zone were given access to free medical and pharmaceutical care over and above care for the injured. At the same time, not long after the earthquake in Abruzzo, numbers of persons moved out of the L'Aquila district and into the Pescara region to obtain accommodation. The local health authority in the Pescara region complained that the workload increased to cope with the victims. Many of them did not have access to transport but were still entitled to the free health care. This required more personnel and vehicles to assist these new patients. The cost of the increased number of personnel were not taken into account by the authorities, and no extra funds to cover them were forthcoming (Sargiacomo, 2015).

Perkiss and Moerman's (2020) article examined the experiences of members of the low socioeconomic community after the impact of Hurricane Katrina on New Orleans in 2005. Although not emphasizing the role of accounting in the actions of the United States Government, through the Federal Emergency Management Agency, they emphasized the importance of "accountability" over the needs of individual persons. Their participants reported the horror of their experience and the impact that an emphasis on efficiency had on their recovery. This paper emphasizes that an individual's experiences are important, if only from the long term affects that person's experiences have on their ultimate welfare and productivity. Accounting does not take into account nonfinancial transactions and the experiences of these individual persons, directly attributable to the rescue, and recovery activities were not recorded.

From an accounting perspective, as the reconstruction takes a long period of time for larger disasters, estimations are required for the work that is not completed, as authorities finalize the collation of data. Downton and Pielke (2005) undertook comparisons between the actual costs of a serious flooding event in the United States and the collected data. They found that estimates of economic losses from larger events were more accurate, with positive and negative errors tending to average out. They also found that there appeared to be a greater risk of error in smaller events, possibly because there is less aggregation of data from differing sources. Their paper makes some suggestions regarding how to interpret and use the estimates to reduce the possibility of misinterpretation.

The research discussed previously, apart from Downton and Pielke (2005), has used a qualitative methodology to investigate the role of accounting in resource allocations during natural disasters. Data were gathered from publicly available documents, semistructured interviews, and direct observation. Accounting is shown as a valuable tool, able to control, record, and report on the expenditure of resources to authorities, where it can be collated for decision-making purposes. The research also shows that this recording is not unproblematic. While it has strength in its ability to accurately report expenditure, accounting defines what is to be recorded and what is not. The research shows that some monetary costs are missing from the data which is to be collated.

Classification, used as a tool to control the resources, is open to gaming and limits what may be assigned to the disaster. Accounting does not record nonfinancial data that may be of relevance.

The previous research is important from an Economics perspective because it brings to light two areas of contention. First, accounting numbers are used as tools of power by both the victims and the bureaucracy and, second, that there are possible gaps in information that is collected by this method.

15.3 SUGGESTIONS FOR FURTHER RESEARCH

The world is changing. According to climatologists, droughts, floods, and storms are likely to produce food shortages. There are concerns about the effect of climate change on the extinction of flora and fauna. In the accounting world, there is a movement toward sustainability reporting, as accountants grapple with finite resources and societal expectations. Businesses are grappling with green energy and low carbon futures. If governments are going to come to grips with this new future, they will need new information.

As Economists produce information about a natural disaster, authorities use it not only to report for this natural disaster but also in decision-making to prepare for tomorrow's. It is important that as much data are collected and used as are possible. Further research is necessary to understand what is not included in the calculations and, subsequently in the collation of the data. Accounting is able to make many things visible, by giving them value. At the same time, it renders other things invisible, because it is not easy to give them worth (Hines, 1992). In this new world the true costs of a natural disaster should include costs that are difficult to trace and identify and losses that are difficult to value. Research is needed to uncover costs presently not included in the final calculation and to render visible that which is, at this point in time, invisible, unseen, and untraced.

15.3.1 SEARCHING FOR MISSING DATA

There is some discussion in the literature and in the accounting profession itself, with regard to how the impact of the natural disaster is measured at the firm level. When assets are lost, should they be measured at the replacement cost or the original cost less depreciation? The debate that argues that the increased income of firms which supply goods and services that are needed in the reconstruction phase, such as construction and hardware supply, is valid; however, this may be difficult to measure. There has been some work that has highlighted long-term changes to business activity, both positive and negative, due to a natural disaster and the question is how to trace the difference.

Sometimes there may be no funds spent at all. The accounting profession has assisted in defining this by appropriately written standards that apply to financial recording when there has been a natural disaster. Although there is no single standard for such an event, information is included in specific standards. Some losses are recorded because they are an asset for an organization running a business. They are recorded as a transaction even though no money changes hands, for instance, the loss of inventory is included in IAS 2: inventories, and the loss of accounts receivable (because clients cannot repay as they were victims too) is included in IFRS 9: financial instruments

(International Financial Reporting Standard 9). There may be a loss of income because the disaster has made normal operations impossible. Many organizations, but not all, are covered by insurance, and the organization recovers this loss through their insurance claim. As an example, actions taken by CERA (Canterbury Earthquake Recovery Authority) after the Christchurch earthquake in 2011 required owners of businesses to move away from the center of the city, without giving them sufficient notice to find and move into alternative accommodation. This would have had a significant financial impact on the owners but was not included in CERA's financial statements and, therefore, was unlikely to have been collated into the data that represented the "cost" of the earthquake, finally reported by the authorities.

The reliance on data supplied by insurance firms does not take into account the underinsured and the uninsured. Wahlquist (2019) commented on this in her news article about the extreme bushfire season in 2019 in Australia. Data from her article indicated that at the time of writing there were more than 2000 claims for bushfire damage in NSW and southeast Queensland. Losses were estimated to be more than $165 million. She notes that many homes were uninsured or severely underinsured. This observation agrees with recent developments in northern Australia (Wainwright & Nothling, 2019) and work by King et al. (2014) who investigated insurance responses in the wake of the earthquakes in Christchurch, New Zealand, 2010 and 2011. This study not only notes that there were occurrences of residential and commercial buildings being uninsured or underinsured but also noted that claimants accepted lower payments than they were entitled to because they required the cash flow. Gaps in claims were paid for over time from operational and personal cash flows. It is concerning that the amount of funds used by individuals and commercial entities, not covered by insurance payments, are not likely to be collected by economists to be included in their data.

The ad hoc accounting system designed for a natural disaster takes into account the tangible assets lost or used and the allocation of resources applied to react, recover, and rebuild the community during and after the crisis. There are many tangible and nontangible assets that are lost. Life is precious and yet it is rarely included in the cost of the disaster [note Gray (2010) where the cost of loss of life has been included in the overall calculations for a bushfire in Australia]. The loss of human life, of course, cannot be valued, but a person's life is not divorced from the economy. An individual contributes to the economy by inputting labor and knowledge. This labor and knowledge is lost and has an economic impact, individually small, but when aggregated across a larger number of dead, injured, or traumatized could possibly be of importance. The loss of animal stock, owned as agricultural assets, is measured and included in the calculated cost of the disaster if it has been insured.

Wildlife and the natural environment belong to no one person and, therefore, it is not a loss as far as the accounting process is concerned. Hines (1992) has argued that without a monetary value, things remain unseen, invisible. The accounting process is based on a commercial system, and it could be argued that the natural environment has no value and it is not included in the calculations. It is possible to place a value on the natural world and to produce a number. This number will enable us to see how much our natural environment is worth to us as a society and enable a valid comparison with the costs of protecting it. Utilizing methodologies suggested by the tourism discipline will give a voice to our natural environment. In Whalquist's (2019) interviews with victims of the severe bushfires in Australia, victims describe their distress as they view the loss of their natural environment. It is obvious that these people place a high value upon the ecosystem in which they lived. There has been some research in the tourism discipline into ways of calculating the value, in monetary terms, of historical sites and our natural environment. Kim, Wong, and Cho (2007) asked people how much they would pay to enter into a world heritage site in Korea. This work has progressed, see Saarikoski et al. (2016), for an outline of more recent work in this area.

Accounting, producing numbers which render things visible, has the ability to change social norms (Vosslamber, 2015). Valuing the ecosystem, including the flora and the fauna, is arguably an important component part of the loss to society when a natural disaster takes place. If we were able to place a value on our natural environment and add this to the costs of natural disasters, authorities would be able to more clearly decide how to allocate resources to preventative practices.

15.4 **CONCLUSION**

The numbers produced by accounting information systems are used by economists to give light to the impact of natural disasters for decision-making and policy development. It is important that the information is as accurate and clear as possible. Accounting systems can check, collate, and confirm the costs of the resources that are used in the reaction, recovery, and rebuilding phases. At this point in time, these costs have been limited by the authorities to the use of specific resources in certain places, at certain times, and for specific uses. Individuals will use the accounting system to claim for normal expenditure and, at other times, find themselves unable to rightfully claim for costs attributable to the consequences of the natural disaster. This arrangement can lead to lower than the actual financial impact on the economy as a whole.

Is this the whole picture? We live in a society that is becoming concerned with our environment as well as the financial outlook. There are areas of data that could enhance the information produced by Economists, as the demands on policymakers change in response to the perspectives of society as a whole change. This is an opportunity for Economists, by engaging in multidisciplinary research, to give greater clarity to how losses from natural disasters impact on ourselves and our society.

REFERENCES

Buis, A. (2019). *Can climate affect earthquakes, or are the connections shaky?* NASA (North American Space Administration). From <https://climate.nasa.gov/news/2926/can-climate-affect-earthquakes-or-are-the-connections-shaky/> Retrieved 23.12.19.

Downton, M. W., & Pielke, R. A. (2005). How accurate are disaster loss data? The case of U. S. flood damage. *Natural Hazards, 35,* 211–228.

Gray, D. (2010). *Black Saturday cost $4.4 billion.* Sydney Morning Herald. From <https://www.smh.com.au/national/victoria/black-saturday-cost-44-billion-20100801-11116.html> Retrieved 27.12.19.

Hines, R. D. (1992). Accounting: Filling the negative space. *Accounting Organizations and Society, 17,* 313–341.

Kim, S. S., Wong, K. K. F., & Cho, M. (2007). Assessing the economic value of a world heritage site and willingness-to-pay determinants: A case of Changdcok Palace. *Tourism Management, 28,* 317–322.

King, A., Middleton, D., Brown, C., Johnston, D., Erri, M., & Johal, S. (2014). Insurance: Its role in recovery from the 2010-2011 Canterbury earthquake sequence. *Earthquake Spectra, 30,* 475–491.

Lai, A., Leoni, G., & Stacchezzini, R. (2014). The socializing effects of accounting in flood recovery. *Critical Perspectives on Accounting, 25,* 579–603.

Merz, B., Kreibich, H., Schwarze, R., & Thieken, A. (2010). Assessment of economic flood damage. *Natural Hazards and Earth System Sciences, 10,* 1697–1724.

Miley, F., & Read, A. (2013). After the quake: The complex dance of local government, national government and accounting. *Accounting History, 18,* 447–471.

Miller, P. (1990). On the interrelations between accounting and the state. *Accounting, Organizations and Society, 15,* 315–338.

Perkiss, S. & Moerman, L. (2020). Hurricane Katrina: Exploring justice and fairness as a sociology of common good(s). *Critical Perspectives on Accounting, 67–68,* https://doi.org/10.1016/j.cpa.2017.11.002.

Saarikoski, H., Mustajoki, J., Barton, D. N., Geneletti, D., Langemeyer, J., Gomez-Baggethun., ... Rui Santos, R. (2016). Multi-criteria decision analysis and cost-benefit analysis: Comparing alternative frameworks for integrated valuation of ecosystem services. *Ecosystem Services, 22 B,* 238–249.

Sargiacomo, M. (2015). Earthquakes, exceptional government and extraordinary accounting. *Accounting, Organizations and Society, 42,* 67–89.

Vosslamber, R. (2015). After the earth moved: Accounting and accountability for earthquake relief and recovery in early twentieth-century New Zealand. *Accounting History, 20,* 518–535.

Wahlquist, C. (2019). In the wake of the bushfires: Stricken residents of north coast NSW face grim job of rebuilding. *The Guardian.* From <https://www.theguardian.com/australia-news/2019/dec/15/in-the-wake-of-the-bushfires-stricken-residents-face-grim-job-of-rebuilding> Retrieved 15.12.19.

Wainwright, S., & Nothling, L. (2019). *Northern Australians ditch insurance as premiums soar in disaster-prone regions.* ABC North Queensland. From <https://www.abc.net.au/news/2019-12-21/high-premiums-driving-uninsured-homes-in-northern-australia/11819814> Retrieved 22.12.19.

Walker, S. P. (2014). Drought, resettlement and accounting. *Critical Perspectives on Accounting, 25,* 604–619.

FURTHER READING

Fahlevi, H., Indirani, M., Mulyany, R., & Nadirsyah, (2019). What role of accounting in disaster recovery and relief? A literature review. *IOP Con. Ser.: Earth Environ. Sci.* 273, 012059.

Sargiacomo, M., Ianni, L., & Everett, J. (2014). Accounting for suffering: Calculative practices in the field of disaster relief. *Critical Perspectives on Accounting, 25,* 652–669.

TYPICAL SOCIAL ADAPTATION MEASURES IN CLIMATE CHANGE PLANNING: A TROPICAL REGION CASE STUDY

16

Allan P. Dale[1], Karen J. Vella[2], Ruth Potts[3], Hurriyet Babacan[4], Alison Cottrell[5], Winn Costantini[6], Meegan Hardacker[2] and Petina L. Pert[7]

[1]*The Cairns Institute, James Cook University, Cairns, QLD, Australia* [2]*Science and Engineering Faculty, Queensland University of Technology, Brisbane, QLD, Australia* [3]*School of Geography and Planning, Cardiff University, Cardiff, NSW, Australia* [4]*Rural Economies Centre of Excellence and The Cairns Institute, James Cook University, Cairns, QLD, Australia* [5]*School of Earth & Environmental Sciences, Centre for Disaster Studies, James Cook University, Townsville, QLD, Australia* [6]*Williams College, Williamstown, MA, United States* [7]*CSIRO Land and Water Flagship and Division of Tropical Environments and Societies, James Cook University, Cairns, QLD, Australia*

16.1 INTRODUCTION

While many regions across the globe are vulnerable to climate change (Hare, Cramer, Schaeffer, Battaglini, & Jaeger, 2011), planning for climate change adaptation has tended to focus its efforts on biophysical and engineering adaption responses (Stanley, 2010). The social dimension of these kinds of adaptation responses is typically framed around inclusive citizen participation (Chu, Anguelovski, & Carmin, 2016) or community-based governance approaches to climate planning (Forsyth, 2013). For vulnerable groups and geographies exposed to the more frequent and severe effects of climate change, however, broader social adaptations are needed to enable them to participate in and take effective leadership of climate adaptation outcomes (Chu et al., 2016; Dodman & Mitlin, 2013; Warrick, 2011).

Despite growing trends toward the use of vulnerability assessments, most provide only tokenistic assessments of socioeconomic and political dimensions of vulnerability (McDowell, Ford, & Jones, 2016). This can obscure the identification of social adaptations needed, such as requiring the transformation of new knowledge, or local and political attitudes, or cultural factors necessary to take action (Adger et al., 2009). It can also overlook social capacities that may need to be built to position communities and geographies to adapt. Social approaches for climate adaptation planning alternatively address the adaptive capacity and resilience needs of vulnerable communities (Kirkby, Williams, & Huq, 2017; Marshall et al., 2010). They cogenerate understandings of community vulnerability and resilience. They can deliver more comprehensive and integrated social adaptation

Economic Effects of Natural Disasters. DOI: https://doi.org/10.1016/B978-0-12-817465-4.00016-9

measures within a region, though their use in climate change adaptation planning approaches is limited.

A review of 14 climate adaptation planning approaches across the globe shows the relative disciplinary focus of the adaptive approach employed and the relative balance of biophysical, engineering, and social adaptation measures established (Table 16.1). This highlights a weak inclusion of social adaptation approaches and measures and demonstrates that socially oriented approaches to climate adaptation are the exception rather than the norm. This result is concerning as climate change adaption needs to be an inherently social process.

This marginalization or compartmentalization of social approaches and the social sciences in climate adaptation is not altogether surprising. Social sciences are typically marginalized in natural resource−related institutions (Dale, Taylor, & Lane, 2001). This is related to the significant disciplinary biases that exist within institutions charged with planning and decision-making for land and natural resource management (Dale et al., 2001). This explains why social considerations generally hold a marginalized role within more biophysical or engineering-oriented approaches to climate adaptation planning. Alternatively, isolated fragments of quality social adaptation effort can occur within these more linear or rationalistic approaches (Warrick, 2011).

As a result of this bias toward biophysical and engineering analyses and solutions within emerging climate change adaptation planning activities unfolding across the globe (Warrick, 2011), this chapter aims to expose readers to a breadth of social adaption measures that can emerge from planners taking a more comprehensive and socially oriented approach to climate change adaption planning. We are particularly interested in exploring social measures typical of vulnerable tropical countries. This focus is timely as we see significant global growth in international and national policies and agreements for climate change adaptation measures that compliment global greenhouse gas mitigation measures (Germanwatch and WWF International, 2010). The chapter does this by exploring the outcomes from a research-based field trial of more socially oriented approaches to climate change planning in Tropical North Queensland (TNQ), a region of tropical Australia that is vulnerable to the potential impacts of climate change that could become typical across the globe's tropical regions.

We begin the chapter with a contextual overview of climate adaptation problems and the social planning approach to climate adaptation in TNQ. This leads into our methods used to examine the social adaptation measures that emerged. Our results outline the social adaptation strategies to emerge in TNQ and our discussion considers the relevance of these strategies to other vulnerable tropical regions.

16.2 METHODS

The socially orientated approach to climate change planning underpinning the analysis in this chapter combined social resilience indicators and participatory knowledge-building processes in regional and remote communities in TNQ (see Fig. 16.1) to underpin a climate adaptation plan. These communities are among Australia's most vulnerable in the face of climate change. They face potential for sea-level rise, more intense dry spells, increasing temperatures, more

Table 16.1 The Relative Focus of Typical Examples of Climate Adaptation Strategies from Global to Local Scale.

Scale	Initiative	Approach Orientation	Relative Focus on Adaptation Measures		
			Social	Biophysical	Engineering
Cross-national	Pacific-Australia Climate Change Science and Adaptation Planning Program (Australian Government Department of the Environment, 2013)	Biophysical with some social adaptation potentially embedded	Weak	Strong	Weak
	European Union Adaptation Package (European Commission, 2013)	Biophysical and engineering focus with some social solutions	Weak	Strong	Strong
National scale	CPACC Project (Caribbean Community Secretariat, 2013)	Predominantly biophysical	Weak	Strong	Weak
	Climate Change Adaptation: A Priorities Plan for Canada (Feltmate & Thistlethwaite, 2012)	Engineering/biophysical oriented with some social adaptation	Moderate	Strong	Strong
Provincial scale	Victorian Climate Change Adaptation Plan (Victorian Government, 2013)	Engineering/biophysical oriented with social adaptation embedded	Moderate	Strong	Strong
	Climate Ready: Ontario's Adaptation Strategy and Action Plan 2011–14 (Ontario Ministry of Environment, 2011)	Biophysical/engineering oriented with social adaptation embedded	Moderate	Strong	Strong
	2009 California Climate Adaptation Strategy (California Natural Resources Agency, 2009)	Predominantly engineering and biophysically oriented	Weak	Strong	Strong
Regional scale	Great Barrier Reef Marine Park Authority Climate Change Action Plan (Great Barrier Reef Marine Park Authority, 2012)	Biophysical with some social adaptation embedded	Moderate	Strong	Weak
	Australian Regional NRM Plans for Climate Change (Department of Sustainability, Environment, Water, Population and Communities, 2013)	Biophysical with social adaptation embedded	Moderate to weak	Strong	Weak
	Future-Proofing Perth's Eastern Region: Adapting to Climate Change (Eastern Metropolitan Regional Council, 2009)	Biophysical focused with some engineering/social adaptation	Weak	Strong	Moderate

(Continued)

Table 16.1 The Relative Focus of Typical Examples of Climate Adaptation Strategies from Global to Local Scale. *Continued*

			Relative Focus on Adaptation Measures		
Scale	Initiative	Approach Orientation	Social	Biophysical	Engineering
Local scale	Copenhagen Climate Change Adaptation Plan (Copenhagen Climate City, 2011)	Predominantly engineering	Weak	Moderate	Strong
	Climate Change Adaptation in New York City: Building a Risk Management Response (Rosenzweig & Solecki, 2010)	Predominantly engineering	Weak	Moderate	Strong
	London Climate Change Adaptation Strategy (Greater London Authority, 2011)	Engineering/ biophysical oriented with some social adaptation	Strong	Strong	Strong
	City of Santa Cruz Climate Adaptation Plan 2012–17 (Santa Cruz City Council, 2012)	Predominantly engineering focus with some social background	Weak	Moderate	Strong

CPACC, *Caribbean Planning for Adaptation to Climate Change;* NRM, *natural resource management.*

extensive coral bleaching, *and* the risk of more intense cyclones and floods (CSIRO Bureau of Meteorology, 2007; Hilbert et al., 2014). Consequently, sociologically and economically diverse subregions like the northern Gulf of Carpentaria, the Torres Strait, Cape York Peninsula, and the Wet Tropics face an uncertain future. These subregions represent a social diversity typical across the wider tropics. The Gulf is a vast and flat pastoral, aboriginal, and mining landscape with extensive low-lying floodplains. The Torres Strait comprises an island-based Melanesian culture. Cape York Peninsula is a remote aboriginal and pastoral domain. Finally, the Wet Tropics represents a developed tourism region (based on the city of Cairns) with an intensive agricultural economy. At the same time, these four subregions contain some of Australia's most significant biodiversity, including existing and proposed World Heritage Area sites (Wet Tropics, Great Barrier Reef, and Cape York Peninsula), wetlands of international significance (Gulf of Carpentaria), and places of marine diversity (e.g., the Torres Strait) (Valentine, 2006; Wet Tropics Management Authority, 2007).

As part of an unfolding experiment to undertake a more socially oriented approach to climate adaptation planning at a regional scale, TNQ has been the location of focused adaptation planning activity involving government, universities, and industry sectors since 2010. Beginning in the Wet Tropics subregion, Dale et al. (2011) piloted the use of social resilience indicators and participatory

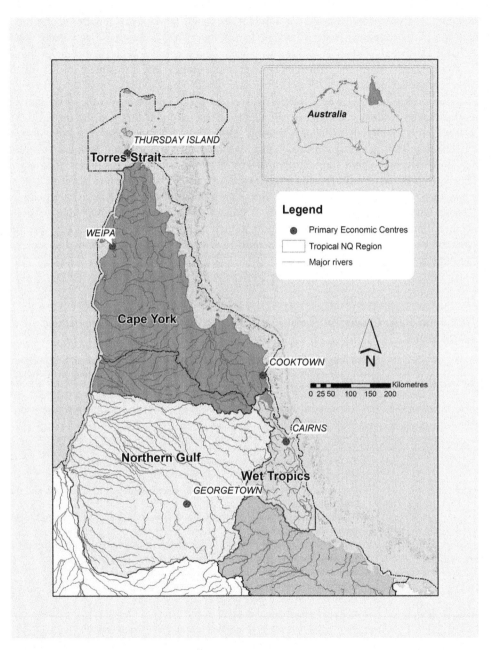

FIGURE 16.1

Map of TNQ, showing four distinct subregions. *TNQ*, Tropical North Queensland.

knowledge-building processes to underpin a social approach to regional climate adaptation planning. In 2012/13 the approach was expanded to all four TNQ subregions (see Dale, Vella, & Cottrell, 2015; Dale et al., 2014). This research-oriented approach has been used to help this region negotiate Queensland (state) and Australian government policy and investment for regional and subregional climate action.

The approach used in TNQ was a hybrid of two dominant approaches to the development and application of social resilience indicators: the technical use of indicators and the use of participatory processes. It was intended that this emerge as a long-term and adaptive social adaptation approach. The approach gathered multiple "lines of evidence" (both formal quantitative indicators and other knowledge sets) into four clusters of key attributes of social resilience. These clusters were refined by Dale et al. (2011) following review of the six key attributes identified by Ross, Cuthill, Maclean, Jansen, and Witt (2010) and later published by Maclean, Cuthill, and Ross (2013). These attribute clusters (see Table 16.2) collectively contain the knowledge necessary to track social resilience at any scale. The indicators gathered were codeveloped with key stakeholders in all four subregions. They were consequently used to devise priority, whole-of-region social adaptation measures at the regional scale.

Table 16.2 Four Basic Clusters of Social Resilience Attributes Useful at Regional and Subregional Scales.

Attribute Cluster	Example Resilience Attributes That Can Be Used to Organize the Gathering of Multiple Lines of Evidence and as a Basis for the Development of Attribute Indicators (Based on Costantini, 2014)
Knowledge, aspirations and capacity	• Regional individual, family, and business awareness of climate change and resource sustainability • Education levels and spread across the region • Skill levels and spread across the region • Regional aspirations for sustainable natural resource management • Regional individual/business leadership/complex problem-solving
Governance	• Connectivity and trust in and among key decision-making institutions and sectors within the region • Adaptive management capacity of key decision-making institutions and sectors within the region • Adaptive use and management of integrated knowledge sets within the region
Economic viability	• Regional diversity and quality of growth in economic activity • Regional vulnerability of natural and energy resource base • Regional inclusiveness and economic fairness/equity • Regional workforce participation and employment
Community vitality	• Regional demographic stability • Well-being/happiness within the general regional community • General regional community health and disparities • Regional community services access, and disparities • Regional measures of housing, accommodation, and accessibility • Regional aspects of built infrastructure vulnerability • Regional community safety and risk

This socially oriented approach to climate change adaptation planning allowed a wide range of social adaptation measures to be identified that are relevant to vulnerable tropical regions. Indeed, it is useful to explore the range of social adaptation measures that emerged from the TNQ approach for two reasons:

1. to further support the TNQ region in developing its own regional social adaptation strategies and to underpin policy and investment negotiations with the Queensland and Australian governments and
2. to explore and more confidently publish the wider range of typical social adaptation measures that might emerge from such an approach, particularly as they might relate to other tropical countries.

Consequently, the methods used in the development of this chapter are simple. Based on the foundation region-wide social adaptation measures or strategies identified in Dale et al. (2014), the research team did the following:

1. They undertook a detailed examination of the literature and experiential knowledge available to further scope the regional context behind the social adaptation measures, to road test the robustness of the narrative concerning the measures, and to determine their specific value within the TNQ region.
2. They engaged regional stakeholders to review these more fully developed measures (see Dale et al., 2014). Key stakeholders included Australian and Queensland government agencies, statutory authorities, regional natural resource management (NRM) bodies, local governments, regional development organizations, industry development bodies, expertise-based research brokerage and delivery institutions and indigenous development, social development, and conservation organizations.
3. They broadly (and qualitatively) evaluated the typical relevance of the types of measures identified to other tropical regions potentially affected by climate change, including developed tropical tourism—agricultural economies, developing tropical economies, and island-based nation-states.

16.3 RESULTS: TYPICAL ADAPTATION MEASURES ARISING FROM THE TNQ SOCIAL RESILIENCE ASSESSMENT

The subregional social resilience assessment processes undertaken in TNQ enabled the identification of 17 potential climate adaptation measures at that scale. These were consolidated into critical measures or strategies that had the potential to address a wider range of social resilience concepts at the whole of TNQ scale. The following subsections outline the 17 strategies that emerged in TNQ in some detail.

16.3.1 ENHANCED REGIONAL AND LOCAL LAND USE PLANNING

The TNQ social resilience assessment considered that a 5-year regional program of land use risk assessment was critical to address identified coastal-related climate risks (e.g., flood, storm surge, and emergency evacuation). Such an assessment would need to integrate biophysical (e.g., surges) and social (e.g., population-wide stress) responses to risk. Regional agreement among key stakeholders would then need to be sought to integrate the consequent spatial and infrastructure implications into next-generation regional land use plans (state) and local government planning schemes. As the key delivery mechanism for this measure, a TNQ social resilience assessment was also proposed to create a special-purpose and regionalized planning capability within the region's two Regional Organisations of Councils. Such an activity would not only build on the strengths of existing regional and local planning mechanisms but also focus on identifying their weaknesses to enable structured negotiation in next-generation land use/infrastructure plans.

16.3.2 **INFRASTRUCTURE RISK IDENTIFICATION**

The TNQ social resilience assessment identified that infrastructure risk identification (physical and social) and suitable planning solutions could be negotiated with the Queensland Reconstruction Authority, resulting in infrastructure risk assessment and retrofitting/redesign estimates of problematic infrastructure predisaster. The proposed resilience assessment could enable more rapid regional negotiation and devolution of Natural Disaster Reconstruction and Relief Arrangement (NDRRA) funds. It was considered that the success of this strategy would rely on higher level policy acceptance of the need to redesign the NDRRA program to enable greater devolution of decision-making powers, more rapid assessment, more flexible delivery, and an ability to replace infrastructure to withstand risks (betterment) rather than just their replacement value.

16.3.3 **CLIMATE INFORMATION SERVICES**

The development of radar facilities increased deployment of monitoring stations, greater use of diffuse data sources (e.g., the Internet of Things) and more effective event-based modeling was another measure identified. It was considered that this would improve community safety in cyclones and intense rainfall events, though the strategic form, location, and cost of these facilities and initiatives would need to be determined. Achieving this would require stronger partnerships between the Bureau of Meteorology and subregional communities, as well as the integration of new arrangements into the standard disaster response practices of households/businesses and farms and within disaster response and management agencies.

16.3.4 **NEW INFRASTRUCTURE FUNDING MODELS**

TNQ has a significant infrastructure deficit due to its limited population and remoteness. The social resilience assessment considered that developing new regional infrastructure funding models and the expansion of an emerging Local Government Roads Alliance Model was essential. This would empower the capacity of local government in asset management of strategic roads to address critical accessibility risks. This approach would require strategic investment to maximize the benefit of regional freight networks, linked to improved port services and development opportunities, and limiting major climate-related market accessibility problems.

16.3.5 **RECASTING RESEARCH MODELS**

Strengthening the current research investment model into the protection of the Great Barrier Reef and Wet Tropics, World Heritage Areas, and wider biodiversity and natural resource assets in TNQ was also considered essential. At the time, the major current investment ($40 million over 5 years) was centrally driven to meet government (vs regional stakeholder) priorities, reducing its adaptive management impact. At least two TNQ subregions had also historically been excluded from this research funding. It was considered that an even stronger research model was needed which fostered more regionally oriented stakeholder and researcher partnerships for the sustainable development of TNQ's natural assets in the face of climate change.

16.3.6 **COLLABORATIVE ALLIANCES AROUND AGRICULTURE AND FISHERIES**

The TNQ social resilience assessment considered that a collaborative research and development alliance focused on tropical agriculture and fisheries was needed to build climate resilience in these industries. Certain agricultural sectors were vulnerable to climate change, including tree crops, forestry, fishing, pastoral, and horticultural sectors. As a result, any emerging approach would need to be focused on being regionally led with key subregional nodes dealing with specific subregional climate problems. Such work would also need to link with productivity and innovation agenda in agriculture and drive educational linkages. A new approach could also become a form of knowledge export and help to explore community innovation and new crops/markets.

16.3.7 **SMALL BUSINESS RESILIENCE**

Strategies to reduce small business and farm debt in vulnerable rural communities were also considered to be important climate adaptation measures. This included farm debt tax write-downs, low-interest loans, farm financial counselors, and longer term small business adjustment capacity and servicing. Rural enterprises were particularly vulnerable in the Gulf and Wet Tropics regions following two Category 4/5 cyclones since 2006. Small businesses across TNQ were also generally vulnerable as a result of intense events. Integrated strategies were required to reduce farm and small business debt across the region, increasing their resilience. Such an approach would require stronger partnerships between governments, NRM bodies, and industry organizations.

16.3.8 **INSURANCE REFORM**

The TNQ social resilience assessment proposed major new partnerships between the insurance industry, governments, and councils that could help improve urban and rural disaster preparedness. Households and businesses in both rural and urban communities were particularly vulnerable in TNQ, most particularly in the Wet Tropics following two Category 4/5 Cyclones since 2006. Some sectors (particularly strata title units, the nursery industry, etc.) and areas (e.g., around Innisfail) were found to be particularly vulnerable to insurance changes. It was considered that a strategy would be required to reduce the impact of increased premiums and the potential for insurance policy exclusions in some sectors and areas. Implementation would require stronger relationships between governments, the regional community, and the insurance industry.

16.3.9 DISASTER COORDINATION AND RESPONSE

Improving disaster coordination and response infrastructure and upgrading cyclone shelter facilities to Category 5 standards were considered important strategies. Recent investment in Category 5 shelters and new disaster response facilities were largely in place in the Wet Tropics, though hospitals and roads for strategic evacuation remained vulnerable. Risk assessment of facilities and housing in Cape York Peninsula communities remained outstanding, and Category 5 shelters and disaster coordination facilities remained underdeveloped. Implementation would require stronger partnerships between local, state, and federal governments/private sector.

16.3.10 CAPITALIZING ON THE REGION'S CLIMATE ADAPTATION SKILLS

The work concluded that developing a Natural Disaster Center of Excellence in TNQ (building on an emerging Energy Remote Service Delivery Capacity Center of Excellence) could also provide real adaptation options for the region. It was considered that large urban centers in the Wet Tropics (with a large proportion of southern migrants) and remote communities with few services were particularly vulnerable in the response phase following major cyclonic events. Implementation would require strong partnerships between the university and education sectors with disaster management agencies and the community. It was also considered that there were opportunities to establish a basis for continuous improvement in disaster response and management and that there was potential to develop an export economy servicing the Pacific and South East Asia.

16.3.11 REGIONAL ENERGY REFORM IN TROPICAL NORTH QUEENSLAND

The TNQ social resilience assessment considered that collective state and commonwealth commitment to the development and negotiated implementation of a whole of NQ energy investment strategy was a critical adaptation measure. The entire NQ region has a very low level of energy security, making it vulnerable to transmission cuts and also presenting a major cost risk in the event of increased energy prices. In TNQ, remote communities were particularly vulnerable due to a reliance on diesel-based generation systems. Implementation will require a high-level policy and targeted financial investment to achieve energy security and affordability in TNQ. Any strategy would need to integrate northern baseload energy generation with a significantly enhanced feed of renewables, particularly in remote communities.

16.3.12 REGIONAL LANDSCAPE MITIGATION

Another important adaptation measure identified included the progression of regional landscape-based options to reduce emissions and to contribute to regional economic diversification. Great opportunities for regional scale greenhouse gas abatement and mitigation in the landscape exist in the Wet Tropics and Cape York Peninsula. A strategy for progressing this opportunity under the Australian government's emissions reduction framework would require innovative partnerships between land managers, NRM bodies, researchers, and carbon emitters.

16.3.13 SPATIAL INFORMATION HUBS FOR PROPERTY SCALE MANAGEMENT AND RESPONSE

The assessment considered that spatial information hubs operating at various scales in the region would provide a basis for better property scale management and local disaster/emergency response. It was considered that the entire region had a low level of available spatial information for adaption planning at the property scale, and for use in a timely manner for disaster planning and response. Implementation strategies would need to support emerging nongovernment, continuously improving spatial information alliances at relevant scales.

16.3.14 HUMAN SERVICE SECTOR CAPACITY

An increase in regional strategic capacity of the region's human services sector was also considered to be an important social adaptation to climate change in the assessment. The whole region suffers significant human service sector capacity limitations, particularly with respect to the capacity of the sector to manage disaster impacts in marginalized communities and groups. The sector would need support to develop a continuously improving human service planning and delivery capacity at the regional and subregional scales.

16.3.15 WATER RESOURCE MANAGEMENT

The assessment considered that regionally managed water assessment for existing water supplies, improved water use efficiency, and new agricultural precincts would be an important regional adaptation. The Wet Tropics region and remote communities have constrained water resources, and the entire region could experience more acute periods of water crisis. An implementation strategy would need to be based on increasing local government water management capacity and increased water development partnership between the private sector and state and Australian governments.

16.3.16 BIOSECURITY AND HEALTH RISKS

A refocus on managing biosecurity and tropical health risks associated with climate change was considered important in the TNQ social resilience assessment. The Wet Tropics community was particularly vulnerable to increased mosquito-borne disease risks, while the Torres Strait and Cape York Peninsula were particularly vulnerable to emerging climate-related food and health security risks in Papua New Guinea (PNG). The Gulf, Cape York Peninsula, and Wet Tropics faced increased weed risks and domestic animal health risks in the face of climate change. A strategic focus would need to include building stronger tropical health research capacity-based in Cairns with strong outreach into communities and PNG.

16.3.17 LONG-TERM ADAPTIVE REGIONAL GOVERNANCE MODELS

An agreed state and federal long-term commitment to enhancing the emerging place-based governance framework for the TNQ region was also considered important in the TNQ social resilience assessment. Such an approach would need to include benchmarking and adaptive management of climate change strategies. The cohesive development of climate adaptation strategies at community, local government, subregional, and regional scales was only just emerging and would need to be progressed on a stable basis. Such an approach, based on strong partnerships between the community and university sector, would enable cohesive adaptations to emerge and the negotiation of appropriate policy and budgetary responses from different levels of government.

16.4 DISCUSSION AND CONCLUSION: THE IMPLICATIONS OF TYPICAL SOCIAL ADAPTATION RESULTS ACROSS THE TROPICS

While the measures identified above are specific to the TNQ region, more generalized characterizations of these measures can be made and their applicability in different tropical regions considered. Table 16.3 outlines the outcomes of our broad review of the potential relevance of these and other generic measures aimed at improving social adaptation across the tropics. For simplicity, we grouped several tropical countries into four broad regions: tropical Asia, tropical Africa, the tropical Americas, and Pacific Island nations. Our overview assessment of these typical social adaptation solutions suggests that they are highly relevant to tropical countries across the globe or are increasingly becoming so. This suggests that some of the approaches taken to identifying, measuring, and building social resilience within this study have potentially wide application across the tropics.

As a result of the subregional and regional engagement emerging from this project, both the research team and key regional and subregional stakeholders are actively working toward the development of a jointly negotiated decision and investment framework for major adaptation in TNQ. Hence, based on our wider lessons, we consider the long-term and stable implementation and evaluation of the impact of this approach is now a future research priority. The potential application of the approach (in partnership with other vulnerable tropical regions) should also be explored. Because of its focus on strong engagement and cost-effective data gathering (i.e., less than AUS $100,000 per benchmark), this approach is applicable in vulnerable regions across the tropical zone, helping to implement international agreements on greenhouse gas abatement. A benchmarking and social adaptation framework of this kind, once in place, has the capacity to start focusing long-term regional attention on very context-specific social adaptation measures. From the experience emerging in TNQ, this at least includes some of the key social resilience building initiatives outlined in the results section.

Table 16.3 The Relevance of Tropical North Queensland Social Resilience Measures to Other Tropical Regions.

Typical Measures/ Tropic Regions	Relevance (Low/Medium/High)				
	Tropical Africa	Tropical Asia	Tropical Pacific	Tropical Americas	Explanatory Comments
Enhanced land use planning	M	H	H	M	All four tropical regions are in the process of diversifying their economies from communal land estates to capital-driven land systems. As a result there is a need for high-level public sector investment in supporting transition planning that can reduce climate change impacts. Land use planning in tropical Asia is a particularly high priority because of the region's rapid urban growth (Tanner, Mitchell, Polack, & Guenther, 2009). Land use planning is also critical to mitigate inundation risks in the short term in low-lying and cyclone-prone areas of tropical Asia and the tropical Pacific (McGranahan, Balkn, & Anderson, 2007; Mimura, 1999).
Spatial information hubs	M	M	M	M	Spatial information and spatial data management/ assessment programs (e.g., GIS) can be used as a decision-support tool by decision-makers before, during, and after natural disasters (Ahmed, 2015; Gunes, Kovel, & Member, 2000). As all of these tropical regions are at risk of more frequent severe weather events, it is critical that decision-makers have evidence to guide decisions on evacuation strategies, the identification of high-risk/vulnerable areas (or areas without sufficient evacuation centers), and planning for emergency services (Zerger & Smith, 2003). Therefore spatial information hubs are relevant to all of these regions.
Improved climate information	M	H	H	M	Communities with good access to weather and climate data have a greater resilience and capacity to prepare for or evacuate dangerous areas during severe weather, and subsequently a greater ability to mitigate some of the negative ramifications of such events (Adger, Hughes, Folke, Carpenter, & Rockstrom, 2005). Improved climate information is particularly critical in tropical Asia, where the high population density and large urban populations make evacuation difficult and time-consuming. The tropical Pacific is also vulnerable because they often have limited ability to retreat to higher ground, lack evacuation infrastructure (Jackson, McNamara, & Witt, 2017), and may need to evacuate to another nation in extreme circumstances (Ronneberg, 2008).

(Continued)

Table 16.3 **The Relevance of Tropical North Queensland Social Resilience Measures to Other Tropical Regions.** *Continued*

Typical Measures/ Tropic Regions	Relevance (Low/Medium/High)				
	Tropical Africa	Tropical Asia	Tropical Pacific	Tropical Americas	Explanatory Comments
Infrastructure risk identification	M	H	H	M	Areas with greater amounts of infrastructure and development are at a higher risk of direct damage in natural disasters compared with less developed and dense areas (Dwyer, Zoppou, Nielsen, Day, & Roberts, 2004). Poor building design and construction also increases risk (Jackson et al., 2017). Based on this, it is critical that densely populated regions are able to identify high-risk infrastructure and minimize risk through redesign, retrofitting, and risk management planning. Many of tropical Asia's megacities are built on deltas (Ericson, Vorosmarty, Dingman, Ward, & Meybeck, 2006; Mirza, Warrick, & Ericksen, 2003), while the tropical Pacific is particularly low-lying, and both regions are consequently highly vulnerable to inundation and destruction of critical infrastructure (Kreimer, Arnold, & Carlin, 2003).
Infrastructure funding models	H	H	H	H	In the Cairns round of G20 Finance Minister Meetings in 2014, priority was given to exploring new models and approaches to infrastructure. Given the significant impact of climate change and sea-level rise on infrastructure, risk management and the need to redesign infrastructure networks (Azevedo de Almeida & Mostafavi, 2016; Nicholls & Cazenave, 2010), this can be easily seen as a genuinely global priority.
Recast research models	M	M	M	M	There is a rapidly increasing international literature concerning the recasting of national research models to overcome power imbalances (Walsh, Brugha, & Byrne, 2016) and move toward better engaged research partnerships with vulnerable regions. This has been identified as an important priority in northern Australia (Babacan et al., 2012; Dobbs et al., 2016; Jones et al., 2008), and these considerations would equally apply to nations across the tropical world. The long-term nature of this cultural change, however, suggests a medium versus high priority.
Agricultural alliances	H	H	M	H	Agricultural research and development alliances provide agricultural producers with support to be innovative and responsive in their land management practices to climatic changes (Brooks & Loevinsohn, 2011; Vermeulen et al., 2012). Such alliances can also provide knowledge and information transfer between landholders and scientists (and vice versa). Tropical Africa, Asia, and Americas all have vast areas of agricultural production to support the needs of their populations (Ramankutty, Evan, Monfreda, & Foley, 2008), while Pacific nations tend to have low agricultural capacity due to poor soil quality (Sharma, 2006). Consequently, agricultural alliances, while still relevant (e.g., see Pelling & Uitto, 2001), are less critical to social resilience in the tropical Pacific, compared with other tropical regions.

Table 16.3 The Relevance of Tropical North Queensland Social Resilience Measures to Other Tropical Regions. *Continued*

Typical Measures/ Tropic Regions	Relevance (Low/Medium/High)				
	Tropical Africa	Tropical Asia	Tropical Pacific	Tropical Americas	Explanatory Comments
Small business resilience	M	M	M	M	Small businesses are likely to be more adversely impacted by the impacts of climate change compared with larger businesses, emphasizing the importance of building the resilience of the small business sector in high-risk regions (Linnenluecke & Griffiths, 2010). The economies of all four tropical regions are still developing and emerging. Subsequently, small businesses represent a substantial portion of their economies, emphasizing the importance of building small business resilience (Liedholm, 2002; Robson, Haugh, & Obeng, 2009).
Insurance reform	M	H	H	M	The increasing frequency and severity of severe weather events is likely to increase the cost or decrease the availability of insurance altogether in high-risk areas (Pryce & Chen, 2011; Stern, 2006). Rising insurance costs also mean that poor communities living in high-risk areas become further marginalized and disadvantaged following severe weather events (Stern, 2006). The effects of this will be particularly felt in tropical Asia (Sawada, 2017) and the tropical Pacific due to the regions' high vulnerability to the impacts of cyclones and flooding/ inundation. In response to these challenges, initiatives such as regional resilience trust funds have been proposed (Keenan, 2017).
Disaster coordination	H	H	H	H	Well-constructed and located disaster facilities in high-risk areas provide both a safe shelter for people during an event and a base for coordination of recovery actions and distribution of basic resources following a disaster (Cho, 2005; Pinkowski, 2008). Developing local (rather than government or top-down) disaster coordination capacities enables communities to mobilize and respond to such events, rather than wait for "help to come" (Kirkby et al., 2017; Pinkowski, 2008). Disaster coordination is critical to the resilience of communities in all tropical regions (Haque, 2003).
Regional energy reform	H	H	H	H	Fossil fuels (coal and diesel) are the primary energy source for communities in tropical Asia and tropical Pacific (Nguyen & Ha-Duong, 2009). Countries in tropical America are highly reliant on hydroelectricity, which is likely to become threatened as rainfall patterns change and Andean glaciers melt (Bradley, Vuille, Diaz, & Vergara, 2006). Tropical Africa draws more equally from a diverse range of energy sources than other tropical regions, including hydroelectricity, coal, natural gas, and biomass (Iwayemi, 2008; Karekezi, 2002). Energy reform is needed in all regions to reduce the vulnerability of their energy systems to climatic shifts and reduce their reliance on nonrenewable energy sources.

(Continued)

Table 16.3 The Relevance of Tropical North Queensland Social Resilience Measures to Other Tropical Regions. *Continued*

Typical Measures/ Tropic Regions	Relevance (Low/Medium/High)				
	Tropical Africa	Tropical Asia	Tropical Pacific	Tropical Americas	Explanatory Comments
Regional landscape change	H	H	M	H	Flexible regional or landscape-scale approaches to managing the impacts of climate change are increasingly being suggested as a key mechanism for managing the critical social, economic, and environmental trade-offs required for regional development (Sayer et al., 2013, 2014). Climate change across these four tropical regions exacerbates the needs for such approaches but is particularly significant in Asia as a result of high land use intensification (van Asselen & Verburg, 2013). Sayer et al. (2014) have already identified the value of greater interchange on these issues and approaches between northern Australia and the wider tropics.
Water resource management	H	H	H	H	Major parts of all four regions experience a wet and a dry season, with alternative water sources needed to support the needs of communities during extended dry seasons. Most islands in the tropical Pacific are highly vulnerable to water crisis because of their reliance on rainfall to support their population (Asian Development Bank, 2013). Tropical regions of Africa, Asia, and America have access to both ground and surface water sources (Shiklomanov, 2000), which can be used based on seasonal water availability. Some of the regions' reliance on these water sources for drinking water and power further emphasizes the importance of using/managing them sustainably (in particular, countries such as India, see The World Bank, 2010a).
Biosecurity and health risks	H	H	H	H	Rising temperatures in all of these tropical regions are likely to increase the spread of invasive species and mosquito- and tick-borne diseases such as dengue fever, malaria, yellow fever, and Lyme disease (Kriticos, 2012; McMichael, Woodruff, & Hales, 2006; Przeslawski, Ahyong, Byrne, Worheides, & Hutchings, 2008). Incidences of malaria have been linked to increased temperatures and ENSO patterns in tropical Africa, tropical America, tropical Pacific, and tropical Asia (McMichael et al., 2006).
Human service sector capacity	H	M	M	H	Disruption of the community and human services sector by extreme weather events can exacerbate the impact of a disaster on already vulnerable populations who are unable to obtain the help or services needed to recover following a disaster without the human service sector (Mallon, Hamilton, Black, Beem, & Abs, 2013). The community and human service sectors in tropical Africa and tropical America are highly underdeveloped and limited in their capacity to provide assistance to communities day to day, let alone during a disaster (Fleury et al., 2013; UNFCCC, 2006). Human services in tropical Asia and tropical Pacific regions are a little more developed, but still evolving.

Table 16.3 The Relevance of Tropical North Queensland Social Resilience Measures to Other Tropical Regions. *Continued*

Typical Measures/ Tropic Regions	Relevance (Low/Medium/High)				
	Tropical Africa	Tropical Asia	Tropical Pacific	Tropical Americas	Explanatory Comments
Improved education	H	H	H	H	Access to education is unequal across/in these tropical regions, with more remote communities having poorer access than those in more populous areas (Carson & McConnel, 2011; Misra, 2006). The results of the OECD Program for International Student Assessment show that tropical African and tropical American countries are performing poorly against other countries in basic education skills (OECD, 2013). Similarly, most countries in tropical Asia scored poorly, with the exception of Vietnam (OECD, 2013). Education in the tropical Pacific is limited by poverty and remoteness, leading to low levels of secondary education (Kidd, 2012).
Capitalizing on disaster skills	L	M	M	M	Tropical Africa has a limited fiscal capacity to invest in disaster adaptation or disaster response and recovery strategies, suggesting limited scope to capitalize on disaster adaptation skills (The World Bank, 2010b). The Asian Disaster Preparedness Center and the Pacific Disaster Center provide nations in tropical Asia and the tropical Pacific regions with decision-support tools for disaster preparation and response (Asian Disaster Preparedness Center, 2013; Pacific Disaster Center, 2013). Countries in tropical America are informed by the Regional Disaster Information Center (CRID) for Latin American and the Caribbean on matters relating to adaptation and disaster risk management. The capacity and connectivity of the many disaster response organizations in tropical regions continues to evolve and grow, presenting new opportunities.
Adaptive governance	H	H	H	H	Responsive and adaptive climate change management governance frameworks involve cohesive strategies, strong partnerships, conflict resolution, and flexibility (Nursey-Bray, 2013). The high vulnerability of tropical regions to extreme weather events and their impacts (Hashim & Hashim, 2016) reinforces the importance and need for strong adaptive governance frameworks to support adaptation and resilience building.

ENSO, *El Niño—Southern Oscillation.*

ACKNOWLEDGMENTS

We would like to thank the Queensland Government's Centre for Social Science Innovation (QCSSI), the Northern Futures Collaborative Research Network, and the Regional Economies Centre of Excellence for their academic and financial support. We also thank the many community stakeholders involved in this work.

REFERENCES

Adger, W., Hughes, T., Folke, C., Carpenter, S., & Rockstrom, J. (2005). Social-ecological resilience to coastal disasters. *The Science of the Total Environment*, *309*, 1036—1039. Available from https://doi.org/10.1126/science.1112122.

Adger, W. N., Dessai, S., Goulden, M., Hulme, M., Lorenzoni, I., Nelson, D. R., ... Wreford, A. (2009). Are there social limits to adaptation to climate change? *Climatic Change*, *93*(3), 335—354. Available from https://doi.org/10.1007/s10584-008-9520-z.

Ahmed, A. (2015). Role of GIS, RFID and handheld computers in emergency management: an exploratory case study analysis. *Journal of Information Systems and Technology Management*, *12*(1), 3—28. Available from https://doi.org/10.4301/s1807-17752015000100001.

Azevedo de Almeida, B., & Mostafavi, A. (2016). Resilience of infrastructure systems to sea-level rise in coastal areas: Impacts, adaptation measures, and implementation challenges. *Sustainability*, *8*(11). Available from https://doi.org/10.3390/su8111115.

Asian Development Bank. (2013). *3 in 4 Asia-Pacific nations facing water security threat: Study.* <http://www.adb.org/news/3-4-asia-pacific-nations-facing-water-security-threat-study>.

Asian Disaster Preparedness Center. (2013). *ADPC at a glance.* <http://www.adpc.net/igo/contents/adpcpage.asp?pid = 2>.

Australian Government Department of the Environment. (2013). *Pacific-Australia climate change science and adaptation planning program.* <http://www.climatechange.gov.au/climate-change/grants/pacific-australia-climate-change-science-and-adaptation-planning-program>.

Babacan, H., Dale, A., Andrews, P., Beazley, L., Horstman, M., Campbell, A., ... Miley, D. (2012). *Science engagement and tropical Australia: Building a prosperous and sustainable future for the north.* Kingston, ACT: Department of Industry, Innovation, Science, Research and Tertiary Education.

Bradley, R. S., Vuille, M., Diaz, H. F., & Vergara, W. (2006). Threats to water supplies in the tropical Andes. *The Science of the Total Environment*, *312*, 1755—1756. Available from https://doi.org/10.1126/science.1128087.

Brooks, S., & Loevinsohn, M. (2011). Shaping agricultural innovation systems responsive to food insecurity and climate change. *Natural Resources Forum*, *35*(3), 185—200. Available from https://doi.org/10.1111/j.1477-8947.2011.01396.x.

California Natural Resources Agency. (2009). *California climate adaptation strategy: A report to the Governor of the State of California in response to executive order S-13-2008.* Sacramento: California Natural Resources Agency. <http://resources.ca.gov/docs/climate/Statewide_Adaptation_Strategy.pdf>.

Caribbean Community Secretariat. (2013). *Caribbean Planning for Adaptation to Climate Change (CPACC) project.* <http://www.caricom.org/jsp/projects/macc%20project/cpacc.jsp>.

Carson, D., & McConnel, F. (2011). Indigenous health and community services employment in remote Northern Territory: A baseline examination of 2006 and 2001 Census data. *The Australian Journal of Rural Health*, *19*(5), 255—258. Available from https://doi.org/10.1111/j.1440-1584.2011.01220.x.

Copenhagen Climate City (2011). *Copenhagen climate change adaptation plan.* Copenhagen: Miljo Metropolen.

Cho, J. (2005). *Smart infrastructures*. UN Chronicle 1, 39, 42.

Chu, E., Anguelovski, I., & Carmin, J. (2016). Inclusive approaches to urban climate adaptation planning and implementation in the Global South. *Climate Policy*, *16*(3), 372−392. Available from https://doi.org/10.1080/14693062.2015.1019822.

Costantini, K. (2014). Improving social resilience in response to climate change in Far North Queensland and Torres Strait. In: *Independent Study Project (ISP) collection*. <https://digitalcollections.sit.edu/cgi/viewcontent.cgi?referer = &httpsredir = 1&article = 2947&context = isp_collection>.

CSIRO Bureau of Meteorology. (2007). *Climate change in Australia: Technical report 2007*. Melbourne: CSIRO.

Dale, A. P., Taylor, N., & Lane, M. (2001). *Social assessment in natural resource management institutions*. Collingwood: CSIRO Publishing.

Dale, A.P., Vella, K., Potts, R., Voyce, B., Stevenson, B., Cottrell, A., ... Pert, P. (2014). *Applying social resilience concepts and indicators to support climate adaptation in Tropical North Queensland, Australia*. JCU, Cairns. <http://elibrary.gbrmpa.gov.au/jspui/bitstream/11017/2966/1/Attributes_Report.pdf>.

Dale, A., Vella, K., & Cottrell, A. (2015). Can social resilience inform SA/SIA for adaptive planning for climate change in vulnerable regions? *Journal of Natural Resources Policy Research*, *7*, 93−104. Available from https://doi.org/10.1080/19390459.2014.963371.

Dale, A., Vella, K., Cottrell, A., Pert, P., Stephenson, B., Boon, H., ... Gooch, M. (2011). *Conceptualising, evaluating and reporting social resilience in vulnerable regional and remote communities facing climate change in tropical Queensland*. Cairns: Reef and Rainforest Research Centre. <http://eprints.jcu.edu.au/21204/>.

Department of Sustainability, Environment, Water, Population and Communities (2013). *Regional NRM planning for climate change fund (Stream 1)*. <http://www.environment.gov.au/cleanenergyfuture/regionalfund/about.html>.

Dobbs, R. J., Davies, C. L., Walker, M. L., Pettit, N. E., Pusey, B. J., Close, P. G., ... Davies, P. M. (2016). Collaborative research partnerships inform monitoring and management of aquatic ecosystems by Indigenous rangers. *Reviews in Fish Biology and Fisheries*, *26*(4), 711−725. Available from https://doi.org/10.1007/s11160-015-9401-2.

Dodman, D., & Mitlin, D. (2013). Challenges For community-based adaptation: Discovering the potential for transformation. *Journal of International Development*, *25*(5), 640−659. Available from https://doi.org/10.1002/jid.1772.

Dwyer, A., Zoppou, C., Nielsen, O., Day, S. and Roberts, S. (2004). Quantifying social vulnerability: A methodology for identifying those at risk to natural hazards. Geoscience Australia, Canberra, <http://www.ga.gov.au/image_cache/GA4267.pdf>.

Eastern Metropolitan Regional Council. (2009). *Future proofing Perth's eastern region: Adapting to climate change*. Perth: Eastern Metropolitan Regional Council. <http://www.emrc.org.au/future-proofing-perth-s-eastern-region-climate-change-adaptation.html>.

Ericson, J., Vorosmarty, C., Dingman, S., Ward, L., & Meybeck, M. (2006). Effective sea-level rise and deltas: Causes of change and human dimension implications. *Global and Planetary Change*, *50*(1−2), 63−82. Available from https://doi.org/10.1016/j.gloplacha.2005.07.004.

European Commission (2013). *EU adaptation strategy package*. <http://ec.europa.eu/clima/policies/adaptation/what/documentation_en.htm>.

Feltmate, B., & Thistlethwaite, J. (2012). Climate change adaptation: A priorities plan for Canada. In: *Report of the Climate Change Adaptation Project (Canada)*. University of Waterloo, Waterloo. <https://uwaterloo.ca/environment/sites/ca.environment/files/uploads/files/CCAP-Report-30May-Final.pdf>.

Fleury, S., Faria, M., Duran, J., Sandoval, H., Yanes, P., Penchaszadeh, V., & Abramovich, V. (2013). *Right to health in Latin America: Beyond universalization*. Chile: United Nations Economic Commission for Latin America and the Caribbean. <http://repositorio.cepal.org/bitstream/handle/11362/35953/S20131047_en.pdf>.

Forsyth, T. (2013). Community-based adaptation: A review of past and future challenges. *Wiley Interdisciplinary Reviews: Climate Change, 4*(5), 439−446. Available from https://doi.org/10.1002/wcc.231.

Germanwatch and WWF International. (2010). *International action on adaptation and climate change: What roads from Copenhagen to Cancun?* Bonn: Germanwatch and WWF International. <http://germanwatch.org/klima/ad-cph-canc.pdf>.

Great Barrier Reef Marine Park Authority. (2012). *Great Barrier Reef Marine Park Authority Climate Change Action Plan.* <http://www.gbrmpa.gov.au/outlook-for-the-reef/climate-change/marine-park-management/climate-change-action-plan>.

Greater London Authority. (2011). *Managing risks and increasing resilience: The Mayor's climate change adaptation strategy.* London: Greater London Authority. <https://www.london.gov.uk/sites/default/files/Adaptation-oct11.pdf>.

Gunes, A., Kovel, J., & Member, P. (2000). Using GIS in emergency management operations. *Journal of Urban Planning, 126*(136−149). Available from https://doi.org/10.1061/(ASCE)0733-9488(2000)126:3(136).

Haque, C. E. (2003). Perspectives of natural disasters in East and South Asia, and the Pacific Island States: Socio-economic correlates and needs assessment. *Natural Hazards, 29*(3), 465−483. Available from https://doi.org/10.1023/A:1024765608135.

Hare, W. L., Cramer, W., Schaeffer, M., Battaglini, A., & Jaeger, C. C. (2011). Climate hotspots: key vulnerable regions, climate change and limits to warming. *Regional Environmental Change, 11*, 1−13. Available from https://doi.org/10.1007/s10113-010-0195-4.

Hashim, J. H., & Hashim, Z. (2016). Climate change, extreme weather events and human health implications in the Asia Pacific region. *Asia-Pacific Journal of Public Health / Asia-Pacific Academic Consortium for Public Health, 28*(2S), 8S−14S. Available from https://doi.org/10.1177/1010539515599030.

Hilbert, D. W., Hill, R., Moran, C., Turton, S. M., Bohnet, I., Marshall, N. A., ... Westcott, D. A. (2014). *Climate change issues and impacts in the Wet Tropics NRM cluster region.* Cairns: James Cook University. <https://publications.csiro.au/rpr/download?pid = csiro:EP14913&dsid = DS3>.

Iwayemi, A. (2008). *Nigeria's dual energy problems: Policy issues and challenges* (pp. 17−21). University of Ibadan, Ibadan, Nigeria: IAEE Energy Forum. <http://www.iaee.org/en/publications/fullnewsletter.aspx?id = 8>.

Jackson, G., McNamara, K., & Witt, B. (2017). A framework for disaster vulnerability in a small island in the Southwest Pacific: A case study of Emae Island, Vanuatu. *International Journal of Disaster Risk Science.* Available from https://doi.org/10.1007/s13753-017-0145-6.

Jones, A., Barnett, B., Williams, A. J., Grayson, J., Busilacchi, S., Duckworth, A., ... Murchie, C. D. (2008). Effective communication tools to engage Torres Strait Islanders in scientific research. *Contintenal Shelf Research, 28*(16), 2350−2356. Available from https://doi.org/10.1016/j.csr.2008.03.027.

Karekezi, S. (2002). Poverty and energy in Africa: A brief review. *Energy Policy, 30*, 915−919. Available from https://doi.org/10.1016/S0301-4215(02)00047-2.

Keenan, J. M. (2017). Regional resilience trust funds: An exploratory analysis for leveraging insurance surcharges. *Environment Systems and Decisions.* Available from https://doi.org/10.1007/s10669-017-9656-3.

Kidd, S. (2012). *Achieving education and health outcomes in Pacific Island countries: Is there a role for social transfers?* Canberra: Australian Agency for International Development. <http://dfat.gov.au/about-us/publications/Documents/education-health-social-transfers.pdf>.

Kirkby, P., Williams, C., & Huq, S. (2017). Community-based adaptation (CBA): Adding conceptual clarity to the approach, and establishing its principles and challenges. *Climate and Development,* 1−13. Available from https://doi.org/10.1080/17565529.2017.1372265.

Kreimer, A., Arnold, M., & Carlin, A. (2003). *Building safer cities: The future of disaster risk.* Washington, DC: The World Bank Disaster Management Facility. <http://www-wds.worldbank.org/servlet/WDSContentServer/WDSP/IB/2003/12/05/000012009_20031205154931/Rendered/PDF/272110PAPER0Building0safer0cities.pdf>.

Kriticos, D. C. (2012). Regional climate-matching to estimate current and future sources of biosecurity threats. *Biological Invasions, 14*, 1533−1544. Available from https://doi.org/10.1007/s10530-011-0033-8.

Liedholm, C. (2002). Small firm dynamics: evidence from Africa and Latin America. *Small Business Economics, 18*, 227−242. Available from https://doi.org/10.1007/978-1-4615-0963-9_13.

Linnenluecke, M., & Griffiths, A. (2010). Beyond adaptation: Resilience for business in light of climate change and weather extremes. *Business & Society, 49*, 477−511. Available from https://doi.org/10.1177/0007650310368814.

Maclean, K., Cuthill, M., & Ross, H. (2013). Six attributes of social resilience. *Journal of Environmental Planning and Management, 57*, 1−13. Available from https://doi.org/10.1080/09640568.2013.763774.

Mallon, K., Hamilton, E., Black, M., Beem, B., & Abs, J. (2013). *Adapting the community sector for climate extremes: Final report*. Gold Coast: National Climate Change Adaptation Research Facility. <https://www.nccarf.edu.au/sites/default/files/attached_files_publications/Mallon_2013_Adapting_community_sector.pdf>.

Marshall, N. A., Marshall, P. A., Tamelander, J., Obura, D., Malleret-King, D., & Cinner, J. E. (2010). *A framework for social adaptation to climate change: Sustaining tropical coastal communities and industries*. Gland: IUCN. Retrieved from <https://www.iucn.org/content/framework-social-adaptation-climate-change-sustaining-tropical-coastal-communities-and-industries>.

McDowell, G., Ford, J., & Jones, J. (2016). Community-level climate change vulnerability research: trends, progress and future directions. *Environmental Research Letters, 11*, 033001.

McGranahan, G., Balkn, D., & Anderson, B. (2007). The rising tide: Assessing the risks of climate change and human settlements in low elevation coastal zones. *Environment and Urbanization, 19*, 17−37. Available from https://doi.org/10.1177/0956247807076960.

McMichael, A. J., Woodruff, R. E., & Hales, S. (2006). Climate change and human health: Present and future risks. *Lancet, 367*, 849−869. Available from https://doi.org/10.1016/S0140-6736(06)68079-3.

Mimura, N. (1999). Vulnerability of island countries in the South Pacific to sea level rise and climate change. *Climate Research, 12*, 137−143. Available from https://doi.org/10.3354/cr012137.

Mirza, M. M. Q., Warrick, R. A., & Ericksen, N. J. (2003). The implications of climate change on floods of the Ganges, Brahmaputra and Meghna rivers in Bangladesh. *Climatic Change, 57*, 287−318. <https://doi-org.ezp01.library.qut.edu.au/10.1023/A:1022825915791>.

Misra, P. K. (2006). E-strategies to support rural education in India. *Educational Media International, 43*(2), 165−179. Available from https://doi.org/10.1080/09523980600641197.

Nguyen, N. T., & Ha-Duong, M. (2009). Economic potential of renewable energy in Vietnam's power sector. *Energy Policy, 37*, 1601−1613. Available from https://doi.org/10.1016/j.enpol.2008.12.026.

Nicholls, R. J., & Cazenave, A. (2010). Sea-level rise and its impact on coastal zones. *Science (New York, NY), 328*(5985), 1517−1520. Available from https://doi.org/10.1126/science.1185782.

Nursey-Bray, M. (2013). *Climate change policy, conflict and transformative governance, Indo-Pacific Governance Research Centre Policy Brief*. Adelaide, SA: The Indo-Pacific Governance Research Centre. <https://www.adelaide.edu.au/indo-pacific-governance/policy/Nursey_Bray_2013_PB1.pdf>.

OECD. (2013). *PISA 2012 Results in focus: snapshot of performance in mathematics, reading and science*. <http://www.oecd.org/pisa/keyfindings/PISA-2012-results-snapshot-Volume-I-ENG.pdf>.

Ontario Ministry of Environment. (2011). *Climate ready: Ontario's Adaptation Strategy and Action Plan 2011/14*. <http://www.ene.gov.on.ca/stdprodconsume/groups/lr/@ene/@resources/documents/resource/stdprod_085423.pdf>.

Pacific Disaster Center. (2013). *Vision and mission: Building disaster resilience*. <http://www.pdc.org/about/vision-and-mission/>.

Pelling, M., & Uitto, J. I. (2001). Small island developing states: Natural disaster vulnerability and global change. *Environmental Hazards, 3*, 49−62. Available from https://doi.org/10.1016/S1464-2867(01)00018-3.

Pinkowski, J. (2008). *Disaster management handbook*. Boca Raton, FL: CRC Press.

Pryce, G., & Chen, Y. (2011). Flood risk and the consequences for housing of a changing climate: An international perspective. *Risk Management, 13*(4), 228−246. Available from https://doi.org/10.1057/rm.2011.13.

Przeslawski, R., Ahyong, S., Byrne, M., Worheides, G., & Hutchings, P. (2008). Beyond corals and fish: The effects of climate change on noncoral benthic invertebrates of tropical reefs. *Global Change Biology, 14*, 2773−2795. Available from https://doi.org/10.1111/j.1365-2486.2008.01693.x.

Ramankutty, N., Evan, A. T., Monfreda, C., & Foley, J. A. (2008). Farming the planet: Geographic distribution of global agricultural lands in the year 2000. *Global Biogeochemical Cycles, 22*(1). Available from https://doi.org/10.1029/2007gb002952.

Robson, P. J. A., Haugh, H. M., & Obeng, B. A. (2009). Entrepreneurship and innovation in Ghana: Enterprising Africa. *Small Business Economics, 32*, 331−350. Available from https://doi.org/10.1007/s11187-008-9121-2.

Ronneberg, E. (2008). *Pacific climate change fact sheet*. Apia, Samoa: Secretariat of the Pacific Regional Environment Programme. <http://www.sprep.org/climate_change/pycc/documents/pacificclimate.pdf>.

Rosenzweig, C., & Solecki, W. (2010). *Climate change adaptation in New York City: Building a risk management response*. New York: Annals of the New York Academy of Sciences. <http://www.nyas.org/publications/annals/Detail.aspx?cid = ab9d0f9f-1cb1-4f21-b0c8-7607daa5dfcc>.

Ross, H., Cuthill, M., Maclean, K., Jansen, D., & Witt, B. (2010). *Understanding, enhancing and managing for social resilience at the regional scale: Opportunities in north Queensland*. Cairns: Reef and Rainforest Research Centre. <http://www.rrrc.org.au/publications/social_resilience_northqueensland.html>.

Santa Cruz City Council. (2012). *City of Santa Cruz Climate Adaptation Plan: An update to the 2007 Local Hazard Mitigation Plan 2012-2017*. City of Santa Cruz: Santa Cruz City Council. <https://www.yumpu.com/en/document/view/31495562/climate-adaptation-plan-city-of-santa-cruz>.

Sawada, Y. (2017). Disasters, household decisions and insurance mechanisms: A review of evidence and a case study from a developing country in Asia. *Asian Economic Policy Review, 12*(1), 18−40. Available from https://doi.org/10.1111/aepr.12154.

Sayer, J., Sunderland, T., Ghazoul, J., Pfund, J. L., Sheil, D., Meijaard, E., ... Buck, L. E. (2013). Ten principles for a landscape approach to reconciling agriculture, conservation, and other competing land uses. *Proceedings of the National Academy of Sciences of the United States of America, 110*(21), 8349−8356. Available from https://doi.org/10.1073/pnas.1210595110.

Sayer, J., Margules, C., Boedhihartono, A. K., Dale, A., Sunderland, T., Supriatna, J., & Saryanthi, R. (2014). Landscape approaches: What are the pre-conditions for success? *Sustainability Science, 10*, 345−355. Available from https://doi.org/10.1007/s11625-014-0281-5.

Sharma, K. L. (2006). *Food security in the South Pacific Island countries with special reference to the Fiji Islands*. Finland: United Nations University - World Institute for Development Economics Research. <http://archive.unu.edu/hq/library/Collection/PDF_files/WIDER/WRP/WRP203.pdf>.

Shiklomanov, I. A. (2000). Appraisal and assessment of world water resources. *Water International, 25*(1), 11−32. Available from https://doi.org/10.1080/02508060008686794.

Stanley, J. (2010). *Promoting social inclusion in adaptation to climate change: Discussion paper*. Report 10/02 Commissioned by the Dept. of Sustainability and Environment, Victoria. <http://library.bsl.org.au/jspui/bitstream/1/1857/1/Promoting_social_inclusion_in_adaptation_to_climate_change_final.pdf>.

Stern, N. (2006). *The economics of climate change: The Stern Review*. London: HM Treasury. <http://fore.yale.edu/climate-change/science/the-stern-review-on-the-economics-of-climate-change/>.

Tanner, T., Mitchell, T., Polack, E., & Guenther, B. (2009). *Urban governance for adaptation: Assessing climate change resilience in ten Asian cities*. Brighton: Institute of Development Studies. <http://www.ids.ac.uk/publication/urban-governance-for-adaptation-assessing-climate-change-resilience-in-ten-asian-cities1>.

The World Bank. (2010a). *Deep wells and prudence: Towards pragmatic action for addressing groundwater overexploitation in India*. Washington, DC: The International Bank for Reconstruction and Dev..

The World Bank. (2010b). *Report on the status of disaster risk reduction in sub-Saharan Africa*. Washington, DC.: Global Facility for Disaster Reduction and Recovery. <http://www.gfdrr.org/sites/gfdrr/files/publication/AFR.pdf>.

UNFCCC. (2006). *Background paper on impacts, vulnerability, and adaptation to climate change in Africa*. Bonn, Germany: UNFCCC Secretariat. <https://unfccc.int/files/adaptation/adverse_effects_and_response_measures_art_48/application/pdf/200609_background_african_wkshp.pdf>.

Valentine, P. S. (2006). *Compiling a case for World Heritage on Cape York Peninsula: Final report for Queensland Parks and Wildlife Service*, June 2006. JCU, Townsville. <http://www.ehp.qld.gov.au/cape-york/pdf/cape-york-world-heritage-case.pdf>.

van Asselen, S., & Verburg, P. H. (2013). Land cover change or land-use intensification: simulating land system change with a global-scale land change model. *Global Change Biology*, *19*(12), 3648–3667. Available from https://doi.org/10.1111/gcb.12331.

Vermeulen, S. J., Aggarwal, P. K., Ainslie, A., Angelone, C., Campbell, B. M., Challinor, A. J., ... Wollenberg, E. (2012). Options for support to agriculture and food security under climate change. *Environmental Science & Policy*, *15*(1), 136–144. Available from https://doi.org/10.1016/j.envsci.2011.09.003.

Victorian Government. (2013). *Victorian climate change adaptation plan*. Melbourne: Victorian Government.

Walsh, A., Brugha, R., & Byrne, E. (2016). The way the country has been carved up by researchers: Ethics and power in north-south public health research. *International Journal for Equity in Health*, *15*(1), 204. Available from https://doi.org/10.1186/s12939-016-0488-4.

Warrick, O. C. (2011). *Local voices, local choices? Vulnerability to climate change and community-based adaptation in rural Vanuatu*. Doctor of Philosophy in Geography. University of Waikato. Retrieved from <http://hdl.handle.net/10289/5828>.

Wet Tropics Management Authority. (2007). *Climate change in the Wet Tropics: Impacts and responses: State of the Wet Tropics Report 2007-2008*. WTMA Cairns. <http://www.wettropics.gov.au/site/user-assets/docs/2008sowt_report_climatechange.pdf>.

Zerger, A., & Smith, D. (2003). Impediments to using GIS for real-time disaster decision support. *Computers, Environment and Urban Systems*, *27*, 123–141. Available from https://doi.org/10.1016/S0198-9715(01)00021-7.

THE IMPACT OF NATURAL DISASTERS AND CLIMATE CHANGE ON AGRICULTURE: FINDINGS FROM VIETNAM

Trong-Anh Trinh, Simon Feeny and Alberto Posso

Centre for International Development, School of Economics, Finance and Marketing, RMIT University, Melbourne, VIC, Australia

17.1 INTRODUCTION

According to the Intergovernmental Panel on Climate Change (IPCC), climate change includes increasing temperatures, changing rainfall patterns, rising sea levels, saltwater intrusion and a higher probability of extreme weather events such as flooding and droughts (Solomon, Qin, Manning, Averyt, & Marquis, 2007). It is believed that the intensified accumulation of carbon dioxide and other greenhouse gases will exacerbate this global issue beyond the next century. The last 100 years have shown an increase in the global mean surface temperature of between 0.4°C and 0.8°C, and it is predicted to increase by 1.4°C−5.8°C over the next 100 years (IPCC, 2014). As a result, this serious environmental problem will affect not only ecosystems by changing the composition of vegetation (as well as plant and animal diversity, and human health) but also global economies through various channels such as agriculture, water resources, energy, and tourism. Climate change is recognized as one of the most significant challenges facing humans in the 21st century and mitigating its impacts has become imperative for policymakers. The populations of developing countries are commonly believed to be more vulnerable to natural disasters and climate change due to resource scarcity, poor infrastructure, and unstable institutions. With low adaptive capacities in these countries, these climatic shocks will worsen inequalities regarding health status and access to food and clean water, and, therefore, hamper progress at reducing poverty.

A report from the World Bank in 2009 indicated that Vietnam is one of five countries predicted to be among the most affected by natural disasters and climate change due to its long coastlines, large population, and economic activity in coastal areas, with heavy reliance on agriculture, natural resources, and forestry (Harvey, 2009). In the past 50 years, climate change in Vietnam has been characterized by an increase in average temperature by 0.5°C−0.7°C; a rise in sea levels by 200 mm; and a higher frequency of storms, floods, and droughts (Harvey, 2009). By the end of the 21st century, climate-change scenarios for Vietnam predict that its annual average temperature will continue to rise by 2°C−3°C and its sea level will rise by 650−1000 mm (Harvey, 2009). Meanwhile, a change in rainfall patterns is expected to increase precipitation during the country's

Economic Effects of Natural Disasters. DOI: https://doi.org/10.1016/B978-0-12-817465-4.00017-0

rainy season and decrease it during the dry season. Consequently, natural disasters and climate change in Vietnam are considered to be serious and present significant challenges to hunger eradication and poverty reduction, as well as the achievement of the United Nations' Sustainable Development Goals (SDGs).

Expectedly, agriculture is one of the sectors most affected by natural disasters and climate change in Vietnam. With both direct effects on crop production and indirect effects through changes in irrigation, water availability and potential evapotranspiration, the effects of disasters and climate change on Vietnam's agriculture could result in a 0.7%−2.4% reduction of total gross domestic product (GDP) by 2050 (Stern et al., 2006). The expansion of the Mekong Delta—one of three most vulnerable deltas in the world alongside the Nile (Egypt) and the Ganges (Bangladesh and India)—is predicted to damage 1.1 million hectares, or 70% of cultivated land in Vietnam's coastal areas by 2030 due to saltwater intrusion (UNDP, 2007).

In recognizing these remarkable effects, there are a rapidly growing number of studies in the economic literature that examine the impacts of climate change and climatic shocks in both developed and developing countries. On the one hand, a large body of literature has focused on the nexus between agriculture and long-term climatic change, measured by temperature and precipitation (Deschênes & Greenstone, 2007; Mendelsohn, Nordhaus, & Shaw, 1994; Schlenker & Roberts, 2009). On the other hand, a growing number of studies examine the relationship between agriculture and natural disasters. These studies recognize that the economic impacts of natural disasters are arguably larger given their magnitude and intensity (Blanc & Strobl, 2016; Keerthiratne & Tol, 2018; Klomp & Hoogezand, 2018; Spencer & Polachek, 2015).

This chapter contributes to the literature by providing evidence of the impact of natural disasters and climate change on agriculture in Vietnam. Vietnam is an interesting case study due to its heavy dependence on agriculture and high vulnerability to disasters and climate change. Vietnam has been subjected to a wide spectrum of disasters which ranks 4th, 10th, and 16th in terms of the absolute number of people exposed to floods, droughts, and other related events, respectively (UNISDR, 2009). At the same time, agriculture has remained a key sector of the economy, with the contribution of approximately 21% of GDP, and accounting for more than 40% of the labor force (MONRE, 2009). It should be noted that most agricultural activities are located in rural areas where households are more susceptible to climate change and weather shocks. Therefore an investigation of the impact of climate change and natural disasters on agriculture in Vietnam is clearly warranted.

The remainder of this chapter is structured as follows. Section 17.2 provides a background of natural disasters and climate change in Vietnam. Section 17.3 presents an overview of agriculture in Vietnam and discusses the impacts of disasters and climate change on agricultural production. Section 17.4 concludes the chapter.

17.2 BACKGROUND: NATURAL DISASTERS AND CLIMATE CHANGE IN VIETNAM

Located in South-East Asia, Vietnam has been recognized as one of the world's best-performing economies over the past few decades. During 1995−2005, its real GDP grew by an average of

7.3% each year, the share of industry rose from 29% to 41% of GDP and per capita income rose from US$260 in 1995 to US$835 in 2007 (Stern et al., 2006). Vietnam achieved significant progress toward the United Nations Millennium Development Goals (MDGs) targets, despite poverty reduction progressing at a slower rate for the country's ethnic minorities. As a result of industrialization and modernization strategies, the industry and service sectors now contribute a large share of Vietnam's annual GDP, while the contribution of the agricultural, forestry, and fishing sectors has declined. Nevertheless, the latter sectors continue to contribute 21% of GDP and employ over 47% of the country's labor force (UNDP, 2015). In terms of agriculture, rice production is considered as key for food security, rural employment, and foreign exchange. It employs two-thirds of the rural labor force and Vietnam is the world's second largest rice exporter. The Red River Delta and Mekong Delta are the two most important rice-growing areas that contribute to more than 70% of the country's food production.

With its long coastlines, geographic location, diverse topography, and climate, Vietnam is one of the most hazard-prone countries of the Asia-Pacific region, with storms and flooding, in particular, responsible for economic and human loss. Climatic change linked to rising average temperature and sea level as well as substantial changes in rainfall patterns has exacerbated these vulnerabilities. Given that a high proportion of the country's population and economic assets (including irrigated agriculture) are in coastal lowlands and deltas, Vietnam has been ranked among the five countries most likely to be affected by climate change (UNDP, 2015). Developing adaptation and mitigation policies for Vietnam is therefore essential.

17.2.1 OVERVIEW OF VIETNAM'S CLIMATE

Situated in the tropical interior zone of the Northern Hemisphere—closer to the tropics than the equator and heavily influenced by the East Sea—Vietnam has a tropical monsoon climate with distinct characteristics of two major seasons: dry season (November–April) and rainy season (May–October).

During the 50 years between 1958 and 2007, the annual average temperature in Vietnam increased significantly by 0.5°C–1°C (Fig. 17.1). The winter temperatures rose faster than those of summer, while temperatures in Northern zones rose faster than Southern zones (Table 17.1). In provincial cities such as Hanoi, Ho Chi Minh City, and Da Nang, the average temperature between 1961 and 2000 was higher than that of the three previous decades (1931–60).

While the geographic increase of temperature is distinct, the change in annual average rainfall varied across regions (see Fig. 17.2 and Table 17.1). On average, the rainfall between 1958 and 2007 decreased by approximately 2% (MONRE, 2009).

In terms of sea level the data from tidal gauges along Vietnam's coasts indicate that the sea level rose at a rate of approximately 2.8 mm/year during the period 1993–2008, which is comparable with the global rise (MONRE, 2009). Although most stations experienced this large rate, there are some provinces (such as Quang Ninh and Quy Nhon) that experienced smaller changes in sea level. Along the coastal zone, the sea level in the Mid-Central coastal region and South-Western region are likely to increase more than the level of the whole coastal zone of approximately 2.9 mm/year (MONRE, 2009).

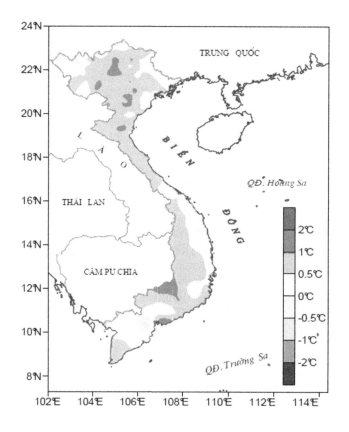

FIGURE 17.1

Changes in annual average temperature (°C) during the last 50 years.

MONRE. (2009). Annual report. *Hanoi: Ministry of Natural Resources and the Environment.*

Table 17.1 Increases in Temperature and Changes in Precipitation During the Last 50 Years in Vietnam's Climate Regions.

Climate Regions	Temperature (°C)			Precipitation (%)		
	January	July	Year	November–April	May–October	Year
North-West	1.4	0.5	0.5	6	− 6	− 2
North-East	1.5	0.3	0.6	0	− 9	− 7
Red River Delta	1.4	0.5	0.6	0	− 13	− 11
North-Central Coast	1.3	0.5	0.5	4	− 5	− 3
South-Central Coast	0.6	0.5	0.3	20	20	20
Central Highlands	0.9	0.4	0.6	19	9	11
South (South-East and Mekong Delta)	0.8	0.4	0.6	27	6	9

MONRE. (2009). Annual report. Hanoi: Ministry of Natural Resources and the Environment.

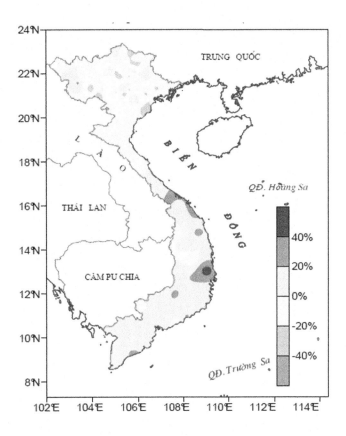

FIGURE 17.2

Changes in precipitation (in percent) during the last 50 years.

MONRE. (2009). Annual report. *Hanoi: Ministry of Natural Resources and the Environment.*

17.2.2 CLIMATE-CHANGE SCENARIOS FOR VIETNAM

According to the Fourth Assessment Report of IPCC, developing climate-change scenarios for Vietnam incorporate greenhouse gas emissions with three different schemes: low-emission scenarios (B1), medium-emission scenarios (B2, A1B), and high-emission scenarios (A2, A1F1) (Solomon et al., 2007). Based on emission predictions, MONRE (2009) publishes Vietnam's official climate-change scenarios, which are also developed for seven climate zones. To assess impact and adaptation planning, the medium-emission scenario (B2) is considered the main scenario, which includes projected changes in temperature, rainfall, and sea level between 2020 and 2100.

The annual mean temperature is expected to increase by approximately 0.4°C−2.8°C from 2020 to 2100 (Table 17.2). The report also indicates that winter temperatures increase faster than summer temperatures and that temperatures in North Vietnam increase faster than those in the South (MONRE, 2009). These observations are all consistent with predicted increases in global average temperatures.

The scenario for precipitation change is more complex due to its seasonal and region-specific patterns. MONRE (2009) projects the annual rainfall to increase 1.5%−3% in Southern regions and 7%−8% in Northern regions by 2100 (Table 17.3). Notably, the dry seasons are predicted to get drier, with higher reductions of precipitation in Southern regions than that in the North. Conversely, the wet seasons are projected to get wetter, with a higher increase in precipitation in the Northern regions than that in the South. As a result, droughts would occur more frequently in the dry season, while floods would occur more regularly in the wet seasons.

Sea level is projected to rise by approximately 30 cm by 2050 and 75 cm by 2100 under the medium-emission scenario (Table 17.4). The sea level is expected to rise most at Hon Dau station

Table 17.2 Projected Increases in Annual Average Temperature (°C).

Climate Regions	2020	2040	2060	2080	2100
North-West	0.5	1	1.6	2.1	2.6
North-East	0.5	1	1.6	2.1	2.5
Red River Delta	0.5	0.9	1.5	2	2.4
North-Central Coast	0.5	1.1	1.8	2.4	2.8
South-Central Coast	0.4	0.7	1.2	1.6	1.9
Central Highlands	0.3	0.6	1	1.4	1.6
South (South-East and Mekong Delta)	0.4	0.8	1.3	1.8	2

MONRE. (2009). Annual report. Hanoi: Ministry of Natural Resources and the Environment.

Table 17.3 Projected Changes in Annual Rainfall (%).

Climate Regions	2020	2040	2060	2080	2100
North-West	1.4	3	4.6	6.1	7.4
North-East	1.4	3	4.7	6.1	7.3
Red River Delta	1.6	3.2	5	6.6	7.9
North-Central Coast	1.5	3.1	4.9	6.4	7.7
South-Central Coast	0.7	1.3	2.1	2.7	3.2
Central Highlands	0.3	0.5	0.9	1.2	1.4
South (South-East and Mekong Delta)	0.3	0.6	1	1.2	1.5

MONRE. (2009). Annual report. Hanoi: Ministry of Natural Resources and the Environment.

Table 17.4 Projected Sea-Level Rise (cm).

	2020	2040	2060	2080	2100
Low scenario	11	23	35	50	65
Medium scenario	12	23	37	54	75
High scenario	12	24	44	71	100

MONRE. (2009). Annual report. Hanoi: Ministry of Natural Resources and the Environment.

(Hai Phong) by approximately 80 cm by 2100 (MONRE, 2009). Generally, the rising sea level in Vietnam is consistent with global forecasts across 2020 and 2100 (Stern et al., 2006). The expected changes in climate variables and sea levels have formed the background for evaluating the effects of climate change on various sectors in Vietnam.

17.2.3 AN OVERVIEW OF NATURAL DISASTERS IN VIETNAM

Located in a tropical monsoon region combined with diverse and complex topography, Vietnam has suffered different types of natural disasters, both hydrometeorological (e.g., flood, storm, and drought) and geophysical (e.g., landslides, earthquakes). According to a report by UNDP (2015), approximately 70% of the population is projected to be exposed to risks from such extreme events. In multiple sectors, from agriculture to industry, from energy to education, the damages of these disasters are expected to be serious. UNDP (2015) also reports an annual economic loss equivalent to 1.3% of GDP or US$3.85 billion in the period 1990−2009. In the coming decades, rapid population growth, socioeconomic development, and urbanization combined with increases in natural hazards will exacerbate the vulnerability of households, especially in rural areas.

Fig. 17.3 summarizes different types of natural hazards that have occurred in the period 1990−2010 in terms of reported events. According to the figure, floods are the most frequent events accounting for 48% of disasters, followed by hailstorms (20%), storms (13%), and flash floods (7%). Cyclones (or typhoons), landslides, and other disasters account for 12% of the reported events.

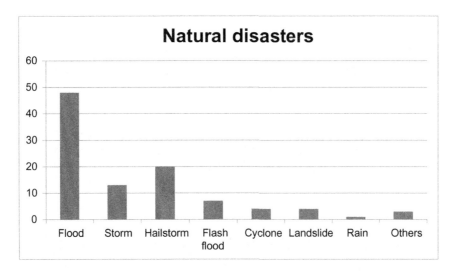

FIGURE 17.3

Proportion of disasters (in percent) in terms of reported events.

Note: Flash flood is defined as an event that occurs within 6 h following the end of the causative event, while the flood is defined as an event that occurs after 6 h following the end of the causative event.

UNDP. (2015). Vietnam special report on managing the risks of extreme events and disasters to advance climate change adaptation.

Hanoi: UNDP.

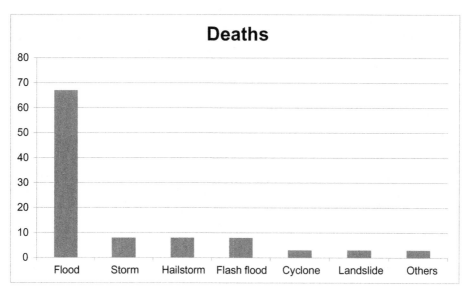

FIGURE 17.4

Proportion of disasters (in percent) in terms of number of deaths.

UNDP. (2015). Vietnam special report on managing the risks of extreme events and disasters to advance climate change adaptation.

Hanoi: UNDP.

In terms of damage, Fig. 17.4 shows that floods account for 67% of deaths while other disasters account for a much smaller proportion. Floods, hailstorm, storm, and flash floods combined are responsible for nearly 90% of the loss of life. Fig. 17.5 indicates that storms cause the most damage to houses (36%), followed by floods (32%) and hailstorms (11%).

Over the period 1990−2010, there is an increasing trend in the number of extreme events recorded, as illustrated in Fig. 17.6. The highest number of reported disaster events has occurred in recent years: 2006 (127), followed by 2008 (123) and 2005 (114).

By region, Fig. 17.7 shows that the most affected provinces by natural hazards are located in Central Vietnam. Inland provinces are more affected by floods while coastal provinces are more affected by storms. Quang Ngai is the province with the highest number of deaths caused by all disaster types over the period, with 924 fatalities. The province with the most houses destroyed is Thanh Hoa (97,383 houses), while the province with the most houses damaged is Quang Binh. Provinces in the Mekong River Delta suffer from a high number of deaths; however, they have a smaller number of reported events than average.

Because of the serious impacts of natural disasters, the Vietnamese government has adopted numerous national policies that focus on improving adaptive capabilities, reflecting its concern and attention to the issue. The National Strategy for Natural Disaster Prevention, Response and Mitigation approved by the Prime Minister in 2007 is considered as a milestone in Vietnam's disaster prevention. The general goal of the strategy is: "Complete the relocation, arrangement and stabilization of the life for people in disaster-prone areas according to the planning approved by

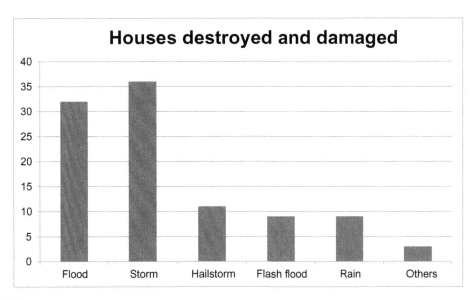

FIGURE 17.5

Proportion of disasters (in percent) in terms of houses destroyed and damaged.

UNDP. (2015). Vietnam special report on managing the risks of extreme events and disasters to advance climate change adaptation.

Hanoi: UNDP.

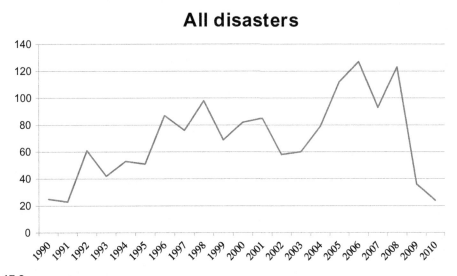

FIGURE 17.6

Number of all disasters per year in the period 1990–2010.

UNDP. (2015). Vietnam special report on managing the risks of extreme events and disasters to advance climate change adaptation.

Hanoi: UNDP.

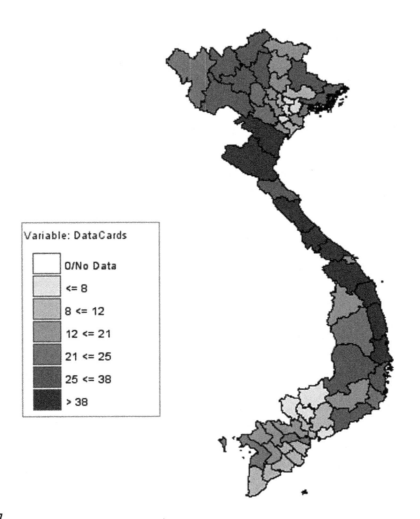

FIGURE 17.7

Spatial distribution of reports for all disaster types.

UNDP. (2015). Vietnam special report on managing the risks of extreme events and disasters to advance climate change adaptation.

Hanoi: UNDP.

authorized government agencies. Up to 2010, manage to relocate all populations from flash flood and landslide high-risk areas and dangerous areas to safe places." However, there are challenges that have impeded the implementation of these programs including corruption and lack of coordination between ministries, or government and local communities (UNDP, 2015).

The Mekong Delta, one of three "extreme" global hotspots in terms of potential population displacement as a result of climate change (Solomon et al., 2007), has received special attention from the government. For example, the program "Living with floods" has been implemented since 2000 to mitigate the risks of natural disasters as well as improving households' livelihoods and the country's economic development. One of the main objectives is to provide safe and permanent residences with access to basic infrastructures such as clean water, schools, and health clinics. These new constructions, in turn, link with initiatives for increased economic development and the urbanization of rural areas.

Recently, mitigating the impact of natural disasters in Vietnam has also received support from international organizations. For example, the ASEAN community has funded and implemented disaster risk reduction projects in 23 provinces by cooperating with local nonprofit organizations. The World Bank has also implemented a community-based disaster risk management program in 12 provinces across Vietnam to build flood and storm mitigation infrastructures, including river dikes, evacuation routes, and drainage systems. To enhance the process of mitigation, the Vietnamese government has cooperated with these organizations by supporting legal frameworks or being a connection between international groups and local communities.

17.2.4 INSTITUTIONAL FRAMEWORK FOR NATURAL DISASTERS AND CLIMATE CHANGE IN VIETNAM

The Vietnamese government has approved several strategies and plans to cope with natural disasters and climate change (Table 17.5). Generally, these frameworks have included vulnerability assessments across sectorial, regional, and community levels; enhancing the role of science and technology for adaptation solutions; increasing public awareness and participation; and integrating climate change into development strategies, plans, and programs in all sectors (Asian Development Bank, 2014).

17.3 IMPACTS OF NATURAL DISASTERS AND CLIMATE CHANGE ON AGRICULTURE IN VIETNAM

17.3.1 OVERVIEW OF VIETNAM'S AGRICULTURE

After 27 years of reform, agriculture has significantly improved and developed into a key economic sector for Vietnam. Despite the rapid rate of industrialization in the last two decades, agriculture remains a major economic sector in Vietnam that generates employment and income for a significant part of the population. Between 1995 and 2005, its contribution to the economy decreased from 27.2% to 20.5%; however, 40% of the total labor force was still employed in the agricultural sector in 2015 (World Bank, 2016).

With its long coastal line and diverse topography, Vietnam grows a wide variety of crops including rice, maize, cassava, sugarcane, coffee, and vegetables. However, as shown in Table 17.6, rice is still considered the most important crop and occupies most of the cultivated land areas across regions. Rice production occurs mainly in two major deltas: about 52% of paddy rice production is from the Mekong Delta, 18% in the Red River Delta and the remaining production in the North-East and North-Central Coast (see Table 17.7). In most regions, irrigated rice is cultivated in two to three harvests each year due to improved irrigation, new rice varieties, new rice technologies, and increased triple-cropping in the Mekong Delta (GSO, 2009).

Table 17.5 Government Strategies and Plans on Climate Change.

Strategy/Plan	Priorities	Implementing Legislation
SEDP for 2011–15	• Emphasize the actions to cope with climate change, increase forest coverage, improve water-supply coverage, improve treatment of industrial waste, improve treatment of solid waste and prosecute pollution violators • Emphasize the response to sea-level rise and vulnerability of low-lying coastal regions	
Strategic Orientation for Sustainable Development in Vietnam (2004) (Vietnam Agenda 21)	• Develop an institutional system supporting sustainable development (e.g., National Council on Sustainable Development) • Develop and implement the local and sector Vietnam Agenda 21: six pilot provinces (Son La, Thai Nguyen, Ninh Binh, Quang Nam, Lam Dong, and Ben Tre) and four pilot sectors (agriculture, fishery, construction, and industry) • Mobilize and allocate the resources for the implementation of Vietnam Agenda 21	
NSEP until 2010 and Visions to 2020 and 5-year action plans	• Promote environmental protection, pollution prevention, area-specific environmental management, and biodiversity conservation	• Land Law 2003 (revised) • Law on Environment Protection 2005 (amended) • Law on Water Resources
National Biodiversity Action Plan to 2010 and Orientations toward 2020	• Conserve terrestrial biodiversity • Conserve biodiversity in wetlands and marine areas • Conserve and develop agricultural biodiversity • Use biological natural resources sustainably • Strengthen state-management capacity on biodiversity and biosafety	• Law on Forest Protection and Development 2004 (implemented through the MARD) • Law on Biodiversity 2009 • Law on Water Resources
National Strategy on Climate Change for 2050 and Visions to 2100; National Target Program to Respond to Climate Change (2008)	• Focus on the construction of coastal breakwaters and irrigation works to combat seawater intrusion and high tides	• Benefits of a Climate Change Act are currently being considered by the National Assembly

Table 17.5 Government Strategies and Plans on Climate Change. *Continued*

Strategy/Plan	Priorities	Implementing Legislation
Second National Strategy and Action Plan for Disaster Mitigation and Management 2001–20	• Require all sectors and local governments to prepare and implement climate-change action plans in their areas of responsibility, as well as report on progress • Stress importance of coexistence with floods in appropriate situations • Establish disaster-forecast centers in the north, center, and south of the country • Construct flood corridors and flood-retention areas in South Vietnam • Design principally to address short-term climate extremes	• Law of Electricity 2004 • Law on Environment Protection 2005 (amended) • Disaster Risk Reduction and Management Law currently under consideration by the National Assembly
Action Plan on Climate Change Response of Agriculture and Rural Development (2011)	• Ensure stable agricultural production and food security, and the maintenance of dyke and infrastructure systems • Focus on the former Cuu Long and Red River Delta, as well as central and mountainous areas • Reduce emissions from deforestation and forest degradation	• Law on Forest Protection and Development 2004
National Target Program on Energy Efficiency and Conservation for 2006–15	• Undertake greenhouse gas mitigation actions • Improve energy efficiency in major industries with an objective to achieve an 8% reduction in energy consumption by 2015, with emphasis on renewable sources	• Law of Electricity 2004
Transport Climate Change Action Plan 2011–15 (2010)	• Assess climate-change effects on transport infrastructure and activities • Identify suitable mitigation and adaptation options for transport projects	
Strategic Environmental Assessment, Environmental Impact Assessment, and Environmental Protection Commitments	• Improve capacity for the review of Strategic Environmental Assessment, Environmental Impact Assessment and Environmental Protection commitments for master plans (socioeconomic development plans), and large and complex projects in Vietnam	• 2011 Environment Decree (Decree 29/2011/ND–CP)

(Continued)

Table 17.5 Government Strategies and Plans on Climate Change. *Continued*

Strategy/Plan	Priorities	Implementing Legislation
Second National Strategyfor Disaster Mitigation and Management in Vietnam to 2020 and the Ordinances on flood and storm control	• Develop capacity of line ministries to review plans and projects for environmental issues • Decentralize responsibility for smaller projects to provincial authorities • Mandate the creation of provincial and other subnational disaster-risk management strategies and plans, and establish subordinate provincial and district committees for flood and storm control	• Draft Law on Disaster Prevention and Fighting pending National Assembly approval in 2012

MARD, *Ministry of Agriculture and Rural Development;* NSEP, *National Strategy for Environmental Protection;* SEDP, *Socio-Economic Development Plan.*
Asian Development Bank. (2014). Viet Nam: Financial sector assessment, strategy, and road map. *Mandaluyong City: Asian Development Bank.*

Vietnam has complex seasonal crop rotations and its crop calendar and patterns vary across agroecological zones. To cope with the uneven distribution of rainfall, its government has invested in irrigation systems to manage water resources for agricultural production. From 1961 to 2002, the percentage of irrigated arable land increased significantly from 18 to 70 (Table 17.8). Currently, most flatland is under irrigation, with a large percentage of crops produced from irrigated land (Stern et al., 2006).

17.3.2 IMPACTS OF NATURAL DISASTERS AND CLIMATE CHANGE ON AGRICULTURAL PRODUCTION

Climate change and natural disasters are expected to have significant effects on crop yields. It reduces agricultural productivity, both directly via the damage inflicted to standing crops and indirectly via their impact on cropping decisions. The most disasters-related agricultural losses in Vietnam are from floods and storms that destroy both standing crops and agricultural infrastructure, including irrigation systems. The increase in these extreme events is also associated with crop diseases which exacerbate further adverse impacts on crop yield. It is estimated that since 2000, Vietnam has lost over 10 million tonnes of rice as a direct consequence of floods and storms, while approximately 120,000 km^2 of rice fields have been submerged (UNDP, 2015). Furthermore, the report indicates that the true losses may be considerably higher due to the difficulty in isolating the impacts of natural hazards on crop yield. Natural hazards also have an indirect impact on productivity by changing agricultural behavior. For example, the delayed planting of summer crops due to a late start to the rains implies that crops are more vulnerable to floods, which typically occur from the autumn.

Table 17.6 Harvested Areas (In Hectare) and Crop Yields (Quintal per Hectare) by Agroecological Zones, 2007.

Regions	Paddy Rice		Maize		Cassava		Sugarcane		Coffee		Vegetables	
	Area	Yield	Area	Yield	Area	Yield	Area	Yield	Area	Yield	Area	Yield
North-West	157.7	3.6	172	3.1	42.9	9.8	12.1	58.1	3.5	1.6	91.1	11.1
North-East	552.5	4.6	236	3.2	55.4	13	13.4	48.5	0	0	–	–
Red River Delta	1111.6	5.7	84.7	4.2	7.5	12	2.3	52.3	0	0	158.6	18
North-Central Coast	683.2	4.7	137.3	3.6	58.9	15.4	63.4	57	7	1.6	68.5	9.8
South-Central Coast	375.8	5.1	42.1	4	65.3	15.6	49.8	48.7	1.6	1.1	44	14
Central Highlands	205	4.2	233.4	4.4	129.9	15.2	33.5	52.5	458.2	2	49	20.2
South-East	431.6	4.2	126.1	4.6	130.8	21.2	49.4	60.8	36.1	1.4	59.6	13
Mekong Delta	3683.6	5.1	36.3	5.6	6.3	11.6	66.9	76.3	0	0	164.3	16.6
Total	7201	5	1067.9	3.8	497	16.1	290.8	59.8	506.4	2	635.1	15.2

GSO. (2009). The 2009 Vietnam population and housing census: Completed results. Hanoi: General Statistical Office.

Table 17.7 Percentage Shares of Crop Production by Agroecological Zone in 2007.

Regions	Paddy Rice	Maize	Cassava	Sugarcane	Coffee	Vegetables
North-West	18	16	13	22	0	12
North-East	54	16	15	14	0	13
Red River Delta	65	4	1	1	0	29
North-Central Coast	36	6	10	41	0	8
South-Central Coast	31	3	17	39	0	10
Central Highlands	11	14	26	23	12	13
South-East	20	6	31	33	1	9
Mekong Delta	70	1	0	19	0	10
Total production	*36*	*4.1*	*8*	*17.4*	*1*	*9.6*

Note: *Crop production is aggregated in quintal.*
GSO. (2009). The 2009 Vietnam population and housing census: Completed results. *Hanoi: General Statistical Office.*

Table 17.8 Typical Seasonal Crop Rotations by Agroecological Zone.

Regions	Spring Crop	Summer Crop	Winter Crop
North-West	Spring rice, maize, soybean, sweet potato	Summer rice, maize, soybean, vegetables	Vegetables
North-East	Spring rice, maize, soybean	Summer rice, soybean	Maize, soybean, sweet potato
Red River Delta	Spring rice, vegetables	Summer rice, vegetables	Winter rice, vegetables, upland crops
North-Central Coast	Spring rice, peanut, upland crops	Summer rice, soybean, upland crops	Vegetables
South-Central Coast	Spring rice, vegetables, cotton	Summer rice, vegetables	Vegetables
Central Highlands	Winter−spring rice, maize, soybean, vegetables, cassava	Summer rice, maize, soybean, cotton, cassava	Winter−spring rice, upland crops
South-East	Spring rice, maize, cotton, vegetables, upland crops	Summer rice, maize	Autumn−winter rice
Mekong Delta	Rice, vegetables	Rice, vegetables	Rice, vegetables

GSO. (2009). The 2009 Vietnam population and housing census: Completed results. *Hanoi: General Statistical Office.*

The impacts of climate change and natural disasters are heterogeneous across regions. By using computable general equilibrium (CGE) models, the World Bank (2010) finds that yield changes vary widely across crops and agroecological zones under climate-change scenarios. As presented in Table 17.9, North-West and Central-Highland regions tend to have the highest decline in crop yields under both high and low scenarios.

Table 17.9 Effects of Natural Disasters and Climate Change on Crop Yields.

Regions	Potential Effects of Climate Change and Natural Disasters Without Adaptation
North-West	Rice yield declines by 11.1%−28.2%; yields of other crops decline by 5.9%−23.5%. Generally, a high scenario results in more yield reduction than a low scenario; a medium scenario has the least yield reduction.
North-East	Rice yield declines by 4.4%−39.6%; yields of other crops decline by 2.7%−38.3%. The largest yield reduction occurs with either high or low scenarios, depending on crops; a medium scenario has the least yield reduction.
Red River Delta	Rice yield declines by 7.2%−32.6%; yields of other crops decline by 4.1%−32.9%. The largest yield reduction occurs with either high or low scenarios, depending on crops; a medium scenario has the least yield reduction.
North-Central Coast	Rice yield declines by 7.2%−32.6%; yields of other crops decline by 4.1%−32.9%. The largest yield reduction occurs with either high or low scenarios, depending on crops; a medium scenario has the least yield reduction.
South-Central Coast	Rice yield declines by 8.4%−27%; yields of other crops decline by 4%−20.9%. Generally, a high scenario results in more yield reduction than a low scenario; a medium scenario has the least yield reduction.
Central Highlands	Rice yield declines by 11.1%−42%; yields of other crops decline by 7.5%−45.8%. The largest yield reduction occurs with either high or low scenarios, depending on crops; a medium scenario has the least yield reduction.
South-East	Rice yield increases by 4.3% in a high scenario, remains the same in a low scenario, and declines by 8.8% in a medium scenario. Yields of other crops decline by 3%−22.7%. The largest yield reduction occurs with any of the three scenarios, depending on crops.
Mekong Delta	Rice yield declines by 6.3%−12%; yields of other crops decline by 3.4%−26.5%. The largest yield reduction can occur under any of the three scenarios, depending on crops.

World Bank. (2010). Vietnam—Economics of adaptation to climate change. *World Bank.*

For rice, the key factors influencing yields are the projected reduction in a run-off in the Mekong Delta and the effect of higher temperatures (especially minimum temperatures). With a 1°C increase in the average temperature, yields are predicted to decline by 0.6 tonnes each hectare. The worst yield reductions (for a high scenario) are approximately 12% in the Mekong Delta and 24% in the Red River Delta. At the country level, rice yields are expected to decrease between 10% and 20% by 2050 (World Bank, 2010).

Generally, "low" scenarios result in lower yield reductions than high scenarios, but there are exceptions in the Red River Delta, which has a greater reduction in yields for both the 2030 and 2050 periods. This is because the wet season has higher increases in minimum and average temperatures during the spring rice season in the Red River Delta, which can shorten the growing period, thus leading to lower yields (World Bank, 2010).

As illustrated in Table 17.10, the effects of climate change and natural disasters on production are mostly negative in the three scenarios. Specifically, climate change and disasters may reduce rice production by between 2 and 7 million tonnes each year by 2050. A "medium" scenario generates the smallest effects on crop production.

Combined with the extensive inundation of cropland in the rainy season and increased salinity intrusion in the dry season, it is estimated that a 30 cm rise in sea level by 2050 will result in a loss

Table 17.10 Effect of Yield Change on Production by Scenario (Million Tonnes).

Year	Scenario	Rice	Maize	Cassava	Sugarcane	Coffee	Vegetables
2030	High	− 6.4	− 1.1	− 1.8	− 3.1	− 0.4	− 1.5
	Low	− 4	− 0.7	− 2.1	− 1.8	− 0.4	− 2.4
	Medium	− 2.2	− 0.2	− 0.4	− 0.8	− 0.1	− 0.5
2050	High	− 6.7	− 1.1	− 1.9	− 3.7	− 0.4	− 1.7
	Low	− 5.8	− 1	− 2.6	− 2.9	− 0.4	− 3.1
	Medium	− 3.4	− 0.3	− 0.6	− 1.4	− 0.1	− 0.9

Note: *To estimate the yield-change effects on production, World Bank (2010) employs the DGVM to compute the effects of climate change and carbon-dioxide fertilization on global yields of major crops at a spatial resolution of 0.5 × 0.5 degrees. Yield simulations are based on process-based implementations of GPP, maintenance respiration, water stress and biomass allocation, dynamically computing the most suitable crop variety and growing period in each grid cell. DGVM, Dynamic global vegetation model; GPP, gross primary production.*
World Bank. (2010). Vietnam—Economics of adaptation to climate change. *World Bank.*

Table 17.11 Effects of Natural Disasters and Climate Change on Production by Scenario (Million Tonnes).

Climate Scenario	Paddy Rice			Maize Yields	Cassava Yields	Sugarcane Yields	Coffee Yields	Vegetable Yields
	Yields	Sea Level	Total					
High	− 6.7	− 2.4	− 9.1	− 1.1	− 1.9	− 3.7	− 0.4	− 1.7
Low	− 5.8	− 2.5	− 8.4	− 1	− 2.6	− 2.9	− 0.4	− 3.1
Medium	− 3.4	− 2.4	− 5.8	− 0.3	− 0.6	− 1.4	− 0.1	− 0.9

World Bank. (2010). Vietnam—Economics of adaptation to climate change. World Bank.

of 193,000 ha of land used to cultivate rice due to inundation and 294,000 ha due to salinity intrusion in the Mekong Delta. At current yields, this loss of land will lead to an approximate 2.6 million tonnes decline in rice production each year—this is more than 13% of today's rice production in the Mekong Delta. Even with adaptation strategies, the total loss of paddy rice due to sea-level rise will be approximately 2−2.5 million tonnes in 2050 (Table 17.11). The same results are found in other research using agronomic simulation models (Nguyen, Vu, & Nguyen, 2007; Phung, 2012).

17.4 CONCLUSION

It is understood that natural disasters and climate change will be serious threats to Vietnam's continued socioeconomic development. An increase in temperature and unpredicted patterns of precipitation, along with a high frequency of natural disasters, will all have remarkable effects across sectors, regions and income groups, particularly on the livelihoods of people residing in rural areas. So far, the Vietnamese government has not yet established the required national climate-change adaptation strategies to ameliorate these impacts. It is therefore necessary to launch adequate and effective policy responses to cope with this phenomenon.

This chapter attempts to provide evidence of the impacts of climate change and natural disasters on agricultural production in Vietnam. On average, there is clear evidence that climate change and natural disasters pose a negative effect on crop yield. The impact is heterogeneous, depending on the type of crop as well as the region. For example, rice production is found to be most affected by climate change, while the Central Highland and North-West regions are most vulnerable. In addition, this chapter investigates the impacts of climate change and natural disasters using different climate scenarios and confirms a significant threat to agricultural production in Vietnam.

While a negative impact of climate change and natural disasters on agriculture has been confirmed, there are several suggestions for future research. Farming involves a continuous process of adaptation, depending on weather, technology, and economic influences. In addition, it is likely that crops are affected differently by climate change, which provides a challenge to estimate the total effects of climate change on agriculture. So far, the literature has used agronomic models (e.g., Nguyen et al., 2007) based on controlled experiments. However, one limitation of this model is that it often ignores the specific characteristics of farmers and changes in price. More recently, other studies (e.g., World Bank, 2010) use a CGE approach, but in turn have difficulties relating to model specification and data consistency in Vietnam. Therefore future studies should apply different approaches to address these shortcomings.

REFERENCES

Asian Development Bank. (2014). *Viet Nam: Financial sector assessment, strategy, and road map.* Mandaluyong City: Asian Development Bank.

Blanc, E., & Strobl, E. (2016). Assessing the impact of typhoons on rice production in the Philippines. *Journal of Applied Meteorology and Climatology, 55*(4), 993−1007.

Deschênes, O., & Greenstone, M. (2007). The economic impacts of climate change: Evidence from agricultural output and random fluctuations in weather. *American Economic Review, 97*(1), 354−385.

GSO. (2009). *The 2009 Vietnam population and housing census: Completed results.* Hanoi: General Statistical Office.

Harvey, D. (2009). Reshaping economic geography: The world development report 2009. *Development and Change, 40*(6), 1269−1277.

IPCC. (2014). Climate change 2014: Synthesis report. In Core Writing Team, R. K. Pachauri, & L. A. Meyer (Eds.), *Contribution of working groups I, II and III to the fifth assessment report of the Intergovernmental Panel on Climate Change* (151 pp.). Geneva: IPCC.

Keerthiratne, S., & Tol, R. S. (2018). Impact of natural disasters on income inequality in Sri Lanka. *World Development, 105,* 217−230.

Klomp, J., & Hoogezand, B. (2018). Natural disasters and agricultural protection: A panel data analysis. *World Development, 104,* 404−417.

Mendelsohn, R., Nordhaus, W. D., & Shaw, D. (1994). The impact of global warming on agriculture: a Ricardian analysis. *American Economic Review,* 753−771.

MONRE. (2009). *Annual report.* Hanoi: Ministry of Natural Resources and the Environment.

Nguyen, H. N., Vu, K. T., & Nguyen, X. N. (2007). Flooding in Mekong River Delta, Viet Nam. *Human Development Report, 2008,* 23.

Phung, P. T. K. (2012). *Impact of climate change on rice production in Vietnam*. Asia Geospatial Digest.

Schlenker, W., & Roberts, M. J. (2009). Nonlinear temperature effects indicate severe damages to US crop yields under climate change. *Proceedings of the National Academy of Sciences of the United States of America*, *106*(37), 15594–15598.

Solomon, S., Qin, D., Manning, M., Averyt, K., & Marquis, M. (2007). *Climate change 2007—The physical science basis: Working group I contribution to the fourth assessment report of the IPCC* (Vol. 4). Cambridge University Press.

Spencer, N., & Polachek, S. (2015). Hurricane watch: Battening down the effects of the storm on local crop production. *Ecological Economics*, *120*, 234–240.

Stern, N., Peters, S., Bakhshi, V., Bowen, A., Cameron, C., Catovsky, S., et al. (2006). *Stern Review: The economics of climate change* (Vol. 30). London: HM Treasury.

UNDP. (2007). *Human development report 2007/8. Fighting climate change. Human solidarity in a divided world*.

UNDP. (2015). *Vietnam special report on managing the risks of extreme events and disasters to advance climate change adaptation*. Hanoi: UNDP.

UNISDR. (2009). *Global assessment report on disaster risk reduction*. Geneva: United Nations International Strategy for Disaster Reduction Secretariat.

World Bank. (2009). *World development report 2009: Reshaping economic geography*. World Bank.

World Bank. (2010). *Vietnam—Economics of adaptation to climate change*. World Bank.

World Bank. (2016). *Vietnam development report 2016: Transforming Vietnamese agriculture: Gaining more from less*. World Bank.

ECONOMIC IMPACT ASSESSMENT AFTER A NATURAL DISASTER USING DEMATEL METHOD

18

Michael Dzator[1,2] and Janet Dzator[2,3]

[1]*School of Access Education, Central Queensland University, Mackay, QLD, Australia* [2]*Australia Africa Universities Network (AAUN) Partner, Newcastle, NSW, Australia* [3]*School of Business, The University of Newcastle, Newcastle, NSW, Australia*

18.1 INTRODUCTION

A disaster is the disruption of the normal functioning of a system or community, which causes a strong impact on people, structures, and environment. Disaster can be grouped as man-made and natural. An example of a man-made disaster is a terrorist attack and examples of natural disaster include earthquake, cyclone, wildfire, storm, and flood. There is an increase in natural disasters in frequency and severity for the past decades. These events often have devastating impacts on human health, causing hundreds of thousands of deaths, illness, or injuries which require some emergency and continuous care. The natural disasters in the last 30 years ago have resulted in economic losses in the tens of billions of dollars (Botzen, Deschenes, & Sanders, 2019). Examples of some of the natural disasters within that period include Northridge earthquake in the United States in 1994, the Kobe earthquake in Japan in 1995, the 2004 Indian Ocean tsunami, Hurricane Katrina in the United States in 2005, the 2011 earthquake and tsunami in Japan, and Hurricane Harvey in the United States in 2017. Specifically, the annual cost of natural disasters between 2000 and 2012 is around 100 billion dollars (Kousky, 2014). Hurricane Harvey alone resulted in flood damage of 85 billion dollars in 2017 (Botzen, Kunreuther, Czajkowski, & de Moel, 2019).

Natural disasters affect a country's economy and temporarily or permanently threaten the economic development in the country. Alexander (1997) broadly defined a natural disaster as an impact of the natural environment upon the socioeconomic system. For example, in Australia, cyclone Debbie damaged farms in Queensland which amounted to more than $100 million in 2017 (ABC, 2017). This affected the supply of crops to the shops which resulted to price increases in the shops. Another natural disaster that affected the community in Northern Queensland recently was the flood in Townsville. This flood destroyed people's houses and livelihood which affected the economic condition of the community (ABC, 2019). There have also been devastating bushfires in Queensland and New South Wales (NSW) in Australia recently and the bushfires continue to have a negative impact in the community leading to a temporary closure of more than 500 schools in NSW (Chrysanthos, 2019).

The economic impacts of natural disasters are diverse which as an example include the destruction of homes, equipment, workplaces, farms, and health outcomes. These impacts are

Economic Effects of Natural Disasters. DOI: https://doi.org/10.1016/B978-0-12-817465-4.00018-2

interrelated, and one impact may affect the other. For example, the destruction of people's home may affect their mental state. A farmer whose crops are destroyed will affect his or her income. It may also lead to starvation in the society if the food that is available is scarce or the price is too high. This will result in an increase of the welfare cost for a government which may affect a country's budget.

The overall economy in a disaster area is affected by different impacts. In this study we would identify the key economic impacts of natural disasters on small- and medium-sized enterprises (SMEs) by detailed literature review and discussion with a number of experts. This study proposes Decision Making Trial and Evaluation Laboratory (DEMATEL) method to develop a structural model for evaluating influential economic impacts. Few papers have examined economic research on natural disasters (Noy, 2009). Moreover, to the best of our knowledge, no paper has evaluated the impact of natural disasters using DEMATEL. The chapter will use DEMATEL method to evaluate the economic impact of a natural disaster.

The primary objective of this research is to:

- identify the direct and indirect economic impacts of natural disasters on SMEs;
- determine key direct and indirect economic impacts from the cause-and-effect perspective; and
- provide policy recommendation of addressing the direct and indirect economic impacts of natural disasters.

This study will identify the key negative direct and indirect economic impacts of a natural disaster. The study will reveal the direct and indirect economic impacts to focus on in order to improve the economic condition of a society after a natural disaster. This study will also help the policymakers to adopt the key impacts which will improve the economy of an area devastated by a natural disaster.

The rest of the chapter is organized as follows. Section 18.2 presents the literature review on the economic impacts of natural disasters on SMEs. The research methodology is presented in Section 18.3. The results, key findings, and discussions are presented in Section 18.4. The conclusion and suggestion for future work are presented in Section 18.5.

18.2 ECONOMIC IMPACTS OF A NATURAL DISASTER ON SMALL- AND MEDIUM-SIZED ENTERPRISES

The economic impact of natural disasters on SMEs is discussed in this section. A comprehensive literature search was conducted using electronic databases (ScienceDirect, Scopus, and EM-DAT). We also conducted literature search using Google Scholar. The search terms included "natural disasters," "economic impact of natural disasters," "the economic impact of natural disasters on small and medium-sized enterprises," and "DEMATEL and disaster."

In 1971 Bolton Committee defined a small firm as an enterprise that has a relatively small share of the market place; managed by owners in a personalized way and it does form part of a large enterprise (Kayanula & Quartey, 2000). Currently, the definition of SME differs from country to country. For example, in the United States, the small business annual gross revenue does not exceed $250,000 while the medium business average annual gross income does not exceed $7 million for the past 3 years (U.S. International Trade Commission, 2010). The definition of SMEs also varies from country to country in terms of employment for manufacturing, construction, and mining sector. In Australia up to 200 people are to be employed, up to 199 in Canada, up to 999 in China, up to 250, and 499 people are to be employed in SMEs in the United States (Alkhoraif, Rashid, & McLaughlin, 2019).

In many countries, SMEs play an important role in their economy in terms of social and local employment (Auzzir, Haigh, & Amaratunga, 2018). They form the core area of the economic activities in the developed and developing countries. For example, in the developed economy, 99% of the economic activities are attributed to SMEs and they account for 66% of all jobs in the private sector (Auzzir et al., 2018). While in the developing countries the SMEs account for 90% of the enterprises and employing 59% of the workforce (Auzzir et al., 2018). The study in Malaysia by Auzzir et al. (2018) shows that the impacts of natural disaster are severe for SMEs. The main natural disaster identified in the study by Auzzir et al. (2018) was flood with loss of sales or production stated as the major impact on SMEs. Gunathilaka (2018) in a similar study in Sri Lanka studied the impacts of natural disaster on SMEs. Interviews were carried with business owners to identify problems they encounter after a natural disaster. The impacts were grouped into four main groups namely capital, labor, logistics, and markets. The SMEs thus play a vital part of an economy and contribute significantly towards the stable and sustainable growth of developing economics (Liu, Xu, & Han, 2013).

A study by Marks and Thomalla (2017) on the impact of natural disasters on SMEs on which they surveyed 26 retail shops in Bangkok stated that the damage of natural disaster to businesses was 12 times higher than the average income and worth more than 1 year of business turnover. The damage caused by the natural disaster has resulted in the deterioration of the economic health of the market community (Marks & Thomalla, 2017). The study shows that there was loss of income for businesses and reduction of sales due to less disposable income for customers. Drought as a natural disaster has also had a devastating effect on SMEs in Australia according to a study by Miles et al. (2007). Some of the impacts according to the study include price increases, staff reduction, cost-cutting, reduced stock, and reduced spending. Wedawatta, Ingirige, and Proverbs (2014) in the study on businesses noted a number of impacts of natural disasters on SMEs after a flood. Eighteen economic impacts were ranked with travel difficulties as the top-ranked economic impact. Other highly ranked impacts include stocks or products damaged or spoiled; the decrease in sales or production; and loss of trading. Equally important impacts include moving to temporary accommodation. An earlier study by Wedawatta, Ingirige, Jones, and Proverbs (2011) on the economic impact of extreme weather on construction work in the United Kingdom investigated the impact on SMEs in the sector. The study listed "nonattendance of employees" as the biggest impact of the extreme weather events of the SMEs followed by loss of sales or production and decrease in turnover or profits. The list of impacts stated in Wedawatta et al. (2014) study is similar to the ones stated in Auzzir et al. (2018).

The literature review reveals that there is lack of studies on the economic impacts of natural disasters. Second, no study has studied the relationships among the various economic impacts. Third, most studies just looked at the various impacts and some of the few studies such as Auzzir et al. (2018), Gunathilaka (2018), and Wedawatta et al. (2014) ranked the economic impacts. Finally, in addition to ranking the economic impacts in terms of importance, it will be good to identify the economic impacts from the cause-and-effect assessment.

18.3 RESEARCH METHODOLOGY

The research methodologies are as follows:

- identifying and selection of economic impacts of natural disasters using literature review and discussion with experts;
- classifying the economic impacts into groups (direct and indirect) and classifying them into matrix form; and
- applying DEMATEL to the data obtained.

DEMATEL is used as a solution methodology because it is best suited for analyzing interrelationships among the economic impacts of natural disasters.

18.3.1 ECONOMIC IMPACTS (E) OF NATURAL DISASTERS

The list of 12 economic impacts of natural disasters on SMEs was obtained using literature review and discussion with four academics in the area of economics and four SME owners who have knowledge about the effect of natural disasters on businesses. Teng (2002) proposed that 5−15 experts are appropriate for a group decision. We therefore used eight experts for the determination of influences among the economic impacts on SMEs. All the 12 direct and indirect impacts on SMEs comprising of 9 indirect impacts are retained and listed as follows:

E1. Loss of sales or production (indirect impact e1, **e1**)
E2. Employees absent from work (indirect impact e2, **e2**)
E3. Damage to property or business premises (direct impact e3)
E4. Damage or loss of stock and equipment (direct impact e4)
E5. Decrease in turnover or profit (indirect impact e5, **e3**)
E6. Reduced access to premises (direct impact e6)
E7. Increase in operating cost (indirect impact e7, **e4**)
E8. Physical or health impacts on employees (indirect impact e8, **e5**)
E9. Disruption to supply chain (indirect impact e9, **e6**)
E10. Increase in insurance cost (indirect impact e10, **e7**)
E11. Increase in unemployment (indirect impact e11, **e8**)
E12. Location and relocation of premises (indirect impact e12, **e9**)

Note: The labels in bold represent indirect economic impacts.

18.3.1.1 Economic impacts description

We describe the various economic impacts in this section:

E1 (e1, e1) Loss of sales or production
    This is when a company's sales or production is reduced. This can be through loss of markets or customers to other competitors or defective products. Loss of sales or production can be caused by a natural disaster that can lead to a reduction in sales and production. Access to the businesses may be affected. Production may also be affected if the materials needed cannot be delivered because of a natural disaster.

E2 (e2, **e2**) *Employees absent from work*

This is a loss in working hours to the company/organization. This may be due to holidays or disaster occurring at the workplace to require an employee to vacate factory/company premises.

E3 (e3) *Damage to property or business premises*

This is a defective building which is unfit to work in. Either part or the whole building is an unsuitable place to work in. For example, if flooding destroyed part of a building of a business.

E4 (e4) *Damage or loss of stock and equipment*

This is a spoiled stock or equipment that is not safe to be kept at the workplace; these are hazards to staffs and the working environment, for example, a broken generator.

E5 (e5, **e3**) *Decrease in turnover or profit*

This is a situation where there is a reduction in sales or operational costs of the business are running faster than sale revenue. For example, increases in electricity, water bills, rent, insurance, repairs while sales revenue remains the same will result in a reduction in the profit.

E6 (e6) *Reduced access to premises*

It is a process of decreasing the number of entrances to premises or narrow the entrance to premises, for example, reduction of the access to business due to a natural disaster.

E7 (e7, **e4**) *Increase in operating cost*

That is, if the variable cost of a business gets higher as a result of variable inputs of production. For example, a rise in wages/salaries, and any other production-related bills such as overheads, distribution expenses, courier fees, and insurance. For a disaster, there will be an increase in operating cost because the business will not run smoothly. For example, the cost of the movement of materials or goods to the business will increase leading to the increase in operating the business.

E8 (e8, **e5**) *Physical or health impacts on employees*

A situation or condition which prevents an employees' ability to work to optimum level or capacity. For example, an employee is on light duty because of health issues or physical look or ability.

E9 (e9, **e6**) *Disruption to supply chain*

There may be a delay in providing orders or receiving items from suppliers. The delay may be due to lack of goods or materials or lack of workforce.

E10 (e10, **e7**) *Increase in insurance cost*

This is a situation when the insurance premium is raised due to claims or a change in policy. For example, claiming insurance after a disaster may result in the increase of the insurance premium.

E11 (e11, **e8**) *Increase in unemployment*

When the number of the labor force who are doing suitable paid jobs decline. This is a fall in the number of people who are working. For example, a whole factory may be closed temporarily or permanently because of a natural disaster.

E12 (e12, **e9**) *Location and relocation of premises*

The movement by the management of staff, materials, machines, equipment, and other company assets from one location to another for a short or long period. For example, a business may have to move temporary to a different location while repair work is done after a natural disaster.

18.3.1.2 Generation of the matrix

Suppose E_k is the $n \times n$ matrix obtained from kth expert. The matrix entry $e_{ij(k)}$ gives the level of influence of economic impact e_i of natural disasters on the economic impact e_j of natural disasters given by the kth expert. The five levels of influence are defined and presented in Table 18.1.

Eight experts completed Tables A.1 and A.2 using Table 18.1 as follows: the values of the levels for e1 influencing e2, e1 influencing e3 up to e1 influencing e12, then e2 influencing e1, e2 influencing e3 up to e2 influencing e12. The process continues from e12 influencing e1, e12 influencing e2 up to e12 influencing e11. That will complete all the rows and columns of Table A.1. For example, if economic impact e1 has a medium influence on economic impact e2, then 2 was written under e2 in row 1. The same procedure was used to complete Table A.2 which is the influences among the indirect impacts. The economic impact on itself is zero therefore the diagonals in Tables A.1 and A.2 are zero.

18.3.2 DECISION-MAKING TRIAL AND EVALUATION LABORATORY METHOD

The DEMATEL method was developed in Switzerland by Gabus and Fontela (1972, 1973) to be used to serve as a tool that would make it possible to solve complex problems and to analyze a variety of causal links. For example, a set of a complex problem has a set of elements which have binary relation with each other. It is a commonly used method for modeling relationship between variables. DEMATEL method has advantages and benefits over other methods such as statistical methods because the method captures causal relationships and not only dependencies inherent in the data (Hiete, Merz, Comes, & Schultmann, 2011).

DEMATEL which involve five steps are described as follows:
The steps of DEMATEL are presented as follows:

Step 1: Generate the direct relation matrix. The matrix represents the aggregate influence scores for various variables or factors. The rating of 0 (no influence), 1 (low influence), 2 (medium influence), 3 (high influence), and 4 (very high influence) among the various factors or variables is done by experts. The average of the matrices which represent the rating from each expert is determined which will result in a square matrix A with zero at the principal diagonal. $A = [a_{ij}]_{n \times n}$ where, a_{ij} represents the degree which factor i impacts factor j. That is

$$A = \frac{1}{N} \sum_{k=1}^{N} a_{ij}^k, \tag{18.1}$$

N is the number of experts.

Table 18.1 Level of Influence.

Number	Level of Influence	Explanation
0	No	Economic impact e_i has no influence on economic impact e_j
1	Low	Economic impact e_i has low influence on economic impact e_j
2	Medium	Economic impact e_i has medium influence on economic impact e_j
3	High	Economic impact e_i has high influence on economic impact e_j
4	Very high	Economic impact e_i has very high influence on economic impact e_j

Step 2: Normalize the direct relation matrix. The normalized direct relation matrix B is obtained as follows

$$B = \frac{A}{\max\left[\max\sum_{i=1}^{n} a_{ij}, \max\sum_{j=1}^{n} a_{ij}\right]}, \quad i,j \in \{1, 2, \ldots, n\} \tag{18.2}$$

That is, divide matrix A by the maximum of the sum of rows and sum of columns.

Step 3: Develop the total relation matrix C from the normalized direct relation matrix B as follows. The total relation matrix C in which I is the identity matrix and B^m is m-indirect influence is obtained by

$$C = [c_{ij}]_{n \times n} = B + B^2 + \cdots + B^m = B(I - B)^{-1} \quad \text{when } m \to \infty \tag{18.3}$$

Step 4: Produce a causal diagram by using the sum of rows D and sum of the columns E

$$D = [d_{ij}]_{n \times 1} = \left[\sum_{j=1}^{n} c_{ij}\right]_{n \times 1} \tag{18.4}$$

$$E = [e_{ij}]_{1 \times n} = \left[\sum_{i=1}^{n} c_{ij}\right]_{1 \times n} \tag{18.5}$$

The horizontal axis $(D + E)$ represents the importance of the variables, whereas the vertical axis $(D - E)$ shows the cause-and-effect relationships. The variables with positive $(D - E)$ values are the cause factors, whereas those with negative are effect factors. The sum of ith row D is the causal influence and the sum of the jth column E is the effect influence of the total relation matrix C.

Step 5: Depict structural relation between variables

The structural relation amongst variables is shown through an inner dependence matrix by retaining only those variables whose effect in the matrix C is greater than the threshold value μ can be given by the experts, based on literature review or obtained by averaging the values of C matrix elements. The use of the threshold to simplify the total relation matrix is to filter out the variables having negligible effects from the total relation matrix C. The matrix obtained by setting the values which are less than the threshold value is known as the inner dependency matrix. There are different methods for deciding the threshold. The common ones are a discussion with experts, finding the average value of matrix C (Awasthi & Gryzbowska, 2014) and adding two standard deviations to the mean (Zhu, Sarkis, & Lai, 2014). We would use the average value of matrix C as the threshold in the chapter.

There are a number of studies with a detailed discussion of the steps that applied DEMATEL to examine disaster problems. A DEMATEL method was used by Khazai, Merz, Schulz, and Borst (2013) to account for the interactive and causal processes between the social and industrial vulnerability of communities affected by a natural disaster. For example, the loss of an industry due to a natural disaster can impact on the unemployment rate in a region. High unemployment in the region can lead to a cycle of problems, which can indirectly magnify the impact of industrial losses in that region, such as lower consumption of industrial goods after a natural disaster. A model was developed by Merz, Hiete, Comes, and Schultmann (2013) based on DEMATEL for risk and vulnerability assessment of an industry due to a natural disaster. This study is similar to that of Khazai et al. (2013). The model was applied to assess the industrial vulnerability of 44 districts in German federal state of Baden-Wuerttemberg.

In facility location problems, Celik (2017) used DEMATEL by a comprehensive literature review and experts' opinion that led to consideration 14 criteria for the location of temporary shelters for disaster management. Celik (2017) found that the optimal distribution of relief is the most important factor for disaster management. Yang, Shieh, Huang, and Tung (2018) used DEMATEL to identify factors that would enable volunteers to respond to inquiries and to notify the public and disaster management systems of flood disasters. The cause-and-effect factors were identified in the study. Trivedi (2018) also used DEMATEL method to determine shelter site location during a disaster by considering relationships among factors for shelter location. The study shows that factors such as favorability of terrain, community infrastructure, transportation infrastructure, and type of ownership have a strong influence on factors such as proximity and safety and security.

The DEMATEL method is used widely because of the following reasons:

- It is flexible (Bouzon, Govindan, & Rodriguez, 2018).
- It allows variations in relationships among factors (Bouzon et al., 2018; Zhu, Sarkis, & Geng, 2011).
- It provides multiple directional relationships (Zhu et al., 2011).
- It has a key advantage in its ability to produce possible results with the least amount of data (Bouzon et al., 2018).

18.4 RESULTS AND DISCUSSION

We present the results of DEMATEL for direct and indirect economic impacts and indirect impacts in this section. The matrices for each step with the impact diagrams are presented.

18.4.1 DIRECT AND INDIRECT IMPACT

Step 1: The average matrix is obtained by adding all the eight matrices from experts (SMEs and economic) and dividing by 8 using Microsoft Excel. Matrix A *(Fig. 18.1)* is obtained using Eq. (18.1).

Step 2: Matrix A is divided by maximum (sum of rows, sum of columns) which is matrix A divided by maximum (34.125, 36.875). That is, matrix A is divided by 36.875. Matrix B *(Fig. 18.2)* is obtained using Eq. (18.2).

Step 3: Multiply matrix B by the inverse of the identity matrix I minus matrix B. Matrix C is obtained using Eq. (18.3).

Step 4: Sum the rows and columns of matrix C to obtained D and E using Eqs. (18.4) and (18.5), respectively. The results are shown in Table 18.2.

Matrix A

0.000	2.375	1.250	1.875	3.875	0.625	3.000	2.125	2.375	1.500	3.625	1.750
2.875	0.000	0.000	0.625	3.125	1.500	2.500	1.875	2.375	1.375	2.625	1.625
4.000	2.750	0.000	3.750	3.750	3.125	3.375	3.000	3.375	3.125	3.000	3.625
3.750	1.875	2.375	0.000	3.750	1.125	2.750	1.875	2.625	2.875	2.125	1.875
3.000	1.375	0.625	0.750	0.000	0.250	2.500	2.125	1.750	1.125	2.625	2.000
3.250	2.875	1.250	1.125	3.250	0.000	2.500	2.375	2.625	1.750	2.625	3.000
2.250	0.875	0.625	0.750	2.875	0.750	0.000	1.000	1.375	2.125	1.250	2.375
2.750	3.250	0.875	1.125	2.500	1.000	1.750	0.000	1.625	1.875	3.750	1.625
3.625	1.625	0.500	1.750	3.000	1.000	2.500	1.125	0.000	2.000	1.125	2.000
0.875	0.750	0.875	0.375	3.000	0.000	3.500	0.875	0.125	0.000	1.125	1.500
3.125	2.375	0.500	0.625	2.750	0.875	1.250	2.000	1.000	0.875	0.000	2.000
2.125	2.000	0.875	1.125	2.250	1.625	2.250	1.250	1.500	2.000	1.625	0.000

FIGURE 18.1

Average matrix *A*.

Matrix B

0.000	0.064	0.034	0.051	0.105	0.017	0.081	0.058	0.064	0.041	0.098	0.048
0.078	0.000	0.000	0.017	0.085	0.041	0.068	0.051	0.064	0.037	0.071	0.044
0.109	0.075	0.000	0.102	0.102	0.085	0.092	0.081	0.092	0.085	0.081	0.098
0.102	0.051	0.064	0.000	0.102	0.031	0.075	0.051	0.071	0.078	0.058	0.051
0.081	0.037	0.017	0.020	0.000	0.007	0.068	0.058	0.048	0.031	0.071	0.054
0.088	0.078	0.034	0.031	0.088	0.000	0.068	0.064	0.071	0.048	0.071	0.081
0.061	0.024	0.017	0.020	0.078	0.020	0.000	0.027	0.037	0.058	0.034	0.064
0.075	0.088	0.024	0.031	0.068	0.027	0.048	0.000	0.044	0.051	0.102	0.044
0.098	0.044	0.014	0.048	0.081	0.027	0.068	0.031	0.000	0.054	0.031	0.054
0.024	0.020	0.024	0.010	0.081	0.000	0.095	0.024	0.003	0.000	0.031	0.041
0.085	0.064	0.014	0.017	0.075	0.024	0.034	0.054	0.027	0.024	0.000	0.054
0.058	0.054	0.024	0.031	0.061	0.044	0.061	0.034	0.041	0.054	0.044	0.000

FIGURE 18.2

Normalized direct relation matrix *B*.

Step 5: The inner dependence matrix is obtained by omitting values which less than the threshold of 0.111 from matrix *C*. The average of the values matrix *C* is used as the threshold for the study.

Table 18.2 shows the degree of influence of each direct and indirect economic impacts over the other impacts using the total relation matrix shown in Fig. 18.3. The economic impacts having the higher value of $D + E$ are among the economic impacts that have the highest degree of relationship with other economic impacts of natural disasters. The importance of the 12 economic impacts based on $D + E$ (the greater the value the more important is the economic important) in Table 18.2 are as follows: e1 > e5 > e3 > e7 > e11 > e2 > e12 > e8 > e4 > e9 > e6 > e10. The most important economic impact (direct and indirect) of natural disaster on SMEs according to the experts is e1 which

Table 18.2 The Sum and the Difference of Influences for the 12 Direct and Indirect Economic Impacts.

Economic Impacts	Row Sum E	Column Sum E	$D+E$	$D-E$	Rank	Cause or Effect
E1(e1)	1.476	1.909	3.385	−0.433	1	Effect
E2(e2)	1.230	1.351	2.581	−0.121	6	Effect
E3(e3)	2.249	0.601	2.850	1.648	3	Cause
E4(e4)	1.664	0.835	2.499	0.829	9	Cause
E5(e5)	1.106	2.074	3.180	−0.968	2	Effect
E6(e6)	1.620	0.705	2.325	0.915	11	Cause
E7(e7)	0.989	1.710	2.699	−0.721	4	Effect
E8(e8)	1.344	1.215	2.559	0.129	8	Cause
E9(e9)	1.238	1.260	2.498	−0.022	10	Effect
E10(e10)	0.782	1.244	2.026	−0.462	12	Effect
E11(e11)	1.076	1.590	2.666	−0.514	5	Effect
E12(e12)	1.144	1.424	2.568	−0.280	7	Effect

Matrix C

```
0.105  0.133  0.064  0.092  0.209  0.054  0.167  0.120  0.127  0.105  0.178  0.122
0.158  0.061  0.027  0.052  0.172  0.068  0.140  0.103  0.116  0.089  0.139  0.105
0.258  0.182  0.051  0.163  0.266  0.136  0.227  0.177  0.190  0.182  0.208  0.209
0.210  0.130  0.098  0.052  0.222  0.071  0.176  0.123  0.144  0.149  0.153  0.136
0.153  0.091  0.040  0.053  0.084  0.035  0.132  0.103  0.095  0.079  0.133  0.108
0.195  0.154  0.067  0.078  0.205  0.041  0.165  0.133  0.141  0.118  0.163  0.160
0.125  0.071  0.038  0.049  0.148  0.044  0.063  0.070  0.080  0.099  0.090  0.112
0.164  0.149  0.051  0.069  0.166  0.060  0.129  0.061  0.103  0.107  0.174  0.111
0.176  0.102  0.042  0.082  0.171  0.056  0.143  0.084  0.057  0.107  0.103  0.115
0.077  0.056  0.039  0.033  0.134  0.020  0.138  0.057  0.039  0.035  0.074  0.080
0.154  0.115  0.036  0.049  0.152  0.050  0.100  0.100  0.077  0.071  0.066  0.106
0.134  0.107  0.048  0.063  0.145  0.070  0.130  0.084  0.091  0.103  0.109  0.060
```

FIGURE 18.3

Total relation matrix C.

is "loss of sales or production" with the value of 3.385 and the least important is e10 which is "increase in insurance cost" with the value of 2.026. Table 18.3 shows the cause-and-effect economic impacts.

The difference between the row sum D and column sum E determines the cause group and the effect group. The direct and indirect economic impacts with positive $D-E$ values are in the cause group while the indirect economic impacts with negative values are in the effect group. The cause direct and indirect economic impacts as shown in Table 18.3 and Fig. 18.4 include e3, e4, e6, and e8 since $D-E$ is positive and the effect direct and indirect economic impacts include e1, e2, e5, e7, e9, e10, e11, and e12 since $D-E$ is negative. The cause-and-effect direct and indirect economic impacts of natural disasters are shown in Table 18.3. The cause direct and indirect economic

Table 18.3 Cause-and-Effect Direct and Indirect Economic Impacts.		
Cause Economic Impacts	**Effect Economic Impacts**	
Damage to properties or business (e3)	Loss of sales or production (e1)	Disruption to supply chain (e9)
Damage or loss of stock and equipment (e4)	Employees absent from work (e2)	Increase in insurance cost (e10)
Reduced access to premises (e6)	Decrease in turnover or profit (e5)	Increase in unemployment (e11)
Physical or health impacts on employees (e8)	Increase in operating cost (e7)	Location and relocation of premises (e12)

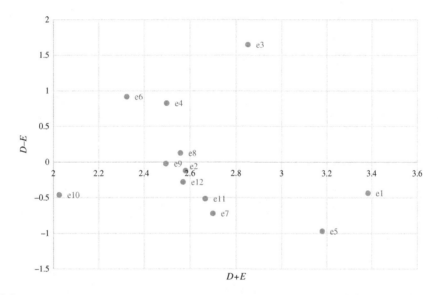

FIGURE 18.4

Prominence—causal diagram for direct and indirect economic impacts.

impacts have a net influence on the effect economic impacts. That is, the effect economic impacts are being influenced by the cause economic impacts. For example, "damage to properties or businesses of SMEs" due to a natural disaster will have an effect on "loss of sales or production."

The four causes of direct and indirect economic impacts influence the other indirect economic impacts; thus they are the main driving factors when SMEs are economically directly and indirectly impacted on after a natural disaster. Loss of sales and production that is in the effect group is the most important indirect economic impact according to the experts and will be influenced by all the other direct and indirect economic impacts of natural disasters in the cause group. Minimizing the cause direct and indirect economic impacts while addressing economic impacts on SMEs due to a natural disaster will minimize the indirect economic impacts in effect group. For example,

0.000	0.133	0.000	0.000	0.209	0.000	0.167	0.120	0.127	0.000	0.178	0.122
0.158	0.000	0.000	0.000	0.172	0.000	0.140	0.000	0.116	0.000	0.139	0.000
0.258	0.182	0.000	0.163	0.266	0.136	0.227	0.177	0.190	0.182	0.208	0.209
0.210	0.130	0.000	0.000	0.222	0.000	0.176	0.123	0.144	0.149	0.153	0.136
0.153	0.000	0.000	0.000	0.000	0.000	0.132	0.000	0.000	0.000	0.133	0.000
0.195	0.154	0.000	0.000	0.205	0.000	0.165	0.133	0.141	0.118	0.163	0.160
0.125	0.000	0.000	0.000	0.148	0.000	0.000	0.000	0.000	0.000	0.000	0.112
0.164	0.149	0.000	0.000	0.166	0.000	0.129	0.000	0.000	0.000	0.174	0.111
0.176	0.000	0.000	0.000	0.171	0.000	0.143	0.000	0.000	0.000	0.000	0.115
0.000	0.000	0.000	0.000	0.134	0.000	0.138	0.000	0.000	0.000	0.000	0.000
0.154	0.115	0.000	0.000	0.152	0.000	0.000	0.000	0.000	0.000	0.000	0.000
0.134	0.000	0.000	0.000	0.145	0.000	0.130	0.000	0.000	0.000	0.000	0.000

FIGURE 18.5

Simplified matrix threshold is 0.111—inner dependence matrix.

minimizing "damage to properties or business premises" will minimize the indirect economic impacts in the effect group such as "loss of sales or production." The remaining direct and indirect economic cost in the cause group will have a similar effect on the impacts in the effect groups.

The row sum in Table 18.2 shows the influence of a particular impact over the other impacts of a natural disaster. The higher the row sum the higher the influence over the other impacts. This confirms the importance of "damage to properties or business premises" with the highest row sum of 2.249 and will have maximum influence over the other impacts. Property damage of a business will have a flow-on effect on the other economic impacts. The column sum in Table 18.2 also shows impacts that receive the highest effect from the other impacts. "Loss of sales or production" is the column sum with the highest value of 1.909 Therefore "loss of sales or production" will have the highest effect from the other economic impacts. The inner dependency matrix from Fig. 18.5 shows that "loss of sales or production" is mainly caused by e3 (damage to properties or business), e4 (damage or loss of stock or equipment), and e6 (reduced access to premises). Damage to properties or business is the most influential economic impact. Hence, it will be useful to minimize that impact to achieve viability of SMEs during a natural disaster. The least influential on "loss of sales or production" as shown in Fig. 18.5 is e10 (increase in the insurance cost) with a value of 0.272 (row sum corresponding to e10).

18.4.2 RESULTS AND DISCUSSION (INDIRECT ECONOMIC IMPACT)

Step 1: The average matrix is obtained by adding all the eight matrices from experts (SMEs and economic) and dividing by 8 using Microsoft Excel. Matrix A *(Fig. 18.6)* is obtained using Eq. (18.1) .

Step 2: Matrix A is divided by maximum (sum of rows, sum of columns) which is matrix A divided by maximum (20.625, 23.000). That is, matrix A is divided by 23.000 matrix B *(Fig. 18.7)* is obtained using Eq. (18.2) .

Step 3: Multiply matrix B by the inverse of the identity matrix I minus matrix B. Matrix C *(Fig. 18.8)* is obtained using Eq. (18.3) .

Matrix A

0.000	2.375	3.875	3.000	2.125	2.375	1.500	3.625	1.750
2.875	0.000	3.125	2.500	1.875	2.375	1.375	2.625	1.625
3.000	1.375	0.000	2.500	2.125	1.750	1.125	2.625	2.000
2.250	0.875	2.875	0.000	1.000	1.375	2.125	1.250	2.375
2.750	3.250	2.125	1.750	0.000	1.625	1.875	3.375	1.625
3.625	1.625	3.000	2.500	1.125	0.000	2.000	1.125	2.000
0.875	0.750	3.000	3.500	0.875	0.125	0.000	1.125	1.500
3.125	2.375	2.750	1.250	2.000	1.375	0.875	0.000	2.000
2.125	2.000	2.250	2.125	1.250	1.625	2.000	1.625	0.000

FIGURE 18.6

Average matrix *A*.

Matrix B

0.000	0.103	0.169	0.130	0.092	0.103	0.065	0.158	0.076
0.125	0.000	0.136	0.109	0.082	0.103	0.060	0.114	0.071
0.130	0.060	0.000	0.109	0.092	0.076	0.049	0.114	0.087
0.098	0.038	0.125	0.000	0.044	0.060	0.092	0.054	0.103
0.120	0.141	0.092	0.076	0.000	0.071	0.082	0.147	0.071
0.158	0.071	0.130	0.109	0.049	0.000	0.087	0.049	0.087
0.038	0.033	0.130	0.152	0.038	0.005	0.000	0.049	0.065
0.136	0.103	0.120	0.054	0.087	0.060	0.038	0.000	0.087
0.092	0.087	0.098	0.092	0.054	0.071	0.087	0.071	0.000

FIGURE 18.7

Normalized direct relation matrix *B*.

Matrix C

0.322	0.322	0.497	0.407	0.289	0.300	0.260	0.415	0.308
0.403	0.207	0.439	0.363	0.261	0.282	0.237	0.355	0.281
0.379	0.247	0.288	0.337	0.252	0.241	0.212	0.332	0.275
0.307	0.193	0.355	0.206	0.183	0.198	0.224	0.242	0.258
0.397	0.336	0.404	0.336	0.187	0.254	0.255	0.384	0.280
0.404	0.254	0.412	0.348	0.218	0.173	0.249	0.280	0.278
0.216	0.157	0.315	0.303	0.152	0.123	0.115	0.201	0.198
0.378	0.280	0.388	0.287	0.246	0.225	0.197	0.227	0.268
0.318	0.246	0.349	0.303	0.201	0.218	0.228	0.269	0.175

FIGURE 18.8

Total relation matrix *C*.

Step 4: Sum the rows and columns of matrix C to obtained D and E using Eqs. (18.4) and (18.5) respectively. The results are shown in Table 18.3.

Step 5: The inner dependence matrix is obtained by omitting values which are less than the threshold of 0.280 from matrix C. The average of the values matrix C is used as the threshold for the study.

Tables 18.4 shows the degree of influence of each indirect economic impact over the other impacts using the total relation matrix shown in Fig. 18.9. The importance of the nine indirect economic impacts based on $D + E$ (the greater the value the more important is the economic impact) in Table 18.4 are as follows: **e1 > e3 > e8 > e2 > e4 > e5 > e6 > e9 > e7**. The most important

Table 18.4 The Sum and the Difference of Influences for the Indirect Economic Impacts.

Economic Impacts	Row Sum D	Column Sum E	$D+E$	$D-E$	Rank	Cause or Effect
E1(e1)	3.120	3.124	6.244	− 0.004	1	Effect
E2(e2)	2.828	2.242	5.070	0.586	4	Cause
E5(e3)	2.563	3.447	6.010	− 0.884	2	Effect
E7(e4)	2.166	2.890	5.056	− 0.724	5	Effect
E8(e5)	2.833	1.989	4.822	0.844	6	Cause
E9(e6)	2.616	2.014	4.630	0.602	7	Cause
E10(e7)	1.780	1.977	3.757	− 0.197	9	Effect
E11(e8)	2.496	2.705	5.201	− 0.209	3	Effect
E12(e9)	2.307	2.321	4.628	− 0.014	8	Effect

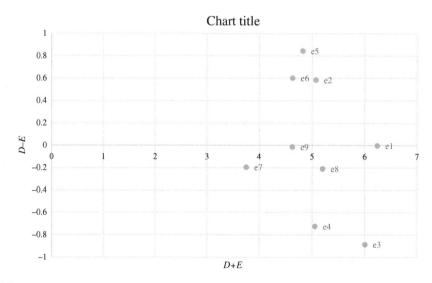

FIGURE 18.9

Prominence causal diagram for indirect economic impacts.

Table 18.5 Cause-and-Effect Indirect Economic Impacts.

Cause Economic Impacts	Effect Economic Impacts	
Employees absent from work (**e2**)	Loss of sales or production (**e1**)	Increase in insurance cost (**e7**)
Physical or health impacts on employees (**e5**)	Decrease in turnover or profit (**e3**)	Increase in unemployment (**e8**)
Disruption to supply chain (**e6**)	Increase in operating cost (**e4**)	Location and relocation of premises (**e9**)

indirect economic impact of natural disaster on SMEs according to the experts is **e1** which is "loss of sales or production" with the value of 6.244 and the least important is **e7** which is "increase in insurance cost" with the value of 3.757.

The indirect economic impacts with positive $D - E$ values are in the cause group, while the indirect economic impacts with negative values are in the effect group. The cause indirect economic impacts as shown in Table 18.5 and Fig. 18.9 include **e2**, **e5**, and **e6** since $D - E$ is positive and the effect indirect economic impacts include **e1**, **e3**, **e4**, **e7**, **e8**, and **e9** since $D - E$ is negative. The cause-and-effect indirect economic impacts of natural disasters are shown in Table 18.5. The cause indirect economic impacts have net influence on the effect indirect economic impacts. For example, "disruption to supply chain" of SMEs due to natural disaster will have an increasing effect on unemployment.

The three cause indirect economic impacts influence the other indirect economic impacts thus they are the main driving factors when SMEs are economically indirectly impacted on after a natural disaster. Loss of sales and production which is in the effect group is the most important indirect economic impact according to the experts and will be influenced by all the other indirect economic impacts of natural disasters. Minimizing the cause indirect economic impacts while addressing economic impacts on SMEs due to a natural disaster will minimize the indirect economic impacts in effect group. For example, minimizing "disruption of the supply chain" will minimize the indirect economic impacts in the effect group such as "loss of sales or production." The remaining indirect economic impact in the cause group will have a similar effect on the impacts in the effect groups.

The row sum in Table 18.4 shows the influence of a particular impact over the other impacts of a natural disaster. The higher the row sum, the higher the influence over the other impacts. This confirms the importance of "loss of sales or production" with the highest row sum and will have maximum influence over the other impacts. The column sum in Table 18.4 also shows impacts that receive the highest effect from the other impacts. "Loss of sales or production" is the column sum with the highest value of 3.124. Therefore "loss of sales or production" will have the highest effect from the other indirect economic impacts. The inner dependency matrix of Fig. 18.10 shows that **e5** (physical or health impacts on employees) has the highest influence on "loss of sales or production" followed by **e2** (employees absent from work). The least influential on "loss of sales or production" as shown in Fig. 18.10 is **e7** (increase in the insurance cost) with a value of 0.618 (row sum corresponding to **e7**).

Inner dependence matrix

0.322	0.322	0.497	0.407	0.289	0.300	0.000	0.415	0.308
0.403	0.000	0.439	0.363	0.000	0.282	0.000	0.355	0.281
0.379	0.000	0.288	0.337	0.000	0.000	0.000	0.332	0.000
0.307	0.000	0.355	0.000	0.000	0.000	0.000	0.000	0.000
0.397	0.336	0.404	0.336	0.000	0.000	0.000	0.384	0.280
0.404	0.000	0.412	0.348	0.000	0.000	0.000	0.280	0.000
0.000	0.000	0.315	0.303	0.000	0.000	0.000	0.000	0.000
0.378	0.280	0.388	0.287	0.000	0.000	0.000	0.000	0.000
0.318	0.000	0.349	0.303	0.000	0.000	0.000	0.000	0.000

FIGURE 18.10

Simplified matrix threshold is 0.280—inner dependence matrix.

18.5 CONCLUSION

We identified and determined the key economic impacts of natural disasters on SMEs in this chapter. The economic impacts were identified by literature review and discussion with experts in the field of economics and owners of SMEs. The determination of the key economic impacts first involves direct and indirect economic impacts and the second involves indirect economic impacts. The top three key economic impacts on SMEs from direct and indirect economic impact include "loss of sales or production," "decrease in turnover and profit," and "damage to property or business premises." The top three key indirect economic impacts include "loss of sales or production," "decrease in turnover or profit," and "increase in unemployment."

These findings support the results obtained from Auzzir et al. (2018), Gunathilaka (2018), and Wedawatta et al. (2014). Ability to determine key economic impacts using causality and prominence by DEMATEL, for the first time to evaluate the economic impacts of natural disasters is the strength of the study. The determination of the key economic impacts of natural disasters will enable policymakers and economic and town planners to target these key economic impacts to minimize the economic impacts of natural disasters.

The proposed study has limitations that need to be investigated for future research. Suggestions for future research include the following:

- Increasing the number of respondents for DEMATEL method.
- More economic impacts could be explored.
- The literature review could be extended.
- The results could be compared with other similar techniques.

ACKNOWLEDGMENT

The authors acknowledge the Australia Africa Universities Network (AAUN) as partners and funder (Grant number G1900649).

CONFLICT OF INTEREST

The authors declare no conflict of interest.

REFERENCES

ABC. (2017). *Cyclone Debbie hits Queensland farmers hard, hundreds of millions in crops lost.* From <https://www.abc.net.au/news/2017-03-31/queensland-farmers-and-crops-hit-hard-by-cyclone-debbie/8405422> Retrieved 10.11.19.

ABC. (2019). *Townsville flooding crises leaves almost 1000 displaced two months on.* From <https://www.abc.net.au/news/2019-04-02/townsville-flooding-leaves-a-thousand-displaced-two-months-on/10959250> Retrieved 10.11.19.

Alexander, D. (1997). The study of natural disasters, some reflections on a changing field of knowledge. *Disasters,* 284−304. From <https://onlinelibrary.wiley.com/doi/epdf/10.1111/1467-7717.00064> Retrieved 14.10.19.

Alkhoraif, A., Rashid, H., & McLaughlin, P. (2019). Lean implementation in small and medium enterprises: Literature review. *Operations Research Perspectives, 6.* From <https://reader.elsevier.com/reader/sd/pii/S2214716018301659?token = 7BBF8F20F78F40C6EE0857F3CE9579E1A0B49A32EE7D439FE2B781192C8C61081EF6BD3-C1C9B8A9A480FC118A097C5C0> Retrieved 22.10.19.

Auzzir, Z., Haigh, R., & Amaratunga, D. (2018). Impacts of disaster to SMEs in Malaysia. *Procedia Engineering, 212,* 1131−1138.

Awasthi, A., & Gryzbowska, K. (2014). Barriers of the supply chain integration process. In P. Golinska (Ed.), *Logistics operations, supply chain management and sustainability. EcoProduction (environmental issues in logistics and manufacturing)* (pp. 15−30). Cham: Springer.

Botzen, W. J., Kunreuther, H., Czajkowski, J., & de Moel, H. (2019). Adoption of individual flood damage mitigation measures in New York: An extension of protection motivation theory. *Risk Analysis.* From <https://onlinelibrary.wiley.com/doi/epdf/10.1111/risa.13318> Retrieved 12.10.19.

Botzen, W. J., Deschenes, O., & Sanders, M. (2019). The economic impacts of natural disasters: A review of models and empirical studies. *Review of Environmental Economics and Policy, 13*(2), 167−188.

Bouzon, M., Govindan, K., & Rodriguez, C. M. T. (2018). Evaluating barriers for reverse logistics implementation under a multiple stakeholders' perspective analysis using grey decision making approach. *Resources, Conservation and Recycling, 128,* 315−335.

Celik, E. (2017). A cause and effect relationship model for location of temporary shelters in disaster operations management. *International Journal of Disaster Risk Reduction, 22,* 257−268.

Chrysanthos, N. (2019). 'Better to be cautious': At least 500 schools to close in Sydney and NSW due to fire danger. *The Sydney Morning Herald.* From <https://www.smh.com.au/national/nsw/better-to-be-cautious-300-schools-to-close-in-sydney-and-nsw-due-to-fire-danger-20191111-p539cs.html> Retrieved 12.11.19.

Gabus, A., & Fontela, E. (1972). *World problems, an invitation to further thought within the framework of DEMATEL.* Geneva: Battelle Geneva Research Centre.

Gabus, A., & Fontela, E. (1973). *Perceptions of the world problematique: Communication procedure, communicating with those bearing collective responsibility (DEMATEL report no. 1).* Geneva: Battelle Geneva Research Centre.

Gunathilaka, S. (2018). The impact of natural disasters on Micro, Small and Medium Enterprises (MSMEs): A case study on 2016 flood event in Western Sri Lanka. *Procedia Engineering, 212,* 744−751.

Hiete, M., Merz, M., Comes, T., & Schultmann, F. (2011). Trapezoidal fuzzy DEMATEL method to analyze and correct for relations between variables in a composite indicator for disaster resilience. *OR Spectrum, 34,* 971−995.

Kayanula, D., & Quartey, P. (2000). *The environment for promoting small and medium-sized enterprises in Ghana and Malawi.* Institute for Development Policy and Management, University of Manchester. ISBN: 1902518675. From <http://www.microfinancegateway.org/sites/default/files/mfg-en-paper-the-policy-environment-for-promoting-small-and-medium-sized-enterprises-in-ghana-and-malawi-2000.pdf> Retrieved 21.10.19.

Khazai, B., Merz, M., Schulz, C., & Borst, D. (2013). An integrated indicator framework for spatial assessment of industrial and social vulnerability to indirect disaster losses. *Natural Hazards,* 1−23. Available from https://doi.org/10.1007/s11069-013-0551-z, Published online 20.01.13.

Kousky, C. (2014). Informing climate adaptation: A review of economic cost of natural disasters. *Energy Economics,* 576−592. From <https://www.sciencedirect.com/science/article/pii/S0140988313002247> Retrieved 15.10.19.

Liu, Z., Xu, J., & Han, B. T. (2013). Small and medium-sized enterprise post-disaster reconstruction management patterns and application. *Natural Hazards, 68,* 809−835.

Marks, D., & Thomalla, F. (2017). Responses to 2011 floods in central Thailand: Perpetuating the vulnerability of small and medium enterprises? *Natural Hazards, 87,* 1147−1165.

Merz, M., Hiete, M., Comes, T., & Schultmann, F. (2013). A composite indicator model to assess natural disaster risks in industry on a spatial level. *Journal of Risk Research, 16*(9), 1077−1099.

Miles, R. L., Hyland, P., Soosay, C., Greer, L., O'Dea, G., Alock, D., et al. (2007). *Effect of drought on small businesses in regional Queensland: Implications for sustainable regional development.* From <https://www.researchgate.net/publication/237719943> Retrieved 04.10.19.

Noy, I. (2009). The macroeconomic consequences of disasters. *Journal of Development Economics, 88*(2), 221−231.

Teng, J. Y. (2002). *Project evaluation: Methods and applications.* Taiwan: National Taiwan Ocean University.

Trivedi, A. (2018). A multi-criteria decision approach based on DEMATEL to assess determinants of shelter site selection in disaster response. *International Journal of Disaster Risk Reduction, 31,* 722−728.

U.S. International Trade Commission (2010). Small and medium-sized enterprises: Overview of participation in U.S exports. (investigation No. 332−508). Washington, DC. From <www.usitc.gov/publications/332/pub4125.pdf> Retrieved 04.10.19.

Wedawatta, G., Ingirige, B., Jones, K., & Proverbs, D. (2011). Extreme weather events and construction SMEs: Vulnerability, impacts, and responses. *Structural Survey, 29*(2), 106−119.

Wedawatta, G., Ingirige, B., & Proverbs, D. (2014). Small businesses and flood impacts: The case of the 2009 flood event in Cockermouth. *Journal of Flood Risk Management, 7*(1), 42−53.

Yang, C. L., Shieh, M. C., Huang, C. Y., & Tung, C. P. (2018). A derivation of factors influencing the successful integration of corporate volunteers into public flood disaster inquiry and notification systems. *Sustainability, 10,* 1973.

Zhu, Q., Sarkis, J., & Geng, Y. (2011). Barriers to environmentally-friendly clothing production among Chinese apparel companies. *Asian Business & Management, 10*(3), 425−452.

Zhu, Q., Sarkis, J., & Lai, K. H. (2014). Supply chain-based barriers for truck-engine remanufacturing in China. *Transportation Research Part E: Logistics and Transportation Review, 68,* 103−117.

APPENDIX

Table A.1 12 × 12 Matrix Showing the Level of Influence (Direct and Indirect Impacts).

	e1	e2	e3	e4	e5	e6	e7	e8	e9	e10	e11	e12
e1	0											
e2		0										
e3			0									
e4				0								
e5					0							
e6						0						
e7							0					
e8								0				
e9									0			
e10										0		
e11											0	
e12												0

Table A.2 9 × 9 Matrix Showing the Level of Influence (Indirect Impacts).

	e1	e2	e3	e4	e5	e6	e7	e8	e9
e1	0								
e2		0							
e3			0						
e4				0					
e5					0				
e6						0			
e7							0		
e8								0	
e9									0

CLIMATE CHANGE AND NATURAL DISASTERS: MACROECONOMIC PERFORMANCE AND SUSTAINABLE DEVELOPMENT

19

Janet Dzator[1,2], Alex O. Acheampong[1,3] and Michael Dzator[2,4]

[1]*School of Business, The University of Newcastle, Newcastle, NSW, Australia* [2]*Australia Africa Universities Network (AAUN) Partner, Newcastle, NSW, Australia* [3]*Faculty of Business and Law, The University of Newcastle, Newcastle, NSW, Australia* [4]*School of Access Education, Central Queensland University, Mackay, QLD, Australia*

19.1 INTRODUCTION

The occurrences of natural disasters continue to pose a threat to economic and human development. Recently, the report from the Centre for Research on the Epidemiology and Disasters (CRED) (2019) indicates that in the year 2018 a total of 315 natural disaster events were recorded with economic damage totally US$ 131.7 billion, while over 68 million people were affected with a total of 11,804 people losing their life. The report further indicates that the cost associated with the occurrence of natural disasters was not shared equally. For instance, Asia suffered the highest impact of disasters, with 45% of natural disasters, 80% of death, and 76% of people affected. Specifically, while Indonesia recorded the highest number of death (47%), India recorded the highest number of people affected (35%). The report further indicates that natural disaster types such as earthquake accounted for the highest number of death (45%), followed by flood (24%). In addition, flood further accounted for the highest number of people affected (58%), while storms accounted for 28% of the number affected.

The rapid occurrences of natural disasters have heightened the desires of international organizations, policymakers, and researchers to better understand the impacts of the natural disasters on economic development. From a theoretical perspective the impact of natural disasters of economic output is ambiguous (Klomp & Valckx, 2014; Panwar & Sen, 2018). For instance, the neoclassical growth theories argue that natural disasters could limit or cause a permanent deviation from previous growth trajectory if disaster leads to the destruction of existing capital stock. Contrarily, the endogenous growth theory indicates the replacement of damaged physical capital caused by disaster with new physical capital increases the existing capital stock with more productive capital, thereby fueling economic growth temporarily (Albala-Bertrand, 1993; Cavallo, Galiani, Noy, & Pantano, 2013; Chhibber & Laajaj, 2008; Hallegatte & Dumas, 2009).

It is further contested that the question on the impact of natural disasters on economic output is an empirical one (Cavallo et al., 2013; Klomp & Valckx, 2014). After the seminal paper of Albala-Bertrand (1993), numerous studies have examined the impact of natural disasters on economic

Economic Effects of Natural Disasters. DOI: https://doi.org/10.1016/B978-0-12-817465-4.00019-4

output; however, the existing empirical literature remains contradictory. For instance, some studies report that natural disasters hinder economic growth (Cavallo et al., 2013; Felbermayr & Gröschl, 2014; Klomp & Valckx, 2014), while another strand of the empirical literature suggests that natural disaster promotes economic growth (Albala-Bertrand, 1993; Skidmore & Toya, 2002). In addition, depending on the type of natural disaster, some studies have also reveale48-els-che_chaiechi-een-1632387_ch019.3d that not all organizations have a significant effect on economic output (Fomby, Ikeda, & Loayza, 2013; Loayza, Olaberría, Rigolini, & Christiaensen, 2012; Panwar & Sen, 2018).

Shabnam (2014) argues that the empirical literature on the economics of natural disasters is still at the infant stage, and together with the contradictory results, Chhibber and Laajaj (2008) argue that further empirical studies are needed to understand the impact of natural disasters on economic output. Therefore this study aims to investigate the short and long-run impact of natural disasters on economic growth, private sector investment, and public sector spending using a comprehensive panel dataset for 61 countries for 1960−2018. In this direction, our study contributes to the literature in the following directions. First, Loayza et al. (2012) argue that the majority of the empirical studies have mainly examined the aggregate economic growth effect of natural disasters with little attention given to the sectorial effect of disasters. These authors argue that disasters could have a different impact on different sectors of the economy, and future studies should rather pay more attention to the sectorial effect of disasters. Our study contributes in this direction by not only estimating the impact of natural disaster on economic growth (GDP per capita) but also further investigate the impact of natural disasters on sectorial growth such as private sector investment and public sector spending. Second, some of the recent literature has indicated that different types of natural disaster could have different effects on aggregate economic growth and other sectors of the economy (Chhibber & Laajaj, 2008; Fomby et al., 2013; Loayza et al., 2012; Panwar & Sen, 2018). We contribute to the literature in this direction by providing new empirical evidence on disaster types such as earthquake, drought, storms, and floods on economic growth, private sector investment, and public sector spending. Finally, it argued that different countries respond differently to the effect of natural disasters.

Our empirical analysis reveals that in the short run the composite indicator for natural disaster and flood significantly reduces economic growth, while other specific types of natural disasters such as drought, storm, and earthquake exert an insignificant negative effect on economic growth. The results also indicate that in the long run, natural disasters have negligible impacts on economic growth. Our findings further suggest that none of the proxies of natural disaster exert a significant effect on public sector spending. On the other hand, the results reveal that in the short run, only flood and earthquake significantly increase private sector investment, but drought retards private sector investment in both short and long run. The remaining sections of the chapter proceed in four steps. In Section 19.2, we provide a literature review on the economics of natural disasters, while methodology and data for the study are described in Section 19.3. The results and discussions are presented in Section 19.4, while conclusions and policy recommendations are presented in Section 19.5.

19.2 LITERATURE REVIEW ON THE ECONOMICS OF NATURAL DISASTERS

Research on natural disaster economics continues to gain momentum among social scientists. In the theoretical literature the impact of natural disasters on economic development remains controversial.

One strand of the theoretical literature argues that natural disaster could hamper economic growth. In the neoclassical economic growth models, both physical and human capitals are fundamental for sustaining higher economic growth (Romer, 1986; Solow, 1956). However, the occurrence of natural disaster could reduce the accumulation of both physical and human capitals, thereby retarding economic growth (Cuaresma, 2010; Klomp & Valckx, 2014). Damages caused by a natural disaster to physical capital could increase depreciation rate (Klomp & Valckx, 2014), which subsequently results in the diversification of resources from planned investment and further increases fiscal deficit and inflation (Benson & Clay, 1998; Chhibber & Laajaj, 2008). Panwar and Sen (2018) argue that damages caused by a natural disaster to physical assets could lead to the loss of potential labor hours and subsequently result in the fall of agricultural, industrial, and the overall economic output.

Contrarily, based on the Schumpeterian creative destruction model, the endogenous growth theory indicates that the destruction of capital stock per worker temporarily increases economic growth after a natural disaster by increasing its marginal return (Loayza et al., 2012; Panwar & Sen, 2018). Thus the replacement of damaged physical capital with new physical capital increases the existing capital stock with more productive capital, thereby fueling economic growth temporarily (Albala-Bertrand, 1993; Chhibber & Laajaj, 2008; Hallegatte & Dumas, 2009). The impact of natural disaster on long-run economic output remains unclear. Chhibber and Laajaj (2008) illustrate four possible scenarios of the impact of natural disaster on economic. They argued that natural disaster may have two types of temporal effect on growth. These could be (1) an initial drop in growth followed by a return to original trend line or (2) a drop followed by an overshoot of the trend line before growth returning to trend line. Alternatively, there could be a permanent shift to a new growth path below the trend line or a strong positive recovery resulting in a permanent shift above the trend line.

With the theoretical argument on the impact of natural disaster on the economic output being contentious, the empirical studies have also resulted in conflicting outcomes. For instance, Albala-Bertrand (1993) examined the impact of 28 natural disasters on economic growth in 26 developing countries between 1960 and 1979. The results from the ordinary least squares (OLS) approach indicated that economic growth increases in most cases after the natural disaster. The authors argued the positive effect is due to the replacement of damaged physical capita with efficient physical assets. Similarly, Skidmore and Toya (2002) investigated the long-run relationship between natural disaster, capital accumulation, total factor productivity, and economic growth. Using a cross-sectional dataset, the results revealed that climatic disasters are associated with the higher rates of human capital accumulation that increases total factor productivity and economic growth but reduces expected rate of return on physical capital.

On the other hand, Klomp and Valckx (2014) used metaregression analysis to examine the impact of natural disaster on economic growth and found that natural disaster retards output per capita; however, the impact differs across disaster type and sample used. Strobl (2012) further investigated the impact of hurricane in the Caribbean and Central America region, and the study found that hurricane reduces economic growth by 0.83% points. In another study, Felbermayr and Gröschl (2014) investigated the impact of geophysical and meteorological events on economic growth for the period of 1979−2010 in more than 100 countries. The study revealed that a disaster in top 1-percentile is associated with 6.83% fall in economic growth, while a disaster in top 5-percentile is associated with 0.33% fall in economic growth. In addition, the results indicated that the smallest 25-percentile retards economic growth at approximately 0.01%. The study concluded that these results vary across income groups with the developing countries bearing the highest cost of the natural disaster. Similarly, Cavallo et al. (2013) demonstrated that extreme disaster hinders economic growth in both short and long run; however, the impact becomes insignificance after controlling for political changes.

The study of Noy (2009) revealed that natural disaster retards economic growth, but with similar magnitude, developing countries experienced major macroeconomic shocks relative to the developed countries. Similarly, Noy and Vu (2010) examined the impact of natural disaster on Vietnam's economic growth. Using provincial data together with system-generalized method of moment (System GMM), the results revealed that lethal disasters contribute to the fall economic growth. However, disasters that destroy physical assets significantly increase economic growth in the short run. Shabnam (2014) investigated the effect of natural disaster on economic growth for 187 countries between 1960 and 2010 and found that total number of people affected by floods significantly reduces economic growth while the death toll from floods exerts an insignificant effect on economic growth

While the majority of the studies have focused on the aggregate economic growth effect of natural disasters, some studies have started investigating the impacts of disasters on different sectors of the economy, and these studies are scarce in the literature. For instance, Loayza et al. (2012) applied the System GMM estimator to examine the impact of natural disaster on economic growth and the sectors of the economy for 1961−2005 for a panel of 94 countries. The results indicated that disaster impacts economic output but the impact is not also negative. For instance, the overall indicator of natural disaster exerts an insignificant effect on economic growth, while flood was found to boost the overall economic growth and the agricultural, service, and the industrial sectors of the economy. In addition, drought was found to reduce economic, agricultural, and industrial growth, while its impacts on the service sector are negligible. Storm was found to significantly reduce agricultural growth, while it significantly increases industrial growth. Earthquake was found to significantly increase industrial growth, but its impact on the remaining sectors of economy and GDP per capita is insignificant. The authors concluded that developing countries are vulnerable to natural disasters relative to developed countries.

Similarly, Panwar and Sen (2018) applied System GMM approach to investigate the impact of natural disaster on economic, agricultural, and industrial growth using a panel of 102 countries. The study revealed that the composite indicator for natural disaster has an insignificant effect on economic growth. However, floods increase GDP per capita and agricultural growth, while it exerts an insignificant positive effect on industrial growth. Contrarily, droughts were found to retard GDP, agricultural, and industrial growth, but the impact is significant on agricultural growth. Storms exert an insignificant effect on GDP, agricultural, and industrial growth, while earthquake significantly increases industrial growth. Interestingly, the authors argue that the impact of natural disaster on economic output is higher for developing countries. Our study contributes to the literature by investigating the impact of natural disasters on economic growth, private sector investment, and public sector spending for a panel of 61 countries over the period between 1960 and 2018.

19.3 METHODOLOGY AND DATA

19.3.1 EMPIRICAL MODEL SPECIFICATION

In estimating the impact of natural disaster on economic growth, private sector investment, and public sector spending, we employed panel dataset that has numerous advantages over the cross-sectional and time series dataset. Thus panel dataset provides a more accurate inference of model parameters as it gives a large number of dataset, has a greater capacity for constructing more realistic behavioral hypothesis, controls for the impact of omitted variables, and simplifies computational and statistical inference (Hsiao, 1985, 2007, 2014). Therefore to investigate the impact of natural disaster on economic output, we adopted the standard reduced-form modeling approach as specified in the following equation:

$$lnY_{it} = a_0 + \beta_1 lnNat_{it} + \beta_n lnX_{it} + \mu_i + \varepsilon_{it}$$

where Y_{it} is the economic output, which is represented using economic growth, private sector investment, and public sector spending, of country i at time t; a_0 is the intercept; Nat represents natural disaster variables, which is the variable of interest; X represents a set of control variables that potentially impact on output per capita; μ and ε represent the country-specific random effect and the stochastic error term, respectively; and $\beta_1 - \beta_n$ are the coefficients to be estimated. Since we are interested in determining both short and long run of natural disaster, two main econometric estimators are used. We employed a dynamic estimator to estimate the short-run effects of natural disasters. Thus we utilized the Blundell and Bond (1998) dynamic System GMM approach, which controls for endogeneity, to estimate the short-run effect of natural disasters. We also employed the fixed effect estimator, which is statistic estimator, to investigate the long-run impact of natural disasters.

19.3.2 DATA

The study used a panel dataset for 61 countries[1] from 1960 to 2018. The total sample consists of 24 low-income countries, 39 lower middle–income countries, 37 upper middle–income countries, and 22 higher income countries. In this study, economic growth, private sector investment, and public sector spending are used as the dependent variables. For the natural disaster variables, we used a composite measure of natural disaster and its disaggregated measures such as drought, storms, flood, and earthquake. Some researchers have used the total economic loss from a natural disaster as a proxy for natural disaster (Cavallo et al., 2013; Toya & Skidmore, 2007). However, it is argued that the use of such measure may be endogenous to economic growth since the disaster loss estimated are mostly recorded based on relief and recovery expenditures incurred flowing natural disaster; therefore using the economic loss as a proxy for natural disaster could result in spurious regression (Panwar & Sen, 2018). Therefore the most appropriate measure for a natural disaster is to use the total number of people affected (see Felbermayr & Gröschl, 2014; Klomp & Valckx, 2014; Loayza et al., 2012; Panwar & Sen, 2018). Furthermore, to account for the potential variables that affect the dependent variables, we control for physical capital, trade openness, government expenditure, foreign direct investment, inflation, education, foreign aid and financial depth, external debt stock, real interest rate, and real exchange rate in our regression. The variables description and the sources of the data are presented in Table 19.1.

---

[1]The countries used for the study are presented in Appendix Table 19.1.

19.4 EMPIRICAL RESULTS

In this section, we report the short- and the long-run impact of the composite natural disaster and its specific types such as earthquake, drought, storms, and floods on economic growth, private sector investment, and public sector spending.

19.4.1 NATURAL DISASTERS AND ECONOMIC GROWTH

Table 19.2 presents the short-run estimates, while Table 19.3 shows the long-run estimates. All the variables are expressed in the natural logarithms; hence, our estimates should be interpreted as elasticity. Focusing on the variables of interest, Table 19.2 shows that the composite indicator for natural disaster and flood significantly reduces economic growth. The results also indicate that specific types of natural disasters such as drought, storm, and earthquake exert an insignificant negative effect on economic growth. The significant negative effect of the composite natural disaster indicator and flood on economic growth in the short run indicates destruction caused by natural disasters reduce the accumulation of both physical and human capitals, thereby retarding economic growth in the short run (Cuaresma, 2010; Klomp & Valckx, 2014). In addition, the destruction caused by the composite natural disaster indicator and flood leads to the loss of potential labor hours and subsequently results in fall overall economic output (Panwar & Sen, 2018).

However, in the long run (see Table 19.3), the composite natural disaster indicator and disaster types such as flood, drought, and earthquake exert an insignificant positive effect on economic growth, while storm retains its negligible negative impact on economic growth. Thus consistent with Chhibber and Laajaj (2008), natural disasters and its kinds have no meaningful impact on economic growth in the long run. Although natural disaster and its types have a negligible effect on economic growth in the long run, most of their signs are positive. This indicates that natural disaster and its kind (except storms) could probably increase economic growth in the long run providing supports to the endogenous growth theory and the Schumpeterian "creative destruction" theory. Thus the replacement of damaged physical capital with new physical capital increases the existing capital stock with more productive capital, thereby fueling economic growth and other sectors of the economy in the long run (Albala-Bertrand, 1993; Chhibber & Laajaj, 2008; Hallegatte & Dumas, 2009).

For the standard determinants of economic growth, physical capital, trade openness, and inflation, most exert an insignificant effect on economic growth in both short and long run. Although the results indicate that education exerts negligible impacts on economic growth in the short run but increases economic growth in the long run. The results further suggest that financial development significantly contributes to economic growth in both short and long run; however, the effect is more profound in the long run. Foreign aid mostly reduces economic growth in both short and long run, while government expenditure significantly reduces economic growth in the long run but not in the short run (Tables 19.4 and 19.5).

19.4.2 NATURAL DISASTERS AND PRIVATE SECTOR INVESTMENT

In the short run, flood and earthquake significantly exert a positive effect on the private sector investment, while drought significantly reduces the private sector investment. But in the long run, it is only drought that significantly reduces the private sector. In both short and long run,

Table 19.1 Variable Description.

Variables	Descriptions	Mean	Sd	Min	Max	Source
Private sector		0.847	1.528	− 13.515	3.292	WDI
Public sector		1.446	2.151	− 5.118	13.884	WDI
Economic growth	GDP per capita (constant 2010 US $)	24.454	1.957	20.364	30.011	WDI
Government expenditure	General government final consumption expenditure (% of GDP)	2.491	0.439	− 0.093	4.308	WDI
Real exchange rate	Real effective exchange rate index (2010 = 100)	4.703	0.442	2.930	8.175	WDI
Real interest rate	Real interest rate (%)	1.822	1.039	− 4.916	7.054	WDI
Terms of trade	Net barter terms of trade index (2000 = 100)	122.247	53.042	39.744	721.053	WDI
External debt stock	External debt stocks (% of GNI)	3.621	0.884	− 1.255	7.117	WDI
Physical capital	Gross capital formation (% of GDP)	3.050	0.420	0.422	4.557	WDI
Trade openness	Trade (% of GDP)	3.828	0.676	− 1.787	6.093	WDI
Foreign direct investment	Foreign direct investment, net inflows (% of GDP)	− 0.009	1.670	− 10.489	4.069	WDI
Inflation	Inflation, consumer prices (annual %)	1.924	1.258	− 3.283	10.076	WDI
Education	School enrollment, secondary (% gross)	3.675	0.849	0.011	5.064	WDI
Foreign aid	Net ODA received (% of GNI)	4.546	6.870	− 0.741	94.946	WDI
Financial depth	Domestic credit to private sector (% of GDP)	3.054	0.994	− 0.711	5.452	WDI
Natural disaster	Total number affected by natural disaster	11.099	3.105	0.000	19.673	CRED (EM-DAT)
Drought	Total number affected by drought	13.738	2.241	6.908	19.615	CRED (EM-DAT)
Storms	Total number affected by storms	10.068	3.470	1.099	18.516	CRED (EM-DAT)
Earthquake	Total number affected by the earthquake	9.270	3.128	0.000	17.666	CRED (EM-DAT)
Flood	Total number affected by floods	10.742	2.896	1.609	19.296	CRED (EM-DAT)

Table 19.2 System Generalized Method of Moment Results for Economic Growth (Short-Run Estimates).

	Model 1	Model 2	Model 3	Model 4	Model 5	Model 6
Economic growth lagged	0.887***	0.931***	0.911***	0.977***	0.874***	0.920***
	(0.037)	(0.036)	(0.129)	(0.025)	(0.099)	(0.046)
Physical capital	0.027*	0.022	0.041	0.034	0.019	0.011
	(0.014)	(0.017)	(0.085)	(0.025)	(0.049)	(0.020)
Trade openness	0.018	0.009	0.003	− 0.000	0.052	0.013
	(0.017)	(0.013)	(0.038)	(0.015)	(0.044)	(0.017)
Government expenditure	0.027	− 0.034	0.084	− 0.110	0.087	− 0.047
	(0.035)	(0.038)	(0.321)	(0.105)	(0.079)	(0.046)
Foreign direct investment	0.011***	0.008***	0.003	0.011**	0.020*	0.012***
	(0.003)	(0.003)	(0.006)	(0.004)	(0.011)	(0.005)
Inflation	0.005	− 0.001	− 0.004	− 0.007	0.014	0.001
	(0.003)	(0.004)	(0.032)	(0.004)	(0.019)	(0.004)
Education	0.030	0.013	0.119	− 0.033	0.050	0.017
	(0.028)	(0.039)	(0.169)	(0.071)	(0.115)	(0.068)
Foreign aid	− 0.010***	− 0.005	− 0.002	− 0.004	− 0.014*	− 0.006*
	(0.003)	(0.003)	(0.025)	(0.004)	(0.008)	(0.004)
Financial development	0.025*	0.028*	− 0.022	0.037	− 0.014	0.032**
	(0.015)	(0.014)	(0.046)	(0.028)	(0.023)	(0.015)
Natural disaster (composite measure)		− 0.003*				
		(0.002)				
Drought			− 0.001			
			(0.018)			
Storm			− 0.002			
			(0.002)			
Earthquake					− 0.001	
					(0.001)	
Flood						− 0.004*
						(0.002)
Constant	0.489***	0.447**	− 0.021	0.402	0.402	0.562**
	(0.178)	(0.219)	(0.790)	(0.287)	(0.277)	(0.242)
Observations	1139	881	106	312	231	643
Hansen	29.198	33.780	3.460	18.791	8.822	32.047
P(Hansen)	0.507	0.290	0.749	0.944	0.786	0.365
AR(1)	0.004	0.002	-	0.050	0.787	0.024
AR(2)	0.834	0.156	-	0.112	0.072	0.735

*Standard errors in parentheses. *P < .10, **P < .05, ***P < .01.*

Table 19.3 Fixed Effect Results for Economic Growth (Long-Run Estimates).

	Model 1	Model 2	Model 3	Model 4	Model 5	Model 6
Physical capital	− 0.043	− 0.029	− 0.121	0.039	0.503***	− 0.042
	(0.066)	(0.081)	(0.169)	(0.116)	(0.118)	(0.099)
Trade openness	0.172*	0.108	0.616	− 0.057	0.211	0.157
	(0.089)	(0.079)	(0.366)	(0.071)	(0.187)	(0.116)
Government expenditure	− 0.114	− 0.129	− 0.432	− 0.410**	− −0.084	− 0.200*
	(0.099)	(0.097)	(0.316)	(0.184)	(0.242)	(0.104)
Foreign direct investment	0.028**	0.021	− 0.035	0.029	− 0.017	0.006
	(0.012)	(0.013)	(0.028)	(0.033)	(0.036)	(0.017)
Inflation	− 0.010	− 0.013	− 0.031	− 0.027	− 0.033	− 0.002
	(0.014)	(0.014)	(0.023)	(0.027)	(0.022)	(0.013)
Education	0.242***	0.392***	0.283	0.760***	0.576***	0.503***
	(0.084)	(0.090)	(0.216)	(0.254)	(0.202)	(0.110)
Foreign aid	− 0.016***	− 0.012*	− 0.011	0.011	− 0.030	− 0.012
	(0.006)	(0.007)	(0.019)	(0.009)	(0.022)	(0.009)
Financial development	0.240***	0.230***	0.319**	0.346***	0.039	0.270***
	(0.053)	(0.042)	(0.123)	(0.081)	(0.043)	(0.039)
Natural disaster (composite measure)		0.003				
		(0.003)				
Drought			− 0.011			
			(0.025)			
Storm				0.001		
				(0.005)		
Earthquake					0.002	
					(0.004)	
Flood						0.002
						(0.003)
Constant	5.747***	5.406***	4.474***	4.520***	3.301**	4.807***
	(0.440)	(0.511)	(1.373)	(1.035)	(1.203)	(0.707)
Observations	1139	881	106	312	231	643
r2	0.527	0.566	0.478	0.637	0.631	0.633
r2_w	0.527	0.566	0.478	0.637	0.631	0.633
r2_o	0.545	0.530	0.561	0.332	0.357	0.518
r2_b	0.556	0.557	0.476	0.424	0.511	0.541
Rho	0.929	0.932	0.881	0.928	0.922	0.934
RMSE	0.209	0.199	0.230	0.197	0.210	0.190

Standard errors in parentheses. *P < .10, **P < .05, ***P < .01.

government expenditure, real exchange rate, and external debt stock significantly reduce the private sector. In addition, terms of trade significantly reduce private sector in the short run but remain insignificant in the long run. On the other hand, the real interest rate significantly increases the private sector investment in both short and long run. It also observed that economic growth mostly reduces private investment in the short run while it increases in the long run.

Table 19.4 System Generalized Method of Moment Results for the Private Sector Investment (Short-Run Estimates).

	Model 1	Model 2	Model 3	Model 4	Model 5	Model 6
Private sector (lagged)	0.324***	0.406***	0.429**	0.132	0.646***	0.220
	(0.069)	(0.112)	(0.191)	(0.163)	(0.213)	(0.299)
Economic growth	− 0.130**	− 0.128**	− 0.303**	0.338**	− 0.609	− 1.264***
	(0.059)	(0.055)	(0.125)	(0.172)	(0.425)	(0.489)
Government expenditure	− 0.310***	− 0.142	− 0.478	− 0.432	0.079	− 0.990
	(0.096)	(0.145)	(0.568)	(0.796)	(0.705)	(0.874)
Real exchange rate	− 0.463***	− 0.532***	− 1.470*	− 0.534	0.987	− 1.979
	(0.169)	(0.191)	(0.763)	(0.529)	(0.918)	(1.705)
Real interest rate	− 0.070	− 0.079	0.103*	0.411***	0.114	− 0.381
	(0.065)	(0.066)	(0.053)	(0.151)	(0.119)	(0.356)
Terms of trade	0.000	− 0.001	− 0.056**	0.017***	− 0.040	− 0.008
	(0.001)	(0.001)	(0.023)	(0.006)	(0.032)	(0.008)
External debt stock	− 0.066	− 0.116*	− 1.012***	0.370	− 1.144	− 0.802
	(0.052)	(0.060)	(0.290)	(0.369)	(0.959)	(0.521)
Natural disaster (composite measure)		− 0.002				
		(0.022)				
Drought			− 0.102*			
			(0.052)			
Storm				0.028		
				(0.032)		
Earthquake					0.112**	
					(0.045)	
Flood						0.180***
						(0.066)
Constant	7.297***	7.470***	26.483***	− 8.895	18.272	48.176**
	(2.011)	(2.137)	(6.942)	(7.434)	(15.253)	(22.689)
Observations	119	98	24	51	32	76
Sargan	121.513	94.320	13.598	17.690	41.934	16.485
P(Sargan)	0.233	0.330	0.192	0.279	0.000	0.170
AR(1)	0.051	0.137	0.306	0.095	0.489	0.503
AR(2)	0.870	0.268	0.321	0.236	0.230	0.131

Robust standard errors in parentheses. $*P < .10$, $**P < .05$, $***P < .01$.

Table 19.5 Fixed Effect Results for the Private Sector Investment (Long-Run Estimates).

	Model 1	Model 2	Model 3	Model 4	Model 5	Model 6
Economic growth	0.014	0.047	0.655***	0.407***	0.307***	0.107
	(0.223)	(0.196)	(0.036)	(0.067)	(0.077)	(0.120)
Government expenditure	− 0.706*	− 0.744*	0.852	− 4.455***	− 2.931**	− 0.735**
	(0.383)	(0.400)	(1.331)	(1.127)	(1.099)	(0.319)
Real exchange rate	− 0.796***	− 0.969***	− 5.133***	− 2.784**	− 2.353***	− 0.811
	(0.249)	(0.294)	(0.700)	(0.862)	(0.528)	(0.492)
Real interest rate	0.115	0.251	0.421***	0.521**	0.647**	0.251
	(0.237)	(0.245)	(0.062)	(0.171)	(0.215)	(0.269)
Terms of trade	0.002	0.000	− 0.017	0.008	− 0.001	− 0.001
	(0.002)	(0.002)	(0.012)	(0.006)	(0.007)	(0.001)
External debt stock	− 0.154*	− 0.305***	− 1.274***	− 0.248	− 0.185	− 0.377***
	(0.073)	(0.097)	(0.148)	(0.307)	(0.288)	(0.112)
Natural disaster (composite measure)		− 0.002				
		(0.030)				
Drought			− 0.390***			
			(0.045)			
Storms				0.030		
				(0.056)		
Earthquake					− 0.060	
					(0.035)	
Flood						0.019
						(0.032)
Constant	5.935	6.404	15.019***	12.374*	10.911**	4.090
	(5.885)	(5.182)	(0.091)	(5.787)	(3.636)	(3.806)
Observations	137	115	24	58	38	88
r2	0.111	0.177	0.812	0.611	0.655	0.165
r2_w	0.111	0.177	0.812	0.611	0.655	0.165
r2_o	0.129	0.067	0.019	0.180	0.052	0.039
r2_b	0.034	0.019	0.134	0.194	0.002	0.034
Rho	0.412	0.583	0.980	0.849	0.838	0.648
RMSE	0.867	0.785	0.300	0.550	0.570	0.792

Robust standard errors in parentheses.$^*P<.10$, $^{**}P<.05$, $^{***}P<.01$.

19.4.3 NATURAL DISASTERS AND PUBLIC SECTOR SPENDING

In both short and long run (see Tables 19.6 and 19.7), natural disaster in general and its specific types (flood, drought, storms, and earthquake) exert an insignificant effect on the public sector spending. Thus natural disaster and its kind will not have any impact on public spending in both the short and

Table 19.6 System Generalized Method of Moment Results for the Public Sector Spending (Short-Run Estimates).

	Model 1	Model 2	Model 3	Model 4	Model 5	Model 6
Public sector (lagged)	0.971**	0.734***	1.023***	0.910***	0.631***	0.911***
	(0.405)	(0.121)	(0.107)	(0.123)	(0.138)	(0.198)
Economic growth	− 4.770	− 1.260	− 0.827	0.891	− 0.313	− 0.463
	(6.771)	(1.477)	(1.178)	(2.161)	(1.875)	(0.467)
Real exchange rate	− 7.344	− 3.192	− 3.021	2.041	− 1.251	− 1.079
	(10.244)	(2.962)	(3.060)	(4.954)	(5.176)	(1.284)
Real interest rate	− 0.804	− 0.428	− 0.309	0.199	0.159	− 0.044
	(1.875)	(0.520)	(0.617)	(0.421)	(0.616)	(0.280)
Terms of trade	0.040	0.005	− 0.024	0.022	− 0.003	− 0.003
	(0.062)	(0.011)	(0.018)	(0.024)	(0.041)	(0.004)
External debt stock	− 5.274	− 0.909	− 1.161	1.640	0.677	− 0.241
	(7.789)	(1.607)	(2.561)	(2.996)	(2.850)	(0.646)
Natural disaster (composite measure)		0.159				
		(0.187)				
Drought			0.055			
			(0.470)			
Storm				0.012		
				(0.187)		
Earthquake					− 0.054	
					(0.135)	
Flood						0.130
						(0.097)
Constant	178.766	48.289	41.920	− 41.455	10.222	16.198
	(252.753)	(54.726)	(44.627)	(89.633)	(84.966)	(16.841)
Observations	378	325	53	150	82	260
Sargan	12.289	26.247	10.919	11.955	26.466	11.375
P(Sargan)	0.266	0.158	0.364	0.288	0.151	0.181
AR(1)	0.057	0.036	0.595	0.144	0.333	0.042
AR(2)	0.723	0.272	0.963	0.121	0.335	0.387

*Standard errors in parentheses. *P<.10, **P<.05, ***P<.01.*

long run. These results contradict the claim that damages caused by natural disasters lead to fiscal deficit (Benson & Clay, 1998; Chhibber & Laajaj, 2008). We argue that when the occurrences of disasters are more localized without causing damages to properties and humans, it will not impact on government expenditure. The controlled variables such as economic growth, real exchange rate, terms of trade, and external debt stock exert an insignificant effect on the public sector spending. However, in the long run, economic growth significantly increases public sector spending.

Table 19.7 Fixed Effect Results for the Public Sector Spending (Long-Run Estimates).

	Model 1	Model 2	Model 3	Model 4	Model 5	Model 6
Economic growth	4.334**	4.493**	5.434***	3.619***	3.905***	4.044**
	(1.697)	(1.587)	(0.614)	(0.393)	(0.702)	(1.448)
Real exchange rate	− 0.207	0.041	− 7.000***	4.164	1.749	− 1.448
	(2.661)	(3.184)	(1.605)	(5.431)	(2.620)	(3.251)
Real interest rate	− 0.234	− 0.845	1.180*	− 1.693	0.777*	0.076
	(0.945)	(0.933)	(0.546)	(1.313)	(0.390)	(0.461)
Terms of trade	− 0.018	− 0.017	− 0.008	− 0.030*	0.011	− 0.005
	(0.011)	(0.012)	(0.018)	(0.016)	(0.038)	(0.011)
External debt stock	0.323	0.731	− 0.206	0.092	− 0.932	0.792
	(1.154)	(1.359)	(0.598)	(1.105)	(1.225)	(1.462)
Natural disaster (composite measure)		− 0.101				
		(0.199)				
Drought			− 0.361			
			(0.346)			
Storm				− 0.076		
				(0.205)		
Earthquake					0.027	
					(0.151)	
Flood						0.052
						(0.203)
Constant	− 111.093**	− 115.989**	− 108.012***	− 109.443***	− 117.603***	− 104.148**
	(44.722)	(43.319)	(14.844)	(34.998)	(28.869)	(40.180)
Observations	382	329	53	153	82	263
r2	0.112	0.133	0.498	0.237	0.285	0.136
r2_w	0.112	0.133	0.498	0.237	0.285	0.136
r2_o	0.022	0.029	0.013	0.097	0.082	0.025
r2_b	0.199	0.204	0.115	0.169	0.057	0.250
Rho	0.884	0.895	0.967	0.911	0.888	0.896
RMSE	5.501	5.171	3.335	4.406	3.985	4.989

Standard errors in parentheses. $^*P<.10$, $^{**}P<.05$, $^{***}P<.01$.

19.5 CONCLUDING REMARKS

The global economy continues to witness an increase in the occurrence of natural disasters of various kinds and magnitudes. Such frequent occurrences have attracted the attention of researchers to empirically understand the impact of these natural disasters in shaping the economic development trajectories of affected countries. However, the empirical literature on the impact of natural disasters on economic growth remains inconclusive. Therefore this study investigates the short- and long-run impact of composite natural disaster, floods, earthquake, storms, and drought on GDP per capita, private sector investment, and public sector spending using a comprehensive panel dataset for 61 countries for 1960–2018.

The empirical results indicate that in the short run the composite indicator for natural disaster and flood significantly reduces economic growth, while other specific types of natural disasters such as drought, storm, and earthquake exert an insignificant negative effect on economic growth. The results also indicate that in the long-run, natural disasters have negligible impacts on economic growth. Our findings further suggest that natural disaster in general and its specific types (flood, drought, storms, and earthquake) exert an insignificant effect on the public sector spending. We argue that when the occurrences of disasters are more localized without causing damages to properties and humans, it will not impact on government expenditure.

The results also suggest that in the short run, only flood and earthquake significantly increase private sector investment, but drought retards private sector investment in both short and long run. The significant positive effect of flood and earthquake on the private sector investment lends some support to the Schumpeterian "creative destruction" theory. Thus as damaged physical capital needs to be replaced, it requires the private sector to replace the damaged physical capital with new physical capital, thereby increasing the private sector investment.

From these findings, we conclude that the impact of natural disaster is not homogenous across different sectors of the economy. We also argue that the effect of natural disaster on the economy is dependent on the specific type of natural disaster. Also, the impact of natural disasters is time specific.

ACKNOWLEDGMENT

The authors are grateful to the Centre for Research on the Epidemiology of Disasters (CRED) for allowing us to use the EM-DAT to carry out this empirical exercise.

CONFLICT OF INTEREST

The authors declare no conflict of interest.

REFERENCES

Albala-Bertrand, J.-M. (1993). Political economy of large natural disasters: With special reference to developing countries. *OUP Catalogue*, Oxford University Press, number 9780198287650.

Benson, C., & Clay, E. (1998). *The impact of drought on sub-Saharan African economies: A preliminary examination*. The World Bank.

Blundell, R., & Bond, S. (1998). Initial conditions and moment restrictions in dynamic panel data models. *Journal of Econometrics*, 87(1), 115−143. Available from https://doi.org/10.1016/S0304-4076(98)00009-8.

Cavallo, E., Galiani, S., Noy, I., & Pantano, J. (2013). Catastrophic natural disasters and economic growth. *Review of Economics and Statistics*, 95(5), 1549−1561.

Chhibber, A., & Laajaj, R. (2008). Disasters, climate change and economic development in sub-Saharan Africa: Lessons and directions. *Journal of African Economies*, 17(Suppl. 2), ii7−ii49. Available from https://doi.org/10.1093/jae/ejn020.

Cuaresma, J. C. (2010). Natural disasters and human capital accumulation. *The World Bank Economic Review*, 24(2), 280−302.

CRED. (2019). *Natural disasters*. Brussels, Belgium: Centre for Research on the Epidemiology of Disasters (CRED) Institute of Health and Society (IRSS) Université catholique de Louvain.

Felbermayr, G., & Gröschl, J. (2014). Naturally negative: The growth effects of natural disasters. *Journal of Development Economics*, 111, 92−106. Available from https://doi.org/10.1016/j.jdeveco.2014.07.004.

Fomby, T., Ikeda, Y., & Loayza, N. V. (2013). The growth aftermath of natural disasters. *Journal of Applied Econometrics*, 3(28), 412−434.

Hallegatte, S., & Dumas, P. (2009). Can natural disasters have positive consequences? Investigating the role of embodied technical change. *Ecological Economics*, 68(3), 777−786. Available from https://doi.org/10.1016/j.ecolecon.2008.06.011.

Hsiao, C. (1985). Benefits and limitations of panel data. *Econometric Reviews*, 4(1), 121−174. Available from https://doi.org/10.1080/07474938508800078.

Hsiao, C. (2007). Panel data analysis—Advantages and challenges. *Test*, 16(1), 1−22.

Hsiao, C. (2014). *Analysis of panel data*. Cambridge University Press.

Klomp, J., & Valckx, K. (2014). Natural disasters and economic growth: A meta-analysis. *Global Environmental Change*, 26, 183−195. Available from https://doi.org/10.1016/j.gloenvcha.2014.02.006.

Loayza, N. V., Olaberría, E., Rigolini, J., & Christiaensen, L. (2012). Natural disasters and growth: going beyond the averages. *World Development*, 40(7), 1317−1336. Available from https://doi.org/10.1016/j.worlddev.2012.03.002.

Noy, I. (2009). The macroeconomic consequences of disasters. *Journal of Development Economics*, 88(2), 221−231. Available from https://doi.org/10.1016/j.jdeveco.2008.02.005.

Noy, I., & Vu, T. B. (2010). The economics of natural disasters in a developing country: The case of Vietnam. *Journal of Asian Economics*, 21(4), 345−354. Available from https://doi.org/10.1016/j.asieco.2010.03.002.

Panwar, V., & Sen, S. (2018). Economic impact of natural disasters: An empirical re-examination. *Margin: The Journal of Applied Economic Research*, 13(1), 109−139. Available from https://doi.org/10.1177/0973801018800087.

Romer, P. M. (1986). Increasing return and long run growth. *The Journal of Political Economy*, 94(5), 1002−1037.

Shabnam, N. (2014). Natural disasters and economic growth: A review. *International Journal of Disaster Risk Science*, 5(2), 157−163. Available from https://doi.org/10.1007/s13753-014-0022-5.

Skidmore, M., & Toya, H. (2002). Do natural disasters promote long-run growth? *Economic Inquiry*, 40(4), 664−687.

Solow, R. M. (1956). A contribution to the theory of economic growth. *The Quarterly Journal of Economics*, *70*(1), 65–94.

Strobl, E. (2012). The economic growth impact of natural disasters in developing countries: Evidence from hurricane strikes in the Central American and Caribbean regions. *Journal of Development Economics*, *97* (1), 130–141. Available from https://doi.org/10.1016/j.jdeveco.2010.12.002.

Toya, H., & Skidmore, M. (2007). Economic development and the impacts of natural disasters. *Economics Letters*, *94*(1), 20–25. Available from https://doi.org/10.1016/j.econlet.2006.06.020.

APPENDIX

COUNTRIES INCLUDED

Algeria, Argentina, Australia, Bangladesh, Brazil, Canada, Chad, Chile, China, Colombia, Congo, Dem. Rep., Cuba, Ecuador, El Salvador, Ethiopia, Fiji, France, Ghana, Guatemala, Haiti, Honduras, Hong Kong SAR, China, India, Indonesia, Iran, Islamic Rep., Jamaica, Japan, Kenya, Korea, Rep., Lao PDR, Madagascar, Malawi, Malaysia, Mexico, Morocco, Mozambique, Myanmar, Nepal, New Zealand, Nicaragua, Niger, Nigeria, Pakistan, Panama, Papua New Guinea, Paraguay, Peru, Philippines, Rwanda, Senegal, South Africa, Sri Lanka, Sudan, Tanzania, Thailand, Uganda, Venezuela, RB, Vietnam, Zambia, Zimbabwe.

NATURAL DISASTERS: MACROECONOMIC IMPLICATIONS AND MEASUREMENT ISSUES

20

Janet Dzator[1,2] and Michael Dzator[2,3]

[1]*School of Business, The University of Newcastle, Newcastle, NSW, Australia* [2]*Australia Africa Universities Network (AAUN) Partner, Newcastle, NSW, Australia* [3]*School of Access Education, Central Queensland University, Mackay, QLD, Australia*

20.1 INTRODUCTION

The Centre for Research on the Epidemiology and Disasters (CRED) (2019) defines natural disasters "as a situation or an event that overwhelms local capacity, necessitating a request at the national or international level for external assistance; an unforeseen and often sudden events that cause great damages, destruction and human suffering." Natural disasters are associated with human and economic loss. For instance, CRED (2019) report suggests that a total of 315 natural disaster events were recorded with economic damage totally US$ 131.7 billion, while over 68 million people were affected with a total of 11,804 people losing their lives in 2018.

From an economic growth theory perspective the impact of natural disasters on economic growth is prior uncertain. For instance, Horwich (2000) argues that natural disasters may not have an effect on the overall economy since the occurrence of natural disasters is often localized. The traditional neoclassical growth theory indicates that natural disasters could be devastating to economic performance since it reduces the accumulation of physical and human capital (Cuaresma, 2010; Felbermayr & Gröschl, 2014; Strobl, 2012). Contrarily, the endogenous growth theory based on the Schumpeterian creative destruction model argues that natural disasters could increase economic growth because of the replacement of damaged physical capital with efficient and new capital stock (Cavallo, Galiani, Noy, & Pantano, 2013; Loayza, Olaberría, Rigolini, & Christiaensen, 2012; Panwar & Sen, 2018).

The macroeconomic effect of natural disasters remains a subject of empirical question (Cavallo et al., 2013). After the seminal paper of Albala-Bertrand (1993a, 1993b), several studies that have been examined have attempted to provide an empirical understanding of the impact of natural disasters on economic performance; however, the findings from the existing empirical literature remain contradictory. Studies presenting a critical review of the literature on the macroeconomic effects of natural disasters are scarce. Our contribution to the broad literature is to provide a critical review of research on the macroeconomic effects of natural disasters. The outcome of this review will provide an understanding of the reasons for the conflicting results and further presents natural

Economic Effects of Natural Disasters. DOI: https://doi.org/10.1016/B978-0-12-817465-4.00020-0

disaster measurements and modeling issues. In addition, the results from this review will present the gaps in the literature, which will inform future research.

Our review results revealed that developing countries are more vulnerable to natural disasters because of poor institutions and governance, poor financial system, weak economic growth, high income inequality, and low level of human capital accumulation. The findings also indicate that the impact of different kinds of natural disaster on the economy is not homogenous. It was further found that hydrometeorological/climatic disasters have a profound effect on the macroeconomy. The Emergency Disaster Database (EM-DAT) remains the mostly use database in the empirical literature; however, like any other database, it suffers from measurement errors. It was also observed from the research that the measurement of natural disasters outcomes tends to focus mainly on the direct effects but neglects indirect/postdisaster complications such as depression, attempted suicide (due to loss of properties), and poorer well-being, and have lower chance of survival of the affected people. From an analytical perspective, panel data studies dominate research on the macroeconomic effects of natural disasters while time series techniques are the least use approach in the empirical literature. The review further indicates that the empirical results on the macroeconomic impacts of natural disasters are inconclusive, and this is caused by the source of natural disasters data, measures for natural disasters, study period, and differences in estimation techniques. The remaining sections of the chapter are organized as follows. In Section 20.2, we provide a theoretical review of the arguments on the macroeconomic effects of natural disasters. Section 20.3 presents empirical studies on the macroeconomic effect of natural disasters, while Section 20.4 stresses on the disaster measurement and modeling issues. Conclusions, policy implications, and direction for future research are presented in Section 20.5.

20.2 ECONOMICS OF NATURAL DISASTERS: THEORETICAL PERSPECTIVE

Theoretically, natural disasters have an ambiguous effect on the economy. One school of thought argues that frequent occurrence of natural disaster events would hinder economic growth. From the classical growth model, physical capital and human capital are the critical factors of production to sustain economic growth. Natural disasters are very disruptive to both physical and human capital as it drops savings rate due to private spending on medical care and emergency expenditure rises (Klomp & Valckx, 2014). In addition, large-scale natural disaster events could increase fatalities and migrations among the labor force, thereby reducing the accumulation of human (Cuaresma, 2010; Felbermayr & Gröschl, 2014; Klomp & Valckx, 2014; Klomp, 2016). Benson and Clay (2004) further argued that natural disasters could limit economic growth as it reallocates expenditure from planned investment, increases inflation through higher fiscal deficit and discourages investors by increasing uncertainty.

On the other hand, the endogenous growth theory based on the Schumpeterian creative destruction model argues that destruction of capital stock per worker temporarily increases economic growth after a natural disaster by increasing its marginal return (Loayza et al., 2012; Panwar & Sen, 2018). Thus the replacement of damaged physical capital with new physical capital increases the efficiency and the effectiveness of the existing capital stock, thereby boosting economic growth (Albala-Bertrand, 1993a, 1993b; Chhibber & Laajaj, 2008; Hallegatte & Dumas, 2009). Skidmore and Toya (2002) further argue that natural disaster could cause a substitution effect between physical capital and human capital such that a fall in physical capital is substituted by investment in human capital, thereby increasing economic growth.

With the ambiguous effect of natural disaster, Chhibber and Laajaj (2008, pp. ii14–ii15) summarize four possible scenarios through which natural disaster could impact on economic output as demonstrated. In the first two scenarios, Chhibber and Laajaj (2008, pp. ii14–ii15) argue that natural disaster does not affect the long-run path of economic growth. However, the natural disaster shocks retard economic growth that is eventually followed by expansion during reconstruction and production returns level returns to its long-run state of equilibrium. In the third scenario, natural disaster permanently reduces capital stock, and the new long-run equilibrium is established at a lower level of economic growth. The establishment of new capital results in technological change that boosts the long-run growth rate of the economy. Different types of natural disaster are associated with different scenarios. For instance, earthquake drives technological change as its devastating effect is more likely to follow by the considerable building of new capital, thereby increasing economic growth. Contrarily, floods do not affect the long-run path of economic growth; however, floods could reduce growth that is eventually followed by expansion during reconstruction and production returns level returns to its long-run state of equilibrium. Also, floods could permanently reduce capital stock, and the new long-run equilibrium is established at a lower level of economic growth. Strulik and Trimborn (2019) added that economic growth is driven above its predisaster level when natural disaster exerts mainly durable consumption goods, but when productive capitals are devastated by disaster natural disaster, it retards economic growth. The authors further argue that when natural disasters exert a neutral effect on economic growth when it destroys durable goods, residential houses, and productive capital.

20.3 THE MACROECONOMIC IMPACTS OF NATURAL DISASTERS: EMPIRICAL EVIDENCE

In this section, we present the empirical works on the macroeconomic effects of natural disasters. Various approaches have been used to evaluate the role of natural disasters on the macroeconomy. As presented in Fig. 20.1, the empirical studies have used various approaches to investigate the impact of natural disasters on macroeconomic variables such as gross domestic product (GDP), agricultural growth, industrial growth, service sector growth, inflation, private consumption, investment, trade, unemployment, and government spending and revenue. In this study, we categorized the empirical under cross-sectional studies, panel data studies, time series studies, and computable general equilibrium (CGE) studies.

20.3.1 CROSS-SECTIONAL STUDIES

The classical work of Albala-Bertrand (1993a, 1993b) examined the impact of natural disasters on economic growth in 26 developing countries between 1960 and 1979, and the results indicated that economic growth increases in most cases after the natural disaster. The authors argued the positive effect is due to the replacement of damaged physical capita with efficient physical assets. Skidmore

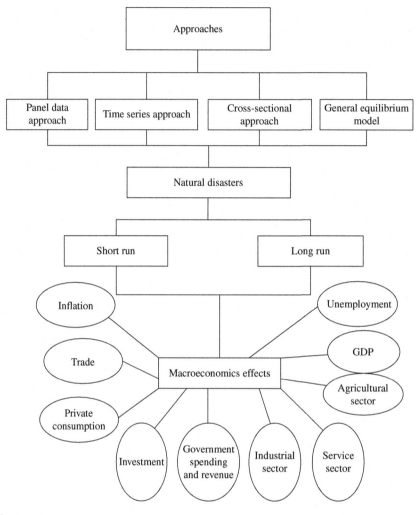

FIGURE 20.1

Approaches for examining the macroeconomic effect of natural disasters.

Authors' construct.

and Toya (2002) further provided an empirical assessment on the long-run impact of climatic and geological disasters on macroeconomic variables such as GDP per capita growth, physical capital growth; human capital and total factor productivity for a cross section of 89 countries. Using natural disaster data from EM-DAT and Davis, the empirical results revealed that total damages caused climatic disasters significantly boost GDP per capita growth while damages caused by geologic disaster retard GDP per capita growth. Although geologic disaster retards GDP per capita growth, it was a geological disaster from EM-DAT that was significant effects the GDP per capita growth.

Further analysis revealed that climatic disasters increase human capital and total factor productivity, while geologic disasters exert an insignificant negative impact on human capital and total factor productivity. Physical capital growth was not significantly affected by both climatic and geological disasters. From the results, Skidmore and Toya (2002) argue that natural disaster causes a substitution effect between physical capital and human capital such that a fall in physical capital is substituted by investment in human capital. The authors further argue that because disasters enhance the adoption of new technologies and replacement of damaged physical capitals, it results in the overall increase in total factor productivity. Cuaresma (2010) further used cross-sectional approach to examine the role of natural disasters on human capital accumulation (secondary school environment) for 80 countries. The results from the Bayesian model averaging reveal that natural disasters reduce human capital accumulation, but the impact of natural geological disasters is severe. The authors argue that their findings are homogenous and are independent of the level of countries economic development and amount of human capital accumulation.

20.3.2 PANEL DATA STUDIES

Using the Hausman–Taylor random effect estimator, Noy (2009) examined the impact of the magnitude of the disaster on GDP per capita in 109 countries for the period of 1970–2003. The results from the benchmark regression revealed that total damages caused by natural disaster significantly reduce GDP per capita, whereas the number of people killed and the number of people affected have a negligible positive effect on GDP per capita. Constructing their measure of disaster magnitude, Noy (2009) revealed that natural disaster retards economic growth, but with similar magnitude, developing countries experienced significant macroeconomic shocks relative to the developed countries. In another study, Noy and Nualsri (2011) employed the panel vector autoregression (PVAR) to examine the response of government spending and revenue to natural disasters in 42 (22 developed and 20 developing countries) countries for the period of 1990Q1–2005Q4. The results indicated that for developed countries a shock to natural disaster initially increases government spending and gradually declines while government revenue drops cumulatively. It was also revealed that a shock to natural disaster increases government payment and outstanding debt while government cash surplus diminishes. On the other hand, for the developing countries, the impulse response results reveal that government spending, government revenue, government payment, and outstanding debt react negatively to natural disasters while government increases cash surplus. Noy and Nualsri (2011) argue that the observed results in the developing countries could be due to the implementation of procyclical fiscal policy in developing countries after the occurrence of natural disasters.

Bergholt and Lujala (2012) further used the fixed effect approach to probe the impact of the number of people affected by climatic disasters (storms and flood) on GDP per capita in 171 countries. The findings indicated that climatic disasters (floods and storms) significantly reduce GDP per capita. Using hurricane data from the North Atlantic Hurricane database and Eastern North Pacific Tracks File, Strobl (2012) investigated the effect of hurricane on economic growth in 30 Central America and Caribbean countries using the bias-corrected least squares dummy variable (LSDV) estimator and found that hurricane retards GDP per capita; however, this depends on timing and control of local economy characteristics. In addition, for robustness check, Strobl (2012) found that hurricane data from the EM-DAT database exert an insignificant negative effect on GDP per capita.

Generally, the majority of the literature has mainly examined the aggregate economic growth effect of natural disaster with little attention given to the sectorial effect of disasters. However, natural disasters could have different implications for different sectors of economy disasters (Loayza et al., 2012). Because of this lacuna, some studies have attempted to extend the empirical analysis to include other segments of the economy. Loayza et al. (2012), for instance, applied the system-generalized method of moment (system-GMM) estimator to examine the short-term impact of natural disaster on economic growth and the sectors of the economy for 1961−2005 for a panel of 94 countries. The results indicated that disaster impact on economic output but the effect is not also negative. For instance, the overall indicator of natural disaster exerts an insignificant impact on economic growth, while floods were found to boost the overall economic growth and the agricultural, service, and the industrial sectors of the economy. In addition, drought was found to reduce economic, agricultural, and industrial growth while its impacts on the service sector are negligible. Storms were found to significantly reduce agricultural growth while it substantially increases industrial growth. Earthquake was detected to boost industrial growth significantly, but its impact on the remaining sectors and the overall economic growth is insignificant. The authors concluded that developing countries are vulnerable to natural disasters relative to developed countries.

Following Loayza et al. (2012), Panwar and Sen (2018) further probed the short-run impact of natural disaster on economic, agricultural, and industrial growth using a panel of 102 countries. Using the system-GMM approach, the findings from their study revealed that the composite indicator for natural disaster has an insignificant effect on economic growth. However, floods increase GDP per capita and agricultural growth while it exerts a negligible positive impact on industrial growth. Contrarily, droughts were found to retards GDP growth, agricultural and industrial growth, but the effect is significant on agricultural growth. Storms exert an insignificant impact on GDP growth, agricultural and industrial growth, while earthquake significantly increases industrial growth. Like Loayza et al. (2012), the authors argue that developing countries bear the highest cost associated with the natural disaster on economic output.

In another study, Felbermayr and Gröschl (2014) investigated the impact of natural disaster on GDP per capita for the period of 1979−2010 in more than 100 countries. Using disaster data from GeoMet, EM-DAT, and NatCatSERVICE database, the study revealed that GeoMet disaster indicator reduces GDP per capita, while EM-DAT and NatCatSERVICE disaster data exert an insignificant effect on GDP per capita. The study concluded that the results vary across income groups with the developing countries bearing the highest cost of the natural disaster. Cavallo et al. (2013) utilized the quasiexperimental approach to examine the effect of natural disaster on GDP per capita in 196 countries for the period of 1970−2008. The study demonstrated that extreme disaster has a neutral impact on GDP per capita.

Fomby, Ikeda, and Loayza (2013) used the PVAR to examine the impact of various disasters (floods, earthquake, storms, and drought) on GDP per capita, agricultural and nonagricultural growth in 84 countries over the period of 1960−2007. The study revealed that droughts have an overall negative effect on GDP growth, agricultural and nonagricultural growth, while floods increase GDP growth, agricultural and nonagricultural growth. Also, earthquake and storms were found to have a weaker effect on GDP growth as well as agricultural and nonagricultural growth. The study concluded that the macroeconomic impact of natural disasters is stronger in developing countries relative to developed countries. Similarly, Cunado and Ferreira (2014) employed the PVAR to examine the macroeconomic impact of floods in 135 countries for the period of 1985−2008. The study revealed that a shock to floods increases GDP per capita growth, and this is limited to developing countries and moderate floods. In addition, the study indicated that the positive effect of floods is more significant in the agricultural sector of developing countries, but the impact of floods on the nonagriculture sector is indirect. Also, using fixed effect estimator, Shabnam (2014) examined the effect of floods on GDP per capita for 187 countries from 1960 to 2010 and found that the total number of people affected by floods significantly reduces economic growth, while the death toll from floods exerts an insignificant effect on economic growth.

Utilizing the PVAR model, Mohan, Ouattara, and Strobl (2018) investigated the response of export, import, government expenditure, investment, private consumption, and GDP to a shock to the hurricane for 21 Caribbean countries for the period of 1970−2011. The study revealed that a shock to hurricane generally increases government expenditure, investment, and import, while it reduces export and private consumption. Using the PVAR and Granger causality approach, Benali, Mbarek, and Feki (2019) investigated the effect of the link between natural disasters, government budget, economic growth, inflation, and government debt in six high- and middle-income countries for the period of 1990−2013. The results revealed that for middle-income countries, natural disaster unidirectionally causes government expenditure and government debt while for the high-income countries, natural disaster unidirectionally to economic growth, government debt, and government expenditure. In another study, Sawada, Bhattacharyay, and Kotera (2019) used the instrumental variable regression to probe the effect of natural disasters (hydrometeorological, geophysical, climatological, technological, and biological floods) on GDP per capita in 189 countries for the period of 1968−2001 and found that in the short run, natural disasters exert more significant adverse effect on GDP per capita while natural disasters increase GDP per capita in the long run.

Results from some of the panel studies revealed that developing countries are vulnerable to natural disasters relative to developed countries (Felbermayr & Gröschl, 2014; Loayza et al., 2012; Noy, 2009; Panwar & Sen, 2018). Attempts have been to provide reasons why economic loss are more pronounced in developing countries compared to the developed economies. Studies by Toya and Skidmore (2007), Kahn (2005), Padli, Habibullah, and Baharom (2010), and Strömberg (2007) suggested that institutions, governance, income inequality, financial system, education attainment, trade openness, income/economic development, population density, and public awareness are among the crucial factors that explain the disparity in economic loss caused by natural disasters between developing and developed economies.

20.3.3 TIME SERIES STUDIES

Using the structural vector autoregression (SVAR) model, Chaiechi (2014) investigated the economic impact of extreme weather (floods and cyclones) events in Queensland (Australia) for the period of 2002−11 and found that a shock to the climatic events initially drops investment and savings but restores to its preshock level after the 1 year. In addition, productivity was found to initially respond positively to a shock to the climatic events, while net exports increase immediately after the disaster before declining and improving again. Also, income distribution and unemployment were found to fluctuate after the disaster. In another study, Doyle and Noy (2015) utilized the VAR model to investigate the impact of the Canterbury earthquake (September 4, 2010 and

February 22, 2011) on New Zealand's economy. The findings from the study indicated that the Canterbury earthquake has a less pronounced effect on New Zealand's economy. Thus the earthquake moderately increased inflation, and the negative impact of GDP per capita was short lived. Shaari, Abd Karim, and Hasan-Basri (2017) also used autoregressive distributed lag (ARDL) to investigate the effects of floods on manufacturing and agricultural sector of Malaysia's economy for the period of 1960–2013 and found that in the long run, flood damage reduces agricultural growth while increasing manufacturing growth. The study indicated that flood size, duration, and frequency have no effect on agricultural and output growth in the long run. In the short run, it was found that flood size and damage increase manufacturing and agricultural growth, whereas flood duration exerts no effect on manufacturing and agricultural growth. In addition, flood frequency was found to significantly increase only manufacturing growth in the short run.

20.3.4 COMPUTABLE GENERAL EQUILIBRIUM STUDIES

Deviating from the econometric analysis, another strand of the empirical studies has utilized the computable general equilibrium (CGE) model to examine the macroeconomic effect of natural disasters. The use of CGE for impact studies stems from the fact that it incorporates all the sectors of the economy into one model that makes it superior over input–output models (Asafu-Adjaye, 2014; Narayan, 2003). Generally, the results from the CGE models reveal that natural disasters are inimical to economic development.

For instance, Selcuk and Yeldan (2001) utilized the CGE model to examine the effect of the August 1999 earthquake in Turkey on key macroeconomic variables such as GDP, consumption, investment, external deficit, and foreign borrowing. The authors conducted their simulation exercise under four scenarios: The first scenario is when there is no policy change, the second scenario is when discretionary adjustments on indirect taxes rates, the third scenario is when there are flexible indirect tax adjustments to recover the capital loss, and the last scenario is when there is foreign aid to recover capital loss. The simulation analysis reveals that when there is no policy change, earthquake was found to reduce GDP, consumption, and investment and increases the external deficit. On the other hand, when there are discretionary adjustments on indirect taxes rates by the government on all sectors to revamp loss capitals, indirect taxes increase the negative impact of earthquakes on GDP, total investment, and consumption, but external deficit declines. Also, when there are flexible indirect tax adjustments to recover the loss capital, GDP does not change significantly, but consumption falls. However, investment, external deficit, and foreign borrowing increase. Finally, when loss capital is financed by foreign aid, consumption initially falls but converges later above the initial path. The authors concluded that using foreign assistance to finance capital loss remains the best outcome to improve consumer welfare after a disaster.

Narayan (2003) further employed the CGE model to investigate the impact of cyclone "Ami" on Fiji's economy. The outcome from the study revealed that cyclones significantly cause a decline in private income, consumption, savings, private investment, government revenue, and GDP per capita, worsening balance of payment and the overall welfare in Fiji's economy. Keen and Pakko (2011) further used the dynamic stochastic general equilibrium model to examine the impact of 2005 Hurricane Katrina on US monetary policy and found that nominal interest rate increases after the hurricane reduces productivity and destroys capital stock.

20.4 **DISASTER MEASUREMENT AND MODELING ISSUES**

Measurements on natural disasters exclude indirect or postdisaster outcomes. The EM-DAT, which is maintained by CRED, remains the most popular source of disaster data used for empirical analysis (see Appendix Table 20.A1). The EM-DAT contains data on hydrometeorological disasters, geophysical disasters, and biological disasters.[1] However, before an event is recorded as a disaster by CRED, it must at least satisfy one the following criteria: (1) 10 or more people reported killed, (2) 100 people reported affected, (3) declaration of state emergency, and (4) call for international assistance. Generally, measurement of natural disasters is reported using either the number of affected people, the number of people killed, or the economic damages caused during the disasters without considering the associated indirect/postdisaster implications. Thus we argue that current measurements on natural disasters are incomplete and problematic since it neglects the postdisaster outcome. Indirectly, people suffered from postdisaster complications such as depression, attempted suicide (due to loss of properties), and poorer well-being and have lower chance of survival. For instance, the CRED criteria for recording the impact of natural disasters overlook these postdisaster complications, hence underreporting the actual or total effects of natural disasters.

Furthermore, EM-DAT data could suffer from measurement errors. For instance, Strobl (2012) argued that information used by CRED to collate the list of natural disaster events comes from multiple sources, and this could cause inconsistency across natural event source, countries, and time. Concerning this, there appear that natural disaster events could be reported over time and the probability that events recorded in earlier periods is more likely to have exceeded the minimum specified criteria in the data (Noy, 2009; Strobl, 2012; Toya & Skidmore, 2007). Strobl (2012) further added that damages reported in the EM-DAT data are unsatisfactory. The author demonstrated that there are several well-known natural disaster events such as hurricane strikes in the Caribbean region, which are missing from the EM-DAT database. Strobl (2012) further claimed that 24 out of 31 Caribbean countries do not have information on damages caused by the hurricane data from the EM-DAT database. In addition, it is also argued that natural disaster damages are overestimated in developing countries to attract international financial aids (Toya & Skidmore, 2007).

The question is how do these measurement issues affect disaster modeling? It must be noted that the measurement errors of the popular EM-DAT and any other disaster database could create attenuation bias when appropriate econometric techniques are not used. Attenuation bias could bias the estimates on the macroeconomic effects of natural disasters. The instrumental variable regressions offer the best approach for solving attenuation bias (Acemoglu, Johnson, & Robinson, 2001). It could be observed from Appendix Table 20.A1 that some of the empirical studies have used an instrumental variable approach to correct for attenuation bias.

---

[1]Hydrometeorological disasters includes floods, droughts, storms, landslides, wave surges; the geophysical disasters include volcanic eruption, earthquake, and tsunami while the biological disasters include the epidemics and insect infestations.

20.5 CONCLUDING REMARKS, POLICY IMPLICATIONS, AND DIRECTION FOR FUTURE RESEARCH

The current study provides both theoretical and empirical review of the macroeconomic effects of natural disasters. The conclusions that emanated from this review are presented as follows: The review indicates that developing countries are more vulnerable to natural disasters because of poor institutions and governance, poor financial system, weak economic growth, high income inequality, and low level of human capital accumulation. The literature indicates that the impact of different kinds of natural disaster on the economy is not homogenous. It was further found that hydrometeorological/climatic disasters have a profound effect on the macroeconomy. The EM-DAT remains the mostly use database in the empirical literature; however, like any other database, it suffers from measurement errors. It was also observed from the research that the measurement of natural disasters outcomes tends to focus mainly on the direct effects but neglects indirect/postdisaster complications such as depression, attempted suicide (due to loss of properties), poorer well-being, and have lower chance of survival of the affected people. From an analytical perspective, panel data studies dominate research on the macroeconomic effects of natural disasters, while time series techniques are the least use approach in the empirical literature.

The review also indicates that the empirical results on the macroeconomic effects of natural disasters are inconclusive. Thus three main empirical findings emerged; some studies report negative effect, while other reports positive impact. The last strand of the empirical literature reports that natural disaster has a neutral effect on the macroeconomy. The inconclusiveness of the results is due to the following factors: First, the source of natural disaster data is one of the critical causes of the disparities in the findings. For instance, Strobl (2012) demonstrated this using natural disaster (hurricane) data from the North Atlantic Hurricane database/Eastern North Pacific Tracks File and EM-DAT database to examine the effect of hurricanes on GDP per capita. The hurricane data from the North Atlantic Hurricane/Eastern North Pacific Tracks File database were found to retard GDP per capita, but the hurricane data from the EM-DAT database were found to have a neutral effect on GDP per capita. Similarly, Skidmore and Toya (2002) further used natural disaster data from EM-DAT and Davis database to provide an empirical assessment on the long-run impact of climatic and geological disasters on macroeconomic variables such as GDP per capita growth. Interestingly, the results revealed that geologic disaster retards GDP per capita growth; it was a geological disaster from EM-DAT that significantly affects the GDP per capita growth. Therefore the use of different disaster data from different databases could potentially lead to conflicting results.

Another potential source of the conflicting results emanates from the indicator/proxies of a natural disaster. In the empirical literature, some studies used total economic damages as a proxy for natural disasters. Other measures such as the number of people killed, people affected by disaster, and the total number of people affected as the proxy for natural disasters. It is argued that the use of total economic damages associated with disaster may be endogenous to economic growth since the disaster loss estimated are mostly recorded based on relief and recovery expenditures incurred flowing natural disaster. Therefore using the total economic loss as a proxy for natural disaster could result in spurious regression (Panwar & Sen, 2018). Some scholars have also argued that the most appropriate measure for a natural disaster to be used is the total number of people affected (see Felbermayr & Gröschl, 2014; Klomp & Valckx, 2014; Loayza et al., 2012; Panwar & Sen,

2018). To buttress our point, Shabnam (2014), for instance, empirically demonstrated that the total number of people affected by floods significantly reduces GDP per capita, while the total death toll caused by floods exert an insignificant impact on GDP per capita. Also, various econometric approach such as static models (fixed effect and ordinary least squares (OLS)), dynamic models (system-GMM, ARDL, VAR, PVAR), and quasiexperimental approaches are used for empirical analysis. Given that these models operate under different statistical assumption, it could yield inconsistency in the literature. Third, the use of different estimation approach also serves another source the conflicting results.

Improving institutions, governance, income inequality, financial system, human capital, and economic development and public awareness are some of strategies and policy measures to mitigate the disastrous effect of natural disasters in developing countries. The literature review presents some knowledge gaps for future research. The review of the literature suggests that research on the macroeconomic effects of disasters are still at the infant stage. However, the majority of the empirical studies have investigated the impact of various disasters on GDP per capita while little attention paid to the other sectors of the economy. Future research evaluating the sectorial effect of different natural disasters are welcome. Also, future research could use a multidisciplinary approach to investigate the psychological implications of natural disasters at the national, global, and regional levels. Furthermore, existing empirical literature assume the impact of natural disaster on the economy is symmetrical. However, it could suggest natural disasters could have an asymmetrical effect on the macroeconomy such that positive and negative shocks to natural disasters could have different implications for the economy. Although panel data analysis presents efficient estimates, the conclusions and policy implications from panel data results may not apply to individual specific countries due to institutional and structural difference among countries. In regards, future research should utilize time series techniques that are less applied in the literature and probe the macroeconomic effect of disasters. Another area of research that deserves much attention is the implications of natural disasters on corruption, political, and civil instability.

ACKNOWLEDGMENT

The authors acknowledge the Australia Africa Universities Network (AAUN) as partners.

CONFLICT OF INTEREST

The authors declare no conflict of interest.

REFERENCES

Acemoglu, D., Johnson, S., & Robinson, J. A. (2001). The colonial origins of comparative development: An empirical investigation. *American Economic Review*, *91*(5), 1369−1401.

Albala-Bertrand, J.-M. (1993a). Political economy of large natural disasters: With special reference to developing countries. In *Oxford University Press (OUP) catalogue*.number 9780198287650

Albala-Bertrand, J. M. (1993b). Natural disaster situations and growth: A macroeconomic model for sudden disaster impacts. *World Development, 21*(9), 1417−1434. Available from https://doi.org/10.1016/0305-750X(93)90122-P.

Asafu-Adjaye, J. (2014). The economic impacts of climate change on agriculture in Africa. *Journal of African Economies, 23*(Suppl. 2), ii17−ii49. Available from https://doi.org/10.1093/jae/eju011.

Benali, N., Mbarek, M. B., & Feki, R. (2019). Natural disaster, government revenues and expenditures: Evidence from high and middle-income countries. *Journal of the Knowledge Economy, 10*(2), 695−710. Available from https://doi.org/10.1007/s13132-017-0484-y.

Benson, C., & Clay, E. (2004). *Understanding the economic and financial impacts of natural disasters*. The World Bank.

Bergholt, D., & Lujala, P. (2012). Climate-related natural disasters, economic growth, and armed civil conflict. *Journal of Peace Research, 49*(1), 147−162. Available from https://doi.org/10.1177/0022343311426167.

Cavallo, E., Galiani, S., Noy, I., & Pantano, J. (2013). Catastrophic natural disasters and economic growth. *Review of Economics and Statistics, 95*(5), 1549−1561.

Chaiechi, T. (2014). The economic impact of extreme weather events through a Kaleckian-post-Keynesian lens: A case study of the state of Queensland, Australia. *Economic Analysis and Policy, 44*(1), 95−106. Available from https://doi.org/10.1016/j.eap.2014.01.002.

Chhibber, A., & Laajaj, R. (2008). Disasters, climate change and economic development in sub-Saharan Africa: Lessons and directions. *Journal of African Economies, 17*(Suppl._ 2), ii7−ii49. Available from https://doi.org/10.1093/jae/ejn020.

Cuaresma, J. C. (2010). Natural disasters and human capital accumulation. *The World Bank Economic Review, 24*(2), 280−302.

Cunado, J., & Ferreira, S. (2014). The macroeconomic impacts of natural disasters: The case of floods. *Land Economics, 90*(1), 149−168. Available from https://doi.org/10.3368/le.90.1.149.

Doyle, L., & Noy, I. (2015). The short-run nationwide macroeconomic effects of the Canterbury earthquakes. *New Zealand Economic Papers, 49*(2), 134−156. Available from https://doi.org/10.1080/00779954.2014.885379.

Felbermayr, G., & Gröschl, J. (2014). Naturally negative: The growth effects of natural disasters. *Journal of Development Economics, 111*, 92−106. Available from https://doi.org/10.1016/j.jdeveco.2014.07.004.

Fomby, T., Ikeda, Y., & Loayza, N. V. (2013). The growth aftermath of natural disasters. *Journal of Applied Econometrics, 3*(28), 412−434.

Hallegatte, S., & Dumas, P. (2009). Can natural disasters have positive consequences? Investigating the role of embodied technical change. *Ecological Economics, 68*(3), 777−786. Available from https://doi.org/10.1016/j.ecolecon.2008.06.011.

Horwich, G. (2000). Economic lessons of the Kobe earthquake. *Economic Development and Cultural Change, 48*(3), 521−542.

Kahn, M. E. (2005). The death toll from natural disasters: The role of income, geography, and institutions. *Review of Economics and Statistics, 87*(2), 271−284.

Keen, B. D., & Pakko, M. R. (2011). Monetary policy and natural disasters in a DSGE model. *Southern Economic Journal, 77*(4), 973−990. Available from https://doi.org/10.4284/0038-4038-77.4.973.

Klomp, J. (2016). Economic development and natural disasters: A satellite data analysis. *Global Environmental Change, 36*, 67−88. Available from https://doi.org/10.1016/j.gloenvcha.2015.11.001.

Klomp, J., & Valckx, K. (2014). Natural disasters and economic growth: A meta-analysis. *Global Environmental Change, 26*, 183−195. Available from https://doi.org/10.1016/j.gloenvcha.2014.02.006.

Loayza, N. V., Olaberría, E., Rigolini, J., & Christiaensen, L. (2012). Natural disasters and growth: Going beyond the averages. *World Development*, *40*(7), 1317−1336. Available from https://doi.org/10.1016/j.worlddev.2012.03.002.

Mohan, P. S., Ouattara, B., & Strobl, E. (2018). Decomposing the macroeconomic effects of natural disasters: A national income accounting perspective. *Ecological Economics*, *146*, 1−9. Available from https://doi.org/10.1016/j.ecolecon.2017.09.011.

Narayan, P. K. (2003). Macroeconomic impact of natural disasters on a small island economy: Evidence from a CGE model. *Applied Economics Letters*, *10*(11), 721−723. Available from https://doi.org/10.1080/1350485032000133372.

Noy, I. (2009). The macroeconomic consequences of disasters. *Journal of Development Economics*, *88*(2), 221−231. Available from https://doi.org/10.1016/j.jdeveco.2008.02.005.

Noy, I., & Nualsri, A. (2011). Fiscal storms: Public spending and revenues in the aftermath of natural disasters. *Environment and Development Economics*, *16*(1), 113−128.

Padli, J., Habibullah, M. S., & Baharom, A. H. (2010). Economic impact of natural disasters' fatalities. *International Journal of Social Economics*, *37*(6), 429−441. Available from https://doi.org/10.1108/03068291011042319.

Panwar, V., & Sen, S. (2018). Economic impact of natural disasters: An empirical re-examination. *Margin: The Journal of Applied Economic Research*, *13*(1), 109−139. Available from https://doi.org/10.1177/0973801018800087.

Sawada, Y., Bhattacharyay, M., & Kotera, T. (2019). Aggregate impacts of natural and man-made disasters: A quantitative comparison. *International Journal of Development and Conflict*, *9*(1), 43−73.

Selcuk, F., & Yeldan, E. (2001). On the macroeconomic impact of the August 1999 earthquake in Turkey: A first assessment. *Applied Economics Letters*, *8*(7), 483−488. Available from https://doi.org/10.1080/13504850010007501.

Shaari, M. S. M., Abd Karim, M. Z., & Hasan-Basri, B. (2017). Does flood disaster lessen GDP growth? Evidence from Malaysia's manufacturing and agricultural sectors. *Malaysian Journal of Economic Studies*, *54*(1), 61−81.

Shabnam, N. (2014). Natural disasters and economic growth: A review. *International Journal of Disaster Risk Science*, *5*(2), 157−163. Available from https://doi.org/10.1007/s13753-014-0022-5.

Skidmore, M., & Toya, H. (2002). Do natural disasters promote long-run growth? *Economic Inquiry*, *40*(4), 664−687.

Strobl, E. (2012). The economic growth impact of natural disasters in developing countries: Evidence from hurricane strikes in the Central American and Caribbean regions. *Journal of Development Economics*, *97*(1), 130−141. Available from https://doi.org/10.1016/j.jdeveco.2010.12.002.

Strömberg, D. (2007). Natural disasters, economic development, and humanitarian aid. *Journal of Economic Perspectives*, *21*(3), 199−222.

Strulik, H., & Trimborn, T. (2019). Natural disasters and macroeconomic performance. *Environmental and Resource Economics*, *72*(4), 1069−1098. Available from https://doi.org/10.1007/s10640-018-0239-7.

Thirawat, N., Udompol, S., & Ponjan, P. (2017). Disaster risk reduction and international catastrophe risk insurance facility. *Mitigation and Adaptation Strategies for Global Change*, *22*(7), 1021−1039. Available from https://doi.org/10.1007/s11027-016-9711-2.

Toya, H., & Skidmore, M. (2007). Economic development and the impacts of natural disasters. *Economics Letters*, *94*(1), 20−25. Available from https://doi.org/10.1016/j.econlet.2006.06.020.

APPENDIX

Table 20.A1 Summary of Disaster Studies.

Authors	Study Sample	Proxies for Natural Disaster	Source of Natural Disaster Dataset	Methodology	Macroeconomic Indicator(s)	Findings
Albala-Bertrand (1993a, 1993b)	26 developing countries	Natural disaster	–	*Study type*: Cross-sectional *Data period*: 1960−79 *Estimation approach*: Pre- and postdisaster analysis	GDP per capita	Economic growth increases in most cases after the natural disaster.
Skidmore and Toya (2002)	89	Total damages by climatic and geologic disasters	Davis and CRED/EM-DAT	*Study type*: Cross-sectional analysis *Data period*:1960−90 *Estimation approach*: OLS	GDP per capita growth	Climatic disasters increase GDP per capita growth, while geological disasters retard GDP per capita growth. However, the EM-DAT geologic disaster indicator remains insignificant. The disaster indicators exert an insignificant negative effect on physical capital growth. Climatic disasters increase human capital indicators, while geologic disasters exert an insignificant negative effect on human capital. Climatic disasters increase total factor productivity, while geologic disasters exert an insignificant negative effect on total factor productivity.
Noy (2009)	109	Number of people, total damages and number of people affected by natural disasters	CRED/EM-DAT	*Study type*: Panel data *Data period*:1970−2003 *Estimation approach*: Hausman−Taylor estimator	GDP per capita	Natural disaster retards economic growth, but with similar magnitude, developing countries experienced significant macroeconomic shocks relative to the developed countries.
Cuaresma (2010)	80	Total damages by climatic and geologic disasters	CRED/EM-DAT	*Study type*: Cross-sectional analysis *Data period*: 1980−2000 *Estimation approach*: Bayesian model averaging	Human capital accumulation (secondary school enrollment)	Geologic disasters significantly reduce human capital accumulation.
Noy and Nualsri (2011)	44	Total damages caused by a natural disaster	CRED/EM-DAT	*Study type*: Panel data *Data period*: 1990Q1−2005Q4 *Estimation approach*: PVAR	Government spending and revenue	In developed countries a shock to natural disaster initially increases government spending and gradually declines, while government revenue drops cumulatively. It was also revealed that a shock to natural disaster increases government payment and

Table 20.A1 Summary of Disaster Studies. *Continued*

Authors	Study Sample	Proxies for Natural Disaster	Source of Natural Disaster Dataset	Methodology	Macroeconomic Indicator(s)	Findings
						outstanding debt, while government cash surplus diminishes. In the developing countries the results show that government spending, government revenue, government payment, and outstanding debt react negatively to natural disasters while increases cash surplus.
Bergholt and Lujala (2012)	171	People affected by climatic disasters (floods and storms)	CRED/EM-DAT	*Study type*: Panel data *Data period*: 1980–2007 *Estimation approach*: Fixed effect	GDP per capita growth	Climatic disaster significantly reduces GDP per capita. Specifically, the number of people affected by floods and storms significant retards GDP per capital growth.
Loayza et al. (2012)	94	Total number of people affected by natural disaster in general and specifically by floods, storms, earthquake and drought	CRED/EM-DAT	*Study type*: Panel data *Data period*: 1961–2005 *Estimation approach*: system-GMM	GDP per capita, agricultural, service and industrial growth	The composite indicator of a natural disaster does not affect GDP per capita, while flood significantly increases GDP per capita; agricultural, service, and industrial growth. Drought reduces GDP per capita, agricultural and industrial growth, while it has no impacts on service growth. Storm significantly reduces agricultural growth, while it substantially increases industrial growth. Earthquake significantly increase industrial growth, but it remains insignificant on GDP per capita and the growth of remaining sectors. Developing countries are more vulnerable to natural disasters relative to developed countries.
Panwar and Sen (2018)	102	Total number of people affected by natural disaster in general and specifically by floods, storms, earthquake and drought	CRED/EM-DAT	*Study type*: Panel data *Data period*: 1961–2005 *Estimation approach*: system-GMM	GDP per capita, agricultural and industrial growth	The aggregate indicator for natural disaster does not influence GDP per capita. Floods increase GDP per capita and agricultural growth, while it does not affect growth. Droughts reduce GDP growth, agricultural and industrial growth, but the impact is significant on agricultural growth. Storms exert an insignificant effect on GDP growth, agricultural and industrial growth, while earthquake significantly increases industrial growth.

(Continued)

Table 20.A1 Summary of Disaster Studies. *Continued*

Authors	Study Sample	Proxies for Natural Disaster	Source of Natural Disaster Dataset	Methodology	Macroeconomic Indicator(s)	Findings
Strobl (2012)	30 Central America and the Caribbean	The composite index for hurricane	North Atlantic Hurricane database and Eastern North Pacific Tracks File	*Study type*: Panel data *Data period*: Since the 1950s *Estimation approach*: LSDV	GDP per capita	Hurricane reduces GDP per capita but depends on timing and control of local economy characteristics.
Cavallo et al. (2013)	196	Number of people killed by disasters	CRED/EM-DAT	*Study type*: Panel data *Data period*: 1970−2008 *Estimation approach*: Comparative case study (quasiexperimental approach)	GDP per capita	The impact of natural disaster on GDP per capita is generally neutral in both short run and long run.
Fomby et al. (2013)	84	The intensity of drought, floods, earthquake and storms	EM-DAT	*Study type*: Panel data *Data period*: 1960−2010 *Estimation approach*: PVAR	GDP growth, agricultural and nonagricultural growth.	Droughts have an overall negative effect on GDP growth as well as agricultural and nonagricultural growth. Floods increase GDP growth as well as agricultural and nonagricultural growth. Earthquake and storms exert a weaker effect on GDP growth as well as agricultural and nonagricultural growth.
Felbermayr and Gröschl (2014)	More than 100 countries	The composite index for natural disaster	GeoMet, EM-DAT and NatCatSERVICE	*Study type*: Panel data *Data period*: 1979−2010 *Estimation approach*: Dynamic fixed effect	GDP per capita	GeoMet disaster indicator reduces GDP per capita, while EM-DAT and NatCatSERVICE disaster data exerts an insignificant effect on GDP per capita.
Shabnam (2014)	187	Floods	EM-DAT	*Study type*: Panel data *Data period*: 1960−2010 *Estimation approach*: Fixed effect	GDP per capita	The total number of people affected by floods significantly reduces GDP per capita, while the total death toll caused by floods exerts a negligible impact on GDP per capita.
Cunado and Ferreira (2014)	135	Floods	Global Archive of Large Flood Events/DFO	*Study type*: Panel data *Data period*: 1985−2008 *Estimation approach*: PVAR	GDP growth, agricultural and nonagricultural growth.	Floods increase GDP per capita, agricultural and nonagricultural sector growth.
Chaiechi (2014)	Queensland (Australia)	Climatic events (floods and cyclones)	Bureau of Meteorology	*Study type*: Time series *Data period*: 2002−11 *Estimation approach*	Investment, savings, productivity, income distribution, net exports and unemployment	A shock to the climatic events initially drops investment and savings but restores its preshock level after the 1 year. In addition, productivity was found to initially respond positively to a shock to the climatic events, while net exports increase immediately after disasters before declining and improving again. Also, income distribution and unemployment were found to fluctuate after the disaster.

Table 20.A1 Summary of Disaster Studies. *Continued*

Authors	Study Sample	Proxies for Natural Disaster	Source of Natural Disaster Dataset	Methodology	Macroeconomic Indicator(s)	Findings
Doyle and Noy (2015)	New Zealand	Earthquake	Statistics New Zealand	*Study type*: Time series *Data period*: September 4, 2010 and February 22, 2011 *Estimation approach*	GDP per capita, inflation, the balance of payment, unemployment rate, interest rate, private consumption, government consumption, performing manufacturing index	The Canterbury earthquake has a less pronounced effect on New Zealand's economy.
Thirawat, Udompol, and Ponjan (2017)	Malaysia	Floods	Department of Irrigation and Drainage Malaysia	*Study type*: Time series *Data period*: 1960–2013 *Estimation approach*: ARDL	Manufacturing and agricultural growth	In the long run, flood damage reduces agricultural growth while increasing manufacturing growth. The study indicated that flood size, duration, and frequency have no effect on agricultural and output growth in the long run. In the short run, it was found that flood size and damage increase manufacturing and agricultural growth, whereas flood duration exerts no effect on manufacturing and agricultural growth. In addition, flood frequency was found to significantly increase only manufacturing growth in the short run.
Mohan et al. (2018)	21 Caribbean countries	Hurricane	Hurricane database HURDAT	*Study type*: Panel data *Data period*: 1979–2011 *Estimation approach*: PVAR	Export, import, government expenditure, investment, private consumption	The study revealed that generally, a shock to hurricane increases government expenditure, investment, import, while it reduces export and private consumption.
Benali et al. (2019)	6 countries	Composite measure for natural disaster	EM-DAT	*Study type*: Panel data *Data period*: 1990–2013 *Estimation approach*: PVAR and Granger causality	Government budget, economic growth, inflation, and government debt	For middle-income countries, natural disaster unidirectionally causes government expenditure and government debt while for the high-income countries, natural disaster unidirectionally to economic growth, government debt, and government expenditure.
Sawada et al. (2019)	189 countries					Hydrometeorological, geophysical, climatological, technological and biological disasters
EM-DAT	*Study type*: Panel data *Data period*: 1968–2001 *Estimation approach*: Instrumental variable (IV) estimator	GDP per capita	In the short run, natural disasters exert a greater negative effect on welfare, while in the long run, natural disasters increase GDP per capita.			

DFO, Darmouth Flood Observatory; EM-DAT, *emergency disaster database*; GMM, *method of moment*; HURDAT, *hurricane research division's hurricane re-analysis project*; LSDV, *least square dummy variable*; PVAR, *panel vector autoregression*.

THE ASIAN TSUNAMI AND TOURISM INDUSTRY: IMPACT AND RECOVERY

21

Maneka Jayasinghe[1], Saroja Selvanathan[2] and E.A. Selvanathan[2]

[1]*Asia Pacific College of Business and Law, Charles Darwin University, Darwin, NT, Australia* [2]*Griffith Business School, Griffith University, Nathan Campus, QLD, Australia*

21.1 INTRODUCTION

The tsunami of December 26, 2004, widely known as the Asian tsunami or the Boxing Day tsunami, which was caused by the fourth largest earthquake of recent times, had a devastating impact on the coastal regions of countries in the Indian Ocean, particularly in Maldives, Thailand, Sri Lanka, Indonesia, and India (Ramalanjoana, 2006). The Asian tsunami not only resulted in a significant death toll but also destroyed infrastructure, transport and communication networks, and the livelihoods of thousands of people (United Nations Office for Disaster Risk Reduction (UNDRR), 2005). The estimated values of the economic, infrastructural, and human development losses to those countries due to the tsunami are more than US$10 billion (Cosgrave, 2005).

Many places affected by the tsunami are popular tourist destinations with beach tourism a staple of the tourism industry in most of the tsunami-affected countries (Sharpley, 2005). However, tourism infrastructure, including many hotels and resorts, in the coastal areas were badly destroyed by the tsunami (Bandara & Naranpanawa, 2007; IMF, 2005; Robinson & Jarvie, 2008). Furthermore, although the beach is often an integral part of tourism and leisure, tourists were too afraid to travel to these coastal areas immediately after the tsunami. These conditions resulted in a significant decline in tourist bookings, arrivals, and revenue following the tsunami, causing a severe downturn in the tourism industry in some of the tsunami-affected countries. Francesco Frangialli, former Secretary-General of the World Tourism Organization, viewed the 2004 tsunami as "the greatest catastrophe ever recorded in the history of world tourism" (BBC News, 2005b).

It is widely accepted that tourism plays a significant role in enhancing economic growth in developing countries. In many developing and emerging economies, tourism is one of the top export categories, representing 40% of service exports (United Nations World Tourism Organization (UNWTO), 2017). Many studies have shown empirical evidence on the presence of tourism-led (economic) growth (see, e.g., Balaguer & Cantavella-Jordà, 2002, for Spain; Durbarry, 2004, for Mauritania; Katircioglu, 2010, for Singapore; Belloumi, 2010, for Tunisia; Kreishan, 2011, for Jordan; Amaghionyeodiwe, 2012, for Jamaica; Obadiah, Odhiambo, & Njuguna, 2012, for Kenya;

Economic Effects of Natural Disasters. DOI: https://doi.org/10.1016/B978-0-12-817465-4.00021-2

Mishra, Rout, & Mohapatra, 2010, for India; Srinivasan, Kumar, & Ganesh, 2012, for Sri Lanka; and Malik, Chaudhry, Sheikh, & Farooqi, 2010, and Hye & Khan, 2013, for Pakistan).

Natural disasters such as tsunami and earthquakes bring about a significant adverse impact on the tourism industry. Mendoza, Brida, and Garrido (2012) analyzed the impacts of three earthquakes; April 2007, November 2007, and February 2010 in the north, south, and center of Chile, respectively, based on inbound arrivals to Chile. Using the seasonal autoregressive integrated moving average (SARIMA) model on monthly visitor arrivals between January 2004 and June 2010, Mendoza et al. (2012) compared the difference between the predicted and the actual number of tourist arrivals (TA) after the three earthquakes. Mendoza et al. (2012) also found that the TA in some regions fully rebound after 4 months, while other regions did not. The study concluded that the impacts of the earthquake on TA depend on the magnitude of the earthquake, the perception toward the safety situation created by the media, and the type of activities that attract tourists to the region where the earthquake has occurred (Mendoza et al., 2012).

Applying autoregressive integrated moving average (ARIMA) models to analyze time series data during the period from January 1996 to November 2012, Wu and Hayashi (2013) analyzed the impact of the Great East Japan Earthquake that hit Tohoku in March 2011, on the volume of inbound TA. This study found that Japan's inbound tourism was adversely affected by the devastating earthquake with inbound tourism not yet fully recovered even 21 months after the disaster. The study also revealed that the number of TA rebound rapidly in the early part of the recovery process and subdued afterward.

Mazzocchi and Montini (2001) examined the economic impact of the earthquake that hit the Umbria region in Italy on September 26, 1997, using time series data. The findings based on the event study method revealed that the region experienced a significant economic downturn between October 1997 and June 1998 followed by a substantial drop in the performance of the tourism industry. Huang and Min (2002) analyzed whether Taiwan tourism had rebounded completely after the September 21, 1999, earthquake in Taiwan, using the SARIMA model. This study established a time series model to forecast Taiwan's inbound tourism demand, in terms of the volume of monthly TA from September 1999 to July 2000. The estimated results revealed that Taiwan's inbound arrivals had not fully recovered even after 11 months of the earthquake.

If a country's economic performance is highly driven by the tourism industry, any impact on the tourism industry resulting from natural disasters could seriously damage their economy. The objective of this chapter is to examine the impact of the 2004 Asian tsunami on TA and GDP in five countries that were severely affected by the Asian tsunami: Maldives, Thailand, Sri Lanka, Indonesia, and India. For this purpose, we first investigate whether tourism causes GDP in these five counties, using long-run and short-run Granger causality tests. Second, we examine the impact of the tsunami on tourism and GDP, using Box—Jenkins ARIMA model to forecast TA and GDP for posttsunami period (2005—10). We then compare the ARIMA model forecasted values and actual out-of-sample values. This exercise enables us to identify whether tsunami adversely affected TA and GDP, quantify the magnitude of the impact and track how TA and GDP have been recovering in the posttsunami period in Maldives, Thailand, Sri Lanka, Indonesia, and India. An analysis of the scale of impact of the tsunami and length of the recovery in the posttsunami period will provide useful insights to identify the overall impact of large-scale disasters, such as tsunami and earthquakes, on the economy in the short run and long run.

This chapter is organized as follows. Section 21.2 presents a preliminary analysis of time series data on international TA and GDP in the five selected countries: Maldives, Thailand, Sri Lanka, Indonesia, and India. The research methodology is presented in Section 21.3, while the estimation results are discussed in Section 21.4. Section 21.5 provides concluding comments.

21.2 PRELIMINARY DATA ANALYSIS

This section presents an overview of the tourism industry in all five countries of interest in this chapter and discusses the trends in TA and GDP in detail. For this study, we use data on the number of international TA and GDP (constant US$). Based on the availability of data, we selected periods 1980−2004 for the Maldives, 1985−2004 for Thailand, 1980−2004 for Sri Lanka, 1990−2004 for Indonesia, and 1981−2004 for India. These data are sourced from the World Bank (2019) open data and the Tourism Department websites of the respective countries.

Table 21.1 presents preliminary statistics on the tourism industry's contribution to the respective economies and the impact of the tsunami on the economy.

Next we discuss the impact of the 2004 Asian tsunami on the local economies of the five countries, with a special emphasis on TA and GDP. The time series trend in tourism growth rates depicted in Figs. 21.1−21.5 clearly indicates a sudden drop in international TA immediately after the tsunami in all five countries. While GDP growth rates show a significant drop in Maldives and Thailand immediately after the tsunami, it reflected a minimal or no impact on GDP growth rate in Sri Lanka, Indonesia, and India.

21.2.1 MALDIVES

The Maldives has experienced about a 10-fold increase in the number of inbound international tourists between 1980 and 2004, with an annual growth rate of above 9% (Ministry of Tourism

			Growth Projection in 2005[b]		
Country	Tourism Earnings (% of GDP)[a]	Damage Caused by Tsunami (US$ Billions)[b]	Pretsunami	Posttsunami	Posttsunami Actual GDP Growth Rate[c]
(1)	(2)	(3)	(4)	(5)	(6)
Maldives	42.0	0.4	6.5	1.0	− 13.1
Thailand	5.4	0.8	6.1	5.5−6.5	4.2
Sri Lanka	4.6	1	6	5.3	6.2
Indonesia	3.9	4−5	5.5	5.25−5.5	5.7
India	2.0	1.7	6.4	6−6.4	7.9

Table 21.1 Economic Impact of Asian Tsunami.

[a]*BBC News (2005a).*
[b]*IMF (2005).*
[c]*World Bank (2019).*

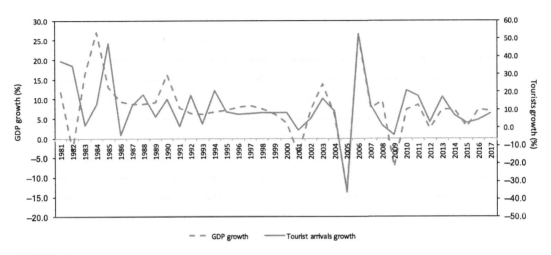

FIGURE 21.1

GDP and international tourist arrivals growth rate, Maldives, 1981–2017.

Authors' compilation using World Bank (2019).

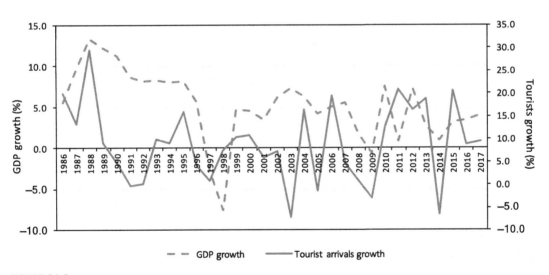

FIGURE 21.2

GDP and international tourist arrivals growth rate, Thailand, 1986–2017.

Authors' compilation using World Bank (2019).

Maldives, 2004). The average contribution of the tourism industry accounted for about 30% of Maldives gross domestic product (GDP) between 1999 and 2003 and contributed to about a 10-fold increase in Maldives' foreign exchange earnings during 1983–2003 (Ministry of Tourism Maldives, 2004). The Maldives was hit hard by the tsunami. The adverse impact on

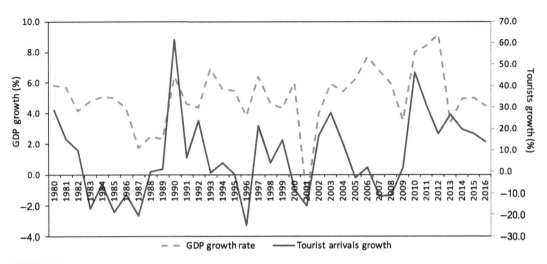

FIGURE 21.3

GDP and international tourist arrivals growth rate, Sri Lanka, 1980–2016.

Authors' compilation using World Bank (2019).

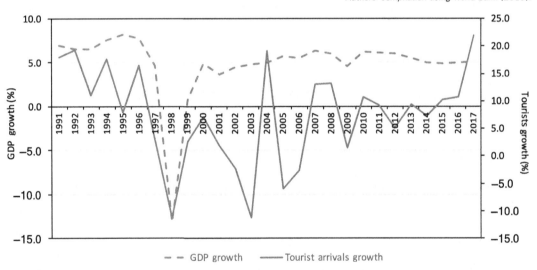

FIGURE 21.4

GDP and international tourist arrivals growth rate, Indonesia, 1980–2017.

Authors' compilation using World Bank (2019).

the economy was substantial as the disaster resulted in a significant reduction in earnings from tourism. United Nations World Tourism Organization (UNWTO) (2006) reported a 39% decrease in TA to the Maldives in 2005. The impact on the overall economy is evident in the growth projections and growth performance of the posttsunami period presented in Table 21.1.

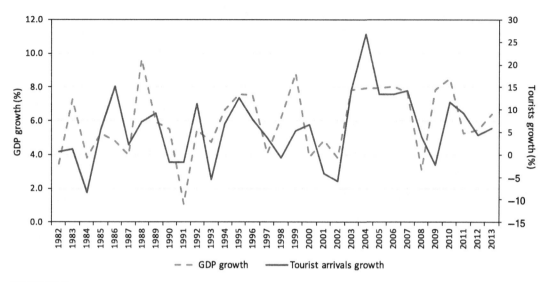

FIGURE 21.5

GDP and international tourist arrivals growth rate, India, 1982−2013.

Authors' compilation using World Bank (2019).

As shown in Fig. 21.1, Maldives' tourism growth rates and GDP growth rates follow a very similar pattern, indicating that GDP is notably dependent on earning from tourism in the Maldives. While before the tsunami, the Maldives was projected to grow at 6.5% in 2005, the tsunami resulted in a significant drop in growth projections (to 1%). Nevertheless, in 2005 Maldives reported a negative growth rate of 13.1%. However, by late 2006 the Maldives regained the momentum of TA, demonstrating the highest growth in TA (52%) in the country's history, after having the highest rate of drop in 2005 (World Bank, 2019). Followed by this growth in TA, as can be seen in Fig. 21.1, the Maldives reported a remarkable GDP growth rate of 26.1% in 2006.

21.2.2 THAILAND

Thailand is one of the key players in the global and Asia Pacific travel and tourism industry. By the time the tsunami hit Thailand in 2004 December, Thailand had been enjoying years of tourism growth (Karatani, 2008). Nearly half of 5,395 of the total tsunami death toll in Thailand were foreign tourists and as a result, the tourism industry experienced a downturn immediately after the disaster (Fig. 21.2). For example, although year-to-year growth in the month of January showed significant growth from about 650,000 (in 1998) to 1.2 million (in 2004), in January 2005, the number of TA dropped by about one-third (to 800,000) (Blanke & Chiesa, 2007). Nevertheless, Thailand TA were restored toward the second half of 2005 and, overall, the country experienced only about 1.4% reduction in TA in 2005 (United Nations World Tourism Organization (UNWTO),

2006). As can be seen in Fig. 21.2, Thailand's GDP appears sensitive to the fluctuations in the tourism industry to a certain extent. For example, during the Asian financial crisis in the late 1990s, the tsunami in 2004, and the global financial crisis in the late 2000s both TA and GDP have experienced a downturn.

21.2.3 SRI LANKA

The tourism industry in Sri Lanka experienced notable growth after 1977, with TA increasing from 153,665 in 1977 to 407,230 in 1982 (Sri Lanka Tourism Development Authority (SLTDA), 2019). However, due to the intensifying civil war in the country, the TA declined to 182,662 in 1987 and were observed to marginally increase thereafter, until the cease-fire agreement took place in 2002. By 2004 the TA to Sri Lanka grew rapidly to 566,202. The annual growth of TA for the period 1980−2016 is shown in Fig. 21.3. Although the tsunami severely affected the tourism industry in Sri Lanka, this was barely visible in terms of TA. The official statistics reported of 549,308 international tourists. Nevertheless, these numbers included a large number of international aid workers who received tourist visas (Robinson & Jarvie, 2008). United Nations World Tourism Organization (UNWTO) (2006) reported a 3% drop in TA to Sri Lanka in 2005 compared to that of 2004. The impact of the tsunami on tourism-related assets lost was quite significant (Bandara & Naranpanawa, 2007). Although the tsunami resulted in revising the growth forecast by about 0.5%−1.0%, mainly due to the disruptions to the fisheries and tourism industries, the Central Bank of Sri Lanka (2005) noted that the extensive reconstruction and rehabilitation programs contributed in offsetting the negative impact of the tsunami on economic growth. As shown in Fig. 21.3, in 2005 the country reported 6.2% of actual growth. Although the TA restored to the pretsunami level by 2006, due to the reescalation of the civil war, the tourism industry continued to suffer until the end of the war in 2009.

21.2.4 INDONESIA

Indonesia claimed the highest number of casualties (94,000) and the physical damage to infrastructure (4−5 US$ billion). However, as earnings from the tourism industry in Aceh, the main area affected by the tsunami, contribute only about 2% of Indonesia's GDP, Indonesia experienced only a minimal impact on the economy as a result of the tsunami (IMF, 2005). Consequently, although TA showed a 6% reduction in 2005 (United Nations World Tourism Organization (UNWTO), 2006), there was no major revision in growth projection. The actual GDP growth rate for the year 2005 was even slightly higher than the pretsunami growth forecast (IMF, 2005). The TA restored by 2006 and demonstrated a steady growth thereafter. World Tourism Organization (WTO) (2005) documented that the Indonesian experience of the impact of the tsunami on tourism and the recovery was different to other tsunami-affected countries. While international tourists avoided Ache, immediately after the tsunami, that of Bali showed an increase throughout 2005. As can be seen in Fig. 21.4, overall, except for the period of the Asian financial crisis of the late 1990s, fluctuations in TA do not have major implications on Indonesia's GDP.

21.2.5 INDIA

The tourism industry represents about 2% of India's GDP. While the tsunami death toll in India reported 10,881 (which is a small proportion in relation to India's 1 billion population), the impact of the tsunami on the economy and the tourism industry was limited. Overall, India reported a 13% growth in TA in 2005 compared to that of 2004. Fig. 21.5 presents the annual growth rates of TA and GDP. Although India experienced a reduction in TA immediately after the tsunami, no significant impact on GDP is evident.

21.3 METHODOLOGY

In this section, we present the econometric and time series models/techniques we use for our analysis in two parts. First, we investigate the causal relationship between GDP and tourism (measured by the number of international TA) using several econometric techniques, such as stationarity of the time series, bounds test for cointegration, and Granger causality test. Second, we use pure time series modeling—univariate Box–Jenkins ARIMA method (Box & Jenkins, 1976) to model the GDP and TA time series individually and forecast each time series to see how the tsunami impacted on these two variables by comparing the out-of-sample forecasts (2005–09) with actual (observed) values.

21.3.1 ECONOMETRIC TECHNIQUES

We use the following long-run model to estimate the relationship with GDP and TA for a country i.

$$\ln GDP_{it} = \beta_0 + \beta_1 \ln TA_{it} + \varepsilon_{it} \qquad i = 1, 2, \ldots, 5; \; t = 1, 2, \ldots, T \qquad (21.1)$$

where GDP_{it} is the GDP for country i (Maldives, Thailand, Sri Lanka, Indonesia, and India) in year t; TA_{it} is the total number of international TA to country i in year t, and ε_{it} is the error term. Hereinafter, both variables are used in natural logarithm form.

To avoid any spurious regression results, first, we investigate whether GDP and international TA data series in Model (1) are stationary. For this purpose, we use three commonly used unit root tests, Augmented Dickey and Fuller (1979) (ADF) test, Phillips and Perron (1988) (PP) test, and KPSS (Kwiatkowski, Phillips, Schmidt, & Shin, 1992) unit root tests. For the first two tests the null hypothesis is H_0: Variable has a unit root (or the series is nonstationary) and for the KPSS test the null hypothesis is H_0: Variable has no unit root (or the series is stationary).

Although the unit root test results conclude that the two variables are either both $I(1)$ or one $I(0)$ and the other is $I(1)$, if the two variables are cointegrated, then the estimation results from Model (1) is still valid. We further investigate whether those variables are mutually cointegrated, using the autoregressive distributed lag (ARDL) error-correction framework (Pesaran, Shin, & Smith, 1996). The unrestricted error-correction version for Model (1) for country i can be written as

$$\Delta \ln GDP_{it} = \beta_0 + \sum_{j=1}^{m} \beta_1 \Delta \ln GDP_{i,t-j} + \sum_{j=0}^{p} \beta_2 \Delta \ln TA_{i,t-j}$$
$$+ \gamma_1 \ln GDP_{i,t-1} + \gamma_2 \ln TA_{i,t-1} + \varepsilon_{it} \qquad i = 1, 2, \ldots, 5; \quad t = 1, 2, \ldots, T \tag{21.2}$$

where Δ is the first difference operator. The right-hand side of Eq. (21.2) has two components: the short run (variables in first difference form) and the long run (variables in level form). The following hypothesis, based on F-test, can be used to test the presence of a long-run relationship (or cointegration).

H_0: $\gamma_1 = \gamma_2 = 0$ (no cointegration or no long-run relationship).
H_1: At least one $\gamma_i \neq 0$ (cointegration or no long-run relationship exists).

The previous F-test is based on a nonstandard distribution (Pesaran & Pesaran, 1997). The critical values that correspond to this distribution are based on three things, (1) the number of regressors, (2) whether the variables included in Model (2) are either $I(0)$ or $I(1)$, and (3) whether Model (2) has an intercept and/or a trend term. Asymptotic critical values are given in Pesaran and Pesaran (1997) and Pesaran, Smith, and Shin (2001) and finite sample critical values are given in Narayan (2005). Two sets of critical values, named the lower bound critical value $I(0)$ and upper bound critical value $I(1)$ need to be used as critical values for the abovementioned F-test. The F-test statistic values that are higher than the $I(1)$ critical value result in a rejection of the null hypothesis, concluding that there is support for cointegration. The F-test statistic values lower than the $I(0)$ critical value lead to the conclusion that there is no support for cointegration and the variables are stationary in their level form. If the value of the F-test statistic is in between $I(0)$ and $I(1)$, the order of integration of the underlying regressors is required to make a conclusion.

If we find evidence for a cointegrating relationship between the selected variables based on the F-test, then for country i, we estimate an ARDL (m,p) model for the long-run relationship in the form

$$\ln GDP_{it} = \beta_0 + \sum_{j=1}^{m} \beta_{1j} \ln GDP_{i,t-j} + \sum_{j=0}^{p} \beta_{2j} \ln TA_{i,t-j} + \varepsilon_{it} \tag{21.3}$$

Model (3) coefficients β_{2j} can be used to test the long-run Granger causality by testing the null hypotheses H_0: $\beta_{20} = \beta_{21} = \cdots = \beta_{2p} = 0$ against an alternative H_1: at least one $\beta_{2p} \neq$ zero.

The required number of lags m and p are determined by using the Akaike information criteria (AIC) and/or the Schwarz Bayesian criteria (SBC). A maximum lag length of 2 is recommended for annual data (Pesaran & Shin, 1999). If the variables are cointegrated then the following error-correction model provides the short-run estimate.

$$\Delta \ln GDP_{it} = \beta_0 + \sum_{j=1}^{m} \beta_{1j} \Delta \ln GDP_{i,t-j} + \sum_{j=0}^{p} \beta_{2j} \Delta \ln TA_{i,t-j} + \phi ECM_{it-1} + \upsilon_{it} \tag{21.4}$$

where the error-correction term ECM_{it-1} is calculated from the long-run estimated results from Model (3) in the form

$$ECM_{it-1} = \ln GDP_{it-1} - \beta_0 - \sum_{j=1}^{m} \beta_{1j} \ln GDP_{i,t-j} - \sum_{j=0}^{p} \beta_{2j} \ln TA_{i,t-j} \tag{21.5}$$

The β_{1j} and β_{2j} coefficients in Model (4) measure the short-run dynamics of the model's convergence to equilibrium and ϕ measures the speed of adjustment or convergence. Model (4) coefficients β_{2j} can be used to test the short-run Granger causality by testing the null hypotheses H_0: $\beta_{20} = \beta_{21} = \cdots = \beta_{2p} = 0$ against an alternative H_1: at least one of β_{2j} coefficient $\neq$ zero.

21.3.2 AUTOREGRESSIVE INTEGRATED MOVING AVERAGE MODELS

To investigate whether the tsunami had an impact on the international TA of the respective countries and GDP, using the univariate Box–Jenkins ARIMA approach, we identify suitable ARIMA models to model the TA and GDP time series. The ARIMA models based on Box–Jenkins approach have been widely used to predict future values of a time series based on previously observed (historical) time series data. These forecasted values are used to investigate the impact of a particular natural disaster, such as an earthquake on TA (e.g., see, Huang & Min, 2002; Mazzocchi & Montini, 2001; Mendoza et al., 2012). We use historical (within-sample) data on international TA and GDP up to the year 2004 (year of the tsunami) to estimate suitable individual ARIMA(p,d,q) models. The sample period is 1980–2004 for the Maldives, 1985–2004 for Thailand, 1980–2004 for Sri Lanka, 1990–2004 for Indonesia, and 1981–2004 for India. We then use these models to forecast the out-of-sample TA and GDP for the following 6 years (2005–10) and compare the actual TA and GDP for the period 2005–10 with the forecast values. If the actual values are less than the forecast values, we consider that tsunami adversely impacted the TA and GDP.

ARIMA analysis involves a four-stage process, identification, estimation, diagnostic checking, and forecasting.

Let X_t is a nonstationary time series that requires d times differencing to get a stationary time series W_t. This means

$$W_t = (1-B)^d X_t \tag{21.6}$$

where B is backward-shift operator and d is the order of differencing.

If the series W_t follows an autoregressive moving average of order p and q [ARMA(p,q)] then X_t follows and ARIMA(p,d,q). That is, the path of W_t can be written as

$$\varphi(B)W_t = \theta_0 + \theta(B)a_t \tag{21.7}$$

where $\varphi(B) = 1 - \varphi_1 B - \varphi_1 B^2 - \cdots - \varphi_2 B^p$ is a polynomial of order p and $\theta(B) = 1 - \theta_1 B - \theta_2 B^2 - \cdots - \theta_q B^q$ is a polynomial of order q, θ_0 is a trend parameter and a_t is a white-noise process. Therefore Eqs. (21.6) and (21.7) can be combined into

$$\varphi(B)(1-B)^d X_t = \theta_0 + \theta(B)a_t \tag{21.8}$$

Eq. (21.8) is the standard ARIMA(p,d,q) model. The ARIMA (p,d,q) models for the five countries of interest in this chapter were estimated using EVIEWS statistical software. EVIEWS uses various model selection criteria to identify a suitable ARIMA(p,d,q) model. The number of lags is selected based on the AIC.

21.4 ESTIMATION RESULTS

In this section, we present the estimation results for the econometric and time series models.

21.4.1 ECONOMETRIC ESTIMATION

The time series plots (Figs. 21.1–21.5) showed irregular patterns, indicating the possibility of non-stationary nature in data. Nevertheless, the ADF, PP, and KPSS unit root test results presented in Table 21.2 indicate that all the variables are stationary in the first differenced form, that is $I(1)$.

Column (2) of Table 21.3 gives the calculated value of the F-test statistics, columns (3) and (4) present the $I(0)$ and $I(1)$ critical values (as discussed in Section 21.3.1), column (5) gives the estimates for the error-correction terms, and column (6) gives the conclusion of the bounds-test hypothesis testing. As can be seen, the F-test statistic values in column (2) are larger than the $I(1)$ critical values in column (4). Therefore we have support for the alternate hypothesis that the variables in the long-run relationship are cointegrated or that a long-run relationship exists between GDP and TA in all five countries. This is further supported by the statistically significant (at the 5% level) error-correction term estimates. The error-correction terms reported in column (5) are negative and less than one in absolute values, as they should be.

The last four columns of Table 21.3 present the Wald test results for the Granger causality. As can be seen, a causal relationship exists in the short run as well as in the long run running from TA to GDP. This indicates that any impact on TA could affect GDP.

21.4.2 AUTOREGRESSIVE INTEGRATED MOVING AVERAGE MODEL ESTIMATION AND FORECASTING

We have identified suitable ARIMA models based on the AIC values using the annual international TA and GDP data up to the year 2004. The estimated results for these ARIMA models are presented in Table 21.4. As can be seen, most of the estimated parameters are statistically significant. Using these estimation results, we forecasted TA and GDP for years 2005–10 for all five countries. These values were then compared against the actual (observed) values for the same period. Figs. 21.6–21.10 plot the actual time series within-sample up to 2004 and the forecasted TA and GDP values for the out-of-sample period 2005–10 for the five countries. These figures depict the deviation of actual values from the forecasted values immediately after the tsunami disaster and recovery trajectory of TA and GDP. Table 21.5 presents the out-of-sample (2005–10) actual and predicted values as well as the percentage forecast error of TA and GDP data series.

As Maldives' GDP is significantly dependent on the tourism industry, tsunami brought about some critical impact on the tourism industry as well as on GDP in the Maldives. As shown in Fig. 21.6, immediately after the tsunami, Maldives experienced a considerable deviation in forecasted TA from actual TA. The tourism industry began to recover by 2007 before getting affected by the global financial crisis thereafter. GDP in the Maldives is very much sensitive to the movements in the tourism industry. This is because earnings from tourism contribute to about 42% of GDP in the Maldives (see Table 21.1). The percentage forecast error values given in Table 21.5 for the Maldives indicate that the impact of the tsunami on the TA is even greater (-51.5%) than the

Table 21.2 Tests for Stationarity.

(1)	(2)	ADF Test			PP Test			KPSS		
		Test Stat	Critical Value (5%)	P-Value	Test Stat	Critical Value (5%)	P-Value	Test Stat	Critical Value (5%)	Conclusion
		(3)	(4)	(5)	(6)	(7)	(8)	(9)	(10)	(11)
Maldives										
Tourist arrivals	Level	−2.676	−2.948	.088	−3.059	−2.943	.039	0.192	0.146	Nonstationary
	First diff	−1.975	−1.951	.048	−5.656	−1.950	.000	0.265	0.463	Stationary, $I(1)$
GDP	Level	−2.119	−3.537	.519	−2.424	−3.537	.362	0.205	0.146	Nonstationary
	First diff	−6.455	−2.946	.000	−6.510	−2.946	.000	0.405	0.463	Stationary, $I(1)$
Thailand										
Tourist arrivals	Level	−2.081	−3.537	.539	−2.219	−3.537	.466	0.741	0.463	Nonstationary
	First diff	−4.378	−2.951	.002	−5.726	−2.946	.000	0.091	0.463	Stationary, $I(1)$
GDP	Level	−1.659	−3.540	.749	−1.210	−3.534	.894	0.167	0.146	Nonstationary
	First diff	−3.212	−2.946	.027	−3.214	−2.946	.027	0.318	0.463	Stationary, $I(1)$
Sri Lanka										
Tourist arrivals	Level	−0.692	−3.537	.966	−1.002	−3.537	.932	0.150	0.146	Nonstationary
	First diff	−4.246	−2.946	.002	−4.299	−2.946	.002	0.263	0.463	Stationary, $I(1)$
GDP	Level	−0.941	−3.537	.940	−1.082	−3.537	.919	0.164	0.146	Nonstationary
	First diff	−4.634	−2.946	.007	−4.634	−2.946	.001	0.256	0.463	Stationary, $I(1)$
Indonesia										
Tourist arrivals	Level	−1.523	−3.587	.796	−1.769	−3.588	.692	0.654	0.463	Nonstationary
	First diff	−4.034	−2.981	.005	−4.001	−2.981	.005	0.223	0.463	Stationary, $I(1)$
GDP	Level	−1.400	−3.588	.838	−1.631	−3.586	.753	0.670	0.463	Nonstationary
	First diff	−3.758	−2.981	.009	−3.732	−2.981	.010	0.102	0.463	Stationary, $I(1)$
India										
Tourist arrivals	Level	−1.840	−3.558	.662	−1.840	−3.558	.662	0.172	0.146	Nonstationary
	First diff	−4.296	−2.960	.002	−4.323	−2.960	.002	0.308	0.463	Stationary, $I(1)$
GDP	Level	−1.966	−3.558	.597	−1.751	−3.558	.705	0.192	0.146	Nonstationary
	First diff	−5.574	−2.960	.001	−5.680	−2.960	.000	0.395	0.463	Stationary, $I(1)$

Note: ADF and PP tests: H_0: series is nonstationary, KPSS: H_0: series is stationary. ADF, Augmented Dickey and Fuller; KPSS, Kwiatkowski, Phillips, Schmidt, and Shin; PP, Phillips and Perron. Authors' compilation using World Bank (2019).

Table 21.3 Bounds Test and Granger Causality Test.

(1)	Bounds Test						Granger Causality			
	F-Test Statistic	Critical Values at 5%		Error-Correction Term	Conclusion	Long Run		Short Run		
		I(0)	I(1)			Wald χ^2 Test Statistic	Conclusion	Wald χ^2 Test Statistic	Conclusion	
(1)	(2)	(3)	(4)	(5)	(6)	(7)	(8)	(9)	(10)	
Maldives	5.57	3.62	4.16	−0.340* (.000)	Cointegrated	47.500* (.000)	TA→GDP	47.070* (.000)	TA→GDP	
Thailand	4.46	3.62	4.16	−0.101* (.006)	Cointegrated	5.290* (.021)	TA→GDP	3.345* (.077)	TA→GDP	
Sri Lanka	5.02	3.62	4.16	−0.003* (.000)	Cointegrated	6.589* (.037)	TA→GDP	6.529* (.011)	TA→GDP	
Indonesia	5.33	4.68	5.15	−0.574* (.001)	Cointegrated	28.436* (.000)	TA→GDP	25.615* (.000)	TA→GDP	
India	61.64	3.62	4.16	−0.001* (.000)	Cointegrated	8.917* (.012)	TA→GDP	8.126* (.004)	TA→GDP	

Note: P-values are given in parenthesis. * Significant at the 5% level. TA, Tourist arrivals.
Authors' compilation using World Bank (2019).

Table 21.4 ARIMA Model Estimation Results.

Variable	Maldives		Thailand		Sri Lanka		Indonesia		India	
	Tourist Arrivals	GDP	Tourist Arrivals	GDP	Tourist Arrivals	GDP	Tourist Arrivals	GDP	Tourist Arrivals	GDP
(1)	(2)	(3)	(4)	(5)	(6)	(7)	(8)	(9)	(10)	(11)
	ARIMA (2,2,2)	ARIMA (0,1,2)	ARIMA (0,1,0)	ARIMA (0,1,1)	ARIMA (1,1,2)	ARIMA (0,1,0)	ARIMA (0,1,0)	ARIMA (0,1,0)	ARIMA (4,1,3)	ARIMA (0,1,0)
C	−0.005*	0.083*	0.086*	0.058*	12.743*	0.044*	0.064*	0.040*	0.042*	0.055*
	(.005)	(.000)	(.000)	(.001)	(.000)	(.000)	(.035)	(.018)	(.000)	(.000)
Y_{t-1}	−0.563*	0.4164**	—	—	0.527*	—	—	—	1.3169**	—
	(.048)	(.068)			(.012)				(.054)	
Y_{t-2}	−0.614*	−0.278	—	—	—	—	—	—	−1.558*	—
	(.004)	(.289)							(.028)	
Y_{t-3}	—	—	—	—	—	—	—	—	0.914**	—
									(.052)	
Y_{t-4}	—	—	—	—	—	—	—	—	−0.690	—
									(.118)	
ε_{t-1}	−1.151	—	—	0.727*	0.576*	—	—	—	−2.891	—
	(.999)			(.000)	(.006)				(.972)	
ε_{t-2}	0.151	—	—	—	0.864*	—	—	—	2.891	—
	(.999)				(.004)				(.978)	
ε_{t-3}	—	—	—	—	—	—	—	—	−0.999	—
									(.988)	
AIC value	−1.806	−2.846	−2.171	−3.843	−0.620	−5.127	−1.526	−2.745	−2.736	−4.930

Note: P-values are given in parenthesis. * Significant at the 5% level; ** significant at the 10% level. ARIMA, autoregressive integrated moving average.

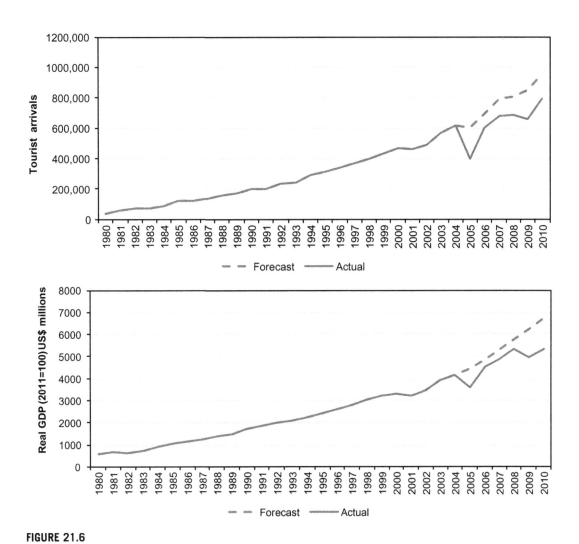

FIGURE 21.6

Actual versus forecast tourist arrivals and GDP, Maldives, 1980–2010.

Authors' compilation using World Bank (2019).

impact of the global financial crisis on TA (-29.6%; -20.5%). However, the tsunami's impact on overall GDP is (-23.93%) slightly less than the impact of the global financial crisis on GDP (-25.7%; -27.3%).

One of the major challenges the Maldives had to face in restoring tourist confidence in traveling to the Maldives followed by the tsunami was to clear the perception that the Maldives was severely impacted by the tsunami and to prevent cancellation of tours. The massive investments on tourism promotional activities and update of (correct) status of the damage around the globe by the Government of Maldives, such as conducting roadshows and promotional workshops, representing

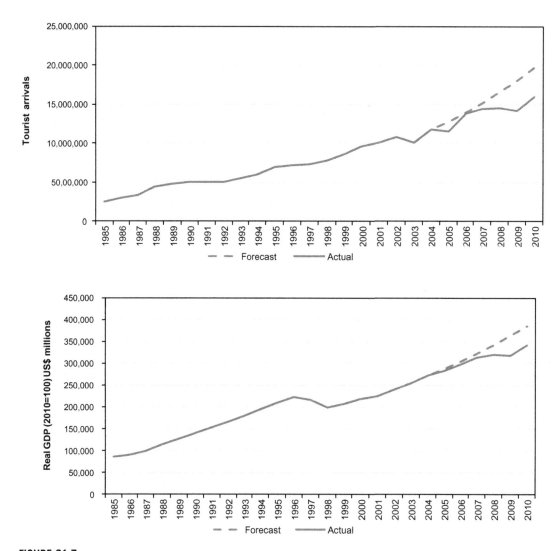

FIGURE 21.7

Actual versus forecast tourist arrivals and GDP, Thailand, 1985–2010.

Authors' compilation using World Bank (2019).

international trade shows and promotional fairs, and advertising on international media significantly contributed to the notable recovery performance in the tourism industry (Carlsen, 2006; Carlsen & Hughes, 2008).

As can be seen in Fig. 21.7, the Thailand tourism industry was visibly interrupted by the tsunami. Nevertheless, by 2006 the industry fully recovered until the period the tourism downturn began due to the global financial crisis. Thailand's GDP was also mildly affected by the tsunami.

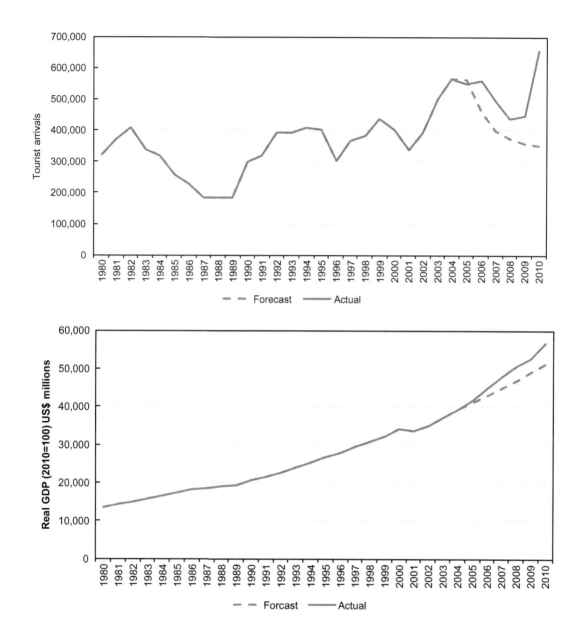

FIGURE 21.8

Actual versus forecast TA and GDP, Sri Lanka, 1980–2010. *TA*, Tourist arrivals.

Authors' compilation using World Bank (2019).

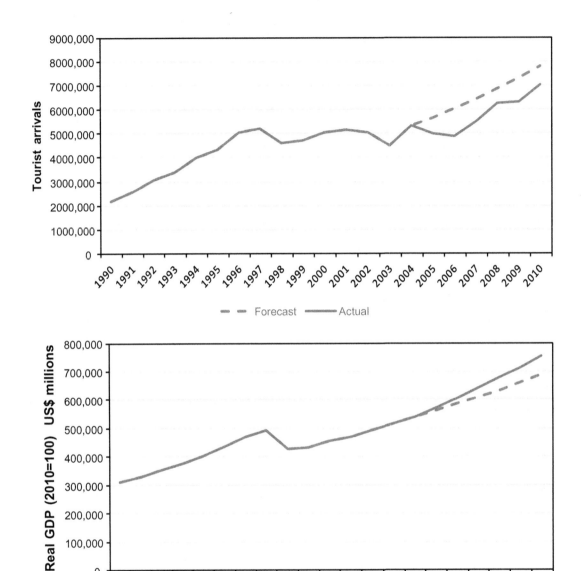

FIGURE 21.9

Actual versus forecast tourist arrivals and GDP, Indonesia, 1990–2010.

Authors' compilation using World Bank (2019).

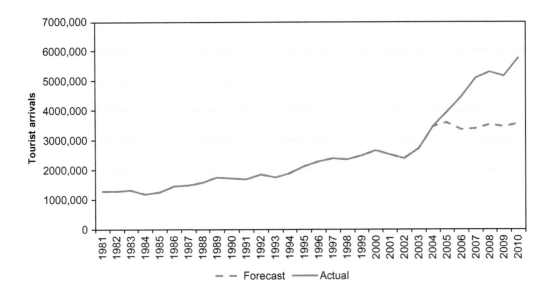

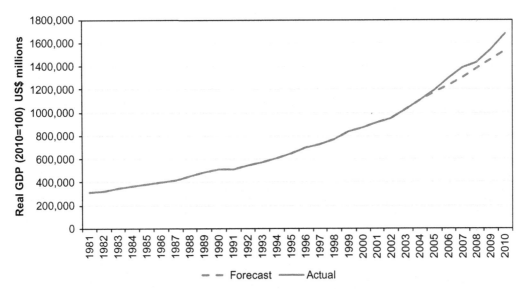

FIGURE 21.10

Actual versus forecast tourist arrivals and GDP, India, 1981–2010.

Authors' compilation using World Bank (2019).

This is clearly reflected in percentage forecast errors of −10.1% and −1.1% for tourism and GDP, respectively. Nevertheless, the global financial crisis had a more severe impact on Thailand's tourism industry and GDP than that of the tsunami. Tourism Authority of Thailand initiated an aggressive marketing campaign followed by the tsunami, such as organizing familiarization trips for news

Table 21.5 Percentage Forecast Error of Tourist Arrivals and GDP, 2005–10.

	Tourist Arrivals			Real GDP US$ Millions		
	Actual	Forecast	Percentage Forecast Error	Actual	Forecast	Percentage Forecast Error
(1)	(2)	(3)	(4)	(5)	(6)	(6)
	Maldives					
2005	395,000	598,421	− 51.5	3583.077	4440	− 23.9
2006	602,000	690,000	− 14.6	4518.672	4849	− 7.3
2007	676,000	794,510	− 17.5	4867.236	5267	− 8.2
2008	683,000	803,463	− 17.6	5328.910	5722	− 7.4
2009	656,000	850,102	− 29.6	4943.692	6215	− 25.7
2010	792,000	954,580	− 20.5	5302.857	6752	− 27.3
	Thailand					
2005	11,567,000	12,791,574	− 10.6	283,767	286,954	− 1.1
2006	13,822,000	13,940,902	− 0.9	297,865	304,234	− 2.1
2007	14,464,000	15,193,498	− 5.0	314,055	322,553	− 2.7
2008	14,584,000	16,558,640	− 13.5	319,474	341,976	− 7.0
2009	14,150,000	18,046,440	− 27.5	317,267	362,568	− 14.3
2010	15,936,000	19,667,920	− 23.4	341,105	384,401	− 12.70
	Sri Lanka					
2005	549,308	561,251	− 2.2	41,633	40,966	1.6
2006	559,603	460,313	17.7	44,826	42,826	4.5
2007	494,008	400,040	19.0	47,873	44,770	6.5
2008	438,475	371,506	15.3	50,721	46,802	7.7
2009	447,890	357,290	20.2	52,516	48,926	6.8
2010	654,476	350,014	46.5	56,726	51,147	9.8
	Indonesia					
2005	5,002,000	5,671,646	− 13.4	571,205	562,351	1.6
2006	4,871,000	6,045,399	− 24.1	602,627	585,150	2.9
2007	5,506,000	6,443,782	− 17.0	640,863	608,873	5.0
2008	6,234,000	6,868,417	− 10.2	679,403	633,558	6.7
2009	6,324,000	7,321,036	− 15.8	710,852	659,244	7.3
2010	7,003,000	7,803,481	− 11.4	755,094	685,971	9.2
	India					
2005	3,918,610	3,601,702	8.1	1,193,873	1,168,586	2.1
2006	4,447,167	3,354,154	24.6	1,290,108	1,234,465	4.3
2007	5,081,504	3,380,831	33.5	1,388,940	1,304,059	6.1
2008	5,282,603	3,509,673	33.6	1,431,813	1,377,576	3.8
2009	5,167,699	3,462,051	33.0	1,544,380	1,455,238	5.8
2010	5,775,692	3,542,207	38.7	1,675,615	1,537,277	8.3

Authors' compilation using World Bank. World Bank open data. (2019).

reporters and travel writers and hosting exhibitions and meetings in tsunami-affected areas. These initiatives resulted in Phuket, one of the severely affected popular tourist destinations in Thailand, to win the Star of Travel Award for the best destination in Southeast Asia in 2005 (Rittichinuwat, 2006).

Since the early 1980s TA to Sri Lanka showed significant fluctuations due to the civil war that erupted in the country. However, the tourism industry was starting to gain momentum as a result of the cease-fire agreement between the Sri Lankan government and the Liberation Tigers of Tamil Eelam during 2002–05 when the tsunami hit in 2004. Therefore the drop in TA immediately after the tsunami was mainly a result of the tsunami rather than the security-related issues in the country.

Although the Sri Lankan tourism industry rebound in 2006, the global financial crisis, breakdown of the cease-fire agreement, and intensified war adversely affected the tourism industry in the late 2000s until the war ended in mid-2009. Due to war-related security concerns, the tourism industry was playing a relatively low key in Sri Lanka's economy (about 4.6% of GDP, see Table 21.1). Therefore the negative impact of the tsunami on tourism did not reflect as severe damage on the country's GDP. The actual GDP reported higher than the forecasted GDP.

As a recovery strategy, Sri Lanka mainly focused on restoring infrastructure that was massively destroyed by the tsunami to facilitate tourists. For example, duty-free import privileges and loan facilities were given to small- and medium-scale enterprises operating in the tourism industry. These initiatives considerably contributed to restoring the infrastructure facilities to serve incoming tourists (Jayasuriya, Steele, & Weerakoon, 2006). In addition, the "Bounce Back Sri Lanka" advertising campaign that highlighted a diverse range of tourist attractions, such as wildlife and archaeological sites in the country and hosting an international surfing competition in 2005 in the Eastern province of Sri Lanka, helped in rebuilding the confidence of tourists to visit the country (Robinson & Jarvie, 2008).

As can be seen in Fig. 21.9, the Indonesian tourism industry was clearly impacted by the tsunami. This is evident from lower actual TA compared to those of forecasted values and negative percentage forecast errors (e.g., −13.4%; −24.1%). The country showed signs of recovery after 2007. Nevertheless, this downturn in TA did not adversely affect the overall GDP in Indonesia. The actual GDP values were even higher than the forecasted values. This could be because earnings from tourism contributed only about 4% of GDP in 2004.

As shown in Fig. 21.10, the actual out-of-sample TA are higher than the forecasted values for the period 2005–10 in India. This indicates that tsunami had little or no impact on the total number of international TA to India. Similarly, the actual GDP values are consistently higher than the forecasted GDP values, indicating the GDP performance was uninterrupted in India due to tsunami. This is as expected as the tsunami hit only some areas of South India, which is a small proportion of the whole of India and earnings from tourism contribute to only about 2% of GDP in India (see Table 21.1). The percentage forecast errors further confirm that neither TA nor GDP were severely impacted by the tsunami in India.

21.5 CONCLUDING COMMENTS

Several countries in South Asia and Southeast Asia were devastated by the giant tsunami waves triggered by a massive earthquake on the east coast of Sumatra on December 26, 2004. This chapter investigated the impact of the 2004 Asian tsunami on TA and GDP in five tsunami-affected Asian countries: Maldives, Thailand, Sri Lanka, Indonesia, and India. For this purpose, we have used several econometric and time series techniques such as unit root tests, bounds test for cointegration, Granger causality, and univariate Box—Jenkins ARIMA method. To gauge tsunami's effect on TA and GDP, we used international TA and GDP time series up to the year 2004 to identify individual ARIMA models. We then used the ARIMA coefficient estimates to forecast the TA and GDP for the out-of-sample period 2005—10. The forecasted values are then compared with the actual out-of-sample TA and GDP values to assess the impact and recovery of TA and GDP following the tsunami.

The results indicated that all countries experienced a notable drop in TA followed by the tsunami. However, smaller nations such as the Maldives and Sri Lanka showed greater responsive in TA due to tsunami compared to geographically dispersed countries, such as Thailand, Indonesia, and India. Almost all countries embarked on massive promotional and advertising campaigns to restore the tourism industry after the tsunami and increased investments in rebuilding tourism-related infrastructure that was severely damaged by the tsunami. These strategies immensely contributed to restoring the tourism industry within a maximum of 24 months in all countries and the upward trend continued till the 2007—08 global financial crisis, except in Sri Lanka. In Sri Lanka the civil conflict continued to adversely affect the tourist confidence in visiting Sri Lanka until 2009 when the civil war came to an end. It was evident in our analysis that the speed of recovery of the tourism industry from a natural disaster such as a tsunami varies from country to country and highly dependent on the efficiency of disaster management strategies in the short run and the effectiveness of postdisaster tourism promotion strategy.

When considering the overall impact of the tsunami on the local economies, GDP in the Maldives showed the most visible sensitivity toward the tsunami. This is because the Maldives GDP is highly dependent on the tourism industry. The 2004 tsunami's impact on GDP in Thailand was minor. However, in Sri Lanka, Indonesia, and India, the actual GDP exceeded the forecasted GDP, indicating that tsunami had limited or no visible impact on GDP. In general, our results suggest that the economies that are highly dependent on one industry such as tourism are more vulnerable to disasters that affect that industry than the economies that are less dependent on that particular industry.

REFERENCES

Amaghionyeodiwe, L. A. (2012). A causality analysis of tourism as a long-run economic growth factor in Jamaica. *Tourism Economics, 18*(5), 125—133.

Balaguer, J., & Cantavella-Jordà, M. (2002). Tourism as a long-run economic growth factor: The Spanish case. *Applied Economics, 34*, 877—884.

Bandara, J. S., & Naranpanawa, A. (2007). The economic effects of the Asian tsunami on the 'tear drop in the Indian Ocean': A general equilibrium analysis. *South Asia Economic Journal, 8*(1), 65—85.

BBC News. *At-a-glance: Tsunami economic impact.* (2005a). Available from <http://news.bbc.co.uk/2/hi/business/4154277.stm-Sri%20Lanka> Retrieved 15.08.19.

BBC News. *Tourism plan agreed after tsunami.* (2005b). Available from <http://news.bbc.co.uk/2/hi/asia-pacific/4226975.stm> Retrieved 11.11.19.

Belloumi, M. (2010). The relationship between tourism receipts, real effective exchange rate and economic growth in Tunisia. *International Journal of Tourism Research, 12*, 550−560.

Blanke, J., & Chiesa, T. (2007). *The travel & tourism competitiveness report*. Geneva: World Economic Forum.

Box, G. E. P., & Jenkins, G. M. (1976). *Time series analysis forecasting and control*. Oakland, CA: Holden-Day.

Carlsen, J. (2006). Post-tsunami tourism strategies for the Maldives. *Tourism Review International, 10*(1), 69−79.

Carlsen, J., & Hughes, M. (2008). Tourism market recovery in the Maldives after the 2004 Indian Ocean tsunami. *Journal of Travel and Tourism Marketing, 23*, 2−4.

Central Bank of Sri Lanka. (2005). *Annual report*. Colombo: Central Bank of Sri Lanka.

Cosgrave, J. (2005). *Tsunami evaluation coalition: Initial findings*. Tsunami Evaluation Coalition.

Dickey, D., & Fuller, W. (1979). Distribution of the estimators for autoregressive time series with a unit root. *Journal of the American Statistical Association, 74*, 427−431.

Durbarry, R. (2004). Tourism and economic growth: The case of Mauritius. *Tourism Economics, 10*(4), 389−401.

Huang, J. H., & Min, J. C. (2002). Earthquake devastation and recovery in tourism: The Taiwan case. *Tourism Management, 23*, 145−154.

Hye, Q. M. A., & Khan, R. E. A. (2013). Tourism-led growth hypothesis: A case study of Pakistan. *Asia Pacific Journal of Tourism Research, 18*(4), 303−313.

IMF. *Preliminary assessment of the macroeconomic impact of the tsunami disaster on affected countries, and of associated financing needs*. (2005). Available from <https://www.imf.org/external/np/oth/2005/020405.htm> Retrieved 25.08.19.

Jayasuriya, S., Steele, P., & Weerakoon, D. (2006). *Post-tsunami recovery: Issues and challenges in Sri Lanka, .* ADB Institute research paper series: (Vol. 71). Tokyo: Asian Development Bank Institute.

Karatani, Y. (2008). Tourism industry losses and recovery process from the Indian Ocean tsunami—A case of the affected tourist destination in Southern Thailand. In *Paper presented at the 14th World Conference on Earthquake Engineering*, Beijing, China.

Katircioglu, S. T. (2010). Testing the tourism-led growth hypothesis for Singapore—An empirical investigation from bounds test to cointegration and Granger causality tests. *Tourism Economics, 16*(4), 1095−1101.

Kreishan, F. M. (2011). Time-series evidence for tourism-led growth hypothesis: A case study of Jordan. *International Management Review, 7*(1), 89−93.

Kwiatkowski, D., Phillips, P. C. B., Schmidt, P., & Shin, Y. (1992). Testing the null hypothesis of stationarity against the alternative of a unit root. *Journal of Econometrics, 54*(1−3), 159−178.

Malik, S., Chaudhry, I. S., Sheikh, M. R., & Farooqi, F. S. (2010). Tourism, economic growth and current account deficit in Pakistan: Evidence from cointegration and causal analysis. *European Journal of Economics, Finance and Administrative Sciences, 22*, 1450−2275.

Mazzocchi, M., & Montini, A. (2001). Earthquake effects on tourism in central Italy. *Annals of Tourism Research, 28*, 1031−1046.

Mendoza, C. A., Brida, J. G., & Garrido, N. (2012). The impact of earthquakes on Chile's international tourism demand. *Journal of Policy Research in Tourism, Leisure and Events, 4*, 48−60.

Ministry of Tourism Maldives. (2004). *Tourism statistics*. Maldives: Statistics Section-Ministry of Tourism.

Mishra, P. K., Rout, H. B., & Mohapatra, S. S. (2010). Causality between tourism and economic growth: Empirical evidence from India. *European Journal of Social Sciences, 18*(4), 518−527.

Narayan, P. (2005). The saving and investment nexus for China: Evidence from cointegration tests. *Applied Economics, 37*(17), 1979−1990.

Obadiah, N. K., Odhiambo, N. M., & Njuguna, J. M. (2012). Tourism and economic growth in Kenya: An empirical investigation. *International Business & Economics Research Journal, 11*(5), 517−528.

Pesaran, H. M., & Pesaran, B. (1997). *Working with Microfit 4.0: Interactive econometric analysis*. Oxford: Oxford University Press.

Pesaran, M. H., & Shin, Y. (1999). An autoregressive distributed lag modeling approach to cointegration analysis. In S. Strom, A. Holly, & P. Diamond (Eds.), *Centennial volume of Rangar Frisch*. Cambridge: Cambridge University Press.

Pesaran, M. H., Shin, Y., & Smith, R. J. (1996). *Testing for the 'existence of a long run relationship'*. Cambridge: University of Cambridge, Department of Applied Economics Working Paper no.9622.

Pesaran, M. H., Smith, R. J., & Shin, Y. (2001). Bounds testing approaches to the analysis of level relationships. *Journal of Applied Econometrics, 16*, 289−326.

Phillips, P., & Perron, P. (1988). Testing for a unit root in time series regression. *Bimetrika, 75*, 335−346.

Ramalanjoana, B. N. (2006). Impact of 2004 tsunami in the islands of Indian Ocean: Lessons learned. *Emergency Medicine International, 2011*, 1−3.

Rittichinuwat, B. N. (2006). Tsunami recovery: A case study of Thailand's tourism. *Cornell Hotel and Restaurant Administration Quarterly, 47*(4), 390−404.

Robinson, L., & Jarvie, J. K. (2008). Post-disaster community tourism recovery: The tsunami and Arugam Bay, Sri Lanka. *Disasters, 32*(4), 631−645.

Sharpley, R. (2005). The tsunami and tourism: A comment. *Current Issues in Tourism, 8*(4), 344−349.

Sri Lanka Tourism Development Authority (SLTDA). (2019). *Tourism growth trends 1970 to 2018*. Colombo: Sri Lanka Tourism Development Authority.

Srinivasan, P. K., Kumar, S., & Ganesh, L. (2012). Tourism and economic growth in Sri Lanka. *Environment and Urbanization, 3*(2), 397−405.

United Nations Office for Disaster Risk Reduction (UNDRR). (2005). *Indian Ocean earthquake—Tsunami 2005*. UNDRR.

United Nations World Tourism Organization (UNWTO). (2006). *Tourism highlights*. Madrid: UNWTO.

United Nations World Tourism Organization (UNWTO). (2017). *Tourism highlights: 2017*. United Nations World Tourism organization (UNWTO).

World Bank. *World Bank open data*. (2019).

World Tourism Organization (WTO). *South Asia: Tsunami recovery—One year on*. (2005). Available from <https://reliefweb.int/report/sri-lanka/south-asia-tsunami-recovery-one-year> Retrieved 01.09.19.

Wu, L., & Hayashi, H. (2013). The impact of the Great East Japan Earthquake on inbound tourism demand in Japan. In *Working paper 21*. Kyoto University.

EFFECT OF DROUGHT ON DEVELOPMENT OF CHILDREN: FIELD OBSERVATIONS FROM THE DROUGHT-PRONE DISTRICT OF DECCAN PLATEAU OF SOUTHERN INDIA

G.S. Srinivasa Reddy, C.N. Prabhu, Lenin Babu Kamepalli and S. Jagadeesh

Karnataka State Natural Disaster Monitoring Centre, Bengaluru, India

22.1 TITLE INTRODUCTION TO CREEPING DISASTER—DROUGHT

The Karnataka state is prone to different kinds of natural disasters such as droughts, floods, cyclones, hailstorms, landslides, and earthquakes. Drought is usually a widespread phenomenon and has at times affected 92% (as in the year 2003) of the blocks in the state. Floods are hydrometeorological disasters that have affected crops and human settlements in a few districts of the Karnataka state. Cyclones and coastal erosion are restricted to coastal districts, while landslides are frequent in the areas of Western Ghats. Coastal and Malnad regions of the state have observed incidents of seismic activity. Of the various types of disasters, drought found to cause maximum damage that also impacts large geographical areas within the state. The extent of the affected area, loss of property, and socioeconomic losses due to different disasters in the state are in the following order: droughts > floods > hailstorms > cyclones > landslides > earthquakes (KSDMA, 2016).

A drought can be short-lived, as few as 15 days or can last for months or years, having a substantial effect on the agriculture, human population, livestock, and ecosystem in the affected region (NDMA, 2010). The small and marginal farmers, who are already facing food and livelihood insecurities, experience the deleterious effects of drought (UNICEF, 2016). Seven types of droughts are recognized in India (DAC&FW, 2016; NDMA, 2010):

1. Meteorological drought—Short-lived but usually precedes all other kinds of droughts.
2. Hydrological drought—If the runoff is less than 75% of the normal and results in a shortage of water supplies in surface and subsurface.
3. Agricultural drought—Meteorological and hydrological droughts usually trigger this type and directly impact the crop yields. Deficiency of soil moisture during the vegetative period leads to wilting and reduced biomass and yields. Agricultural drought is defined as a period of 4 consecutive weeks (of severe meteorological drought) with a rainfall deficiency of more than

Economic Effects of Natural Disasters. DOI: https://doi.org/10.1016/B978-0-12-817465-4.00022-4

50% of the long-term average or with a weekly rainfall of 5 cm or less from June to September. About 80% of India's total crop is planted during this period or six such consecutive weeks during the rest of the year (called as Rabi season).

4. Soil moisture drought—Water supply to soil profile is lower, and the evaporative demand is more due to high temperature and winds with low relative humidity.

5. Socioeconomic drought—Reduction in the availability of food/income loss due to the failure of crops. This affects the socioeconomic fabric of the society and leads to compulsive migrations from village to cities in search of greener pastures for their food and livelihood.

6. Ecological drought—This affects the ecosystem, eco-environment, and its production system significantly as a consequence of distress-induced environmental damage.

7. Famine: If droughts prolong for more extended periods, there will be substantial scale deficiency of food/feed/water that leads to significant scale mass starvation causing deaths.

A drought year is defined by the Indian Meteorological Department (IMD) as a year in which the overall rainfall deficiency is more than 10% of the Long Period Average value and if more than 20% of its area is affected by drought conditions, either moderate or severe or combined moderate and severe. When the spatial coverage of drought is more than 40%, it is known as a severe drought year (DAC&FW, 2016).

Karnataka is considered the most drought-prone state after Rajasthan State and entire rainfed area in the state regularly affected by drought. Nearly 80% of blocks in the state are drought-prone (Fig. 22.1). According to the Ministry of Agriculture and Farmers Welfare (MoAFC&W), Government of India (GoI) 16 districts of the state, the majority of which are from north interior Karnataka, experienced drought for a period of 10 years during the last 15 years (2001−15). Drought causes crop loss, drinking water scarcity, fodder scarcity, and unemployment to rural agricultural laborers. Increased incidence of droughts reduces crop productivity, affecting nutrition and consequently, resistance to infections.

Drought has its implications on several aspects such as infant mortality, maternal mortality, malnutrition among children and women, high incidence of childhood diseases, and inadequacies in water supply and sanitation and ultimately results in a low Human Development Index (HDI) ranking (PPMSD, 2014; Shivashankar & Ganesh Prasad, 2015). Due to a very low HDI, some districts are categorized as Aspirational Districts by the GoI (NITI, 2018a,b) with some of these identified by UNICEF (2016) for their intervention in the areas of health, education, nutrition, water, sanitation, and child protection as well. An in-depth understanding about direct and indirect impacts of drought on children and their health, nutrition, protection, education, and other related issues may provide insights for concerned stakeholders to devise an effective recovery, mitigation measures, and intervention for achieving long-term climate and disaster resilience.

22.1.1 THE STUDY

The present study is envisaged to develop a knowledge base that could help the government or other agencies address drought and other climate-related risks in a more proactive, prepared manner to help minimize and prevent immediate and long-term hardship for the most vulnerable, that is, children, through better planning. Specific objectives include the following:

- Assess the impact of drought/drought-like situations in Yadgir district to understand its impact and related manifestations on communities, particularly children and women, in terms of access to essential services and coping mechanisms.

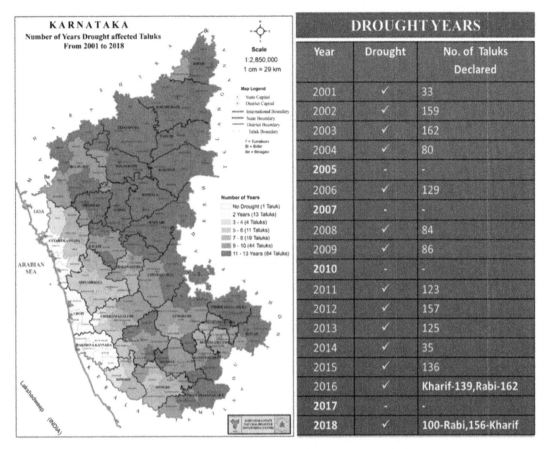

	DROUGHT YEARS	
Year	Drought	No. of Taluks Declared
2001	✓	33
2002	✓	159
2003	✓	162
2004	✓	80
2005	-	-
2006	✓	129
2007	-	-
2008	✓	84
2009	✓	86
2010	-	-
2011	✓	123
2012	✓	157
2013	✓	125
2014	✓	35
2015	✓	136
2016	✓	Kharif-139,Rabi-162
2017	-	-
2018	✓	100-Rabi,156-Kharif

FIGURE 22.1

Incidences of drought in Karnataka.

KSNDMC.

- Assess the impact of drought/drought-like situations in Yadgir district and its related manifestation on five major program priorities: health, education, nutrition, water and sanitation, and child protection.
- Critical appreciation of governance systems at district level to identify system strengthening requirements in terms of planning, climate adaptation, preparedness, and response.
- Provide recommendations to the administration for its current routine and future programing.

22.1.2 **METHODOLOGY**

Yadgir district is selected as the study area as data from Karnataka State Natural Disaster Monitoring Centre and Government of Karnataka indicate that among all other districts, this district is prone to repetitive droughts. Further, various indicators of the HDI place this district at the bottom of the human development scale.

Of the three blocks in the district, Shahapur, Surapura, and Yadgir, indicators such as the extent of area prone to drought, percentage of population of Scheduled Castes and Scheduled Tribes, lower level of social infrastructure, and extent of out-migration indicate that Yadgir block is the most vulnerable to drought and other natural disasters. Hence, Yadgir block was selected for the study of the impact of drought. Selective random selection of villages was adopted for the survey to collect baseline information (UNICEF, 2016). Assuming drought vulnerability is dependent on several factors, such as economic status, social infrastructure level, and diversification of livelihood options, information was collected on two levels:

- social infrastructure level [Anganwadi Centers (AWCs), schools, drinking water facilities, etc.] and
- at household level for details such as economic status, livelihood options, household assets, and demographic details.

Assuming that vulnerability is dependent on economic status, a random selection of the household was made, based on their economic status, for instance, landless, marginal, and small farmers in the village. Assistance of focal points of Integrated Child Development Schemes—AWCs was taken to identify the households in villages. At the household level, information was collected on the impact of drought on various aspects of day-to-day life, for instance, changes in the food basket and availability of water (Rajpoot & Kumar, 2019; Roy & Hirway, 2007; UNISDR, 2018; WMO, 2017). Efforts were made to elicit responses from female members of the household as they are typically more involved in managing these issues.

Sample size: Randomly selected 30 villages formed the study area in across the Yadgir block. Different schedules, seven in number, were used to collect information on vulnerability and coping measures in each village, as given in Table 22.1.

Table 22.1 Sampling Details.

Sl. No.	Schedule/ CheckList	Information Sought	Mode of Information Collection	Total Number
1	Village information	Physical and social infrastructure of village	FGD and key informant interaction	30 Villages
2	Anganwadi center information	Details of ICDS services provided and beneficiairies	Questionnaire	30 AWC
3.	School information	Availability of water for drinking and WASH	Questionnaire	30 Schools
4.	Household information	Impact of drought on children	Questionnaire	30 Households
		Impact of drought on adolescent girls	Questionnaire	30 Households
		Impact of drought on pregnant women	Questionnaire	30 Households
		Impact of drought on lactating women	Questionnaire	30 Households
Total schedules				**210**

ICDS, *Integrated Child Development Services.*

22.1.3 DESCRIPTION OF STUDY VILLAGES

Of the district population, about 23.3% and 12.5% belong to Scheduled Castes and Scheduled Tribes, respectively, and around 87% of the total district population is rural, indicating the importance of the farm economy and its vulnerability to drought. The child population constitutes about 16% of the total district population, and in rural areas, the proportion of the child population is slightly higher than that in urban areas. Of the three blocks, Shahapur, Surapura, and Yadgir, Yadgir has the lowest HDI and also lower percentage of irrigation facilities compared to the other two blocks.

In all 30 selected study villages, rainfed cultivation is the primary occupation. Villages have some infrastructure facilities, such as electricity, borewell-based piped water supply (PWS) system, preliminary education, and Anganwadi Centers (childcare centers). In some villages, primary health centers, flour mills, retail shops, and post offices are available. Interactions with key informants (representatives of local self-government agencies) indicate that there has been a steady decline in groundwater table over the years. Availability of drinking water in summer months is a common concern across all villages. Outer migration for manual work is the most preferred option during the drought years.

22.1.4 KEY OBSERVATIONS

Drought has a devastating impact on the economy in general and more particularly in regions where the first-order economic activities such as agriculture constitute the primary source of livelihoods, as is the case with study villages. During drought seasons, agricultural activities do not take place in full capacity. They cause a decline in the demand for farm labor and, consequently, income for those households, the farm labor of which is an important source of income. A decline in income coupled with drought conditions triggers several chain reactions at the household level, as described in the following sections.

22.1.5 DROUGHT AND LIVELIHOOD

Agriculture is the primary source of income in all the villages studied. About 75% of sampled households own agricultural land, with an average landholding of 0.5 Ha. Income from farm labor also constitutes an important source for the sampled households. Only 35% of them have irrigation facilities, indicating the importance of having a normal monsoon for their livelihood. In the absence of alternate employment avenues during drought, migration to urban centers remains a viable option. Instances of the drop-in farm yield up to 51% and reduced demand for farm labor up to 60% were attributed to drought conditions in the sampled villages (Table 22.2). Besides, livestock operations (both for milch animals, sheep and goat rearing) were also severely affected by drought, due to fodder shortages. Fodder banks established by the state government were of little use because of the distance and help from friends and relatives is primary means of assistance to tide over the drought.

22.1.6 NUTRITION

Impact of drought on nutrition has a double-edged impact, that is, (1) reduced household income and consequently reduced purchasing power and (2) drought-induced reduction in the availability

Table 22.2 Gainful Employment From Farm Operations.

No. of Households	Gainful Employment (In Months)
6	0
48	6
50	10

Primary survey.

of food materials in villages. About 7% of households were immune to drought-induced changes while all others felt the impact of drought at varying degrees. The impact was moderate for 26% of sampled households and extreme for 2% of respondents. Majority of the respondents were aware of the potential long-term adverse impacts of inadequate nutrition on children's health, migration is considered as a viable option, though it affects education [CD1]. On gender aspects, 58% of respondents opined that drought affects women folk more than the male members of the family. Electronic media was an important source of information about drought, followed by word of mouth. Agricultural extension centers and other such institutions were not considered for drought-related information. Only 53% of households have participated in drought relief programs initiated by the government such as Mahatma Gandhi Rural Employment Guarantee Scheme (MGNREGS), while 35% respondents preferred to take help from their friends and relatives to tide over the drought.

Cutting down the expenses through a shift to low-cost food items followed by distress sale of livestock were consequences of drought. Reducing food intake and exploring the avenues of child labor were also considered by the respondents. About their perceptions regarding various drought mitigation measures, sinking of new borewells for drinking water and desiltation of tanks were welcomed by the majority. Services by agricultural extension services were appreciated by only 46% of respondents only.

22.1.7 HEALTH

In Yadgir, at the district level, infant mortality rate (IMR) is 48, which is higher than the IMR rate of Karnataka at 35. Similar is the maternal mortality rate (MMR) at 186, higher than the state average of 144. In terms of health infrastructure also, Yadgir district is relatively less developed. It has the lowest number of health subcenters at 169 in the state, while Belagavi district has highest at 620. The district has 42 primary health centers, ranking third from the bottom, while Belagavi, Mysore, and Tumkur districts have more than 140 PHCs. Some of the villages do not even have bare minimum health infrastructure, and villagers are forced to go to neighboring villages for medical help. In such context, drought and its associated effects such as inadequate nutrition and shortage of water for personal hygiene are bound to have repercussions.

22.1.7.1 Water and sanitation and child protection

Borewell-based water supply system is the primary source of water even in a typical year for the majority (98%) of the respondents with public point is the source of water collection. However, in

Ramasamudra, Jenikere, Kaulur, Nagalapura, and Venkateshapura villages, tanker-based water supply was required even in normal monsoon year. During a drought year the borewell-based PWS was able to cater to only 40% of dependents. Thus water availability and access becomes a concern for about 60% of respondents and are forced to source another borewell. Failure of borewell that was closer to home has resulted for 46% of households to spend additional time up to 30 minutes every day, and about 3% of households spend more than half an hour additional time every day for collection of water (Fig. 22.2). For only 6% of respondents, there was no need to spend additional time on water collection even in a drought year. Such a ground situation is not conducive and certainly a concern for achieving norms recommended by National Rural Drinking Water Programme (75 L per capita per day by 2021). It was also found that even during a normal monsoon year, water availability falls short by about 18% and during the drought year, the situation becomes far worse with the gap between demand and supply widening to the tune of 31%, that is, about 100 L shortage for a family of five members.

Due to inadequate availability of water during drought conditions, households were forced to reduce their demand for water. As the demand for drinking, cooking, and other functions are inelastic, a compromise has to be made in the quantity of water used for personal hygiene. During the survey, it was enquired about the extent of forced reduction in water use for WASH activities. About 16% of households responded that drought conditions had forced them to reduce water for WASH activities, but the extent was marginal only. For 58%, reduction of water for WASH activities was considerable while the significant reduction was reported from about 18% households. Health issues such as skin diseases due to poor personal hygiene observed in 17 households (Table 22.3). Several households have reported problems related to the digestive system due to poor quality of drinking water.

Another significant impact of water shortage observed was a reversal to open defecation practice. Due to inadequate water availability for the use of individual household toilets, households have preferred to revert to the open defecation practices. Among the respondents, about 29% of households have stopped using the toilets facility and about 45% of households have drastically reduced the extent of toilet use. For 21% households, there was a marginal impact on the use of the toilet facility (Table 22.4).

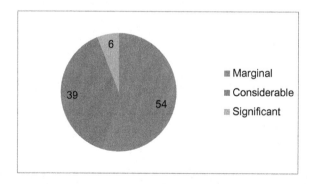

FIGURE 22.2

Households affected by a decline in water quality (in %).

Table 22.3 Water for WASH/Personal Hygiene.

Water for Hygiene	Marginal	Considerable	Significant
No. of households	19	69	22
Percentage of households	16	58	18

Primary survey.

Table 22.4 Impact of Inadequate Water Availability on Reversal to open defecation practice .

Reversal to ODF	No Impact	Marginal	Considerable	Significant
No. of households	3	25	54	35
Percentage of households	3	21	45	29

Primary survey.

22.1.8 EDUCATION

Regarding the levels of literacy, Yadgir district stands at the bottom with the lowest literacy rank in the state. Table 22.5 presents a comparison of literacy levels in different blocks of Yadgir with the state average. A comparison of literacy in various blocks of Yadgir along with district and state average is given in Table 22.5. Drought conditions may have played a significant role in the literacy levels of the district.

Integrated Child Development Services (ICDS) has a strong presence in the district. AWCs functions as an outlet for a package of six ICDS services that aim for holistic child development in a life cycle mode. Starting from pregnant women to lactating women to children from birth to 6 years of age, and adolescent girls, this package broadly includes health, education and nutritional support, community mobilization, and nonformal preschool education for children in the age-group of 3−6 years. From the child education perspective, AWC has a critical role in children readiness and dimension of school readiness (Fig. 22.3). During the survey an attempt was made to study the ground reality of AWC functioning and its infrastructure. For those AWCs working out of school premises, water is not a problem, but for other AWCs, collection of water from the public post is a concern and storage of collected water also needs improvement. In addition to water availability, lack of adequate playground is also an issue that forces children to remain indoors only during their stay in AWC.

In addition to AWCs, schools were also surveyed in the sample villages to find out the status of facilities such as toilets, playground, and drinking water. In 30 surveyed schools, average teacher to pupil ratio found to be 38. The number of classrooms varies from just a single room (Bachavara Tanda) with a single teacher to a maximum of 13 rooms with 15 teachers in Bandalli village. The average distance that a student has to travel to reach school and from school was less than 500 m.

Table 22.5 Literacy Levels.

Block	Male (%)	Female (%)	Total (%)
Shahapur	61.63	41.25	51.50
Shorapur	66.42	43.07	54.82
Yadgir	58.47	39.78	49.06
District average	62.25	41.38	51.83
State average	82.47	68.08	75.36

Primary survey.

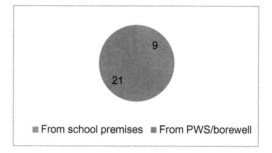

FIGURE 22.3

Source of water for Anganwadi Centers.

Regarding the condition of physical structures, in 9 villages, the structures were in good conditions, but in about 12 villages, the structures need immediate attention (according to the opinion of teachers). Of the 30 schools, 4 schools found to be in requirement of toilet facilities for students as these schools have toilets for only teachers. For 7 schools out of the 30 surveyed schools, PWS was the source of water supply. For nine schools, dug wells were the source of water supplies. Borewell within the premises of schools was the source for about 13 schools. It was opined during the survey that the existing sources of water were sufficient to meet the water demands of the school and that recently there were no incidences of water shortages. However, apprehensions were made that if schools were to function in full swing during summer months, water availability would become a major issue. An effort made to list the perceptions of community that influence child education.

22.1.9 WAY FORWARD

Children are the building blocks of future societies and nations, and they are highly vulnerable. Drought affects livelihood systems in rural economies. Despite various measures and support by the government, requirements of an affected community cannot be met completely and force affected communities to suffer from the scarcity of water, fodder for livestock, credit/finances, etc. These factors trigger migrations from villages to urban areas for their livelihood, which in turn, affect the education, safety, and security of the children. Results indicate that drought is adversely affecting nutrition, health, and education. It is resulting in stress and psychosocial illness in the study villages. Since all the elders are busy and preparing to face the disaster situation, the children are usually neglected or overlooked regarding food, medication, sanitation, and cleanliness.

Separation of the children during a disaster from the family causes insecurity, trauma, and increased vulnerability. Children are separated either by force or with the consent of the family members and taken to other places, which may not be congenial for living. Girl children are to be treated with utmost care since they are more vulnerable to abuse. During the disaster the children are treated differently based on gender even in the family as regards food, health care, sanitation, and clothing. Children with disabilities need special care during such stressed conditions. To reduce the risks of disasters on children, suitable measures that address their safety and security should be undertaken.

22.1.10 MEASURES TO ADDRESS NEEDS UNDER DROUGHT RISK REDUCTION

To save vulnerable children group the programs that address all the issues of the children are to be undertaken by the government, philanthropic, private health, and academic institutions during disaster. These programs need to consider the following points.

22.1.10.1 Water and sanitation

Access to safe drinking water and improved sanitation to safeguard the children from the spread of waterborne, vector-borne, and sanitation-based diseases (Department of Public Works, Panchayat Raj, and Department of Health and Family Welfare, Directorate of Social Security).

22.1.10.2 Education

Usually in drought/disaster, the children are not sent to schools by parents. Hence, proper counseling is to be given to the parents or elders of the family (schools and educational institutions).

22.1.10.3 Drought awareness program

Mock drills are to be arranged to bring awareness in the children about the measures to be taken during the emergency [NDRF, Department of Fire and Schools, National Institute of Disaster Management (NIDM)].

22.1.10.4 Capacity building

Through training the children should be prepared both mentally and physically to face the stress of drought, that is, building resilient groups (DM, SDMA, NIDM, CC-DRR, and UNICEFF and schools).

22.1.10.5 Training to the teachers

To create a group of ambassadors in schools for the implementation of measures under disaster emergencies (NIDM and schools).

22.1.10.6 Capacity building (resilience) in society

As an outreach program, teachers, students, and civil societies/NGOs should educate the communities about disaster management to create disaster-resilient groups (civil society organizations, NGOs) in the community.

22.1.10.7 Food and nutrition

Access to healthy and balanced nutritious food through midday meal in the schools. It may be extended even to the summer holiday period either through schools or AWC (AAY—Antyodaya Anna Yojana, PHCs, Nutritional Rehabilitation Centers).

22.1.10.8 Psychosocial care of the children

During drought, children will be under trauma due to fear and insecurity in the schools and outside the schools. Psychosocial care should be provided to children in the affected community to help them to overcome the trauma of drought and its implications.

22.1.10.9 Micronutrients, malnutrition (hidden hunger)

Affects one-third of the global population. Children health should be monitored regularly and measures to be taken as per the child's need.

22.1.10.10 Child protection

Children are left behind with elders or relatives as parents migrate to cities in search of livelihood and may stay away for extended periods. At such times, neither these children nor the aged people protect themselves from exploitation and abuse, namely, abduction, trafficking, and forced labor. Circumstances may even force parents to consider early child marriages as they feel the burden of adolescent girls in the family. Hence, they need secured environment through child protection services. Proper care and timely counseling to the parents are to be ensured by local organizations.

Checklist of the safety and security measures that are child-friendly under emergency may be prepared with consultation of local, state, national organizations (KSDMA, NIDM, and Institutional Organizations, like UNICEF).

22.1.10.11 Department-wise measures

Considering the need of the children in drought conditions and measures to be taken for their security and safety at different places, for instance, home schools and outside schools could be broadly divided into five categories:

1. nutrition;
2. education;
3. capacity building;
4. health, hygiene, and sanitation; and
5. protection and security.

If the concerned authorities take these measures, children can be saved from the adverse impacts of drought and other disasters. The measures that are to be taken up by different departments during drought are as described in Table 22.6.

Table 22.6 Suggested Initiatives in CC-DRR.

Sl. No	Measures	Departments/Organizations
1.	Inclusion of DM in schools as a part of their curriculum	Department of Education, SDMA, NIDM
2.	Preparation of school safety plans (DMP) and DM	Department of Education, SDMA, Rural Development and Panchayat Raj, Department of Health, Department of Police and Fire Services
3.	Promotion of safety health care education to mitigate the effects of disaster/droughts	Department of Primary and Secondary Education, Department of Health
4.	Safe and sufficient water supply to children, maintenance of water quality (WASH)	Department of Public Health, Rural Development and Panchayat Raj
5.	Promoting preventive health care, health camps for children	Department of Education and Department of Health
6.	Capacity building in children to build resilience under emergency due to disaster	Schools and NIDM
7.	Checking of minimum standards in school such as sanitation, hygiene, and cleanliness	Department of Education, NIDM, District Administrations, Swachh Bharat scheme, Department of Health
8.	Ensure the children to practice safety measures even in outside the schools during emergency	Department of Education, Department of Police and Public Safety
9.	Ensure social safety to children during disaster	Department of Primary and Secondary Education, Department of Police and Public Safety
10.	Main streaming the children issues in the policy document	Department of Education, Department of Police and Public Safety
11.	Monitoring of safety measures to assess effective implementation	Department of Education, civil society organizations
12.	Ensure the security of children from forced labor, trafficking, mental, physical abuse, and exploitation	Department of Police and Public Safety, NGOs

DM, *Disaster management.*

22.1.11 DROUGHT RISK REDUCTION: GAPS IN GOVERNANCE

Drought Management starts well before the actual occurrence of drought. Based on the meteorological data, weather forecasting is done, which, in turn, indicates the probability of drought incidence. In the event of possible drought, the institutional mechanism should be geared up to prepare the people to face the drought. Adequate administrative measures should help in planning and monitoring in the affected area regarding the needs such as water, food, sanitation, health, and social security of the people, especially in the drought-affected areas. Concerned departments should ensure that the needs of the affected community are met through effective implementation and monitoring of drought mitigation strategies in these areas. These drought management activities should be taken up continuously and systematically in war footing to save people, livestock, environment, and other affected components in the drought-affected areas. It is observed that right from the declaration of droughts till the implementation of drought relief measures, there are inordinate delays.

To alleviate adverse drought effects on the affected communities, the gaps or lapses in the drought management, mentioned next may be looked into for good governance:

- For drought preparedness, monitoring, and management, the state should have a drought management policy with specific guidelines to each department dealing with the drought management.
- The drought management involves many organizations, national, state, and local levels; at times they lack proper guidelines to handle the emergencies. Coordination and cooperation of these organizations is a must to handle the droughts effectively.
- Need for a proper location-specific assessment of the drought indicating the extent, severity, and possible drought-proofing measures for the drought vulnerable areas. Different institutes involved tackling the same activity that creates operational problems, so the goals of each institute are to be specified.
- Drought management is to be dealt with a humanistic approach with a positive attitude of saving people, livestock, biota, and environment. Drought management measures such as water, food, fodder, sanitation, health, and social security should be implemented meticulously and with sincerity without any delay.
- Local organizations find very low importance in drought management, but they will have site-specific information on the impact of droughts and the existing indigenous technologies, so they should be involved and entrusted the responsibility to solve the problems.
- The paucity and timely availability of funds and manpower. Provision of funds through microcredit or under drought relief without any delay would save the needy effectively. Measures are to be taken in capacity building of the different vulnerable sections (women) of the affected communities through training in traditional skill development.

Climate change projections indicate a higher incidence and intensity of warm conditions in several regions in India and also increased water stress due to conflicting demands from different sectors. Hence, to equip with changing climate conditions, we need to initiate suitable, long-term measures as soon as possible, and all these plans should consider child-centered risk reduction aspects.

REFERENCES

DAC & FW. (2016). *Manual for drought management 2016.* New Delhi: Department of Agriculture, Cooperation & Farmers Welfare, Ministry of Agriculture & Farmers Welfare, Government of India.

Indian Meteorology Department (IMD). Available from: <http://imd.gov.in/section/nhac/wxfaq.pdf>.

Karnataka State Disaster Management Authority (KSDMA). (2016). Karnataka State Disaster Management Plan 2016, Revenue Department, Disaster Management, Government of Karnataka, Bengaluru. Available from: <http://www.dm.karnataka.gov.in/files/District%20Disaster%20Management%20Plans/State%20DM %20Plan%202016-17.pdf>.

NDMA. (2010). *National disaster management guidelines: Management of drought. A publication of the National Disaster Management Authority.* New Delhi: Government of India, ISBN 978-93-80440-08-8, September 2010.

National Institution for Transforming India (NITI). (2018a). DEEP DIVE insights from champions of change — The aspirational districts dashboard. Available from: <http://championsofchange.gov.in/assets/docs/Deep%20Dive%20-%20V1%20-%201st%20Delta%20Ranking%20(June%202018).pdf>.

National Institution for Transforming India (NITI). (2018b). Transformation of aspirational districts — Baseline ranking and realtime monitoring dashboard 2018. Niti Aayog, Government of India. Available from: <http://championsofchange.gov.in/assets/docs/Aspirational%20Districts%20-%20Baseline%20Ranking%20-%20March%202018.pdf>.

Planning, Programme Monitoring and Statistics Department (PPMSD). (2014). Millennium development goals report of Karnataka, 2014. Government of Karnataka. Available from: <https://www.karnataka.gov.in/spb/Reports/Draft%20MDG-Report-Karnataka.pdf>.

Rajpoot, P. S., & Kumar, A. (2019). Impact assessment of meteorological drought on rainfed agriculture using drought index and NDVI modeling: A case study of Tikamgarh district, M. P., India. *Applied Geomatics*, *11*, 15–23. https://doi.org/10.1007/s12518-018-0230-6.

Roy, A. K., & Hirway, I. (2007). *Multiple impacts of droughts and assessment of drought policy in major drought prone states in India project report submitted to the planning commission.* New Delhi: Government of India Available from. Available from http://planningcommission.nic.in/reports/sereport/ser/stdy_droght.pdf.

Shivashankar, P., & Ganesh Prasad, G. S. (2015). Human Development Performance of Gram Panchayats in Karnataka — 2015, Abdul Nazir Sab State Institute of Rural Development and Panchayat Raj and Planning, Programme Monitoring and Statistics Department, Government of Karnataka. Available from: <http://www.sirdmysore.gov.in/GPHDI/IntroductiontoHDI.pdf>.

UNICEF. (2016). *When coping crumble — Drought in India 2015-16. A rapid assessment of the impact of drought on children and women in India.* New Delhi. India: UNICEF India Country Office.

UNISDR (2018). Economic losses, poverty and disasters 1998–2017. Centre for Research on the Epidemiology of Disasters (CRED), UNISDR.

WMO/GWP. (2017). Benefits of action and costs of inaction: Drought mitigation and preparedness — a literature review prepared by Nicolas Gerber and Alisher Mirzabaev; World Meteorological Organization (WMO) and Global Water Partnership (GWP) Geneva, Switzerland and GWP, Stockholm, Sweden, p. 24.

FLOOD DISASTERS IN ABA NORTH LOCAL GOVERNMENT AREA OF ABIA STATE, NIGERIA: POLICY OPTIONS

23

Christopher Onyemaechi Ugwuibe[1], Francisca N. Onah[2] and Charles Nnamdi Olise[1]

[1]*Department of Public Administration & Local Government, University of Nigeria, Nsukka, Nigeria*
[2]*Social Science Unit, School of General Studies, University of Nigeria, Nsukka, Nigeria*

23.1 INTRODUCTION

Floods remain the most frequent environmental hazards that occur in Nigeria, resulting in property damage. Floods can occur at both rural and urban areas. Flooding has become a major disaster in developing countries including Nigeria in recent years. According to Oriola (2000), the persistent occurrence of floods in strategic locations in Nigeria in recent times has been a great concern and challenge to the people and governments. These stakeholders are increasingly concerned about the threats of flooding to communal safety, economic implications, and national development.

The recent flooding in Nigeria that affected the 36 states including the federal capital territory (Abuja) could not only be attributed to the poor drainage system but partly to climate change. A public opinion agreed that it was due to the water released from Cameroun dam, for which many lives and millions of properties got destroyed. One distinguished characteristic about flooding is that it does not discriminate against any individual or location but marginalizes whosoever refuses to prepare for its occurrence.

In Nigeria flooding has forced millions of people from their homes, destroyed businesses, polluted water resources, and increased the risk of diseases (Akinyemi, 1990; Baiye, 1988; Edward-Adebiyi, 1997; Nwaubani, 1991). According to World Bank (2012), developing countries are presently characterized by rising flooding events, this is one of the effects of increasing urbanization that ultimately alters the fluvial process of the area, because urbanization increases surface run-off due to increase in relative proportion of impervious surfaces. Flooding is a situation that results when land that is usually dry is covered with water of a river overflowing or heavy rain. Flooding occurs naturally on the flood plains which are prone to disaster. It occurs when water in the river overflows its banks or sometimes results from a poorly constructed dam or erosion control channel. It happens without warning but with a surprise package that always delivers to unprepared community like the ones that occurred in some communities in Aba North in recent times such as Eziama, Uratta, Umuola Egbelu, Umuogo Osoke, Ariaria, Ogbor, and Asa Okpulo. The News Agency of Nigeria (Obichie, 2019) reported that a heavy downpour on Wednesday night (June 12) in Aba, the

Economic Effects of Natural Disasters. DOI: https://doi.org/10.1016/B978-0-12-817465-4.00023-6

commercial hub of Abia, has left many homes, shops, and markets submerged in flood. NAN further reported that goods and property estimating at over N15 million were also destroyed due to it.

According to the United Nations population estimates as on July 14, 2019, Nigeria is the most populous country in Africa with the largest economy and with a population of 200,663,734 (United Nations, 2019). Studies on Nigerian flooding appear to be scanty in many important global flood studies and documents. It is on this premise that this study attempts to assess flood disasters in Aba North Local Government Area of Abia State, Nigeria with a view of making policy recommendations. In this chapter, we shall attempt some fairly comprehensive and reasonable overview of flooding scenario in Nigeria as well as in our study area. To this end, issues to be discussed in this chapter is organized around the following major subthemes: method of the study, description of the study area, the concept of flooding, theoretical framework, major causes of flooding in Aba North Local Government, discussion of findings, and concluding remarks.

23.2 METHOD OF STUDY

The study adopted both primary and secondary data. The primary data were obtained through questionnaire, direct observation, and oral interview of key informants. A total of 107 questionnaires were purposively distributed to the existing nine communities within the study area. The sample population includes traders, public servants, and students. These key informants were selected based on their knowledge of relevant information required for the study.

23.3 DESCRIPTION OF THE STUDY AREA

The study covered Aba North local government area. The area covers about 60% of the urban area of Aba and it is located in Southeastern Nigeria, on latitude 507°N and longitude 70220°E. It has an estimated population of 6446 in 2009 (https://en.wikipedia.org/wiki/Aba_North). Aba North Local Government Area has its headquarters at Eziama, it covers a total area of about 3440 ha. It has nine communities—Eziama, Uratta, Usoke Amato, Umuola Egbelu, Umuogo Osoke, Umunneato Ariaria, Ogbor, Asa and Okpulo. According to Akintoye et al. (2015) Aba North has physical features, that is, the Aba River, popularly known as "water side," and a high land in Ogbor Hill. In landmarks, it is the smallest town in Abia State; it is bounded on the South by Aba South Local Government Area, in the East by Obingwa and on the West by Ugwunagbo and Ukwa West Local Government Area, respectively. Fig. 23.1 presents a map of Aba North Local Government Area, which is the study area.

23.4 THE CONCEPT OF FLOODING

Flooding in the cities and the towns is a recent phenomenon caused by increasing incidence of heavy rainfall in a short period of time, indiscriminate encroachment of waterways, inadequate capacity of drains, and lack of maintenance of the drainage infrastructure. In Nigeria, reports on flood disaster are always on the high side especially during the raining season ranging from July and

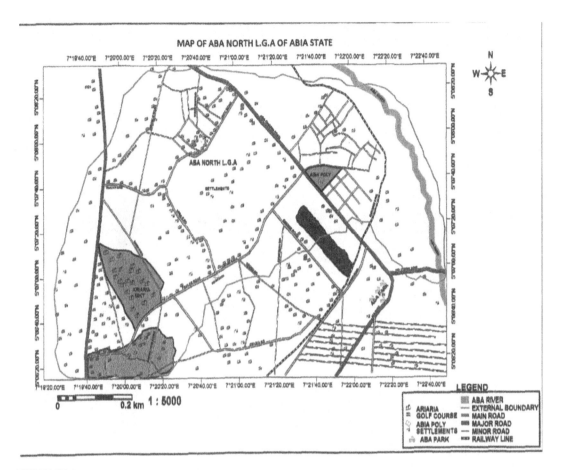

FIGURE 23.1

Map of Aba North Local Government Area.

Nwoko, A.U. (2013). Flooding in Nigerian cities: Problems and prospects. A case study of Aba Urban, Abia State, Nigeria. *Available from <www.academia.edu/3669614/>.*

September. Between these peaks is the "little dry season" commonly known as "August Break." The rainy or wet season starts early March and lasts till October or early November of each year. The length of the wet season is at least 7 months including the period of August-break. The dry season lasts between November and March.

Flooding is defined as a large amount of water covering a particular area or community that was usually dry (Olajuyigbe, 2012). Nwafor (2006) defined flood as a natural hazard like drought and desertification which occurs as an extreme hydrological (runoff) event. Flood can simply be defined as a flow of water above the carrying capacity of erosion channel. In the past, in Nigeria, hundreds of lives and properties worth millions of Naira have been destroyed directly or indirectly from flooding every year. It is an environmental disaster occurring from time to time in all rivers and natural drainage systems. Flooding and rainstorm apart from causing destructions to lives and properties often causes significant damage to the livelihood system of the victim. This causes untold hardship to the individuals. According to Ologunorisa (2004), the negative impact of floods has been on the increased due to a number of factors, with rising sea level and increased developments on flood plain among other issues. The worst affected areas by flood in Aba North include Ariaria International Market, Uratta, Umuogo, and Asa Okpulo.

23.5 THEORETICAL FRAMEWORK

The study adopted the institutional theory propounded by North (1990). Peters et al. (2000) define an institution as a formal or informal, structural, societal, or political phenomenon that transcends the individual level, which is based on more or less common values, has a certain degree of stability and influences behavior.

The institutional theory is found credible in interrogating the causes of flooding disaster in Aba North Local Government. Guy (2000) defines institution as a formal or informal, structural, societal, or political phenomenon that transcends the individual level, based on more or less common values, with a certain degree of stability and influences behaviors. It considers the processes by which the structures, including schemes, rules, norms, and routines, become established as authoritative guidelines for social behavior. The key idea behind institutionalism is that much organizational action reflects a pattern of doing things that evolves over time and becomes legitimate within an organization and environment. Most metropolises in Nigeria are just like any other municipal area councils in a developing country which has not benefited from the contributions made by regional development plans because management is via planning authorities, using policies made for them, to achieve different goals that relate to physical planning. Master planning as a policy tool and technique to solve physical planning problems in a regional arena has failed.

This institutional restriction becomes a constraint by not supporting and facilitating the infrastructure delivery to promote sustainable city planning as observed by absence of adequate master planning frameworks to address flooding disaster. In addition, the approval of a plan by government may have to go through lengthy procedures and could then be incorporated into the relevant section of a plan. Aba North Local Government is the fastest growing cosmopolitan town in Abia State but lacks an institutional framework for the effective implementation of any plan to control development. All these are the justification for the relevance of the institutional theory in understanding the place of flood disaster in Aba North Local Government Area.

23.6 MAJOR CAUSES OF FLOODING IN ABA NORTH LOCAL GOVERNMENT AREA

23.6.1 FLASH FLOODING

Flash flooding is the major cause of flood in Aba North. Flash floods are defined as those that occur as a result of excessive rise in water resulting from either during or within a few hours of a heavy downpour that produces the rise. Therefore flashfloods occur within small catchments areas as in the case of Aba North Local Government Area, where the response time of the drainage basin is short. The rainfall regime is bimodal with peaks in July and September. Between these peaks is the "little dry season" commonly known as "August Break." The rainy or wet season of the area begins about March and lasts till October or early November. The length of the wet season is at least 7 months including the period of August-break. The dry season lasts from November to March. Many hydrological factors have relevance to the occurrence of a flash flood. Besides, other factors constituting flash flooding include topography of the land, types of vegetative obtainable, human factor, and antecedent rainfall. Flash flood is also caused by concatenation of both meteorological and hydrological activities on the soil. Most flash floods associated with rainfall are produced by thunderstorms, that is, deep, moist convection. A single thunderstorm cell is unlikely to produce enough rainfall to cause a flash flood (Doswell, 2003).

23.6.2 GROUNDWATER FLOODING

Flood, as a form of environmental disaster, occurs when soil and vegetation cannot absorb all the water in a particular area. This then runs off the land in quantities that cannot be absorbed in stream channels or retained in natural ponds and constructed reservoirs. According to Parker (2000), groundwater is a major source of water supply globally. It is the source of about one-third of global water withdrawals and provides drinking water to a reasonable population of the world.

For example, the Aza River that rises at Abayi and flows south through Aba Township into Imo River at a point near Okpontu is a major source of recharge to groundwater in the underlying aquifers. Both the river and groundwater are important sources of water to the city of Aba North Local Government and neighboring communities. The impacts of climate change are being witnessed, in the form of high intensity of rain leading to flooding on one part and drought in the other, all revealing groundwater-level fluctuations.

According to Zbigniew and doll (2009), groundwater recharge is projected to increase in the warming world, but many semiarid areas that suffer from water stress already may face decreased groundwater recharge. However, in areas with a land surface elevation of a few meters or more, groundwater availability is more strongly impacted by changes in groundwater recharge.

23.6.3 POOR DRAINAGE SYSTEM

The urban nature in Aba North Local Government is characterized by high population concentration. Within the area, there exist poor public transportation network, inadequate basic infrastructure, high level of environmental pollution, and flooding. The drainage system in Aba North is poorly constructed, and as a result, it leads to blockage whenever there is a heavy downpour of rain. According to Tucci (2001), one of the major implications of poor drainage system include increase in peak flows and in frequency owing to the higher runoff capacity through conduits and canals, increased sediment production from unprotected surfaces and production of solid waste (refuse); and deterioration in the quality of surface and groundwater owing to the transport of solid material and clandestine sewage and stormwater connections. The abuse of the Master Plan poses a great challenge to the Abia State Government and Abia State Environmental Agency in meeting up with the provision of urban infrastructure that meets the demand of its populace. The after effect of this

FIGURE 23.2

Flood hits Ariaria International Market.

Field Survey (2019).

environment disaster is excessive flow exceeding the hydraulic capacities of the drains and subsequent flooding with disastrous consequences on urban transportation that leads to either Ariaria International Market, Ahia-Ohuru, Old Park and other strategic commercial and residential areas in the local government (Fig. 23.2).

Floodwater can seriously and suddenly disrupts one's personal activity by disrupting the mean of transportation and channel of communication. Floods disrupt normal drainage system in both urban and rural areas, and sewage spills are common, which represents a serious health hazard, along with standing water and wet materials in people's domain.

Data in Table 23.1 show the views of key informants from the nine communities prone to flooding disaster in Aba North Local Government Area. The results show that 23 key informants from the sampled population representing 21.4% indicated that flooding disaster led to destruction of property. This consisted of both individual residential belongings such as houses, shops, and religion/worship centers. 38 key informants representing 35.6% posit that flooding seriously disrupted their economic activities, thereby imposing untold hardship to the victims. 29 key informants representing 27.1% indicated that indiscriminate disposal of solid waste was as a result of the environmental disaster.

Table 23.1 Effects of Flooding Disaster in Aba North Local Government Area.

S. No.	Statement	Frequency	Percentage
1.	Destruction of property	23	21.4
2.	Disruption of economic activities	38	35.6
3.	Indiscriminate disposal of solid waste	29	27.1
4.	Harboring of pest and diseases	17	15.9
		107	100

Field Survey (2019).

These emerge from the neglect of the building master plan by the Aba North Urban Planning Unit. While 17 key informants representing 15.9% indicated that flooding disaster in Aba North led to harboring of pest and diseases among the inhabitant. The result in the table established that flooding had a major effect on disruption of economic activities in the study area.

23.7 DISCUSSION OF FINDINGS

The result of the key informants' opinion in the study area revealed that flooding in Aba North is caused by flash flooding resulting in poor drainage systems, as well as prolonged rainfall, which affects most parts of Aba North. These when combined with indiscriminate waste disposal by individuals worsen the flood situation, especially in densely populated areas such as Ariaria, Faulks road, Uratta road, and Umule vicinity.

Responses during oral interviews also revealed that the presence of a very large population in a limited area of land (population pressure) is one of the causes of flooding. Increase in population of the inhabitants invariably leads to increase in waste disposal, blocked culverts, grates, scuppers are blocked with dirt. The major loss that is experienced from flood occurrence in the study area is the destruction of both public and private property, dilapidation, and collapse of buildings especially in Uratta road, Aba.

In order to proffer solution to the recurrent occurrences of flood disaster in Aba North Local Government Area, key informants were of the view that there the Ministry of Environment and Abia State Environmental Agency should construct sustainable drainage systems, proper waste disposal system, as well as legislation against the disruption of the State Build Master Plan.

23.8 CONCLUDING REMARKS

Flooding disasters has been a recurrent environmental problem affecting humanity. Its frequency and intensity have continued to increase on yearly bases. The residents of Aba North are devising ways in which they can permanently adapt to address the problem. Some individuals now resort to occasional removal of sand in blocked drainages, construction of water channels, and fumigation of stagnant floodwater.

The study recommends a preventive and sustainable flood management options. That is, there should be a close and working relationship between local people and agencies such as the Town Planning Authorities who are saddled with the responsibility of tackling building plans for new houses and flooding in the area. The study further recommends adequate public enlightenment campaign. The study observed that environmental attitude of most residents of Aba especially regarding drainage channels and waste disposal is very irritating. There is a need for elaborate public enlightenment and civil education on the use and maintenance of drains, as well as the health implications of waste disposal. Residents should be adequately sanctioned on any drain across their individual environment.

Besides, there is a dire need for effective partnership by local/state/federal governments and stakeholders in Aba North Local Government Area to support town planning, and State

Environmental Agencies' efforts targeted at combating flooding. The compulsory every last Saturday monthly environmental sanitation in Abia State should be reform with greater emphasis on solid waste control measures.

REFERENCES

Akintoye, O. A., Digha, O. N., Uzochukwu, D. C., & Harrison, U. E. (2015). The consequences of flood on occupational trips duration and flood risk assessment of Aba North Local Government Area of Abia State, Nigeria. *Journal of Environment and Earth Science, 5*(11), 153−165. Assessed from <www.iiste.org>.

Akinyemi, T. (Friday, July 20, 1990). Stemming the tide of Lagos floods. *The Guardian*, pp. 7.

Baiye, E. (Thursday, October 8, 1988). Numan in the throes of floods. *The Guardian*, pp. 9.

Doswell, C. A. (2003). *Flooding*. Norman, OK: University of Oklahoma.

Edward-Adebiyi, R. (Saturday, May 17, 1997). The story of Ogunpa. *The Guardian*, pp. 9.

Guy, B. P. (2000). Institutional theory in political science: The new institutionalism. London, UK: Continuum.

North, D. C. (1990). *Institutions, institutional change, and economic performance*. Cambridge, MA: Cambridge University Press.

Nwafor, A. N. (2006). Flood extent mapping in Bayesa State, Nigeria. *Journal of Geography, 8*(33), 18−29.

Nwaubani, C. (October 21, 1991). Ogunpa River leaves bitter aftertaste in tragic course through Abeokuta. *The Guardian*, pp. 9.

Nwoko, A. U. (2013). *Flooding in Nigerian cities: Problems and prospects. A case study of Aba Urban, Abia State, Nigeria*. Available from: <www.academia.edu/3669614/>.

Obichie, B. (2019). Flood destroys over N15m worth of goods in Aba, renders many homeless (New Agency of Nigeria). Available from: <https://www.legit.ng/1243363-flood-destroys-n15m-worth-goods-aba-renders-homeless.html>.

Olajuyigbe, E. A. (2012). Mapping and analysis of 2012 flood disaster in Edo State using geospatial technic. *Journal of Environmental sciences, 6*(5), 32-44.

Ologunorisa, E. T. (2004). An assessment of flood vulnerability zones in the Niger Delta, Nigeria. *International Journal of Environmental Studies, 61*(1), 31−38.

Oriola, E. O. (2000). Flooding and flood management. In H. I. Jimoh, & I. P. Ifabiyi (Eds.), *Contemporary issues in environmental studies* (pp. 100−109). Ilorin: Haytee Press & Publishing Coy.

Parker, D. J. (Ed.), (2000). *Floods* (Vol. 1, pp. 116−127). London & New York: Routledge.

Peter, B. G. (2000). Institutional theory in political science: The new institutionalism: London, UK: Continuum.

The News Agency of Nigeria. (2019). Breaking news report.

Tucci, C. E. M. (ed.) (2001) *Urban drainage in specific climates: Urban drainage in humid tropics*, Vol.1. Technical reports in Hydrology, No. 40, Vol. 1, UNESCO Press, Paris, France.

United Nations. (2019). *Population of Nigeria*. Available from: <https://worldometers.info/world-population/Nigeria-population>. Retrieved July 14.

World Bank. (2012). *Thai flood 2011: Rapid assessment for resilient recovery and reconstruction planning: Overview (English)*. Washington, DC: World Bank. <http://documents.worldbank.org/curated/en/677841468335414861/Overview>.

Zbigniew, K., & Doll, P. (2009). Will groundwater ease freshwater stress under climate change? *Hydrological Sciences−Journal−des Sciences Hydrologiques, 54*(4; Special issue: Groundwater and Climate in Africa).

INSTITUTIONAL RESPONSE AND MECHANISMS TO THE MANAGEMENT OF NATURAL DISASTER IN NIGERIA

Chukwuma Felix Ugwu[1], Chioma S. Ugwu[2] and Kalu T.U. Ogba[3]

[1]*Department of Social Work, Faculty of the Social Sciences, University of Nigeria, Nsukka, Nigeria* [2]*Department of Political and Administrative Studies, University of Port Harcourt, Port Harcourt, Nigeria* [3]*Department of Psychology, University of Nigeria, Nsukka, Nigeria*

24.1 INTRODUCTION

Globally, about half of the worldwide population were potentially exposed to natural disasters in 2018 [Centre for Research on the Epidemiology of Disasters (CRED), 2018a, 2018b]. Natural disasters are triggered by hydrometeorological, climatological, geophysical, and biological phenomena (Mata-Lima, Alvino-Borba, Pinheiro, Mata-Lima, & Almeida, 2013). According to United Nations (2002), there are notable volcanic eruptions in Indonesia, Montserrat, Ecuador, and the Philippines; earthquakes in Japan, Turkey, El Salvador, Indonesia, India, and Peru; and floods in Bangladesh, Ethiopia, Guinea, India, Mozambique, Nigeria, Sudan, Thailand, Venezuela, Vietnam, and Algeria. In the 21st century, floods have affected more people than any other type of natural hazard (CRED, 2017/2018; Mulugeta et al., 2017).

Among these forms of natural disaster, flood is the most prominent across countries. Flood is the accumulation and overflow of water bodies such as sea, river, lake, sea and rainfall, dam breakdown, and snow melt (Leister, 2009). As a result of greenhouse warming, flood is projected to be an unprecedented event in modern times (Omotosho, Baloguun, & Ogunjobi, 2000). The frequency of occurrence of floods grows faster than other natural disasters. The expected consequences include death, economic losses of planted crops, damage to infrastructure, and disruptions in the electric system, of drinking water, and loss of sewage disposal facilities (Caruso, 2017). For example, about 27 million people were reportedly affected by flooding in India, Nepal, and Bangladesh, with 450 million people living in potentially exposed area. Of those potentially exposed, 18% were directly affected in Bangladesh, 6% in Nepal, and 2.5% in India (CRED, 2017/2018).

Natural disaster is usually unpredictable and affects human lives in many different ways. Over the years, natural hazards such as droughts, wild land fires, storms and tropical cyclones, earthquakes, volcanic eruptions, and floods have been detrimental to human lives and livelihoods. In most cases the impact has created widespread social, economic, and environmental damages (United Nations, 2002). Majorly at the macroeconomic dimension, there is the likelihood that

Economic Effects of Natural Disasters. DOI: https://doi.org/10.1016/B978-0-12-817465-4.00024-8

natural disasters destroy infrastructure affecting industries, growth, and employment, while at a microeconomic dimension, they can destroy assets and affect nutrition and access to education and health services. Again, natural disasters often times generate outbreaks of illness and provoke casualties in the affected areas (Caruso, 2017).

Climate change, which is the major cause of floods in the world (Olanrewaju, Chitakira, Olanrewaju, & Louw, 2019), has countless financial implications in all spheres of life. Counting the cost, between 2003 and 2013, damages from natural disaster amounted to about USD 550 billion and affected 2 billion in developing countries. Such disasters weaken national economic growth and development goals, as well as sustainable development of the agriculture sector and its subsector (FAO, 2015). CRED (2017) noted that underreporting climate-related natural disasters is majorly a challenge to other continents such as Europe, Asia, Oceania, and the Americas, but most acute in Africa, where economic cost is available for less than 14% of all disasters.

In Africa the people are prone to natural and human-induced disasters such as floods, hurricanes, earthquakes, tsunamis, droughts, wildfires, pest plagues, and air and water pollution with losses to livelihoods and property (Mulugeta et al., 2017). The human-induced is attributed to productive ventures through the use of technology among other engagements or lifestyle of man to survive in the environment. Some of the activities of man impact on climate conditions, which in turn threaten human and natural existence around the world. According to Ellis (2008), research shows that climate change do not evenly affect all countries, but most likely to have greater impact in equatorial regions such as sub-Saharan Africa. This may likely affect food production both in the present and future agriculture in the continent of Africa. It implies that climate change and the growth of agriculture are inseparably interconnected. Although agriculture is dependent on the weather to thrive, climate change has a negative impact on agriculture in many parts of the world because of increasingly severe weather patterns. Due to the worsening situation of weather, it is projected that climate change will continue to cause floods, worsen desertification, and disrupt growing seasons (Ellis, 2008). In the same vein, the Intergovernmental Panel on Climate Change (IPCC) (2007) projected that by 2020 agriculture yields in some African countries could be reduced by up to 50%.

In developing countries, including Nigeria, flooding costs not only affect agriculture but include other spheres of lives in terms of the impact on the economy, property and human lives and livelihoods. The damage and losses to infrastructures and public facilities costs huge sums of money. In addition, people face health challenges, especially women and children in the rural areas are the worst affected with cholera and malaria due to contaminated and stagnated water (Agwu & Okhimamhe, 2009). For example, the identified immediate public health risks of the displaced population by World Health Organization (WHO), in Nigeria's 2012 floodwaters, are numerous. The health risks are overcrowding in designated camps or with host families or communities, leading to high risk of meningitis, measles, and incidence of acute respiratory infections; interruption of safe water and sanitation supplies, which may lead to high risk of outbreaks of water- and foodborne diseases, such as cholera; and increased vector breeding sites with risk of yellow and malaria fever; among other health setbacks (World Health Organization, 2012). The health cost is highly undesirable for sustainable living. Besides the effect of flood on health, the hazard also extends to immediate-, medium-, and long-term concerns for national growth and development.

Flood in Nigeria is characterized by immediate-, medium-, and long-term effects. The immediate impact in Nigeria is basically destruction of homes and displacement of the people; destruction of livelihood such as farming, poultry, and fishing. Included in this category are the small-, medium-, and large-scale businesses. The medium- and long-term impacts include food shortages, high food prices, and food insecurity (OlajuyIgbe & Ajayeoba, 2012). Communities that primarily depend on natural resources such as agriculture and forest for livelihood are likely to be most affected by climate change (Thapa, Joshi, & Joshi, 2015).

Across disciplines, natural disasters are believed to be naturally occurring events, the consequences of which are often heightened by man-made actions that exceed the capability of man to contain (Mata-Lima et al., 2013). To this end, the poor capacity for self-recovery of the affected areas attracts the needed assistance from external bodies (Noy & Nualsri, 2010), and local environment. The significance of building the resilience of nations and communities to disasters is paramount to any Sustainable Development Goals (SDGs). The previous Hyogo Framework for Action 2005–15 that is the key instrument for implementing disaster risk reduction sought for building the resilience of nations and communities to disasters. Currently the Sendai Framework for Disaster Risk Reduction 2015–30's three strategic goals appeal for investing in disaster risk reduction, covering resilience for almost all aspects of life [International Federation of Red Cross and Red Crescent Societies (IFRC), 2016].

This current framework is in line with the 17 SDGs, globally agreed at the end of 2015 as are placements to the 2000–15 Millennium Development Goals, basically set to build infrastructure, promote inclusive and sustainable industrialization, and foster innovation and make human settlements inclusive, safe, resilient, and sustainable (IFRC, 2016). Hence, there is a need to appraise institutional response and mechanisms to the management of natural disaster in Nigeria. The broad objective of the chapter is to examine the institutional response and mechanisms to the management of natural disaster with regards to risk-reduction options in Nigeria. Specific efforts are made to explore the relationship between climate change and flood disaster, impact of flood on livelihood, adaptation and community coping mechanisms or resilience, and postdisaster needs assessment in Nigeria.

24.2 CLIMATE CHANGE AND FLOOD DISASTER IN NIGERIA

Climate change has been a subject of discourse in recent years. Climate change is already affecting the natural environment and its inhabitants in the whole world. Despite 2015 Paris climate change treaty, countries signatory to the treaty are yet to make substantial implementation strategies to mitigating the effect of ozone layer depletion. The emission of greenhouse gases into the atmosphere such as carbon dioxide, methane, nitrous oxide, and other gaseous substances greatly contributed to natural disasters of varying degrees across nations of the world. Environmental changes are not only caused by natural phenomenon; however, human-induced natural disaster contributes immensely to natural disasters leading to rise in the sea levels leading to floods. Floods have been noted as the most occurring natural disaster, especially in those countries along coastal lines. In the coastline areas flooding is caused as a result of tsunamis, hurricanes—cyclonic storms, and high tides. In some other areas it is caused by long-term processes such as subsidence and rising sea level as a consequence of global warming can lead to the advancement of the sea on to the land (Nelson, 2016). Flood is a major risk to riverside population and floodplains with associated impacts on aquatic fauna and flora (Agbonkhese, Yisa, & Daudu, 2013).

Gwary (2008) observed that in Nigeria flooding occurs in three main forms, river flooding, urban flooding, and coastal flooding. The river floods may be climate change-induced with attendant floods becoming more frequent and large than normal in some places or become smaller and less frequent in other places [U.S. Environmental Protection Agency (EPA), 2016]. Mallakpour and Villarini (2015) rightly pointed out that warmer temperature causes more water to evaporate from the land and oceans, and subsequently changes in the size and frequency of *heavy precipitation* events, which may affect the size and frequency of river flooding. The urban flooding can pollute water supplies and intensify the spread of waterborne diseases such as diarrhea, typhoid, scabies, cholera, malaria, and dysentery(Agbonkhese et al., 2013). Typically in Nigeria flooding may be attributed to both natural and human-induced flooding, but mostly attributed to nature.

Most floods in Nigeria are associated with climate change. According to Fitchett, Grant, and Hoogendoorn (2016), floods associated with rising sea levels are related to climate change. In an examination of the impact of climate change on the coastal areas in Nigeria, Obot, projected that the sea level is anticipated to increase in Lagos, Bayelsa, Port-Harcourt, Warri, and Calabar and will face increase in sea level of 6 m due to rising temperature. In 2010 residents in Lagos and Ogun states of southern Nigeria were engulfed by floods occasion by heavy rainfall leading to the release of overflowing water from the Ovan dam into the rivers, causing riverbanks overflow. This single event of river overflow displaced about 1000 residents (Okonkwo, 2013). In addition, there is a notable progression in floods in the country far above the recorded event in the preceded years. In 2012 the flooding that resurged in some parts of the country incidentally consumed lives, peoples' property, and sources of livelihood. In the history of Nigeria an unprecedented 2012 flood affected 25 out of 36 states, displaced about 3,871,063 persons, injured about 5871, killed 363 persons, and destroyed 597,400 homes [United Nations Development Programme (UNDP), 2019].

Conversely, population explosion and human activities, which include the location of infrastructures, rapid industrialization and urbanization, and exploitation of natural resources intensify the occurrence of floods [United Nations Human Settlements Programme (UN-Habitat), 2011]. Furthermore, Agbonkhese et al. (2013) noted that in spite of heavy rains, human activities in the environment and poor drainage system in most cities in Nigeria caused the deaths of hundreds of people and rendered some homeless. Again, such flooding in cities may contaminate water supplies and intensifies the spread of diarrhea, cholera, typhoid, malaria, scabies, and other diseases (Agbonkhese et al., 2013).

Like most cities across coastline areas in many countries, in recent times, the phenomenon of flood in Nigeria is a common and known occurrence among the people. That is to say that the effect of climate in terms of flooding has come to stay if urgent measures across national, regional, and global communities do nothing about its adaptation and mitigation. For example, some geographical locations of Nigeria are prone to flood due to its proximity and low-lying nature. This therefore agrees with the classification of Nigeria into three major climatic regions. The country's three major climatic regions are characterized by tropical rainy season in the south, wet and dry in the central area, and semiarid in the far north. These climatic conditions that are associated with ecological zones in the country predispose the zones to natural disasters such as coastal erosion, landslides, and floods in the southern, while desertification, drought and occasional flooding are common in the northern Nigeria (United Nations Economic Commission for Africa, 2015). According to NEST (1991a, 1991b), specific geographical areas in Nigeria are bedeviled with flood disaster than other parts. The areas are:

1. low-lying areas in the southern parts of the nation where annual rainfall is very heavy;
2. the Niger Delta zone;
3. the floodplains of the larger rivers of the Niger, Benue, Taraba, Sokoto, Hadeja, Cross River, Imo, Anambra, Ogun, Kaduna, etc.; and
4. flat low-lying areas around and to the south of Lake Chad, which may be flooded during and for a few weeks after the rain NEST.

NEST (1991a, 1991b) observed that primarily, river floods in the country are caused by water coming from the Fouta Djallon Highlands in Guinea, and the runoff from the previous year's rainfall on these highlands, which permeates gradually to Nigeria. The first casualties are the Niger Delta areas. These include towns, villages, and agricultural lands, particularly in the Sagbama and Yenagoa areas in Bayelsa state and other riverine states of Edo, Delta, Cross River, and Akwa-Ibom states. According to United Nations Economic Commission for Africa, climate-related disaster development solutions should essentially include:

- pursuing disaster reduction, adaptation, and sustainable development as mutually supportive goals;
- considering risk reduction as an essential investment in sustainable development, not as an additional cost; and
- corrective development planning that ensures development does not generate risks (United Nations Economic Commission for Africa, 2015, p. 10).

24.3 IMPACT OF FLOOD ON LIVELIHOOD

Nigeria nation has battled the menace of flood for several decades. The most pronounced of them all is the 2012 flood in the country, which incidentally affected virtually all the states of the federation. The flood is accompanied by human and material losses. However, the most affected 12 states include, Bayelsa, Rivers, and Anambra (most affected); Delta, Kebbi, Kogi, Taraba, Adamawa, Nasarawa (second-most affected); and Benue, Edo, and Jigawa (third-most affected). These most affected states were linked with attendant damage and losses.

> Evidently, the most affected in terms of destroyed assets is that of housing (damage of over N1 trillion). The sector of agriculture, which comprises crop production, livestock raising, and fishery, was the second-most affected, wherein the most important features of disaster effects are production losses (amounting to N380 billion) and the destruction of physical assets, at a cost of a further N100 billion. Third-most affected was the commerce sector (including wholesale and retail activities), which sustained destruction of its assets (premises and stocks of goods to sell) of N18.6 billion, as well as very large losses in sales (N376 billion). The oil sector sustained damages to oil wells and suffered other associated infrastructure and production losses worth a combined total of N230 billion.
>
> **FGN (2013, p. xxII)**

Although flood has occurred in previous years in Nigeria, it is believed that the 2012 flood disaster remains the most substantial of all in the history of Nigeria. Countless damage and losses

Table 24.1 Summary of Damage and Losses Caused by the 2012 Floods in Nigeria's Most Affected States.

Sector	Subsector	Damage	Disaster Effects, Million Naira Losses	Total
Social		1,256,299.3	73,557.9	1,329,857.2
	Education	82,134.6	15,211.2	97,345.8
	Health	18,204.8	9476.8	7681.7
	Housing	1,155,959.9	48,869.9	204,829.7
Productive		147,996.5	1,037,070.0	1,185,066.5
	Agriculture	101,008.2	380,520.8	481,528.9
	Manufacture	21,795.2	74,425.0	96,220.2
	Commerce	18,693.1	357,124.2	375,817.3
	Oil industry	6500.0	225,000.0	231,500.0
Infrastructure		54,019.6	8013.6	62,033.2
	Water and sanitation	12,902.2	–	12,902.2
	Electricity	329.0	8013.6	8342.6
	Transport	40,788.4	–	40,788.4
Cross-sectoral		23,840.2	17,167.0	41,007.2
	Environment	23,840.2	17,167.0	41,007.2
Total		1,482,155.6	1,135,808.5	2,617,964.0

FGN. (2013). Data from Nigeria post-disaster needs assessment 2012 floods. *Abuja: Government Press, xxvii (27).*

are huge in the context of social living, productive ventures, infrastructure, and the environment. The collaborative assessment by federal government of Nigeria, with technical support from the World Bank, European Union, United Nations, and other development partners succinctly summarized the major impact of 2012 flood, including substantial and quantifiable damage and losses in major sector and subsector in Nigeria. Table 24.1 reveals as follows.

24.4 DISASTER ADAPTATION AND RESILIENCE/COMMUNITY COPING MECHANISMS IN NIGERIA

The concepts of disaster adaptation and resilience are intertwined and inseparable. In this context, disaster adaptation underscores the need to adjust to new conditions arising from climatic variations, while resilience connotes the ability to manage environmental stressors or pressures relating to natural disaster. Although the definition of resilience is based on disciplinary bias or orientation and or the nature of hazard. The central idea is toward making an adjustment or positive shift toward a better living. Resilience in the direction of natural disaster is crucial in every disaster-ridden or -prone areas. Disaster resilience encompasses the ability of individuals and communities, organizations, and states to adapt to and recover from hazards, shocks, and/or stresses considering long-term planning toward uncertainties (Combaz, 2014).

Ozor et al. (2012) described adaptation as a concept likened to system resilience expressed in the capacity of a system, organization, community, and or individuals to create, alter, or implement multiple adaptive actions for sustenance. Due to extreme weather events such as flooding, which have interrupted economic activities, caused huge losses, displaced people, and threatened the sustainability of communities, resilience building among the affected is far and wide recognized as important for disaster risk reduction in many nations (Mallakpour and Villarini, 2015; Montz, 2009; Oladokun, Proverbs, & Lammond, 2017). This new insight corroborates the stance of Hammond, Chen, Djordjević, Butler, and Mark (2015) and Park, Seager, Rao, Convertino, and Linkov (2013), who emphasize paradigm shift from relying solely on flood defense and structural systems to an approach that focuses on the concept of community resilience as an aspect of flood risk management.

Since impact of flood varies, so do community-based adaptation strategies depend on cultures and societies. Researchers are of the opinion that adaptation practices should be community-based interest to encourage sustainable development. The adaptation strategies is more successful when it is localized, in terms of the affected communities making substantial input based on preferences and priorities, norms and values (Newton, Paci, & Ogden, 2005). This implies that community-based adaptation recognizes existing environmental knowledge, peculiarities, believes, and interests.

Environmental vulnerabilities and peculiarities vary according to geographical zones of the country. Equally, the communities' adaptation to flood hazards, which is mostly a product of climate change, and coping mechanisms may largely dependent on their knowledge and awareness of the hazardous situation. The people's ability to cope is hinged on existing adaptive capacities in all spheres of life. In view of the communities' ability, Sayne (2011) asserts that most households and individuals are grounded with local knowledge that can be improved for effective responses to flood disasters. More so, the inclusion of the communities by government in planning and implementation of adaptive measures would improve opportunities to sustain the existing social fabrics and strengthen resilience among the people (Sayne, 2011).

Focusing on experiences of the people, Nwaogu and Ezekwe (2018) study in Ndoni, Rivers State, Nigeria, on communities adaption on extreme flood event shows that flood is not seen as preventable issue but something to live within their community. The study further revealed that the community adapt to flood hazard following early warning signs on the media (radio and television), and by clearing drainage systems and dredging of rivers within and around the community, fortification of houses, among other safety networks to assist the most disadvantaged individuals and households. Another study in Rivers State by Igwe and Wordu (2016) revealed that although the Orashi community was not privileged to have preflood awareness, early warning, and preparedness before the 2012 flood in the community, the community resilience was based on self-help efforts to addressing the challenges of the hazard that accompanied the flood disaster. For communities that had the privilege to and those that did not know about early warning for adequate preparation for evacuation, the import here is that communities have in-built coping mechanisms to adapt and cope with the expected and unanticipated flood disaster, even unaided.

A major setback to adaptive functioning is poor socioeconomic background and attitudes toward the environment. Research findings in six states representing the six geopolitical zones of Nigeria, by Daramola, Oni, Ogundele, and Adesanya (2016), revealed that attitudes such as the dumping of solid waste on unapproved sites, individuals with poorly or inferior roof construction materials, and those in rural areas are the most affected. Furthermore, this findings revealed that those who do not have personal savings and those whose livelihood is based on agriculture and a source of income increased likelihoods of experiencing severe negative effects of natural disaster. Notwithstanding, the overall objectives of disaster adaptation and resilience are to save lives; protect infrastructure and livelihoods, social systems, and the environment; and support broader resilience in contexts of violent conflict or fragility (Combaz, 2014).

24.5 INSTITUTIONAL RESPONSE AND MECHANISMS TO FLOOD MANAGEMENT IN NIGERIA

Many nations are working to address the effects of climate change with regards to impacts varying in extent and nature (Onu & Ikehi, 2016). Institutional response and sustainable mechanisms are crucial in any system of government to prevent and protect citizens' livelihoods. In particular, this is one of the rights and privileges enjoyed by people, irrespective of color, status, or creed. It goes to say that a government that does not protect, promote, and conserve the rights of its citizens may be confronted with agitations and resistance to governmental authority. Protecting the people is similar to protecting the environment. Environmental quality depends on its management in terms of process of control and organization. Environmental management aims at reducing and/or eliminating the chances of the vulnerability of the environment to disaster preparedness, prevention, and mitigation (Adeniran, 2013). To this effect, many nations, including Nigeria, have institutionalized risk-reduction and management agency to address the plight of the people.

The National Emergency Management Agency (NEMA) of Nigeria is a risk-management body in the country. The agency is an offshoot of the defunct National Emergency Relief Agency (NERA) of 1976. Prior to the establishment of NEMA, NERA had the mandate to collect and distribute relief materials to disaster victims. Specifically the agency dealt with drought of 1972−73, which then devastated the Northern Nigeria, leading to socioeconomic consequences, loss of lives and property (UNDP, 2019). NEMA established via Act 12, as amended by Act 50 of 1999. According to UNDP, the primary objective of NEMA is to "coordinate and facilitate disaster management efforts aimed at reducing the loss of lives and property and protect lives from hazards by leading and supporting disaster management stakeholders in a comprehensive risk-based emergency management program of mitigation, preparedness, response and recovery" (UNDP, 2019). The mandate of NEMA also includes:

> The systematic process of using administrative decisions, organization, operational skills and capacities to implement policies, strategies and coping capabilities of the societies and communities to lessen the impacts of natural hazards and related environmental and technological disasters. This comprises all forms of activities, including structural and non-structural measures to prevent or limit (mitigation and preparedness) adverse effects of hazards.
>
> **NEMA (2011, p. 243)**

Flood is an environmental problem, the destructive tendency of which can be prevented or minimized through policy and planning by the government and background information on flood risk is a prerequisite to managing the impacts of flooding and making informed decisions in addressing such impacts (Adebayo, 2014). In its commitment to environmental degradation and natural disasters, the Nigeria government had entered and ratified several global treaties on the protection and conservation of the environment. Notably among the treaties include the Ramsar Convention of 1971, Stockholm Declaration on the Human Environment 1972, the Vienna Convention for the Protection of the Ozone Layer of 1988, the Basel Convention on the Control of Transboundary Movements of Hazardous Wastes and their Disposal, the Biodiversity Convention of 1994, Convention on Climate Change of 1992, and the Rio Declaration 1992. All these intergovernmental treaties which Nigeria is a signatory have their mission to promote and protect both national and transnational environment. It may be concluded that impacts motivated an attempt to develop the National Disaster Response Plan (NDRP) for Nigeria. NDRP establishes a process and structure for the systematic, coordinated, and effective delivery of federal assistance, to address the consequences of any major disaster or emergency declared by the President of the Federal Republic of Nigeria. This national plan is supported by a management agency (NEMA), which was established in March 1999.

NEMA is supported by the Nigerian Meteorological Agency (NIMET). However, NEMA at the national government is replicated at the states and local government areas (LGAs). The organogram in Fig. 24.1 depicts the system of operation in levels of government.

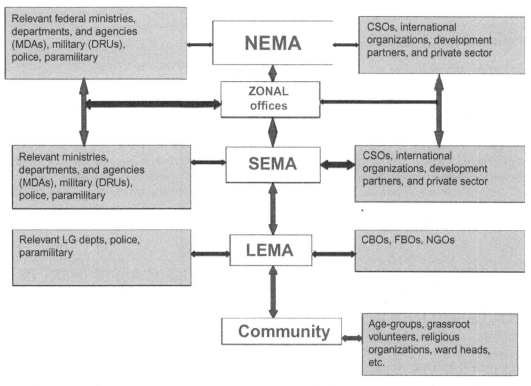

FIGURE 24.1

Systemic structure of the DRM in Nigeria. *DRM*, Disaster risk management.

Adapted from Nigeria National Disaster Framework (2010, p. 21).

NEMA, state emergency management agency, and local emergency management authority are managed by nation, state, and LGA, respectively. The agencies are structured to collaborate with relevant ministries, departments, nongovernmental agencies (local and international), international agencies, security, and others to build resilience and reduce vulnerabilities to disasters. It is common knowledge that disasters can hardly be avoided; however, research findings show that effective disaster management plan is a precondition to mitigating the impacts. In effect, national and international governments are partnering at the regional and global levels to implement disaster management cycle involving mitigation, preparedness, response, and recovery (James et al., 2013). As a follow-up to national commitment to addressing disaster, NEMA has preliminary contingency plan known as the NDRP. The contingency plan is structured in a systematic and coordinated delivery of national assistance to addressing the consequences of disaster, or as directed by the President of Nigeria (James et al., 2013). Above all, according to Sani (2012), the framework for disaster management activities in Nigeria is grouped into six:

1. national development planning and disaster management,
2. disaster prevention,
3. disaster mitigation,
4. disaster preparedness,
5. disaster response, and
6. disaster recovery.

All these frameworks of activity are channeled toward the overall well-being of the affected population. It takes into consideration disaster prevention and mapping, budgeting, and provision of relief materials, and other contingencies. Although the framework of action is supposedly decentralized in structure and operation across states and LGAs, evidently the actual functioning is domiciled with the national government. Considering the prevailing and fluctuating economic conditions of Nigeria, the task of disaster management and reduction may be enormous to embark on at all times. There is need for concerted effort at other levels of government to stem the tide natural disaster in Nigeria. Improved funding and technical input to the responsible government agencies will create a robust and patterned process in disaster mapping, postdisaster needs assessment, and provision of necessary support to the needy.

24.6 DISASTER MAPPING AND POSTDISASTER NEEDS ASSESSMENT IN NIGERIA

Flood is one of the biggest threats to lives and livelihoods. The havoc is unprecedented and rages as climate change effects intensify. An understanding of the potentials of flooding is the combined adoption of meteorological forecasts and land-surface hydrological models to predict riverine and surface runoff flooding (Pappenberger et al., 2013). It implies that the skill of weather forecasting has steadily improved over the last decades (Simmons & Hollingsworth, 2002), and the application of weather forecasts is in anticipation of disasters (Pappenberger et al., 2013). For instance, in line with its mandate to monitor weather and climate, and water information for sustainable development and safety, the NIMET produces the seasonal rainfall prediction that borders on onset and cessation dates, length of the rainy season, annual amount of rainfall, expected changes in the predictions of these conditions, and socioeconomic implications of the predictions for Nigeria (NIMET, 2013). This is critical in determining and cushioning the hazardous effects of flood when it eventually occurs. Developing Nigeria involves not only the improvement on political and socioeconomic aspects but also include the importance of taking into account technological and meteorological hazards and weather prediction (Ojo, Olanusi, Akinnubi, & Akinnubi, 2011). This will enhance early warning systems for environmental hazard preparedness.

Disaster mapping is channeled toward identifying areas prone or vulnerable to disasters, with attendant precautions and preparedness to mitigating the effects. According to Kaoje and Ishiaku (2017), flood vulnerability mapping is a simple presentation of areas that are at risk of flood events on maps. In Nigeria, those communities living close to coastal areas, such as Lagos, Akwa-Ibom, Rivers, Delta, Jigawa, Cross River, Benue, and Kogi states, are most likely to flood vulnerability. Disaster mapping enables forecast organization to engage the natives on early warnings, and best strategies to avert casualties among other concerns. In the event of floodwaters overflowing riverbanks, it is expected that the socioeconomic patterns of living will be disrupted for a while. Considering the consequences of flood hazard in the country, it becomes pertinent to assess the impact on the overall well-being of the affected people for inclusiveness and optimal living. While disaster mapping in Nigeria is an independent and official responsibility of the government, the postdisaster needs assessment in Nigeria is not only an independent practice but also takes the form of collaboration or interface between the government and community members and development partners locally and beyond national boundaries.

With technical assistance from European Union, United Nations, World Bank, and relevant partners, the Federal Republic of Nigeria (FRN) (2013) published a report on the 2012 flood in Nigeria. The report detailed that disaster in the country was occasioned by a combination of three major factors, high-intensity rain between June and September 2012, land-use practices that obstructed drainage system, which eventually increased the water flows, and the release of excess water from river dams within and outside Nigeria's land border into rivers Niger and the Benue. According to the 2012 report, Cameroon released enormous amounts of water from its Lagdo dam into the Benue river, thereby adding to the releases from Nigeria's Kainji, Shiroro, and Jebba dams. Furthermore, the report indicated that of 36 states of the country, 254 LGAs and 3870 communities in 28 states of the country were affected by flood. In addition, 685,501 and 321,866 houses were damaged and partially damaged, respectively. More so, 82 schools, 20 markets, and over 3 million hectares of farmlands were decimated. On human impact, over 2 million people were displaced, 7 million people were affected either directly or indirectly, and there were 18,422 injured and 381 dead (FGN, 2013). UNICEF (2013) opined that although NEMA and NIMET reported that toward the end of 2012, flood receded throughout the country; however, the impact of the flood on affected communities remains critical due to loss of livelihood and poor access to water, hygiene and sanitation, and health services. In addition, a high percentage of water sources are likely to be contaminated or damaged due to flooding of the communities. WHO assessment of the 2012 flooding in Nigeria revealed that flood has affected rural communities heavily reliant on subsistence agriculture. WHO further noted that basic services were severely overstretched and many communities totally isolated by the floodwaters (World Health Organization, 2012).

Among other postdisaster needs assessment in Nigeria, the United Nations Economic Commission for Africa (2015), in reaction to 2012 floods in the country, explains that the postdisaster needs assessment in November and December 2012 objectives relied on its assessment in terms of losses and damages; determining the overall impact of floods on socioeconomic development at the national, state, and community levels; formulating a recovery and reconstruction framework that captures the costs, short-, medium-, and long-term needs; and defining a strategy for disaster risk management in the country. Postdisaster assessment of flood hazard may have direct or indirect impact on the people. The direct effects include damages such as displacement of the population, injuries, infrastructures and interference with socioeconomic lives of the people (Adelekan, 2010; IFRCS, 2012; Jha, Bloch, & Lamond, 2012), and loss of farmland and farm produce, and outbreak of epidemics that threatens the health of survivors (Bello & Ogedegbe, 2015). On the other hand, survivors of flood disaster face the indirect impacts of flood hazard, which include but not limited to damage to the environment and psychosocial effects (Kayode, Yakubu, Ologunorisa, & Kola-Olusanya, 2017), and symptoms of posttraumatic stress disorder, depression, and anxiety (Mason, Andrews, & Upton, 2010).

The shortcoming of this laudable effort by the United Nations Economic Commission for Africa is bedeviled with Nigeria's institutional approach. The commission's findings noted that a major factor hindering optimal disaster risk reduction mainstreaming into development plans is the sluggish progress by state and local government authorities to fully establish emergency management agencies and emergency management committees. In addition, it places additional burden on NEMA, which in turn overstretches the agency's capacity and lessens the effectiveness of its responses. Among other issues observed as factors hindering mainstreaming risk-reduction plan into national development agenda include insufficient funding for the implementation of programs (United Nations Economic Commission for Africa, 2015). Although the government of Nigeria, through NEMA, oversees the management of disasters in the country, there is no substantial evidence indicating robust budget for disaster management in Nigeria. Again, Nasiru (2012) furthermore explains that lack of political will and lack of provision of facilities such as working equipment and communication system are some challenges impeding the effective and efficient management of disasters by NEMA. This may be due to the circumstance that it is an annual event, or an occurrence within a short period of time. Obviously, this may contribute to inadequate attention to the recruitment of qualified personnel for disaster management in the country.

The United Nations Economic Commission for Africa (2015) noted inadequate level of trained personnel in addressing disasters in the country. This observed anomaly in the workforce agreed with the findings of the Global Network of Civil Society Organizations for Disaster Reduction (GNDR) in 2018. A survey in Nigeria by GNDR (2018) showed that public officials responsible for the implementation of Sendai Framework do not clearly understand the framework of action. This is another major setback in alleviating the challenges of the victims of flood disaster, especially in rescue operation, needs assessment, and the nature of relief materials to the victims of flood disaster in Nigeria.

Nasiru (2012) observed that disaster relief remains an important factor in disaster management, but not enough to reduce human needs and an environmental impact of future disaster. An appraisal of the Nigerian government responses to flooding has been abysmal in operation and coordination of relief materials compared to the developed societies where government institutions are strong and reliable. According to Adefisoye (2015), poor coordination of relief and rescue activities worsens the situation of the affected people, noting that the people resign to faith as they depend on their effort to rebuild their houses and businesses. Flooding in the country is mostly managed by the victims. The people count their losses and take to self-help measures in terms rebuilding and reconstructing their homes, devise means of livelihoods after the flood is over.

Across the states and local government levels, emergency responses and management are more or less nonexistent. The states depend solely on the national government for rescue operations and management. The implication is that more of the budget and funding are provided by the national government. According to Onwabiko (2012), disaster response initiatives are worse off and emergency services dysfunctional due to the inability of state governors to establish institutional structures at the grassroot level of the 36 states and 774 local government councils of the country. In addition, Adefisoye (2015) explains that lack of legal backing, in addition to nonconformity and compliance to local government provisions for emergency situations another major challenge to relieving the conditions of those affected by natural disaster. Sustainable emergency management requires proper understanding of disaster risk so as to mitigate vulnerability and enhance coping capacities in the face of disaster (Okoli, 2014).

24.7 CONCLUSION

Climate change resulting to flooding remains one of the most common natural disasters experienced across nations of the world, Nigeria inclusive. Climate change, which has both natural and human-induced causes, mostly results in rising sea levels in Nigeria, leading to flood. The consequences of flood are enormous constituting social, economic, and environmental dimensions. The impact of flood affects every sphere of human lives, including but not limited to agriculture, economy, property and infrastructures, human health, and loss of live stocks. The most disastrous of them all is loss of human lives. The economic and environmental impact of flood is most notably on land as observed through erosions, thus limiting agricultural production that is the main source of livelihood for vast population in Nigeria. The need to overcome the problem of flooding cannot be overemphasized. Thus, to achieve SDGs, adequate measures are required for adaptation and mitigation of natural disaster (flood). Institutional response and sustainable mechanisms are crucial to prevent and protect citizens' livelihoods, building resilience to disasters, engaging in disaster mapping; sufficient funding and prudent management of disaster management body are panacea to the problem of flooding in Nigeria.

REFERENCES

Adebayo, W. A. (2014). Environmental law and flood disaster in Nigeria: The imperative of legal control. *International Journal of Education and Research*, 2(7), 447–468.

Adefisoye, T. (2015). An assessment of Nigeria's institutional capacity in disaster management. *Scientific Research Journal (SCIRJ)*, 3, 37–48.

Adelekan, I. O. (2010). Vulnerability of poor urban coastal communities to flooding in Lagos, Nigeria. *Environment and Urbanization*, 22, 433–450.

Adeniran, A. J. (2013). Environmental disasters and management: Case study of building collapse in Nigeria. *International Journal of Construction Engineering and Management*, 2(3), 39–45. Available from https://doi.org/10.5923/j.ijcem.20130203.01.

Agbonkhese, O., Yisa, G. L., & Daudu, P. I. (2013). Bad drainage and its effects on road pavement conditions in Nigeria. *Civil and Environmental Research, 3*(10), 7–15.

Agwu, J., & Okhimamhe, A. A. (2009). *Gender and climate change in Nigeria: A study of four communities in north-central and south-eastern Nigeria.* Lagos: Heinrich Böll Stiftung (Hbs).

Bello, I. E., & Ogedegbe, S. O. (2015). Geospatial analysis of flood problems in Jimeta Riverine Community of Adamawa State, Nigeria. *Journal of Environment and Earth, 5*(12), 32–45.

Caruso, G. D. (2017). The legacy of natural disasters: The intergenerational impact of 100 years of disasters in Latin America. *Journal of Development Economics, Elsevier, 127*(C), 209–233.

Centre for Research on the Epidemiology of Disasters (CRED). (2018a). *Natural disasters 2018.* Brussels: Author.

Centre for Research on the Epidemiology of Disasters (CRED). (2018b). *2018 review of disaster events.* Brussels: Author.

Combaz, E. (2014). *Disaster resilience: Topic guide.* Birmingham: GSDRC, University of Birmingham.

CRED. (2017). *Economic losses, poverty and disasters 1998-2017.* Brussels: Author.

CRED. (2017/2018). *Natural disasters 2017.* Brussels: CRED.

Daramola, A. Y., Oni, O. T., Ogundele, O., & Adesanya, A. (2016). Adaptive capacity and coping response strategies to natural disasters: A study in Nigeria. *International Journal of Disaster Risk Reduction, 15,* 132–147. Available from https://doi.org/10.1016/j.ijdrr.2016.01.007.

Ellis, S. (2008). *The changing climate for food and agriculture: A literature review.* Minnesota: Institute for Agriculture and Trade Policy.

FAO. (2015). *The impact of disasters on agriculture and food security.* Italy: FAO, /Prakash Singh.

FGN. (2013). *Nigeria post-disaster needs assessment 2012 floods.* Abuja: Government Press.

FGN. (2010). *Nigeria: National disaster framework.* Abuja: FGN.

FGN. (2013). *Nigeria post-disaster needs assessment 2012 floods.* Abuja: FGN.

Fitchett, J. M., Grant, B., & Hoogendoorn, G. (2016). Climate change threats to two low-lying South African coastal towns: Risks and perceptions. *South African Journal of Science, 112,* 1–9.

GNDR. (2018). *Local voices for resilience.* London: Global Network of Civil Society Organisations for Disaster Reduction (GNDR).

Gwary, D. (2008). Climate change, food security and Nigeria agriculture. In *Paper presented at the workshop on the challenges of climate change for Nigeria.* NISER 19th–20th May, 2008.

Hammond, M. J., Chen, A. S., Djordjević, S., Butler, D., & Mark, O. (2015). Urban flood impact assessment: A state-of-the-art review. *Urban Water Journal, 12,* 14–29.

IFRCS. (2012). *Emergency appeal operation updates.* Retrieved from <http://reliefweb.int/sites/reliefweb.int/files/resources/MDRNG01401.pdf>.

Igwe, M. C., & Wordu, S. A. (2016). Community-based resilience to the 2012 flood disaster in Orashi Region of Rivers State, Nigeria. *International Journal of Novel Research in Humanity and Social Sciences, 3*(6), 51–66.

International Federation of Red Cross and Red Crescent Societies (IFRC). (2016). *Resilience: Saving lives today, investing for tomorrow.* Geneva: IFRC.

IPCC. (2007). *From climate change 2007.* Available from <www.ipcc.ch/pdf/assessment-report/ar4/wg3/ar4-wg3-chapter8.pdf>.

James, G., Shaba, H., Zubair, O., Teslim, A., Yusuf, G., & Nuhu, A. (2013). *Space-based disaster management in Nigeria: The role of the international charter "space and major disasters".* Abuja: Environment for Sustainability.

Jha, A. K., Bloch, R., & Lamond, J. (2012). *Cities and flooding: A guide to integrated urban flood risk management for the 21st century.* Washington, DC: World Bank.

Kaoje, I. U., & Ishiaku, I. (2017). Urban flood vulnerability mapping of Lagos, Nigeria. *International Journal of Science and Technology, 3*(1), 224–236, *Special Issue.*

Kayode, J. S., Yakubu, S., Ologunorisa, T. M., & Kola-Olusanya, A. (2017). A post-disaster assessment of riverine communities impacted by a severe flooding event. *Ghana Journal of Geography*, *9*(1), 17–41.

Leister, P. (2009). *Flooding in Wales*. Retrieved from <http://www.environmentagency.gov.uk/static/documents/Research/ENV0005_Flooding_in_Wales_ENGLISH_AW_LR(1).pdf>.

Mallakpour, I., & Villarini, G. (2015). The changing nature of flooding across the central United States. *Nature Climate Change*, *5*, 250–254.

Mason, V., Andrews, H., & Upton, D. (2010). The psychological impact of exposure to floods. *Psychology, Health & Medicine*, *15*(1), 61–73.

Mata-Lima, H., Alvino-Borba, A., Pinheiro, A., Mata-Lima, A., & Almeida, J. A. (2013). Impacts of natural disasters on environmental and socio-economic systems: What makes the difference? *Ambiente & Sociedade*, *XVI*(3), 45–64.

Montz, B. (2009). Emerging issues and challenges: Natural hazards. *Journal of Contemporary Water Research and Education*, *142*, 42–45.

Mulugeta, G., Durrheim, R., Konare, A., Amponsah, P. E., Ayonghe, S. N., Benouar, D., et al. (2017). *Natural and human-induced hazards and disasters*. Pretoria: ICSU Regional Office for Africa.

Nasiru, G. (2012). *Transformation agenda for disaster management in Nigeria*. Available from <http://sahar-areports.com/articles/transformation-agenda-disastermanagement-Nigeria>.

Nelson, S.A. (2016). *River systems and causes of flooding*. Tulane University. Retrieved on 25th October, 2019 at <https://www.tulane.edu/~sanelson/Natural_Disasters/riversystems.htm>.

NEMA. (2011). *2010 annual report*. Abuja: National Emergency Management Agency.

NEST. (1991a). *Nigeria's threatened environment: A national profile*. Ibadan: Intec Printers.

NEST. (1991b). *Nigeria's threatened environment: A national profile*. Ibadan: Environmental Study/Action Team.

Newton, C., Paci, D., & Ogden, A. E. (2005). Climate change adaptation and regional forest planning in Southern Yukon, Canada: Mitigation and Adaptation Strategies. *Global Change*, *13*(8), 833–861.

NIMET. (2013). *Seasonal rainfall prediction*. Abuja: NIMET.

Noy, I., & Nualsri, T. B. (2010). The economics of natural disasters in a developing country: The case of Vietnam. *Journal of Asian Economics*, *21*, 345–354.

Nwaogu, N. R., & Ezekwe, I. C. (2018). Lessons in resilience: A rural community's adaption to an extreme flood event in northern Rivers State of Nigeria. *International Journal of Environmental Sciences & Natural Resources*, *9*(3), 001–008. Available from https://doi.org/10.19080/IJESNR.2018.09.555764.

Ojo, M. O., Olanusi, O. B., Akinnubi, R. T., & Akinnubi, R. T. (2011). Application of meteorology and weather prediction in the sustainable environmental quality in Nigeria. *Journal of Environmental Issues and Agriculture in Developing Countries*, *3*(3), 100–105.

Okoli, A. C. (2014). Disaster management and national security in Nigeria: The nexus and the disconnect. *International Journal of Liberal Arts and Social Science*, *2*(1), 21–59.

Okonkwo, I. (2013). *Effective flood plain management in Nigeria: Issues, benefits and challenges*. Retrieved from <http://transparencyng.com/index.php/contributions/60-guest/8548-effective-flood-plain-management-in-nigeria-issues-benefits-and-challenges>.

Oladokun, V. O., Proverbs, D. G., & Lammond, J. (2017). Measuring flood resilience: A fuzzy logic approach. *International Journal of Building Pathology and Adaptation*, *35*, 470–487.

OlajuyIgbe, O., & Ajayeoba, A. (2012). *Floods in Nigeria*. Abuja: Actionaid.

Olanrewaju, C. C., Chitakira, M., Olanrewaju, O. A., & Louw, E. (2019). Impacts of flood disasters in Nigeria: A critical evaluation of health implications and management. *Journal of Disaster Risk Studies*, *11*(1), 1–9.

Omotosho, J. B., Baloguun, A. A., & Ogunjobi, K. O. (2000). Predicting monthly and seasonal rainfall, onset and cessation of the rainy season in West Africa using only surface data. *International Journal of Climatology*, *20*, 865–880.

Onu, F. M., & Ikehi, M. E. (2016). Mitigation and adaptation strategies to the effects of climate change on the environment and agriculture in Nigeria. *IOSR Journal of Agriculture and Veterinary Science (IOSR-JAVS)*, *9*(4), 26–29.

Onwabiko, E. (2012). *Human rights writers association of Nigeria*. Available from <www.huriwa.Blogspot.com>.

Ozor, N., Madukwe, M. C., Enete, A. A., Amaechina, E. C., Onokala, P., Eboh, E. C., et al. (2012). A framework for agricultural adaptation to climate change in Southern Nigeria. *International Journal of Agriculture Sciences*, *4*(5), 243–251.

Pappenberger, F., Wetterhall, F., Dutra, E., Giuseppe, F. D., Bogner, K., Alfieri, L., et al. (2013). Seamless forecasting of extreme events on a global scale. Climate and land surface changes in hydrology. In *Proceedings of H01, IAHS-IAPSO-IASPEI assembly*. Gothenburg, Sweden, July 2013 (IAHS Publ. 359, 2013).

Park, J., Seager, T. P., Rao, P. S., Convertino, M., & Linkov, I. (2013). Integrating risk and resilience approaches to catastrophe management in engineering systems. *Risk Analysis*, *33*, 356–367.

Sani, M. (2012). *Nigeria: NEMA develops national contingency plan on disaster management*. Retrieved 14/06/2013 from <allafrica.com/stories/201206180409.html>.

Sayne, A. (2011). *Climate change adaptation and conflict in Nigeria*. Washington, DC: United States Institute of Peace.

Simmons, A. J., & Hollingsworth, A. (2002). Aspects of the improvement in skill of numerical weather prediction. *Quarterly Journal of the Royal Meteorological Society*, *128*(580), 647–677.

Thapa, S., Joshi, G. R., & Joshi, B. (2015). Impact of Climate Change on Agricultural Production in Nepal. *Nepalese Journal of Agricultural Economics*, *2–3*, 64–78.

UNICEF. (2013). *UNICEF Nigeria situation report*. Available from <https://www.unicef.org/appeals/files/UNICEF_Nigeria_SitRep_February_2013.pdf>.

United Nations. (2002). *Natural disasters and sustainable development: Understanding the links between development, environment and natural disasters*. Geneva: United Nations International Strategy for Disaster Reduction.

United Nations Development Programme (UNDP). (2019). *Baseline study on disaster recovery in Africa: Transitioning from relief to recovery*. New York: Author.

United Nations Economic Commission for Africa. (2015). *Assessment report on mainstreaming and implementing disaster risk reduction measures in Nigeria*. Addis Ababa: United Nations Economic Commission for Africa.

United Nations Human Settlements Programme (UN-Habitat). (2011). *Cities and climate change global report on human settlements 2011*. London: Author.

U.S. Environmental Protection Agency (EPA)]. (2016). *Climate change indicators in the United States* (4th ed.). Retrieved from <https://www.epa.gov/sites/production/files/documents/climate_ind>.

World Health Organization. (2012). *Public health risk assessment and interventions: Flooding disaster Nigeria, November 2012*. Geneva: World Health Organization.

CHALLENGES OF ADDRESSING NATURAL DISASTERS IN NIGERIA THROUGH PUBLIC POLICY IMPLEMENTATION: AN EXAMINATION OF ISUIKWUATO EROSION AND THE ECOLOGICAL FUND

25

Emeh Ikechukwu Eke[1] and Kalu T.U. Ogba[2]

[1]*Department of Public Administration, University of Nigeria, Nsukka, Nigeria* [2]*Department of Psychology, University of Nigeria, Nsukka, Nigeria*

25.1 INTRODUCTION

Handicapped by the disastrous gully erosion threats confronting them, the Isuikwuato Local Government Council resorted to caution the users of Uturu-Isuikwuato road of the danger of the road split. This caution became necessary given the danger to lives and properties the ditch portends as many unsuspecting commuters have fallen into the ditch leading to life-threatening injuries, destruction of vehicles, and loss of valuables and lives. Indeed, the people of Isuikwuato in particular and Abia North Senatorial District in general have been subjected to untold ecological, economic, and psychological ordeals to erosions, which have swept off farmlands and business outfits and threatened their ancestral homes. Of the adverse effects of this gully erosion (natural disaster), the ripping off and shutting down of the only commutable access to their ancestral homes (villages) is gruesome and the only option of diverting into the bush to meander around the terrible spots has gone bizarre; hence, people are now exposed to armed robbery attacks, kidnaping, raping, and herdsmen attacks. These security challenges are reminiscent of Igwe and Fukuoka's (2010) assertion that erosion has affected nearly every part of the country but more aggressive in the southeastern part where it has killed people; torn roads in shreds; destroyed homes, schools, and farmlands; and displaced people from their homes.

Unfortunately, extant literature from the West has not seem to accord due recognition to erosion as a natural disaster with life-threatening capacity. But, in Nigeria, an avalanche of literature has shown that erosion is a natural disaster with unmatched capacity to wreak havoc on both the environment and the economy of the people; hence, almost every state in Nigeria is currently threatened by soil erosion (see Figs. 25.1–25.4), especially the southeastern states, with Abia, Imo, and

FIGURE 25.1

The straight view of the road ruptured by erosion gully at Mgbelu-Umunnekwu, Isuikwuato.

FIGURE 25.2

The right hand side view of erosion gully at Mgbelu-Umunnekwu from Uturu and the right hand side view of erosion gully at Mgbelu-Umunnekwu from Ovim-Isuikwuato.

FIGURE 25.3

Electric pole pushed down and destroyed by erosion gully at Mgbelu-Umunnekwu.

FIGURE 25.4

Road breakage as a result of erosion gully menace at Mgbelu-Umunnekwu, Isuikwuato.

Anambra topping the list. In these states, evidence abounds that erosions have caused severe damages to structures and systems as buildings have collapsed and properties swept away. The effect is felt more during raining seasons when flooding increases and crops and farmland are swept away. During this period also, fragile buildings yield to the pressures of increased flooding and are uprooted, same goes to roads. Unfortunately, the movement of debris (e.g., remains of buildings and roads uprooted) also causes extensive damage to farmlands.

While Igwe (2012) had asserted the gradual but constant dissection of the landscape by soil erosion, which threatens settlements and scarce arable land as the greatest threat to the environmental settings of southeastern Nigeria, Ofomata (1975) remarked that more than 1.6% of the entire land area of eastern Nigeria is occupied by gullies. This figure is very significant for an area that has the highest population density 500 persons per km^2 in Nigeria then, because it portends high vulnerability and susceptibility to extinction. This is because before the 1980s the classical gully sites in the region were the Agulu, Nanka, Ozu-item, Oko in Aguata area, Isuikwuato, and Orlu. Unfortunately, since the 1980s till now, the situation has not abated, instead exacerbated such that the southeast has about 2800 active erosion gullies currently (Igwe & Fukuoka, 2010; Nwankwo, 2018).

It is this kind of circumstance that necessitated the establishment of Ecological Fund in 1981 as an intervention fund for the amelioration of ecological problems such as soil erosion, flood, drought, desertification, oil spillage, pollution, and general environmental pollution. Unfortunately, this establishment that currently receives 3% of the federation earnings monthly has been characterized by gross mismanagement (Tables 25.1 and 25.2). This gross mismanagement of the Ecological Funds spurred Senator Mohammed Hassan to lead a debate on the proper usage of the Ecological Funds. The essence of the debate, as it seems, was to instill discipline, transparency, and accountability in the administration of the Ecological Fund because according to the senator, while erosion has ravaged the southeast (where Isuikwuato is), the north is threatened by dessert encroachment,

Table 25.1 Revenue Sharing Formula for the Three Tiers of Government.

S/No	Tier of Government	Derivation	Ecology	Total
1	Federal	0.49	0.97	1.46
2	State	0.24	0.48	0.72
3	Local	0.20	0.40	0.60
Total				2.78

https://www.osgf.gov.ng/storage/app/media/uploaded-files/background%20on%20ecological%20fund.pdf.

Table 25.2 Adjusted Special Funds Under the Federal Government.

Special Funds	Before 2004 (%)	2004 to Date (%)
FGN Share of Derivation and Ecology	1.46	1.00
Development of Natural Resources	3.00	1.68
Stabilization Fund	0.725	0.5

https://www.osgf.gov.ng/storage/app/media/uploaded-files/background%20on%20ecological%20fund.pdf.

yet, instead of putting the funds where they are statutorily meant to go into, they are diverted into other things not captured by the mandates and objective of the Ecological Fund.

These points raised by that bill and total exclusion of Isuikwuato in the publication from the Ecological Fund Office (EFO) on the projects approved between May 2015 and September 2019 are the main concerns of this chapter. Thus, given the dreaded situation of Isuikwuato erosion, a total exclusion from the Ecological Fund distribution within the 5 years published only calls for a reexamination of public policies implementation toward the vulnerability of the people of the country whose lives and properties are threatened by natural disasters. This chapter, therefore, examined the challenges of addressing the erosion menace in Isuikwuato in Abia North senatorial zone, Abia State, Nigeria, through the Ecological Fund administration.

25.2 THE NATURE OF SOIL AND GULLY EROSION MENACE IN ISUIKWUATO, ABIA STATE, NIGERIA

Anytime tale of erosion menace in southeast Nigeria is told, a mention of Isuikwuato, Abia State, is imminent. While Isuikwuato is not the only local government area of Abia State ripped apart by erosion (see Table 25.3), its peculiarity stems from the volatile and corrosive nature of erosion in the area. For instance, not only is Isuikwuato geologically prone to soil washing that leads to and further exacerbates gully erosions, the bordering communities of Uturu and Bende also have threatening erosion gullies. Apart from these threats from outside Isuikwuato, virtually every community in Isuikwuato is heavily threatened by erosion such that it is difficult, if not impossible to assert with profound level of exactitude the areas that are erosion flashpoints and erosion-free zones. This is the reason communities in Isuikwuato featured prominently in Uluocha and Uwadiegwu's (2015) categorization of towns at risk of erosion in Abia State (see Table 25.4). In the study, communities in Isuikwuato such as Amanta and Amaiyi featured in the severe risk zone category, while Umunnekwu was enlisted in the moderate risk zone, but as at November 4, 2019, Umunnekwu has migrated from the moderate risk zone to a death trap as shown in Figs. 25.1−25.3. These threats start from Mgbelu-Umunnekwu part of Ohukabia to Amaiyi Obinohia. The situation of Ohukabia today used to be the story of Umunnekwu Agbo some 10 years ago but even though it has not returned to its old self, it is manageably motorable without recourse to diversion into the bush path (see Figs. 25.2−25.5). Even Amaokwe-Elu community of Isuikwuato that was completely displaced due to the expansion of gully sites in 2011 is still rated among the high risk zone (Nwilo, Olayinka, Uwadiegwu, & Adzandeh, 2011; Uluocha & Uwadiegwu, 2015).

While erosion menace in Isuikwuato is a precarious sight, Okey (2019) sees the situation as hellish. He compared it to that of Umunwanwa clan of Umuahia South Local Government Area of Abia State where not less than 10 houses were swallowed at once by erosion in 2018 and asserted that the erosion menace in Isuikwuato is horrendous. He decried the devastation natural disaster has done to a serene and beautiful landscape that was hitherto, a natural beauty to behold. The narrative of a commercial driver, identified as Chidiebere Eke, revealed more about the state of the gully erosion when he asserted, "I have been plying this road for the past 20 years and we (the motorists) have not had it this bad. When the first erosion site started, we were managing to find our way to Uturu but, with the second gully on the road, going to Uturu is hellish and, accessing Abia State

Table 25.3 Status of Projects Awarded by the Ecological Fund Office (May, 2015 to September, 2019).

S/No	Project Titles	Location	Status
1	Public Convenience	Oyo State	Completed
2	Model Village Development in Adamawa and Gombe	Adamawa and Gombe	Completed
3	Remediation of Kaduna Refinery	Kaduna	Completed
4	Soil Erosion Control Works at Usho/Ise Road, Ondo	Ondo State	Completed
5	Soil Erosion Control Works at Zuba	FCT	Completed
6	Kaltungo Gully Erosion Control Works, Gombe State	Gombe	Completed
7	Reconstruction of Yola-Mubi Bridge at Kaa'a Shiwa and Mail 30—Kwambla	Adamawa	Completed
8	Oghere Ahmed Erosion Control Works, Idimu, Lagos	Lagos state	Completed
9	Kudeti Flood Control Works, Ibadan Oyo State	Oyo State	Completed
10	Gully Erosion Control Works at Veritas University, Bwari, Abuja	FCT	Completed
11	Daura Township Erosion and Flood Control Project, II	Katsina	Completed
12	Daura Township Erosion Control, Phase I, Daura	Katsina	Completed
13	Gully Erosion Control Works at Magodo, Lagos State	Lagos	Completed
14	Soil Erosion Control Works at Nigerian Police Housing Estate Kurudu	FCT	Completed
15	Lohmak-Kumkwum Erosion Control Works, Langtang North LGA	Plateau	Completed
16	Doma Flood Erosion Control Works, Nasarawa State	Nassarawa	Completed
17	Umuoki-Umuowasi Erosion Control and Road Improvement Works	Imo	Completed
18	Umudiato Flood Control Works, Orlu LGA, Imo State	Imo	Completed
19	Owalla Avuvu Erosion Control Works	Imo	Completed
20	Reconstruction of Maska Earth Dam, Maska Funtua	Katsina	Completed
21	Remediation of Warri Refinery Phase II	Delta	Completed
22	Kwoi Erosion and Flood Control Project at Jaba	Kaduna	Completed
23	Mubi Flood/Erosion Control Works and River Yalzaram Training, Phase I	Adamawa	Completed
24	Gadan Toro to Market Road Erosion Control Works, Toro LGA	Bauchi	Completed
25	Adiabo Okutikang Beach Gully Erosion Control Works	Cross River	Completed
26	Kumo Gully Erosion Site, Anguwan Jauro Musa, Akko LGA	Gombe	Completed
27	Ganye Yelwa Road Erosion and Flood Control Works, Ganye LGA	Adamawa	Completed
28	Azare Township Flood and Erosion Control Works	Bauchi	Completed
29	Tsoho Gwaram Erosion and Flood Control Project, Gwaram LGA	Jigawa	Completed
30	Kumo Gully Erosion Site, Anguwan Jauro Sabo, Akko LGA	Gombe	Completed
31	Channelization and Flood Control of Ponj-Afo River, Ikirin	Osun	Completed
32	Ganye-Dabora-Daksam Road Erosion and Flood Control Works, Ganye LGA and Reconstruction of Jada Water Works System	Adamawa	Completed
33	Erosion and Flood Control Work in Bogoro LGA	Bauchi	Completed

Table 25.3 Status of Projects Awarded by the Ecological Fund Office (May, 2015 to September, 2019). *Continued*

S/No	Project Titles	Location	Status
34	Erosion Control Works at Nda, Nnobi and Agbor, Idemili LGA	Anambra	Completed
35	Sabke/Dutsi and Mashi Water Supply Project	Katsina	Completed
36	Erosion/Flood Control Project of Rice Field, Suru LGA	Kebbi	Completed
37	Erosion/Flood Control Project of Rice Field, Shanga LGA	Kebbi	Completed
38	Gully Erosion Control Works at Federal Government College, Otobi	Benue	Completed
39	Erosion/Flood Control Project of Rice Field, Bagudu LGA	Kebbi	Completed
40	Channelization and Desilting of Ogbagba and Okoko Rivers in Osogbo I & II	Osun	Completed
41	Ecological Challenges at the Federal University, Ndufu Alike Ikwo	Ebonyi	Completed
42	Erosion Control on the Drainage Basin of Ahmadu Bello University, Zaria Lake/Dam	Kaduna	Completed
43	Abaji Gully Erosion and Flood Control Works, Phase II, FCT	FCT	Completed
44	Rijau, Dugge, Dukku, T/Magajiya and Genu, Rijau Local Government Erosion and Flood Control Project	Niger	Completed
45	East of Kumo General Hospital to Jauro Musa Stream Gully Erosion Control	Glombe	Completed
46	Amper Flood Hazards and Erosion Control Works, Tarka LGA	Benue	Completed
47	Gully Erosion Control and Flood Improvement Works at Awlaw, Oji-River	Enugu	Completed
48	Erosion Control Works at Odo Alaamo, Ogbomoso	Oyo	Completed
49	Gully Erosion and Road Improvement Works, Nasarawa Town: Phase I	Nassarawa	Completed
50	Nkari Erosion Control and Road Improvement Works, Ini LGA	Akwa Ibom	Completed
51	University of Calabar Erosion & Flood Control Works, Calabar	Cross River	Completed
52	Erosion Control and Channelization of Okoko River, Oshogbo	Osun	Completed
53	Flood Control and Desilting of Ogbagba in Oshogbo Township	Osun	Completed
54	Mitigation of Flooding into Uburu Salt Lake and Over 20,000 HA of Rice Farm at Asu-Umunaga	Ebonyi	Completed
55	Soil Erosion Control and Road Improvement Works, Yalenguruza	Gombe	Completed
56	Soil Erosion and Flood Control Project at Agenebode/Fugar Erosion Site, Edo	Edo	Completed
57	Magama LGA Erosion and Flood Control Project	Niger	Completed
58	Erosion & Flood Control at Ifedayo/Boluwaduro/Ila in Osun State	Osun	Completed
59	Erosion and Flood Control Works at Bulangu Township and collector Drains Linking Ponds, Kafin Hausa	Jigawa	Completed
60	Erosion control Works at makinde way and Environs Idimu	Lagos	Completed
61	Amachara Flood and Erosion Control Works, Amachara Umuahia	Abia	Completed
62	Tambuwal-Dogondaji Erosion Works, Sokoto State	Sokoto	Completed
63	Odolu Community Soil Erosion and Flood Control Project	Kogi	Completed

(Continued)

Table 25.3 Status of Projects Awarded by the Ecological Fund Office (May, 2015 to September, 2019). *Continued*

S/No	Project Titles	Location	Status
64	Erosion and Flood Control Project Aiyeotoro Community	Ondo	Completed
65	Erosion Control Works, Miango, Bassa	Plateau	Completed
66	Erosion Control Project at State House Presidential Villa, Abuja	FCT	Completed
67	Bagole-Kofare Erosion Control Works	Adamawa	Completed
68	Erosion Control Works in Uvu, Askira	Borno	Completed
69	Gully Erosion and Flood Control, Idumu Osigbudu Street	Delta	Completed
70	Gully Erosion and Flood Control Works at Akpene Eket	Akwa Ibom	Completed
71	Flood, Channelization and Reconstruction of Failed Structures Along Ijebu-Ode/Ilesc Road Via Owa River	Ogun	Completed
72	Abaji Gully Erosion and Flood Control, FCT, Abuja	FCT	Completed
73	Erosion and Flood Control Works in Eruwa, Ibarapa East LGA	Oyo	Completed
74	Erosion and Flood Control Works Medical School, Bauchi Town	Bauchi	Completed
75	Erosion and Flood Control Works in Tarka and Guma LGA	Benue	Completed
76	Jalingo Metropolis (Kofar Fada) Shetima and Kasuwan Yelwa Areas, Turaki	Taraba	Completed
77	Okwe-obioha Obi-Ebere Road Erosion Control Project, Ikwuano	Abia	Completed
78	Erosion and Flood Control at Dumbulun, Tsanyawa LGA	Kano	Completed
79	Flood and Erosion Control of Damaturu-Balmari-Gashua Road, Damaturu	Yobe	Completed
80	Gully Erosion/Flood Control & Road Improvement Works at Sokale Community, Dutse Alhaji, Bwari Area Council, FCT	FCT	Completed
81	Erosion Control Works Along Bauchi-Ningi Road Junction to the Main Campus of Abubakar Tafawa Balewa University, Bauchi	Bauchi	Completed
82	Jikwoyi Erosion and Flood Control Works, AMAC	FCT	Completed
83	Stream Erosion Control at Chief Ogungbo Road One River Along Olabisi Onabanjo University Road, Ago Iwoye, Ijebu North LGA, Phase 1, Ogun State	Ogun	Completed
84	Gully Erosion/Flood Control and Road Improvement Works at Durumi, Mpampe	FCT	Completed
85	Emergency Works for the Damaged Culvert and Road Improvement Works at Saburi, AMAC, Abuja	FCT	Completed
86	Burga—Gwammadaji Flood and Erosion Control Works, Bauchi	Bauchi	Completed
87	Ondo Township Erosion Control Works, Ondo State	Ondo	Ongoing
88	Contract for Shoreline Protection, Aleibiri	Bayelsa	Completed
89	Erosion Control and Road Improvement Works of OORA/AAYO River, Oke-Ayo Area, Ilesha, Osun State	Osun	Completed
90	Soil Erosion and Flood Control Works at Project Development Institute Enugu (PRODA) Federal Ministry of Science & Technology, Enugu	Enugu	Completed
91	Flood and Erosion Control Works at Karishi Town, Abuja	FCT	Completed

Table 25.3 Status of Projects Awarded by the Ecological Fund Office (May, 2015 to September, 2019). *Continued*

S/No	Project Titles	Location	Status
92	Gandu, Tukuntawa and Zoo Road Erosion and Flood Control Project	Kano	Completed
93	Tundun Magayaki/Cemetary/Gully Erosion and Flood Control, Sokoto South	Sokoto	Completed
94	Gully Erosion Control at Okwohia Obowo/Ihitte Uboma LGA, Imo	Imo	Completed
95	Erosion and Flood Control Project at Jimeta Yola North LGA	Adamawa	Completed
96	Construction of Road and Drainage Along Amba-Bassa Road, Kokona LGA	Nasarawa	Completed
97	Flood and Erosion Control in Gashua and Nguru LGAs	Yobe	Completed
98	Erosion/Flood Control and Road Improvement Works Along kampani-Kogo Village, Bogoro LGA	Bauchi	Completed
99	Erosion/Flood Control at Odo-Owa in Oke-Ero LGA	Kwara	Completed
100	Bera-Gokana Stadium-Kpor Town Flood/Canalization Project, Gokana LGA	Rivers	Completed
101	Erosion Control Works at Sarius Palmetum & Botanic Garden in Maitama, AMAC, Abuja	FCT	Completed
102	Evacuation/Reclamation & Establishment of Integrated Solid Waste Management Facility Including Auxillary Works in Kurudu	FCT	Completed
103	Reconstruction of Drainage System in Mai'dua LGA, Katsina State	Katsina	Completed
104	Ikenne Town Flood/Erosion Control Project, Ikenne Town, Ogun	Ogun	Completed
105	Auchi-Fuga Agenebode Road Gully Erosion Control Works	Edo	Completed
106	Flood and Erosion Control at Jan Bako in Maradun LGA, Zamfara	Zamfara	Ongoing
107	Ote River Channelization, Erosion Control and Access Road, Osin Ekiti, Oye	Ekiti	Completed
108	Idye Basin Flood Control Measures, Makurdi, Benue State	Benue	Ongoing
109	Flood and Erosion Control Around Ochanja Market, Onitsha	Anambra	Completed
110	Police Signpost-FHA Estate Erosion Control and Road Improvement Works, Lugbe, AMAC	FCT	Completed
111	Umunze Erosion Control/Flood Control and Road Improvement Works, Orumba South LGA, Anambra State	Anambra	Completed
112	Erosion Control Project at State House Presidential Villa, Phase II	FCT	Ongoing
113	Public Service Institute of Nigeria, Abuja	FCT	Completed
114	Gully Erosion Control and Road Improvement Works at Agbozu, Umueze Amaba and Methodist Church Compound, Obiohia Uzuakoli	Abia	Completed
115	Erosion Control and Rehabilitation of Hong Garaha Road	Adamawa	Ongoing
116	Ogbeche Community Farms Road Erosion/Flood Control, Otukpo	Benue	Completed
117	Erosion Menace Control at the Federal Road Safety Corps Academy, Udi, Udi	Enugu	Completed
118	Erosion and Flood Control Along Abacha Road, Old Karu	Karu	Completed

(Continued)

Table 25.3 Status of Projects Awarded by the Ecological Fund Office (May, 2015 to September, 2019). *Continued*

S/No	Project Titles	Location	Status
119	Flood Control Project at Enhwe, Isoko LGA	Delta	Completed
120	Ramat Polytechnic Erosion and Flood Control Project	Borno	Completed
121	Erosion/Flood Control and Road Improvement Works at Federal University	Jigawa	Completed
122	Gishare-Talbushi Flood and Erosion Control Project at Dengi, Kanam	Plateau	Ongoing
123	Erosion Control Bridges/Road Improvement Works at Ashara-Wako, Kwali Area Council, Phase I	FCT	Completed
124	Dredging and Channelization of Mosafejo Canal, Surulere, Lagos	Lagos	Completed
125	Erosion and Flood Control in Dilapidated Building Infrastructure in Colleges Kaduna Polytechnic	Kaduna	Completed
126	Gully Erosion Control Project at Madufa Town, Zaki LGA	Bauchi	Completed
127	Ibadan Flood and Erosion Control Works	Oyo	Completed
128	Ibiaku-Unit-Uran Route Erosion Control Works, Uruan/Uyo	Akwa Ibom	Completed
129	Execution of Dutsi Water Supply Project, Phase II, Section II	Katsina	Completed
130	Execution of Sabke Water Supply Project, Phase II, Section I	Katsina	Ongoing
131	Additional Works on the Erosion Control and Rehabilitation of Hong Garaha	Adamawa	Ongoing
132	Construction of 3-Cell Box Culvert and Road Improvement Works at Abacha Road Karu	Nasarawa	Completed
133	Erosion Control in Ekwetekwe/Ogbunyagu Community, Ezza North	Ebonyi	Ongoing
134	Bera-Gokana Stadium-Kpor Town Flood/Canalization Project, Gokana LGA, Rivers State (Phase 2)	Rivers	Ongoing
135	Erosion Control and Road Improvement Works at Gora-Roguwa Road, Karu	Nasarawa	Ongoing
136	Gully Erosion and Flood Control Works at Fededral Government Boys College, Wuye	FCT	Completed
137	Road Improvement works at Uruata by Port Harcout Road and Soil Erosion Control	Abia	Ongoing
138	Ngomari Airport Erosion Control Works Jere LGA	Borno	Ongoing
139	Jakada Road Flood/Road Improvement Measures, Maddakia Zonkwa, Zango Kataf	Kaduna	Completed
140	Gully Erosion Control Works Within Bida Town	Niger	Completed
141	Emergency Channelization and Soil Erosion Reclamation at Paraku Crescent, Wuse II	FCT	Ongoing
142	Procurement of Incinerator 3nos (18−200 Containerized Medical Incinerator) at NAFDAC Office	FCT	Completed
143	Gully Erosion Control and Road Improvement Works at Army Post Service Estate, Phase 5 Road, Kurudu, AMAC	FCT	Completed
144	Onireke Flood Channelization and Road Improvement Measures, Ile-Ife	Osun	Completed

Table 25.3 Status of Projects Awarded by the Ecological Fund Office (May, 2015 to September, 2019). *Continued*

S/No	Project Titles	Location	Status
145	Erosion Control, Construction of Bridge and Road Improvement Works, Ashara-Wako, Kwali Area Council, Phase 2	FCT	Completed
146	Umuma Isiaku Erosion Control Works, Ideato South LGA	Imo	Ongoing
147	Construction of Jetty and Shoreline Protection Facilities at Underwater Warfare School (UWWS), Lagos	Lagos	Ongoing
148	Road Improvement Works and Construction of Bridge at Dutse Saki Village in Bogoro LGA	Bauchi	Ongoing
149	Erosion Control of Flooded Areas/Road Improvement Works in Owo	Ondo	Ongoing
150	Flood and Erosion Control in Oyo Town	Oyo	Ongoing
151	Uyanga-Ojor-Ifumkpa-Owai Erosion Control and Road Improvement Works, Akampka LGA	Cross River	Ongoing
152	Uyanga-Ojor-Ifumkpa-Owai Erosion Control and Road Improvement Works, Akampka LGA	Niger	Ongoing
153	Tulu-Tama Flood and Erosion Control Project, Tulu-Tama, Toro LGA	Bauchi	Ongoing
154	Ecological Fund Intervention at the Main Campus Phase II Site, ABU, Zaria	Kaduna	Ongoing
155	Uromi Road Rehabilitation and Erosion Control Works at Okhele Esan North-East LGA, Edo State	Edo	Ongoing
156	Erosion Control and Road Improvement Works at Bangshika to Gashaka Road, Adamawa State	Adamawa	Ongoing
157	Flood and Erosion Control Works at Kazaure, Kazaure LGA	Jigawa	Ongoing
158	Construction of Reinforced Concrete Drains and Land Reclamation Works at Suleja, Niger State	Niger	Ongoing
159	Gully Erosion and Road Rehabilitation at Dutsen Makaranta, Bwari Area Council, FCT	FCT	Completed
160	Road and Storm Water Drainage at the Federal College of Education, Yola	Adamawa	Ongoing
161	Erosion Control Works at Odugbo Community Along Opkokwu Stream, Apa LGA, Benue State	Benue	Ongoing
162	Erosion Control and Rehabilitation Works Along Rimi Zayam, Poichi and Gwalfadako River Bridge, Toro LGA, Bauchi	Bauchi	Ongoing
163	Construction of Drainage Structures and Road Improvement Works at Okemesi, Ekiti West LGA, Ekiti State	Ekiti	Ongoing
164	Establishment of Integrated Industrial Pollution Management Facility in Chalawa, Sharada and Bompai Industrial Areas of Kano, Kano State	Kano	Ongoing
165	Flood and Erosion Control and Road Improvement Works in Parts of Nnobi, Alor and Umudiaokka Towns in Idemili/Dunukofia LGA	Anambra	Ongoing
166	Ecological Challenges at the Federal University, Wukari, Taraba State	Taraba	Ongoing

(Continued)

Table 25.3 Status of Projects Awarded by the Ecological Fund Office (May, 2015 to September, 2019). *Continued*

S/No	Project Titles	Location	Status
167	Flood/Erosion Control Works at Federal Teaching Hospital, Abakaliki	Ebonyi	Completed
168	Construction of Culverts along Layin Gebedado Farm in Saminaka Ward, Yola	Adamawa	Completed
169	Erosion Control Works at Dekina Town, Dekina LGA, Kogi State	Kogi	Ongoing
170	Repairs of Embakment Failures at Leadership Training Centre, Apapa, Lagos (Federal Ministry of Youth and Sports Development)	Lagos	Ongoing
171	Riverbank Erosion Control at Maru, Zamfara State	Zamfara	Ongoing
172	Construction of Hydraulic Structures and Road Improvement Works on Eroded Tudun Wada to Karshi Road (Phase 1), FCT	FCT	Ongoing
173	Erosion Control and Rehabilitation Works Along Rimi Zayam, Polchi and Gwalfadako River Bridge, Toro LGA, (Phase II)	Bauchi	Ongoing
174	Erosion Control Works at Faruruwa Bridge, Faruruwa Ward Road, Takai LGA	Kano	Ongoing
175	Gully Erosion Control Works at the National Institute of Construction Technology (NICT) Uromi, (Phase I), Edo State	Edo	Ongoing
176	Special Intervention on Integrated Waste Management System for Katsina Township, Katsina State	Katsina	Ongoing
177	Flood Control and Swamp Upgrading/Road Improvement Works at Ibbi Road, GRA, Wukari, Taraba State	Taraba	Ongoing
178	Erosion and Flood Control Works/Reinforcement of Threatened Bridge Base at FUTO, Owerri, (Phase I), Imo State	Imo	Ongoing
179	Desertification Control and Small Scale Afforestation Scheme in Yusufari and Karasuwan LGA	Yobe	Ongoing
180	Erosion Control and Road Improvement Along Trademoor e/Voice Of Nigeria Road, Sabon Lugbe, FCT	FCT	Ongoing
181	Gully Erosion in Wuro Patuji (GindinKurna), Mubi South LGA	Adamawa	Ongoing
182	Erosion Control/Channelization Works at Ayekale/Olorunda to Lake 264, Oshogbo, Osun State	Osun	Ongoing
183	Proposed Access Road Gurku IDP Camp, Karu LGA, Nasarawa State	Nasarawa	Ongoing
184	Ecological Challenges of Siltation, Flooding and Pollution of Ferry Channels and It's Adjoining Communities in Lagos (Phase I), Lagos	Lagos	Ongoing
185	Erosion Control Works at Sabon Gida Village, Langtang South LGA	Plateau	Ongoing
186	Ecological Control and Road Improvement Works a Kafe District, FCT	FCT	Ongoing
187	Erosion Control and Road Improvement Works Along 3.8 km Enugu Eke-Ogul Eke-Eke Market ObodoAmankwo-Oma Eke Road Project Udi LGA	Enugu	Ongoing

Table 25.3 Status of Projects Awarded by the Ecological Fund Office (May, 2015 to September, 2019). *Continued*

S/No	Project Titles	Location	Status
188	Gully Erosion Control/Township Roads and Drainage Improvement Works a Sabon Garin Sirika and Shargalle, Dutsi LGA, Katsina State	Katsina	Ongoing
189	Construction of Drainage Retaining Walls and Erosion Control at the University of Abuja Main Campus, Airport Road, FCT	FCT	Ongoing
190	Flood Control/Road Improvement Works at Jega LGA, Kebbi State	Kebbi	Ongoing
191	Erosion Control at Federal College of Horticulture, Dadin Kowa	Gombe	Ongoing
192	Erosion and Flood Control Works at Patigi Town, (Phase I), Kwara	Kwara	Ongoing

FCT, *Federal Capital Territory.*
From The Ecological Fund Office. Status of Ecological Fund Office projects executed between May 2015 and September 2019. (2019). <https://ecologicalfund.gov.ng/wp-content/uploads/2019/09/EFOPROJECTS-MAY-2015-September-2019.pdf>. Projects awarded by the Ecological Fund Office from May 2015 to September 2019.

University, Uturu from the Isuikwuato axis now, is like going to the land of the dead (Figs. 25.1, 25.2 and 25.5)."

This horrible state of Isuikwuato erosion spurred Okey (2019) to assert that, in a matter of months, the entire Abia North Senatorial District may be cut off from the rest of the state if nothing reasonable is done urgently to checkmate both the existing deadly Isuikwuato gullies and the advancing gully erosion before the first Abiriba junction on the Umuahia/Ohafia highway.

It is instructive and essential to assert that the Isuikwuato erosion menace has lingered for a very long time with successive administrations not able to address the menace on account of lean resources of the state and local governments (Nwankwo, 2018; Umahi, 2016), hence the calls on relevant agencies of the federal government, to come and perform their constitutional duties to the people of Isuikwuato by declaring the area and others in the state similarly threatened by ecological mishaps, as ecological disaster zones requiring urgent critical intervention to avert impending ecological and economic dooms and/or compounding the existing ecological and economic challenges (Nwankwo, 2018).

25.3 EROSION-INDUCED ECOLOGICAL AND ECONOMIC CHALLENGES OF THE PEOPLE OF ISUIKWUATO

The people of Isuikwuato in Abia State have been exposed to various challenges as a result of the erosive character of their land, which the people depend on for their survival. With the current trend of erosion menace, lives have been lost, houses swallowed, farmlands swept, business outfits (such as petrol stations and restaurants) have been closed down, and the environment of the people has been negative tampered with. This tampering has had negative economic connotations on the people, which have also led to psychological traumatic experiences. The import of these challenges and impacts on the survivability of the people is in keeping with Lal's (2010)

Table 25.4 State-Based Analysis of the Projects Awarded by the Ecological Fund Office from May, 2015 to September, 2019.

S/No	State	Number of Projects
1	Abia	4
2	Adamawa State	13
3	Akwa Ikbom State	3
4	Anambra State	4
5	Bauchi State	12
6	Bayelsa State	1
7	Benue State	6
8	Borno State	3
9	Cross River State	3
10	Delta State	3
11	Ebonyi State	4
12	Edo State	4
13	Ekiti State	2
14	Enugu State	4
15	FCT	27
16	Gombe State	7
17	Imo State	6
18	Jigawa State	4
19	Kaduna State	6
20	Kano State	5
21	Katsina State	9
22	Kebbi State	4
23	Kogi State	2
24	Kwara State	2
25	Lagos State	7
26	Nasarawa State	6
27	Niger State	5
28	Ogun State	3
29	Ondo State	4
30	Osun State	8
31	Oyo State	6
32	Plateau State	4
33	Rivers State	2
34	Sokoto State	2
35	Taraba State	3
36	Yobe State	3
37	Zamfara State	2

FCT, *Federal Capital Territory.*
Compiled by the authors.

FIGURE 25.5

The first Isuikwuato erosion site at Umunnekwu Agbo, Isuikwuato.

assertion that soil erosion has severe adverse economic and environmental impacts on the people due to direct effects on crops or plants; pollution of natural waters and adverse effects on air quality due to dust and emissions of radioactive gases. There are also damages to civil structure; siltation of water ways and reservoirs, and additional costs involved in water treatment. For the purpose of this study, we identified and detailed these challenges under ecological and economic perspectives.

25.4 THE EROSION-INDUCED ECOLOGICAL CHALLENGES OF THE PEOPLE OF ISUIKWUATO

The ecological challenges of the people of Isuikwuato due to erosion menace are discussed under: flooding, landsliding, deforestation, and road destruction.

25.4.1 FLOODING

Erosion initially starts with flooding, which is the large volume of water rising and its uncoordinated spreading across a region. It is the most common natural disaster in southeastern Nigeria where heavy rains are prevalent, thus, leading to the removal of soil surface necessary for agriculture and making soil inhabitable for other microorganisms. In fact, flooding in Isuikwuato has destroyed agricultural lands with crops submerged in the water. This has caused a lot of economic loss to farmers and the community at large. In Isuikwuato, flooding has wrecked houses, schools, hospitals, businesses, and civic centers. Even transportation means, road networks, bridges, sewerage systems, and communication facilities have equally been disrupted by flooding. People of Isuikwuato have in recent times witnessed severe damage to power transmission and power generation leading to outright loss of power supply. Epidemics and waterborne diseases due to waterlogging have infested the people and typhoid, cholera, malaria, and diarrhea are common signs of the adverse effect of lack of access to good drinking water. Ultimately, human beings, animals, and plants are debilitated by flooding. Besides, the natural estheticism in cultural settlement and communal settlement is destroyed and carried away by flooding. When flooding has successfully and deeply eroded the soil surface, landsliding is imminent.

25.4.2 LANDSLIDING

Beyond flooding is landsliding, which involves any downslope movement of soil and flooding occurs in a sloppy area. Therefore landslides in Isuikwuato is usually initiated in slopes already on the verge of movement by rainfall, stream erosion, changes in ground water, disturbance by human activities, or any combination of these factors. Some of the landslides in Isuikwuato occur slowly over time due to gradual effect of flooding while most destructive ones happen suddenly after a heavy rainfall. Landslide like flooding usually left the ecosystem, schools, hospitals, businesses, transportations, roadways, canals, bridges, sewerage systems, communication, etc. damaged. In Isuikwuato, roads will be motorable today, by next week, it will fall to erosion gullies because the soil had already eroded and therefore became weak and prone to landsliding. One of the major effects of this landslide is deforestation.

25.4.3 DEFORESTATION

Due to deforestation, trees are forcefully removed either by nature or by human plot for the purpose of agriculture, grazing, or for firewood and timber for construction and manufacturing. Unfortunately, this act by erosion does not leave any affected portion of land useful any more. This is the case with Isuikwuato. Ordinarily, forestry is huge source of revenue to the government. For instance, forested areas like the Isuikwuato areas can provide food, medicine, and fuel. This definitely adds value to the economy of the area. In the same vein, a good percentage of terrestrial biodiversity, containing a wide array of trees, plants, animals, and microbes, is accommodated by forests. It also provides home for a diverse collection of living organisms that are also an important resource, but erosion destroys all these benefits made available by forests. Trees also absorb carbon dioxide, mitigating greenhouse gas emissions produced by human activity; hence, as climate change continues, trees play important role in carbon seizure, or the capture and storage of excess carbon dioxide. Gibbs, Harris and Seymour (2018) observed that trees alone are estimated to provide about 23% of the climate mitigation that is required to offset climate change. Sadly, deforestation not only removes vegetation that is important for removing carbon dioxide from the air, it also produces greenhouse gas. This assertion was confirmed by the Derouin (2019) that deforestation is the second leading cause of climate change. Down to the people of Isuikwuato, the effects are not different.

Deforestation in Isuikwuato also altered and damaged roads and footpaths. This has also endangered and prevented inter/intracommunal sense of interactions and coexistence, which exemplify the typical Isuikwuato. There are roads and footpaths that locally linked one village/clan and the other, thus, geographically binding the clans as one community in one assumed atmosphere. This is more psychological than ecological. However, when the ecological atmosphere is ruptured, the ripple effects spill down to the other aspects of human existence. Hence, it is a challenge that the normal communal interaction and coexistence has been affected by erosion.

In the same vein, Isuikwuato that used to be cold almost all the year round is now becoming hot. This is as a result of the direct rays of the Sun on the bare surface of the soil that was exposed by erosive deforestation. Agriculturally, it has affected the type of crops presently grown. For instance, perishables are grown out of seasons unlike now they grow only crops that can survive the hotness of the Sun. As such, foods are very scarce. In addition to the effects of global warming, deforestation by erosion has created its own erosion-induced local warming in Isuikwuato. According to Ezeigwe (2016), most societies under erosion menace suffer from this erosion-induced local warming because the environment has been subjected directly to the intense rays of sunlight.

25.4.4 ROAD DESTRUCTION

Road destruction and breakage seem to be the end point of erosion. When roads are destroyed and broken, most of the economic activities of the area become stalled. For instance, Isuikwuato is strategically and geographically positioned that it has roads (that are broken by erosion) linked to other neighboring states of Imo, Ebonyi, and Enugu, which thrive in agricultural productions, commercial engagements, and corporate engagements. As the road is destroyed, goods and services from these states are no longer accessible to or are difficult to access by the people of Isuikwuato in particular.

25.5 THE EROSION-INDUCED ECONOMIC CHALLENGES OF THE PEOPLE OF ISUIKWUATO

Economic challenges due to erosion menace explain how erosion has affected lives, finances, goods and services, farming industry, etc. of these people. While some of these challenges may have been mentioned in the section of ecological challenges perhaps in passing, here, the erosion-induced economic challenges of the people of Isuikwuato are discussed under agriculture, properties, and human life subthemes.

25.5.1 LOSS OF AGRICULTURAL PRODUCE AND AGRO-RELATED SOURCES OF LIVELIHOODS OF THE PEOPLE

Isuikwuato is the major food-producing community in Abia State, Nigeria. Different kinds of food and economic trees are grown. The economic challenges of people whose environment has been destroyed by erosion are mostly felt on agricultural productions of that society and the challenges seem to be the same across erosion-ridden areas (Kendall & Pimentel, 1994). First, erosion makes land very bad and infertile (mostly impossible) for agricultural productions. The major means of livelihood of this people is farming. When the land is rendered infertile and unusable, at first, the people will be faced with the difficulty of feeding not to talk of the generating money through agricultural proceeds that sustains them.

This particular effect of erosion seems to be general with every erosive environment and scholars such as Osadebe and Enuvie (2008), Poesen, Nachtergaele, Verstraeten, and Valentin (2003), and Dave (2003) have reported same. Specifically, Pimentel et al. (1995) noted that erosion creates water deficiency together with shortage of basic plant nutrients, such as nitrogen, phosphorus, potassium, and calcium, which are essential for crop production and that when this occurs, plant growth is stunted and crop yield declines, invariably, affecting the buying and selling of agricultural produce as their means of livelihood is farm. Means of survival, however, becomes very tough and hard. More worrisome is that the erosion had broken the roads and footpath through which other neighboring communities ply to exchange agricultural goods and services. Motorists, commuters, and other road users are threatened. For instance, most able bodied men and women do go to other nearby clans for menial agricultural labor. Second, Isuikwuato has a very large hectare of cashew plantation. The economic importance of cashew plantation to the host community, Abia State and Nigeria, at large is beyond measure. But erosion is gradually pulling down and destroying these economic trees in this area (Fig. 25.6). The nuts of this cashew that are usually transported to other states have been hampered by the roads that were destroyed by erosion, and it becomes extremely difficult for the conveyance of these nuts for economic purposes. Other states linked by Isuikwuato roads now hardly ply through the road. Those who ply the road are now forced to divert into the bush path that has become a safe haven for armed robbers and kidnapers. As a result, the local community mounted a local toll gate at the very point of diversion where commuters, motorists, and other road users are taxed, causing more hardship to the people. The toll gate was initiated to serve as a security front at the bush path (Fig. 25.7).

FIGURE 25.6

The erosion gullies falling cashew trees.

FIGURE 25.7

Local toll gate at the Mgbelu-Umunnekwu, Isuikwuato diversion into the bush path due to erosion gully menace.

25.5.2 BREAKDOWN OF PROPERTIES OF THE PEOPLE OF ISUIKWUATO DUE TO EROSION MENACE

One of the yardsticks to measure growing economy is good houses; hospitals; good road networks; businesses (such as fuel station, restaurants, and hotels); and infrastructural facilities such as pipelines and utility cables. Therefore the impact of erosion on schools, for instance, reduces the quality of education, which in turn affects the developmental span and economy of the society. Buildings of the schools have been cracked by erosion and most of the students are afraid of staying in their class rooms for classes. Most worrisome is that Abia State University that is a state-owned university is few kilometers away from the erosion site. It is even speculated that in the nearest future, say 5 years, if urgently it is not checked, half of the university premises would have gone down into erosion gullies.

In the same vein, many lives are threatened and lost due to destruction of hospitals and health centers. Just as businesses like filling stations have been deserted due to erosion, it is the economy that is suffering. What seems to be the worst is the destruction of infrastructural facilities and utility cables such as electricity cables (Fig. 25.8). In fact, it was reported that they cannot remember when last they saw electricity in their community. In this situation nothing can thrive successfully because no economic endeavor thrives in a total blackout environment. As stated before, there is a

FIGURE 25.8

The front and side view of abandoned fuel station at Nkwonta in Umunnekwu, front view of another abandoned fuel station due to erosion at Umunnekwu Agbo.

massive destruction of cash crops like cashew along Ama Mmiriukwu community in Uturu all through Umunnekwu Agbo to Amaiyi Obinohia. This destruction also directly kills the economy.

25.5.3 THREAT TO HUMAN LIVES DUE TO EROSION MENACE

Of course, the major challenge of erosion menace in Isuikwuato is destruction of lives. Many lives have been lost to erosion challenges in this community as many people have been reportedly drowned. Consequently, the inhabitants now live in perpetual fear and anxiety. The affected families reported having recurring memories of the drowning scene and when such individuals are not psychologically treated, posttraumatic stress disorder is more likely to occur, leading to frequent flashbacks of negative or sad experience of an individual into the person's memory and as such the flashback may not allow the person to function effectively. Most worrisome is that the breakage of the road due to erosion has led to diversion from the main road into the bush to enable meandering of the eroded paths (Figs. 25.7 and 25.9). At this diversion a local toll gate authorized by the local community is mounted, (Fig. 25.7) where they task every motorist between #50 and #100, thereby creating another difficulty for the road users. The meandering nature of the diversion has made the road a dead trap for road users, hence a preying point for armed robbers and kidnapers subjecting the motorists to intense fear and tension as they now feel very insecure. These fears and tensions negatively affect business transactions that are supposed to boost the economy of the community. An inhabitant of the community in charge of the toll gate recalled their ordeal in the hand of armed robbers in these dreaded diversion paths. The diversionary paths are very scary and whenever under such tensions, anything can happen (see Fig. 25.7). Many vehicles are abandoned along the diversionary pathways for the fear of victimization in the hands of hoodlums when these vehicles develop faults especially at odd hours (Fig. 25.9).

FIGURE 25.9

The bush path diversion at Mgbelu-Umunnekwu due to road breakage by erosion gully.

25.5.4 THE ESTABLISHMENT AND ADMINISTRATION OF ECOLOGICAL FUND IN NIGERIA

The establishment and administration of Ecological Fund in Nigeria resonates the premium placed on land in Nigeria as a cherished resource perhaps because it is not just a symbol of identity and a means of cultural affiliation, but also a means of social and economic survival of a people (Ifesinachi, Adibe, & Wogu, 2015). Despite the importance and relevance of this resource, natural disasters such as flooding, erosion, landsliding, desert encroachment, and deforestation have continuously threatened this single vital resource. These natural threats are not peculiar to any particular region. They traverse the entire country in different forms, shades, and shapes (see Tables 25.1 and 25.2), collectively known as ecological problems.

The protection of lives and properties of the people that is the exclusive preserve of government made the administration of Shehu Shagari, on the recommendation of the Okigbo Commission of 1981, to establish a special fund through the Federation Account Act of 1981 called the Ecological Fund to serve as an intervention measure by the Federal Government of Nigeria to address the multifarious ecological challenges in various communities across the country.

Under the act that established the Ecological Fund, 1% from the federation account was allotted to it. While this 1% was considered adequate at that time given the scope of its operations, years later, because of the growing ecological challenges in the country and the creation of states and more states up to 36 states and the Federal Capital Territory (FCT), it became necessary to pump more money into the fund and both Decrees 36 of 1984 and 106 of 1992 raised the percentage from 1 to 2. A further modification to this was effected in the federation account order of July 8, 2002. This raised the total fund percentage to 3. Out of the 3%, 2% is from the consolidated revenue of the federation and the remaining 1% is from the derivation allocation. It is however instructive to assert here that the state and local governments receive their share of the Derivation and Ecology Fund as part of their monthly allocation at the monthly meetings of the Federal Account Allocation Committee and the remaining fund is administered on the instruction of the president through the EFO headed by a permanent secretary (Agbese, 2017; Isah & Ayado, 2018).

But for clarity purposes, the fund is shared among the three tiers of government along existing revenue sharing formula as follows:

The residue of the two funds (3.00% − 2.78% = 0.22%) was transferred to the stabilization account. To implement the excess crude savings for 2004, the government approved a grant of 2% of the Federation Account to be added to the states' share of revenue allocation, thus the special funds under the federal government had to be adjusted as follows:

According to Agbese (2017), the fund had at its inception its core mandates of:

1. reducing ecological problems nationwide to the barest minimum,
2. facilitating quality and effective implementation of the projects,
3. utilizing the fund judiciously and equitably, and
4. managing the Ecological Fund projects effectively.

These mandates are in tandem with the prime objective of the act, which is to have a pool of fund that would be solely devoted to the funding of ecological projects to ameliorate serious ecological problems nationwide. The beneficiaries of these funds are the EFO, the National Emergency Management Agency (NEMA), the National Agency for the Great Green Wall, and the North East Development Commission (https://ecologicalfund.gov.ng/about-us/fund-sources-allocations/). However, added to the list of beneficiaries of this fund is "direct assistance to the Governments or/and any other projects on the approval of Mr. President."

Since then, the fund has been operative and interventions have happened in many parts of the country (see Tables 25.1 and 25.2) with the needed funds made available. For instance, while Blueprint Editorial (2019) documented that the Secretary to the Government of the Federation (SGF), Boss Gida Mustapha has reiterated that over 160 communities throughout the federation have benefitted from the federal government accrued share of Ecological Fund, Owete (2016) documented that between 2007 and 2011, the Fund was allocated a total of $217 billion, according to data from the Central Bank of Nigeria. More explicitly, the fund got $32,698 billion in 2007; $43,284 billion in 2008; $39,032 billion in 2009; $43,568 in 2010; and $58,876 billion in 2011. While the funding data may not adequate in terms of span of coverage in this instance, it is meant to demonstrate factually that funds were indeed released for ecological interventions.

However, the question that begs for answer is "how come the situation is this bad when there are mapped out funds to salvage the situation and the funds have been made available for this purpose especially given the fact that the fund has for long been instituted as a public policy aimed at solving existing and impending ecological problems such as soil erosion, flood, drought, desertification, oil spillage, pollution, and general environmental pollution at least since 1981 (Aliyu, Nura, & Abiodun, 2017)?"

Even before then, there have been official and formal courses of action attending to ecological problems such as the establishment of a soil conservation scheme financed by the Colonial Welfare and Development Fund. The project was to serve as a model for checking gully growth and as an antierosion demonstration for other agencies working in areas suffering similar problems. By 1950 the erosion control unit had worked on some 134 gullies and built 805 dams, 24 mil (384 km) of contour ridges, and 33 mil (53 km) of path with 4336 sumps (Udo, 1971). According to Ajaero and Mozie (n.d.), the Eastern Region of Nigerian Government in 1964 also declared its intention to fight the gully expansion. The government was just about to come to grasps with the problem when the pogrom and the Nigeria-Biafra civil war stopped every plan on the projects targeted at the abovementioned ecological problems is executed on behalf of the federal government by the EFO. In 1974 the federal government awarded a multimillion naira contract to an Italian firm, Technosynesis S.P.A., to study the erosion phenomenon in Nigeria, produce a soil erosion map of the country, and present a battery of measures required to check erosion in each of gully erosion zones (Eze, 1979), and others. It suffices therefore that these efforts were unsuccessful; hence, the launching of the Ecological Fund in 1981, which was necessary because of the myriad ecological threats the nation faced that seemed to have defied every effort at apprehending the menace.

Unfortunately, since the inception of the fund in 1981, whereas interventions have been made, the core mandates of the fund have not been met. This inability of the fund to meet its core mandates is what has kept Isuikwuato in this bad light. The implication is that while the Ecological Fund is a splendid and well-thought out public policy to remedy identified public problems, its implementations have been severely constrained. The identification of these constraining factors and proffering of tenable measures toward addressing these challenges are the cardinal purposes of this chapter described in the next section.

25.6 CHALLENGES OF ADDRESSING NATURAL DISASTERS IN NIGERIA THROUGH ECOLOGICAL FUND POLICY

Myriads of challenges have been identified as impediments to effective utilization of the Ecological Fund as curative measure to ecological problems in Nigeria. These challenges are more visible during policy implementations than policy formulation stages and according to Ugboma (2015), these challenges are better understood within the context of Nigeria's social, political, historical, and economic development. With these points in mind, we identify and treat the challenges next under the following subheadings.

25.6.1 MISMANAGEMENT OF THE ECOLOGICAL FUND AT THE FEDERAL GOVERNMENT LEVEL

Literature on the challenges of addressing ecological problems through Ecological Fund administration is ripe with concepts such as abuse, misuse, and misappropriation of funds. Some even use the word corruption (Adelana, 2016; Agbese, 2017; Emorinken, 2018; Olurounbi, 2018). In all, these concepts are aimed at explaining the mismanagement that has trailed the administration of the Ecological Fund.

In the words of Senator Mohammed Hassan, the utilization of the Ecological Fund, particularly since the return to civilian rule in 1999, has been entangled in endless controversies that in most cases are hinged on gross mismanagement hence rather than serve as a veritable tool for addressing the myriads of the country's ecological problems, the utilization of the fund has been derailed as funds are sometimes diverted to ecologically nonrelated issues such as election funding. Ifesinachi et al. (2015) also identified that the utilization of the Ecological Fund was characterized by nonconformity to stipulated disbursement guidelines, incomplete remittance of the fund to relevant agencies, as well as the unauthorized diversion of the fund. In some instances the fund was utilized to serve private interests. According to Owete (2016), while the total allocation to the fund from statutory allocation between 2007 and 2011 was $164 billion, representing 76% of total allocation, and allocation from the Excess Crude Account was $53 billion, representing 24% of the total allocation, Nigeria Extractive Industries Transparency Initiative (NEITI), in one of its reports, acknowledged that some of the disbursements were not utilized for the purpose for which the funds were established as some of the funds were released to the Consolidated Revenue Fund, FCT Authority, state governments, the military and ministries, departments, and parastatals. It reported that while $6.75 billion was released to the FCT Development Authority, $93.76 billion was released to the federal government; $10.46 billion to ministries, departments, and agencies (MDAs) of the federal government; $10.06 billion to the military; and only $23.77 billion to NEMA; and $15.40 billion to state governments to solve ecological problems. While $56.91 billion was released to the Office of the SGF Ecological Fund Account for execution of various ecological projects in different locations across the country, the sum of $93,768,951,164 was released as loan to the federal government during the years under review for funding of budget deficits and advances to state and local governments to meet shortfalls in their revenue. Owute (2016) averred that the Ecological Fund has lately become a cesspool of corruption.

Corroborating these assertions are few detailed cases of misuse and abuse of Ecological Funds as documented encapsulated in Senator Mohammed Hassan's proposed bill. According to the senator in his lead debate for an act to establish the Ecological Fund for the purpose of funding certain federal agencies dealing with environmental and ecological problems, in 2002, the fund gave a grant of $728 million to the Presidential Research and Communication Unit and disbursed $928 million for nonecological projects. In 2003 $1.9 billion was disbursed for nonecological programs out of which $800 million was given to the Ministry of Aviation for renovation of the Aminu Kano Airport and $150 million to Kaduna State government to manage sectarian crisis in the state. In 2009 $93.7 billion was illegally transferred from the Ecological Fund to the Consolidated Revenue Fund to fund the acceleration of capital budget and advances to state and local governments to meet revenue shortfalls. In December 2009 the National Economic Council (NEC) reported that about $200 billion belonging to the Ecological Fund had been spent on questionable projects that were either nonexistent or had no relation to the fight against ecological disasters or problems. The next year, by June 2010 precisely, the House of Representatives raised queries over curious withdrawals and loans to agencies and individuals totaling $146.594 billion, mainly to irrelevant objectives of the Fund. By December 2011 it was reported that an illegal deduction totaling $20.1 billion was made from the Ecological Fund for debt servicing. In 2013 the government of Goodluck Jonathan gave $2 billion to certain states of the federation but excluded other states. In the same year, some $22 billion was withdrawn and shared out to some states and local governments, while $2.078 billion was withdrawn toward the building of the Second Niger Bridge, although there is nothing on ground to justify the purported expenses. He concluded that in the last 15 years from 2002, the federal and state governments have reportedly misused about $500 billion meant for addressing ecological problems. Even the Senate Committee on Public Accounts Report uncovered a massive abuse and misappropriation of the fund, which included $154.9 billion projects not related to the environment.

These and more may have spurred Senator Mohammed Hassan to lead the debate for an act to established the Ecological Fund for the purpose of funding certain federal agencies dealing with environmental and ecological problems, so that questions of stewardship as regards the management of ecological disasters and the funds provided will be properly addressed to a particular office.

25.6.2 UNHEALTHY POLITICS OF WHO CONTROLS THE ADMINISTRATION OF THE ECOLOGICAL FUND

In 2016 the presidency was engulfed in a fierce battle over who controls the fund within the frame of a bill proposed by the presidency for proper administration and management of ecological issues in the country. Recall that in 2015, the Attorney General of the Federation and Minister of Justice, Abubakar Malami, and the Minister of Environment, Amina Mohammed, were engaged in a battle over the control of the Ecological Fund. According to Owete (2016), a bill proposed by Mr. Malami putting the management of the Fund under the Office of the SGF was opposed by Mrs. Mohammed who demanded that the fund be placed under her ministry. The Minister of Environment said the management of the fund should be placed under her ministry instead of the Office of the SGF as spelt out in Section 25.3 of the draft bill because the exigencies of the work of the Federal Ministry of Environment, including the increasing incidences/complexities of ecological issues exacerbated by the impacts of climate change, as well as the need for Nigeria to achieve the targets and compliments approved by the president under our Internal Nationally Determined Contribution necessitates that the proposed Ecological Fund Management Authority be located under and within the purview of the Federal Ministry of Environment. She concluded that crosscutting functions of other MDAs notwithstanding the bulk of ecological problems serviced by the fund is environmental in nature. Again, the need to locate the authority in the Federal Ministry of Environment is in keeping with common practice where other similar agencies are located within the purview of their supervisory ministries. the Industrial Training Fund supervised by the Federal Ministry of Trade and Investment; the River Basin Development Authority supervised by Federal Ministry of Water Resources; and the Tertiary Education Trust Fund supervised by the Federal Ministry of Education are few examples.

The Premium Times documented that the Minster of Justice and Attorney General of the Federation, Mr. Malami, and the Chief of Staff to the President, Abba Kyari, worked concertedly to forestall placing the authority under the Minister of Environment, Mrs. Mohammed, because placing such huge allocation to the fund under her would make her uncontrollable. This high level politics of who runs the Ecological Fund may be because of the funds' abuse, misuse, and misappropriations as documented previously.

25.6.3 POSSIBLE SHORTAGE OF FUND TO ATTEND TO THE CRITICAL ECOLOGICAL PROBLEMS

Nigeria is a large country with numerous challenges emanating from her large coverage. While some people may not be at home with this fact given their view of issues of governance at the central level, but with 774 local governments, each having up to 30 communities submerged under wards who are represented by councilors at the local government level, it becomes clear that Nigeria is a large entity. However, it becomes clearer when seen from the perspective of Nigeria being referred to as the "Giant of Africa," owing to her large population of about 186 million inhabitants or the recent United Nations estimation of 200 million inhabitants. The implication is that Nigeria is the most populous country in Africa and the seventh most populous country in the world (Holmes, 1987).

Consequently, there are numerous ecological disasters in the country and there is hardly any state that has no serious ecological threat. Erosion is a serious case that has traversed every state with high impact on the ecology and economy. Whereas the World Bank (2012) claimed that nature bequeathed Nigeria with not less than 3000 erosion gullies that grow wider and deeper with each rainfall, in 2018, the Premium Times documented the World Igbo Environmental Foundation as saying that the five southeastern states of Nigeria have over 2800 active erosion sites. While Anambra State alone has over 1000 active erosion sites, in Imo there are about 300, Abia 500, Enugu 500, and Ebonyi 500.

This massive number of ecological problems needing Ecological Fund intervention has reduced the fund available to the office to attend to qualified ecological disasters and this is one of the challenges of addressing ecological issues in the country. This is perhaps, the reason Nnodim (2018) documented the EFO as saying that the amount being requested by various communities, groups, and politicians for the control of soil erosion and flooding in their respective surroundings from the EFO is now over $1.1 trillion. The office further stated that despite the increased demand for ecological attention in communities across the country, it was only getting between $12 billion and $15 billion quarterly to address erosion and flood problems in different parts of the country. She confirmed that as in 2017, the office visited about 2890 erosion and flooding sites in Nigeria that are demanding for intervention. This dearth of fund to attend to the peoples' demands on the office for intervention may have propelled the senate to move for upward review of the allocation to the Ecological Fund from the existing 3% to something reasonable and considerate to cater for the high increase in ecological cases in the country (Olurounbi, 2018).

25.6.4 THE ABSENCE OF NATIONAL ASSEMBLY OVERSIGHT ON THE MANAGEMENT OF THE ECOLOGICAL FUND

While the National Assembly has, through a bill sponsored by Senator Mohammed Hassan, proposed increase in the allocation to Ecological Fund for resources adequacy and effective and efficient control of ecological problems in Nigeria, many commentators are perplexed about the nonchalance of the National Assembly toward the abuse and misuse of the Ecological Funds as well as other corrupt practices that allegedly trailed the administration of the Ecological Fund for about two decades now.

That the administration of the Ecological Fund has been enmeshed in what Owete (2016) called corruption cesspool is not in contention as the lead debate for the bill sponsored by Senator Mohammed Hassan gave detailed facts of high-profile abuse and misuse of the Ecological Fund that has hindered the expected successes of the Office of the Ecological Fund. Other studies such as Isah and Ayado (2018), Adelana (2016), and Agbese (2017) also confirmed the mismanagement of the fund.

What is amiss, however, is the fact that the senate, instead of probing the mismanagement of these funds that have denied the EFO the opportunity to fulfill its mandates, proposed an increase in financial allocation to the fund. This position of the senate and the manner in which the Chairman of the Senate Committee on Ecology and Climate Change, Senator Bukar Ibrahim handled enquires from the press regarding the true position of the senate on the mismanagement of the fund and the inability of the fund to meet its mandate, citing scarce fund as the reason for their inability to meet their core mandates, has left many commentators especially Olurounbi (2018) to postulate and pontificate that the senate is culpable in the Ecological Fund misappropriations.

Indeed, the insinuation that upward review of Ecological Fund has shown the National Assembly's culpability in mismanagement of Ecological Fund may not need to be wished away given the fact that in 2010, the House of Representatives had raised queries over curious withdrawals and loans to agencies and individuals totaling $146.5 billion, mainly for matters irrelevant to the objectives of the fund. Recall that on the March 15, 2018, allafrica.com documented that in December 2009, the NEC reported that about $200 billion belonging to the Ecological Fund had been spent on questionable projects that were either nonexistent or had no relations to the fight against ecological disasters. Again, the federal government on March 4, 2003 approved the sum of $10 million through the Ecological Fund account of the SGF to Sitbabs Engineering Limited for the construction of an abattoir in Bida, Niger State, with Lower Niger River Basic Development Authority (LNRBA). Even the senate revealed that for 15 years, the federal and state governments have diverted about $500 billion Ecological Funds meant to tackle environmental problems of communities in the country. Little wonder, Senator Mohammed Hassan asserted that that over the years, the Ecological Funds was operated like "slush" money and rather used for varying unrelated purposes. Undeniably, successive governments have been serially accused of diverting and enriching themselves with the Ecological Funds; hence, the operation of the funds had been largely discretionary, with zero guidelines on the disbursement and utilization of the monies. And one wonders why the senate, armed with these useful information, instead of scrutinizing the misappropriation of the funds, rather proposed an increase. It becomes difficult to assert the position of the senate in this matter: whether they are for efficient and effective administration of the Ecological Fund, or for enhanced sharing of the Ecological Fund.

This lack of legislative oversight for the administration of the Ecological Fund is perhaps one of the greatest challenges facing the addressing of natural disasters in Nigeria through public policy administration.

25.6.5 STATE GOVERNMENTS' MISMANAGEMENT OF THEIR SHARE OF ECOLOGICAL FUND

The state and local governments have not been very helpful in the administration of Ecological Fund and this is a terrible obstacle for the EFO to achieve its core mandates. Recall that the various state governments and local governments enjoy certain percentage of the Ecological Funds in their statutory monthly allocations. However, it is on record that successive administrations, especially that of former president Goodluck Jonathan, have severally shared bulk of monies from the Ecological Fund to them for purposes outside the purview and mandate of the commission, yet these state governors and local government chairmen conceal the receipt of such funds from the people of the states and still make demands on the federal government to come to the aid of the people of affected communities in their states. One wonders what the governors and local government chairmen do with the Ecological Funds they receive every month (Adelana, 2016).

Therefore saying that the Nigerian state governors are having field day with the Ecological Fund amounts to stating the obvious. The case of former governor of Plateau state, Joshua Dariye who acknowledged sanctioning the disbursement of the said $1.2 billion Ecological Fund by the now defunct All States Trust Bank, which was diverted to sponsoring the 2003 presidential election of the People's Democratic Party (PDP) and also paid to companies some of which were found to be owned by the former state governor, is still fresh given the recent turn of events in the court courtesy of Economic and Financial Crime Commission (EFCC). Dariye is not alone on this. Former governor of Benue State, Gabriel Suswam is also facing the same hurdle. According to the EFCC, Suswam also illegally diverted $6 billion of Ecological Funds during his two terms as Benue State governor (Adelana, 2016).

PMnewsnigeri.com on Thursday, November 1, 2018 documented the Permanent Secretary, EFO, Dr. Habiba Lawal, urging state governments to judiciously use Ecological Funds to complement their efforts in addressing the multifarious ecological challenges in various communities across the country. According to her, the state governments should ensure that the shares of their Derivation and Ecological Funds are judiciously expended and by doing so, the huge cost of intervention projects will be far reduced and the savings thereof will be harnessed to address other national developmental needs.

This call became necessary as the Deputy Chief Whip of the House of Representatives, Hon. Nkeiruka Onyejeocha alleged of state governors' non-utilization of Ecological Fund. Nkeiruka Onyejeocha. The lawmaker lamented that the federal government releases Ecological Funds to states monthly, but the state governors rarely tackle ecological problems in their states. She emphatically stated that governors are not deploying Ecological Funds to tackle erosion menaces bedeviling their states and therefore called for the separation of Ecological Fund from federal allocations accruing to states to avoid abuse of the fund by state governors, so that Ecological Fund will be strictly deployed to addressing ecological challenges devastating many communities in Nigeria (Alaribe, 2019).

These allegations of mismanagement of the Ecological Fund meant for salvaging communities ravaged by ecological problems that include erosion did not stop at the table of the state governors. The Office of the Managing Director of Niger Delta Development Commission (NDDC) has also come under heated criticism for allegedly squandering the sum of $400 billion released to the NDDC, through spurious and fictitious contracts awarded by the Acting Managing Director, Prof. Nelson Braimbaifa and his son, Christopher, who is the Special Assistant to the acting Managing Driector (MD), on technical matters. An investigation by Pointblanknews.com revealed that while $300 billion from the $1.5 trillion Ecological Fund owed the agency by the federal government was released to the NDDC in February 2019, bulk of the sum was spent on fake contracts awarded to loyalists of the All Progressive Congress (APC), ahead of the elections. Another $100 billion was released to the agency in April 2019, out of which, over $10 billion was stolen through such shady awards, inflated contracts, and fake certificates under the supervision of the acting NDDC boss. These details were made available by a senior management staff of NDDC. According to this source, few weeks to the general elections in January 2019, the NDDC office in Port Harcourt became a Mecca of sort with APC loyalists trooping in and out with contract papers. He stated that Christopher (the son of the acting MD) gave over $10 billion worth of fake water hyacinth contracts to people. The source further revealed that during the tenure of Nsima Ekere as MD of NDDC, between December 2017 and January 2018, he awarded about $3 billion fake water hyacinth contracts to 1000 APC loyalists at $3 million per person (Pointblanknews.com, 2019).

25.6.6 POLITICIZATION OF ECOLOGICAL FUND–BASED PROJECT AWARDS BY THE STATE GOVERNMENT

Just as should be expected in country of mixed tribes, language, culture, and personalities, politics prevails and pervades the distribution of national project. The phenomenon is crystal in the award of ecological contracts to the deserving states/communities in Nigeria. Take for instance, sometime in November 2018, precisely on the Wednesday 28th, the Federal Executive Council (FEC) presided over by Vice President Yemi Osinbajo approved $9.6 billion for 11 ecological intervention projects across the country. Somehow, across the country means Lagos, Oyo, Ondo, Cross River, Adamawa, Bauchi, Jigawa, Kaduna, Niger State, and Abuja. Our emphasis is that, though it was reported by Mr. Femi Adesina, the Special Adviser to the President on Media and Publicity as 11 ecological intervention projects across the country, no southeast states of Anambra, Enugu, Imo, Abia, and Ebonyi got a share notwithstanding the terrible threat of erosion with global acknowledgment in that region (Chafe, 2018).

Again, sometime in August, 2018, the Nigerian government also approved over $12 billion for the execution of ecological projects across the country. This time around, out of the five southeast states, one was favored as Anambra, Lagos, Oyo, Akwa Ibom, Adamawa, Bauchi, Borno, Jigawa, Kaduna, Plateau, as well as the FCT and the State House are the beneficiaries. In these two instances, while Abia State did not get even one, states such as Oyo, Bauchi, Jigawa, Adamawa, Kaduna, and Abuja (the FCT) got two Ecological Fund–sponsored projects apiece (Choji, 2018). What is even more surprising is the fact that the government house got one project. One wonders the difference between Abuja (FCT) and the government house.

It is on record that Abia State is highly threatened by erosion such that across the 17 local government areas, erosion menace has footprint there (see Table 25.4). Yet of these two instances Abia State was not recognized as threatened by any ecological problem let alone erosion. This may well explain why out of 192 projects awarded from 2015 to September 2019, Abia has only 4 (with Anambra, Enugu, and Ebony) projects and Imo has 6 perhaps because of the then APC governor, Rochas Okorocha (see Tables 25.1 and 25.2).

This politics and politicking of ecological intervention projects awards has been a serious spanner in the wheel of possible amelioration of the arrays of dreaded ecological problems such as flooding, erosion deforestation, and waste disposal management in Nigeria. Imagine Abuja with 27 projects while the combined 5 southeast states got 22 projects only.

25.6.7 AWARD AND ABANDONMENT OF ECOLOGICAL FUND–SPONSORED PROJECTS

Across the spectrum of Nigeria, the stories of award of ecological projects and later abandoned for years before completion while some are left fallow for life are prevalent. In October 25, 2017 Jide

Ojo wrote that he learnt, through a revelation made by a commissioner in the state, that there were over 500 abandoned federal government projects in Akwa Ibom state alone. He went further to cite the Daily Trust report of Sunday, June 28, 2015 that there were approximately 56,000 abandoned government projects across the country according to the Director of Administration of Chartered Institute of Project Management of Nigeria, Mr. David Godswill Okoronkwo, who listed the abandoned project according to geopolitical zones, viz., southeast 15,000; southwest, 10,000; South−South, 11,000; northwest, 6000; North−Central, 7000; northeast, 5000; and Abuja, 2000. He further cited another report in the Guardian of Sunday, July 2, 2017, saying that Akwa Ibom Integrity Group published over 300 projects abandoned by the NDDC in the state. Leader of the group, Chief Okon Jim, alleged that the commission abandoned a total of 121 rural roads, 75 classroom blocks, 69 rural water schemes, and 43 mini-electrification projects among others across the state (Ojo, 2017).

The story has not changed today as litany of abandoned projects is visible in every state of the federation but the one that serves as a great challenge to addressing ecological problems through the administration of Ecological Fund is abandonment of Ecological Fund projects awarded by the EFO under the presidency. Take for instance, On May 16, 2007, a contract worth $82,253,500 was awarded to Hazardous Waste Management Engineering Limited for the procurement, supply, and installation of 1 unit 150 kg/h rotary kiln medical waste incinerators at the Federal Medical Centre, Makurdi, Benue State, to the delight of the medical community. What is however not delightful is that 9 years after approval was given and 80% of the contract sum for the project paid to the contractor, the facility is far from being functional.

A similar trend was also observed in the execution of projects in Kogi state when, in July, 2013, the government of former president, Goodluck Jonathan approved a sum of $105 million for erosion control in Okaito to Messers Total Unique Nigeria Limited. However, 3 years after the contractor reported that it had achieved 100% completion, the problem of erosion still lingers. The Amutu of Eko in Okaito, Chief Ayodele Abass was enraged when he was informed that the erosion control contract has been certified completed by the EFO. Chief Abass hinted the likelihood of monitoring and evaluation officers from the government and the contractor conniving to give a misleading report that a good job was done. The same scenario was observed in Idah, the headquarters of Igala Kingdom where 6 years after a $150 million contract was awarded to ETA Associates Limited to control erosion in the area, residents still live in fear whenever it rains as the job was done halfway and the contractor claimed to have been paid to do the work halfway.

The situation in Ilorin, the state capital of Kwara, is identical from what holds sway in other places. The pathetic situation in Amilengbe, Isale Koko, and Aduralere axis of the state capital where residents have to vacate their houses whenever it rains gives strong credence to this. The lynchpin in this case is the Asa River that suffers terrible pollution from operations within the riverbed. For the purpose of these areas a $1.2 billion contract was awarded to Ambico Sendirian Nigeria Limited for the channelization of Asa River and its tributaries in 2013. The 2.3 km channelization is from Unity Bridge to Emir Bridge and then Amilengbe. Unfortunately, according Adelana (2016), residents said that attempts made so far by the contractor have not ameliorated their sufferings, adding that the contract expected to be executed within nine dry months has lingered for far too long and indeed, the project was far from being completed and residents of the area especially those in Amilegbe and Isale Koko have been witnessing massive flooding when it rains considering the snail speed of work by the contractor.

25.6.8 THE ECOLOGICAL FUND OFFICE WORKING WITH OUTDATED DATA ECOLOGICAL PROBLEM SITES

Correct information depends usually on the currency of our data. Use of unupdated data cannot only be counterproductive but can mar set goals and targets. But the greatest challenge of the use of or working with outdated information is the risk of marginalization of a people. In the case of ecological problems such as erosion and flooding, up-to-date information is nonnegotiable.

On July 18, 2019, representing Abia North Senatorial District, Senator Orji Uzor Kalu wrote a letter to the SGF, Mr. Boss Mustapha on the terrible situation of roads in his constituency, which have been heavily devastated by erosion. The people of Isuikwuato, which is heavily battered by erosion gullies, got sight of that letter and found that Isuikwuato was not among the affected areas the senator mentioned to Mr. Boss Mustapha and they are angry with the senator to the extent of accusing him of marginalizing the people of Isuikwuato.

A closer look at the letter revealed that it was a reminder of a letter already sent to the Office of the SGF, perhaps when the Isuikwuato erosion gully menace has not reached this ugly present state. However, one may want to look at it, it seems inexcusable to omit isuikwuato in any erosion disaster intervention in Abia State as a whole, let alone Abia North Senatorial District.

The adverse implication of the letter of the senator will be more pronounced when the EFO acts with the data it has, which was captured and compiled before the recent wreak of the Okigwe-Isuikwuato road by erosion. Recall that by March 26, 2018, Okechukwu Nnodim reported the Permanent Secretary of the EFO, Mrs. Habiba Lawal as saying that the EFO has the data of requests from various groups and the last was about $1.1 trillion, and that is specifically for soil erosion and flood control. She stated that based on the instruction of President Muhammadu Buhari, her office carried out a comprehensive national survey of ecological zones, variations, scope, and severity of sites across the country. She noted that the survey updated the federal government's data bank on ecological problems in all parts of the country, adding that it was aimed at providing a holistic, pragmatic, and reliable source of information on ecological challenges in Nigeria. Unfortunately for the people of Isuikwuato and others who got prone to ecological issues lately, the exercise that was conducted to ensure an evidence-based and strategic project selection and management process was carried out in three phases from August to December 2017 with some 2890 sites visited during the exercise (Nnodim, 2018). This kind of data that the EFO relies on is no longer reliable and tenable given the nature of both erosion and flooding devastating the major paths of the country as revealed in the literature, hence the need to update and save lives and livelihoods.

25.6.9 DISREGARD FOR THE SANCTITY OF HUMAN LIFE AND THE PROVISIONS OF THE CONSTITUTION ON THE PROTECTION OF LIVES AND PROPERTIES

One glaring fact is that life has lost its sanctity in Nigeria and the constitution that provided for the protection of lives and properties has been grossly disregarded. If not, how can a public servant or political office holder, who swore to the constitution to protect it and uphold its dictates, watch fellow human beings lose their lives and means of livelihood to ecological problems such as flooding and erosion, while the funds legally provided for guarding against such mishaps are either diverted to other issues not related to the life-threatening ecological problems or basically shared among those who swore by the constitution to protect lives and properties of Nigerians. According to an International Centre For Investigative Reporting (ICiR) report of May 28, 2016, on August 2014 at the famous Makera iron fabrication market in Gombe State, an elderly man sat in his shop perhaps waiting to go home after the rain lost his life in dreaded floods. The report further stated that in Herwa Gana, which extends to Kasuwan Makera, the $352,317,046 contract awarded in 2008 to ZEF Nigeria Limited, which was to be completed in June 2009 was only completed in 2014 after people had suffered so many losses, including collapsed buildings. Furthermore, in four northeast states— Adamawa, Bauchi, Borno, and Gombe, several projects that were listed by the EFO as completed and fully paid for were in reality either uncompleted or shoddily done while some could not be located, meaning that they probably did not exist. Again, icirnigeria.org uncovered at least eight projects, six of which were listed by the EFO as completed that symbolized the wasteful utilization of money because, even with $2.2 billion paid for the projects, they were anything but completed.

The human cost of uncompleted and abandoned projects is unimaginable. According to icirnigeria.org, in Baraza and Kardam in Bauchi State, villagers were shocked to hear that their villages benefitted from over $400 million contract for the construction of erosion control projects in 2007. The contract, valued at $442,379,368, was awarded to Anbeez Services Limited and was expected to be completed in 2008. It was for four villages, Baraza, Kardam, and Bazil in Dass Local Government Area and Rimin Zayam in Toro Local Government Area. Even though the EFO listed the project as completed, the reporter could not locate Bazil, as no one in the two local governments remembered any village with that name. However, in the three other villages, residents, including farmers, are battling with erosion and flood. Baraza and Kardam are separated by a big river that overflows its banks during rainy season but dries up during dry season. When it rains heavily, farmlands, houses, schools, and roads are submerged in water, with people opting to stay indoors. The contractor abandoned the work after constructing the culverts, leaving the drainages undone. As a result, erosion has continued to wash farmlands and houses its way into people's farms and houses, leaving the victims fearing for their lives. In one instance, erosion has washed inches close to an electricity pole and it is a question of when, rather than if, the nearby house gives way. The owner of the land, Garba Aminu, said this gives him sleepless nights.

In Umuahia, Okey (2019) observed that several buildings have, over the years, been submerged, while many more are on the verge of being washed away, even as communities become islands overnight, having been cut off from the rest of the state by ferocious and rampaging soil erosion. Just last year (2018), no fewer than 10 houses were swallowed in one fell swoop, while many people were rendered homeless by erosion in three villages, Umuekwaa, Umuokwasa, and Mgbarakuma, in the Umunwanwa clan of Umuahia South Local Government Area of the state. Yet, these areas are not fixed rather more devastated. If not for disregard of lives, how can leaders lead like this?

25.6.10 LACK OF KNOWLEDGE OF THE PROCESSES AND PROCEDURES OF ASSESSING ECOLOGICAL FUND

While the people of the communities infested by ecological mishaps may say that politics takes center stage and even upper hand in the awarding of ecological intervention projects, but the truth of the matter is that there is no harm in trying. It is therefore essential that the people whose ancient land, means of livelihoods, lives, and ancestral homes are threatened by ecological problems should at least know the processes and procedure of assessing the Ecological Intervention Fund from the EFO. This may be the reason for creating a link in the website of the EFO process of assessing the intervention fund from the federal government.

It has often been said that no knowledge is a waste as it is the key that unlocks peoples' destinies. Therefore the lack of knowledge of the processes and procedures of assessing the Ecological Fund intervention is one of the biggest challenges in addressing ecological issues in Nigeria through the administration of Ecological Fund. According to Nnodim (2018), various communities, groups, and politicians have flooded the database of the EFO with requests for intervention in their areas/communities prone to ecological challenges such as erosion and flooding. The implication is that those people (including the people of Isuikwuato) who thought it is only the governor or the senator who can talk to the federal government concerning Ecological Fund intervention have missed a key point as groups in a given community, a political party, or even a single politician irrespective of level can process or assess Ecological Fund. When this knowledge is missing, the politicians tend to make themselves God over the people and it vitiates the principles of felt need theory, which Onyenemezu and Olumati (2013) see as changes deemed necessary by people to correct the deficiencies they perceive in their community.

25.6.11 ADDRESSING ISUIKWUATO EROSION MENACE THROUGH THE ECOLOGICAL FUND ADMINISTRATION

The people of Abia State are evident on the list of the projects awarded by the Office of the Ecological Fund between May 2015 and September 2019, but what is amiss is that despite the dreaded nature of erosion gullies in Isuikwuato, they were not captured in the Abia State's four slots that went to Umuahia, Ikwuano, Bende, and Aba. While it may be attributed to the politics that goes with award and citation of such projects, we assume, the challenge may not be outside the 10 challenges listed and discussed previously. Consequently, we discuss the possible and feasible ways of addressing the Isuikwuato erosion menace under the following points:

25.6.12 ENGAGING THE EROSION AND GULLIES IT CREATES EARLY TO COMBAT IT WHEN IT IS STILL SMALL

The truth, which cannot be excused, is that this hellish erosion menace in Isuikwuato, just like every other places in Nigeria started very small. At that level the EFO will not recognize it, but if the people recognize its impending effect, they will start handling it. At that level the local government can reasonably fix it. Even community self-help approach can salvage it but the general problem has been the tendency to wait for our politicians, especially those in the government house at the state level or those at Abuja to come and fix, what can and should be fixed at the elementary level. Indeed, the politicians will not come to fix it either because, at that level, it will not give them the media hype they need to score some political points or because there are existing situations that are more compelling because of their high threat level or because those are the type the EFO will recognize to sponsor given its level of devastation of the ecology and the economy.

But generally speaking, it is a terrible attitude of ours and the government to wait and watch our environment dilapidate before it could be recognized for fixing and that time, it will gulp huge amount of money and may have consumed lives and livelihoods, instead fixing it when it was small, and requires small resources with lives and livelihoods saved.

25.6.13 HOLDING BOTH THE STATE AND LOCAL GOVERNMENTS ACCOUNTABLE FOR THEIR SHARE OF ECOLOGICAL FUNDS COLLECTIBLE EVERY MONTH AS BUILT INTO THEIR MONTHLY ALLOCATIONS

Although the trend has been that the people inflicted with ecological mishaps such as erosion, in the instance of Isuikwuato, to blame the government and consequently call on them (the government) to come and rescue them (the people affected by the natural disaster) with emphasis of the federal government, but the primary responsibility to protect the people and their environment belongs to the local and state governments. If not for any other reason, at least, because they (the state and local governments) are closer to the people and should have quicker access to the people and their challenges. In this case the government of Nigeria as a collective entity has already foreseen these ecological mishaps and made provisions for checkmating them. These provisions are the percentages of the Ecological Fund accruing to the state (0.72) and local (0.60) governments.

Unfortunately, many people from Isuikwuato local government area do not know that the government of Isuikwuato has for many years been receiving certain amount of money specifically for the purpose of controlling and/or preventing erosion menace in Isuikwuato. In the same vein, they may also not have known that the government of Abia State (including that of Orji Uzor Kalu, who is now the senator representing them at the National Assembly) was collecting certain amount of money from the federal government specifically designated to control and/or prevent erosion menace in every nooks and cranny of the state. According to Hon. Nkiruka Onyejeocha, the federal government releases Ecological Funds to states every month, which is lumped together in the federal allocations, yet state governors rarely tackle ecological problems in their states. Ecological Funds should be strictly deployed to addressing ecological challenges.

Anytime the menace hit a crisis stage, both the local government and the state will go to the venue and tell the people how handicapped they are in terms of resources to fix the problems given the slim resources at their disposal and given the magnitude of the problem. But the reality is, if the local governments and, sometimes, the state governments have been proactive when the erosion was small and relatively insignificant, the erosion would not have gotten to this stage and perhaps, the loss of lives and livelihood would not have arisen in the first place. Therefore the people of Isuikwuato and others affected by ecological problems should confront their governments and let them know that they are aware that they receive certain amount of money every month for the purpose of controlling or/and preventing these menace so that both the state and local governments should stop exonerating themselves from the menace because, in the real sense, they are the main architect of this menace because, if they had utilized the fund accruing to them monthly on that erosion, the situation would not have been what it is now. Therefore that role the local and state governments were supposed to play which they have not been playing is still very much needed now because there are still very many areas where these erosions are budding and with such existing funds, those areas should be checkmated now to avoid more of what is at hand now.

25.6.14 ENGAGING THE SENATOR REPRESENTING ABIA NORTH SENATORIAL DISTRICT AND THE HONORABLE MEMBER OF THE HOUSE OF REPRESENTATIVE FOR ISUIKWUATO AND UMUNNEOCHI FEDERAL CONSTITUENCY

The people of Isuikwuato have a man who has ample experience in the issues of governance and politics in the person of Dr. Orji Uzor Kalu as the senator, representing Abia North Senatorial District, having served two tenures as governor of Abia State and currently as an APC senator, which means that he has or is expected to have good alignment with the presidency more than the governor who is of PDP extraction.

Even though, the people of Isuikwuato are not happy with the senator for not including Isuikwuato in his letter to the Secretary of the Federal Government, Boss Mustapha, for federal intervention in communities ripped apart by erosion in his constituency, with the letter as an evidence, the people of Isuikwuato should engage the senator in discussions through courtesy calls and other means, including but not limited to letter writing, soliciting his help and support for federal government intervention on the erosion menace. This process will lead to the senator explaining the circumstances surrounding that letter and with time will definitely yield to helping the people in pressurizing the federal government to include Isuikwuato in the schedule of interventions in earnest.

In the same vein, they should engage the member of House of Representatives representing Isuikwuato and Umunneochi, Hon. Onyejeocha Nkeruka who have asserted her enduring efforts to getting Isuikwuato and Umunneochi off the map of erosion devastated zone in Nigeria. While she decried the erosion menace bedeviling her constituency, she assured that the Ecological Office would soon intervene as the various erosion sites in the constituency had been captured in the database of the office and were undergoing the process of design preparatory to contract award. She further disclosed that she has also drawn the attention of the NDDC, and other relevant agencies to the erosion menace in her constituency (Amoye, 2019).

With Dr. Senator Orji Uzor Kalu as Senate Chief Whip and former governor and Hon. Onyejeocha as Deputy Chief Whip of the House of Representatives, representing Isuiklwuato, we think with proper engagement that these two giants can and will drag federal intervention to Isuikwuato, thereby achieving their long-time dream of getting Ecological Fund intervention on the erosion devastated Uturu-Isuikwuato road.

25.6.15 ENSURING THAT THE ECOLOGICAL FUND OFFICE IS KEPT ABREAST OF THE CURRENT STATE OF EROSION THREAT IN ISUIKWUATO AS PROJECTS ARE AWARDED ON PRIORITY BASES

It is said that information is very vital for the success of any undertaking, but what is more critical is the accuracy and currency of such information. In this regard, whereas the Deputy Chief Whip of the

House of Representatives Hon. Nkeiruka Onyejeocha has ensured that the erosion sites in Isuikwuato have been captured in the EFO database, it is only necessary to ensure that important update on such information is maintained because, not just that the office work with the information available to them, such information inform their decision on which problem to attend to bearing in mind the priority level of such ecological problems. To do this, our former point on engagement with the senator and the member of House of Representatives suffice and perhaps the next point.

25.6.16 ACQUAINTANCE WITH THE PROCESSES AND PROCEDURE OF ASSESSING ECOLOGICAL FUNDS

An old adage says, if you want to do something well, do it yourself. Here, emphasis is on the fact that you can do it yourself and if you can, then you should. It has been revealed by the EFO that communities, groups, and individuals can assess this intervention and so what is left is for the people of Isuikwuato, to know first of all that they can assess this intervention by themselves and second how they can go about it.

Due to the criticality of this information, the EFO, in their website, created a section that dealt with this all-important processes and procedure of assessing the Ecological Funds Intervention. According to the EFO, there are simple seven steps to follow, and they are:

1. Requests are made to the EFO for intervention from several stakeholders, including communities, government agencies at local, state, and federal government levels.
2. Requests are submitted with designs and cost estimates.
3. Requests are stored in the EFO's data bank, collated, and analyzed technically.
4. Consultants are engaged by the EFO for further appraisal of the proposed projects, conduct studies at site, and preparing engineering designs/drawings and adequate cost estimates.
5. The EFO seeks the presidential approval through SGF for projects recommended by EFO.
6. The approved projects are forwarded to the Federal Ministry of Finance for the release of the approved fund to SGF's Ecological Account.
7. The EFO award contracts through FEC and EFO Tenders Board depending on the threshold.

Following the information, the people of Isuikwuato are, therefore, advised to take a hard look at the previous adumbrated steps of assessing the Ecological Intervention Fund so that even without the governor, senator, or member of House of Representatives, they can possibly secure this intervention given the seriousness they accord to this procedure and the processes.

25.6.17 THE LOCAL GOVERNMENT CAN LOOK OUT FOR INTERNATIONAL DONOR AGENCIES ASSISTANCE

Apart from the 5 above, the people of Isuikwuato, especially the Isuikwuato local government can get help from outside. It is on record that the World Bank has collaborated with the Nigerian government to initiate efforts at combating erosion in Nigeria. The effort resulted in new project in seven states of Abia, Anambra, Cross River, Ebonyi, Edo, Enugu, and Imo as pilot project. The bank responded through the US$500 million IDA-financed Nigeria Erosion and Watershed Management Project and has mobilized a strong coalition at national and international levels to tackle and reduce soil erosion on a war footing. The project became effective on September 16,

2013 and has received $3.96 million from the Global Environment Facility and $4.63 million from the Special Climate Change Fund. Therefore, if the local government teams are willing, they can get such assistance where our government is not forthcoming.

25.7 CONCLUSION

While it is correct to assert that Abia State has been battered by erosion, Isuikwuato is the heart of this sad story in recent times. This does not preclude assertions that there are many other areas of the state that are equally ravaged by erosion, but with the cutting of the Isuikwuato/Uturu highway into two, the economic life of Isuikwuato has been denigrated to the point that it can be asserted that Abia food supply is highly threatened by erosion menace as Isuikwuato, which is a major food-producing community in Abia State, has her farmlands washed away and cut off from their neighboring states of Ebonyi, Imo, and Enugu. This erosion menace has led to untold hardship to the people of the area.

As pitiable as this situation is, there is program of the government instituted since 1981 and has been receiving funding and requisite attention especially by this administration. Part of this funding has been going to both the Abia State government and Isuikwuato local government on monthly basis, with the remaining residing with the EFO for the singular purpose of ensuring that Isuikwuato and other affected areas do not experience the kind of psychological, ecological, and

Table 25.5 Soil Loss (Tons/Acre) for Each Local Government Area in Abia State.

Local Government Name	Soil Loss Per Tons/Acre in 1986	Soil Loss Per Ton/Acre in 2003
Aba North	269.19	266.73
Aba South	219.09	224.01
Arochukwu	242.84	258.44
Bende	525.73	547.54
Ikwuano	154.98	164.27
Isiala Ngwa North	258.56	262.37
Isiala Ngwa South	140.06	154.39
Isuikwuato	568.45	594.65
Obioma Ngwa	298.06	371.89
Ohafia	544.89	538.90
Osisioma Ngwa	232.06	236.86
Ugwunagbo	142.23	138.57
Ukwa East	208.31	222.41
Ukwa West	194.83	185.30
Umu-Nneochi	1082.58	1120.59
Umuahia North	412.95	720.89
Umuahia South	304.67	304.22

Modified from Uluocha, N. O, & Uwadiegwu, I. (2015). Mapping gully erosion in Abia State, Nigeria using geographic information systems and remote sensing techniques. Journal of Soil Science and Environmental Management, 6(10), 284–300. Mapping gully erosion in Abia State, Nigeria using geographic information system (GIS) and remote sensing techniques.

Table 25.6 Soil Loss (Ton/Ha) and Risk Rate of Towns Affected by Soil Erosion in Abia State.

Risk Rate

Severe		High		Moderate		Low		Very Low	
Town Name	Soil Loss	Town Name	Soil Loss	Town Name	Soil Loss	Town Name	Soil Loss	Town Name	Soil Loss
UmuOru	915.23	UraNtaUmuarandu	447.39	UmuUvo	269.33	Umuoru	89.21	Umuzomgbo	21.24
UmuNwaNwa	878.39	Amaokwe-Elu	527.7	UmuUhie	267.84	UmuomayiUku	135.75	Umuozuo	25.91
Umukalika	858.16	UmuOmei	607.69	Umuosu	173.94	UmuOkohia	124.06	Umuosi	5.93
Umuiroma	1053.11	UmuOkoroUku	386.78	UmuosoOnyoke	231.82	UmuOcha	55.19	Umuokpe	8.73
Umuasua	839.61	UmuOko	334.58	Umuopia	272.36	Umunmachi	90.66	Umuokorola	5.5
Owaza	1325.88	UmuOjimaOgbu	311.14	Umuokoro	231.84	UmuMba	138.65	UmuOhia	23.18
OnichaNgwa	925.86	Umuode	554.36	UmuOkahia	185.76	UmuEzeUku	106.59	UmuOcham	8.43
Okwu	864.89	UmuObiakwa	349.86	UmuohuAzueke	245.29	UmuDosi	90.66	Umumba	49.84
OhuhuNsulu	1152.95	UmuNkpe	469.29	Umuodo	234.11	UmuAla	135.75	UmuikuUko	25.91
Ohafiafgn	819.43	UmuNkiri	305.63	Umuocheala	156.27	UmuAjuju	135.75	Umuememike	9.43
Ndiudumaukwu	971.95	Umuihi	630.21	Umuobiala	151.09	UmuAda	90.66	Umuellem	6.57
NdiOrieke	1455.25	Umuhu	542.45	UmuNta	218.41	Ubani	54.64	Umuawa	0
MgbedeAla	751.25	Umuezu	379.09	Umunnekwu	243.81	Ubaha	124.06	Umuakwu	6.57
Ekenobizi	1101.37	UmuevuOloko	491.33	Umulehi	275.39	Ovuoku	58.93	Umuabia	6.14
Asaga	821.5	Umueteghe	345.28	Umuko	190.19	Okpo	64.13	Umuabanyi	5.93
Amuzukwu	1154.74	UmuEnyere	425.16	UmuEzegu	272.36	Okoko	105.78	UkwaNkasi	7.37
Amiyi	797.79	Umuchiakuma	429.36	UmuEgwu	211.4	OkahiaUga	88.35	Ukpo	4.45
AmaUru	1200.78	Umuawa	336.5	Umudike	156.27	OgoOmerenama	113.61	Ozu-Akoli	9.64
AmaUke	1093.34	Umuanughu	349.11	Umuanyi	260.14	Oduenyi	60.14	Onuasu	0
Amanta	1556.38	Umuanya	352.91	Obor	251.29	Obiohia	51.57	Okwe	4.95
Akoli	1029.01	Umuamachi	309.45	Umuchima	216.91	Amuma	93.69	Okpuala	30.3
Abala	1371.49	Amaoku	478.8	UmuAvo	216.91	Obinto	79.66	Okopedi	16.99
Umuakwu	713.38	UmuAkpara	468.65	Oboro	209.46	Obieze	92.83	Okon	45.63
Umuopara	680.7	Umu-Aja	561.47	Obuohia	242.86	Obete	106.59	Okoloma	5.93
Uturu	1060.8	Umuahia	374.44	UmuAkwumeke	222.71	Egbelu	119.13	Ndiachinivu	6.32

Modified fromUluocha, N. O., & Uwadiegwu, I. (2015). Mapping gully erosion in Abia State, Nigeria using geographic information systems and remote sensing techniques. Journal of Soil Science and Environmental Management, 6(10), 284–300. Mapping gully erosion in Abia State, Nigeria using geographic information system (GIS) and remote sensing techniques.

economic suffering they are experiencing now. Unfortunately, while the Ecological Funds are flowing into the bank accounts of these three designated beneficiaries, ecological problems such as erosion in the southeast (where Isuikwuato is situated) has also kept on creeping closer to the ancestral homes of the people, having overrun their farmlands and uprooted their business outfits as well as schools, hospitals, civil centers, and electric poles.

Sadly, the people are ignorant of this flowing of Ecological Fund into their local and state government treasuries; hence, at the instance of outburst of any ecological crisis, the leaders of both the state and local governments cry wolf of their financial incapacitation to ameliorate the situation, whereas for years, these governmental units have been misappropriating the funds legally meant for this purpose.

Apart from this attitude of the local and state governments of the affected areas, those in charge of this fund at the federal government level has been very culpable in the mismanagement of funds meant for management of ecological crisis in the country. Perhaps, because of this large sum available for misappropriation at the federal level, intense politicking of the management of the fund is another great challenge for achieving the purpose of Ecological Fund in Nigeria. As a result of this unhealthy abuse and misuse of fund, the remaining has obviously become inadequate to attend to ecological crisis, even as the senate's lack of oversight on the mismanagement of this fund has not helped solving issues. Then again, politics of who gets what project, when, and how has also negative implication to the success of Ecological Fund Act as abandonment of ecological funded contracts. This abandonment speaks volume of the disregard for the sanctity of lives and the constitution in this country. But what seems to be the worst is the lack of knowledge of the process and procedure of accessing this fund and the possible usage of outdated data by the EFO to administer these funds.

Having identified these challenges of using the Ecological Fund to address ecological crisis such as erosion in Isuikwuato, this chapter proffered six possible measures that can address the Isuikwuato erosion menace through the use of Ecological Fund (Tables 25.5 and 25.6).

REFERENCES

Adelana, O. (2016). *Ecological funds: A tale of corruption and waste.* <https://www.icirnigeria.org/main-headline-the-ecological-fund-of-corruption-and-misuse/>.

Agbese, D. (2017). *The ecological fund and the cans of worms.* <https://guardian.ng/opinion/the-ecological-fund-and-the-cans-of-worms/>.

Ajaero, C. K., & Mozie, A. T. (n.d.). *The Agulu-Nanka gully erosion menace in Nigeria: What does the future hold for population at risk?.*

Alaribe, U. (2019). *State governors not utilizing ecological fund—Hon. Onyejeocha.* <https://www.vanguardngr.com/2019/10/state-governors-not-utilizingecological-fund-%E2%80%95-hon-onyejeocha/> Retrieved on 18.11.19.

Aliyu, H. I., Nura, A. Y., & Abiodun, O. A. (2017). Assessing the socio-economic impact of gully erosion in Chikun Local Government Area, Kaduna State, Nigeria. *Science World Journal, 12*(1), 42−47.

Amoye, T. (2019). *State governors not utilizing ecological fund—Hon. Onyejeocha. Retrieved from* <https://www.vanguardngr.com/2019/10/state-governors-not-utilizing-ecologicalfund-%E2%80%95-hon-onyejeocha/#gsc.tab = 0on21/11/2019>.

Blueprint Editorial (2019). 160 communities benefit from Ecological Fund − SGF. https://www.blueprint.ng/160-communities-benefit-from-ecological-fund-sgf/

Chafe, I. (2018). *Government approves N9.6bn for 11 ecological projects in Abuja, nine states.* <https://www.environewsnigeria.com/government-approves-n9-6bn-for-11-ecological-projectsin-abuja-nine-states/>.

Choji, T. (2018). *Cabinet approves N12bn for ecological projects.* <https://www.von.gov.ng/cabinet-approves-n12-billion-for-ecological-projects/>.

Dave, F. M. (2003). *Effects of soil erosion.* <http://soilerosion.net/doc/onsitehtml>.

Derouin, S. (2019). Deforestation: Facts, Causes & Effects. Retrieved from https://www.livescience.com/27692-deforestation.html (6/6/2020).

Ecological Fund Office. *Background on ecological fund.* (2019). <https://www.osgf.gov.ng/storage/app/media/uploaded-files/background on ecological fund.pdf>.

Ecological Fund Office. *Procedure of assessing the ecological fund.* (2019). <https://www.osgf.gov.ng/storage/app/media/uploadedfiles/Procedure%20for%20Accessing%20Ecological%20Fund.pdf>.

Emorinken, M. (2018). *How N28.2b ecological fund was looted, by AGF.* <https://thenationonlineng.net/how-n28-2b-ecological-fund-was-looted-by-agf/>.

Eze, U. (1979). *Niger — Techno (1978) Soil Erosion control in Imo and Anambra State* Summary Report.

Ezeigwe, P. C. (2016). Evaluation of the socio-economic impacts of gully erosion in Nkpor and gully erosion in Nigeria: Causes, impacts and possible solutions. *Journal of Geosciences Research*, 5(1), 24—37.

Gibbs, D., Harris, N., & Seymour, F. (2018). By the numbers: The value of tropical forests in the climate change equation. Retrieved from https://www.wri.org/blog/2018/10/numbers-value-tropical-forests-climate-change-equation (5/6/2020).

Holmes, P. (1987). *Nigeria: Giant of Africa.* <https://www.amazon.com/Nigeria-Giant-Africa-PeterHolmes/dp/0950849812>.

Ifesinachi, K., Adibe, R., & Wogu, C. (2015). The management of ecological fund and natural-resource conflicts in northern Nigeria, 2009-2013. *European Scientific Journal*, 11(25), 115—128.

Igwe, C.A. (2012). Gully Erosion in Southeastern Nigeria: Role of Soil Properties and Environmental Factors, Research on Soil Erosion, Danilo Godone, Silvia Stanchi, IntechOpen, https://doi.org/10.5772/51020. Available from: https://www.intechopen.com/books/research-on-soil-erosion/gully-erosion-in-southeastern-nigeria-role-of-soil-properties-and-environmental-factors.

Igwe, O., & Fukuoka, H. (2010). Environmental and socio-economic impact of erosion in Nigeria, West Africa. *International Journal of Erosion Control Engineering*, 3(1).

Isah, A., & Ayado, S. (2018). *FG, states diverted N500bn ecological fund in 15 years senate.* <https://leadership.ng/2018/03/15/fg-states-diverted-n500bn-ecological-fund-in-15-yearssenate/>.

Lal, R. (2010). Soil erosion impact on agronomic productivity and environment quality. *Critical Reviews in Plant Sciences*, 17(4).

Kendall, H. W., & Pimentel, D. (1994). Constraints on the expansion of the global food supply. *Ambio*, 23, 198—205.

Nnodim, O. (2018). *Ecological Fund Office receives N1.1tn requests for erosion, flooding.* <https://punchng.com/ecological-fund-office-receives-n1-1tn-requests-for-erosion-flooding/>.

Nwankwo, S. (2018). *Erosion threatens Abia food supply.* <https://thenationonlineng.net/erosion-threatens-abia-food-supply/retrievedonthe6/11/2019>.

Nwilo, P. C., Olayinka, D. N., Uwadiegwu, I., & Adzandeh, A. E. (2011). An assessment and mapping of gully erosion hazards in Abia State: A GIS approach. *Journal of Sustainable Development*, 4(5), 196—211.

Ofomata, G. E. K. (1975). *Soil erosion. Nigeria in maps, eastern states.* Benin City: Ethiope Publishing House.

Ojo, J. (2017). *FG's criminal abandonment of government projects.* <https://punchng.com/fgs-criminal-abandonment-of-government-projects/>.

Okey, S. (2019). Abia battered by erosion. Retrieved from https://www.sunnewsonline.com/abia-battered-by-erosion/ (5/6/2020).

Olurounbi, R. (2018). *Upward review of Ecological Fund questions N/Assembly's culpability in mismanagement.* <https://tribuneonlineng.com/upward-review-of-ecological-fund-questions-nassemblys-culpability-in-mismanagement/>.

Onyenemezu, E. C., & Olumati, E. S. (2013). The imperativeness of felt-needs in community development. *Journal of Education and Practice, 4*(2), 156−159.

Osadebe, C. C., & Enuvie, G. (2008). Factor analysis of soil spatial variability in gully erosion area of southeastern Nigeria: A case study of Agulu-Nanka-Oko area. *Scientia Africa, 7*(2), 45.

Owete, F. (2016). *Buhari's ministers battle over control of Ecological Fund.* From <https://www.premiumtimesng.com/news/top-news/207989-exclusive-buharisministers-battle-control-ecological-fund.html> Retrieved on 11.06.16.

Pimentel, D., Harvey, C., Resosudarmo, P., Sinclair, K., Kurz, D., McNair, M., ... Blair, R. (1995). Environmental and economic costs of soil erosion and conservation benefits. *Science, 267*(1), 1117−1123.

Poesen, J., Nachtergaele, J., Verstraeten, G., & Valentin, C. (2003). Gully erosion and environmental change: Importance and research need. *Catena, 50*(2−4), 91−133.

Pointblanknews.com. *Acting NDDC boss, son, squander N400billion on fake contracts.* (2019). <http://pointblanknews.com/pbn/exclusive/acting-nddc-boss-son-squander-n400billion-onfake-contracts/>.

The Ecological Fund Office. *Status of Ecological Fund Office projects executed between May 2015 and September 2019.* (2019). <https://ecologicalfund.gov.ng/wp-content/uploads/2019/09/EFOPROJECTS-MAY-2015-September-2019.pdf>.

The World Bank. *Nigeria − Erosion and watershed management project.* (2012). <http://documents.worldbank.org/curated/en/728741468334143813/pdf/679830PAD0P1240osed0401901200simult.pdf>.

Udo, R. K. (1971). *Geographic regions of Nigeria.* Ibadan: Heinemann Publishers.

Ugboma, P. P. (2015). Environmental degradation in oil producing areas of Niger delta region, Nigeria: the need for sustainable development. *STECH, 4*(2), 75−85.

Uluocha, N. O., & Uwadiegwu, I. (2015). Mapping gully erosion in Abia State, Nigeria using geographic information systems (GIS) and remote sensing techniques. *Journal of Soil Science and Environmental Management, 6*(10), 284−300.

Umahi, H. (2016). *Erosion swallows billions of naira.* <https://www.sunnewsonline.com/erosion-swallows-billions-of-naira/>.

ECOLOGICAL AND ECONOMIC COSTS OF OIL SPILLS IN NIGER DELTA, NIGERIA

26

Chukwuma Felix Ugwu[1], Kalu T.U. Ogba[2] and Chioma S. Ugwu[3]

[1]*Department of Social Work, Faculty of the Social Sciences, University of Nigeria, Nsukka, Nigeria*
[2]*Department of Psychology, University of Nigeria, Nsukka, Nigeria* [3]*Department of Political and Administrative Studies, University of Port Harcourt, Port Harcourt, Nigeria*

26.1 INTRODUCTION

Oil industry exploration and production is harming the ecosystems that are supportive of agriculture business across the oil-rich countries. The refinery's oil production products in the country are liquefied petroleum gas, premium motor spirit, kerosene (aviation and domestic), automotive gas oil (diesel), low pour point fuel oil, high pour point fuel oil, and unleaded gasoline (UNEP, 2011). Some of the oil facilities are located close to homes, water sources, and farmlands. The volatile properties of petroleum have wiped out large areas of vegetation. When there are spills around and within the drainage basin, the hydrologic force of the river and tides pushes spilled petroleum to shift into regions of vegetation (Adelana, Adeosun, Adesina, & Ojuroye, 2011). For decades, oil spills and oil-related pollution have damaged the soil, water, and air quality and expose living organisms to extinction. It regularly occurs due to theft, accidents, human miscalculations, and procedural discharges of petroleum hydrocarbon to the environment. The environmental hazard obstructs the maximum functioning of plants and animals and creates environmental states incompatible for a healthy living.

One of the largest and oldest oil producers in Africa is Nigeria, with over 50 years of commercial extractive activity (Malden, 2017). It joined the Organization of Petroleum Exporting Countries (OPEC) in 1971. The country has OPEC-proven crude oil reserves of 36.97 (3.1%) at the end of 2018 (OPEC's Annual Statistical Bulletin, 2019). It has huge deposits of natural resources such as natural gas, petroleum, tin, iron ore, coal, limestone, niobium, lead, zinc, and arable land in its territory (Central Intelligence Agency, 2019). In sub-Saharan Africa, Nigeria is the leading economy with oil as the central source of its foreign earnings and revenues (Central Intelligence Agency, 2019). The oil exploration and production and distribution activities are major businesses of Nigeria and located in the region geographically designated as the Niger Delta region. This region is vulnerable to oil exploration and exploitation. It implies that the unwarranted consequences of oil production cause harm to the indigenous people and their environment.

Similar to other oil-producing nations, oil spill is peculiar in the Niger Delta region of Nigeria. It is the world's largest wetland and Africa's largest delta covering some 70,000 km^2 (Ogbe, 2011;

Economic Effects of Natural Disasters. DOI: https://doi.org/10.1016/B978-0-12-817465-4.00026-1

Ohimain, 2012; World Bank, 1995). This area extends from Apoi to Bakassi and from Mashin creek to the Bight of Benin. It has a coastline of 560 km and about two-thirds of the entire coastline of Nigeria. The region's population is estimated to about 20 million, with over 40 ethnic groups, with about 300 communities and 250 spoken divergent languages (Ugbomeh & Atubi, 2010), and the population continues to upsurge at about 3% annually (Federal Ministry of Environment, 2010). The Niger Delta region comprises mainly states in South-South (Akwa Ibom, Bayelsa, Cross River, Rivers, Delta, and Edo) and partly fractions South-East (Abia and Imo) and South-West (Ondo) of Nigeria. This region of the country that is known for its wetlands and water bodies embodies large mangrove forests and a network of creeks and rivers (Omorede, 2014) and remains an essential source of livelihood to the indigenous people, as well as to the various organisms that inhabit these ecosystems (Adelana et al., 2011). The grave impact of oil spills in the area suggests the impoverishment of the indigenous inhabitants and species of organisms in the environment.

Over the years, oil spill is one of the utmost environmental hazards in the oil region of Nigeria. Most environmental hazards in the region are consequences of oil spills, gas flaring, and oil pollution. The environmental hazard has led to the descend of local fishing and farming, the loss of habitat, and biodiversity, as well as acid rain damage and health impacts of pollution (Reed, 2009). These environmental hazards have deep impacts on the people who depend majorly on the environment for livelihood. This growing environmental hazard has continued to generate tension between the Nigerian government, multinational oil companies (MOCs) through the Shell Petroleum Development Company (SPDC), and the communities affected by oil pollution in the Niger Delta (Yakubu, 2017). This incidence is likely to persist decades and threatens any hope for a sustainable living (Ejiba, Onya, & Adams, 2016).

In spite of huge gross earnings for the country, the region is constantly subjected to hazardous challenges ranging from economic and social to negligence of the environment by MOCs (Omorede, 2014). Also, people in the affected areas are engulfed with health-related challenges such as breathing problems and skin lesions, coupled with basic human rights issues (Adelana et al., 2011).

The major oil spills that are uncontrolled create an atmosphere of public hazards and uncertainties. Nwilo and Badejo (2005) gave an account of major oil spills in Nigeria from 1978 to 1998. According to them, the major spills in the coastal zone include the GOCON's Escravos spill in 1978 of about 300,000 bbl, SPDC's Forcados Terminal tank failure in 1978 of about 580,000 bbl, and Texaco Funiwa-5 blowout in January 17, 1980 of about 400,000 bbl. This was closely followed by the incidents of the Abudu pipeline in 1982 of about 18,818 bbl. More than a decade later, the Jesse Fire Incident and the Idoho Oil Spills of January 1998 spilled about 40,000 bbl in the environment. According to Nwilo and Badejo (2005), within these periods, the 1980 oil spills to the Atlantic Ocean, and the spills that damaged 340 ha of mangrove, is the largest offshore spills in the country. Further records show that in the region, between 1976 and 1996, oil spills on land, swamp, and offshore environments were approximately 6%, 25%, and 69%, respectively (Egberongbe, Nwilo, & Badejo, 2006). In recent times, similar trend of oil spills continues unabated in the region.

A recent study by Ejiba et al. (2016) found that about 65.13% of oil spilled in 2014 resulted from sabotage; 17.38% causes were unascertained; 14.35% was through natural accidents, corrosion, equipment collapse, and human miscalculation; and while 3% was due to unknown circumstances. Also in 2014, a total of 1087 oil spills averaging 91 spill incidences occurred monthly. The last 5 years witnessed an average of 733 spill occurrences with 23,000 bbl annually. However, Ejiba et al. (2016) further noted that these estimates are constantly been disputed by the MOCs who argue that the bulk of oil spills that transpire in the region results through sabotage or vandalism. Notwithstanding, what is common is evidence of oil spills on the land and coastal environment of the Niger Delta community. Incidentally, this continuous negative development in the area has attracted both local and international concerns.

The World Bank (1995) assessment of the region in 1995 concluded among other things that the environmental loss of biodiversity and valuable timber and tree crop species indicates that indices of development in the region were below globally acceptable standards. The Rio Declaration on Environment and Development in Rio de Janeiro from June 3 to 14, 1992 states among other things that to achieve sustainable development, state parties to the treaty shall ensure environmental protection that shall compose an essential part of the development process. The Rio Declaration is the reaffirmation of the Declaration of the United Nations Conference on the Human Environment, endorsed in Stockholm on June 16, 1972 (United Nations, 1992). However, within the period under review, many mitigating measures have not produced positive outcomes in the oil-rich region.

As a consequence, the people have over the years engaged in different forms of agitations and conflicts against the government of Nigeria to press home the demand for inclusive development and socioeconomic emancipation of the region (Ugboma, 2015). According to Sagay (2005), in the midst of poverty and deprivation, environmental abuse and degradation seem to be the greatest threats to the continued existence of the people in the oil-bearing region. In this direction, studies showed that increasing oil spills is dangerous to the environment, which proves difficult to clean up when contaminated by oil and which constitutes loss of biodiversity and environmental degradation (Ugboma, 2015). It implies therefore that as long as petroleum resource is explored and exploited, spills will still take place recurrently (Udoh & Ekanem, 2011). In this discourse the major objective of this chapter is the ecological and economic costs of oil spills in Niger Delta in Nigeria. The specific objectives are essentially to:

1. understand the inherent status of ecosystem in the Niger Delta region of Nigeria;
2. examine the environmental hazards of oil spills to the land and coastal region with relatedness to human health, diverse species of organisms(animal and plants) in the land and coastal region;
3. determine the socioeconomic cost of oil spills to environment of the region;
4. highlight the spillover effects of oil spills to the ecosystem of the region; and
5. examine the impact of oil spill on economic growth and development of the region and Nigeria at large.

26.2 STATUS OF ECOSYSTEM IN THE NIGER DELTA REGION OF NIGERIA

The ecosystem and biodiversity are an integral element of human survival. A significant number of world populations rely on plants and animals for the cure and management of several disease conditions (Izah & Seiyaboh, 2018). For instance, in the process of pollination, insects in the environment are essential. Wildlife and plant resources are also resourceful in the environment. The former is a of protein constitute with hides and skin, and while the product of the later is used as timber for construction, while several sicknesses, diseases, and infections are treated using shrubs and herbs (Izah, Angaye, Aigberua, & Nduka, 2017). These assertions apply to the people of the region. The indigenous inhabitants traditionally are fishermen, farmers, and hunters exploiting available resources as sources of livelihood from land, water, and forest. However, there are evidence of distortions in the environment, health, social, and economic activities of the people largely due to environmental degradation resulting from environmental exploitative activities of MOCs (Ayanlade, 2014; Ikemeh, 2015). According to the federal government of Nigeria, the country has good records in biodiversity and rich in the diversity of plant and animal species. There are about 7895 plant species within 338 families and 2215 genera. About 22,000 vertebrates and invertebrates species are recorded. These species include about 20,000 insects, 1000 birds, 1000 fishes, 247 mammals, and 123 reptiles. About 0.14% of these animals are under threat while 0.22% are endangered (Federal Republic of Nigeria, 2006).

More so, about 1489 species of microorganisms are recorded in Nigeria. These animal and plant species are found in great quantity in the vegetation that ranges from the mangrove along the coast in the south to the Sahel in the north and usually sufficient for the rural economy (Federal Republic of Nigeria, 2006) and contributes to national earnings. However, the continued conservation and sustenance of the biodiversity raises fundamental questions in the present condition of the country. For instance, Table 26.1 delineates an estimated overall status of biodiversity in the ecosystem, especially for plant and animal species in Nigeria.

Specifically, the region remains the most diverse areas of West Africa due to its endowment with biodiversity (Obot, 2003; Phil-Eze & Okoro, 2009). For instance, several studies have indicated that species-rich reptile and amphibian communities are commonly seen in the deltaic swamp forests of the area (Akani, Aifesehi, Petrozzi, Amadi, & Luiselli, 2014).

Insects are commonly used in the region as food. A survey conducted in the region showed species of insects used as food, which cut cross six orders: Isoptera, Orthoptera, Coleoptera, Lepidoptera, Hemiptera, and Diptera (Okore, Avaoja, & Nwana, 2014). These insects emanate from the wild at different times of the year and are eaten by all ages in the population (Okore et al., 2014).

Since the beginning of oil exploration and production, the foremost environmental hazards confronting the Niger Delta of Nigeria are environmental degradation resulting in change in land use, forest degradation, and environmental pollution from oil and gas production and biodiversity loss (Ayanlade, 2014). The identified hazards on both land and coastal region have consequences to human health, animal, and plants. The next subtheme delves into oil spills and pollution contaminated environment and associated peculiarities in the ecosystem.

26.3 ENVIRONMENTAL HAZARDS OF OIL SPILLS IN THE NIGER DELTA REGION

In the direction of the interest of discourse, this subsection shall examine the impact of environmental hazards of oil spills to the ecosystem with relatedness to human health, socioeconomic living, and diverse species of organisms in the land and coastal region. Having explored the abundant resources in the ecosystem in preceding section, it may be needful to dissect different dimensions

Table 26.1 Threatened Plant and Animal Species and Their Uses.

Species	Main Uses	Status
A. Plants		
Milicia excelsa	Timber	Endangered
Diospyros elliotii	Carving	Endangered
Triplochiton scleroxylon	Timber	Endangered
Mansonia altissima	Timber	Endangered
Masilania accuminata	Chewing	Endangered
Carcina manni	Chewing stick	Endangered
Oucunbaca aubrevillei	Trado-medical	Almost extinct
Erythrina senegalensis	Medicine	Endangered
Cassia nigricans	Medicine	Endangered
Nigella sativa	Medicine	Endangered
Hymenocardia acida	General	Endangered
Kigelia africana	General	Endangered
B. Animals		
Crocodylus niloticus	Food/medicine/leather	Endangered
Osteolaemus tetraspis	Food/medicine	Endangered
Struthio camelus	Food/medicine	Endangered
Psittacus erithacus	Medicine/pet	Endangered
Cercopithecus erythrogaster	Food	Endangered
Loxodonta africana	Food/ivory	Endangered
Trichechus senegalensis	Food	Endangered
Giraffa camelopardalis	Food/medicine	Endangered
Python sebae	Bags	Endangered
Gazella dorcas	Food	Endangered

Adapted from Federal Ministry of Environment (2010). Fourth national biodiversity report. *Abuja: Federal Ministry of Environment.*

of environmental-related contamination encumbering growth and development of the people, other natural habitats, and ecosystem regeneration.

26.3.1 HUMAN HEALTH

Oil spills, flaring, and pollution have severe environmental and health hazards on the people of the Niger Delta region of Nigeria. For example, a survey in Rivers State of the region revealed a range of illnesses connected with the pollution, such as gastrointestinal problems, skin diseases, cancers, and respiratory ailments (Baumüller, Donnelly, Vines, & Weimer, 2011). It is observed that the continuous activities connected with petroleum exploitation and production and consequent release of hydrocarbon chemical wastes have led to soils, groundwater pollution, and ecosystem degradation (Ite, Ibok, Ite, & Petters, 2013). The surface and groundwater toxic chemicals released due to oil production include benzene, toluene, ethylbenzene, and xylene and other toxic chemicals such as toxic polyaromatic hydrocarbons (PAHs) (Ana, Mynepalli, & Bamgboye, 2009). In the region, studies indicated that over the years, over 75% of produced gases have been flared, which is about 45 million tons of carbon dioxide daily (Yakubu, 2017). Annually, approximate of 35 million metric tons of carbon dioxide and methane flares in the atmosphere of the region (Nriagu, 2011). More complicated is the incomplete combustion of the gas with resultant effect of various organic products that include volatile organic compounds (VOCs) and PAHs, and among other inorganic pollutants to the environment (Kostiuk, Johnson, & Thomas, 2004). According to Agbola and Olurin (2003), the heat emitted from gas flaring is estimated to 45.8 billion kilowatts, averaging 1.8 million cubic feet of gas daily.

Consequent to gas flaring and pollution of the environment, tens of thousands of people have been affected by oil spills and pollution. The people in the environment are worried about their health, including the consumption of fish or drinking water from streams or rainwater (Amnesty International, 2011). In addition, this has detrimental consequences on the people. The deposits of potentially dangerous toxic substances to human population are associated with long-term chronic and terminal diseases, especially if not treated or managed. It is common knowledge among the people of the region that the emission of hydrocarbons among other noxious materials into the atmosphere may have an effect on fertility and result to abnormal births. More so, such release of dangerous substances may cause skin-related cancer and respiratory diseases especially chronic restrictive lung conditions (Atubi, 2015). For greater understanding of the health challenges of oil production, Table 26.2 shows various impacts of discharges of hydrocarbon to the environment of the oil-rich region of the country.

In Table 26.2, considering some of the human health-related effects of gas flaring pollutants is associated with damaging effects. It is obvious that this may likely determine the life expectancy of the people in the area. For instance, findings by United Nations Environment Program (UNEP) (2016) indicated that Ogoni community in Rivers State is unprotected in the air they breathe and drinking water and are also exposed to skin-related diseases due to contamination of the environment. Some inhabitants migrated to other areas of the country they consider safer than areas where the environmental hazards are less severe. This has implications for not only health but also includes the socioeconomic fabrics and emotional situations of the people. According to Yakubu (2017), the government should review and update the existing laws to address the prevailing environmental hazards in the region of the Niger Delta of Nigeria. For further details on the impact of oil spills and gas flaring on human health, see Agency for Toxic Substances and Diseases Registry ASTDR (2016): https://www.atsdr.cdc.gov/toxfaqs/tfacts175.pdf and Gobo, Richard, and Ubong (2010): https://www.ajol.info/index.php/jasem/article/view/55348.

While the discussion of human related-effects of oil spills may be unending at the moment, it is crucial to understand the role of plants and animals in the ecosystem. Human beings have similar functions with plants and animals and coexist in a way that is interdependent in the ecosystem. That is to say that plants and animals receive similar adverse effects to the point of extinction. While the preceding subtheme advanced the existing effects oil spills, gas flaring, and pollution in the region of Niger Delta, the subsequent subtheme will focus on the peculiar effect of oil production to plants and animals in the region.

Table 26.2 Some Human Health Effects of Gas Flare Pollutants in the Niger Delta.

Serial No.	Chemical Name	Human Health Effect
1.	Alkanes: methane, ethane, and propane	Low levels: can result in swelling, itching, and inflammation. High levels: may cause skin infections such as eczema and acute lung swelling.
2.	Alkenes: ethylene and propylene	May result in weakness, nausea, and vomiting.
3.	BTEX	They are toxic and are either carcinogenic or probable carcinogens. Targets on exposure are usually the nervous system and blood-forming organs.
4.	Carbon monoxide	Low levels: can cause permanent damage to the heart and brain. May harm the mental development of fetus and children. High levels: can lead to miscarriage and death.
5.	Hydrogen sulfide	Low levels: nausea, headaches, delirium, disturbed equilibrium, tremors, convulsions, and skin and eye irritation. High levels: respiratory tract and mucous membrane irritation; may cause immediate or delayed pulmonary edema. May result in extremely rapid unconsciousness and death.
6.	Nitrogen dioxides (NO, NO_2)	Low levels: cause irritation of eyes, nose, throat, and lungs. Cough, shortness of breath, tiredness, and nausea may also occur. Buildup of fluid in the lungs 1 or 2 day(s) after exposure is also possible. High levels: may result in rapid burning, spasms, and swelling of the upper respiratory tract and throat tissues. Reduced oxygenation of body tissues, a buildup of fluid in lungs and death.
7.	Sulfur dioxide	Low levels: asthmatics are very sensitive to respiratory effects. High levels: burning sensation of the nose and throat, breathing difficulties, and severe airway obstructions were observed in miners exposed to a copper mine explosion A 100 ppm in 100 parts of air has been demonstrated to be immediately dangerous to human health and life.

BTEX, *Benzene, toluene, ethylbenzene, and xylene.*
Modified from Yakubu, O. H. (2017). Addressing environmental health problems in Ogoniland through implementation of United Nations Environment Program recommendations: Environmental management strategies. Environments, 4(28), 1–19. doi:10.3390/environments4020028, p. 6.

26.3.2 PLANTS AND ANIMALS

The risk and uncertainties associated with oil spills are likely to remain at the lower levels of river, ocean, and sea for many years and harmful to aquatic lives (Saleh, Ashiru, Sanni, Ahmed, & Muhammad, 2017). Aquatic resources primarily are the traditional source of enterprise and livelihood of most communities in the Niger Delta (Baumüller et al., 2011). Recent studies examined spatial variations in the mortality of fish species with regards to marine ecosystems particularly in the Arctic region. Although varying results indicate the resistance of fish species to the toxicity of the environment, oil spill on marine ecosystem is a major contributor to environmental hazard that limits fish production (Carroll et al., 2018; Langangen et al., 2017). This explains why the agricultural interventions such as Agricultural Credit Guarantee Scheme Fund failed to substantially improve the fishing agricultural subsector in Nigeria (Osuagwu, & Olaifa, 2018). To this end, Osuagwu and Olaifa (2018) suggest an enhanced social protection policy rather than credit facility for the inhabitants of the Niger Delta since there is short supply of arable land for cultivation and clean water for the continued existence of aquatic ecosystem. In essence, this will drastically reduce the high loan default among the populace.

Soil is the bedrock of almost all plant species. Plants germinate and grow through the support of water, air, and nutrients that equally support agricultural products. Repeated oil spills on agricultural soil and the harmful effect on all forms of life expose the soil to toxic substances which consequently renders the active surface layer unproductive for plants and crops (Abii & Nwosu, 2009; Essien & John, 2010). For example, a study by Ahmadu and Egbodion (2013) on the effect of oil spills on cassava production in the Niger Delta region of Nigeria found that the most significant effect on the farmland was the increased soil temperature/toxicity. The study also revealed other effects that included reduction of soil fertility, degradation of the farmland, lowland productivity, destruction of soil structure, poor soil aeration, and destruction of soil microorganisms. A similar study by Uquetan et al. (2017) focused on the germination of tree crops such as cocoa, cashew, pawpaw, mango in a state of Niger Delta revealed that seed sprouting began in 10 and 16 days, after planting in the control and planting in the polluted samples, respectively. Further, the results revealed that the unpolluted soil was higher than soil samples treated with crude oil and spent lubricating oil (contamination levels on seeds). It is discovered that the effect increased as concentration of treatment increases. The reduced germination of seeds is attributed to elements of oil in the seed coat that prompts unfavorable conditions for seeds germination. Also soils polluted with crude oil and spent lubricating oil show poor wetability, reduced aeration and compaction, and increased propensity to heavy metal accumulation.

Pollution of the environment occurs when environmental changes negatively affect the quality of human and animal life, including microorganisms and plants (Abosede, 2013). Abosede (2013) investigated the effect of crude oil pollution on some soil physical properties at Effurun, Warri, Delta State, Nigeria. The findings arising from polluted sites revealed slight differences compared to unpolluted soil profile due to oil pollution in the area. In summary, major findings associated with the effect of oil spills on plants include the following:

1. Inhibit the germination of plants.
2. Delay germination by inducing stress, which prolongs lag phase.
3. Inhibit the uptake of water and nutrients by the root of the plant, hence causing deficiency to other parts as the leaves.
4. Affect regeneration of stumps.
5. Affect anatomical features of leaves.
6. Cause cellular and stomatal abnormalities.
7. Disruption of the plant water balance, which indirectly influences plant metabolism.
8. Cause root stress, which reduces leaf growth via stomata conductance.
9. Cause chlorosis of leaves.
10. Enlargement of cells in various tissues due to oxygen starvation (FiriAppah, Okujagu, & Bassey, 2014, p. 12).

26.3.2.1 SOCIOECONOMIC COST OF OIL SPILLS TO ENVIRONMENT

An important feature of the people of Niger Delta is their socioeconomic life. The social and economic activities are intertwined and inseparable. The region is rich in aquaculture and farmland that are sources of livelihood in the area. The people depend on the soil, water resources, and forest for subsistence and survival by engaging in farming and fishing. The traditional occupation of the people includes canoe carving, palm oil and garri processing, mat making, thatch roof making, and fish smoking. In all these industries, fishing and farming from water and land remain the mainstay of the people's lives (Ipingbemi, 2009).The economic decline in the area presents unimaginable frustration among the people.

Frustration occasioned by oil spills and pollution leads to violence and militancy, reduction in tourism and hospitable industries (Okonkwo, 2014). It is pertinent to note that the consequential effect of the environmental hazards on the ecology and ecosystem in the region accumulated to the birth and uprise of the Movement for the Survival of the Ogoni People (MOSOP) targeting environmental justice. This movement of the people through agitation against the violation of rights and privileges attracted both national and international concerns. The Ken Saro-Wiwa MOSOP–led agitations became the mother of all other agitations in the region and further gave rise to resource control impasse between the Nigerian government and the Niger Delta people. Despite the justice-driven peaceful-led approach by Ken Saro-Wiwa, he was convicted and executed in the struggle for resource control by the military regime for his doggedness and assertiveness toward the protection of the environment. The radical pursuit by the people ushered in the derivation formulae, which is a percentage of the oil sales accrued to the oil-producing areas for compensation. Despite the allocation resulting from oil exploration and exploitation, the emission of gaseous substances to the environment remains an obstacle to the ecosystem, economic growth, and development over the years.

Generally, the economic impacts of oil spills include but not limited to the cost of cleanup and compensation, destruction to agricultural lands, fishery, vegetation, and wildlife (Okonkwo, 2014). The effects of oil spill on crops leave much to be desired. The damages to crops are enormous. Oil spills permeate the soil thereby inhibiting germination, growth, and performance of crops such as cassava, pepper, and tomatoes (Osuoke & Emeka-Okpara, 2014). Farmers have been compelled to abandon their hitherto land to search for alternative means of livelihood. The increased soil sterility due to the devastation of soil microorganisms led to dwindling agricultural productivity (Agbogidi & Ayelo, 2010). The consequential spread of oil spillage destroys the farmlands, contaminates drinkable water, and causes shortcomings in fishing of coastal waters (Atubi, 2015). In addition, around the coastal and inland water areas of Niger Delta, aquaculture and fishing are major sources of food, employment, and economic benefits to human populations. However, hazardous environmental impact on aquaculture ventures has led to the loss of substantial investment (Akankali & Nwafili, 2017). The implications of oil spill on economic activities of the people have impelled some serious thoughts to changing their means of survival (Ipingbemi, 2009). It may be concluded that oil production in the area has diminished the natural resources thereby incapacitating the inhabitants impoverished and undermined their social and economic capabilities.

26.4 SPILLOVER EFFECTS OF OIL SPILLS TO THE ECOSYSTEM OF NIGER DELTA REGION

In this context, the highlight of spillover effects of oil spills to the ecosystem of the Niger Delta region is conceived in two categorical dimensions. First, oil spills and pollution have short-term implications for ecosystems. It is expressed in human health and economic livelihood, plants and animal survival, and the environment. As a subject of serious concern, effort to mitigate the incursion oil spill impact is overwhelmingly not attainable at the moment. Second, oil spills and the associated pollution manifest with long-term effects.

26.4.1 SHORT-TERM EFFECTS

Oil pollution has both short-to-medium- and long-term consequences. In the short-term, oil pollution would continue to deny the people of their immediate means of sustenance in farming and fishing that have been the traditional means of survival; hence individuals and households resort to other means of livelihood. Also, individuals exposed to pollution may suffer from debilitating diseases such as asthma, headaches, nausea, fatigue, and drowsiness.

26.4.2 LONG-TERM EFFECTS

Oil pollution exposes the current and future generations to health and environmental hazards.

The presence of petroleum hydrocarbons varies according to the environment. In effect, the dose of hydrocarbon emission and duration in the environment defines the degree of toxicity present in the water, air, soil, density, food, and other media people interact within their environment. People are vulnerable to petroleum that disperses into the air, including those petroleum-contaminated surface water or groundwater when used for bathing, washing, cooking, and drinking (UNEP, 2011). In this vein, Eykelbosh (2014) noted that oil spills may result in toxic and other physical health, mental health, and community health impacts. Toxic effects may take place when residents are exposed to the compound mixture of dangerous composite in petroleum, which includes VOCs (e.g., benzene, toluene, ethylbenzene or xylene), PAHs, heavy metals, and, in the case of controlled or uncontrolled burning of spilled oil, particulate matter, and other combustion products.

These hazardous compounds mixtures in petroleum as well as their short- and long-term effects are well documented in the region. The consequences are further exemplified in Table 26.3.

In Table 26.3, it shows a good number of much of expected health challenges confronting the inhabitants of the Niger Delta. The coping and treatment of the aforementioned effects are associated with financial implications at the expense of the affected. In addition, the economic status of the people may not guarantee adequate access to health care due to the poor economic earnings. Besides the health hazards, it may rejuvenate militancy and conflict against the state in an already restive region.

Oil spills and pollution of the environment, if not addressed, can lead to degradation of the Niger Delta mangrove forests, destruction of ecosystems, and deterioration in the fish and agricultural harvests that are dominant to the livelihoods of local communities (Ite et al., 2013). The

Table 26.3 Oil-Derived Contaminants and Short- and Long-Term Effects on Human Health.

Compound	Short-Term Effects	Long-Term Effects
Particulate matter		
PM$_{2.5}$ (particles measuring less than 2.5 μm) PM$_{10}$ (particles measuring less than 10 μm)	Respiratory effects (exacerbation of asthma, decreased function, inflammation)	Cardiovascular and respiratory disease, premature death
VOCs		
Benzene	Hematopoietic, nervous, and immune effects	Carcinogenic to humans; reproductive and developmental effects in animals
Toluene	Nervous effects (headaches, nausea, fatigue, drowsiness)	Upper respiratory symptoms; nervous effects; developmental effects
Ethylbenzene	Eye and throat irritation, dizziness	Possible human carcinogen; developmental effects in animals
Xylene	Nervous effects and nose, eye, and throat irritation; skin irritation and vasodilation	Developmental effects in animals; not classifiable as to its carcinogenicity to humans
PAHs (polyaromatic hydrocarbons) (as mixtures)	Headaches, nausea, vomiting, loss of appetite, skin irritation (itching, burning, edema), eye irritation	Liver damage; hematological effects; reproductive and developmental effects in animals; known or suspected carcinogens
Hydrogen sulfide	Respiratory effects (sore throat, cough, shortness of breath, and impaired lung function in asthmatics), nervous effects (loss of consciousness), eye irritation	Central nervous effects
Dispersant components c		
2-Butoxyethanol	Headache, irritation of the nose and throat, vomiting, metallic taste	Developmental and reproductive effects in animals; not classifiable as to its carcinogenicity to humans
Heavy metals		
Cadmium (Cd)	Respiratory effects at extremely high exposures	Kidney damage, respiratory disease and decreased lung function, carcinogenic to humans
Mercury (Hg)	Various nervous and respiratory effects ranging from low to high exposures	Nervous and respiratory effects; not classifiable as to its carcinogenicity to humans
Nickel (Ni)	Respiratory effects (inflammation, atrophy of the nasal epithelium)	Chronic lung inflammation; carcinogenic to humans, developmental effects in animals

VOCs, *Volatile organic compounds.*
Modified from Eykelbosh, A. (2014). Short- and long-term health impacts of marine and terrestrial oil spills. *Available from* <*https://pdfs.semanticscholar.org/20f6/2a334fade4ecb11afd11d00a17f6654bbce1.pdf?* _ga = 2.175862694.581943255.1573117965-286584850.1566786334>*, pp. 2−3.*

phenomenon of oil production requires active and effective prevention and controls measure to mitigate the threat envisaged to consume the environment and deplete the ecosystem in the region and other parts of the country (Saleh et al., 2017). Improved cooperation of oil explorers, national toward global standards to mitigate the implications of oil spills among oil-rich areas and promote climate change regeneration, is fundamental. To deal with the situation of environmental hazards in the region, concerted efforts are required to curb the menace through institutional support.

26.5 THE IMPACT OF OIL SPILL ON ECONOMIC GROWTH AND DEVELOPMENT OF THE NIGER DELTA REGION

The Niger Delta region of Nigeria is abundantly endowed with rich natural resources producing oil and gas that account for over 85% of the nation's gross domestic product, over 95% of the national budget, and over 80% of the national wealth. The region's oil production has contributed over 90% of the nation's export earnings since 1975 (Ebegbulem, Ekpe, & Adejumo, 2013). Despite these immense contributions to the nation's economic growth and development, the oil industry has negatively affected the economic progress of the host region. The region that consists of a number of different ecological zones, mangrove swamps, coastal ridge barriers, forests, lowland rain forest, and freshwater swamps is inhibited by rural communities that depend mainly on the natural environment for survival (Ebegbulem et al., 2013). The major source of livelihood for the rural communities is agriculture consisting primarily of fishing and farming. According to the United Nations Development Programme (UNDP) (2006), more than 70% of the people depend on natural environment for their livelihood. However, consistent oil exploration and exploitation in the region over the years by oil industries accompanied by spillage have degraded the environment and negatively impacted on the ecosystems. Destruction of farmland, fishery and aquatic resources, and mangrove ecosystem are consequences of oil spillage (Orubu, Odusola, & Ehwarieme, 2004). Oil spills that occur on land destroy crops and damage the quality and productivity of soil utilized for farming while those that occur in water damages fisheries and other aquatic animals as well as contaminate water used for drinking (Amnesty International, 2011). These hazardous situations undermine the economic progress and development of the inhabitants of the region.

A report documented by Amnesty International (2011) has it that "In June 2005, a pipeline surveillance company, contracted by Shell, discovered an oil spill from a high-pressure pipeline operated by Shell, in Oruma, Bayelsa state. The oil spread into many fish ponds and killed fish and shellfish on which the community relied for livelihood and food." Incidences as this often rob owners of fishponds, their economic source of survival, and adversely reduce fish availability for sale with attendant consequences on market price rise. More so, Emmanuel, Olayiwola, and Babatunde (2009) documented a study carried out in the Niger Delta in 1995, which revealed that between 1992 and 1993 the total area under major food crop production in Bayelsa, Rivers and Delta States, decreased by 41.7 and 15% in 1995. The decrease was attributed to the increasing incidents of oil spillage that destroyed farmlands. The study went further disclosed that about 2,185,000 ha that is the land area of Rivers and Bayelsa states estimated half of it is swamp, and in Delta State out of 1,769, 800 representing its total land area estimated one-third is swamp. These swamp areas are not favorable for agricultural production. Hence, persistence loss of fertile lands through oil spillage is to paralyze agricultural production and majority of inhabitants' source of economic power.

It is evidenced that oil spillage cripples the economic activities of affected areas. Complimenting this assertion, Orubu et al. (2004) in their work highlighted how oil spills that occurred in the region between 1976 and 2005 affected the lives of inhabitants. The authors reported that in 1986 about eight major creeks and villages were affected by oil spills resulting in loss of barrels of oil and crippling of economic activities in the areas, including damages on fishnets, ponds, and traps of the farmers worth over 2 million naira. The net effect of such incidences is economic destabilization of affected individuals that may have psychological health implications on the people.

The economic growth and development of the nation as well as the Niger Delta region is limited by oil spillage that occurs with continuous production. The nation's sole reliance on the oil sector as its main source of foreign earnings coupled with occurrences of oil spills in the region has undermined the agricultural sector's contribution to economic growth and development. The nondiversification of the economy may have a negative implication on the nation's economic status on the long run when oil becomes limited. The literature has it that oil contributes over 79.5% of total government revenue while other sectors, including agriculture, contribute less than 20% of the nation's income (Omamuyovwi & Akpomuvire, 2017). This may not be healthy for the nation's future economy. The resultant oil spills have negative impact on agricultural production as it limits the quantity of agricultural products available for both consumption and income generation. The Niger Delta region to an extent depends on neighboring states for agricultural produce as the fertility of their lands has been reduced due to oil spills.

The poverty index of the region is high, and the reason for that can be deduced from the negative impact of oil spillage on the lands of the region, insensitivity of government and oil multinational companies to develop the area and ensure spill cleanup. World Bank Institute (2003) asserts that much of the social conflicts that occur in the Niger Delta are as a result of pervasive poverty and lack of development. The region is bedeviled with lack of access to clean pipe-born water, good roads, schools, health-care services, poor education, and unemployment. In some areas where some of these amenities exist, they are only to enhance further exploitations. The Niger Delta Human Development Report (2006) has it that the region's human development index score, encompassing longevity of life, knowledge, and a decent standard of living, remains at a low value of 0.564 (with 1 being the highest score). According to the report, the region rates far below regions with comparable oil and gas resources such as Saudi Arabia with 0.800 in 2000, while United Arab Emirates, Kuwait, Libya, Venezuela, and Indonesia were placed at of 0.849, 0.844, 0.799, 0.772, and 0.697 in 2003, respectively.

The abovementioned report underscores the development level of the region in spite of enormous wealth it contributes to the nation. Government and Oil multinational response to the region's plight is eminent to enhance economic growth and development of both the region and the nation.

26.6 CONCLUSION

Oil, no doubt, is a precious gift of nature. When effectively and efficiently explored and managed, it leads to the growth and development of a nation, otherwise, the nation suffers. Oil spill results from a number of reasons/causes ranging from sabotage, natural accidents, erosions, equipment failure, and human error to even unknown circumstances. For whatever reason, it is obvious that oil spill is an aberration and has done more harm. Most worrisome is that much has not been done to ensure proper cleanups (in the event of oil spill) and proper management in terms of exploration, exploitation, adequate social responsibilities, and compensation to the affected communities. Consequently, the affected community through to the nation at large suffers ecologically and economically. For instance, oil spills in Niger Delta in Nigeria have heavily impoverished and damaged human health and existence and undermined plant and animal survival as well as other agricultural (land, fishery, vegetation, and wildlife) habitat and activities that would have boosted the economy and living standard of the nation. Hence, the ecological and economic cost of oil spills in Niger Delta in Nigeria is unquantifiable, immeasurable but lethal to human, animal, and plant interrelationships as well as mutual coexistence.

REFERENCES

Abii, T. A., & Nwosu, P. C. (2009). The effect of oil-spillage on the soil of Eleme in Rivers State of the Niger Delta area of Nigeria. *Research Journal of Environmental Sciences*, *3*(3), 316−320.

Abosede, E. E. (2013). Effect of crude oil pollution on some soil physical properties. *Journal of Agriculture and Veterinary Science (IOSR-JAVS)*, *6*(3), 14−17.

Adelana, S. O., Adeosun, T. A., Adesina, A. O., & Ojuroye, M. O. (2011). Environmental pollution and remediation: Challenges and management of oil spillage in the Nigerian coastal areas. *American Journal of Scientific and Industrial Research*, *2*(6), 834−845.

Agbogidi, O. M., & Ayelo, E. (2010). Germination of African oil bean (*Pentaclethra macrophylla*, Benth.) seeds grown in crude oil polluted soil. *Bioresearch Bulletin*, *3*, 147−156.

Agbola, T., & Olurin, T. (2003). *Land use and land cover: Change in the Niger Delta*. Abuja: Centre for Democracy and Development.

Agency for Toxic Substances and Diseases Registry (ASTDR). *Nitrogen oxides*. (2016). Available from <https://www.atsdr.cdc.gov/toxfaqs/tfacts175.pdf>.

Ahmadu, J., & Egbodion, J. (2013). Effect of oil spillage on cassava production in Niger Delta Region of Nigeria. *American Journal of Experimental Agriculture*, *3*(4), 914−926.

Akankali, J. A., & Nwafili, S. A. (2017). An assessment of the socioeconomic impact of crude oil pollution on aquaculture in Gokana Local Government Area Rivers State, Nigeria. *Nigerian Journal of Fisheries and Aquaculture*, *5*(1), 87−94.

Akani, G. C., Aifesehi, P. E. E., Petrozzi, F., Amadi, N., & Luiselli, L. (2014). Preliminary surveys of the terrestrial vertebrate fauna (mammals, reptiles, and amphibians) of the Edumanon Forest Reserve, Nigeria. *Tropical Zoology*, *27*, 63−72.

Amnesty International. (2011). *The true 'tragedy': Delays and failures in tackling oil spills in the Niger Delta*. United Kingdom: Amnesty International.

Ana, G. R. E. E., Mynepalli, S., & Bamgboye, E. A. (2009). Environmental risk factors and health outcomes in selected communities of the Niger delta area, Nigeria. *Perspectives in Public Health*, *129*(4), 183−191. Available from https://doi.org/10.1177/1466424008094803.

Atubi, A. O. (2015). Effects of oil spillage on human health in producing communities of Delta State, Nigeria. *European Journal of Business and Social Sciences*, *4*(8), 14−30.

Ayanlade, A. (2014). *Remote sensing of environmental change in the Niger Delta, Nigeria. Thesis is submitted to King's College London, University of London, in partial fulfilment of the requirements for the degree of*

Doctor of Philosophy. Department of Geography, School of Social Sciences and Public Policy, King's College London University of London.

Baumüller, H., Donnelly, E., Vines, A., & Weimer, M. (2011). *The effects of oil companies' activities on the environment, health and development in Africa*. Belgium: European Parliament.

Carroll, J., Vikebo, F., Howell, D., Broch, O. J., Nenstad, R., Augustine, S., et al. (2018). Assessing impacts of simulated oil spill on the Northeast Arctic cod fishery. *Marine Pollution Bulletin, 126*, 63−73.

Central Intelligence Agency. *The world fact book*. (2019). https://www.cia.gov/library/publications/the-world-factbook/geos/ni.html Retrieved 14.08.19.

Ebegbulem, J. C., Ekpe, D., & Adejumo, T. O. (2013). Oil exploration and poverty in the Niger Delta region of Nigeria: A critical analysis. *International Journal of Business and Social Science, 4*(3), 279−287.

Egberongbe, F. O. A., Nwilo, P. C., & Badejo, O. T. (2006). *Impacts and management of oil spill pollution along the Nigeria coastal areas: Administering marine spaces*. Retrieved from <https//www.fig.net/pub.fig>.

Ejiba, I. V., Onya, S. C., & Adams, O. K. (2016). Impact of oil pollution on livelihood: Evidence from the Niger Delta Region of Nigeria. *Journal of Scientific Research & Reports, 12*(5), 1 12.

Emmanuel, A. O., Olayiwola, J. B., & Babatunde, A. W. (2009). Poverty, oil exploration and Niger Delta crisis: The response of the youth. *African Journal of Political Science and International Relations, 3*(5), 224−232.

Essien, O. E., & John, I. A. (2010). Impact of crude-oil spillage pollution and chemical remediation on agricultural soil properties and crop growth. *Journal of Applied Sciences and Environmental Management, 14*(4), 147−154.

Eykelbosh, A. (2014). *Short- and long-term health impacts of marine and terrestrial oil spills*. Available from <https://pdfs.semanticscholar.org/20f6/2a334fade4ecb11afd11d00a17f6654bbce1.pdf?_ga = 2.175862694. 581943255.1573117965-286584850.1566786334>.

Federal Ministry of Environment. (2010). *Fourth national biodiversity report*. Abuja: Federal Ministry of Environment.

Federal Republic of Nigeria. (2006). *National oil spill detection and response agency (establishment) act, 2006 as amended*. Abuja: Federal Republic of Nigeria.

FiriAppah, C., Okujagu, D. C., & Bassey, S. E. (2014). Effects of crude oil spill in germination and growth of *Hibiscus esculentus* (Okra) in Bayelsa state Niger Delta Region of Nigeria. *International Journal of Engineering and Science (IJES), 3*(6), 30−40.

Gobo, A., Richard, G., & Ubong, I. (2010). Health impact of gas flares on Igwuruta/Umuechem communities in Rivers State. *Journal of Applied Sciences and Environmental Management, 13*, 27−33.

Ikemeh, R. A. (2015). Assessing the population status of the critically endangered Niger Delta Red Colobus (*Piliocolobus epieni*). *Primate Conservation* (29), 87−96.

Ipingbemi, O. (2009). Socio-economic implications and environmental effects of oil spillage in some communities in the Niger delta. *Journal of Integrative Environmental Sciences, 6*(1), 7−23.

Ite, A. E., Ibok, U. J., Ite, M. U., & Petters, S. W. (2013). Petroleum exploration and production: Past and present environmental issues in the Nigeria's Niger Delta. *American Journal of Environmental Protection, 1*(4), 78−90. Available from https://doi.org/10.12691/env-1-4-2.

Izah, S. C., Angaye, T. C. N., Aigberua, A. O., & Nduka, J. O. (2017). Uncontrolled bush burning in the Niger Delta region of Nigeria: Potential causes and impacts on biodiversity. *International Journal of Molecular Ecology and Conservation, 7*(1), 1−15.

Izah, S. C., & Seiyaboh, E. I. (2018). Challenges of wildlife with therapeutic properties in Nigeria; a conservation perspective. *International Journal of Avian & Wildlife Biology, 3*(4), 252−257.

Kostiuk, L., Johnson, M., & Thomas, G. (2004). *University of Alberta flare research project: Final report November 1996−September 2004*. Edmonton, AB: University of Alberta, Department of Mechanical Engineering.

Langangen, O., Olsen, E., Stige, L. C., Ohlberger, J., Yaragina, N. A., Vikebo, F. B., et al. (2017). The effects of oil spills on marine fish: Implications of spatial variation in natural mortality. *Marine Pollution Bulletin*, *119*(1), 102−109.

Malden, A. (2017). *Nigeria's Oil and Gas Revenues: Insights from new company disclosures.* <https://resourcegovernance.org/sites/default/files/documents/nigeria-oil-revenue.pdf> Retrieved 14.08.19.

Niger Delta Human Development Report (2006) . Abuja: United Nations Development Programme (UNDP) UNDP Nigeria.

Nriagu, J. (2011). Oil industry and the health of communities in the Niger Delta of Nigeria. In O. N. Jerome (Ed.), *Encyclopedia of environmental health* (pp. 240−250). Burlington, NJ: Elsevier.

Nwilo, P. C., & Badejo, O. T. (2005). *Oil spill problems and management in the Niger Delta*. Miami, Florida, USA: International Oil Spill Conference.

Obot, E. A. (2003). *Niger delta bioscope: Integration of important faunal zones of the delta.* Gland: World Wildlife Fund.

Ogbe, M. G. (2011). Managing the environmental challenges of the oil and gas industry in the Niger Delta, Nigeria. *Journal of Life Science*, *1*(1), 1−17.

Ohimain, E. I. (2012). Environmental impacts of petroleum exploration dredging and canalization in the Niger Delta. In A. S. Akpotoe, S. H. Egboh, A. I. Ohwona, C. O. Orubu, S. B. Olabaniyi, & R. O. Olomo (Eds.), *Five decades of oil production in Nigeria: Impact on the Niger Delta* (pp. 391−405). Abraka, Nigeria: Centre for Environmental and Niger Delta Studies.

Okonkwo, E. C. (2014). Oil spills in Nigeria: Are there social and economic impacts?. *International Oil Spill Conference Proceedings*, *2014*(1), 300289.https:/doi.org/10.7901/2169-3358-2014-1-300289.1.

Okore, O., Avaoja, D., & Nwana, I. (2014). Edible insects of the Niger Delta area in Nigeria. *Journal of Natural Sciences Research*, *4*(5), 1−9.

Omamuyovwi, A. I., & Akpomuvire, M. (2017). The impact of oil exploitation on the socio-economic life of orogun community, an oil producing community in Delta State, Nigeria. *American Journal of Environmental and Resource Economics*, *2*(2), 73−79.

Omorede, C. K. (2014). Assessment of the impact of oil and gas resource exploration on the environment of selected communities in Delta State, Nigeria. *International Journal of Management, Economics and Social Sciences (IJMESS)*, *3*(2), 79−99.

OPEC's Annual Statistical Bulletin. *OPEC share of world crude oil reserves 2018* (2019). <https://www.opec.org/opec_web/en/data_graphs/330.htm> Retrieved 14.08.19.

Orubu, C. O., Odusola, A., & Ehwarieme, W. (2004). The Nigerian oil industry: Environmental diseconomies, management strategies and the need for community involvement. *Kamla-Raj Journal of Human Ecology*, *16*(3), 203−214.

Osuagwu, E. S., & Olaifa, E. (2018). Effects of oil spills on fish production in the Niger Delta. *PLoS One, 13* (10), 1−14.

Osuoke, G. O., & Emeka-Okpara, F. O. (2014). Evaluation of the impact of crude oil spillage on Izombe community and their productivity implications. *International Journal of Research in Engineering and Technology*, *3*(10), 208−211.

Phil-Eze, P. O., & Okoro, I. C. (2009). Sustainable biodiversity conservation in the Niger Delta: A practical approach to conservation site selection. *Biodiversity and Conservation*, *18*, 1247−1257.

Reed, K. (2009). *Crude existence: Environment and the politics of oil in Northern Angola.* CA: University of California Press.

Sagay, I. E. (2005). The Niger Delta and the case for resource control. A public lecture delivered on honour of Honourable Justice Adolphus Karibi-Whyte at the University of Ife.

Saleh, M. A., Ashiru, M. A., Sanni, J. E., Ahmed, T. A., & Muhammad, S. (2017). Risk and environmental implications of oil spillage in Nigeria (Niger-Delta Region). *International Journal of Geography and Environmental Management*, *3*(2), 44−53.

Udoh, J. C., & Ekanem, E. M. (2011). GIS based risk assessment of oil spill in the coastal areas of Akwa Ibom state, Nigeria. *African Journal of Environmental Science and Technology*, *5*(3), 205−211.

Ugboma, P. P. (2015). Environmental degradation in oil producing areas of Niger Delta region, Nigeria: The Need for sustainable development. *International Journal of Science and Technology*, *4*(2), 75−85.

Ugbomeh, B. A., & Atubi, A. O. (2010). The Role of the oil Industry and the Nigerian State in defining the future of the Niger Delta Region of Nigeria. *African Research Review*, *4*(2), 103−112.

United Nations Development Programme (UNDP). (2006). *Human development report, beyond scarcity: Power, poverty and the global water crisis*. New York: UNDP.

United Nations Environment Program (UNEP). (2016). Environmental assessment of Ogoniland. Available at: http://postconflict.unep.ch/publications/OEA/UNEP_OEA.pdf.

UNEP. (2011). *Environmental assessment of Ogoniland*. Nairobi: UNEP.

United Nations. *Rio declaration on environment and development*. (1992). Available from <https://www.cbd.int/doc/ref/rio-declaration.shtml>.

Uquetan, U. I., Osang, J. E., Egor, A. O., Essoka, P. A., Alozie, S. I., & Bawan, A. M. (2017). A case study of the effects of oil pollution on soil properties and growth of tree crops in Cross River state, Nigeria. *International Research Journal of Pure and Applied Physics*, *5*(2), 19−28.

World Bank. (1995). *Defining an environmental development strategy for the Niger Delta. Vols. I & II*. West Central Africa: Industry and Energy Operation Division.

World Bank Institute. (2003). *Nigeria poverty-environment linkages in the natural resource sector empirical evidence from Nigerian case studies with policy implications and recommendations*. Africa Environment and Social Development Unit, World Bank. Retrieved from <http://documents.worldbank.org/curated/en/115331468775788053/pdf/multi0page.pdf>.

Yakubu, O. H. (2017). Addressing environmental health problems in Ogoniland through implementation of United Nations Environment Program recommendations: Environmental management strategies. *Environments*, *4*(28), 1−19. Available from https://doi.org/10.3390/environments4020028.

SOCIOECONOMIC VULNERABILITY TO URBAN FLOODS IN GUWAHATI, NORTHEAST INDIA: AN INDICATOR-BASED APPROACH

27

Shrutidhara Kashyap[1] and Ratul Mahanta[2]

[1]*Department of Economics, Arya Vidyapeeth College, Guwahati, India*
[2]*Department of Economics, Gauhati University, Guwahati, India*

27.1 INTRODUCTION

An effective disaster risk reduction requires quantification of vulnerability. Vulnerability of hazard indicates the risk associated with the social and economic liability, the ability to cope with it, and the degree to which a system may react adversely in a hazardous situation (Proag, 2014). The increased instances of urban floods worldwide with notable socioeconomic vulnerability enhance the importance of measuring socioeconomic vulnerability and incorporating in flood management plans in urban centers (Armas & Gavris, 2016; Birkmann, 2006; Cutter, Boruff, & Shirley, 2003). The experiences of Bangladesh and Nepal show that successful implementation of flood-protection measures depends on identification of vulnerable clusters with specific socioeconomic impacts (Dewan, 2015). Not even flood, study on Bucharest, Romania reports that socioeconomic factors contribute toward mitigation of other natural hazards. It is possible through integrating societal impacts in monitoring process to enrich administrative tools for hazard management (Armas & Gavris, 2016). Guwahati, the gateway to Northeast India, is rapidly urbanizing with a population of 9.69 lakhs (Government of India, 2011). The economic scenario of the city has changed with the establishment of new industrial and residential expansions induced by the urbanization process at the beginning of the 21st century (Borthakur & Nath, 2012). The haphazard land-use−land-cover conversions and encroachments upon ecosensitive areas are responsible for frequent floods in Guwahati. Deficiency in basic urban infrastructure has made the flood situation more worse causing havoc in the city. The urban flood vulnerability in the form of human and animal losses, damages to infrastructures such as schools, public health centers in Guwahati Metropolitan Area (GMA) becomes irreversible (DDMA, 2017). The new areas submerged into floods prove the increased exposure and vulnerability to floods. In order to resolve a long-term solution to this growing problem, the need of the hour is to quantify socioeconomic vulnerability posed by urban flooding. Such analysis provides greater insight into solving urban disarray and wretchedness.

The increased trends and exposures to human intervened disasters receive considerable attention from research communities across the globe in recent years (Antwi et al., 2015). Expanded urban activities have led to reduction in urban greenery and increased urban disasters (Bhaskar, 2012). The increased frequency and vulnerability of floods have proved it as the most distressing disaster all over the world (Dankers et al., 2014; Hall et al., 2014; Svetlana, Radovan, & Jan, 2015). Higher concentration of assets and infrastructure with complexity involved in management results in higher flood vulnerability in urban hubs (Diakakis, Deligiannakis, Katsetsiadou, & Meleki, 2017). India has been experiencing an increasing exposure to urban floods in major urban centers of the country. The spatial expansion of urban areas during 2001–11 has resulted in increased flood risks (Centre for Science and Environment, 2016). Almost 56% of Indian smart cities are vulnerable to urban flooding. In recent decades, urban India has experienced some notable flood incidences causing havoc in urban hubs Hyderabad (2000), Ahmedabad (2001), Delhi (2002, 2003, 2009), Chennai (2004, 2015), Mumbai (2005), Surat (2006), Kolkata (2007), Jamshedpur (2008), Guwahati (2010, 2012, 2014), and Srinagar in 2014 (NDMA, 2016).

In the advent of frequent urban floods in Guwahati, NE India, the chapter intends to make an analysis of socioeconomic vulnerability to urban floods in the city. The chapter has been designed to give an idea about the socioeconomic dimensions of urban floods and the extent of vulnerability in terms of those dimensions as faced by the exposed population. The chapter has been organized into seven sections. The first section introduces the idea followed by a discussion on socioeconomic vulnerability impacts of urban floods. The third section discusses the theoretical background of vulnerability assessment and the fourth section deals with the assessment procedure of flood vulnerability in Guwahati. The results from the assessment of socioeconomic vulnerability to urban floods are depicted in the fifth section. The sixth section illustrates the limitations and future implications of the study with concluding observations in the seventh section.

27.2 URBAN FLOOD AND SOCIOECONOMIC VULNERABILITY

The interrelated factors such as climate change, land use change, increased urban expansion, deterioration in open spaces, improper urban infrastructure and lack of maintenance, and absence of drainage system are primarily responsible for the occurrences of urban floods (Jha et al., 2011). Both natural and flash floods have been experienced in urban locations. It depends upon the nature of occurrences and their impacts. However, flash floods pose higher threats in cities due to its own characteristics of the rapidity and suddenness at the inception and higher intensity. Urban areas being congested with higher population density and asset concentration in a smaller geographic area, flash floods cause massive damages to physical and human systems (Jonkman, 2005; Jonkman & Vrijiling, 2008; Boloque et al., 2016). Flash flood is a swiftly occurring flood caused by heavy rains and it is different from natural flood regarding short-time scale, quick evolution, and occurrence due to a few hours of heavy rainfall (Naulin, Payrastre, & Gaume, 2013).

The urban flood vulnerability indicates the extent of damages, direct or indirect to the exposed population caused by urban floods. Petry (2002) identifies that direct and indirect damages can be subgrouped into tangible and intangible damages. Direct tangible damages are physical damages to property and the expenditure incurred for rehabilitation of exposed population. Direct intangible damages are in the form of the sufferings to living beings (casualties, illness, accident etc.). The indirect tangible damages include distortion of trade and commerce and reduced purchasing power, while indirect intangible damages are increased hazard vulnerability and emigration. The tangible and intangible losses to floods incorporate physical and socioeconomic vulnerability to floods respectively. The physical vulnerability of urban floods may be realized in terms of damages to assets, institutions, and infrastructure (Dewan, 2015; Debionne, Ruin, Shabou, Lutoff, & Creutin, 2016; Dorbot & Parker, 2007). The social vulnerability of urban floods indicates the intangible damages to human system with the associated effects of social inequalities in terms of education, health, social security, etc. in the postflood situation. There are ample of studies (Antwi et al., 2015; Pouya, Nouri, Monsouri, & Lashaki, 2017; Proag, 2014; Rufat, Tate, Burton, & Maroof, 2015; Saxena, Geethalakshmi, & Lakshmanan, 2013; Sherbinin & Bardy, 2015; Tapsell, Rowsell, Tunstall, & Wilson, 2002) discussing the components and indicators of social vulnerability. The economic vulnerability refers to incapability of exposed population in accessing resources to withstand adverse shocks (UNISDR, 2009). Among various economic vulnerability indicators, working status, dependency ratio, financial deprivations, and livelihood disruptions caused by floods receive prime attention in flood vulnerability studies (Antwi et al., 2015; Karagiorgos, Thaler, Heiser, Hube, & Fuchs, 2016; Proag, 2014; Saxena et al., 2013). The recent improvements in hazard literature advocate the application of livelihoods approaches for measuring vulnerability with socioeconomic components. Livelihood approach intends to capture social and economic dimensions of flood vulnerability through assessing sensitivity and adaptive capacity following IPCC (2014).

27.3 **VULNERABILITY ASSESSMENT: THEORETICAL PERSPECTIVE**

Vulnerability, a multifaceted concept related to hazards and shocks, has been operationalized using various approaches—engineering, geographic, hydrological, and social science approaches. Physical vulnerability to floods has been studied using hydrological models by different researchers (Birhanu, Kim, Jang, & Park, 2016; Bodoque et al., 2016; Garrote, Alvarenga, & Herrero, 2016; Yin, Yu, Yin, Liu, & He, 2016). The tangible damages from floods including economic losses of households' assets and properties, infrastructure damages, and road network disruptions are assessed through this approach. The effects of floods upon spatiotemporal changes have been analyzed in studies applying geographical approaches (Daungthima & Kazunori, 2013; Elalem & Pal, 2015; Halounova & Holubee, 2014; Mahmood, Khan, & Ullah, 2016; Setyani & Saputra, 2016; Shields, 2016). Quantitative techniques like econometric analysis have been performed to measure flood vulnerability by the researchers (Debionne et al., 2016; Grames, Prskawetz, Grass, Viglione, & Blosch, 2016; Lakhdar, Tayeb, & Boudersa, 2015; Mustafa, Marzuki, Ariffin, Salleh, & Rahaman, 2014; Renald, Tjiptoherijanto, & Suganda, 2016; Spitalar et al., 2014). The social science approach incorporates human component of flood vulnerability with the application of indexing method. The researchers (Antwi et al., 2015; Pouya et al., 2017; Saxena et al., 2013) aim to capture the intangible damages from floods through the construction of a composite index of different vulnerability components. Integration of physical and social components is required to bridge the gap between natural sciences approach and social science approaches (Karagiorgos et al., 2016). Social science researchers advocate indexing methods incorporating a set of socioeconomic indicators that reflect flood exposure, sensitivity, and adaptive capacity as defined in IPCC (2001) (Armas & Gavris, 2016; Birkmann, 2006; Cutter et al., 2003). The IPCC (2014) report advocates "paradigm shift" in defining vulnerability stating vulnerability not being dependent on exposure. A vulnerable system is one which is exposed to hazard with adverse shocks. The new paradigm calls for vulnerability assessment on the basis of hazard-specific indices of sensitivity and adaptive capacity without considering exposure indicators (IPCC, 2014; Sharma & Ravindranath, 2019). IPCC (2001) considers "vulnerability" as the end point a keen to "outcome vulnerability" (Kelly & Adger, 2000) which has been redefined in IPCC (2014) as starting point or "contextual vulnerability" (Kelly & Adger, 2000; O'Brien, Eriksen, Schjolden, & Nygaard, 2007).

The increasing concern about livelihood security threats caused by hazard vulnerability has led to the analysis of vulnerability though livelihood approaches. Households' ability in terms of natural, social, financial, physical, and human capital to survive after a shock has been assessed using sustainable livelihoods Approach in several studies (Chambers & Conway, 1992; Serrat, 2017; Twigg, 2001). However, the opponents consider this approach as incomplete being addressed only the sensitivity and adaptive capacity components of vulnerability while exposure left unaddressed (Haan, Riederer, & Foster, 2009). The navigation toward a complete vulnerability assessment with livelihood vulnerability as the focal point results in the construction of livelihood vulnerability index (LVI). The LVI approach developed by Haan et al. (2009) has been used in some earlier studies (Adu, Kuwornu, Somuah, & Sasaki, 2017; Etwire, Al-Hassan, Kuwornu, & Osei-Owusu, 2013; Laitae, Praneetvatakul, Vijitsrikamol, & Mungkung, 2013) involving the assessment of vulnerability to climate changes.

27.4 ASSESSMENT OF SOCIOECONOMIC URBAN FLOOD VULNERABILITY IN GUWAHATI: LVI APPROACH

The assessment of urban flood vulnerability in GMA has been conducted using the LVI approach. The application of LVI is expected to provide the basis for capturing the flood vulnerability element in its entirety. The LVI–IPCC complying with IPCC (2014) definition of vulnerability may bring a fruitful framework for flood vulnerability assessment in GMA. The area-specific assessment through the construction of localized LVIs paves the way for designing and implementing flood management plans on the basis of local resources and local needs. The LVI consists of nine major components—flood exposure, sociodemographic characteristics, social network and security indicators, effects upon livelihood, health, education, transportation, and amenity services. The major components and subcomponents of LVI are presented in Table 27.1.

Both natural and flash floods have been experienced in GMA over the years. Natural floods are caused by the excessive rainwater in the Brahmaputra in the close proximity areas and low lying nearby reservoirs and wetlands. Flash floods become regular yearly events primarily in central areas of the city. Table 27.2 reflects the areas under urban flood, both natural and flash one in 2017 in Kamrup (Metro) district where Guwahati is located.

The study attempts to make an area-specific flood vulnerability assessment on the basis of 2017 flood in GMA. The lesser time gap between data collection and event year may be expected to be helpful in getting the adequate data in terms of better remembrance for the respondents. The primary data has been collected through extensive field visits during August–September of year 2017. Four areas from GMA [two from Guwahati Municipal Corporation Area (GMCA) and two

Table 27.1 Major Components and Subcomponents of Livelihood Vulnerability Index.

Major Components	Subcomponents	Explanation of Subcomponents	Vulnerability Effects	No. of Subcomponents	Source
Sociodemographic profile	Dependency ratio	population (less than 14 years + greater than 65)/population (15–64)	+	4	Antwi et al. (2015), Haan et al. (2009)
	Housing ownership Structure	Percentage of population having own houses	−		Cutter and Emrich (2006), Karagiorgos et al. (2016)
	Income	Average household income	−		Karagiorgos et al. (2016), Rufat et al. (2015), WHO (2010)
	Educational attainment	Percentage of households with highest level of education below HSLC in the family	+		WHO (2010), Rufat et al. (2015)
Flood exposure	Intensity	Percentage of households witnessing flood intensity to increase than previous years	+	5	Parker (1995), Diakakis et al. (2017), Spitalar et al. (2014)
	Frequency	Percentage of households experiencing flood more than twice a year	+		Diakakis et al. (2017)
	Height	Height index = height of flood/height of base of house in relation to ground level	+		Pouya et al. (2017)
	Damage	Percentage of households experiencing house damage	+		Diakakis et al. (2017), Mahmood et al. (2016)
	House structure	Percentage of households not having adequate housing Structure (concrete/ semiconcrete + pucca)	+		WHO (2010), Bodoque et al. (2016)
Livelihood	Income	Percentage of households facing loss in income-generating activity	+	4	Proag (2014), Saxena et al. (2013), Karagiorgos et al. (2016)
		Percentage of households alter occupation	+		
		Average income loss	+		
	Evacuation	Percentage of households leaving properties	+		Karagiorgos, et al. (2016)
Health	Human health	Percentage of households facing injury/illness	+	3	Haan et al. (2009), Karagiorgos et al. (2016), Tapsell et al. (2002)
		Avg. treatment cost	+		
	Livestock	Percentage of households facing illness/injury/ death of livestock	+		Proag (2014), Saxena et al. (2013)
Education	Education disruption	Percentage of households facing education disruption due to flood	+	1	Haan et al. (2009)

(Continued)

Table 27.1 Major Components and Subcomponents of Livelihood Vulnerability Index. *Continued*

Major Components	Subcomponents	Explanation of Subcomponents	Vulnerability Effects	No. of Subcomponents	Source
Transportation	Traffic problem	Percentage of households having experienced traffic congestion caused by flooding	+	2	Debionne et al. (2016)
	Transportation cost	Percentage of households experiencing increased transportation cost	+		Dorbot and Parker (2007), Yin et al. (2016)
Amenity	Sanitation	Percentage of households waterlogged in toilets	+	4	Proag (2014)
		Percentage of households defecated outside home due to flood	+		
	Drinking water	Percentage of households buy or collect drinking water from outside during flood	+		Pilot survey
	Drainage	Percentage of households stating inadequate drainage provision	+		Pilot survey
Social network	Length of residency	Avg. no. of years resided	−	5	Karagiorgos et al. (2016)
	Local association	Percentage of households any member having associated with clubs or NGOs	−		
	Community participation In relief	Percentage of households accessing institutional structural relief	−		Bodoque et al. (2016), Petry (2002), Renald et al. (2016)
		Percentage of households accessing institutional nonstructural relief	−		Petry (2002)
		Percentage of households accessing noninstitutional relief	−		Petry (2002)
Security indicators	Individual preparedness	Percentage of households raising floor of house	−	4	Kurosaki (2017)
		Percentage of households moving children and elderly persons away from home	−		
		Percentage of households keeping own savings for coping	−		Cutter (1996), Cutter & Emrich (2006), Cutter et al. (2006)
		Expenditure on coping	−		

Authors on the basis of reviewed literature.

Table 27.2 Areas Submerged From Urban Flood in Kamrup (Metro) District, 2017.

Revenue Circle	Type of Flood	No. of Affected Sites	Affected Areas
Dispur	Flash	12	Anil Nagar, Nabin Nagar, Shree Nagar, Tarun Nagar, Zoo Road, Udalbakra, VIP road, Narengi, Joya Nagar, Hatigaon, Dakhin Gaon, Kahilipara
Guwahati	Flash	9	Lachit Lane, Rajgarh Road, B.T. College Road, Pandu, Chatribari, Lamb Road, Chandmari, Fatasil, Pub Sarania
Azara	Natural and flash	9	Dharapur, VIP, SOS Road, Charmajuli, Charnihali, Garal, Agchia, Borjhar, Majir Gaon
Chandrapur	Natural	31	Chandrapur Bagisha, Tintukura, Ram Singhsapori, Rajabari, Kherbari, Bengenabari, Chapaidang, Hatisila, Hajongbari, Tamulbari, Kalimandir, Belguri, Ghuligaon, Lahapara, Pachim Mayang Mouza, Panikhaiti, Tatimara, Khankor, No-2 Kharguli, No-2 Chandrapur, Kamarpur Gaon, Kamarpur Pahar, No-1 Dhamkhunda, No-2 Dhamkhunda, Dhipuji pathar, Dhipujijanpam, Gobhali, Ghoramara, Kajolichowki, Hatibagora, Gobardhon
Sonapur	Natural	14	MaloibariNC, Maloibari Jungle, MalaibariGaon, Malobaripathar, Pub Maloibari, Gaon Bimoria, NizDimoria, 1 No. Kachutoli, 2 No. Kachutoli, Sonapurpathar, Juboi, Jarikuchi, Bajram, Parabuma

DDMA (2017). Report on urban flood in Guwahati, District Disaster Management Authority, Kamrup(Metro), Government of Assam.

from GMA excluding GMCA] have been chosen for the vulnerability assessment. The areas are selected on the basis of vulnerability rankings in terms of exposed population and exposed geographical area (ASDMA, 2014). Anil-Nabin Nagar (highly vulnerable) and Maligaon (low and moderately vulnerable) are chosen from GMCA and Panikhaiti (highly vulnerable) and Azara (low and moderately vulnerable) are from GMA excluding GMCA. The location of study area has been presented in the following Fig. 27.1.

A multistage sampling method has been followed while conducting field survey in the selected study areas of GMA. GMA being purposively selected, in the second stage, two areas (Anil-Nabin Nagar and Maligaon) from GMCA and Panikhaiti and Azara from GMA excluding GMCA are selected. In the next stage, two wards are randomly selected from the selected areas from GMCA and two villages from outgrowth areas. In the final stage, in order to capture the variations in social, income, and gender classes, households in each ward and village are selected using stratified random sampling procedure. The distribution of total sample size of 281 among the wards and villages is presented in Table 27.3.

27.4.1 CALCULATING THE LIVELIHOOD VULNERABILITY INDEX AND LIVELIHOOD VULNERABILITY INDEX—IPCC

The first step toward construction of LVI involves the normalization of subcomponents (Table 27.1) measured in different scales. There are ample of methods of normalization of

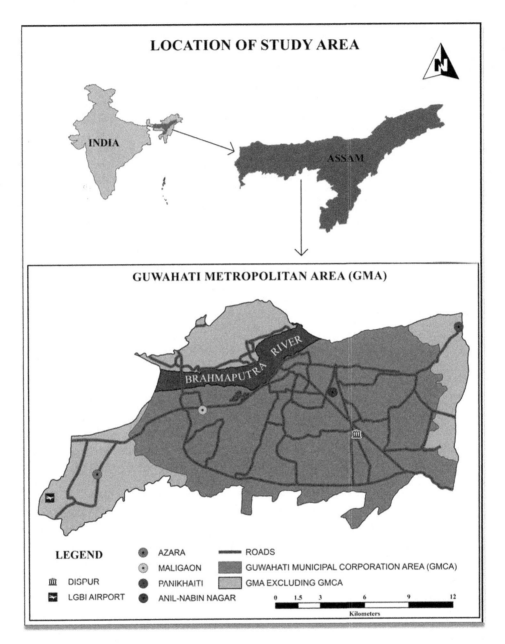

FIGURE 27.1

GMA map with location of study areas. *GMA*, Guwahati Metropolitan Area.

GIS Mapping, Satellite map is generated from United States Geological Survey (USGS), Google Earth Explorer.

Table 27.3 Sampling Design for Assessing Flood Vulnerability in Guwahati Metropolitan Area (GMA).

GMA			
GMCA		**GMA Excluding GMCA**	
Highly vulnerable	Low and moderately vulnerable	Highly vulnerable	Low and moderately vulnerable
Anil-Nabin Nagar	Maligaon	Panikhaiti	Azara
Ward No. 20A, 20B	Ward No. 5A, 5B	Two villages—Rajabari, Bengenabari	Two villages—Barbori, Mirzapur
No. of sample = 50	No. of sample = 48	No. of sample = 95	No. of sample = 88

GMCA, *Guwahati Municipal Corporation Area.*
ASDMA (2014). Review of studies on urban floods in Guwahati: From flood knowledge to urban action, Assam State Disaster Management Authority, Government of Assam; GMDA (2009). Master Plan—2025, GMA, Guwahati Metropolitan Development Authority, Assam.

indicators, however, the study follows the method as used in the construction of Physical Quality of Life Index (PQLI) by Morris D. Morris (1969); Human Development Index (UNDP, 1990) and adapted for the construction of LVI (Adu et al., 2017). Some subcomponents are positively to vulnerability while a few are negatively related. Two separate equations have been applied for the two cases. Eq. (27.1) is used for positive indicators and Eq. (27.2) for negative indicators.

$$\text{Index}_{sai} = \frac{S_a - S_{min}}{S_{max} - S_{min}} \tag{27.1}$$

$$\text{Index}_{sai} = \frac{S_{max} - S_a}{S_{max} - S_{min}} \tag{27.2}$$

where S_a is observed subcomponent of indicator of area a and S_{max} and S_{min} are the maximum and minimum values, respectively.

In the next step, index for each major component has been calculated taking averages of standardized subcomponents using:

$$M_a = \sum_{(i=1)}^{n} \frac{\text{index}_{sai}}{n} \tag{27.3}$$

where M_a represents indices for major components—sociodemographic profile (SDP), flood exposure (FE), livelihood (L), health (H), education (E), transportation (T), amenity (A), social network (SN), and security indicator (SI); n is the number of subcomponents in each major component. Then LVI for each of the four study areas is constructed following:

$$\text{LVI} = \frac{W_{SDP}\text{SDP}_a + W_{FE}\text{FE}_a + W_L L_a + W_H H_a + W_E E_a + W_T T_a + W_A A_a + W_{SN}\text{SN}_a + W_{SI}\text{SI}_a}{W_{SDP} + W_{FE} + W_L + W_H + W_E + W_T + W_A + W_{SN} + W_{SI}} \tag{27.4}$$

The number of subcomponents in each major component has been used as the weights for the specific major components. A balanced weighted average method (Adu et al., 2017; Haan et al.,

2009; Sullivan, Meigh, & Fediw, 2002) has been followed in order to regard equal importance or weight of subcomponents in resource-poor setting and unaware mass respondents in developing economies (Haan et al., 2009). The LVI for the study areas ranges from 0 representing "the least vulnerable" to 0.6 as "the most vulnerable."

27.4.2 LIVELIHOOD VULNERABILITY INDEX—IPCC

LVI—IPCC has been applied to capture the contributing factors—exposure, sensitivity, and adaptive capacity to vulnerability in accordance with the IPCC working definition of vulnerability, 2001 and 2007 in some earlier studies (Adu et al., 2017; Haan et al., 2009). The IPCC (2014) has depicted, exposure being a preexisting situation, sensitivity, and adaptive capacity as internal properties of a vulnerable system (IPCC, 2014; Sharma & Ravindranath, 2019). The present study has made an attempt to assess vulnerability on the basis of IPCC (2014) with better reflection of sensitivity and adaptation of flood-affected population in the study areas. The organization of major components into two vulnerability components is presented in Table 27.4.

However, the same standardization procedure for subcomponents and major components followed in case of calculation of LVI has been tried out to obtain LVI—IPCC. The vulnerability components or contributory factors to LVI—IPCC have been calculated applying the formula:

$$CF_a = \frac{\sum_{(i=1)}^{n} W_{M_i} M_{ai}}{\sum_{(i=1)}^{n} W_{M_i} M_{ai}} \tag{27.5}$$

where CF_a is an IPCC defined contributing factor, M_{ai} are major components indexed by i, W_{M_i} is weight of each major component, n is the number of major components in each contributing factor.

Then the final index is obtained as weighted aggregates of sensitivity and adaptive capacity scores computed applying (27.5):

$$LVI-IPCC = \frac{W_s S + W_a A}{N} \tag{27.6}$$

where LVI—IPCC is the LVI incorporating IPCC vulnerability framework, S the sensitivity score, A adaptive capacity score, W_s and W_a are the weights of sensitivity and adaptive capacity, respectively, in terms of number of subcomponents as assumed in this study and N is the sum of weights of contributing factors to vulnerability. The range of LVI—IPCC varies from 0 (least vulnerable) to 1 (most vulnerable) (IPCC, 2014; Department of Science and Technology, GOI, 2019).

Table 27.4 Organization of Major Components Into Vulnerability Components of IPCC (2014).

Components	Major components
Sensitivity	Livelihood, health, education, transportation, amenity
Adaptive capacity	Sociodemographic profile, social network, security indicator

IPCC (2014), Climate Change, 2014. Impacts, adaptation and vulnerability, contribution of working group II to the fifth assessment report, New York.

27.5 **RESULTS AND DISCUSSIONS**

The household data for the subcomponents collected from the four study areas—Anil-Nabin Nagar, Maligaon, Panikhaiti, and Azara—have been compiled to get the percentages, ratios, and averages for each study area. These results are standardized applying Eqs. (27.1) and (27.2) depending upon their nature of relationship with flood vulnerability. The indexed subcomponents of LVI, hence-forth, obtained for the four study areas—Anil-Nabin Nagar, Maligaon, Panikhaiti, and Azara—are reported in Table 27.5.

The results from location-specific flood vulnerability assessment reveal that the four areas are vulnerable to floods. Differential intensities in regards to vulnerabilities of subcomponents have been experienced for the study areas. It signifies the need to separate local area-centric flood man-agement plans for GMA. However, the flood exposure for Anil-Nabin Nagar is 0.614, for Maligaon 0.290, Panikhaiti and Azara exhibit 0.767 and 0.509, respectively. Anil-Nabin Nagar is quite favor-able in terms of SDP (0.401), which is highly influenced by lower dependency ratio and higher educational attainment as indicated by the indices 0.148 and 0.120, respectively. Anil-Nabin Nagar exhibits a high flood exposure index (0.614). The higher flood exposure along with higher housing damage index (0.860) enhances the extent of vulnerability. Income generation has been found to be disrupted for the private sector employees and those working in urban informal sectors with adverse effects upon their livelihood. Education disruption index is quite higher (0.820) revealing floods have higher negative impacts upon education of children due to water in their houses, roads, or in schools. Transportation (0.970) is another area of concern for Anil-Nabin during flood days with waterlogged roads, and resulting traffic congestion for longer hours or even for a few days. Amenity provisions are not well off due to waterlogged toilets and problems in drinking water facilities. Social network, an important component in hazard arena, is weak for Anil-Nabin Nagar (0.612). The leakages in undertaking adaptation measures have made the area vulnerable toward security indicators (0.653). The overall LVI score for Anil-Nabin Nagar has been calculated as 0.510 (highly vulnerable).

Maligaon, although having experienced with lower physical vulnerability to floods, results show some noticeable socioeconomic vulnerability issues. The loss in income-generating activity (0.354) enhances the miseries of the households. Evacuation index is low (0.042) as only homesteads are affected by floods in most cases. Education is again seriously disrupted as indicated by higher scores for education major component (0.729). In terms of transportation (0.709), increased traffic congestion in waterlogged roads and higher fares cause trouble in road mobility. However, amenity provisions such as sanitation, drinking water are less affected. Maligaon is found to be more vulner-able toward social network (0.766) and security indicators (0.732). The overall LVI for Maligaon has been found as 0.415 (moderately vulnerable).

Panikhaiti and Azara are located in outgrowth areas of GMA. The nature and extent of urbani-zation as well as flood vulnerability issues are different from the city center. The households are found to be highly exposed to floods which are intensified by damages of houses and lack of ade-quate housing structure. Kaccha houses are very common for the residents. However, previous flood experiences help the households to keep the base of their houses at a very high level and to keep safe themselves to some extent. The mass households engaged in unorganized sector are found to be vulnerable in terms of earning livelihoods even for 1−2 months. Alternative livelihood

Table 27.5 Indexed Subcomponents, Major Components, and Livelihood Vulnerability Index (LVI) for Four Study Areas.

Major Components	Subcomponents	Anil-Nabin Nagar	Maligaon	Panikhaiti	Azara
SDP	Dependency ratio	0.148	0.151	0.172	0.131
	Percentage of population having own houses	0.460	0.542	0.021	0
	Average household income	0.876	0.860	0.959	0.905
	Percentage of households with highest level of education below HSLC in the family	0.120	0.188	0.832	0.307
SDP index		0.401	0.435	0.496	0.336
FE	Percentage of households witnessing flood intensity to increase than previous years	0.920	0.575	0.989	0.966
	Percentage of households experiencing flood more than twice a year	0.960	0.520	1.00	0.557
	Height index	0.269	0.082	0.149	0.172
	Percentage of households experiencing house damage	0.860	0.104	0.737	0.466
	Percentage of households not having adequate housing structure (concrete/semiconcrete + pucca)	0.060	0.167	0.958	0.386
FE index		0.614	0.290	0.767	0.509
Livelihood	Percentage of households facing loss in income-generating activity	0.340	0.354	0.474	0.307
	Percentage of households alter occupation	0	0.021	0.200	0.034
	Average income loss	0.081	0.023	0.107	0.137
	Percentage of households leaving properties	0.340	0.042	0.274	0.068
Livelihood index		0.190	0.110	0.264	0.137
Health	Percentage of households facing injury/illness	0.480	0.188	0.516	0.489
	Avg. treatment cost	0.069	0.009	0.030	0.087
	Percentage of households facing illness/injury/death of livestock	0.020	0	0.189	0.523
Health index		0.190	0.066	0.245	0.366
Education	Percentage of households facing education disruption due to flood	0.820	0.729	0.737	0.545
Transportation	Percentage of households having experienced traffic Congestion caused by flooding	0.960	0.750	0	0.011
	Percentage of households experiencing increased transportation cost	0.980	0.667	0.011	0.011
Transportation index		0.970	0.709	0.006	0.011
Amenity	Percentage of households waterlogged in toilets	0.780	0.021	0.737	0.375
	Percentage of households defecated outside home due to flood	0.040	0	0.747	0.307
	Percentage of households buy or collect drinking water from outside during flood	0.740	0.042	0.295	0.057
	Percentage of households stating inadequate drainage provision	0.300	0.500	0	0.200

Table 27.5 Indexed Subcomponents, Major Components, and Livelihood Vulnerability Index (LVI) for Four Study Areas. *Continued*

Major Components	Subcomponents	Anil-Nabin Nagar	Maligaon	Panikhaiti	Azara
Amenity index		0.465	0.141	0.445	0.235
SN	Avg. no. of years resided	0.820	0.890	0.422	0.404
	Percentage of households any member having associated with clubs or NGOs	0.460	0.042	0.905	0.443
	Percentage of households accessing institutional structural relief	0.280	0.896	0.347	0.682
	Percentage of households accessing institutional nonstructural relief	0.720	1.00	1.00	1.00
SN index		0.612	0.766	0.731	0.692
Security	Percentage of households accessing noninstitutional relief	0.780	1.00	0.979	0.932
	Percentage of households raising floor of house	0.740	0.750	0.842	0.693
	Percentage of households moving children and elderly persons away from home	0.880	0.896	0.789	0.932
	Percentage of households keeping own savings for coping	0.120	0.292	0.347	0.216
	Expenditure on coping	0.870	0.991	0.967	0.950
Security index		0.653	0.732	0.736	0.698
LVI		0.510	0.415	0.523	0.415

FE, *Flood exposure;* SDP, *sociodemographic profile;* SN, *social network.*
Authors' computations.

options like working as a daily labor and fishing are undertaken by the households which provide them with temporary relief to the sufferings from income and employment losses. The evacuation index for Panikhaiti is 0.274 indicating that households leave their houses during flood days and shelter in relative's houses or in the camps at nearby schools. The injuries and illness caused by floods to the households give rise to health vulnerability index to be 0.516. Education disruption index is higher (0.737) with evidence of closure of schools even for months. The use of schools for rehabilitation purposes is found to be a considerable reason behind longer hindrance toward education. The transportation is lesser affected due to availability of boating facility during flood days. But sanitation facility is found to be adversely affected on account of waterlogged toilets (0.737) for months. Difficulty in collecting drinking water has been emerged due to floodwater exposed tube-wells and wells of households. Social network component of LVI indicates that the area is completely vulnerable (0.731). Insufficient institutional structural relief measures, non-availability of nonstructural measures for flood control and mitigation has made the situation worse. Households are reported to get some relief from Down-Town University located at Panikhaiti. The households are found to be vulnerable toward flood security measures as

Table 27.6 (Livelihood Vulnerability Index–IPCC) in Four Study Areas.

LVI–IPCC Components	Anil-Nabin Nagar	Maligaon	Panikhaiti	Azara
Sensitivity	0.425	0.239	0.767	0.225
Adaptive capacity	0.560	0.654	0.660	0.584
LVI–IPCC	0.490	0.439	0.715	0.398

Authors' computations.

indicated by security index 0.736. The overall LVI score for Panikhaiti is computed as 0.523 (highly vulnerable).

Azara, a center for urban agglomeration in GMA, has experienced increased intensity of urban floods than previous years. Although flood exposure is not so higher yet vulnerability in terms of income and employment loss is substantial. Some households are found to be earning bread and butter through fishing during flood days. The evacuation index is low (0.068) due to lower exposure of houses to floodwater. Health issue is quite remarkable in terms of injuries and illness of human beings and live stocks. The deaths and illness of cows, goats, and hens due to water-borne diseases and unhygienic conditions are caused by dirty floodwater. It hampers the economic conditions of those households depending on livestock rearing business. Education becomes adversely affected owing to the use of schools for evacuation purposes. Transportation is found to be less vulnerable while few households are reported to face trouble for waterlogged toilets. Social network and security measures are found to be weak at Azara. Institutional relief measures are insufficient (only foods for live stocks) adding to the vulnerability. The overall LVI for Azara is 0.415.

The LVI computed on the basis of IPCC framework (LVI–IPCC) shows (Table 27.6) that the four study areas—Anil-Nabin Nagar, Maligaon, Panikhaiti, and Azara—are vulnerable to floods. Being grouped into the IPCC components of flood vulnerability-sensitivity and adaptive capacity, with sensitivity 0.425 and adaptive capacity 0.560 for Anil-Nabin Nagar, LVI–IPCC becomes 0.490 (highly vulnerable). Maligaon reveals moderate vulnerability as reflected in LVI–IPCC score (0.439), sensitivity 0.239, and adaptive capacity 0.654.

The higher sensitivity index (0.767) for Panikhaiti with adaptive capacity index of 0.660 results in a highly vulnerable LVI–IPCC score (0.715). Azara seems to be lesser vulnerable with LVI–IPCC score 0.398. The adaptive capacity (0.584) score is higher at Azara as compared to its sensitivity score (0.225). The lesser LVI–IPCC of 0.398 has been influenced by better adaptation found in the surveyed households.

27.6 LIMITATIONS AND FUTURE IMPLICATIONS

Despite having immense help in capturing livelihood vulnerability caused by floods, LVI and LVI–IPCC framework suffers from certain limitations. The subjectivity involved in the selection of subcomponents and determination of the nature of relationship with flood vulnerability may be studied further. Some other limitations of this approach include the use of mean values for standardization and much simpler method of weighting with weights being the number of subcomponents in each major component. However, such simpler approach may be helpful for examining the effectiveness of flood management policies in a flood-prone area. It is made possible through substituting the values of the indicators expected to change and recompute the LVI in a post-implementation phase of a flood control plan (Haan et al., 2009). The use of LVI in vulnerability assessment could provide a basis for estimating future vulnerability projections under changing flood exposures. The use of primary data eliminates the difficulties involved in using secondary data, data being collected in different temporal and spatial scales. The future research needs a careful study of flood vulnerability implications across the households, genders, educational attainment levels, etc. Better absorption of shocks in a flood-prone system needs enhancement of institutional provisioning and individual preparedness. Urban planning in such a system should be carefully designed incorporating the infrastructure development and legislation strengthening to minimize the vulnerability. System-oriented actions to cope up with post outcome effects and resettlement of vulnerable population require research effort on flood vulnerability, on a microscale, with the involvement of government, NGOs, and local community. More research emphasis should be given on social network component bearing a great role in enhancing social capital for locally managed flood vulnerability reduction programs.

27.7 CONCLUSION

The location-specific vulnerability assessment in GMA under LVI and LVI–IPCC framework may be of assistance toward designing flood management plans on the basis of the needs of different areas. It helps to identify the key areas to be given attention while allocating government funds for urban development activities. Drainage and solid waste management facilities are inadequate in GMA (Master Plan, GMA, 2009). The policymakers may give proper attention toward building structural measures such as drains and sewerage management provisions in order to prevent water-logging and reduce the extent of flood exposure. The study paves the way for rethinking of redesigning flood control programs with inclusion of nonstructural measures such as enact of law, flood warning (Amoateng, Finlayson, Howard, & Wilson, 2018) having long-term effects upon reducing flood damages. Intangible damages causing social inequalities in terms of health and education (Antwi et al., 2015; Sherbinin & Bardy, 2015) could be met up with the development of community development centers (Madan & Routray, 2015; Tanwattana, 2018), well-equipped with the arrangements to provide shelter to flood victims. The flood vulnerability assessment with emphasis on livelihood vulnerability is expected to be useful for the city planners to develop flood mitigation plans compatible with local needs and minimizing livelihood vulnerability.

REFERENCES

Adu, D. T., Kuwornu, J. K. M., Somuah, H. A., & Sasaki, N. (2017). Application of Livelihood vulnerability index in assessing smallholder maize farming households' vulnerability to climate change in Brong-Ahafo region of Ghana. *Kasetsart Journal of Social Sciences (2017)*, 1−11. Retrieved from <https://doi.org/10.1016/j.kjss.2017.06.009>.

Amoateng, P., Finlayson, C., Howard, J., & Wilson, B. (2018). A multi-faceted analysis of annual flood incidences in Kumasi, Ghana. *International Journal of Disaster Risk Reduction, 27*, 105−117.

Antwi, E. K., Danqah, J. B., Owusu, A. B., Loh, S. K., Mensah, R., Boafo, Y. A., et al. (2015). Community Vulnerability Assessment Index for Flood Prone Savannah Agro-ecological Zone: A Case Study of Wa West District, Ghana. *Weather and Climate Extremes, 10*, 56−69.

Armas, I., & Gavris, A. (2016). Census-based social vulnerability assessment for Bucharest. *Procedia Environmental Sciences, 32*, 138−146. Retrieved from <www.sciencedirect.com>.

ASDMA. (2014). *Review of studies on urban floods in Guwahati: From flood knowledge to urban action.* Assam State Disaster Management Authority, Government of Assam.

Bhaskar, P. (2012). Urbanisation and changing green spaces in Indian cities. *International Journal of Geology, Earth and Environmental Sciences, 2*(2), 148−156.

Birhanu, D., Kim, H., Jang, C., & Park, S. (2016). Flood risk and vulnerability of Addis Ababa city due to climate change and urbanization. *Procedia Engineering, 154*, 696−702. Retrieved from <www.sciencedirect.com>.

Bodoque, J. M., Amerigo, M., Herrero, A. D., Garcia, J. A., Cortes, B., Canovas, J. A. B., et al. (2016). Improvement of resilience of urban areas by integrating social perception in flash flood risk management. *Journal of Hydrology, 541*, 665−676.

Borthakur, M., & Nath, B. K. (2012). A Study of Changing Urban Landscape and Heat Island Phenomenon in Guwahati Metropolitan Area. *International Journal of Scientific and Research Publications, 2*(11). Retrieved from <www.ijsrp.org>.

Birkmann, J. (2006). *Measuring vulnerability to natural hazards − Toward disaster resilient societies.* Tokyo: UNU-Press.

Centre for Science and Environment. Government of India (2016). *Why urban India floods: Indian cities grow at the cost of their wetlands.* Available at <http://csestore.cse.org.in>.

Chambers, R. & Conway, G. (1992). Sustainable Rural Livelihoods: Practical Co for the 21st century, IDS Discussion Paper 296, Brighton: IDS.

Cutter, S. L. (1996), Vulnerability to environmental hazards. Progress in Human Geography, *20*(4), 529−539.

Cutter, S. L., Boruff, B. J., & Shirley, W. L. (2003). Social vulnerability to environmental hazards. *Social Science Quarterly, 84*(2), 242−261.

Cutter, S. L., & Emrich, C.T. (2006). Moral hazard, social catastrophe: The changing face of vulnerability along the hurricane coasts. Annals of American Academy, *604*(1), 102−112.

Cutter, S.L., Emrich, C.T., Mitchell, J.T., Boruff, B.J., Gall, M., Schmidtlein, M.C., & Melton, G. (2006). The long road home: Race, class and recovery from Hurricane Katrina. Environment, *48*(2), 8−20.

Dankers, R., Arnell, N. W., Clark, D. B., Falloon, P. D., Fekete, B. M., Gosling, S. M., et al. (2014). First look at changes in flood hazard in the inter-sectoral impact model inter-comparison project ensemble. *Proceedings of the National Academy of Sciences of the United States of America, 111*, 3257−3261. Retrieved from <https://doi.org/10.1073/pnas.1302078110>.

Daungthima, W. & Kazunori, H. (2013). Assessing the flood impacts and the cultural properties vulnerabilities in Ayutthaya, Thailand. *Procedia Environmental Sciences, 17*, 739−748. Retrieved from <www.sciencedirect.com>.

DDMA. (2017). *Report on urban flood in Guwahati.* District Disaster Management Authority, Kamrup (Metro), Government of Assam.

Debionne, S., Ruin, I., Shabou, S., Lutoff, C., & Creutin, J. D. (2016). Assessment of commuters' daily exposure to flash flooding over the roads of the Gard Region, France. *Journal of Hydrology, 541*, 636−648.

Department of Science and Technology. (2019). *Government of India Climate vulnerability assessment for the Indian Himalayan Region using a common framework*. Available at <https://dst.gov.in>.

Dewan, T. H. (2015). Societal impacts and vulnerability to floods in Bangladesh and Nepal. *Weather and Climate Extremes, 7*, 36−42.

Diakakis, M., Deligiannakis, G., Katsetsiadou, K. N., & Meleki, M. (2017). Mapping and classification of direct flood impacts in the complex conditions of an urban environment. The case study of the 2014 flood in Athens, Greece. *Urban Water Journal, 14*(10), 1065−1074.

Dorbot, S., & Parker, D. J. (2007). Advances and challenges in flash flood warning. *Environmental Hazards, 7*(3), 173−178.

Elalem, S., & Pal, I. (2015). Mapping the vulnerability hotspots over Hindu-Kush Himalaya Region of flooding disasters. *Weather and Climate Extremes, 8*, 46−58.

Etwire, P. M., Al-Hassan, R. M., Kuwornu, J. K. M., & Osei-Owusu, Y. (2013). Application of livelihood vulnerability index in assessing vulnerability to climate change and variability in northern Ghana. *Journal of Environment and Earth Sciences, 3*(2), 157−170.

Garrote, J., Alvarenga, F. M., & Herrero, A. D. (2016). Quantification of flash flood economic risk using ultra-detailed stage-damage functions and 2-D hydraulic models. *Journal of Hydrology, 541*, 611−625.

GMDA (2009). Master Plan- 2025, GMA, Guwahati Metropolitan Development Authority, Assam.

Grames, J., Prskawetz, A., Grass, D., Viglione, A., & Blosch, G. (2016). Modelling the interaction between flooding events and economic growth. *Ecological Economics, 129*, 193−209.

Government of India. (2011). *Reports on census of India*. Available at: <https://censusindia.gov.in>.

Haan, M. B., Riederer, A. M., & Foster, S. O. (2009). The livelihood vulnerability index (LVI): A pragmatic approach to assessing risks from climate variability and change—A case study in Mozambique. *Global Environmental Change, 19*, 74−88.

Hall, J., Arheimer, B., Borga, M., Brazdil, R., Claps, P., Kiss, A., et al. (2014). Understanding flood regime changes in Europe: A state-of-the-art assessment. *Hydrology and Earth System Sciences, 18*(7), 2735−2772.

Halounova, L., & Holubee, V. (2014). Assessment of flood with regards to land cover changes. *Procedia Economics and Finance, 18*, 940−947. Retrieved from <www.sciencedirect.com>.

IPCC (2001). *Climate change 2001. Impacts, adaptation and vulnerability. Contribution of working group II to the third assessment report*. Geneva: IPCC.

IPCC (2014). *Climate change 2014. Impacts, adaptation and vulnerability. Contribution of working group II to the fifth assessment report*. New York: IPCC.

Jha, A., Lamond, J., Bloch, R., Bhattacharya, N., Lopez, A., Papachristodoulou, N. et al. (2011). *Five feet high and rising: Cities and flooding in the 21st century (Working paper no. 5648)*. Washington, DC: The World Bank. Retrieved from <http://eprints.uwe.ac.uk/16002>.

Jonkman, S. N. (2005). Global perspectives on loss of human life caused by floods. *Natural Hazards, 34*(2), 151−175.

Jonkman, S. N., & Vrijiling, J. K. (2008). Loss of life due to floods. *Journal of Flood Risk Management, 1*(1), 43−56.

Karagiorgos, K., Thaler, T., Heiser, M., Hube, J., & Fuchs, S. (2016). Integrated flash flood vulnerability assessment: Insights from East Attica, Greece. *Journal of Hydrology, 541*, 553−562.

Kelly, P. M., & Adger, W. N. (2000). Theory and practice in assessing vulnerability to climate change and facilitating adaptation. *Climatic Change, 47*, 325−352.

Kurosaki, T. (2017). Household level recovery after floods in a tribal and conflict- ridden society. *World Development, 94*, 51−63.

Laitae, C., Praneetvatakul, S., Vijitsrikamol, K., & Mungkung, N. (2013). Assessment of vulnerability index on climate variability in the East of Thailand. In *Conference proceedings, conference on intuitional resilience management and rural development, Stuttgart, Germany, Sept. 17−19, 2013*. University of Hohenheim.

Lakhdar, B., Tayeb, S., & Boudersa, G. (2015). Vulnerability's evaluation front of risk in urban space precaution as a principle of sustainable development. Case of Ghardia Town. *Energy Procedia, 74*, 1000−10006. Retrieved from <www.sciencedirect.com>.

Madan, A., & Routray, J. K. (2015). Institutional framework for preparedness and response of disaster management institutions from national to local level in India with focus on Delhi. *International Journal of Disaster Risk Reduction, 14*, 545−555.

Mahmood, S., Khan, A. H., & Ullah, S. (2016). Assessment of 2010 flash flood causes and associated damages in Dir valley, Khyber, Pakhtunkhwa, Pakistan. *International Journal of Disaster Risk Reduction, 16*, 215−223.

Morris, M. D. (1979), Measuring the condition of the world's poor: The physical quality of life index. New York: Pergamon Press for the Overseas Development Council.

Mustafa, C. S., Marzuki, N. A., Ariffin, M. T., Salleh, N. A., & Rahaman, N. H. (2014). Relationship between social support, impression management and well-being among flood victims in Malaysia. *Procedia Social and Behavioural Sciences, 155*, 197−202. Retrieved from <www.sciencedirect.com>.

Naulin, J. P., Payrastre, O., & Gaume, E. (2013). Spatially distributed flood forecasting in flash flood prone areas: Application to road network supervision in Southern France. *Journal of Hydrology, 486*, 88−99.

NDMA (2016), National disaster management plan. Government of India.

O'Brien, K., Eriksen, S., Schjolden, A., & Nygaard, L. P. (2007). Why different interpretations of vulnerability matter in climate change discourses. *Climate Policy, 7*, 73−88.

Parker, D. J. (1995). Floods in cities: Increasing exposure and rising impact potential. Built Environment, *21* (2/3), 114−125.

Petry, B. (2002). Coping with floods: Complementarity of structural and non-structural measures. In B. Wu, et al. (Eds.), *Flood Defence*. New York: Science Press. (Keynote lecture).

Proag, V. (2014). The concept of vulnerability and resilience. *Procedia Economics and Finance, 18*, 369−376. Retrieved from <www.sciencedirect.com>.

Pouya, A. S., Nouri, J., Monsouri, N., & Lashaki, A. K. (2017). An indexing approach to assess flood vulnerability in the Western Coastal Cities of Mazandaran, Iran. *International Journal of Disaster Risk Reduction, 22*, 304−316.

Renald, A., Tjiptoherijanto, P. & Suganda, E. (2016). Toward resilient and sustainable city adaptation model for flood disaster prone city: Case study of Jakarta Capital Region. *Procedia Social and Behavioural Sciences, 227*, 334−340. Retrieved from <www.sciencedirect.com>.

Rufat, S., Tate, E., Burton, C. G., & Maroof, A. S. (2015). Social vulnerability to floods: Review of case studies and implications for measurement. *International Journal of Disaster Risk Reduction, 14*, 470−486.

Saxena, S., Geethalakshmi, V., & Lakshmanan, A. (2013). Development of habitation vulnerability assessment framework for coastal hazards: Cuddalore Coast in Tamil Nadu, India—A case study. *Weather and Climate Extremes, 2*, 48−57.

Serrat, O. (2017). *The sustainable livelihoods approach. Knowledge solutions*. Singapore: Springer. Retrieved from <https://doi.org/10.1007/978-981-10-0983-9_5>.

Setyani, R. E., & Saputra, R. (2016). Flood-prone areas mapping at Semarang City by using simple additive weighting method. *Procedia Social and Behavioural Sciences, 227*, 378−386. Retrieved from <www.sciencedirect.com>.

Sharma, J., & Ravindranath, N. H. (2019). Applying IPCC 2014 framework for hazard- specific vulnerability assessment under climate change. *Environmental Research Communications, 1*. Retrieved from <https://doi.org/10.1088/2515-7620/ab24ed>.

Sherbinin, A., & Bardy, G. (2015). Social vulnerability to floods in two coastal megacities: New York and Mumbai. *Vienna Yearbook of Population Research, 13*, 131−165.

Shields, G.M. (2016). Resiliency planning: Prioritizing the vulnerability of coastal bridges to flooding and scour. *Procedia Engineering, 145*, 340−347. Retrieved from <www.sciencedirect.com>.

Spitalar, M., Gourely, J. J., Lutoff, C., Kristetter, P. E., Brilly, M., & Carr, N. (2014). Analysis of flash flood parameters and human impacts in the US from 2006 to 2012. *Journal of Hydrology, 519*, 863−870.

Sullivan, C., Meigh, J. R., & Fediw, T. S. (2002). *Derivation and testing of the water poverty index phase 1, final report*. Wallingford, Oxon: Department for International Development.

Svetlana, D., Radovan, D., & Jan, D. (2015). The economic impact of floods and their importance in different regions of the world with emphasis on Europe. *Procedia Economics and Finance, 34*, 649−655. Retrieved from <www.sciencedirect.com>.

Tanwattana, P. (2018). Systematizing community-based disaster risk management (CBDRM): Case of urban flood-prone community in Thailand upstream. *International Journal of Disaster Risk Reduction, 28*, 798−812.

Tapsell, S. M., Rowsell, E. C. P., Tunstall, S. M., & Wilson, T. L. (2002). Vulnerability of flooding: Health and social dimensions. *Philosophical Transactions; Mathematical, Physical and Engineering Sciences, 360* (1796), 1511−1525.

Twigg, J. (2001). Sustainable livelihoods and vulnerability to disasters. In *Disaster management working paper 2/2001*. Benfield Greig Hazard Research Centre.

UNDP (1990). Human Development Report, 1990, Oxford University Press, New York, USA.

UNISDR. (2009). *Terminology on disaster risk reduction*. Geneva: United Nations International Strategy for Disaster Risk Reduction.

WHO (2010). Protecting health from climate change: Vulnerability and adaptation assessment. Geneva: World Health Organisation.

Yin, J., Yu, D., Yin, Z., Liu, M., & He, Q. (2016). Evaluating the impact and risk of pluvial flash flood on intra-urban road network: A case study in the city centre of Shanghai, China. *Journal of Hydrology, 537*, 138−145.

AGRICULTURAL PRODUCTION AND INCOME IN A DISASTER YEAR: FINDINGS FROM THE STUDY OF MELALINJIPPATTU VILLAGE AFFECTED BY CYCLONE THANE

28

T.P. Harshan

School of Habitat Studies, Tata Institute of Social Sciences, Mumbai, India

28.1 INTRODUCTION

Disasters disrupt the normal functioning of society and create multidimensional impact in the society. Economic losses created by disaster are increasing from the second half of the 20th century mainly because of the increase in assets and population at risk as well as the concentration of economic activities in disaster-prone regions (Shabnam, 2014). These economic fluctuations due to disasters and lack of resources with the people to cope with a disaster are one of the causes of persistent poverty and poverty traps among different households. Many studies have shown that relative economic impacts of disasters will be higher in the poor and developing countries (Mendelsohn, Basist, Kurukulasuriya, & Dinar, 2007). A primary reason for this is the low level of income in households that limits the response of households, regions, and nation to the disaster.

The study of disasters using microlevel household data is not widespread. Most of the studies focus on the macrounits (e.g., district and provinces). Detailed assessments of various variables, which create the impact of disaster in a village, in various years, are important to understand comprehensively the impact of disaster among various socioeconomic categories in the affected area. This helps the state to develop appropriate measures for reducing the impact and improve the recovery process in the affected area.

This study tries to understand the economic impacts of the disaster across different occupation groups. The economic impact of disasters includes the destruction of physical and financial resources, which in turn affects the income, consumption, production, and employment in the local economy. This study is based on a cyclone affected village (cyclone Thane, one of the significant weather shocks of 2011 in India) and explores how agricultural incomes are affected by the disaster. Cyclone Thane hit the Tamil Nadu coast in India on December 29–30, 2011 and damaged rather destroyed houses and standing crops and led to loss of human life and livestock.

Economic Effects of Natural Disasters. DOI: https://doi.org/10.1016/B978-0-12-817465-4.00028-5

28.1.1 NATURE OF DISASTERS

Disasters have been broadly classified into two—natural and technological (industrial accident, transport accident, miscellaneous accident). The frequency of disasters that occurred over a period of 50 years (1951–2000) is almost as same as the number of events reported between 2001 and 2016 [Centre for Research on the Epidemiology of Disasters (CRED), 2000]. These changes are mainly because of the increase in the number of reported cases of disaster at the regional level. Table 28.1 shows different types of disasters and their frequencies. It indicates that hydrometeorological disasters such as floods, droughts, cyclones, avalanches, heat waves, and cold waves are more frequent than the others. That is, 79% of the total disaster types are hydrometeorological in nature. From 1900 to 2018, globally, floods represented 47% of natural hazard events, storms 39%, and droughts 9% (see Table 28.2).

28.1.2 IMPACT OF DISASTERS ON ECONOMY

The impacts of environmental and weather shocks are unevenly distributed among nations, regions, communities as a result of difference in exposure,[1] coping ability, and resilience (Clark, Moser, & Ratick, 1998). These impacts depend on the relative size of the economy and the magnitude and

Table 28.1 Different Types of Natural Disasters From 1900 to 2018.

Disaster Types	1900–49	1950–99	2000–09	2010–18	Total	Percentage
Hydrometeorological	348	5003	3529	2705	11,585	79
Geological	190	829	354	332	1705	12
Biological	29	634	612	179	1454	10
Total	567	6466	4495	3216	14,744	100

EM-DAT.

Table 28.2 Different Types of Hydrometeorological Disasters Between 1900 and 2018.

Disasters	1900–49	1950–99	2000–18	Total	Percentage
Drought	33	608	317	934	9
Extreme temperature (heat and cold wave)	2	162	405	533	5
Flood	43	1888	3044	4721	47
Storm	158	2079	1929	3942	39
Total	236	4737	5695	10,130	100

EM-DAT.

---

[1]Exposure: risk to be affected from disaster; cope: capacity to grasp effect of disaster and function; resilience: capacity to return back to normalcy after the disaster.

depth of the disaster. Different members of the society are affected differently and this differential impact is largely on account of existing inequalities and disparities in society rather than on account of the accidental and unexpected geophysical nature of disasters. The inability of a substantial section of the world's poor to escape from the vicious cycle of poverty makes them more vulnerable to different disasters (Guha-Sapir, Hargitt, & Hoyois, 2004).

Economic losses due to extreme climate events have increased across the global landscape. The reported global cost of disasters increased 14-fold between 1950 and 1990 (Guha-Sapir et al., 2004). The Asia-Pacific region covers 26% of the world's land area and 60% of the world's population lives in this region. It is estimated that 50% of the world's disasters occur in the same region. India and Bangladesh are included among the 10 worst disaster-affected countries. Average losses from disaster, as a proportion of GDP, are 20% higher in developing countries in comparison with the losses sustained by the developed countries (Kreimer, 2001).

28.1.3 DISASTER PROFILE OF INDIA

According to the National Institute of Disaster Management (NIDM) (2012), 58.6% of the total geographical area is prone to earthquakes, 12% of the land is vulnerable to floods and river erosion, 75% of the coastal line is under threat of cyclone and tsunami, while 68% of the cultivable area is prone to droughts (see Fig. 28.1). From 1980 to 2018, India was affected by 1291 disaster events, as recorded in the CRED International Disaster Database, in which 192,053 people were killed and 1.94 billion people were affected. Total economic losses for the period 1980−2018 were estimated at $98.65 billion (in current dollars). For India the average economic damage per year due to disaster is $2.53 billion.

As per the Food and Agriculture Organization of the United Nations, in India, 60.5% of the total land is utilized for agriculture and 51.1% of the people are employed in agriculture. Extreme climate events pose serious challenges on livelihood of farmers in India. Farmers are often forced to use a significant portion of their revenue to reduce the impact of the disaster (Swaminathan & Rengalakshmi, 2016). As per the CRED database, 39.3% of the total disasters occurred in India between 2000 and 2018 are floods and 21.1% are cyclones and storms. The IPCC reports on projected temperature changes indicate that there will be an increase of 2°C−4.7°C in the temperature by 2100 (Solomon, Qin, Manning, Alley, & Berntsen, 2007). Also, it is expected that the rainfall will increase between 10% and 40% by the end of the 21st century from the baseline period of 1961−90 (Kumar, Sahai, Kumar, Patwardhan, & Mishra, 2006).

28.1.4 CURRENT STUDY

One of the prominent methods to assess the impact of the disaster is vulnerability[2] indicator−based studies. A notable weakness of such indicator-based studies is the large number of indicators that are repeatedly included in the index. The actual economic impacts of the disasters are measured

---

[2]Vulnerability is the degree to which a system susceptible to and is unable to cope with adverse effects of extreme event (McCarthy, Canziani, Leary, Dokken, & White, 2001).

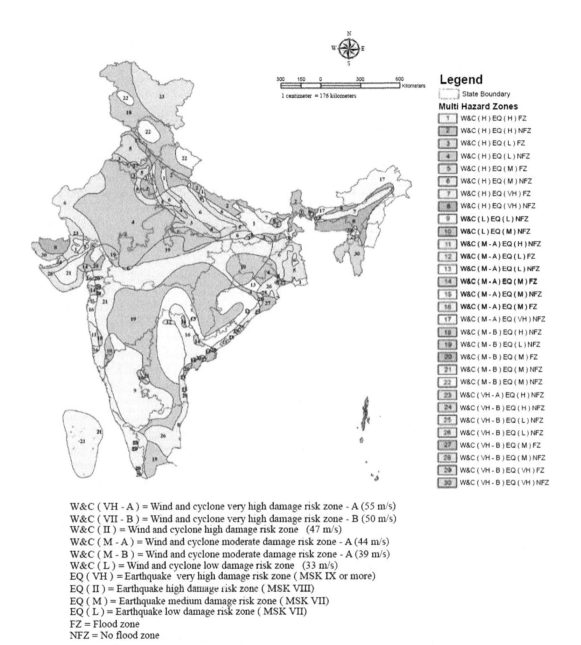

W&C (VH - A) = Wind and cyclone very high damage risk zone - A (55 m/s)
W&C (VII - B) = Wind and cyclone very high damage risk zone - B (50 m/s)
W&C (II) = Wind and cyclone high damage risk zone   (47 m/s)
W&C (M - A) = Wind and cyclone moderate damage risk zone - A (44 m/s)
W&C (M - B) = Wind and cyclone moderate damage risk zone - A (39 m/s)
W&C (L) = Wind and cyclone low damage risk zone   (33 m/s)
EQ (VH) = Earthquake very high damage risk zone (MSK IX or more)
EQ (II) = Earthquake high damage risk zone (MSK VIII)
EQ (M) = Earthquake medium damage risk zone (MSK VII)
EQ (L) = Earthquake low damage risk zone (MSK VII)
FZ = Flood zone
NFZ = No flood zone

FIGURE 28.1

Map showing multihazard zones in India.

Based on Vulnerability Atlas of India, BMTPC. Buiding Material and Promotion Council , Minstry of Housing and Urban Affairs,

Government of India.

rarely in these literatures and there are very few studies that relate the outcome from the disaster with the pre-disaster socioeconomic conditions in the disaster-affected population. Most often, studies focus on aspects such as the spatial dimension of the disaster, the outcome of the disaster as a function of physical hazards as well as socially constructed variables, vulnerability as a function of exposure to disaster and adaptive capacity to cope with these disasters (Balica, Douben, & Wright, 2009; Brooks, Adger, & Kelly, 2005; Cutter, Mitchell, & Scott, 2000; Eakin & Luers, 2006; Ford & Smit, 2004; Lindell, 2011).

The current study focuses on the longitudinal data to measure the impact of the disaster. Repeated observation of variables (e.g., income, production, labor)—that are influenced by a disaster—in different time intervals helps in gaining a comprehensive character of the impacts by that disaster. This study assessed the impact of disaster in rural society using 3 years' data on land, production, and income in a village (pre-disaster, disaster, and post-disaster year).

28.2 METHOD

The study was conducted using the village study method. Census survey was conducted twice in the village, and data pertaining to three time periods were collected (during the first survey, details on pre-disaster and disaster years were collected; the resurvey includes the post-disaster year data which was collected 2 years after the cyclone). The focus of the survey was to measure the economic conditions of the household in different years.

28.2.1 CYCLONE THANE

Cyclone Thane developed over the Bay of Bengal in December 2011 with a wind speed of 35 m/s. It started off as a depression on December 25, 2011 where a storm surge of about 1 m height inundated in the low-lying coastal areas of Tamil Nadu (Vinod, Soumya, Tkalich, & Vethamony, 2014). The landfall of cyclone Thane with a wind speed 140–150 km/h occurred very close to Cuddalore and Puducherry regions. The cyclone was accompanied by heavy rains and high-speed wind that uprooted the trees and knocked down the electric poles due to which the electricity supply was disrupted. According to the official estimates, 53 people were killed in the cyclone—41 in Tamil Nadu and 12 in Puducherry (NIDM, 2012). Apart from the loss of human and animal life, there was extensive agricultural damage. It was estimated that in Cuddalore district alone 1,000,000 coconut trees and 28,000 ha of cashew plantations were destroyed. Apart from the coconut and cashew, paddy, banana, mango, jackfruit, and sugarcane fields were also affected by the cyclone. Almost 80% of standing crops were damaged. The cyclone also caused heavy damage to buildings and an estimated 75,000 thatched, tiled and asbestos-sheeted houses were damaged by the cyclone (NIDM, 2012). Cuddalore, Kanchipuram, Villupuram, Tiruvallur, and Panruti in Tamil Nadu and entire Puducherry were the most affected areas from cyclone Thane.

28.2.2 SELECTION OF STUDY VILLAGE

Cuddalore and Puducherry were the most affected by cyclone (National Institute of Disaster Management, 2012) and hence considered for the current study. A pilot study was conducted by taking sample data from four villages (Karanappattu, Ullerippattu, Karaimedu, and Melalinjippattu) in Cuddalore and two villages (Irulansandy and Kaduvanur) in Puducherry. These villages were selected based on the following criteria—different kinds of cultivation, tenure system, impact of cyclone and the condition of agricultural laborers, tenants, and intervention from the government in the post—cyclone period. Based on the information from panchayat authorities and the discussions with the villagers, Melalinjippattu was selected for the study (289 households, 50.6% SC population). Census type of survey has been used for the data collection since it can provide a comprehensive understanding of the impact of external shock among different strata of the village as well as the recovery methods. The survey was conducted in the month of July—August 2012. The survey was done based on a detailed questionnaire. A resurvey was done in the same village in 2014. The major purpose of the resurvey was to find out the changes in the socioeconomic situations of the different households and the recovery processes adopted by the different households.

28.2.3 MELALINJIPPATTU: THE STUDY VILLAGE

Melalinjippattu village is situated in Cuddalore district of Tamil Nadu. Melalinjippattu village is spread over 141.04 ha. There are around 289 households and the average household size of a family is 5—7 (average household size in the state is 3.9 and the district is 4.1). More than half of the population (50.60%) in the village belongs to scheduled castes. The nearest town of Melalinjippattu is Cuddalore and Puducherry, which is 12 km away from the village. Majority of the households are engaged in agriculture. Out of 141 ha of land, 106.13 ha is used for cultivation. Paddy, brinjal, groundnut, millet, tapioca, coconut, and banana are the major crops in the village.

The total population in the village is 1222 (51.6% men). There are 289 households in the village of which 57% belong to the scheduled caste and 37% belong to the most backward communities category. Table 28.3 describes the primary occupations of the people in the village. Around 43% of the population depends on agriculture for their livelihood. Nonagricultural manual labor including MGNREGS (Mahatma Gandhi National Rural Employment Guarantee Scheme) constitutes the most critical occupation category in the village and 47% of the population was engaged in these activities. Nonagricultural labor opportunities are more due to the closeness of this village to the city. Major nonagricultural labor is in the construction activities.

Table 28.3 Distribution of Population by Primary Occupation.

Primary Occupation	Number of Workers	Share in Total
Cultivator	68	12
Agricultural labor	187	31
Nonagricultural manual labor	281	47
Salaried work	48	8
Business[a]	12	2
Total	596	100

[a]*In the Census, this category includes all work other than cultivation, agricultural labor, and home-based industry.*
Government of India 2011a, 2011b.

28.3 **ASSESSING DAMAGE AND LOSSES DUE TO CYCLONE THANE**

According to the United Nations Office for Disaster Risk Reduction, "losses" and "damages" are used interchangeably to measure the impact of disaster on society, economy, and environment. Losses can be divided into two—one is the direct tangible losses such as damage to public and private buildings and infrastructure, commercial units and the other is indirect losses that arise as a result of the disruptions in the flow of goods and services to the affected area (UNISDR, 2017). According to United Nations General Assembly Seventy First Session Agenda item 19C, direct loss is due to the partial or fully damage of physical assets in the affected area. It usually occurs at the time of disaster or during first few hours after the disaster and is relatively easy to measure. Indirect loss is considered as the decline in the economic value due to direct economic loss as well as human and environmental impacts. It includes the decline in income due to business disruption (macroeconomic impact), destruction of physical and natural assets, temporary unemployment, price increase, etc.. This study is an attempt to assess the direct and indirect losses as a result of cyclone Thane and to see how people cope with these losses. For the study, damage to buildings, livestock as well as crops were considered as direct loss, whereas indirect losses were assessed by explaining the changes in income and labor days.

28.3.1 **DIRECT LOSSES**

This section describes the damages (loss of life, property damage, crop damage) due to cyclone at the regional level as well as village level.

28.3.1.1 *Loss and Damage of Lives, Houses, and Buildings*

The deaths during cyclone Thane were mainly due to collapsing of houses. Majority of the deaths (41) were reported in Cuddalore Taluk; however, no deaths were reported in Melalinjippattu village. According to the state records, 343 livestock succumbed to the cyclone. In Melalinjippattu the loss of livestock due to cyclone was lesser in comparison to other kind of losses. Most of the livestock losses were due to the collapse of the cattle sheds.

The NIDM data shows that more than 3.5 lakh houses were damaged during the cyclone. Among all the taluks in Tamil Nadu, exceptional damage to houses was recorded in Cuddalore. Fifty-one percent of the total houses were partially or fully damaged due to the cyclone (see Appendix). Melalinjippattu village was also severely affected. Table 28.4 shows the damages to buildings in Melalinjippattu. Thatched houses are more vulnerable in the cyclone compared to concrete houses as the thatched roofs are not strong enough to confront the cyclone winds and can easily blow away. Houses with asbestos sheets or tiled roofs were also affected by the cyclone. In Melalinjippattu, there were thatched (44%), tiled (10%), sheet (14%), and concrete roofed houses (30%). Due to the cyclone, 45.8% of the thatched roof houses, 44.4% of the tiled roof houses, and 50% of the asbestos sheet roofed houses were damaged. In addition, 34.3% of the concrete houses were also partially damaged. There was a total loss of Rs. 16.52 lakhs as a result of damage to the buildings in the village. The estimation of damages was noted based on the assessments made during the household survey by the household members and the researcher.

Table 28.4 Damages to Houses and Buildings.

Type of Roof	Total Number	Damaged Houses	Percentage of Damaged Houses to Total Percentage	Value of Damages
Thatch	153	70	45.8	1,356,000
Tiled	36	16	44.4	23,200
Sheet	48	24	50.0	40,550
Concrete	105	36	34.3	233,000
Thatch and sheet	1	0	0.0	
Concrete and sheet	1	0	0.0	
	344	146	42.4	1,652,750

28.3.1.2 Losses and Damages to Crop

Paddy, brinjal, groundnut, millet, tapioca, coconut, banana were the major crops in the village. The source of irrigation is tube well. Crops are cultivated in three seasons—Samba (September–January), Kuruvai (February–April), and Navarai (May–July). There cultivated land is of two types—nanjai (wet) and punjai (dry). Both types of land are irrigated. The normal practice in Melalinjippattu is that paddy is cultivated in two seasons, that is, Samba and Kuruvai; and groundnut, brinjal, and spinach are cultivated in Navarai.

As the cyclone hit Tamil Nadu at the time of harvest, standing crops were damaged widely across the region. Table 28.5 shows magnitude of the destruction of crops in Tamil Nadu. Of the total cultivated area in the state, 54.49% was damaged and paddy was the most affected crop (72.9% of the paddy area was destroyed).

In order to understand the extent of damage on cultivation and production, survey data collected on various crops in the village for 3 years (pre-disaster year: 2010–11; disaster year: 2011–12; post-disaster year: 2014–15) were considered (see Table 28.6 for more details). Both the extent of cultivation as well as production (45% decline) suffered severely during the cyclone. Though the production went up in post-disaster year, it was only 85% of the pre-disaster year production. Though the extent of paddy cultivation increased in the post-disaster year, the production did not do so as paddy was used mainly for household consumption. As can be seen from the table, the yield as well as the gross production declined in the village indicating a lack of enthusiasm of households to engage in cultivation. Significant decline in the income in disaster year is the major reason for the changes in the production pattern.

28.3.2 INDIRECT DAMAGES AND LOSSES

Impact of disaster on land, income, and labor are the major three variables that were considered to measure the indirect damages as well as losses. The impact was measured based on the changes over the years on these variables among different landholding categories in the village. A repeated measures ANOVA with time period (pre-disaster, disaster, and post-disaster year) as within-subject factor and landholding

Table 28.5 Details of Agricultural Damage in Tamil Nadu.

Crops	Total Area Cultivated (acre)	Area Damaged (acre)	Percentage	Remarks
Paddy	96,391	70,271	72.9	Flowering and harvesting stage
Groundnut	9394	2274	24.2	Sown and vegetative stage
black gram	14,132	7000	49.5	Sown and vegetative stage
Sugarcane	29,700	6833	23.0	Planting and growth stage
Coconut	2363	939	39.7	20 years of yielding stage
Cotton	7452	110	1.5	Boll stage
Oil palm	1100	46	4.2	4 years old and yielding palm
	160,532	87,473	54.5	

NIDM (2012).

Table 28.6 Crop-Wise Production and Extent of Cultivation.

Crop	Extent (acre)			Production (kg)		
	Pre-disaster Year	Disaster Year	Post-disaster Year	Pre-disaster Year	Disaster Year	Post-disaster Year
Brinjal	24.8	3.4	9.1	49,380	350	11,229
Green chilly	4.5	0.0	0.7	9000	0	1369
Green gram	39.1	0.0	6.6	8675	0	1440
Groundnut	39.7	38.8	17.3	34,503	7575	12,026
Spinach	4.4	0.0	4.8	6846	4185	8850
Millet	7.9	1.9	11.3	13,339	2500	18,815
Paddy	126.2	118.2	154.6	211,620	75	208,765
Black gram	0.1	0.0	0.0	25	0	0
Tapioca	5.0	0.6	5.8	29,300	0	29,300
Ladies finger	1.3	0.5	0.0	4660	3250	0
Banana	1.0	1.0	0.0	500	350	0
Bitter guard	0.0	0.0	0.8	0	0	100
Marigold flower	0.0	0.0	0.3	0	0	1600
Sugarcane	0.0	0.0	6.5	0	0	5000
	254	164.5	218	367,638	18,285	298,494

category (landless, marginal, small and large landholders) as between-subject factor was also carried out on the income from cultivation and manual labor to understand in detail, the impact of the cyclone.

28.3.2.1 Impact of Cyclone on Land Holding and Cropping Pattern in the Village

Table 28.7 shows that there were 72 households who owned land in the village during the first survey and the number reduced to 65 during the resurvey period. Similarly, during the first survey time, the number of households that were engaged in cultivation was 93, while the same dropped to 69 at the time of resurvey. Twenty-four households moved out of cultivation in 2 years after the disaster. Two years after the disaster, half of the households (tenant households) gave away the lands back to their landlords because the income generated from cultivation was not enough for these tenants to pay rents to landowners, and hence, they moved to manual labor for their livelihood.

28.3.2.2 Impact of Cyclone on Income From Cultivation

During the pre-disaster year, an inequality in income from cultivation existed among different landholding categories (Table 28.8). The income per household for large landholders (4−10 acres of land) was 33 times higher than that of marginal landholders (0.01−0.99 acre of land). Around 34% of the net income from cultivation was earned by 9% of the total households (large landholders), though the small land-holders (1−3.99 acres of land) were five times more than the large landholders. Thus, as expected, large landholders incurred more loss during the disaster year. The results from ANOVA indicates that there was a main effect of time with net income from cultivation being the highest in the pre-disaster year ($M = 62{,}350.99$, $SD = 5934.05$) compared to post-disaster ($M = 44{,}195.70$, $SD = 6553.56$) and disaster year ($M = -17{,}024.17$, $SD = -21{,}633.29$), $F(1.79, 152.47) = 82.08$, $P < .001$. The households had a negative income in the disaster year. The interaction between time and landholding category indicates that the cyclone impacted different landholding categories differently, $F(4, 170) = 25.78$, $P < .001$. However, small and large landholders were affected to a great extent. Though the large landholders could return back to normalcy (there was no significant difference between net income from cultivation in the pre-disaster and post-disaster time period), the small landholders could not and were the most vulnerable group.

Table 28.7 Land Characteristics 2011, 2014.

	2011	2014
Total household	289	289
Land owning households	72	65
Land owning household to total households (%)	24	22
Cultivating households	93	69
Cultivating households to total households (%)	32	24
Tenants households	53	34
Percentage of tenants to cultivating HH	57	49

Table 28.8 Income From Cultivation by Landholding Size for 3 Years.

Year	Landholding Category	No. of HH	Area of Cultivation (acre)	GI	GI per HH	NI from Cultivation	Income per HH	Income per Acre
Pre-disaster	Marginal	40	36.5	677,184	16,930	161,694	4042	4430
	Small	42	142.2	2,848,569	67,823	1,713,328	40,794	12,049
	Large	8	75.3	1,507,677	188,460	997,535	124,692	13,247
Disaster	Marginal	34	18.5	54,338	1598	−222,727	−6551	−12,039
	Small	39	89.5	28,806	739	−509,889	−13,074	−5697
	Large	8	56.5	47,900	5988	−321,070	−40,134	−5683
Post-disaster	Marginal	26	26.7	566,247	21,779	150,560	5791	5639
	Small	32	106.6	2,326,889	72,715	847,688	26,490	7952
	Large	8	85.5	2,146,816	268,352	965,684	120,711	11,295

HH, Households; *GI*, gross income; *NI*, net income.

28.3.3 IMPACT ON WAGE EMPLOYMENT

28.3.3.1 Agricultural Wage Employment

The disaster badly hit agricultural labor (number of labor days, labor days per worker, income per worker; see Table 28.9). As mentioned earlier, agriculture faced the maximum damage due to the cyclone. Within agriculture, agricultural labor days were the most affected compared to other factors pertaining to agriculture. Landless households were more involved in agricultural labor, and so, they were the most vulnerable group in terms of income decline from agricultural labor. There was a significant reduction in the number of labor days following the disaster. This was largely because cultivation in the Kuruvai season did not occur as a result of electricity disruption immediately after the disaster. Few households in the marginal landholding category (0.01−0.99 acre) also faced difficulties due to the decline in agricultural labor days.

28.3.3.2 Income From Nonagricultural Manual Work

Following the disaster, the number of workers engaged in manual labor reduced significantly as most of the women workers migrated to (jobs in shops, factories, etc.). Also, there was a decline in the workers from landless and marginal landholding category. Since landless households depended mostly on manual labor, they were the most affected category during the cyclone (see Table 28.10). Main effects of time $[F(2, 494) = 1.38, P = .25]$ as well as group $[F(3, 247) = 0.95, P = .41]$ were absent indicating that households were able to cope with the disaster due to their income from nonagricultural manual work. However, there was an interaction between time and group, $F(6, 494) = 6.22, P < .001$. Pairwise analysis indicated that landless and marginal land holding categories were the most affected by the cyclone. Although this was the case, landless households went back to normalcy in the post-disaster year (their income was significantly higher in the post-disaster year compared to the pre-disaster year). However,

Table 28.9 Agricultural Labor Days in 3 Different Years Based on Landholding Category.

Year	Landholding Category	Number of Workers	Labor Days	Income	Income/ Worker	Labor Days/ Worker
Pre-disaster	Landless	96	6639	712,465	7422	69
	Marginal	15	596	49,125	3275	40
	Small	10	315	32,500	3250	32
Disaster	Landless	103	1488	180,050	1748	14
	Marginal	8	155	19,300	2413	19
	Small	5	59	6900	1380	12
Post-disaster	Landless	138	3331	510,900	3702	24
	Marginal	6	218	26,850	4475	36
	Small	4	168	27,300	6825	42

Table 28.10 Manual Labor Days in the Village Based on Landholding Size Across 3 Years.

Year	Landholding Category	Number of Workers	Labor Days	Income	Income/ Worker	Labor Days/ Worker
Pre-disaster	Landless	276	32,305	7,169,880	25,978	117
	Marginal	59	7039	1,952,930	33,101	119
	Small	51	6017	1,511,270	29,633	118
	Large	4	800	296,000	74,000	200
Disaster	Landless	275	26,907	6,070,710	22,075	98
	Marginal	63	5783	1,658,790	26,330	92
	Small	50	4825	1,337,420	26,748	97
	Large	4	680	262,200	65,550	170
Post-disaster	Landless	231	32,897	11,097,270	48,040	142
	Marginal	35	4706	1,211,950	34,627	134
	Small	20	2694	732,160	36,608	135
	Large	8	850	466,450	58,306	106

small and large landholders were not affected by the cyclone. Thus the results indicate that the marginal landholders were the most affected category. Taking into consideration the different sources of income, it was found that compared to other livelihoods (cultivation and agricultural labor), the condition of manual labor improved significantly in the post-disaster year. Geographical particularities of the village (situated very close to urban areas) enhanced the accessibility of the workers to manual labor thus helping them to cope with the impacts of cyclone.

Though the labor days in the village reduced significantly following the disaster, the income per worker was not much affected. Thus increase in income even at the time of declining labor days indicates that the wage rate in the region increased. The increase in the wage rate did not have any correlation with the external environmental event. But this helped the landless and marginal households to improve their living conditions (post-disaster year). A small dip in the labor days per worker for the large landholding category was due to their migration to salaried as well as self-employed jobs.

28.4 ADAPTATION STRATEGIES

Three agencies (ADRA, CASA, and Red Cross) were present during the aftermath of the cyclone (George, 2012). Due to the lack of financial resources, the role of panchayat and MLAs were also limited. The cyclone shelter in Cuddalore played an important role in reduction of casualties during the cyclone. MGNREGS is social security mechanism through which rural households are involved in unskilled manual work and it guarantees 100 days of wage employment in a financial year. The MGNREGS was an important livelihood opportunity for a large number of households in the village. The timely increase in MGNREGS wages was a proactive strategy for disaster relief. Households also received compensation from the State. Below, we will take a look at its details.

28.4.1 COMPENSATION

The village economy and households expected to recover from the cyclone by combining their resources and aids from government and other agencies. State in varying degrees took the responsibility of post-disaster relief for the affected population from the cyclone. Collective loss sharing at the state level occurred through compensation and disaster recovery processes.

Compensation for the crop loss was given by the Tamil Nadu government through panchayat. As per the Tamil Nadu government report (Agriculture Department Policy Notes Demand No 5—agriculture 2002–13), Rs. 214.637 crore was sanctioned for crop damage assistance following the cyclone (Rs. 10,000 per hectare for paddy crops; Rs. 7500 for other irrigated crops; Rs. 4000 for the rain-fed crops; 1 ha = 2.5 acres). But the farmers received only Rs. 2000 per acre for all the crops in the village (data obtained from farmers). Table 28.11 gives the details of the compensation based on different land sizes.

Out of 91 households whose crops were destroyed in the cyclone, only 41 (45%) received the compensation. The total loss from the cyclone was Rs. 2,480,835 and the compensation they

Table 28.11 Compensation to Landholding Size.

Landholding Category	Number of HH	Number of HH got Compensation	Loss	Compensation	Proportion of Compensation to Loss
Marginal	43	8	277,335	9000	3
Small	40	25	1,326,000	89,900	7
Large	8	8	877,500	69,000	8
	91	41	2,480,835	167,900	7

HH, Households.

received was only Rs. 167,900. Among the marginal landholders (owning less than 1 acre land), only eight households got the compensation. So, the majority of the marginal landholders were omitted from the compensation (Table 28.11).

28.5 CONCLUSION

This paper tried to assess the impact of a natural disaster on household incomes and employment in a village. Melalinjipattu village in Tamil Nadu was one of the severely affected villages by the Cyclone Thane in December 2011, especially in terms of damage to agriculture production and income. The cyclone destroyed the Samba crop in the village. Due to the disruption of electricity, agricultural production in the subsequent seasons was also affected.

The study showed that all households made net losses in cultivation in the disaster year. The loss per acre was higher for smaller farm sizes, though the total volume of loss varied directly with the size of the farm. To put it simply, the relative losses (loss per acre or loss as a percentage of normal year incomes) are higher for smaller landholders, and they are more vulnerable to the impact of income shocks caused by disasters. Agricultural laborers were also seriously affected by the disaster as they could not find employment during harvesting in the Samba season and also in the subsequent seasons when crop production declined due to disruption of electric supply. Melalinjipattu village had the advantage of easy availability of nonagricultural employment in nearby urban centers. Nonagricultural employment opportunities played an important role in the mitigation of the impact of disaster. However, there were households, particularly female-headed households, who could not access nonagricultural employment and hence were the most vulnerable. For these households, MGNREGS played a significant role in reducing the impact of cyclone as they could cope with the income loss through this scheme. Thus disasters affect different population groups and regions with differing resource base and conditions of production differently. Disaster management strategies, particularly ex-poststrategies need to take cognizance of such differences.

Occurrence of disasters disrupts the normal economic function in the society. Livelihood that is based on natural resources especially agriculture, fisheries, etc. gets affected the most due to disasters. Generation of employment by the state, especially nonagricultural livelihood opportunities, will help the affected society to reduce the economic burden in the aftermath of the disaster. The state should provide sufficient compensation to the affected society; also, compensation should be based on the damage and losses of the properties. Livelihood based on agriculture will sustain following a disaster only with the adequate support from the state in the form of compensation as well as support for the cost of input for cultivation immediately after the disaster. Support should be provided more for the marginal and small peasants since they require more support for their recovery in the aftermath of the disaster.

APPENDIX

Around 3.6 lakhs of houses were damaged (Table 28.A.1). An exceptional damage to the houses was seen in Cudaalore Taluk. 73.5% of the total houses were partially or fully damaged in the cyclone.

Table 28.A.1 Damage to the Houses in Cuddalore District.

| Taluk | Huts | | Tiled Houses | Total Damaged Housed | Total Houses | Percentage of Damaged Houses to Total Houses |
	Fully Damaged	Partially Damaged				
Cuddalore (rural)	49,511		26,835	76,346		
Cuddalore (town)	302	16,094	15,833	32,229		
Panruti	18,545	50,650	26,153	95,348		
Kurinjpadi	3058	39,799	10,916	53,773		
Chidambaram	1362	82,556	16,458	100,376		
Virudachalam	413	6883		7296		
Kattumanarkoil	101	175	10	286		
Tittagudi		228		228		
Total	73,292	196,385	96,205	365,882	717,587	51

Government of Tamil Nadu memorandum submitted to Central Team-India disaster report NIDM (National Institute of Disaster Management).

REFERENCES

Balica, S. F., Douben, N., & Wright, N. G. (2009). Flood vulnerability indices at varying spatial scales. *Water Science and Technology, 60,* 2571−2580. Available from https://doi.org/10.2166/wst.2009.183.

Brooks, N., Adger, W. N., & Kelly, P. M. (2005). The determinants of vulnerability and adaptive capacity at the national level and the implications for adaptation. *Global Environmental Change, 15*(2), 151−163.

Clark, G. E., Moser, S. C., Ratick, S. J., et al. (1998). Assessing the vulnerability of coastal communities to extreme storms: The case of Revere, MA., USA. *Mitigation and Adaptation Strategies for Global Change,* 3−59. Available from https://doi.org/10.1023/A:1009609710795.

Centre for Research on the Epidemiology of Disaster. (2000). *EM-DAT: The OFDA/CRED International Natural Disaster Database 1900-1999.* Louvain, Belgium: Universite Catholique de Louvain.

Cutter, S. L., Mitchell, J. T., & Scott, M. S. (2000). Revealing the Vulnerability of People and Places: A Case Study of Georgetown County, South Carolina. *Annals of the Association of American Geographers, 90*(4), 713−737. Available from https://doi.org/10.1111/0004-5608.00219.

Eakin, H., & Luers, A. L. (2006). Assessing the vulnerability of social-environmental systems. *Annual Review of Environment and Resources, 31*(1), 365.

Ford, J., & Smit, B. (2004). A Framework for Assessing the Vulnerability of Communities in the Canadian Arctic to Risks Associated with Climate Change. *Human Dimensions of the Arctic System, 57*(4), 389−400. Available from https://doi.org/10.14430/arctic516.

George, A. (2012). *Cyclone Thane—Disaster preparedness and response.* Retrieved from <http://trinet.in/?q = node/824>.

Government of India. (2011a). *District Census Handbook Cuddalore. Series 34, Part XII A.* Tamil Nadu: Directorate of Census Operations.

Government of India. (2011b). *District Census Handbook Cuddalore. Series 34, Part XII B.* Tamil Nadu: Directorate of Census Operations.

Guha-Sapir, D., Hargitt, D., & Hoyois, G. (2004). *Thirty years of natural disaster 1974-2003: The numbers*. Centre for Research on Epidemiology. <http://www.unisdr.org/eng/library/Literature/8761.pdf>.

Kreimer, A. (2001). Social and economic impacts of natural disasters. *International Geology Review, 43*(5), 401−405.

Kumar, K. R., Sahai, A. K., Kumar, K. K., Patwardhan, S. K., Mishra, P. K., et al. (2006). High-resolution climate change scenarios for India for the 21st century. *Current Science, 90*, 334−345.

Lindell, M.K. (2011). *Disaster studies*. Sociopedia.isa, Available from: https://doi.org/10.1177/2056846011111O.

McCarthy, J., Canziani, O., Leary, N., Dokken, D., & White, K. (2001). Climate change 2001: impacts, adaptation, and vulnerability. *Contribution of working group ii to the fourth assessment report of the intergovernmental panel on climate change.*

Mendelsohn, R., Basist, A., Kurukulasuriya, P., & Dinar, A. (2007). Climate and rural income. *Climatic Change, 81*(1), 101−118.

National Institute of Disaster Management (NIDM). (2012). *India disaster report 2011 by K.J Anandha Kumar, Ajinder Walia, Shekhar Chaturvedi*. New Delhi: National Institute Disaster Management. <https://nidm.gov.in/PDF/pubs/IndiaDisasterReport2011.pdf>.

Shabnam, N. (2014). Natural disasters and economic growth: A review. *The International Journal of Disaster Risk Science, 5*(2), 157−163.

Solomon, S., Qin, D., Manning, M., Alley, R. B., Berntsen, T., et al. (2007). *Climate change 2007: The physical science basis. Contribution of working group I to the fourth assessment report of the Intergovernmental Panel on Climate Change* (pp. 19−71). Cambridge and New York: Cambridge University Press.

Swaminathan, M. S., & Rengalakshmi, R. (2016). Impact of extreme weather events in Indian agriculture: Enhancing the coping capacity of farm families. *Mausam, 67*(1), 1−4.

UNISDR (United Nations International Strategy for Disaster Reduction). (2017). *Terminology on disaster risk reduction*. Geneva: UNISDR.

United Nations. (2015). *Report of the open-ended intergovernmental expert working group on indicators and terminology relating to disaster risk reduction*, note by the *secretary-general seventy-first session of the general assembly, agenda item 19(c), A/71/644*. New York: United Nations.

Vinod, K. K., Soumya, M., Tkalich, P., & Vethamony, P. (2014). Ocean—Atmosphere interaction during Thane Cyclone: A numerical study using WRF. *Indian Journal of Marine Sciences, 43*(7), 1230−1235. <https://doi.org/10.1007/springerreference_28992>.

FURTHER READING

CRED. (2007). *Annual disaster statistical review: Numbers and trends 2006*. Brussels: Centre for Research on the Epidemiology of Disasters (CRED), School of Public Health, Catholic University of Louvain.

Eakin, H., & Bojotquez-Tapia, L. A. (2008). Insights into the composition of household vulnerability from multicriteria decision analysis. *Global Environmental Change, 18*(1), 112−127.

SOCIO-ECONOMIC VULNERABILITIES TO NATURAL DISASTERS AND SOCIAL JUSTICE

29

Yetta Gurtner and David King

Centre for Disaster Studies, College of Science and Engineering, James Cook University, Townsville, QLD, Australia

29.1 INTRODUCTION

Modeling of climate change impacts predicts an increase in natural hazards as a consequence of global warming [Intergovernmental Panel on Climate Change (IPCC), 2007, 2012]. Droughts (with associated heat waves and bushfires) and floods are predicted to increase in severity as the planet warms. While some locations will become drier and others on average wetter, there will also be an increase in extreme climatic cycles, alternating between longer, more severe droughts, and periods of heavier rainfall and consequent floods. While climate change has anthropogenic causes, other human actions such as clearing vegetation, and altering landscapes and drainage systems are quite independently contributing to increases in both droughts and floods, and the severity of their impacts on human structures and lives (Cinner et al., 2018; Kelman, Gaillard, & Mercer, 2015). Community vulnerability to the hazards of drought, bushfire and flood is increased simultaneously through both human landscape modification and climate change.

This chapter will review the impacts of floods on communities. The focus is on community vulnerability and experiences of flood disasters, as well as suggesting analogs for future scenarios under climate change. Vulnerability to all natural hazards and specifically to the flood hazard is expressed in the risk formula—risk = hazard × vulnerability, or the interaction between exposure, sensitivity and resilience/vulnerability. Communities in especially hazardous locations are significantly more vulnerable to disaster (Anderson-Berry & King, 2005; Granger, Jones, Leiba, & Scott, 1999). However, in communities generally, and especially in the most hazard prone locations, impacts occur unequally across the population. Poorer, more disadvantaged members and groups in society experience relatively greater damage and loss to life and safety, property, livelihoods, and well-being. Vulnerability of the socially disadvantaged extends from limited community infrastructure and lifelines through wealth disparity, to reduced levels of personal, household and community resilience and capacity to recover from disaster.

Disaster risk reduction (DRR) stresses the need for sustainability, but the focus on sustainability has tended to be toward care and protection of the environment. The twin pillar of ecologically

Economic Effects of Natural Disasters. DOI: https://doi.org/10.1016/B978-0-12-817465-4.00029-7

sustainable development (ESD), social justice, is neglected for its structural politico-ecological complexities, while a neo liberal economic agenda has redistributed risk to vulnerable communities under the guise of building resilient communities. The DRR strategies of building resilient communities purport to empower and partner local initiatives but are limited in their capacity to reduce the structural vulnerability of households and communities. Drawing on a recent case study and research this chapter will identify issues of household and community vulnerability in flooded communities in order to identify the social justice and environmental justice issues of floods, both as analysis of past events and as analogs of future climate change scenarios.

This chapter examines community vulnerability to disaster through perspectives of how we construct vulnerability, the complexity of vulnerability issues, and causes and measures of vulnerability. Issues of local knowledge, human rights, social justice, and environmental justice are placed in a disaster vulnerability context. We illustrate these perspectives with reference to the severe monsoonal floods that inundated parts of Townsville in North Queensland early in 2019.

29.2 CONSTRUCTING DISASTER VULNERABILITY

Vulnerability to disaster is predicated on inequalities between people, households, groups, and communities. However, all the terms we use—disaster, community, vulnerability, and resilience are human constructs that exemplify the complexity of human environment relationships. A natural phenomenon, such as a flood, is perceived as a hazard when it threatens human beings and their infrastructure, and it becomes a disaster when it overwhelms communities. The capacity for a potential hazard to overwhelm people may be exacerbated or reduced according to the circumstances, capacities, and resources of the community. Very broadly vulnerability is defined as a susceptibility to harm that is defined by sets of attributes that contribute to potential loss (Anderson-Berry & King, 2005; Blaikie, Cannon, Davis, & Wisner, 1994).

In constructing the idea of vulnerability to disaster or to a hazard we first construct nature and environment as something separate from human society (Hilhorst & Bankoff, 2004). Nature and society are portrayed as in conflict. This separation is reinforced by social values and resource use, whereby we deem the environment to be an object for our use or a threat to our well-being. The most commonly held cultural attitude to nature and the environment is stewardship, whereby we look after the environment and its resources for our ultimate benefit. ESD expresses this same dichotomy by envisaging two pillars of environment and social justice, looking like silo's or opposites, and more importantly being addressed as separate entities (Hilhorst & Bankoff, 2004). The IPCC extends this separation (IPCC, 2007; Kelman, Gaillard, Lewis, & Mercer, 2016). The environment itself is also socially constructed and mediated by society (Cardona, 2004). We anthropomorphize nature, we name its elements, and assign values or threats according to our world views and culture. The hazard in nature is something to be controlled or overcome, thereby justifying physical engineered protection, and otherwise portrayed negatively. However, the environmental movement has generated something of a shift from a confrontation with the environment to acceptance of human society as a part of nature (Hilhorst & Bankoff, 2004), but this is less evident in emergency management and disaster practice. Emergency management comes from a top-down command and control perspective, often drawing personnel from the military and emergency services, who perceive the hazard as enemy and the focus of their intervention to protect or defend people. Inevitably this defines people and community as potential victims who must be saved because they are at risk, or vulnerable to the hazard.

Furthermore, as Cardona (2004) points out, there is no coherent theory of disaster. Disaster researchers and scholars come from and derive perspectives that reflect a range of disciplines spanning the social and physical sciences (Faas, 2016). This has tended toward a maintenance of the idea of society being separate from, and threatened by, the extremes of nature. As the idea of climate change adaptation is increasingly embedded in planning for potential hazards, there is a slight shift to working with nature (Visser, Petersen, & Ligtvoet, 2014).

The International Decade for Natural Disaster Reduction (IDNDR) shifted the focus on DRR from the physical or the hazard toward community vulnerability reduction (Anderson-Berry & King, 2005). As emergency managers and their advisors attempted to translate vulnerability reduction into policy, the problem that emerged was a lack of control over societal vulnerability, because most attributes of vulnerability are structural (Wisner, 2004) and outside the capacity of either communities or emergency managers to instigate change, at least in the short term. Resilience is a concept running alongside vulnerability which emphasizes a different but interrelated set of variables that define community strengths rather than weaknesses. Thus building resilience emerged as a substitute for vulnerability reduction and was easier to build into policies and strategies of DRR. In Australia, building resilient communities is official DRR policy. It also suited a neoliberal economic agenda of smaller government by passing responsibility for community safety back on to the communities themselves. Society bears the cost of preparing for disasters, rather than government providing protective infrastructure, such as levees which anyway are deemed by emergency managers to be unsustainable in the long term.

The problem for DRR is that while resilience building, and subsequently climate change adaptation, are valued strategies for governments and communities, vulnerability remains unmitigated, thereby reducing the effectiveness of both resilience and adaptation activities. Resilience and adaptation build on strengths, whereas vulnerability identifies weaknesses in the community or individual, which by being structurally embedded into society, culture, and economy are frequently insurmountable, or wicked problems. Inevitably many resilience building strategies are constrained or undermined by attributes of vulnerability.

Resilience building for DRR both links to and overlaps with human rights and social justice issues (Heijmans, 2004). Both resilience and vulnerability are also related to development, especially in the poorer developing countries (Delica-Willison & Willison, 2004; Kelman et al., 2016). Resilience also contains within this concept the idea of bouncing back or returning to some predisaster state (Kelman et al., 2016). Neither the disaster recovery nor, in particular, climate change adaptation can take us back to a pre-event "normal state." Human cultures construct an idea of a normal past place, which can distort or thwart both attempts at recovery as well as resilience building and vulnerability mitigation themselves (Kelman et al., 2016). Resilience might better be defined as bouncing forward, but its trajectory must pass through or over the limitations imposed by vulnerability.

Vulnerability is linked to reduced capacity in the face of risk and avoidance of danger (Davis, 2004). Human capacity, including social capital, is reduced when people move away from a hazard or a disaster zone. Global rural to urban migration on top of rapid population increase has reduced people's local knowledge and their capacity to reduce household and community risk (Delica-Willison & Willison, 2004). People lose local knowledge of a known risk on entering strange cities where the risk is unknown to them. Some flows of urban migration have involved people who are fleeing from environmental crisis such as drought, or disasters such as floods, with a mindset of escaping disaster for an unknown "safer" place. Outmigration following a disaster changes the vulnerability status of individuals and households (Graif, 2016). For some people vulnerability is reduced and status improved, while others may carry their vulnerable attributes to a new destination. Also, as many out migrants are young and employable, the community of origin may also become more vulnerable as those left behind face reduced capacity and resources (Boon, Cottrell, King, Stevenson, & Millar, 2012; King et al., 2014; Visser et al., 2014). Furthermore, cities themselves drive hazard vulnerability (Jones, 2017), as most of them lie in coastal areas, thereby being more susceptible to climate change which is further compounded by a rapidly growing population (Jones, 2017).

29.3 ISSUES OF DISASTER VULNERABILITY

In the past few decades too much emphasis of vulnerability mitigation has focused on socio-economic and demographic characteristics of the population (Faas, 2016). Vulnerability is multilayered. Issues surrounding vulnerability include unequal development, vulnerability of infrastructure including places and structures of cultural significance, the narrowness of the definition of vulnerability, climate change adaptation (Bankoff, 2004; Cinner et al., 2018; Faas, 2016; Hilhorst & Bankoff, 2004; Kelman et al., 2015; Pica 2018), as well as structural inequalities of class, race, and wealth inequality (Cardona, 2004). Global inequality, emerging out of colonialism prompted strategies to develop the economies and human capital of newly independent and developing countries. Bankoff (2004) cites the inequality of both development projects and aid which further contribute to the global inequality of vulnerability. It further points to the underlying premise of development being based upon growth that is essentially unsustainable (Bankoff, 2004; Kelman et al., 2015). It should be noted that there is an extensive literature on theories and practice of development which addresses inequality and social justice. There is not enough space in this chapter to discuss these ideas other than to point out the impact on vulnerability to disaster. Development policy has embraced DRR as an essential component of funded development projects [Benson, 2009; Lewis & CICERO (Center for International Climate and Environmental Research—Oslo), 2012]. Heijmans (2004) cites as additional issues that shape the causes of vulnerability, the frequency of disasters that impact on communities, local knowledge, and politics. These areas are often underrepresented in vulnerability assessments. Hilhorst and Bankoff (2004) cite climate change, globalization and migration as complicating vulnerability issues, but they also refer to developing regions or countries that are viewed as unbankable, or not worth investing in, such that their lack of development is further exacerbated. In the developing world, development itself increases vulnerability. Faas (2016), Cinner et al. (2018), Visser et al. (2014), and Kelman et al. (2015) all stress the narrowness of both vulnerability and climate change adaptation measures. Climate change is only one factor in a complex array of issues and attributes that drive vulnerability.

29.4 POTENTIAL CAUSES AND MEASURES OF HOUSEHOLD AND COMMUNITY VULNERABILITY

It is against this background of the complex, multilayered nature of vulnerability, that scholars have suggested causes of vulnerability and measures that may contribute to emergency management policies and strategies aimed at DRR practice at the community level (Davis, 2004). Emergency managers need information on vulnerable sectors of the community in order to be able to respond effectively. Applied research addresses this need in order to inform the professionals and policy makers. At the practitioner level, Williams and Webb (2019) cite the need of emergency managers to define vulnerability in order to be able to respond to the needs of communities. Williams and Webb's (2019) study with emergency managers identified social vulnerability as driven by culture and poverty, moral imperatives, lack of security, and lack of knowledge and awareness. Ndah and Onu Odihi (2017) used a recently modified vulnerability based disaster risk assessment model premised on three stages—causes, dynamic pressures, and unsafe conditions. Extending from Crichton's risk triangle, Anderson-Berry and King (2005) categorized vulnerability characteristics as community, individual and societal in order to identify areas for emergency management intervention. Following from Cutter's extensive methodology (Cutter & Emrich, 2006), Chakraborty, Rus, Henstra, Thistlethwaite, and Scott (2020) use similar socio-economic and demographic characteristics, at small scales to identify vulnerable localities. However, Cutter warned that one size does not fit all when carrying out vulnerability assessments (Cutter & Emrich, 2006). Identification and mapping of vulnerability has continued throughout the 20 years since IDNDR because of the availability of census data at local levels and the usefulness of quickly identifying vulnerable localities (Anderson-Berry & King, 2005; Cutter & Emrich, 2006; Granger et al., 1999; Ndah & Onu Odihi, 2017; Wisner, 2004). Socio-economic and demographic attributes have been used extensively to define vulnerability because the data are easily available. The data have driven the types of assessments and measures and have consequently defined vulnerability.

However, all of these researchers (Anderson-Berry & King, 2005; Cutter & Emrich, 2006; Granger et al., 1999; Ndah & Onu Odihi, 2017; Wisner, 2004) and many more (including Cardona, 2004; Heijmans, 2004; Hilhorst & Bankoff, 2004) point to the complexity of vulnerability that derives from its underlying causes. Understanding the causes of vulnerability is a more meaningful route to its definition. Wisner (2004) identifies four approaches to vulnerability assessment as being demographic, taxonomic, situational (all structuralist), and self-evaluation, which also encompasses politics. Cardona (2004) states that vulnerability is not limited to poverty (or necessarily class and race) but is primarily the inability of individuals to protect themselves. Vulnerability is thus neither limited to demography nor to the severity of the hazard, but he does believe that risk can still be assessed objectively. Benson (2004) and Faas (2016) on the other hand, refer to unreported costs—direct as opposed to indirect impacts—and intangible impacts, causes, and even characteristics.

Cardona (2004) places vulnerability as originating in three areas: physical fragility or exposure, socio-economic fragility—marginalized, segregated and disadvantaged—and a lack of resilience. One must be careful to avoid a linear relationship between vulnerability and resilience. They operate as parallel or interlinked sets of attributes of individuals and households. In identifying both sets of attributes Boon et al. (2012) proposed social ecological model of realms inhabited or interacted by each individual.

In all attempts to list attributes of vulnerability and resilience in order to understand needs as well as causes, there are gaps in understanding that are a consequence of the complexity of hazard vulnerability.

29.5 CONTRIBUTIONS OF LOCAL KNOWLEDGE, HUMAN RIGHTS, SOCIAL AND ENVIRONMENTAL JUSTICE TO THE MITIGATION OF DISASTER VULNERABILITY

There is an observed DRR policy shift from vulnerability to empowerment (Heijmans, 2004). Building resilient communities puts responsibility on communities working with emergency managers and support personnel from a wide range of government, NGO, and private enterprise organizations. It is classical community development as shown by Arnstein (1969) to engage and empower people and their communities to enhance their strengths as they prepare themselves to reduce the risk of hazards and potential disaster. As suggested earlier it has also suited a neoliberal agenda in government to reduce government expenditure, to cut back on services and infrastructure, and to privatize community safety. Basic services are present, but their role is to support communities in their preparations and hazard awareness. Building resilient communities amounts to a privatization of community safety. It is a primary responsibility of the community to protect itself and reduce risk, rather than government responsibility. Vulnerability mitigation is not part of the neoliberal agenda, although obviously a social democratic government, whether of the right or left, will supply basic social services, support, and infrastructure to the most vulnerable members of the community including disadvantaged communities. But this is about meeting basic needs and does not mitigate vulnerability. The social welfare system maintains socio-economic disadvantage, even though it alleviates the worst impacts of poverty. Consequently communities set out to build hazard resilience despite remaining vulnerable. As long as vulnerability remains unaddressed, resilience, as well as climate change adaptation, is likely to be constrained or held back.

Yet the literature already cited in the discussion around causes and measures of vulnerability makes clear the complexity of vulnerability and critiques the excessive reliance on socio-economic and demographic characteristics of people and communities. However, as pointed out above, we use these data because they are so easily available from censuses, and because there is no doubt they are good indicators, albeit a small part of the big picture of attributes of vulnerability.

As we consider at least equally as important factors, like local knowledge we are immediately confronted by difficulties of gathering information, affording it a realistic weighting, quantifying qualitative characteristics and giving shape to the intangible (Faas, 2016). The intangible factors may include beliefs, spirituality, and world views that are optimistic or pessimistic, personality strengths and weaknesses, risk appetite, and many more factors that contribute to community vulnerability and resilience. Local capacity and social capital are clearly attributes that derive from these factors and shape vulnerability (Wisner, 2004), but qualitative research shows us differences between articulate and inarticulate people and communities—narratives or stoicism—that may not necessarily reflect real resilience.

Local knowledge and indigenous knowledge are intangible factors that are particularly susceptible to articulation, interpretation, and understanding. Bankoff (2004) examines local knowledge from a historical geographical perspective in which he places it alongside the emergence of the noble savage concept at the beginning of the industrial period. In many ways we have not moved very far from this patronizing acknowledgment of "primitive" people who live close to nature. The idea has transformed to indigenous knowledge, which is afforded some respect but is clearly secondary to mainstream science. Anthropology inadvertently feeds into this recognition of the "other" as a separate state of knowledge (Bankoff, 2004; Faas, 2016) that is relevant in specific societies but has limited wider application. The growth driven modernization process of development was constructed to remove societies out of the "primitive" close to nature state into the globalized economy. Thus development in initiating social change and modernization erodes local and indigenous knowledge (Heijmans, 2004).

Emergency management is confronted by two types of local knowledge—indigenous knowledge which is attributed to people of nonmainstream cultures, and local knowledge which may be owned by members of the mainstream culture but is specific to environments and landscapes that are occupied and known by defined communities. For the purposes of DRR there is not a practical difference between these two types of knowledge.

Whether the community is indigenous, or a cultural minority, or part of the mainstream culture, the knowledge people possess about their own local environment and its potential hazards is a powerful factor that reduces their vulnerability to disaster. People have access to a local, often oral, history (in both literate and preliterate societies) that is highly specific to their observed landscape processes and patterns. They have a depth of understanding of their local environment. This is more easily observed in communities that live off the land—farming—but it is also the case for long-term urban residents. Networks of experience and knowledge of both the local environment and the skills and personalities of community members transfer informal information across the community and across generations. IT has added to that information flow and is in many places utilized by indigenous as well as mainstream communities. At the community level many people will recognize and define their local knowledge as commonsense—knowledge that is both formal and informal, built on shared experience. We suggest that local knowledge is social capital which contributes to resilience and counteracts or lessens some of the constraints of vulnerability.

Disaster vulnerability encompasses environmental justice issues (Ryder, 2017). Our environment is both the natural and the physical world from which we obtain resources and within which we create and modify landscapes. The environment is everything: it is where we live, work, play, and learn (Ryder, 2017). Vulnerability of the environment and unequal access to environments and resources are both an environmental justice and a social justice issue. Colonialism and oppression of colonized peoples imposed environmental inequality (Bankoff, 2004; Oliver-Smith, 2004; Ryder, 2017). An example of such an impact is the situation of Australian indigenous people who lost land and livelihoods and were subsequently relocated into centralized settlements (often missions) in many cases, co-resident with other cultural and language groups, which further increased their disadvantaged status. Their vulnerability was created by the environmental and social injustice of the colonial process.

As the environmental movement has emerged, the disadvantaged members of society have been excluded (Ryder, 2017) to the extent that vulnerability of the natural environment is given a greater political significance than its human occupants. However, the general usage of environmental justice is more directly related to environmental disadvantages that fall upon social economically vulnerable people and communities. These include air, water and soil pollution, blighted landscapes, poor quality agricultural land, lack of space, lack of access, environmental diseases, and an unequal share of climate change impacts such as drought, fire, and flood.

Natural hazards and disasters are part of the range of environmental vulnerability that impacts unequally those who are already socio-economically vulnerable. Reale and Handmer (2011) outline the specific inequality of land tenure as an environmental and social justice issue. Different types of tenure add to inequality and vulnerability through insecurity, land shortage, lack of access, loss of livelihoods, and homelessness. These are often made worse after a disaster (Reale & Handmer, 2011) where some households and communities may absolutely lose land and likelihood through erosion damage. Predisaster spatial inequalities are compounded by disasters such as flood (as well as bushfires and cyclones). Environmental justice is directly linked with social justice (Chakraborty et al., 2020). Ryder (2017) argues the need to integrate both environmental and social justice into DRR and vulnerability reduction. Wisner (2004) has long argued that disasters are a product of socio-economic inequality and vulnerability.

Relocation from a hazard prone area or a disaster involves extensive migration globally (King et al., 2014) including flows of refugees, some of whom are beginning to define themselves as climate change refugees. Unfortunately, international agencies, especially UN, do not recognize flight from a hazard as constituting refugee status unless life is directly endangered (Suhrke, 2013). Environmental refugees tend to be kept within a country's borders adding to the vulnerability of already impoverished regions and countries (Noll, 2003).

Emergency management attempts to reduce vulnerability by building resilience in people and communities, without addressing the causes and impacts of vulnerability, as the sector has no authority or resources to intercede in areas that are way beyond its control. Both resilience and climate change adaptation as policies and strategies are inevitably counteracted by the constraints of vulnerability. Social vulnerability is grounded in socio-economic and demographic attributes, as well as social class, race, politics, environment, and culture. Such vulnerability is counter to environmental justice, social justice, and human rights. Governments and international agencies sign up to human rights and justice but maintain unequal social systems. Thus efforts at DRR are unsustainable as long as they merely recognize but otherwise ignore the complex multilayered problems of social vulnerability. Building resilient communities and enhancing climate change adaptation constitute one step forward, while vulnerability takes us one step back.

UN sustainability goals provide an agenda for gradually tackling socio-economic inequality and vulnerability. DRR has a role to play in these goals but remains dependent of the activities of a complex range of agencies. On the positive side, however, land use planners have taken responsibility in many countries (King, Gurtner, Firdaus, Harwood, & Cottrell, 2016) to reduce hazard risk through development controls. More proactively, planners and designers have played significant parts in disaster recovery by designing places for people and communities which give meaning to their loss while at the same time reducing the potential for future disaster—building back better (Donovan, 2013; Pica, 2018).

29.6 CASE STUDY OF TOWNSVILLE'S MONSOONAL FLOODS 2019

While socio-economic and demographic profiles are useful for emergency managers (particularly in post event response and recovery), a case study of the Townsville "North and North Queensland

Monsoon Trough" rainfall and flood event of 2019 demonstrates how issues of exposure, sensitivity, vulnerability, and resilience, remain dynamic, and context dependent. Such experience of the complex and multifaceted interface between hazard and traditional constructs of vulnerability elucidates the challenges in achieving effective and sustainable DRR strategies for communities. As climate change is likely to affect variability in both the frequency and severity of extant natural hazards it becomes imperative that broader perspectives and approaches be considered for future events.

Townsville is a regional city in North Queensland with a population of approximately 195,000 people (Australian Bureau of Statistics, 2019). Located on a natural floodplain in the lower reaches of the Bohle and Ross Rivers it has a high natural exposure to flood risk. Significant historical flood occurrences include 1881, 1892, 1946, 1953, 1960, and in more recent memory during Cyclone Althea in 1971 and "The Night of Noah" in 1998 (Green Cross Australia, 2011; National Library of Australia − Trove, 2020). As the built environment and population of Townsville has expanded, environmental degradation, alteration to physical landscapes, water diversion, changes in natural drainage systems, resource management with subsequent development, and construction has only served to increase levels of both flood exposure and risk. In 2019 the Insurance Council of Australia (ICA), rated Townsville as Australia's most flood prone area. While the circumstances and impacts of the 2019 monsoon trough event have been consistently characterized as "unprecedented" the flood hazard narrative is not new for communities within this region.

Recognized as the largest urban center north of the Sunshine Coast, the tyranny of distance, isolation, and the scale of Townsville has resulted in relatively high business and living costs comparative to the rest of Australia [Townsville City Council (TCC), 2019]. In 2016, the Australian Bureau of Statistics (ABS) Index of Relative Socio-economic Disadvantage (IRSD) positioned Townsville lower than the national and state average, indicative of higher levels of disadvantage and potential inequality (id The Population Experts, 2019) (refer to Table 29.1). As a composite of variables such as income levels, employment, educational attainment and skilled occupations, IRSD recognizes "people's access to material and social resources, and their ability to participate in society" (Australian Bureau of Statistics, 2018).

While 2016 socio-economic census data (id The Population Experts, 2019) reveals that Townsville had higher than both state and national medians for labor force participation and trade qualifications, it also had a higher level of unemployment at 8.9%. Household rental rates and the percentage of Aboriginal and Torres Strait residents were similarly higher than Australian averages. In contrast, Townsville had lower than national medians for demographic indicators such as age and the diversity of residents born overseas and/or speaking a language other than English in the household. Although indicative of adaptive capacity, socio-economic well-being, and issues of communicating hazard warnings and information, in the context of the 2019 monsoon and flood event such measures were of limited direct value in determining levels of flood exposure, susceptibility or impacts.

Preceded by years of hot dry weather, limited rainfall, and declared drought conditions, the monsoon flood event of early 2019 is illustrative of the weather extremes faced by the tropical savannah region of Townsville. From January 25 to February 14, 2019, the city experienced periods of sustained heavy rainfall as a result of the convergence of a slow-moving monsoon trough and tropical low pressure systems. Over 2000 mm was recorded within 10 days, exceeding all previous records and the city's average annual rainfall. A significant amount of this rain fell within the local

Table 29.1 Townsville City Council's Index of Relative Socio-economic Disadvantage Small Areas and Benchmark Areas.

Area	2016 Index	Percentile	2019 Flood Inundation
Australia	**1001.9**	**46**	
Queensland	**996.0**	**43**	
Townsville City Council	**989.00**	**39**	
Annandale	1077.0	90	✓
Douglas	1071.0	87	
Bohle Plains—Rangewood—Shaw	1070.1	87	
Idalia—Cluden—Oonoonba	1056.7	80	✓
Townsville City	1055.0	79	✓
Mount Low—Burdell—Beach Holm	1045.1	74	
Cosgrove—Mount Louisa	1037.2	69	
Castle Hill—North Ward	1030.9	64	
Central Business District	1029.5	63	
Alligator Creek and District	1027.0	62	
Black River—Alice River—Hervey Range	1024.3	60	
Belgian Gardens—Rowes Bay	1002.2	47	✓
Thuringowa Central	991.0	40	
Mundingburra	990.0	39	✓
Kirwan	988.0	38	
Hyde Park—Mysterton	973.2	31	✓
South Townsville	973.0	31	✓
Balgal Beach—Rural West	968.8	28	
Woodstock District	968.3	28	
Railway Estate	967.0	28	✓
Cranbrook	959.0	25	
West End	959.0	25	✓
Hermit Park	958.0	24	✓
Magnetic Island	951.4	22	
Currajong	949.0	21	
Wulguru	948.0	21	
Rosslea	948.0	21	✓
Deeragun—Jensen	944.7	20	
Condon	940.0	18	
Aitkenvale	940.0	18	
Kelso	936.0	17	
Gulliver	929.0	15	
Julago—Stuart	919.4	13	✓
Rasmussen	911.0	12	
Pimlico	896.0	9	✓
Heatley	895.0	9	✓
Murray—Roseneath and District	875.0	7	
Vincent	873.0	7	
Shelly Beach—Garbutt and surrounds	868.6	6	

Adapted from id The Population Experts (2019).

catchment area with the Ross River Dam receiving over 850,000 ML peaking at 43.00 m and a capacity of 247% (TCC, 2019). Although the Council had publicly available high-resolution hazard maps identifying established flood inundation zones, developing conditions meant that officials were forced to fully open the spillway gates of the Ross River Dam on February 4, 2019, creating a previously unidentified risk for suburbs such as Idalia and Oonoonba. The release of this water into the city generated rapidly rising, dangerous, and high-velocity flows causing extensive additional damage and disruption (refer to Figure 29.1).

While the loss of human life and injury was considered relatively low given the scale of the event (three indirect fatalities), preliminary data estimates from April 2019 reveal the significant social and economic impacts (Deloitte Access Economics, 2019). During the peak of the floods, six evacuation centers were operational providing temporary accommodation for more than 800 people (over 1100 residents were evacuated). With 3369 residential dwellings impacted, over 1700 people required emergency housing support. Official assessments determined that 1255 (37%) of the affected dwellings were uninhabitable. In addition, 1655 commercial and 777 public buildings were also impacted. Affected public infrastructure and critical lifelines included roads, bridges, rail, water and waste, port, electricity, and telecommunications. Health, social, and community services such as medical centers, schools, child care, and shopping centers were also interrupted. Beyond natural flood degradation and erosion, environmental damage included the build-up of debris, waste, and rubbish. Direct and indirect socio-economic costs related to the event have been

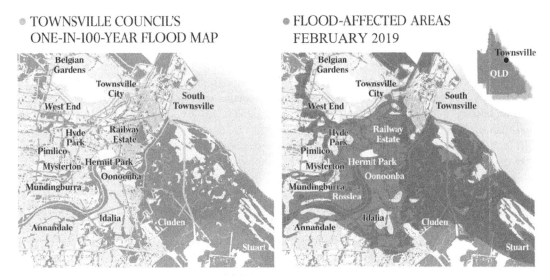

FIGURE 29.1

A comparison of Townsville City Councils 1 in 100 flood map and areas impacted in the 2019 monsoon trough event.

Adapted from The Australian. (February 7, 2019). Townsville flood maps reviewed as more homes go under. The Australian. [Online] <https://www.theaustralian.com.au/nation/politics/townsville-flood-maps-reviewed-as-more-homes-go-under/news-story/ c15de089467702d1aa4a14bede751ce9> (The Australian, 2019).

estimated to exceed $5.6 billion [Deloitte Access Economics, 2019; Queensland Reconstruction Authority (QRA), 2019].

Established flood hazard maps and small area IRSD for the Townsville City Council (TCC) geographic region exhibit limited correlation between direct hazard exposure, traditional measures of community vulnerability, and the subsequent adversity experienced in this event (refer to Fig. 29.1 and Table 29.1). Heavily impacted suburbs such as Idalia, Oonoonba, Cluden, Rosslea, Railway Estate, Annandale, Hermit Park, Hyde Park, Mundingburra, and South Townsville are representative of both high and low measures of inequality and socio-economic disadvantage across the city. According to the QRA more than 116,000 people were identified as experiencing hardship post event, and 62,200 were assisted with psychological first aid (Deloitte Access Economics, 2019; QRA, 2019). It has been estimated that approximately 40% of recovery costs (over $2.25 billion) have been directed to emergency response and clean-up, health, social, and community concerns, including mental health, alcohol misuse, family violence, gambling, and chronic disease. Loss of power, premises, vehicles, contents, equipment, agriculture, livestock, and physical access further impacted local businesses and industries disrupting finance, employment and livelihoods throughout the region (Deloitte Access Economics, 2019; QRA, 2019).

As existing hazard research and DRR strategies traditionally focus on people from poor and disadvantaged socio-economic backgrounds, and in regional and remote areas, the experience of this flood event demonstrates the need for further investigation into related root causes, dynamic pressures, and unsafe conditions which can affect individual, household, and community levels of exposure, sensitivity, and resilience (Wisner, 2004). Post event research in Townsville identified particular issues of community hazard risk perception, coping and adaptive capacity, and macro forces based in established political and institutional systems.

An online social survey of impacted Townsville residents conducted by the Centre for Disaster Studies (2019) between February and March 2019 (unpublished) focused on issues of hazard preparedness, information, and response. While the basic demographic profile of the 705 participants in this survey demonstrates a bias of predominantly females over 30 years old with either vocational or tertiary qualifications—the purpose of the study was to appreciate direct experiences rather than socio-economic characteristics.

In terms of individual and household preparation, respondents were asked to consider pre-event levels of preparedness and particulars such as a comprehensive disaster kit and evacuation plans. With only half of those surveyed indicating they felt adequately prepared, identified constraints included, lack of familiarity with local hazards—especially floods (some residents were new to the region, while others only considered cyclone threats), complacency due to prior local experience and precedence, and poor awareness of the evolving risks. Almost 65% of respondents had disaster kits; yet many lacked sufficient food, water, and supplies to sustain all household members for any longer than 2−3 days, with some completely reliant on mobile phone technology/battery duration for information and communication once power supplies were interrupted. The majority of impacted people indicated that evacuation plans and decisions were only deliberated as the event developed due to the perceived safety and security of localities (or dwelling), fears of looting, and personal concerns that pets would not be accommodated in local evacuation centers.

Consistent with reports from the TCC and ICA (2019) results from the Centre for Disaster Studies (CDS) survey showed that more than 25% of residents and businesses had issues regarding adequate insurance coverage for the flood event. Many community members and businesses were uninsured or underinsured (i.e., had undervalued property replacement costs, only purchased building or contents insurance, or simply did not have appropriate flood damage cover). Flood insurance reparations were further complicated and delayed for some over disputes regarding the specific cause of damage, preexisting issues, and proof of regular maintenance. Individual company repair policies and statements of "like for like" provided limited flexibility to modify preexisting building design and structures or utilize alternative building materials for improved flood proofing. As the costs of insurance premiums for the region become increasingly prohibitive, some respondents indicated that they were willing to accept the financial risks of remaining uninsured, instead investing the financial resources elsewhere. These institutional policies and subsequent decision-making effectively undermines levels of resilience and coping capacity for future hazard events.

A key facet of sustainable DRR is the effective communication of relevant hazard information. This encompasses elements of awareness, comprehension/understanding, information distribution and access, and trust for effective decision-making. CDS survey results and similar studies (Bergin, 2019; TCC, 2019) indicate that many Townsville residents did not sufficiently understand the flood risk. Modeling for flood hazard mapping and the overlays used for new development approvals in Townsville have been based on a 1% (1 in 100) annual exceedance probability. It is evident that this expression of probability was poorly understood and that projections of damage from significant and sustained rainfall and flooding were inadequate for this event. Prior to the monsoonal rainfall and flooding, many people also believed that the engineered structure of the Ross River Dam provided sufficient mitigation and protection from any significant flood impacts. Such misunderstandings produced a false sense of security within the community evidenced in the extent of ground-level development and construction within highly vulnerable flood prone areas. Historically, any dwellings located in these zones were built as high-set or had an elevated habitable floor level to reduce the risk of direct flood damage.

While a formal government review described the TCC's response to the 2019 weather event as "well-coordinated and effective in keeping the community informed, protecting life and property, and managing the supply of essential services" [Inspector General of Emergency Management (IGEM), 2019; TCC, 2019], the CDS survey results did identify some issues in relation to the communication process. The majority of respondents indicated that they accessed and received information regarding the event electronically either directly through official websites such as the Bureau of Meteorology and Townville City Council, through the Townsville Disaster Dashboard, via social media, and/or through text messages—reflecting a heavy reliance on internet access and technology. Census data from 2016 indicates that only 78.3% of Townsville households had internet access (Australian Bureau of Statistics, 2018; id The Population Experts, 2019). Electricity supplies were disrupted to many suburbs during the 2019 rainfall and flood event, mobile phones have limited independent battery life, and AM/FM battery operated radio ownership was not universal. Many were left unable to directly access available information. Participants additionally expressed a high level of frustration regarding the style, content, contradictions, and terminology used in warnings and information updates, creating a sense of confusion and uncertainty in an already high stress context. Related recommendations were proposed in the IGEM flood event review (IGEM, 2019).

As the media highlighted flood-related evacuations, emergency response efforts, and personal experiences, the residents surveyed by the CDS indicated a desire for better coordination between response agencies, a greater commitment of government resources and funding, and the provision of more timely, relevant information. Community cohesion, connectivity, support, altruism, and volunteering, consistent with Aldrich and Meyer (2015) definition of social capital was evident in the post event clean-up, but so was an expectation and demand for both military support and government grants, reflecting a developing culture of dependency. Subsequent weeks and months revealed increasing mental health issues associated more with community dislocation and changing social relations than income or direct loss (Deloitte Access Economics, 2019). Blatman-Thomas (2019) highlighted further marginalization in the predicament of "invisible victims" of the event including indigenous, homeless, and those who were in temporary accommodation, with no physical address to access mainstream support and services. In the response and recovery process, it is evident that institutional and political systems have demonstrated a greater influence on community resilience and adaptive capacity than demographic or socio-economic dynamics.

29.7 CONCLUSION

Lessons from the Townsville monsoon rainfall and flood event of 2019 have implications which extend well beyond this case study. Considerable issues have been identified in terms of increasing physical and community hazard exposure, the vulnerability of the social, economic, and built environment due to planning development, rebuilding and policy decisions, and concerns about the resources, adaptive capacity, and political commitment to sustainable DRR strategies in the face of climate change uncertainty and variability. Each potential hazard is complex, dynamic, and context dependent. Bergin (2019) proposes greater investment in "soft" self-protection measures such as improved community education, training, and the development of real-time situational hazard awareness and understanding for effective decision-making. While beneficial in developing community risk profiles, resilience for extreme climate-induced weather events should not be centered entirely on traditional socio-economic indicators and measures of inequality. Evolving research and hazard experience suggest that effective and sustainable protection of life and property requires greater resources, leadership, commitment, and risk prioritization from all stakeholders.

REFERENCES

Aldrich, D. P., & Meyer, M. A. (2015). Social capital and community resilience. *American Behavioral Scientist*, 59(2), 254–269.

Anderson-Berry, L., & King, D. (2005). Mitigation of the impact of tropical cyclones in Northern Australia through Community Capacity Enhancement. *Mitigation and Adaptation Strategies for Global Change*, 10, 367–392.

Arnstein, S. R. (1969). A ladder of citizen participation. *Journal of the American Planning Association*, 35(4), 216–224.

Australian Bureau of Statistics. (2018). *2033.0.55.001—Census of population and housing: Socio-economic indexes for areas (SEIFA), Australia, 2016.* <https://www.abs.gov.au/ausstats/abs@.nsf/Lookup/2033.0.55.001main + features100002011>.

Australian Bureau of Statistics. (2019). *3218.0—Regional population growth, Australia, 2017-18.* <https://www.abs.gov.au/AUSSTATS/abs@.nsf/mf/3218.0>.

Bankoff, G. (2004). The historical geography of disaster: 'Vulnerability' and 'local knowledge' in Western discourse. In G. Bankoff, G. Frerks, & D. Hilhorst (Eds.), *Mapping vulnerability: Disasters, development and people*. Oxford & New York: Earthscan Routledge.

Benson, C. (2004). Macro-economic concepts of vulnerability: Dynamics, complexity and public policy. In G. Bankoff, G. Frerks, & D. Hilhorst (Eds.), *Mapping vulnerability: Disasters, development and people*. Oxford & New York: Earthscan Routledge.

Benson, C. (2009). *Mainstreaming disaster risk reduction into development: Challenges and experience in the Philippines*. The ProVention Consortium, The International Federation of Red Cross and Red Crescent Societies.

Bergin, A. (2019). *Townsville floods demonstrate need for better disaster planning*. ASPI, Australian Strategic Policy Institute. <www.aspi.org.au/opinion/townsville-floods-demonstrate-need-better-disaster-planning>.

Blaikie, P., Cannon, T., Davis, I., & Wisner, B. (1994). *At risk: Natural hazards, people's vulnerability, and disasters*. London: Routledge.

Blatman-Thomas, N. (2019). Reciprocal repossession: Property as land in Urban Australia. *Antipode, 51*(5), 1395–1415.

Boon, Helen J., Cottrell, A., King, D., Stevenson, R. B., & Millar, J. (2012). Bronfenbrenner's bioecological theory for modelling community resilience to natural disasters. *Natural Hazards, 60*, 381–408.

Cardona, O. D. (2004). The need for rethinking the concepts of vulnerability and risk from a holistic perspective: A necessary review and criticism for effective risk management. In G. Bankoff, G. Frerks, & D. Hilhorst (Eds.), *Mapping vulnerability: Disasters, development and people*. Oxford & New York: Earthscan Routledge.

Centre for Disaster Studies. (2019). *Understanding community preparedness and response to the North Queensland Major Flood Event 2019*. Centre for Disaster Studies.

Chakraborty, L., Rus, H., Henstra, D., Thistlethwaite, J., & Scott, D. (2020). A place-based socioeconomic status index: Measuring social vulnerability to flood hazards in the context of environmental justice. *International Journal of Disaster Risk Reduction, 43*.

Cinner, J. E., Adger, W. N., Allison, E. H., Barnes, M. L., Brown, K., Cohen, P. J., ... Morrison, T. H. (2018). Building adaptive capacity to climate change in tropical coastal communities. *Nature Climate Change, 8*, 117–123, February 2018.

Cutter, S. L., & Emrich, C. T. (2006). Moral hazard, social catastrophe: The changing face of vulnerability along the hurricane coasts. *The Annals of the American Academy of Political and Social Science, 604*, 102–112.

Davis, I. (2004). Progress in analysis of social vulnerability and capacity. In G. Bankoff, G. Frerks, & D. Hilhorst (Eds.), *Mapping vulnerability: Disasters, development and people*. Oxford & New York: Earthscan Routledge.

Delica-Willison, Z., & Willison, R. (2004). Vulnerability reduction: A task for the vulnerable people themselves. In G. Bankoff, G. Frerks, & D. Hilhorst (Eds.), *Mapping vulnerability: Disasters, development and people*. Oxford & New York: Earthscan Routledge.

Deloitte Access Economics. (2019). *The social and economic cost of the North and Far North Queensland Monsoon Trough*. Deloitte Australia Economics. <https://www2.deloitte.com/au/en/pages/economics/articles/social-economic-cost-north-far-north-queensland-monsoon-trough.html>.

Donovan, J. (2013). *Designing to heal: Planning and urban design response to disaster and conflict*. Melbourne, VIC: CSIRO Publishing.

Faas, A. J. (2016). Disaster vulnerability in anthropological perspective. *Annals of Anthropological Practice, 40*(1), 14–27.

Graif, C. (2016). Un)natural disaster: Vulnerability, long-distance displacement, and the extended geography of neighbourhood distress and attainment after Katrina. *Population and Environment, 37*, 288–318.

Granger, K., Jones, T., Leiba, M., & Scott, G. (1999). *Community risk in cairns: A multi hazard risk assessment*. Canberra, Australia: Australian Geological Survey Organisation.

Green Cross Australia. (2011). Harden up chronological history of flooding 1857-2010. In *Harden up Queensland*. <http://hardenup.org/media/347511/queensland_flood_history.pdf>.

Heijmans, A. (2004). From vulnerability to empowerment. In G. Bankoff, G. Frerks, & D. Hilhorst (Eds.), *Mapping vulnerability: Disasters, development and people*. Oxford & New York: Earthscan Routledge.

Hilhorst, D., & Bankoff, G. (2004). Introduction: Mapping vulnerability. In G. Bankoff, G. Frerks, & D. Hilhorst (Eds.), *Mapping vulnerability: Disasters, development and people*. Oxford & New York: Earthscan Routledge.

id The Population Experts. (2019). *City of Townsville community profile*. <https://profile.id.com.au/townsville/seifa-disadvantage-small-area>.

Inspector General of Emergency Management (IGEM). (2019). *The 2019 monsoon trough rainfall and flood review report 3: 2018-19*. Office of the Inspector-General Emergency Management. <https://www.igem.qld.gov.au/sites/default/files/2019-12/IGEM%20MTRF%20Review%28lowres%29.pdf>.

Insurance Council of Australia (ICA) (2019). *Insurers brief deputy premier on Townville catastrophe recovery*. <https://www.insurancecouncil.com.au/media_release/plain/500>.

Intergovernmental Panel on Climate Change (IPCC). (2007). *Climate change (2007)—Impacts, adaptation and vulnerability*. Cambridge, UK: Cambridge University Press.

Intergovernmental Panel on Climate Change (IPCC). (2012). *Managing the risks of extreme events and disasters to advance climate change adaptation. A special report of working groups I and II of the Intergovernmental Panel on Climate Change*. Cambridge, UK: Cambridge University Press.

Jones, B. (2017). Cities build their vulnerability. *Nature Climate Change, 7*, News and Views 237.

Kelman, I., Gaillard, J. C., Lewis, J., & Mercer, J. (2016). Learning from the history of disaster vulnerability and resilience research and practice for climate change. *Natural Hazards, 82*, S129−S143.

Kelman, I., Gaillard, J. C., & Mercer, J. (2015). Climate change's Role in disaster risk reduction's future: Beyond vulnerability and resilience. *The International Journal of Disaster Risk Science, 6*, 21−27.

King, D., Bird, D., Haynes, K., Boon, H., Cottrell, A., Millar, J., ... Thomas, M. (2014). Natural disaster mitigation through relocation and migration: Household adaptation strategies and policy in the face of natural disasters. *International Journal of Disaster Risk Reduction, 8*, 83−90.

King, D., Gurtner, Y., Firdaus, A., Harwood, S., & Cottrell, A. (2016). Land use planning for disaster risk reduction and climate change adaptation: Operationalizing policy and legislation at local levels. *International Journal of Disaster Resilience in the Built Environment, 7*(2), 158−172.

Lewis, J., & CICERO (Center for International Climate and Environmental Research—Oslo). (2012). The good, the bad and the ugly: Disaster risk reduction (DRR) versus disaster risk creation (DRC), version 1. *PLoS Currents, 4*. Available from https://doi.org/10.1371/4f8d4eaec6af8, e4f8d4eaec6af8.

National Library of Australia − Trove. (2020). Search: Townsville, floods. *Digitised Newspapers and More*. [Online] <https://trove.nla.gov.au/newspaper/result?q = %28townsville%2C + flood%29 + + + + + + + + + + + + +&s = 20>.

Ndah, A. B., & Onu Odihi, J. (2017). A systematic study of disaster risk in Brunei Darussalam and options for vulnerability-based disaster risk reduction. *The International Journal of Disaster Risk Science, 8*, 208−223.

Noll, G. (2003). States, refugees and international law. In E. Newman, & J. van Selm (Eds.), *Refugees and forced displacement: International security, human vulnerability and the state*. United Nations University Press.

Oliver-Smith, A. (2004). Theorizing vulnerability in a globalized world: A political ecological perspective. In G. Bankoff, G. Frerks, & D. Hilhorst (Eds.), *Mapping vulnerability: Disasters, development and people*. Oxford & New York: Earthscan Routledge.

Pica, V. (2018). Beyond the Sendai Framework for Disaster Risk Reduction: Vulnerability reduction as a challenge involving historical and traditional buildings. *Buildings, 8*, 50.

Queensland Reconstruction Authority (QRA). (2019). *North and Far North Queensland Monsoon Trough – State recovery plan 2019–2021*. <https://www.qra.qld.gov.au/publications-resources/plans-policies-and-strategies>.

Reale, A., & Handmer, J. (2011). Land tenure, disasters and vulnerability. *Disasters, 35*(1), 160–182.

Ryder, S. S. (2017). A bridge to challenging environmental inequality: Intersectionality, environmental justice, and disaster vulnerability. *Social Thought and Research, 34*, 85–115.

Suhrke, A. (2013). Human security and protection of refugees. In E. Newman, & J. van Selm (Eds.), *Refugees and forced displacement: International security, human vulnerability and the state*. United Nations University Press.

The Australian. (February 7, 2019). Townsville flood maps reviewed as more homes go under. *The Australian.* [Online] <https://www.theaustralian.com.au/nation/politics/townsville-flood-maps-reviewed-as-more-homes-go-under/news-story/c15de089467702d1aa4a14bede751ce9>.

Townsville City Council (TCC). (2019). *Submission to the northern Australia insurance inquiry, Australian Competition & Consumer Commission May 2019*. Townsville City Council. <https://www.accc.gov.au/system/files/Townsville%20City%20Council.pdf>.

Visser, H., Petersen, A. C., & Ligtvoet, W. (2014). On the relation between weather-related disaster impacts, vulnerability and climate change. *Climatic Change, 125*, 461–477.

Williams, B. D., & Webb, G. R. (2019). Social vulnerability and disaster: Understanding the perspectives of practitioners. *Disasters.* Available from https://doi.org/10.1111/disa.12422.

Wisner, B. (2004). Assessment of capability and vulnerability. In G. Bankoff, & D. J. M. Hilhorst (Eds.), *Mapping vulnerability: Disasters, development and people*. Oxford & New York: Earthscan Routledge.

FURTHER READING

Meyer, M. A. (2017). Elderly perceptions of social capital and age-related disaster vulnerability. *Disaster Med Public Health Preparedness, 11*, 48–55.

ALTERNATIVE SCENARIOS FOR LOW-CARBON TRANSPORT IN NIGERIA: A LONG-RANGE ENERGY ALTERNATIVES PLANNING SYSTEM MODEL APPLICATION

30

Michael O. Dioha[1], Atul Kumar[1], Daniel R.E. Ewim[2] and Nnaemeka V. Emodi[3]

[1]*Department of Energy and Environment, TERI School of Advanced Studies, New Delhi, India* [2]*Department of Mechanical Engineering, Mechatronics and Industrial Design, Tshwane University of Technology, Pretoria, South Africa* [3]*Future Energy Research Group, Tasmanian School of Business and Economics, University of Tasmania, Hobart, TAS, Australia*

30.1 INTRODUCTION

The fifth assessment report of the Intergovernmental Panel on Climate Change (IPCC) (2014) established the fact that GHG emissions from anthropogenic activities negatively impact the climate. The recent special report of the IPCC (2018) on global warming of 1.5°C further highlights that warming of the earth due to historical anthropogenic GHG emissions will continue to remain for centuries, and this will also contribute to long-term changes in the climate system. Therefore to reduce the adverse effects of climate change, there is a serious need to mitigate GHG emissions (Dioha, Emodi, & Dioha, 2019). The transport sector catalyzes socioeconomic development but contributes to GHG emissions as the current transport system is highly dominated by fossil fuels such as gasoline and diesel (Prasad & Raturi, 2018). The share of transport sector CO_2 emissions in total global emissions has increased from about 13.5% in 2005 (Anderson, Fergusson, & Valsecchi, 2007) to about 25% in 2016 (IEA, 2016). In fact, some studies have suggested that emissions from the transport sector by 2050 could become about twice the levels in 2010 (Marchal et al., 2011). Thus if global CO_2 emissions from the transport sector are to be curtailed and steered toward realizing a low-carbon future, it will be necessary for all regions of the world to align their transport policies in the direction of sustainability.

The Nigerian transport system mainly runs on fossil gasoline and diesel (Gujba, Mulugetta, & Azapagic, 2013). The demand for fossil fuels in the Nigerian transport sector has also been growing over the years at a compound annual growth rate (CAGR) of 3.68%, from 166 PJ (1990) to 368 PJ (2012)[1] (Fig. 30.1). This demand is also expected to continue growing in the same trend as population, income,

---

[1]https://www.iea.org/statistics/index.html?country = NIGERIA&year = 2016&category = Energy% 20supply&indicator = TPESbySource&mode = table&dataTable = BALANCES.

Economic Effects of Natural Disasters. DOI: https://doi.org/10.1016/B978-0-12-817465-4.00030-3

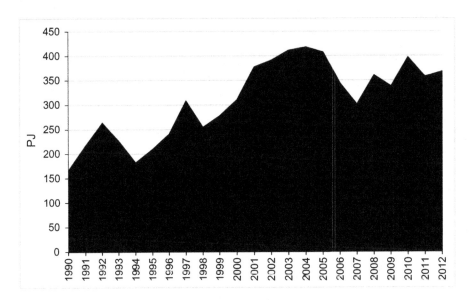

FIGURE 30.1

Time trend of Nigerian transport sector energy consumption.

and the demand for mobility increases in the future. Accordingly, it is expected that the resultant emissions from the Nigerian transport sector will continue growing in the future except vigorous mitigation policies are introduced. However, the design of such policies needs to be informed by scientific studies, which takes into account the impact of different policy pathways on energy demand and the resultant GHG emissions.

Energy systems models are tools that can be used to explore future low-carbon scenarios of a region, country, or sector in order to inform policy decisions (Dioha, 2018). The models are generally classified as bottom-up and top-down models (Herbst, Toro, Reitze, & Jochem, 2012). The bottom-up models are rich in technology representation but lack macroeconomic feedback system. Typical examples include the MARKAL/TIMES, MESSAGE, and Long-range Energy Alternatives Planning (LEAP) System. In contrast, the top-down models represent macroeconomic feedbacks richly but are weak in technology representation. Typical examples include E3ME, GEM-E3, and GTAP. For more information on energy system models, see the work of Dioha (2017). In the literature, there have been a plethora of studies applying a series of these models to study different low-carbon scenarios for the transport sector (Ahanchian & Biona, 2014; Chunark, Thepkhun, Promjiraprawat, Winyuchakrit, & Limmeechokchai, 2015; Dhar & Shukla, 2015; Gül, Kypreos, Turton, & Barreto, 2009; He & Chen, 2013; He & Qiu, 2016; Hong, Chung, Kim, & Chun, 2016; Limanond, Jomnonkwao, & Srikaew, 2011; Martínez-jaramillo, Arango-aramburo, Álvarez-uribe, & Jaramillo-álvarez, 2017; Nealer, Matthews, & Hendrickson, 2012; Peng, Du, Ma, Fan, & Broadstock, 2015; Peng, Ou, Yuan, Yan, & Zhang, 2018; Prasad & Raturi, 2018; Selvakkumaran & Limmeechokchai, 2015; Seo, Park, Oh, & Park, 2016; Shabbir & Ahmad, 2010; Tsita & Pilavachi, 2017; Valderrama, Monroy, & Valencia, 2019; Yan & Crookes, 2009; Zhang, Chen, & Huang, 2016; Zhang, Fujimori, Dai, & Hanaoka, 2018).

However, a critical analysis of these literatures suggests that most of the studies on transport energy system have been focused in developed countries and few developing countries. Thus there is an urgent need to evaluate possible pathways toward decarbonizing the transport system of developing countries such as Nigeria. Studies applying quantitative methods to examine low-carbon pathways for the transport sector in Nigeria are limited. A research has earlier been conducted to examine the life cycle environmental and economic implications of different scenarios for the Nigerian passenger transport system for the period 2003−30 (Gujba et al., 2013).

In this chapter, we applied the Nigerian LEAP system model to evaluate different scenarios for low-carbon transport in Nigeria. We establish three scenarios based on fuel substitution, namely: business-as-usual (BAU), countermeasure 1 (CM1), and countermeasure 2 (CM2). We then compare the energy consumption and environmental emissions across the scenarios. We believe that our study will be of very high importance value for the Nigerian government in its effort to make policies relevant to the decarbonization of the country's transport sector. The remaining part of this chapter is arranged thus: Section 30.2 presents the methodology employed in the study. Section 30.3 describes the study results and analysis. Section 30.4 concludes the study with some recommendations for policymakers.

30.2 METHODOLOGY

30.2.1 THE LONG-RANGE ENERGY ALTERNATIVES PLANNING MODELING TOOL

The LEAP model is a scenario-based modeling package developed by the Stockholm Environment Institute. The modeling tool is very common, and it has been applied for different energy and climate change studies in different parts of the world (especially in developing countries). For examples, on a national level, researchers have applied the LEAP model to study different energy transition scenarios in Mozambique (Mahumane & Mulder, 2019). At a local level, LEAP has been used to study the energy and environmental impacts of different energy access scenarios for the Nigerian household sector (Dioha & Emodi, 2019). Thus the results derived from a LEAP model run can easily be understood by other researchers. LEAP is an integrated accounting framework model that covers the different processes involved in energy resource extraction, conversion, and consumption in all sectors of the economy. Moreover, the LEAP model could be applied for optimization in the power sector that tends to minimize the total cost of the system. LEAP is also fitted with a technology emission database (TED) that consists of the IPCC Tier 1 GHG emission factors and standard emission factors for local air pollutants of energy technologies. However, the TED in LEAP accounts for only direct emissions and does not account for life cycle emissions. This feature allows LEAP to account for environmental implications of energy production and utilization. The model can also be used to perform cost−benefit analysis for different scenarios of energy demand and GHG emissions. As a bottom-up simulation model, LEAP also requires a certain level of data for its analysis. The data used in LEAP is inputted in a hierarchical or tree order, which consists of modules for key assumptions on socioeconomic and demographic variables, energy demand, statistical differences, transformation, stock change, resources, and nonenergy sector (Fig. 30.2) (Heaps, 2002). For a detailed description of LEAP modeling framework, see the user guide (SEI, 2008).

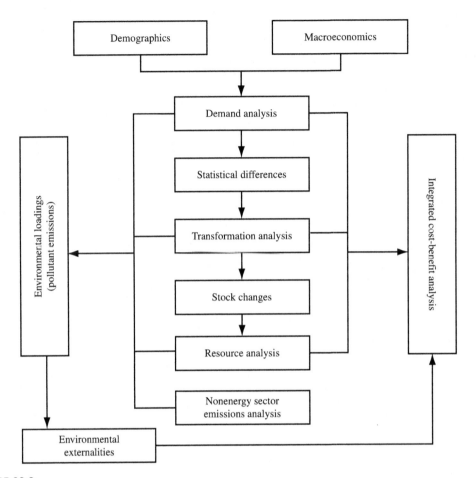

FIGURE 30.2

Calculation flow in LEAP (Heaps, 2002). *LEAP*, Long-range Energy Alternatives Planning.

In this chapter, we used the transport demand module of the Nigerian LEAP model to estimate the final energy demand and associated emissions of the Nigerian land transport system for the period 2015—40. The Nigerian LEAP model was developed by Emodi, Emodi, Murthy, and Emodi (2017) to study different scenarios for low-carbon transitions in Nigeria. The model is an economy-wide model that covers the processes of energy production, transformation, and consumption. The time period of the model is from 2010 to 2040. The main driving factors of the model are GDP and population growth. The supply side of the model contains the various steps involved in primary energy production (both conventional and nonconventional). The supply side also consists of secondary energy production (electricity and petroleum products). The demand side of the model consists of agriculture, commercial, industry, residential, and transport sectors. Our focus in this chapter is using the transport demand module of the model for energy projections. The transport module of the Nigerian LEAP model consists of only the land transport system. The module consists of six categories of vehicles: motorcycles, cars, light goods vehicles, heavy good vehicles, urban buses, and long-distance coaches. The vehicles were further subcategorized according to their fuel technologies. Fig. 30.3 shows the transport module of the Nigerian LEAP model.

For energy demand estimation, LEAP endogenously computes the final energy demand as the product of an activity level and a given end-use energy intensity (SEI, 2008). In general, LEAP calculates energy demand using four different methods: final energy demand, useful energy demand, stock, and transport analysis. The specific method to be applied depends on the level of data and expertise available, as well as the research question. Energy demand can be estimated in LEAP using Eqs. (30.1)–(30.4). For detailed description of the Nigerian LEAP model, see the work of Emodi et al. (2017).

$$\text{Final energy analysis}, E = \sum_{i=1}^{n} Q_I \times I_i \tag{30.1}$$

where E represents the energy demand, Q_I activity level, and I_I energy intensity.

$$\text{Useful energy analysis}, E = Q \times \left(\frac{u}{n}\right) \tag{30.2}$$

where u represents the useful energy intensity and n efficiency.

$$\text{Stock analysis}, E = S \times D \tag{30.3}$$

where S represents the Stock and D device intensity.

$$\text{Transport analysis}, E = S \times \frac{M}{FE} \tag{30.4}$$

where M represents the vehicles miles and FE fuel economy, GHG emissions are estimated in LEAP according to the following equation.

$$G = \sum E \times Ef \tag{30.5}$$

where G represents the total GHG emission, E energy demand of a fuel type, and Ef emission factor of the fuel.

30.2.2 SCENARIOS

Scenarios are self-consistent assumptions or story lines describing how an energy system might unfold over a modeling timeframe, under a given set of conditions (Zhang, Feng, & Chen, 2011). It is worthwhile to mention that scenarios do not attempt to predict the future, but they try to show the plausible changes that may be required if an energy system evolves via a particular pathway. In this study, we have established a BAU scenario as a benchmark for comparison and two other alternative scenarios, that is, the CM1 and CM2 scenarios. In the BAU scenario, we portray a baseline case in which there is no extra policy intervention to decarbonize the Nigerian transport sector, and thus the current status of fossil fuels in the sector remains up to 2040. In CM1 scenario, we depict a scenario in which there is little effort to decarbonize the Nigerian transport sector, while the CM2 scenario reflects a situation in which there is additional effort to decarbonize the sector. Our alternative scenarios are simply based on fuel substitution in the transport sector. As earlier noted

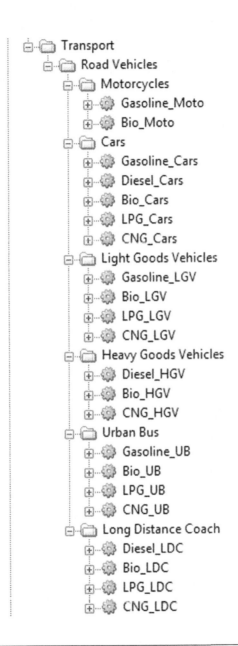

FIGURE 30.3

Tree structure of the transport module of Nigeria LEAP model. *LEAP*, Long-range Energy Alternatives Planning.

in Section 30.1, about 99% of the Nigerian vehicles fleet run on gasoline and diesel. While we acknowledge that there are other ways of decarbonizing the transport sector beyond fuel substitution, we have decided to focus on fuel substitution in this study because of the abundant cleaner alternative fuels in the country, which remains untapped. Here, we consider a shift from fossil gasoline and diesel to compressed natural gas (CNG) and biofuels.

Nigeria has four petroleum refineries with a cumulative capacity for crude distillation at around 10.7 million bbl/day (EIA, 2015). Despite this, Nigeria still imports petroleum products and this is due to the fact that the local demand for petroleum products is more than the supply from the local refineries. This situation occurs because of the low capacity utilization of the local refineries. Thus to reduce the country's reliance on imported oil products, it is only necessary for Nigerian government to increase the local refineries utilization capacities as well as diversify to low-carbon fuels that will also help to improve the country's energy security as well as reduce environmental pollution. Nigeria has around 182 trillion cubic feet of natural gas reserve potential which comprises 46.5% and 53.5% nonassociated and associated gas, respectively (ECN, 2014). At the global level, Nigeria is ranked the 25th largest producer of natural gas and number 7 in terms of gas reserves. Thus the availability of resources for using CNG in Nigeria is not a problem. Nigeria also has abundant biofuel feedstocks that can be utilized for making transport fuels. With respect to biofuels, Nigeria is blessed with abundant arable land that can be used to grow energy crops such as cassava, corn, sugarcane, and sweet sorghum (Ben-Iwo, Manovic, & Longhurst, 2016). In order to avoid competition with food and to improve the sustainability, they can be grown on degraded or marginal lands in the country. Given the huge potential for fuel substitution in the transport sector, our scenarios assume that the potentials of these alternative fuels will be exploited in the country. The detailed assumptions characterizing each of the scenarios are presented in Table 30.1.

30.2.3 COBENEFITS

In this study, the cobenefits of low-carbon development refer to the additional benefits beyond energy demand and GHG reductions resulting from a shift to cleaner transport fuels. It is important

Table 30.1 Share of Vehicle Types for All Scenarios by 2040 (%).

		Motorcycles	Cars	LGV	HGV	UB	LDC
BAU	Gasoline	100	98	100	–	100	–
	Diesel	–	2	–	100	–	100
	CNG	–	–	–	–	–	–
	Biofuel	–	–	–	–	–	–
CM1	Gasoline	–	50	60	–	60	–
	Diesel	–	–	–	70	–	70
	CNG	–	30	20	15	20	15
	Biofuel	10	15	10	10	10	10
CM2	Gasoline	–	25	30	–	30	–
	Diesel	–	–	–	50	–	50
	CNG	–	30	20	15	20	15
	Biofuel	25	25	30	25	35	25

CM1, *Countermeasure 1*; CM2, *countermeasure 2*; CNG, *compressed natural gas*; HGV, *heavy good vehicles*; LDC, *long-distance coaches*; LGV, *light goods vehicles*; UB, *urban buses*.

that researchers and policy analysts acknowledge the importance of cobenefits in low-carbon transition studies. The cobenefits are seen as the immediate benefits of climate mitigation as they support the realization of other sustainable development goals such as good health and sanitation. When cobenefits are well communicated, they can also serve as additional reasons for policymakers to inculcate low-carbon strategies in broader economic development plans. In this chapter, we examined the cobenefits of low-carbon transport from the perspective of air pollution abatement. We analyzed the mitigation of carbon monoxide (CO) and nitrogen oxides (NO_x) due to the alternative scenarios. These local air pollutants cause a lot of health problems such as headache, vomiting, dizziness, nausea, and even death. Thus the mitigation of these air pollutants is needed to avoid such dire health consequences.

30.3 NUMERICAL RESULTS AND ANALYSIS

30.3.1 FINAL ENERGY DEMAND

Fig. 30.4 illustrates the results of the final energy demand for each scenario. Driven by increase in population and the corresponding demand for mobility, the final energy demands of all scenarios are expected to increase steadily up to 2040 but with different CAGRs. In the BAU scenario, final energy demand is expected to grow at a CAGR of 3.0%, from 269 PJ in 2015 to around 563 PJ in 2040; this is around 110% increase within 25 years. However, due to gradual efficiency penetration in the alternative scenarios, the CAGR of final energy demand will be lower under CM1 scenario relative to the BAU scenario, that is, final energy demand by 2040 will be 462 PJ with a CAGR of 2.2%. This is equivalent to

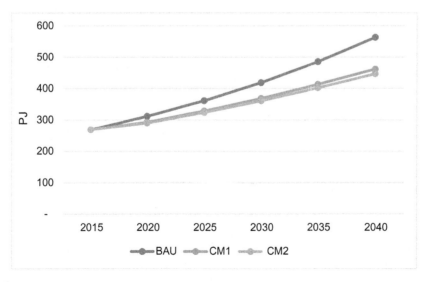

FIGURE 30.4

Final energy demand projections for all scenarios.

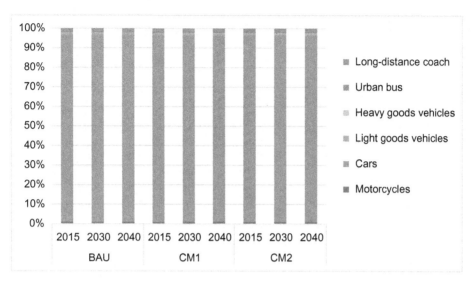

FIGURE 30.5

Share of final energy demand by vehicle categories.

an energy demand reduction of around 18% when compared with the BAU scenario. With additional decarbonization efforts in CM2 scenario, final energy demand is expected to increase at a CAGR of 2.0% to 446 PJ by 2040. This is equivalent to an energy demand reduction of around 21% when compared with the BAU scenario. Thus the current result suggests that a transition from fossil transport fuels to cleaner fuels could reduce the future energy demand of the Nigerian transport system. However, it remains to be seen how the alternative scenarios could be attained in the future. With respect to shares in final energy demand, our analysis (Fig. 30.5) indicates that cars contribute over 90% to final energy demand of the Nigerian transport sector. To advance our understanding of the results, we analyzed the per capita energy demand for all scenarios (Fig. 30.6). In the BAU scenario, per capita energy demand is expected to increase at a CAGR of 0.58%, from 1.47 GJ/person in 2015 to about 1.70 GJ/person in 2040. In the alternative scenarios, our analysis shows that per capita energy demand will fall by 18% and 21% in CM1 and CM2 scenarios, respectively, by 2040, when compared with the BAU scenario. Certainly, the CM2 scenario has the largest capacity for energy demand reduction.

30.3.2 GHG EMISSIONS

Fig. 30.7 illustrates the GHG emission trajectories under the three scenarios analyzed for the period 2015−40. Under the BAU scenario, GHG emissions are expected to grow from 19 $MtCO_2eq$ in 2015 to around 39 $MtCO_2eq$ in 2040 with a CAGR of 2.9%. It could be observed that the rate of GHG emissions is not always correlated exactly with the rate of energy demand as observed in Fig. 30.4. The slightly higher growth rate of energy demand (3.0%) is due to the efficiency levels of the CNG and biofuel vehicles. CNG and biofuel vehicles are not significantly more efficient

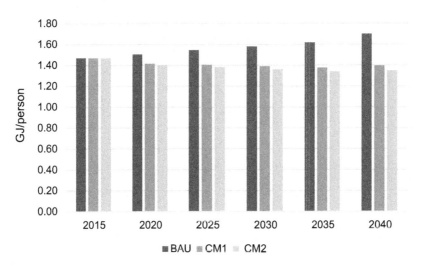

FIGURE 30.6

Per capita energy demand for all scenarios.

than gasoline and diesel vehicles. Under CM1 scenario, GHG emissions will increase to about 27 MtCO$_2$eq in 2040 with a significantly lower CAGR of 1.4%, which is also not in tandem with energy demand growth rate owing to the reason earlier outlined. In the CM2 scenario, due to the vigorous efforts paid toward removing fossil fuels from the Nigerian transport system, we observed that GHG emissions will rise to 24 MtCO$_2$eq in 2040 with a lower CAGR of 0.9%. In sum, our analysis suggests that GHG emissions by 2040 will drop by 31% and 38% in CM1 and CM2 scenarios, respectively, when compared with the BAU case. With respect to shares in total GHG emissions, our analysis (Fig. 30.8) indicates that cars contribute over 90% to total GHG emissions from the Nigerian transport sector. The per capita GHG emissions under the three scenarios are presented in Fig. 30.9. With respect to this measure, our results indicate that it will increase at a CAGR of 0.73%, from 0.1 tCO$_2$eq/person in 2015 to about 0.12 tCO$_2$eq/person in 2040. In the alternative scenarios, our analysis shows that per capita GHG emissions will reduce by 33% and 42% in CM1 and CM2 scenarios, respectively, by 2040 when compared with the BAU case. As anticipated, the CM2 scenario proves to be the case with the largest GHG mitigation potential.

30.3.3 AIR QUALITY

An important aspect of low-carbon development of the transport sector is the air quality benefit it delivers. Our analysis of different air pollutants suggests that the alternative scenarios will reduce the local air pollutants emissions from the transport sector. As can be seen in Fig. 30.10A, emissions of CO are expected to grow from 2074 kt in 2015 to about 4343 kt in 2040, that is, over 100% increase in 25 years. However, in the alternative scenarios, emissions of CO are expected to decline in 2040 by 36% and 41% in CM1 and CM2 scenarios, respectively. Similarly, in Fig. 30.10B, our analysis on NO$_x$ emissions shows that it will grow from 163 kt in 2015 to about

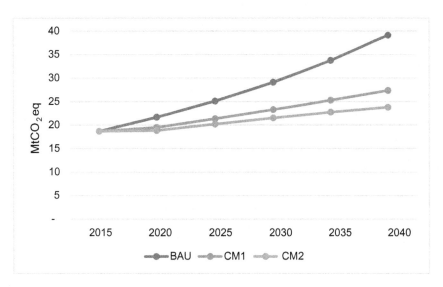

FIGURE 30.7

GHG emissions projections for all scenarios.

342 kt in 2040 (over 100% increase in 25 years). Predictably, NO_x emissions will fall in 2040 in CM1 and CM2 scenarios by around 27% and 37%, respectively. While our results are in terms of numbers, these actually reflect better quality of air for Nigerians. Therefore policymakers need to promote cleaner fuels as the health implications due to local air pollutants from the current transport system are severe.

30.4 CONCLUSION AND OPTIONS FOR THE FUTURE

In spite of the fact that most studies on low-carbon transport systems have been for the developed countries, in this chapter, we have made an attempt to examine different scenarios for low-carbon transitions in one of the least developed countries of the world. We have used the Nigerian LEAP model to understand the implications of different low-carbon policy pathways on the country's transport energy requirements and environmental emissions. We explored a BAU scenario in which there is no effort toward decarbonizing the transport sector and two countermeasure scenarios in which there are different levels of efforts toward decarbonizing the transport sector. We found that there is a significant scope to reduce the GHG emissions of the sector via fuel substitution alone. Our analysis reveals that the future energy and emissions trajectories of the Nigerian transport sector will depend on the many choices that Nigerians make today. Thus various stakeholders need to be properly informed about the opportunities and challenges facing Nigeria toward decarbonizing its transport system. Our analysis showed that cars are the hot spots for reducing energy demand and GHG emissions in the country. Beyond energy demand and GHG emissions reductions, our study indicated that the alternative low-carbon scenarios will also deliver better air quality that is

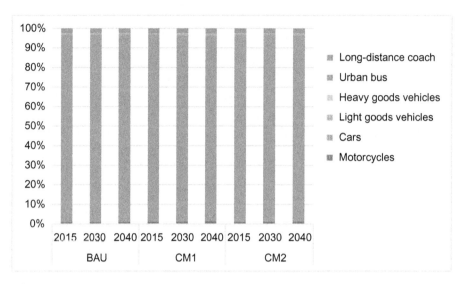

FIGURE 30.8

Share of total GHG emissions by vehicle categories.

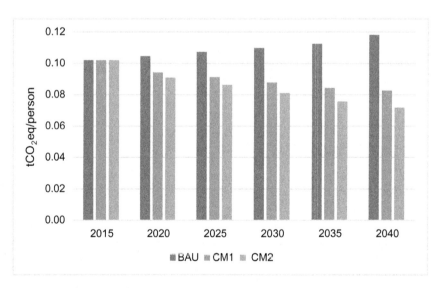

FIGURE 30.9

Per capita GHG emissions for all scenarios.

needed for the health and well-being of Nigerians. While the benefits of the low-carbon scenarios are significant, realizing these scenarios will be challenging in Nigeria given the current socioeconomic situation.

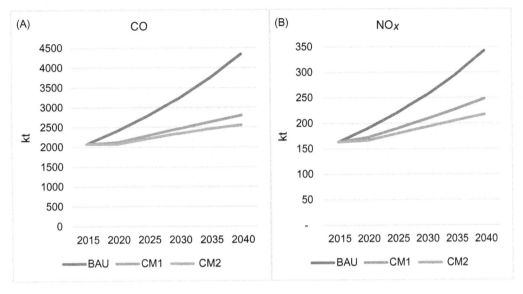

FIGURE 30.10

Air pollutants emissions projections for all scenarios.

The scenarios analyzed in this chapter mainly focused on fuel substitution (i.e., shifting from gasoline and diesel to CNG and biofuel vehicles). However, the central question to be addressed here is if the Nigerian government can provide the necessary infrastructures required to implement a fuel substitution program in the country's transport sector. There will be political, technological, and financial hurdles that the government needs to overcome before it can promote cleaner transport fuels. From the political front, government needs to first make gasoline and diesel fuels unattractive for motorists. This can be achieved in the country by completely removing the current fuel subsidy while redirecting these subsidies to CNG and biofuels. The technological challenge involves the availability of the technologies in the country. There will be a serious need to encourage the manufacture of flex-fuel vehicles in the country. In addition, the current gasoline and diesel stations need to be upgraded in order to provide fuel services for the new types of vehicle. From the financial dimension the aforementioned strategies will involve significant doses of financial investments in new infrastructures such as the refineries for biofuels production. The government alone cannot provide the needed investment. Thus there will be a serious need to mobilize private sector capital. In this case, what the government needs to do is to provide the enabling environment that will allow the private investors to flourish.

As could be seen from Figs. 30.4 and 30.7, none of the low-carbon scenario is enough to hold energy demand and emissions at the base year levels by 2040. In all the scenarios, energy demand and emissions are expected to be increasing compared to the base year values. This suggests that one policy pathway (i.e., fuel substitution) alone is not enough to decarbonize the Nigerian transport sector. Thus there is a need for triangulation of low-carbon policies for the Nigerian transport sector. Looking ahead, Nigeria could adopt other low-carbon strategies such as modal shift, improved vehicle efficiency, and road-pricing schemes. However, these strategies will also have several policy implications.

Reducing transport energy demand and emissions in Nigeria can be realized through the promotion of other modes of transport (passenger and freight) that are less energy intensive. However, this will require serious local and national urban planning policies and programs. In addition, implementation of this strategy will require significant diversification and improvements in the quality of public transport systems. For instance, to implement a modal shift from road-based to rail-based transport, the government needs to provide adequate number of trains and rail infrastructures for both intra- and intercity movements. In addition, there will be a need to create dedicated freight corridors in the country and the expansion of the existing rail network to cater to the new industrial, mining, and port areas in the country. The government needs to provide good spatial map for citizens and visitors of the city to support easy accessibility of different locations as well as to improve traffic safety and the entire quality of urban living. Push and pull measures as well as extensive car-sharing options are also other strategies that the Nigerian government could consider to facilitate modal shift. In terms of improving vehicle fuel efficiencies, the main option available today is to keep improving the conventional vehicle technologies, and this will involve sustained efforts. Improving the efficiency of vehicle engines will help to reduce fossil fuel consumption and the resultant emissions. To achieve this in Nigeria, policies that are aimed at promoting the adoption of state-of-the-art technologies and practices will be needed while respecting the national circumstances, for example, introducing taxes for vehicles that are older than 10 years for a designated income category.

For road-pricing scheme, this involves the billing of road users in a manner that actually reflects the cost of using the road (i.e., paying more when there is a heavy traffic and less when there is light traffic). This strategy seeks to reduce the number of vehicles plying the road and in turn reduces emissions, noise, and accidents. The strategy tends to encourage people to adopt more sustainable means of transportation especially during peak traffic hours. However, this strategy can only be well implemented if coupled with other measures such as dedicated pedestrian and cycling roads as well as efficient public transport systems. Road-pricing scheme has been implemented in many cities such as Singapore, Stockholm, London, and Milan, and it has led to a significant reduction in the total vehicle-km traveled and substantial reduction of CO_2 emissions in these cities. Therefore Nigeria can learn from the experience of these cities and start a road-pricing scheme policy in major cities such as Lagos, Port Harcourt, and Abuja.

Finally, we suggest going further on this study by modeling the impacts of these other low-carbon policy pathways that are suggested. In addition, future studies should go a step further to analyze the economic implications of these scenarios and should consider, including maritime and air transport as well. Beyond the aforementioned low-carbon policies, we also recognize that there are other strategies that could be explored to decarbonize the Nigerian transport and, thus, open the window for further research on this subject.

REFERENCES

Ahanchian, M., & Biona, J. B. M. (2014). Energy demand, emissions forecasts and mitigation strategies modeled over a medium-range horizon: The case of the land transportation sector in Metro Manila. *Energy Policy*, *66*, 615−629. Available from https://doi.org/10.1016/J.ENPOL.2013.11.026.

Anderson, J., Fergusson, M., & Valsecchi, C. (2007). *An overview of global greenhouse gas emissions and emissions reduction scenarios for the future (No. PE400.994)*.

Ben-Iwo, J., Manovic, V., & Longhurst, P. (2016). Biomass resources and biofuels potential for the production of transportation fuels in Nigeria. *Renewable and Sustainable Energy Reviews*, *63*, 172−192. Available from https://doi.org/10.1016/J.RSER.2016.05.050.

Chunark, P., Thepkhun, P., Promjiraprawat, K., Winyuchakrit, P., & Limmeechokchai, B. (2015). *Low carbon transportation in Thailand: CO_2 mitigation strategy in 2050*. SpringerPlus (Vol. 4). Springer International Publishing. <https://doi.org/10.1186/s40064-015-1388-6>.

Dhar, S., & Shukla, P. R. (2015). Low carbon scenarios for transport in India: Co-benefits analysis. *Energy Policy*, *81*, 186−198. Available from https://doi.org/10.1016/J.ENPOL.2014.11.026.

Dioha, M. O. (2018). Modelling the impact of Nigeria household energy policies on energy consumption and CO2 emissions. *Engineering Journal*, *22*(6), 1−19. Available from https://doi.org/10.4186/ej.2018.22.6.1.

Dioha, M. O. (2017). Energy system models for sub-Saharan African countries—A systematic review. *Journal of Sustainable Energy*, *8*(4), 159−168.

Dioha, M. O., & Emodi, N. V. (2019). Investigating the impacts of energy access scenarios in the Nigerian Household Sector by 2030. *Resources*, *8*(3), 127. Available from https://doi.org/10.3390/resources8030127.

Dioha, M. O., Emodi, N. V., & Dioha, E. C. (2019). Pathways for low carbon Nigeria in 2050 by using NECAL2050. *Renewable Energy Focus*, *29*, 63−77. Available from https://doi.org/10.1016/j.ref.2019.02.004.

ECN. (2014). *Energy implications of vision 20: 2020 and beyond*. From <www.energy.gov.ng/index.php?option = com_docman&task = doc_download&gid = 112&Itemid = 49> Retrieved 15.10.17.

EIA. (2015). *Nigeria − International energy data and analysis 2015*. U.S. Energy Information Administration. From <http://www.eia.gov/beta/international/analysis.cfm?iso = NGA> Retrieved 18.04.17.

Emodi, N. V., Emodi, C. C., Murthy, G. P., & Emodi, A. S. A. (2017). Energy policy for low carbon development in Nigeria: A LEAP model application. *Renewable and Sustainable Energy Reviews*, *68*, 247−261. Available from https://doi.org/10.1016/j.rser.2016.09.118.

Gujba, H., Mulugetta, Y., & Azapagic, A. (2013). Passenger transport in Nigeria: Environmental and economic analysis with policy recommendations. *Energy Policy*, *55*, 353−361. Available from https://doi.org/10.1016/J.ENPOL.2012.12.017.

Gül, T., Kypreos, S., Turton, H., & Barreto, L. (2009). An energy-economic scenario analysis of alternative fuels for personal transport using the Global multi-regional MARKAL model (GMM). *Energy*, *34*(10), 1423−1437. Available from https://doi.org/10.1016/J.ENERGY.2009.04.010.

He, L.-Y., & Chen, Y. (2013). Thou shalt drive electric and hybrid vehicles: Scenario analysis on energy saving and emission mitigation for road transportation sector in China. *Transport Policy*, *25*, 30−40. Available from https://doi.org/10.1016/J.TRANPOL.2012.11.006.

He, L.-Y., & Qiu, L.-Y. (2016). Transport demand, harmful emissions, environment and health co-benefits in China. *Energy Policy*, *97*, 267−275. Available from https://doi.org/10.1016/J.ENPOL.2016.07.037.

Heaps, C. (2002). *Integrated energy environment modeling and LEAP*. Stockholm Environment Institute.

Herbst, A., Toro, F., Reitze, F., & Jochem, E. (2012). Introduction to energy systems modelling. *Swiss Journal of Economics and Statistics*, *148*(2), 112.

Hong, S., Chung, Y., Kim, J., & Chun, D. (2016). Analysis on the level of contribution to the national greenhouse gas reduction target in Korean transportation sector using LEAP model. *Renewable and Sustainable Energy Reviews*, *60*, 549−559. Available from https://doi.org/10.1016/j.rser.2015.12.164.

IEA. (2016). *CO2 emissions from fuel combustion. OECD/IEA*. Paris, France: OECD Publishing. Available from https://doi.org/10.1787/co2_fuel-2016-en.

Intergovernmental Panel on Climate Change (IPCC). (2014). *Synthesis report: Contribution of working groups I, II and III to the 5th assessment report of the Intergovernmental Panel on Climate Change (Core Writing Team, R.K. Pachauri and L.A. Meyer)*. Intergovernmental Panel on Climate Change (IPCC): Geneva, Switzerland. Retrieved from <ipcc.ch/pdf/assessment-report/ar5/syr/AR5_SYR_FINAL_All_Topics.pdf>.

Intergovernmental Panel on Climate Change (IPCC). (2018). T. W. V. Masson-Delmotte, P. Zhai, H. O. Pörtner, D. Roberts, J. Skea, P. R. Shukla, A. Pirani, W. Moufouma-Okia, C. Péan, R. Pidcock, S. Connors, J. B. R. Matthews, Y. Chen, X. Zhou, M. I. Gomis, E. Lonnoy, T. Maycock, & M. Tignor (Eds.), *Global warming of 1.5°C: An IPCC special report on the impacts of global warming of 1.5°C above pre-industrial levels and related global greenhouse gas emission pathways, in the context of strengthening the global response to the threat of climate change*.

Limanond, T., Jomnonkwao, S., & Srikaew, A. (2011). Projection of future transport energy demand of Thailand. *Energy Policy, 39*(5), 2754–2763. Available from https://doi.org/10.1016/J.ENPOL.2011.02.045.

Mahumane, G., & Mulder, P. (2019). Expanding versus greening? Long-term energy and emission transitions in Mozambique. *Energy Policy, 126*, 145–156. Available from https://doi.org/10.1016/J.ENPOL.2018.10.056.

Marchal, V., Dellink, R., van Vuuren, D., Clapp, C., Château, J., Lanzi, E., & van Vliet, J. (2011). Chapter 3: Climate change. In: *OECD environmental outlook to 2050*. OECD.

Martínez-jaramillo, J. E., Arango-aramburo, S., Álvarez-uribe, K. C., & Jaramillo-álvarez, P. (2017). Assessing the impacts of transport policies through energy system simulation: The case of the Medellin Metropolitan Area, Colombia. *Energy Policy, 101*, 101–108. Available from https://doi.org/10.1016/j.enpol.2016.11.026.

Nealer, R., Matthews, H. S., & Hendrickson, C. (2012). Assessing the energy and greenhouse gas emissions mitigation effectiveness of potential US modal freight policies. *Transportation Research Part A: Policy and Practice, 46*(3), 588–601. Available from https://doi.org/10.1016/j.tra.2011.11.010.

Peng, B., Du, H., Ma, S., Fan, Y., & Broadstock, D. C. (2015). Urban passenger transport energy saving and emission reduction potential: A case study for Tianjin, China. *Energy Conversion and Management, 102*, 4–16. Available from https://doi.org/10.1016/J.ENCONMAN.2015.01.017.

Peng, T., Ou, X., Yuan, Z., Yan, X., & Zhang, X. (2018). Development and application of China provincial road transport energy demand and GHG emissions analysis model. *Applied Energy, 222*, 313–328. Available from https://doi.org/10.1016/J.APENERGY.2018.03.139.

Prasad, R. D., & Raturi, A. (2018). Low-carbon measures for Fiji's land transport energy system. *Utilities Policy, 54*, 132–147. Available from https://doi.org/10.1016/J.JUP.2018.08.001.

SEI. (2008). *Stockholm Environment Institute. User guide, LEAP: Long Range energy alternative planning system*. Boston: Stockholm Environment Institute.

Selvakkumaran, S., & Limmeechokchai, B. (2015). Low carbon society scenario analysis of transport sector of an emerging economy—The AIM/enduse modelling approach. *Energy Policy, 81*, 199–214. Available from https://doi.org/10.1016/J.ENPOL.2014.10.005.

Seo, J., Park, J., Oh, Y., & Park, S. (2016). Estimation of total transport CO_2 emissions generated by medium- and heavy-duty vehicles (MHDVs) in a sector of Korea. *Energies, 9*(8). Available from https://doi.org/10.3390/en9080638.

Shabbir, R., & Ahmad, S. S. (2010). Monitoring urban transport air pollution and energy demand in Rawalpindi and Islamabad using LEAP model. *Energy, 35*(5), 2323–2332. Available from https://doi.org/10.1016/J.ENERGY.2010.02.025.

Tsita, K. G., & Pilavachi, P. A. (2017). Decarbonizing the Greek road transport sector using alternative technologies and fuels. *Thermal Science and Engineering Progress, 1*, 15–24. Available from https://doi.org/10.1016/J.TSEP.2017.02.003.

Valderrama, E. M., Monroy, C. Á. I., & Valencia, B. E. (2019). Challenges in greenhouse gas mitigation in developing countries: A case study of the Colombian transport sector. *Energy Policy*, *124*, 111−122. Available from https://doi.org/10.1016/J.ENPOL.2018.09.039.

Yan, X., & Crookes, R. J. (2009). Reduction potentials of energy demand and GHG emissions in China's road transport sector. *Energy Policy*, *37*(2), 658−668. Available from https://doi.org/10.1016/J.ENPOL.2008.10.008.

Zhang, H., Chen, W., & Huang, W. (2016). TIMES modelling of transport sector in China and USA: Comparisons from a decarbonization perspective. *Applied Energy*, *162*, 1505−1514. Available from https://doi.org/10.1016/J.APENERGY.2015.08.124.

Zhang, L., Feng, Y., & Chen, B. (2011). Alternative scenarios for the development of a low-carbon city: A case study of Beijing, China. *Energies*, *4*(12), 2295−2310. Available from https://doi.org/10.3390/en4122295.

Zhang, R., Fujimori, S., Dai, H., & Hanaoka, T. (2018). Contribution of the transport sector to climate change mitigation: Insights from a global passenger transport model coupled with a computable general equilibrium model. *Applied Energy*, *211*, 76−88. Available from https://doi.org/10.1016/J.APENERGY.2017.10.103.

DO NATURAL DISASTERS REDUCE FOREIGN DIRECT INVESTMENT IN SUB-SAHARAN AFRICA?

31

Ben Katoka

Hankuk University of Foreign Studies, Yongin-si, Republic of Korea

31.1 INTRODUCTION

Foreign direct investment (FDI) decisions—decisions related to foreign entry and expansion by multinational corporations (MNCs)—are determined by an evaluation of expected risks and returns in existing or potential host countries (White and Fan, 2006). Risks range from political instability and policy uncertainty to terrorist attacks and natural disasters (Perrow, 2011), which may negatively affect the activities of companies in the short- and long-run. Natural disasters, in particular, are directly and indirectly associated with substantial costs for businesses—for example, damages to fixed assets and capital, damages to raw materials, and interruptions in the production of goods and services (Cole, Elliott, Okubo, & Strobl, 2015). Unlike other types of risks, natural disasters are more arduous to control. For these reasons, past or present experience with disasters may deter a firm's entry and expansion in a foreign market (Oetzel & Oh, 2013), thereby reducing FDI.[1]

In Sub-Saharan Africa (SSA),[2] the past five decades have witnessed a dramatic growth in natural disasters such as floods, landslides, and drought (Voigt et al., 2016). As will be discussed in greater detail later, the region has experienced more than 2000 natural disasters, with more than half occurring in the past two decades (World Bank, 2019). Demographic distribution, climatic conditions, and other geographical conditions make some SSA countries more susceptible to major natural disasters (Di Baldassarre et al., 2010). For example, while disaster risks increase with urban growth (Fraser, Leck, Parnell, & Pelling, 2017), countries located in coastal areas of Eastern and Southern Africa are more prone to flooding due to storm surge. Similarly, countries located in regions such as Southern Africa, the Sahel, and the Horn of Africa are increasingly exposed to drought (Shiferaw et al., 2014).

In this context, this chapter addresses the question of whether experiences with natural disasters reduce the inflow of FDI into SSA. Research in this area is vital for three reasons. First, the rising

---

[1]White and Fan (2006, p. 21) note that portfolio analysis "rests on the availability of detailed information about past behaviour," notably regarding the level of risk.

[2]Here, SSA is defined as the region comprising 48 countries situated south of the Saharan desert [United Nations Environment Programme (UNEP), 2002].

Economic Effects of Natural Disasters. DOI: https://doi.org/10.1016/B978-0-12-817465-4.00031-5

occurrences of natural disasters can have and do have severe economic consequences in the region. Second, for many of SSA economies FDI is increasingly becoming the largest source of international capital. Finally, although FDI to SSA are highly influenced by a whole range of economic, policy, political, and geographical factors (Asiedu, Jin, & Kanyama, 2015; Katoka & Kwon, 2018; Onyewu & Shrestha, 2004), no existing studies seem to have investigated the connection between FDI and the natural disasters that struck the region.[3]

There is a large number of studies that investigate the economic impact of natural disasters (Loayza, Olaberria, Rigolini, & Christiaensen, 2012; Skidmore & Toya, 2002; Strobl, 2012; Toya & Skidmore, 2007). Other studies have looked at the extent to which the occurrences of natural disasters can affect MNCs' decisions related to their interest to enter a foreign market and whether they should expand their activities in existing host countries (Hosono et al., 2016; Oh & Oetzel, 2011). Some studies have examined the response of various international financial flows—for example, foreign aid, remittances, equity, and FDI—to disasters (Becerra, Cavallo, & Noy, 2014; David, 2010; Naudé & Bezuidenhout, 2014; Yang, 2008). Empirical studies that focus on the nexus between natural disasters and FDI are more limited. A few exceptions include Anuchitworawong and Thampanishvong (2015) and Escaleras and Register (2011).[4] However, these two studies are respectively based on a single country case—Thailand—and include both developed and developing countries.

The current study focuses on SSA countries. Natural disasters in this study are restricted to four disaster types occurrences—floods, drought, storms, and epidemics. The focus on these four disaster types is dictated by the experience of SSA countries indicating that the majority of disasters consist particularly of floods, droughts, and tropical cyclones—which account for more than 80% in disaster-related economic losses—alongside epidemics (van Niekerk & Nemakonde, 2017; World Bank, 2016).[5]

The remainder of the paper is structured as follows. Section 31.2 develops the theoretical framework to understand the connection between natural disasters and the inflow of FDI. Section 31.3 provides a brief overview of the trends in natural disasters and FDI inflows in SSA. Section 31.4 describes the data and variables to examine the links between natural disasters and FDI, as well as presents the estimation model. Section 31.5 presents the estimation results and discusses their implications, while Section 31.6 is the conclusion.

31.2 UNDERSTANDING THE NATURAL DISASTERS AND FOREIGN DIRECT INVESTMENT NEXUS

The literature has established the destructive effects that natural disasters exhibit in the short and long-run.[6] Disasters can affect gross domestic product (GDP), trade, productivity, poverty reduction, household income and saving, the activities of firms, as well as the international capital

---

[3]See Olatunji and Shahid (2015) for a review of the evidence on the determinants of FDI in SSA countries.

[4]Based on publicly available studies that focus on FDI and natural disaster as the dependent variable and main dependent variable, respectively.

[5]Flooding can lead to landslides—particularly due to land degradation—and epidemics (Gemeda & Sima, 2015). Similarly, drought reduces the availability of potable water and, thus, can increase the likelihood of diseases [United Nations International Strategy for Disaster Reduction (UNISDR), 2014].

[6]For surveys of the literature, see Cavallo et al. (2011), van Bergeijk and Lazzaroni (2015), and Noy and duPont (2016).

flows—such as remittances, foreign aid, equity, and FDI. This section starts by succinctly discussing this literature and then builds on this discussion to elaborate on the ways in which FDI can respond to natural disasters.

31.2.1 NATURAL DISASTERS AND ITS ECONOMIC EFFECTS

Existing studies have placed considerable emphasis on macroeconomic conditions in the aftermath of major natural disasters. For example, disasters commonly damage roads, railways, and container terminals, thereby raising trade costs and reducing trade flows (Osberghaus, 2019). Countries or regions affected by natural catastrophes, especially large earthquakes and meteorological disasters, can experience a severe decline in economic growth (Felbermayr & Gröschl, 2014). Natural disasters can reduce economic growth by discouraging new business creation, particularly entrepreneurship start-up activity in the short run (Boudreaux, Escaleras, & Skidmore, 2019).

Other studies have shown that destructions resulting from disasters can yield favorable economic conditions in the post−disaster period—a process often referred to as "creative destruction."[7] For example, a cross-country study of 89 countries established that countries that were hit by disasters experienced faster economic growth (Skidmore & Toya, 2002). This is because disasters can heighten the return to human capital relative to investment through the adoption of newer and more productive technologies (Crespo, Hlouskova, & Obersteiner, 2008). Also, large earthquakes and cyclones are associated with increased wages in the agricultural sector (Kirchberger, 2017) and reduced household income inequality (Keerthiratne & Tol, 2018a).[8]

However, it should be noted that the consequences of natural disasters may vary depending on disaster types and intensity, economic sectors, and the level of development of the affected country or region (Loayza et al., 2012).

31.2.2 NATURAL DISASTERS AND MNCs

The nature of a particular disaster might determine whether an MNC exits or expands in a host country in the aftermath of a natural catastrophe. For example, Oh and Oetzel's study (2011) of foreign subsidiaries of 71 large European MNCs between 2001 and 2006 found that firms with greater levels of experience operating in countries with a specific type of hazard, including risks related to natural disasters and others—such as terrorist attacks and technological disasters—are more likely to enter and expand into other countries experiencing the same type of risk.

In contrast, Benson and Clay (2004, p. 23) point out that "all major types of disaster can severely damage capital assets and other resources, thereby reducing the productivity of undamaged capital and labor." Such adverse effects, in turn, can disrupt longer-term investment plans for both physical and human capital. Examining how firms responded to the earthquake which occurred in the city of Kobe in 1995, Tanaka (2015) finds that the plants that survived in Kobe's most devastated areas faced severe adverse impacts from the catastrophe, particularly on employment and

---

[7]Creative destruction—coined by economist Joseph Schumpeter in 1943—refers to process whereby natural disasters create an opportunity to update the capital stock and promote capital growth (Leiter, Oberhofer, & Raschky, 2009).

[8]Primarily because natural disasters lead to a shift of the labor force from the labor force from the agricultural sector into the construction sector, thereby raising the marginal product of labor in agriculture.

value-added growth, during the subsequent 3 years. Also, Hosono et al. (2016) find that natural disasters shrink the lending capacity of banks located in affected areas, thereby reducing investment of firms situated both inside and outside of the affected zones. These findings are in line with Klomp (2016), whose analysis of 160 countries between 1997 and 2010 found that large-scale natural disasters adversely affect the solvency of commercial banks.

These disaster impacts on firms led White and Fan (2006, p. 3) to note that "the disaster events could initiate or trigger changes in attitudes or perception of disaster risks by raising the level of risk aversion and therefore the likelihood of an avoidance response." In the same vein, Kato and Okubo (2018) point out that MNCs invest large amounts of capital in foreign markets, which in turn makes them more sensitive to adverse shocks such as those originating from natural disasters. Earlier, Bernard and Jensen (2007) showed that MNCs are more likely to exit a host country than local firms, in response to negative shocks. For these reasons, an MNC might find it profitable to exit a host country struck by a natural disaster as such an event could damage its local upstream suppliers or negatively affect customers.[9] For example, Haraguchi and Lall (2015) show that approximately 15% of firms decided to shut down their factories in Thailand's aftermath of the flood event in 2011.

That said, unless a firm is better equipped to deal with future catastrophes, it is unlikely that it would engage in the settlement or economic operations in environments with high-risk for natural disasters. In their study of 106 European firms operating across 109 countries between 2001 and 2007, Oetzel and Oh (2013) found that experience in high-impact disasters is only strongly related to expansion but not new business entry. In other words, though MNCs can identify business opportunities after a disaster has occurred in countries where they currently operate, they refrain from entering new countries experiencing significant natural hazards. Therefore Oetzel and Oh (2013, p. 736) concluded, "the more dramatic and severe the disaster, the more likely it is to be a deterrent to future investment."

Another way firms can minimize risks is by building resilience (Neumayer, Plümper, & Barthel, 2014). For instance, MNCs that encounter severe high-impact disasters may have a transformative experience that translates into the development of experientially based capabilities for managing future disasters (Lampel, Shamsie, & Shapira, 2009). At the same time, whether a firm decides to continue to operate in a market following a natural disaster might depend on what McKnight and Linnenluecke (2016) refer to as "business continuity management" and "corporate philanthropy." The former refers to whether a natural disaster does not significantly affect a firm's capability to deliver products and service.[10] Corporate philanthropy, on the other hand, refers to when firms offer donations—such as cash and in-kind materials to the community affected by the disaster—in an effort to contribute to disaster relief, with the aim of enhancing their post−disaster reputational and financial benefits (Muller & Kraussl, 2011).[11] As a result, MNCs might choose to continue operating in a region regardless of whether it has been hit by a disaster.

---

[9]Disasters could lead to human losses and reduced household income, thereby negatively affecting existing or potential customers for MNEs.

[10]Such capability essentially depends on the firm's resilience, that is, its ability to identify and mitigate risks.

[11]As Godfrey (2005, p. 779) put it, "If firms suffer losses in firm value during times of crisis, managers may expect donations to generate the reputational capital." Donations therefore are believed to "help [firms] recover shareholder value lost during the disaster" (Muller & Kraussl, 2011, p. 913).

The discussion, so far, indicates that there is no conclusive answer to the question of whether natural disasters reduce MNCs' investment. On the one hand, risks and potential losses associated with disasters can deter investment, while on the other, factors such as corporate philanthropy and post—disaster economic opportunities can induce MNCs to remain and even expand their operations in existing host countries that have been struck by a natural catastrophe. MNCs can eventually contribute to post—disaster reconstruction.

31.2.3 NATURAL DISASTERS AND INTERNATIONAL CAPITAL FLOWS

The literature has also paid attention to the response of various types of international financial flows to natural disasters. For example, Becerra et al. (2014) show that the severity of a disaster—measured in terms of damage—and the economic status of the affected country or region determine the post—disaster surge in foreign aid. Specifically, a low-income country sharply hit by a disaster is likely to receive more foreign assistance relative to its counterpart with a better economic status—for example, upper-middle-income—or that experiences a lower-intensity natural hazard. Keerthiratne and Tol (2018b) observe that the aftermath of natural catastrophes in poor countries tends to be characterized by increased and more diversified types of foreign aid flows, as well as a diversification of the number of donors. In the same way, the flow of remittances toward low- and middle-income countries increases drastically after a natural disaster (Bettin & Zazzaro, 2018).

David (2010) notes that the response to natural disasters varies across different types of capital flows. He finds that the magnitude of remittances and foreign aid received by affected countries increase after a disaster. He also finds that other private capital flows—such as bank lending, equity, and FDI—decrease in response to natural catastrophes. Yang (2008) notes that disasters lead to declines in expected rates of return or increases in risk perceptions, which in turn reduce the attractiveness of private asset sales—such as portfolio investment—and FDI.

A limited number of studies have focused on the nexus between natural disasters and FDI. One main issue investigated has been whether large-scale disasters reduce FDI inflows. Escaleras and Register (2011) examine the impact of various types of natural disasters on FDI in 74 countries during the period 1997—2007. They found that although major natural disasters reduce FDI inflows, not all types of natural disasters affect FDI with the same magnitude. Specifically, they found that windstorms and volcanoes have the strongest negative effects on FDI.

Anuchitworawong and Thampanishvong (2015) study the connection between FDI inflow and the natural disaster that struck Thailand between 1970 and 2012. They observe that a higher degree of severity associated with natural disaster, including the number of people affected by disaster and frequency of occurrence and the amount of financial damages, significantly reduces FDI inflows.

Overall, three key lessons can be drawn from the above discussion. First, damages from natural disasters vary greatly across countries. Not only are some countries more likely to experience natural disasters, the effect on FDI may also vary depending on such factors as the level of economic development, population size, the share of the urban population, and the size of the affected country, among others.

Accordingly, the analysis in this chapter differs in two ways from the existing studies investigating the link between natural disasters and FDI. First, it focuses on the SSA region during the period between 1970 and 2018. In contrast, studies by Escaleras and Register (2011) and Anuchitworawong and Thampanishvong (2015) are based on countries from many regions and a single country case, respectively. Second, the current analysis takes into account the possibility that the effects of natural disasters on FDI are conditional on other variables.

31.3 NATURAL DISASTERS AND FOREIGN DIRECT INVESTMENT TRENDS IN SUB-SAHARAN AFRICA

Almost all countries in SSA are susceptible to some types of natural hazards and the associated risks of disasters. In particular, disasters linked to droughts, storms, floods, and epidemics, are the most common in the region (van Niekerk & Nemakonde, 2017). This section provides an overview of natural disasters across SSA, describing their types and frequency, as well as how FDI inflows in the region have evolved since 1970.

Data for natural disasters are obtained from the Emergency Events Database (EM-DAT) maintained by the Center for Research on the Epidemiology of Disasters (CRED) at the Catholic University of Louvain, Belgium (http://www.emdat.be/). The EM-DAT includes detailed information on different types of natural disasters, including drought, epidemic, flood, landslide, wildfire, earthquake, storm, extreme temperature, mass movement, and volcanic activity. The analysis focuses on four of the most prevalent natural disasters in SSA, namely drought, flood, storms, and epidemics (World Bank, 2016).

Based on these data, it can be observed that the frequency of natural disasters reported in SSA has markedly increased since the 1970s (see Table 31.1). SSA experienced 2039 disasters associated with drought, flood, storms, and epidemics. There were 464 disaster events between 2010 and 2017, compared to just 87 in the 1970s. Floods were the most frequent disasters—790 occurrences—followed respectively by epidemics (785), droughts (246), and storms (218), during the same period.

It has been argued that the impact of natural disasters in SSA tends to be amplified by a number of structural factors (Adelekan et al., 2015; World Bank, 2016). These factors include the rapid and unplanned growth of cities, the lack of solid and good quality infrastructure and services, concentration of population and assets in zones vulnerable to natural hazards, as well as the environmental degradation due to deforestation, among many others. These structural factors, added to high levels of poverty, place SSA countries among the most vulnerable to natural disasters in the world.

Table 31.1 Number of Reported Disasters in Sub-Saharan Africa, 1970–2017.

Period	Total	Drought	Flood	Storms	Epidemics
1970–79	87	24	28	16	19
1980–89	213	59	54	31	69
1990–99	400	50	128	33	189
2000–09	875	65	347	85	378
2010–17	464	48	233	53	130
Total	2039	246	790	218	785

Author's calculations based on data from EM-DAT (2019).

In addition, geographical conditions can determine the frequency and levels of vulnerability to different disaster types. On the one hand, countries with coastal exposure are more likely to suffer losses from storms in any given year, whereas the frequency of epidemics tends to be higher in countries located on or near the equator.

Table 31.2 reports the frequency with which individual SSA countries were affected by drought, flood, cyclone, and epidemics, between 1970 and 2017. The highest number of reported natural disasters was registered in the DR Congo, with a total of 116 natural disasters, including 72 epidemics, 5 storms, 37 floods, and 2 droughts. Nigeria (111) and Kenya (100) are respectively the second and third most hit by natural disasters. Furthermore, 12 other countries reported an average of 1–2.4 disasters per year during the same period. These countries include Mozambique, Ethiopia, Tanzania, Sudan, South Africa, Niger, Madagascar, Uganda, Angola, Malawi, Burkina Faso, and Chad. In comparison, countries such as Equatorial Guinea, Cabo Verde, Gabon, Comoros, Eswatini, Botswana, Guinea-Bissau, Lesotho, Mauritius, Liberia, and Gambia experienced less frequent natural catastrophes during the observation period.

In SSA, the disaster impact on economic growth has been shown to largely reflect the damage to infrastructure and human capital [International Monetary Fund (IMF), 2016]. Also, damage and losses from natural disasters can reduce the FDI attractiveness of the affected countries. The question that arises therefore is whether countries that report more natural disasters receive less FDI inflows compared to countries with less frequent disaster events.

Table 31.3 compares the differences in FDI inflows between two country categories, based on the frequency of natural disasters in the period from 1970 to 2018. The data of the observation period are disaggregated into 5 10-year periods (1970–79, 1980–89, 1990–99, 2000–09, and 2010–18) to show differences between the two country categories. For each period, the first group consists of 10 countries that experienced the highest frequency of disaster events, whereas the second group includes 10 countries with less frequent natural catastrophes. Data about the FDI inflow are taken from the United Nations Conference on Trade and Development (UNCTAD) database.[12]

The data presented in Table 31.3 indicate substantial variations across the two country categories. In particular, countries that exhibit a high frequency of natural disasters receive lower overall FDI inflows relative to the low-frequency countries. Average FDI inflows as a percentage of GDP for countries with a high frequency of natural disasters were consistently lower during the observation period. This is suggestive evidence that natural disasters have negative effects on FDI inflows in SSA countries. It is therefore hypothesized that FDI to SSA countries decrease in response to natural disasters. In other words, countries where natural disasters are more frequent receive less FDI relative to countries where the frequency of natural disasters is lower.

Accordingly, one would intuitively expect that FDI responsiveness to natural disasters would be stronger in countries that have had the highest frequency of disaster events since 1970—specifically, DR Congo, Nigeria, Kenya, Mozambique, and Ethiopia. Table 31.4 presents the number of natural disasters and FDI inflows as a percentage of GDP in selected periods—1970–74, 1975–79, 1980–84, 1985–89, 1990–94, 1995–99, 2000–04, 2005–09, and 2010–14 for those five countries. The data reported in Table 31.4 suggest that the frequency of natural disasters varies greatly over time within and across countries. However, there is no clear indication that increases in the number of natural disasters reduce the inflows of FDI within individual countries.

---

[12]https://unctadstat.unctad.org/wds/ReportFolders/reportFolders.aspx?sCS_ChosenLang = en

Table 31.2 Sub-Saharan Africa Countries: Frequency of Natural Disasters by Type, 1970−2017.

Country	Drought	Flood	Storms	Epidemics	Total Natural Disasters
DR Congo	2	37	5	72	116
Nigeria	1	46	6	58	111
Kenya	14	50	1	35	100
Mozambique	14	35	22	28	99
Ethiopia	14	53	0	25	92
Tanzania	9	38	5	32	84
Sudan	9	34	2	35	80
South Africa	9	34	28	7	78
Niger	10	22	2	41	75
Madagascar	8	7	52	7	74
Uganda	8	20	4	36	68
Angola	8	37	0	21	66
Malawi	7	34	4	13	58
Burkina Faso	8	18	1	23	50
Chad	8	16	3	22	49
Burundi	6	22	5	15	48
Mali	8	23	0	17	48
Benin	2	19	1	25	47
Zimbabwe	7	11	5	22	45
Ghana	3	19	1	22	45
Cameroon	5	16	0	23	44
Zambia	5	18	0	19	42
Senegal	6	18	3	11	38
Central African Republic	1	15	10	11	37
Mauritania	9	16	3	6	34
Guinea	2	13	1	16	32
Rwanda	6	13	1	12	32
Sierra Leone	0	9	3	17	29
Namibia	8	14	0	7	29
Togo	3	12	0	12	27
Côte d'Ivoire	1	11	0	14	26
Congo	1	9	0	15	25
Gambia	6	9	4	5	24
Liberia	1	6	2	12	21
Mauritius	1	0	17	2	20
Djibouti	8	6	1	5	20
Lesotho	5	5	7	3	20
Guinea-Bissau	3	4	1	10	18
Botswana	4	10	1	3	18

Table 31.2 Sub-Saharan Africa Countries: Frequency of Natural Disasters by Type, 1970–2017. *Continued*

Country	Drought	Flood	Storms	Epidemics	Total Natural Disasters
Comoros	1	2	6	6	15
Eswatini	6	3	3	3	15
Gabon	0	2	3	8	13
Cabo Verde	5	1	3	2	11
Equatorial Guinea	0	0	0	1	1

Author's calculations based on data from EM-DAT (2019).

Table 31.3 Natural Disasters and Foreign Direct Investment Inflows (% GDP), Average 1970–2018.

Period	High-Frequency Countries	Low-Frequency Countries
1970–79	3.93	24.62
1980–89	1.74	20.74
1990–99	9.24	34.14
2000–09	35.71	44.09
2010–18	53.41	55.96

Note: For the period 1970–79 the high-frequency countries include Mozambique, Madagascar, Tanzania, Ethiopia, Rwanda, Mauritius, Kenya, Benin, Ghana, and Senegal. The low-frequency countries include Angola, Central African Republic, Comoros, Congo, Equatorial Guinea, Gabon, Guinea, Liberia, Namibia, and Eswatini. For the period 1980–89 the high-frequency countries include Ethiopia, Nigeria, South Africa, Madagascar, Tanzania, Mozambique, Sudan, Mali, Burkina Faso, and Mauritius. The low-frequency countries include Liberia, Central African Republic, Gabon, Cote d'Ivoire, Namibia, Zimbabwe, Burundi, Gambia, Congo, and Equatorial Guinea. For the period 1990–99 the high-frequency countries include Tanzania, Sudan, Nigeria, Ethiopia, Kenya, DR Congo, Mozambique, South Africa, Cameroon, and Niger. The low-frequency countries include Guinea, Gambia, Cote d'Ivoire, Namibia, Cabo Verde, Swaziland, Gabon, Botswana, Comoros, and Equatorial. For the period 2000–09 the high-frequency countries include DR Congo, Nigeria, Kenya, Mozambique, Ethiopia, Uganda, Angola, Madagascar, South Africa, and Sudan. The low-frequency countries include Guinea-Bissau, Djibouti, Swaziland, Botswana, Comoros, Lesotho, Gabon, Mauritius, Cabo Verde, and Equatorial Guinea. Finally, for the entire period after 2010, the high-frequency countries include DR Congo, Kenya, Nigeria, Niger, Mozambique, Angola, Ghana, Madagascar, South Africa, and Malawi. And the low-frequency countries are Lesotho, Botswana, Liberia, Comoros, Swaziland, Djibouti, Mauritius, Cabo Verde, Guinea-Bissau, and Equatorial Guinea.
Author's calculations based on data from EM-DAT (2019).

Building on these preliminary findings, the next section turns to regression analysis to investigate the following hypotheses:

- First (*H1*), FDI to SSA countries decrease in response to natural disasters.
- Second (*H2*), FDI to SSA countries do not significantly respond to natural disasters.

Table 31.4 DR Congo, Nigeria, Kenya, Mozambique, and Ethiopia: Foreign Direct Investment (FDI) Inflows and Natural Disasters in Selected Years.

Year	DR Congo		Nigeria		Kenya		Mozambique		Ethiopia	
	Natural Disasters	FDI Inflows (% GDP)	Natural Disasters	FDI Inflows (% GDP)	Natural Disasters	FDI Inflows (% GDP)	Natural Disasters	FDI Inflows (% GDP)	Natural Disasters	FDI Inflows (% GDP)
1970–74	0	0.24	0	0.83	1	0.47	1	0.05	2	0.47
1975–79	3	0.86	1	0.34	3	0.69	7	0.01	3	0.15
1980–84	1	0.18	1	0.09	2	0.28	5	0.02	4	0.01
1985–89	2	0.03	9	1.05	1	0.29	4	0.09	9	0.003
1990–94	3	0.002	5	2.68	5	0.14	5	0.95	4	0.09
1995–99	14	0.06	14	2.8	13	0.17	11	3.47	15	1.63
2000–04	34	2.57	35	1.84	20	0.34	24	5.6	16	4.07
2005–09	25	5.42	20	2.55	29	1.82	19	4.08	25	1.64
2010–14	18	7.9	18	1.31	19	2.25	17	30.41	12	2.28

Author's calculations based on data from EM-DAT (2019).

31.4 **MODEL SPECIFICATION**

This section describes the data and estimation strategy.

31.4.1 **DATA AND VARIABLES**

As noted earlier, data on natural disasters are taken from the EM-DAT database. This data source provides information on the number of disaster events, the number of people affected, and the total amount of monetary damage per year since 1900. The current analysis, however, uses data covering the period 1970–2018 due to availability considerations. Many SSA countries included in the sample lack relevant data for the pre-1970 years.

On the other hand, the analysis focuses on the number of disaster events as the primary covariate for two reasons. First, it is difficult to estimate the severity—for example, damages—of natural disasters in many SSA countries. The EM-DAT database provides relevant data for only a handful of countries. Second, and more importantly, this analysis is concerned with the possibility that a higher frequency of natural disasters is a deterrent to foreign investment. As noted in the previous section, the analysis is restricted to four types of natural disasters—droughts, floods, storms, and epidemics. The study therefore investigates whether an increase in the frequency of those four types of events is associated with reduced FDI inflows in SSA countries. The dependent variable of interest is the FDI inflows in current US$, taken from the UNCTAD database.[13]

Furthermore, the empirical specification accounts for some variables that might influence causality between natural disasters and the FDI inflows. One control variable of primary interest is the land size measured in square kilometer and taken from the World Bank World Development Indicators (WDI). This variable is included because the effects of natural disasters may vary depending on the size of the affected area.

Other variables relate to a country's urban population as a share of total population, a country's total population, and GDP per capita and GDP growth, all taken from the WDI. First, the share of the urban population in total population is included to capture the potential adverse consequences of the unplanned urbanization in SSA. It has been suggested that intensive and unplanned urbanization, particularly in flood-prone areas in SSA, magnifies the negative impacts of floods (Di Baldassarre et al., 2010), and a broad spectrum of infectious and parasitic diseases (Adelekan et al., 2015: 35). Second, the total population of a country is included to control for the population effect. This is because disaster damages are larger in the most populous and poor SSA countries. Third, both GDP per capita (constant 2010 US$) and GDP growth are included to control for the socioeconomic conditions of the host countries.

Finally, a dummy variable is included to capture the coastal exposure and takes the value 0 if a country is landlocked, and 0 otherwise.

It should be pointed out that FDI inflows are determined by a wider array of factors—policy, socioeconomic, geographical. The interest is in the current analysis is to analyze the most parsimonious model, while retaining only those variables that can potentially influence the relationship between natural disasters and FDI inflows.

---

[13]FDI as a percentage of GDP is used as the alternative dependent variable.

31.4.2 ESTIMATION STRATEGY

The aim is to estimate a regression equation that accounts for the fact that prior disaster occurrences might influence FDI decisions in existing or potential host countries. As discussed earlier, foreign investors might decide to cancel their investment projects or relocate them to other destinations based on ex ante—both very recent events and longer term trends—experience with natural disasters.

Accordingly, the current analysis is based on three variables that capture the frequency of natural disasters occurring during three time periods prior to the measurement of FDI.[14] These include (1) the year prior to the measurement of FDI; (2) the 2 years prior to the measurement of FDI; and (3) the 3 years prior to the measurement of FDI. On the other hand, FDI is measured as follows: $t_1 = 1972$; $t_2 = 1974$; $t_3 = 1976$; $t_4 = 1978$; $t_5 = 1980$; $t_6 = 1982$; $t_7 = 1984$; $t_8 = 1986$; $t_9 = 1988$; $t_{10} = 1990$; $t_{11} = 1992$; $t_{12} = 1994$; $t_{13} = 1996$; $t_{14} = 1998$; $t_{15} = 2000$; $t_{16} = 2002$; $t_{17} = 2004$; $t_{18} = 2006$; $t_{19} = 2008$; $t_{20} = 2010$; $t_{21} = 2012$; $t_{22} = 2014$; $t_{23} = 2016$; and $t_{24} = 2018$.

Although the choice for these time periods is somewhat arbitrary, it allows for adequately addressing the question of whether prior occurrences of natural disasters reduce the FDI inflows in subsequent periods.

To estimate the effect of natural disasters on FDI, the study considers the hybrid random effects model proposed by Allison (2009). The model is well suited for estimating and comparing the relationship between two variables within and between countries. The analysis is based on the estimation of the following model:

$$\text{FDI}_{it} = \alpha + \beta_1(\text{ND}_{ij} - \overline{\text{ND}}_i) + \beta_2\overline{\text{ND}}_i + \beta_3(W_{ij} - \overline{W}_i) + \beta_4\overline{W}_i + \beta_5Z_i + (u_i + e_{ij}) \tag{31.1}$$

where i and t denote the countries and years, respectively. FDI is the net inflows of foreign direct investment for country i in year t. ND represents the total number of disasters to be experienced by country i in the prior year(s)—$t-1$, $t-2$, and $t-3$. W is a vector of the time-variant control variables described above. Z is a matrix of time-invariant variables (in this case, land size and coastal exposure). β_1 and β_3 are the within-country effect of time-variant independent variables(s) on FDI.[15] This is computed by deviating each observation (of each time-varying independent variable) from its country-specific mean. Hence, for the ND variable, for example, β_1 represents how on average a within-country change in the number of natural disasters is associated with a within-country change in FDI (FDI$_{it}$). β_2 and β_4 are the between-country effect of time-varying variables, that is, the country-specific mean of the time-varying independent variable(s) relative to the sample mean. Thus β_2 can be interpreted as how a country's average in the time-varying variable(s) is associated with a change in the average number of natural disasters. u and e denote the between and within error terms, respectively. FDI inflows in current US $, GDP per capita, total population, and land size, are expressed in natural logarithm. If hypothesis 1 (H1) holds, the coefficients on ND will be statistically significant and negative. In contrast, if hypothesis 2 (H2) holds, the coefficients on ND will be statistically insignificant and either positive or negative.

---

[14]This analytical approach follows Escaleras and Register (2011) who generate four variables that consider the total number of disasters occurring during four time periods prior to the measurement of FDI: (1) the prior year; (2) the prior 5 years; (3) the prior 10 years; and (4) the prior 25 years.

[15]Note that within-country effects in the hybrid model are identical to the fixed-effects estimates (see Bell & Jones, 2015).

31.5 EMPIRICAL RESULTS AND IMPLICATIONS

The results of the estimation are reported in Table 31.5. Column (1) presents the estimates using the total natural disasters at $t-1$, column (2) presents the estimates using the total natural disasters at $t-2$, and column (3) shows the estimates using the total natural disasters at $t-3$. It can be observed that the estimated effect of prior natural disaster occurrences on FDI inflows is consistently insignificant when estimated as a between-country effect. Specifically, the estimated between-country effects of the total natural disaster occurrences at $t-1$ (0.295), $t-2$ (0.188), and $t-3$ (0.103) are statistically insignificant. In other words, the average net inflows of FDI for SSA countries that experienced natural disasters in prior years and for countries that did not experience natural disasters do not statistically differ.

Table 31.5 Results of the Hybrid Model.

Variable	(1)	(2)	(3)
Between			
Total natural disasters in the prior year	0.295 (0.389)		
Total natural disasters in the prior 2 years		0.188 (0.216)	
Total natural disasters in the prior 3 years			0.103 (0.141)
Log population	0.568*** (0.149)	0.548*** (0.156)	0.103*** (0.159)
Urban population (% total population)	0.015 (0.0112)	0.015 (0.012)	0.013 (0.012)
Log GDP per capita	0.572*** (0.159)	0.566*** (0.157)	0.585*** (0.158)
GDP growth	0.111* (0.054)	0.11* (0.054)	0.103* (0.054)
Within			
Total natural disasters in the prior year	0.129* (0.060)		
Total natural disasters in the prior 2 years		0.083* (0.037)	
Total natural disasters in the prior 3 years			0.075** (0.029)
Log population	4.198*** (0.348)	4.118*** (0.358)	4.101*** (0.381)
Urban population (% total population)	−0.063*** (0.016)	−0.06*** (0.016)	−0.062*** (0.017)
Log GDP per capita	1.367*** (0.231)	1.372*** (0.231)	1.385*** (0.243)
GDP growth	0.034*** (0.009)	0.033*** (0.009)	0.032** (0.009)
Time-Invariant			
Land area	−0.304 (0.27)	−0.332 (0.277)	−0.295 (0.272)
Landlocked	−0.228 (0.287)	−0.237 (0.277)	−0.236 (0.293)
Observations	1014	1014	980
Number of countries	44	44	44

Standard errors in parentheses. GDP, Gross domestic product.
***$P<.01$.*
**$P<.05$.*
$P<0.1$.

In contrast, the estimated effect of prior natural disaster occurrences on FDI inflows is consistently significant and positive when estimated as a within-country effect. In column (1), for example, the within-country effect is estimated to be 0.129. That is, for a given SSA country, FDI inflows augment by 0.13 percentage points in response to the total number of natural disaster events that occurred in the previous year. Similarly, the coefficients in columns (2) and (3) suggest that for a given SSA country, FDI inflows increase by 0.08 percentage points in response to the total number of natural disasters that occurred in the previous 2 and 3 years.

Overall, the between-country effects of natural disasters are in favor of hypothesis (2): FDI to SSA countries do not significantly respond to natural disasters. In contrast, the within-country effects are in contradiction with hypothesis (1): FDI inflows in SSA countries decrease in response to natural disasters.

A Wald test is performed to test for equality of between- and within-country estimates. In the present case the test statistics suggests that the null hypothesis of equivalence for within and between estimates should not be rejected ($t-1$: Wald chi2 = 0.13; $t-2$: Wald chi2 = 0.13; and $t-3$: Wald chi2 = 0.01). This can be considered evidence in favor of random effects model.[16]

The results of the Wald test suggest that the within-country effects overestimate the effects of disaster occurrences on the inflows of FDI. This is because natural disasters are unlikely to be correlated with other country-specific unobserved factors that adversely impact FDI inflows. Nevertheless, meaningful interpretations of the within-country estimates could be achieved by performing a comparison between the within-country and the between-country effects (between-effect minus within-effect). Such exercise is essential to evaluate the size of the observed impact of natural disaster occurrences on FDI inflows that is determined by differences between countries that experience a higher frequency of natural disasters and their counterparts with less frequent natural catastrophes. Computed differences between the within-country and between-country coefficients are consistent with hypothesis (H2), that is, prior occurrences of natural disasters do not significantly determine the contemporaneous FDI inflows in SSA countries.[17]

Three primary reasons that might explain the insignificant causal effects of natural disasters on FDI in SSA are worth noting. First, although risks and potential losses associated with disasters can deter new investment projects, they do not necessarily induce MNCs to exist from existing host countries. In fact, MNCs can remain and even expand their operations in existing host countries that have been struck by a natural catastrophe. As discussed previously, corporate philanthropy and other business opportunities can cause MNCs to remain in countries hit by natural disasters and even contribute to the post–disaster reconstruction efforts. Future research can explore the extent to which MNCs operating in SSA undertake corporate philanthropy in the aftermath of major natural disaster events.

Second, the kind of FDI received by many countries across the SSA region might be crucial in explaining the impact of natural disasters. For instance, FDI in extractive industries might respond differently to floods and droughts relative to FDI in the agricultural sector. On this particular matter, future research can investigate how FDI in different sectors respond to the natural disasters occurring in SSA.

---

[16]This test was performed following Schunck (2013).

[17]The results of these calculations are not reported due to space considerations but are available upon request.

Finally, the effects of natural disasters might be conditioned by such factors as the land size, population size, and urban population size of the affected countries. For instance, many natural disasters that hit SSA countries may affect only geographically limited areas. Individual disaster events are therefore less susceptible to adversely affect economic activities in larger countries, whereas their impacts in small countries may be countrywide. Similarly, the effects of natural disasters might be conditional on urban growth and population size. In the case of the present study, there are reasons to believe that the causal impact of natural disasters on FDI is conditioned by other variables. Future research can gain leverage over this issue by estimating empirical models that specify this conditional effect explicitly.

31.6 CONCLUSION

The past five decades have witnessed a dramatic growth in natural disasters such as floods, droughts, windstorms, and epidemics in SSA. Such events are associated with human and economic losses, which can reduce the investment attractiveness of the region as a whole. In this context, understanding the impacts of natural disasters on foreign investment is essential because FDI constitutes the largest external financial flow available to SSA countries for economic development.

This study addressed the question of whether experiences with natural disasters reduce FDI into SSA. It examined the extent to which within- and between-country changes in the frequency of natural disasters impact FDI in SSA. An empirical model is developed in which prior occurrences of natural disasters affect the subsequent FDI inflows. Changes in the frequency of natural disasters were considered for three different periods: (1) the year before the measurement of FDI; (2) the 2 years before the measurement of FDI; and (3) the 3 years before the measurement of FDI. The study used hybrid models for a panel of 44 SSA countries in the period from 1970 to 2018.

The results suggest that the changes in the frequency of natural disasters have a little causal effect on the FDI inflows in SSA. Specifically, the net inflows of FDI in SSA countries are not significantly affected by the total number of natural disaster events that have occurred during the preceding one to 3 years, all else being constant.

While discussing possible explanations to these results, the study highlighted a few key areas that need serious consideration for further research on the currently limited literature about the nexus between natural disasters and FDI.

REFERENCES

Adelekan, I., Johnson, C., Manda, M., Matyas, D., Mberu, B., Parnell, S., ... Vivekananda, J. (2015). Disaster risk and its reduction: An agenda for urban Africa. *International Development Planning Review*, *37*(1), 33−43.

A.G. van Bergeijk, P., & Lazzaroni, S. (2015). Macroeconomics of natural disasters: Strengths and weaknesses of meta-analysis versus review of literature. *Risk Analysis*, *35*(6), 1050−1072.

Allison, P. D. (2009). *Fixed effects regression models* (Vol. 160). London: SAGE Publications.

Anuchitworawong, C., & Thampanishvong, K. (2015). Determinants of foreign direct investment in Thailand: Does natural disaster matter? *International Journal of Disaster Risk Reduction*, *14*(2015), 312−321.

Asiedu, E., Jin, Y., & Kanyama, I. K. (2015). The impact of HIV/AIDS on foreign direct investment: Evidence from Sub-Saharan Africa. *Journal of African Trade*, *2*(1−2), 1−17.

Becerra, O., Cavallo, E., & Noy, I. (2014). Foreign aid in the aftermath of large natural disasters. *Review of Development Economics*, *18*(3), 445−460.

Bell, A., & Jones, K. (2015). Explaining fixed effects: Random effects modeling of time-series cross-sectional and panel data. *Political Science Research and Methods*, *3*(1), 133−153.

Benson, C., & Clay, E. (2004). *Understanding the economic and financial impacts of natural disasters*. Washington, DC: The World Bank.

Bernard, A. B., & Jensen, J. B. (2007). Firm structure, multinationals, and manufacturing plant deaths. *The Review of Economics and Statistics*, *89*(2), 193−204.

Bettin, G., & Zazzaro, A. (2018). The impact of natural disasters on remittances to low-and middle-income countries. *The Journal of Development Studies*, *54*(3), 481−500.

Boudreaux, C. J., Escaleras, M. P., & Skidmore, M. (2019). Natural disasters and entrepreneurship activity. *Economics Letters*, *182*(2019), 82−85.

Cavallo, E., & Noy, I. (2011). Natural disasters and the economy—a survey. *International Review of Environmental and Resource Economics*, *5*(1), 63−102.

Cole, M. A., Elliott, R., Okubo, T., & Strobl, E. (2015). Natural disasters, industrial clusters and manufacturing plant survival. *RIETI Discussion Paper Series 15-E-008*. Research Institute of Economy, Trade and Industry.

Crespo, C. J., Hlouskova, J., & Obersteiner, M. (2008). Natural disasters as creative destruction? Evidence from developing countries. *Economic Inquiry*, *46*(2), 214−226.

David, A. (2010). How do international financial flows to developing countries respond to natural disasters?. *IMF Working Paper WP/10/166*.

Di Baldassarre, G., Montanari, A., Lins, H., Koutsoyiannis, D., Brandimarte, L., & Blöschl, G. (2010). Flood fatalities in Africa: From diagnosis to mitigation. *Geophysical Research Letters*, *37*(22), 1−5.

EM-DAT, 2019. CRED / UCLouvain, Brussels, Belgium. http://www.emdat.be

Escaleras, M., & Register, C. A. (2011). Natural disasters and foreign direct investment. *Land Economics*, *87* (2), 346−363.

Felbermayr, G., & Gröschl, J. (2014). Naturally negative: The growth effects of natural disasters'. *Journal of Development Economics*, *111*(2014), 92−106.

Fraser, A., Leck, H., Parnell, S., & Pelling, M. (2017). Africa's urban risk and resilience. *International Journal of Disaster Risk Reduction*, *26*(2017), 1−6.

Gemeda, D. O., & Sima, A. D. (2015). The impacts of climate change on African continent and the way forward. *Journal of Ecology and the Natural Environment*, *7*(10), 256−262.

Godfrey, P. (2005). The relationship between corporate philanthropy and shareholder wealth: A risk management perspective. *Academy of Management Review*, *30*(4), 777−798.

Haraguchi, M., & Lall, U. (2015). Flood risks and impacts: A case study of Thailand's floods in 2011 and research questions for supply chain decision making. *International Journal of Disaster Risk Reduction*, *14* (2015), 256−272.

Hosono, K., Miyakawa, D., Uchino, T., Hazama, M., Ono, A., Uchida, H., & Uesugi, I. (2016). Natural disasters, damage to banks, and firm investment. *International Economic Review*, *57*(4), 1335−1370.

International Monetary Fund (IMF). (2016). *Enhancing resilience to natural disasters in Sub-Saharan Africa. African regional economic outlook—October 2016*. Washington, DC: IMF.

Kato, H., & Okubo, T. (2018). The impact of a natural disaster on foreign direct investment and vertical linkages. *Keio-IES Discussion Paper Series 2017-018*. Institute for Economics Studies, Keio University.

Katoka, B., & Kwon, H. J. (2018). Business regulations and foreign direct investment in Sub-Saharan Africa: Implications for regulatory reform. In R. U. Efobi, & S. Asongu (Eds.), *Financing sustainable development in Africa* (pp. 63−91). Cham: Palgrave Macmillan.

Keerthiratne, S., & Tol, R. S. (2018a). Impact of natural disasters on income inequality in Sri Lanka. *World Development*, *105*(2018), 217−230.

Keerthiratne, S. & Tol, R.S. (2018b). Foreign aid concentration and natural disasters. *University of Sussex Working Paper Series No. 02−2018.*

Kirchberger, M. (2017). Natural disasters and labor markets. *Journal of Development Economics*, *125*(2017), 40−58.

Klomp, J. (2016). Economic development and natural disasters: A satellite data analysis. *Global Environmental Change*, *36*(2016), 67−88.

Lampel, J., Shamsie, J., & Shapira, Z. (2009). Experiencing the improbable: Rare events and organizational learning. *Organization Science*, *20*(5), 835−845.

Leiter, A. M., Oberhofer, H., & Raschky, P. A. (2009). Creative disasters? Flooding effects on capital, labour and productivity within European firms. *Environmental and Resource Economics*, *43*(3), 333−350.

Loayza, N. V., Olaberria, E., Rigolini, J., & Christiaensen, L. (2012). Natural disasters and growth: Going beyond the averages. *World Development*, *40*(7), 1317−1336.

McKnight, B., & Linnenluecke, M. K. (2016). How firm responses to natural disasters strengthen community resilience: A stakeholder-based perspective. *Organization & Environment*, *29*(3), 290−307.

Muller, A., & Kraussl, R. (2011). Doing good deeds in times of need: A strategic perspective on corporate disaster donations. *Strategic Management Journal*, *32*(2011), 911−929.

Naudé, W. A., & Bezuidenhout, H. (2014). Migrant remittances provide resilience against disasters in Africa. *Atlantic Economic Journal*, *42*(1), 79−90.

Neumayer, E., Plümper, T., & Barthel, F. (2014). The political economy of natural disaster damage. *Global Environmental Change*, *24*(2014), 8−19.

Noy, I., & duPont, W. (2016). The long-term consequences of natural disasters − A summary of the literature. *Working Papers in Economics and Finance*. School of Economics and Finance, Victoria Business School.

Oetzel, J., & Oh, C. H. (2013). Learning to carry the cat by the tail: Firm experience, disasters, and multinational subsidiary entry and expansion. *Organization Science*, *25*(3), 732−756.

Oh, C. H., & Oetzel, J. (2011). Multinationals' response to major disasters: How does subsidiary investment vary in response to the type of disaster and the quality of country governance? *Strategic Management Journal*, *32*(6), 658−681.

Olatunji, L. A., & Shahid, M. S. (2015). Determinants of FDI in Sub-Saharan African countries: A review of the evidence. *Business and Economic Research*, *5*(2), 23−34.

Onyewu, S., & Shrestha, H. (2004). Determinants of foreign direct investment in Africa. *Journal of Developing Societies*, *20*, 89−106.

Osberghaus, D. (2019). The effects of natural disasters and weather variations on international trade and financial flows: A review of the empirical literature. *Economics of Disasters and Climate Change*, *3*(3), 305−325.

Perrow, C. (2011). *The next catastrophe: Reducing our vulnerabilities to natural, industrial, and terrorist disasters*. Princeton University Press.

Schunck, R. (2013). Within and between estimates in random-effects models: Advantages and drawbacks of correlated random effects and hybrid models. *The Stata Journal*, *13*(1), 65−76.

Shiferaw, B., Tesfaye, K., Kassie, M., Abate, T., Prasanna, B., & Menkir, A. (2014). Managing vulnerability to drought and enhancing livelihood resilience in sub-Saharan Africa: Technological, institutional and policy options. *Weather and Climate Extremes*, *3*, 67−79.

Skidmore, M., & Toya, H. (2002). Do natural disasters promote long-run growth? *Economic Inquiry*, *40*(4), 664−687.

Strobl, E. (2012). The economic growth impact of natural disasters in developing countries: Evidence from hurricane strikes in the Central American and Caribbean regions. *Journal of Development Economics*, *97*(1), 130−141.

Tanaka, A. (2015). The impacts of natural disasters on plants' growth: Evidence from the Great Hanshin-Awaji (Kobe) earthquake. *Regional Science and Urban Economics, 50*(1), 31−41.

Toya, H., & Skidmore, M. (2007). Economic development and the impacts of natural disasters. *Economic Letters, 94*(1), 20−25.

United Nations Environment Programme (UNEP). (2002). *Regionally based assessment of persistent toxic substances: Sub-Saharan regional report.* Switzerland: UNEP Chemicals.

United Nations International Strategy for Disaster Reduction (UNISDR). (2014). *Disaster risk reduction in Africa: Status report on implementation of Africa regional strategy and Hyogo framework for action.* Nairobi, Kenya: UNISDR Regional Office for Africa.

van Niekerk, D., & Nemakonde, L.D. (2017). Natural hazards and their governance in sub-Saharan Africa. *Natural Hazard Science,* http://dx.doi.org/10.1093/acrefore/9780199389407.013.230.

Voigt, S., Giulio-Tonolo, F., Lyons, J., Kučera, J., Jones, B., Schneiderhan, T., ... Li, S. (2016). Global trends in satellite-based emergency mapping. *Science, 353*(6296), 247−252.

White, C., & Fan, M. (2006). *Risk and foreign direct investment.* Springer.

World Bank. (2016). *Striving toward disaster resilient development in Sub-Saharan Africa: Strategic framework 2016−2020.* Washington, DC: World Bank.

World Bank. (2019). *This is what it's all about: Building resilience and adapting to climate change in Africa.* Retrieved from <https://www.worldbank.org/en/news/feature/2019/03/07/this-is-what-its-all-about-building-resilience-and-adapting-to-climate-change-in-africa>.

Yang, D. (2008). Coping with disaster: The impact of hurricanes on international financial flows, 1970−2002. *The BE Journal of Economic Analysis & Policy, 8*(1), 1−45.

INTEGRATING CLIMATE CHANGE ADAPTATION AND VULNERABILITY REDUCTION FOR SUSTAINABLE DEVELOPMENT IN SOUTH ASIA AND AFRICA

32

Achiransu Acharyya

Department of Economics, Visva-Bharati (A Central University), Santiniketan, India

32.1 INTRODUCTION

Although the South Asian region has experienced economic growth in the past few years, the region still has predominantly a large section of population that is poor and dependent completely on agriculture. More than half the present population in India is dependent on agriculture. This number is 75% for Bangladesh. High population and less land size imply very low man-to-land ratio. Similarly, in sub-Saharan Africa, more than 60% of the populations are small landholding farmers.

Studies show that the impact of climate change will be most severe on the poor rural population of South Asia and Africa. For instance, in South Asia, global warming is predicted to increase the frequency of cyclones, steady rise in sea level, and average temperature as also rainfall. In sub-Saharan Africa the impact of climate change will have different effects. These include erratic rainfall, regular incidence of extreme heat events, and rising aridity of land. This means that the poor farmers in the region will be the worst sufferers (Acharyya, 2014).

Under the circumstances, it is extremely crucial therefore to understand the vulnerability of poor households due to climate change and their capability in adapting to changing climate scenarios with their traditional farming practices for sustained livelihood and development.

Agricultural production is affected by climate change in different ways. Various effects due to climate change such as variation in rainfall, persistent natural calamities such as hurricanes, droughts, typhoons, and rising temperatures have an impact on agriculture in terms of productivity and output [Chambwera & Stage, 2010; International Panel on Climate Change (IPCC), 2001]. From different research works on agriculture and its relation to climate change, it is observed that the output and productivity in agriculture is significantly negatively affected from rising temperatures (Acharyya, 2012; Lobell, Sibley, & Ortiz-Monasterio, 2012), fluctuations in occurrence and degree of extreme climatic scenarios, examples being floods and droughts (Brida & Owiyo, 2013; Singh, Mishra, Singh, & Parmar, 2013), and variation in rainfall patterns (Mall, Gupta, Singh, &

Ratbore, 2006; Prasanna, 2014). The impact of future climate change on cereal crops show that the yield loss could be very high, especially for staple food crops such as wheat and rice in developing and poor countries (Porter et al., 2014). The variation in crop cultivation suitability along with decrease in agricultural biodiversity could lead to lowering of the efficiency of the use of agricultural inputs and preponderance of diseases as also rise in pests. These are the few main effects on agriculture of climate change in different countries of the world (Norton, 2014).

Many smallholder farmers in West Africa have to be satisfied with meager and uncertain crop output and incomes, as also persistent food insecurity. These challenges are very severe in the drylands. Situations such as reduced fertility of soil, land degradation, variation in climatic conditions, and ever-rising cost of fertilizers and water stress reduce crop productivity (Zougmore, Jalloh, & Tioro, 2014). Moreover annual cycles of rainfall are decided by the location of intertropical convergence zone, creating a regional climatic condition that is most unpredictable and erratic in the world (Thomas, Twyman, Osbahr, & Hewitson, 2007) and forecasting of subsequent climate changes, highly ambivalent.

Spatial and temporal transition of temperature along with scarcity of water could lead to crucial ramification for agriculture, notably for reduced crop productivity. Although there will be regional variation in the impact of climate change on agriculture, the general forecast is that the productivity of cereals will fall to a large extent (Asian Development Bank, 2014).

So far as effect of change in climate in agriculture in India is concerned, a rise of 0.4°C in average annual surface air temperature has been recorded within the country over the last century. It is anticipated that by 2050 there will be an in general increase in average temperature varying between 2°C and 4°C [Ministry of Environment and Forests (MEF), 2004]. Moreover, it is also predicted that seasonal and spatial fluctuation in rainfall will also rise in future (IPCC, 2014). Trends historically show a sharp rise in average temperature and big fluctuation in pattern of rainfall in the monsoon season in India (IPCC, 2001). Also, the change in climate in future, under all circumstances, will lead to reduction in most crop yields in the long term (Rao et al., 2013). The rise in climatic variation can lead to strong fluctuations in agricultural production in the near future. Contemporary research on global and local simulation models show that a small rise in temperature in effect will have important negative impact on productivity of rice, maize, and wheat in the region (Aggarwal et al., 2009; Parry, Rosenzweig, Iglesias, Livermore, & Fischer, 2004). Under the abovementioned circumstances, it is extremely important for underdeveloped and developing countries of the region to adapt measures so that the production of agriculture can be sustainably increased to feed the rising population of the countries.

Climate-smart agriculture (CSA) or climate-smart farming (CSF), as defined by Food and Agriculture Organization (FAO) (2013), tries the integration of various dimensions of sustainable development by mutually addressing both the problems of food security and climate challenges. These dimensions of sustainable development are based on mainly the three pillars of community, economics, and the environment. The key is to collectively achieve sustainability in agriculture along with increasing production and income using the three mentioned pillars (FAO, 2013).

CSA is now looked upon as a pathway for the development of scientific oriented policy, backed by investment conditions to ensure the security of food under climate change conditions in a sustainable manner. The significance, concurrence, and expansive ambit of the impact of change in climate on an agricultural arrangement can make an imperative requirement for an all-inclusive integration of these impacts into the future planning of agriculture of the countries backed by programs and investments. This chapter is an attempt to look at the possibilities of CSF adaptation and practices in poor countries of Asia and Africa that would be affected most from climate change. Section 32.2 tries to analyze the possibilities of CSF in agriculture in developing countries with poor farmers and small size of landholding. Section 32.3 looks at the ground realities of CSF, especially in terms of adaptation to scenarios of changing climate by farmers. Section 32.4 is the concluding section.

32.2 CLIMATE-SMART FARMING: POTENTIALS AND PROBLEMS

The main income of an estimated 69% of the rural poor of the world depends on agriculture (World Bank, 2016). Moreover, about 1.3 billion landless workers and smallholders are employed in the agricultural sector (World Bank, 2016). The forecast is that variation in climate will lead to a negative impact on the livelihood of people who are dependent wholly on agriculture. It will also have an adverse impact on food security of the poor farmers both in developing and underdeveloped countries. This situation will be more aggravated with rising food and energy prices, land degradation, and reduced support of investment (Smit, McNabb, & Smithers, 1996). The poorest smallholder farmers will be the worst affected due to a drive-up in poverty and food insecurity. Not only that climate change effects will spread unevenly as farmers situated in environments that are most fragile, such as drylands, mountains, and conflict-prone stretches as also zones along the coast would be more susceptible to stresses and shocks (Chandra, McNamara, Dargusch, Caspe, & Dalabajan, 2017).

CSF emboldens through the use of different programs and drives the sustainable development of various agricultural schemes (FAO, 2010a, 2010b). The purpose of CSF is to achieve under the situation of climate change, low-carbon discharge from development, and greater security of food for the global population (FAO, 2013).

Climate mitigation is any action to lastingly phase out long-term risk from change in climate. The IPCC (2001) in its definition of mitigation defines it as "an anthropogenic intervention to reduce the sources or enhance the sinks of greenhouse gases." So far as climate adaptation is concerned, it is "adjustment in natural or human systems to a new or changing environment" (IPCC, 2001).

Thus the CSA approach tries to integrate the adaptation and mitigation to change in climate to ensure holistic planning of agricultural development. CSA has achieved ample limelight, more so in poor countries, because of the capacity to increase the security of food and provide flexibility in agricultural systems while reducing emission of greenhouse gases (GHGs) (Grainger-Jones, 2011). Such features are very relevant to the countries of South Asia and Africa, since, in these countries, the development of economic and human is dependent on the growth of agriculture, which is also the most susceptible sector so far as the influence of climate change is concerned (Vermeulen, Campbell, & Ingram, 2012).

Numerous technologies and agricultural practices under CSA exist. One such practice is minimum tillage, which is a soil control system with the aim of very little soil handling that is just required for a good crop output and opposed to intensive tillage that disrupts the structure of soil by using plows. Other practices include nutrient and irrigation management, various methods of crop formulation, and residue embodiment. All these can raise crop yields, nutrient and water use efficiency, and benefit in the shrinkage of emissions from GHG that is created due to production in agriculture (Sapkota, Jat, Aryal, Jat, & Khatri-Chhetri, 2015a, 2015b). Likewise, Information and communication technology-related agro-advisories, livestock and crop insurances, adoption of improved seeds, and rainwater harvesting can benefit farmers to lessen the adverse impact of change in climate and weather variations (Acharyya, 2019; Acharyya & Ghosh, 2017; Altieri & Nicholls, 2017).

CSF attempts to assimilate innovative and traditional habits, services, and technologies, appropriate for a specific area for adoption to variability and change in climate (CTA, 2014). The implementation of CSF technologies in a combination or individually has tremendous possibilities to curtail climate change effects on agricultural production. For example, simple adaptation measures such as adoption of irrigation technologies and changes in crop sowing dates can give greater yields with less variation than without adaptation (Finger, Hediger, & Schmid, 2011). Under different climate scenarios, metaanalysis of crop simulation shows that adaptation planning at the microlevel of farming can raise the productivity of crops in the aggregate as in comparison to without adaptation strategies (Challinor et al., 2014). Research also tells us that welfare from adaptation to climate change varies with temperature, with crop type, as also with variation in rainfall pattern (Easterling et al., 2007). Likewise, research shows that CSA technology adoption in all likelihood raises input use efficiency and improve crop yields, reduces emissions of GHGs, and leads to a net income rise for the population (Gathala et al., 2011; Khatri-Chhetri, Aryal, Sapkota, & Khurana, 2016; Sapkota et al., 2014).

Although CSA is an important step in order to accomplish adaptation of agriculture to climate change as also land-related mitigation issues, the application of CSA in reality is difficult. There are serious doubts whether a transition forward to CSA can indeed take place (FAO, 2013). It also involves different types of organizations and new kinds of institutions (FAO, 2013). A change with regard to CSA involves innovation techniques dependent on learning within society as the consequence of change in climate is uncertain as also dynamic in nature.

Many scholars highlight the social and contextual type of education as also change in traditional practices against new risks from climate change. It must be mentioned here though that based on the theoretical foundations, the theories give different kinds of importance to collective and individual learning (Blackmore, 2007). The nature of knowledge creation as also understanding of others' opinions and others' underlying objectives and values are also particularly important (Muro and Jeffrey, 2008). Community natural resource management literature postulates collective action as a significant result of social learning. This idea of collective action at the beginning meant institutional arrangements that would accelerate the coordination in the control and use of resources that are common pool in nature (Ostrom & Gardner, 1994). Later, both ideas of collective action and social learning were used for other common goals. For example, the formation of group by farmers' and cooperation among them were given much significance to obtain access to important resources (Markelova, Meinzen-Dick, Hellin, & Dohrn, 2009).

There are evidences to show that CSA can address the complicated interactions between food production, climate, and social factors at the level of local community. There exists a large volume of CSA literature that asks for a commitment to research methods to harmonize behavior of mitigation and adaptation in the practices of traditional production of food and farming arrangements owned by small and marginal farmers (Harvey et al., 2013; Lipper et al., 2014; Scherr, Shames, &

Friedman, 2012). There is no unique approach that exists for CSF since institutional factors, technological innovations, geography, capacity, and temporal scale, disrupt the synthesis of climate change targets with regard to agriculture. Hence, so far as governments and scientists are concerned, the key challenge remains in the framing of policies dependent on narrow scientific information that affects traditional knowledge practices and culpability of marginal farmers in underdeveloped and poor countries (IPCC, 2014). Practitioners and scientists are facing a tremendous stress to understand how the interventions of CSA would operate at the grassroots level, which can in turn help in the scientific effectiveness of the intervention. Under the circumstances, research on community-based actions that lead to increase in synergies and positive benefits, while reducing conflicts with regards to mitigation and adaptation methods, becomes extremely important. However, it is worth mentioning that although this approach has been frequently voiced (Smith & Olesen, 2010), it is unclear how in reality the mechanism of participation would essentially operate in this scenario, since it is dependent on different conjectures of the process through which smallholder farmers interact with agricultural landscapes and experience climate shocks.

It is extremely important to promote empirical research that makes it possible to inform the smallholder farmer of risks related to change in climate as also help them in the adoption of practices in agriculture that are suitable to the local landscape, available technologies and systems. In such a scenario the task of local institutions becomes extremely important for the scheme and application of policies related to climate change as also programs of farmer households, and there exists a good amount of literature in support of this perspective. To site one example, where private, local, public, and civic institutions have come together to invent and practice the diversity of rural livelihood systems is that of the National Adaptation Programmes of Action (Agrawal & Agrawal, 2008) that has been successful in many developing and underdeveloped countries. In the next section, we try to look at the various CSF practices that have been initiated by bigger organizations such as the FAO as also by the poor farmers.

32.3 CLIMATE-SMART FARMING PRACTICE: A REALITY CHECK

Although technologies related to CSA bring different gains, unfortunately, farmer's acceptance of such technologies in underdeveloped countries is not that high, rather it is much less than expected (Palanisami et al., 2015). Multiple reasons affect the degree of acceptance of CSA technologies such as social and economic attributes of farmers, environmental situation of a typical area, and nature of technologies that have been invented or innovated (Below et al., 2012; Campbell, Thornton, Zoigmoré, van Asten, & Lipper, 2014; Deressa, Hassan, & Ringler, 2011). Even then, an effort has been made in various places across the globe, most importantly in underdeveloped countries of Africa and Asia by the farmers (sometimes with the help of NGOs as also multilateral agencies) to mitigate and accommodate the influence of change in climate through various modes of farming practices (FAO, 2013).

Some of these behaviors of farmers toward CSF practices are worth studying to understand the degree of adaptability. One of these is the "Kihamba" agroforestry system of Mount Kilimanjaro's southern slopes. Here, an 800-year-old traditional agroforestry system exists, which has been an exemplary case of remarkably sustainable form of upland farming, and this form of sustainable farming has been supporting locations with high agrarian population densities in Africa without impairing nature and maintaining sustainability but also arranging livelihoods for a huge population (FAO Sourcebook, 2013).

In the Qinghai province of China, a pilot project by the name of "Three Rivers Sustainable Grazing Project" attempts for the recovery of degraded grasslands through the method of sustainable development and better management of grassland and cutback in pressure due to grazing in overloaded places, along with the sowing of better pastures and improved management of pastures. The aim of the project is to increase the water-holding scope of the soil by locking in more carbon in soils and biomass and, through this, augment grassland biodiversity.

There have been other instances also, where the farmers among themselves or with the help of various institutions have tried at CSF. For example, in Zambia, a landlocked country in south–central part of Africa, various agricultural practices for the improvement of food security have been promoted by the Government of Zambia. The input subsidy program is one of the most important of these policies with the declared aim of enhancing food security and thereby lowering poverty. CSF with respect to legume intercropping and crop rotation as also residue retention is one kind of practice that has gained government priority from the 1990s, especially from the Agriculture and Livestock Ministry with support from various international institutions. Arslan, McCarthy, Lipper, Asfaw, and Cattaneo (2013), in their detailed study of CSF practices in Zambia, show that adoption of a system of agricultural practices that have a CSA potential can benefit farmers adapt to climate change and also produce greater yields.

In the region of Mindanao in the Philippines, the marginal and small farmers have been educated on CSF techniques and applications in various Climate-resiliency Field Schools set up by the government. Information collected from field observations from various municipal corporations of the area demonstrate that various activities such as local planning and management, forums with all the stakeholders as also municipal budgeting strategies ensure the adaptation and mitigation behavior to shift in climate among marginal and poor farmers (Chandra et al., 2017). Moreover, by analyzing data using matrix and also looking at responses of the stakeholders, synergies, conflicts, interactions, and positive benefits can be identified between adaptation and mitigation in food output practices (Chandra et al., 2017). Field Schools based on resiliency to climate have bolstered multiple systems that include rice intensification, the process of organic farming, as also the building of seed banks for local communities. Other processes include reforestation, conservation of soil, and agroforestry. Carbon stocks have been maintained while raising crop production in the area. Field Schools give direction of a multilevel institutional platform so that poor farmers could get information about the climate that then can be used to develop planning of the farm, including timing of preparation of land for agriculture, timing of harvest, and making choices on which crop to grow. Here, it must be mentioned that climate-smart interventions have various technical aspects that must be done meticulously involving knowledge-intensive processes that are highly location specific in nature. Such behavior is motivated by the education and scope of local farmers.

In the drylands of West Africa, infertility of soil and water scarcity are the two most binding elements that pose difficulty in achieving higher agricultural productivity and growth, leading to greater emphasis on water and soil conservation and adoption of smart farming practices in areas that suffer from water stress (Acharyya, 2014). Rivaldo et al. (2016) use a joint analysis framework, by taking into consideration 11 on-farm climate-smart practices adopted by around 500 farmers in the West African Sahelian zone to understand the various climate change–related decisions on adoption of new technology. Their analysis show that although many farmers are adopting techniques and practices related to CSF on a widespread scale in the West African drylands, there is still scope to improve the adoption rates through introduction of more varied technology.

Chhetri, Aggarwal, Joshi, and Vyas (2017), in a study in the four dryland districts of Rajasthan in India, selected 16 villages for primary survey to evaluate the preferences and eagerness of farmers' to pay for adoption of various technologies in agriculture that are climate smart in nature. The villages were chosen based on various rainfall zones, greater reliance on rain-fed agriculture, and also having high possibility of drought predominance. After comprehensive meetings with community service organizations, government officials, and various stakeholders in the villages, the study observed that farmer's desire for technologies that are climate smart in nature is pronounced by some common features as well as variation corresponding to the rainfall zones and socioeconomic features of the farmers. So far as local farmers are concerned, their desired climate-smart technologies included harvesting of rainwater, location-specific management of crop nutrition, insurance of their crop, agro-advisories in weather-based crop management, laser land leveling, and contingent crop planning.

Similarly, a study on 3-year (2008−10) panel sample of marginal and small farmers across 15 rural districts in Zimbabwe based on efficiency and productivity of maize under CSA technologies by Ndlovu, Mazvimavi, An, and Murendo (2013) reveals that farmers produce 39% more in CSF in contrast to conventional farming, although technical efficiency is almost same for both. Interestingly, the results of the study reveal high yield gains in CSF practices and greater output of food.

From the abovementioned discussion, we understand that farmers in underdeveloped countries in South Asia and Africa are making efforts to promote CSF practices. However, such practices require the use of technology as well as education among farmers. Investment is also a bottleneck for the poor farmers in the region. Moreover, given that many farmers in Africa and Asia are poor, it is important therefore that international organizations along with the government try to help the farmers to adopt CSF practices.

32.4 **CONCLUSION**

The endorsement of CSA and CSF by the use of local institutional effectiveness can give boost to a synthesized action toward climate-resilient agricultural practices as also promote coherence between climate- and agricultural-related objectives. They can also connect agriculture- and climate-related investment for a combined path toward sustainable agricultural development through diversification of technologies, crop and livestock output processing systems and through building of networks across the agricultural sector. A multipolicy approach is necessary that facilitates and supports the proper use of CSF technology. What is necessary is the identification of the dominant mechanisms that hinder or threaten sustainable implementation of the technology and under the circumstances, select the most appropriate actions in the poor countries. This will ensure better access to capital in some cases and to specific support of finances in others. Farm-monitoring technology based on cooperative and financing in education and training should also boost the sustainable use of CSA technologies. It must be mentioned though that in most of the scenarios the policy environment should have schemes that are clear in legal and other terms so that it allows for user rights and effective ownership.

One of the arguments given to show that CSF is not "business as usual" is the importance given to implementation of adaptive, context-specific solutions, using systems and methods that assess the pros and cons between the mainstays of sustainable increase in agricultural output, food security, and adaptability to change in climate at the grassroots stage as also at macrostage thereby leading to opportunities that allow the reduction of GHG emissions from agriculture.

Climate smartness cannot be delivered overnight. It is first contextual and localized in nature in the sense that some farming practices may be similar to climate-smart practices in one sense but not in the other. Under the circumstances, what is necessary is information and data that can help farmers in taking decision at the farm level and also policymakers in national governments to prioritize investment in CSF for a sustainable better future.

REFERENCES

Acharyya, A. (2012). Sustainability of groundwater in South Asia: Need for management through institutional change. *NITTE Management Review*, *6*(2), 28–38.

Acharyya, A. (2014). Groundwater, climate change and sustainable well-being of the poor: Policy options for South Asia, China and Africa. *Procedia—Social and Behavioral Sciences*, *157*, 226–235.

Acharyya, A. (2019). IoT and big data analytics in agriculture: A review of theory and practice. In Gupta, & Angadi (Eds.), *Digital India for agricultural and rural development: Scope and limitations* (pp. 70–99). Gaurang Publishing Globalize Pvt. Ltd..

Acharyya, A., & Ghosh, M. (2017). Climate change smart farming practices: A review. In Gupta, & Angadi (Eds.), *Smart farming: Problems and prospects* (pp. 135–155). Mumbai: Gaurang Publishing.

Agrawal, A., & Agrawal, N. (2008). *Climate adaptation, local institutions, and rural livelihoods. IFRI Working Paper # W08I-6.* Michigan:: University of Michigan.

Aggarwal, P. K., Singh, A. K., Samra, J. S., Singh, G., Gogoi, A. K., Rao, G. G. S. N., & Ramakrishna, Y. S. (2009). Introduction. In P. K. Aggarwal (Ed.), *Global climate change and Indian agriculture*. New Delhi: Indian Council of Agricultural Research.

Altieri, M. A., & Nicholls, C. I. (2017). The adaptation and mitigation potential of traditional agriculture in a changing climate. *Climatic Change*, *140*(1), 33–45.

Arslan, A., McCarthy, N., Lipper, L., Asfaw, S., & Cattaneo, A. (2013). Adoption and intensity of adoption of conservation farming practices in Zambia. *ESA Working Paper No. 13-01*. Rome: Food and Agriculture Organization of the United Nations. <http://www.fao.eg/3/a-aq288e.pdf> Accessed 26.07.14.

Asian Development Bank. (2014). *Climate change in South Asia*. Tokyo: Asian Development Bank.

Below, T. B., Mutabazi, K. D., Kirschke, D., Frankc, C., Siebert, S., Siebert, R., & Tsherning, K. (2012). Can farmers' adaptation to climate change be explained by socio-economic household level variables? *Global Environmental Change*, *22*(1), 223–235.

Blackmore, C. (2007). What kinds of knowledge, knowing and learning ate required for addressing resource dilemmas? A theoretical overview. *Environmental Science & Policy*, *10*, 512–525.

Brida, A. B., & Owiyo, T. (2013). Loss and damage from the double blow of flood and drought in Mozambique. *International Journal of Global Warming*, *5*(4), 514–531.

Campbell, B. M., Thornton, P., Zoigmore, R., van Asten, P., & Lipper, L. (2014). Sustainable intensification: What is its role in climate smart agriculture? *Current Opinion in Environmental Sustainability*, *8*, 39–43.

Challinor, A. J., Watson, J., Lobell, D. B., Howden, S. M., Smith, D. R., & Chhetri, N. (2014). A meta-analysis of crop yield under climate change and adaptation. *Nature Climate Change, 4*, 287−291.

Chambwera, M., & Stage, J. (2010). *Climate change adaptation in developing countries: Issues and perspectives for economic analysis.* Available from <www.iied.org/pubs/display.php?o = 15Sl7IIED>.

Chandra, A., McNamara, K. E., Dargusch, P., Caspe, A. M., & Dalabajan, D. (2017). Gendered vulnerabilities of smallholder farmers to climate change in conflict-prone areas: A case study from Mindanao. Philippines. *Journal of Rural Studies, 50*, 45−59.

Chhetri, A. K., Aggarwal, P. K., Joshi, P. K., & Vyas, S. (2017). Farmers' prioritization of climate-smart agriculture (CSA) technologies. *Agricultural Systems, 151*, 184−191.

CTA, 2014. Leading the Way. Annual Report 2013. Technical Centre for Agricultural and Rural Cooperation ACP-EU (CTA). The Netherlands.

Deressa, T. T., Hassan, R. M., & Ringler, C. (2011). Perception of and adaptation to climate change by farmers in the Nile basin of Ethiopia. *Journal of Agricultural Science, 149*(1), 23−31.

Easterling, W., Aggarwal, P., Batima, P., Brander, K., Erda, L., Howden, M., ... Tubiello, F. (2007). Food, fibre and forest products. In M. L. Parry, O. F. Canziani, J. P. Palutikof, P. J. van der Linden, & C. E. Hanson (Eds.), *Climate change (2007): Impacts, adaptation and vulnerability. Contribution of working group II to the fourth assessment report of the intergovernmental panel on climate change* (pp. 273−313). Cambridge University Press.

Finger, R., Hediger, W., & Schmid, S. (2011). Irrigation as adaptation strategy to climate change—A biophysical and economic appraisal for Swiss maize production. *Climatic Change, 105*(3 & 4), 509−528.

Food and Agriculture Organization (FAO). (2010a). *Climate-smart agriculture: Policies, practices and financing for food security.* Rome: Food and Agriculture Organization.

Food and Agriculture Organization (FAO). (2010b). *Adaptation and mitigation.* Rome: Food and Agriculture Organization.

Food and Agriculture Organization (FAO). (2013). *Climate-smart agriculture sourcebook.* Rome: Food and Agriculture Organization of the United Nations (FAO).

Gathala, M. K., Ladha, J. L., Kumar, V., Saharawat, Y. S., Kumar, V., Sharma, P. K., ... Pathak, H. (2011). Tillage and crop establishment affects sustainability of South Asia Rice-Wheat system. *Agronomy Journal, 103*(4), 961−971.

Grainger-Jones, E. (2011). Climate-smart smallholder agriculture: What's different? (no. 3). *IFAD Occasional Paper. International Fund for Agricultural Development*, Rome.

Harvey, C. A., Chaco, M., Donatti, C. I., Garen, E., Hannah, L., Andrade, A., et al. (2013). Climate-smart landscapes: Opportunities and challenges for integrating adaptation and mitigation in tropical agriculture. *Conservation Letters, 7*(2), 77−90.

International Panel on Climate Change (IPCC). (2001). *Climate change (2001): Impacts, adaptation, and vulnerability.* Cambridge: Cambridge University Press.

International Panel on Climate Change (IPCC). (2014). *Climate change (2001): Mitigation of climate change.* Cambridge: Cambridge University Press.

Khatri-Chhetri, A., Aryal, J. P., Sapkota, T. B., & Khurana, R. (2016). Economic benefits of climate-smart agricultural practices to smallholders' farmers in the Indo-Gangetic Plains of India. *Current Science, 110* (7), 1251−1256.

Lipper, L., et al. (2014). Climate smart agriculture for food security. *Nature Climate Change, 4*, 1068−11072.

Lobell, D., Sibley, A., & Ortiz-Monasterio, J. I. (2012). Extreme heat effects on wheat senescence in India. *National Climate Change, 2*, 186−189.

Mall, R. K., Gupta, A., Singh, R. S., & Ratbore, L. S. (2006). Water resources and climate change: An Indian perspective. *Current Science, 90*(12), 1610−1626.

Markelova, H., Meinzen-Dick, R., Hellin, J., & Dohrn, S. (2009). Collective action for small holder market access. *Food Policy, 34,* 1−7.

Ministry of Environment and Forests (MEF). (2004). *India's initial national communication to the United Nations framework convention on climate change.* New Delhi: Ministry of Environment and Forests.

Muro, M., & Jeffrey, P. (2008). A critical review of the theory and application of social learning in participatory natural resource management processes. *Journal of Environmental Planning and Management, 51*(3), 325−344.

Ndlovu, P. V., Mazvimavi, K., An, H., & Murendo, C. (2013). Productivity and efficiency analysis of maize under conservation agriculture in Zimbabwe. *Agricultural Systems, 124,* 21−31.

Norton, R. (2014). Combating climate change through improved agronomic practices and input-use efficiency. *Journal of Crop Improvement, 28*(5), 575−618.

Ostrom, E., & Gardner, R. (1994). *Rules, games and common-pool resources.* Ann Arbor, MI: The University of Michigan Press.

Palanisami, K., Kumar, D. S., Malik, R. P. S., Raman, S., Kar, G., & Monhan, K. (2015). Managing water management research: Analysis of four decades of research and outreach programmes in India. *Economic and Political Weekly, 26/27,* 33−43.

Parry, M. L., Rosenzweig, C., Iglesias, A., Livermore, M., & Fischer, G. (2004). Effects of climate change on global food production under SRES emissions and socio economic scenarios. *Global Environmental Change, 14*(1), 53−67.

Porter, J. R., et al. (2014). Food security and food production systems. In Field, et al. (Eds.), *Climate change 2014: Impacts, adaptation, and vulnerability. Part A: Global and sectoral aspects. Contribution of working croup II to the fifth assessment report of the intergovernmental panel on climate change* (pp. 485−533). Cambridge and New York: Cambridge University Press.

Prasanna, V. (2014). Impact of monsoon rainfall on the total food grain yield over India. *Journal of Earth System Sciences, 123*(5), 1129−1145.

Rao, C. A. R., et al. (2013). *Atlas on vulnerability of Indian agriculture to climate change.* Hyderabad: National Initiative on Climate Resilient Agriculture (NLCRA), ICAR.

Rivaldo, A., Kpadonoua Baba, O., Tom, B. B., Fatima, D., Franck, R., & Kiema, A. (2016). Advancing climate-smart-agriculture in developing drylands: Joint analysis of the adoption of multiple on-farm soil and water conservation technologies in West African Sahel. *Land Use Policy, 61,* 196−207.

Sapkota, T. B., Aryal, J. P., Jat, M. L., & Bishnoi Dalip, K. (2015a). On-farm economic and environmental impact of zero-tillage wheat: A case of northwest India. *Experimental Agriculture, 51*(1), 1−16.

Sapkota, T. B., Jat, M. L., Aryal, J. P., Jat, R. K., & Khatri-Chhetri, A. (2015b). Climate change adaptation, greenhouse gas mitigation and economic profitability of conservation agriculture: Some examples from cereal systems of Indo-Gangetic Plains. *Journal of Integrative Agriculture, 14*(8), 1524−1533.

Sapkota, T. B., Majumdar, K., Jat, M. L., Kumar, A., Bishnoi, D. K., Mcdonald, A. J., & Pampolino, M. (2014). Precision nutrient management in conservation agriculture based wheat production of northwest India: Profitability, nutrient use efficiency and environmental footprint. *Field Crops Research, 155,* 233−244.

Scherr, S. J., Shames, S., & Friedman, R. (2012). From climate-smart agriculture to climate-smart landscapes. *Agriculture and Food Security, 1*(12), 1−15.

Singh, G., Mishra, D., Singh, K., & Parmar, R. (2013). Effect of rainwater harvesting on plant growth, soil water dynamics and herbaceous biomass during rehabilitation of degraded hills in Rajasthan, India. *Forest Ecology and Management, 310,* 612−622.

Smit, B., McNabb, D., & Smithers, J. (1996). Agricultural adaptation to climate variation. *Climate Change, 33,* 7−29.

Smith, P., & Olesen, J. (2010). Synergies between the mitigation of and adaptation to climate change in agriculture. *Journal of Agricultural Science, 148,* 543−552.

Thomas, D. S. G., Twyman, C., Osbahr, H., & Hewitson, B. (2007). Adaptation to climate change and variability: Farmer responses to intra-seasonal precipitation trends in South Africa. *Climate Change, 83*, 301–322.

Vermeulen, S. J., Campbell, B. M., & Ingram, J. S. I. (2012). Climate change and food systems. *Annual Review of Environmental Resources, 37*, 195–222.

World Bank. (2016). World development report. In *Digital dividends*. Washington, DC: World Bank.

Zougmore, R., Jalloh, A., & Tioro, A. (2014). Climate-smart soil water and nutrient management options in semiarid West Africa: A review of evidence and analysis of stone bunds and zai techniques. *Agriculture and Food Security, 3*(16), 1–8.

ASSESSING THE SOCIAL VULNERABILITY TO FLOODS IN INDIA: AN APPLICATION OF SUPEREFFICIENCY DATA ENVELOPMENT ANALYSIS AND SPATIAL AUTOCORRELATION TO ANALYZE BIHAR FLOODS

33

Rupak Kumar Jha[1], Haripriya Gundimeda[1] and Prakash Andugula[2]

[1]*Department of Humanities and Social Sciences, Indian Institute of Technology Bombay, Mumbai, India*
[2]*Centre of Studies in Resource Engineering, Indian Institute of Technology Bombay, Mumbai, India*

33.1 INTRODUCTION

This chapter aims to assess the social vulnerability to floods in Bihar, India. Biophysical realization of disaster in the aftermath of an extreme climatic event is not solely driven by the hazard intensity. The etiology of hazard events to disaster impacts is complex, and a chain of events is involved from the hazard genesis to disaster losses (Blaikie, Cannon, Davis, & Wisner, 1994 and Wisner, Blaikie, Cannon, & Davis, 2004). Ex-post hazard, apparently in most of the cases, it is found that the disaster consequences are differential. The much-evolved research on vulnerability to climate stress has realized that the difference in damage extent is mostly attributed to the differential social vulnerability conditions (Adger, 2006). In the hazard context, social vulnerability highlights those demographic and socioeconomic attributes that can increase or decrease the degree of disaster consequences (Cutter, Emrich, Webb, & Morath, 2009). However, vulnerability as a concept is a complex subject to be empirically assessed. On the one hand, vulnerability in the realm of climate science is understood as the likelihood of the occurrence and impacts of weather events such as floods (Nicholls, Hoozemans, & Marchand, 1999), while on the other, social scientists view vulnerability as a state, independent of any hazard events (Adger & Kelly, 1999; Kelly & Adger, 2000). Vulnerability in this context is the existing structural system that makes communities prone to damages as a result of the external events. However, concerning a particular environmental extreme event, it is the interaction between the extreme environmental events and the existing socially vulnerable conditions that result in various losses and damages outcomes. According to Brooks and Adger (2003), social vulnerability may be considered as one of the crucial determinants of biophysical vulnerability. For example, in the event of flooding, location differences are one of the crucial aspects of the differential disaster intensity. However, the population dwelling at various locations is determined by the existing structural factors in the society. Thus vulnerability essentially becomes a social product (Few, 2003). In addition, given a community or a place facing the hazard of similar scale, the affected place and population experiencing the differential outcomes are mainly attributed to the differential social vulnerability conditions (Cannon, 2008; Ribot, 2010).

Appraising differences in social vulnerability is usually guided by two reasons: (1) understand the magnitude and scale of the threat; and (2) employ a host of remedial measures to downscale the impacts (Adger & Kelly, 1999). Therefore to design and reinforce suitable hazard prevention and disaster mitigation policies, identifying socially vulnerable groups and places would help in identifying targeted solutions and strategies (Tapsell, McCarthy, Faulkner, & Alexander, 2010). A potential problem-resolving approach can be to focus on comprehending the process of vulnerability to a particular hazard and accordingly formulate the policies to attempt obviating these vulnerabilities. Concomitantly, vulnerability mapping can provide a significant visualization tool to the policy makers by constructing a composite vulnerability index using a group of indicators and spatially reference them to illustrate the spatial distribution (Eakin & Luers, 2006). Research guided by the hazard(s) of place model (Cutter, 1996) has followed social vulnerability mapping. Some of the notable examples are (1) Cutter, Mitchell, and Scott (2000) for Georgetown County in the United States; (2) Cutter, Boruff, and Shirley (2003) for the counties of the United States; (3) Chen, Cutter, Emrich, and Shi (2013) for the Yangtze delta region in China; (4) Zhou, Li, Wu, and Wu (2014a) for the various provinces of China; and (5) Gautam (2017) for the districts of Nepal. These literature either consider independent indicators or use principal component (PC) analysis (PCA) as a dimension reduction technique to construct composite indices and geospatially reference them to illustrate the vulnerability distribution on various choropleth maps. In this line, recent attempts have also been made on depicting the temporal and spatial variations. Cutter and Finch (2008) and Zhou, Li, Wu, Wu, and Shi (2014b) are the recent contributions at the subnational-scale social vulnerability mapping for the United States and China, respectively. These studies additionally employed exploratory spatial data analysis (ESDA) to analyze the distribution characteristics of the vulnerability of their respective study areas. We can find a similar approach in some of the research contributions from India in the last decade and half focusing on various vulnerability attributes. At the subnational scale, Chakraborty and Joshi (2016) for the district-wise vulnerability mapping, Mazumdar and Paul (2016) for the coastal districts of India, and Mazumdar and Paul (2018) for the administrative blocks of Odisha are some of the recent examples in this regard.

The crucial issues of importance for mapping vulnerability are the choice of indicators, the weighting process, and the reflections they generate toward the final results. Due to the embedded complications in the vulnerability concept, indicators chosen across the scales are different. Besides, the relative contribution of variables is not symmetric in the vulnerability assessment process. Indicators selection, assigning weights, and vulnerability group classifications are very sensitive issues, and apparently there seem to be no consensus within the scientific community from the published papers. We can find a variety of approaches in this regard. For example, Zhang and Huang (2013) and Chakraborty and Joshi (2016) have used the analytic hierarchy process method to assign weights at the subindices level. Most of the studies on social vulnerability index (SoVI) assumed equal weights across the indicators (e.g., Chen et al., 2013; Cutter et al., 2003; Mavhura, Manyena, & Collins, 2017; and Mazumdar and Paul, 2016). Mazumdar and Paul (2018) have used

PCA for the dimension reduction. However, this method has been also used to assign weights for the household survey−based vulnerability index construction (Opiyo, Wasonga, & Nyangito, 2014). Thus it remains inconclusive at the intermediary stage of aggregation whether the individual indicator should receive weights. Rygel, O'Sullivan, and Yarnal (2006) opine that in highly constrained circumstances, assigning weights to individual indicators is possible as well as desirable. Weights can also be assigned when the vulnerability is well understood. However, in a more complex, highly dynamic, and fuzzy boundary definition of biophysical and socioeconomic space, assigning weights tends to lose its significance as the causes of vulnerability vary temporally. In this regard, they employ Pareto-ranking method (Goldberg, 1989) which overcomes the limitation associated with a priori assigning the weights to the chosen indicators. Data envelopment analysis (DEA) is similar in classifying the concerned units of the study although not so congruent to the Pareto-ranking method. Arguably, it improves upon the traditional methods on account of the weight selection (Huang, Liu, Ma, & Su, 2013; Wei, Fan, Lu, & Tsai, 2004).

Flood disasters have significant direct and indirect economic losses and residual impacts due to the loss of lives and suffering experiences by the affected people in the area. Thus a basic assumption is that in the face of the occurrence of a similar hazard, vulnerability is proportional to the loss and damage; vulnerability would essentially increase if there is an increase in loss and vice versa (Hou, Lv, Chen, & Yu, 2016; Huang et al., 2013). Analogous to measuring the production efficiency of firms, the DEA method considers the spatial units as decision-making units (DMUs). This method can be used to measure the social vulnerability by having various disaster losses as outputs and various environmental and socioeconomic indicators as the inputs.

In the aforementioned background, this chapter attempts to map the district-wise social vulnerability to floods in the Indian state of Bihar that experiences floods of varying intensity every year. We consider the Disaster Management Department (DMD) defined flood-prone districts of Bihar as our DMUs and employ the superefficiency DEA model to rank all the DMUs, including the efficient ones. To do so, we consider a host of indicators reflecting the flood damage outcomes in the output domain and the relevant topographic and socioeconomic indicators in the input domain. We further employ spatial autocorrelation technique to explore the similarity and/ or varying social vulnerability pattern across the districts. The use of DEA as a tool to assess social vulnerability is less explored in the literature assessing social vulnerability to floods in India, and thus our attempt in this study is novel in this regard. The remainder of this chapter is as follows. Section 33.2 explains the study area profile. In Section 33.3, for identifying indicators, we briefly review literature focusing on vulnerability assessment using DEA method. Section 33.4 explains the research design, whereas, Section 33.5 provides the results and discussion. Finally, in Section 33.6, we summarize and conclude the study.

33.2 **STUDY AREA PROFILE**

Bihar is located between $24°20'10''$ to $27°31'15''$N and $83°19'50''$ to $88°17'40''$E (Fig. 33.1). It is an eastern Indian state that geographically constitutes a saucer shape valley. The river Ganges flows from west to east and virtually bisects the state's geography into two. North Bihar that adjoins Nepal is drained by a number of rivers. Ghaghra, Gandak, Burhi Gandak, Adhwara group of rivers,

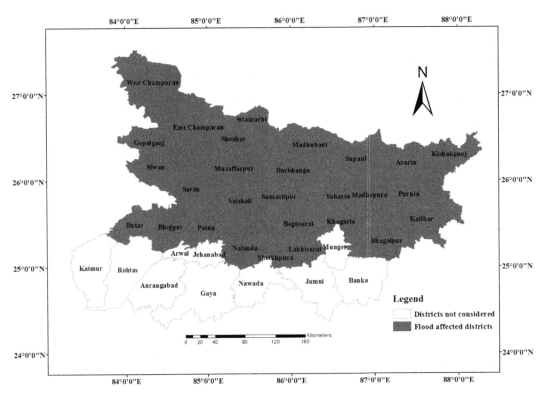

FIGURE 33.1

Study area location.

Bagmati, Kamla—Balan, Kosi, and Mahananda are the major rivers that flow through this terrain, which have their catchments in the Himalayas (http://fmis.bih.nic.in/history.html), whereas, in the south, Karmnasa, Sone, Punpun, Phalgu, Kiul, Chandan, and Sakri are the major rivers that flow toward the river Ganges. Rivers descending from the upper and foothills of Himalayas carry high discharge and very heavy sediment load before their confluence with the river Ganges. In the relatively flat north Bihar plains, these sediments get settled and cause widespread flooding in the area. Rivers in the south, on the other hand, cause seasonal flooding due to the advent of southwest monsoon (http://bsdma.org/images/global/SDMP.pdf). All these rivers function as the tributaries to the river Ganges and divide the states into seven geo-cultural zones. As shown in Table 33.1, districts under first four geo-cultural zones are located in the north Bihar, whereas the districts under rest of the geo-cultural zones are located in the south Bihar.

Bihar is India's most flood-prone state; about 68,800 km^2 out of the total geographical area of 94,160 km^2, constituting 73.06%, is flood affected (http://fmis.bih.nic.in/history.html). According to the DMD, Government of Bihar (GoB), 28 out of 38 districts are flood prone. These districts are shaded in Fig. 33.1.

Table 33.1 Rivers Forming Geo-Cultural Zones in Bihar.		
Sr. No.	**Geo-Cultural Zones**	**Districts**
1	Ghaghra−Gandak	West Champaran, East Champaran, Gopalganj, Siwan, Saran
2	Gandak−Bagmati	Sheohar, Sitamarhi, Muzaffarpur, Vaishali, Samastipur Begusarai
3	Bagmati−Kosi	Darbhanga, Madhubani, Supaul, Saharsa, Khagaria
4	Kosi−Mahananda	Madhepura, Araria, Purnia, Kishanganj, Katihar
5	Karmnasa−Sone	Buxar, Kaimur, Bhojpur, Rohtas
6	Sone−Punpun	Patna, Jehanabad, Arwal, Gaya, Nalanda, Aurangabad, Nawada
7	Punpun−Sakri	Sheikhpura, Lakhisarai, Jamui, Banka, Munger, Bhagalpur

State Disaster Management Plan, Bihar State Disaster Management Authority (BSDMA), Government of Bihar (GoB).

33.3 USING DATA ENVELOPMENT ANALYSIS TO CONSTRUCT VULNERABILITY INDEX: IDENTIFICATION OF THE INDICATORS

There has been a substantial research to develop a composite index to assess vulnerability, but the DEA has not been widely used for constructing vulnerability index, its classification, and the mapping. Index construction using this method is recent—the mainstream antecedent literature can only be traced back to Clark et al. (1998). One of the earliest attempts in this regard comes from Wei et al. (2004) who estimated vulnerability to various disasters at different scales in China. They used two input variables, population and gross domestic product (GDP), and two output variables, people affected and cost of damage, to construct vulnerability index at the provincial scale in China for the period 1989−2000. We find an extension of this method in the work of Zou and Wei (2009). They selected a host of input variables to reflect the economic structure and resilience conditions to examine the residual impacts in terms of death, affected people, and economic damages for southeast Asian countries for the period 1995−2005. Multidimensional aspects of flood vulnerability have also been assessed using DEA in Huang et al. (2012). They explored four vulnerability dimensions: (1) population vulnerability as measured by the affected population due to disaster, (2) death vulnerability measured as death outcome due to disaster, (3) agricultural vulnerability measured as crop area affected due to disaster, and (4) economic vulnerability measured as direct economic loss due to disaster. This study considered 10-year average data during 2001−10 at the provincial scale in China. For all these four models, the percentage of cropped area was used as the common proxy for the hazard. However, the total population and population density were the other common input indicators for assessing the population vulnerability and death vulnerability. For the agricultural vulnerability, the other input indicators were the gross area sown of crops, whereas, for the assessment of economic vulnerability, the input indicators are total population, total GDP, population density, GDP per km^2, and GDP per capita. Values were averaged for 10 years to estimate the final vulnerability index score and illustrated on the choropleth maps. In another study, by Huang et al. (2013), the assessment and classification of regional vulnerability to natural hazards is approached by extending the method proposed by Rygel et al. (2006) while using DEA method. In the output class, disaster-affected area, area with total loss of harvest, proportion of farmland with total loss of harvest, affected population, number of deaths, population with drinking water problem, number of collapsed or damaged building units, and direct economic losses were considered as indicators, while in the input category, regional total population, GDP, per capita GDP, population density, cultivated areas, GDP per km^2, regional urbanization level, and the total fixed investment in construction represented the socioeconomic indicators. To represent the dangerousness of regional hazards, the frequency and magnitude of each type of hazard were used. Further, correlation analysis in the outcome vulnerability score and the socioeconomic indicators showed a positive relationship meaning that socioeconomic aspects shape the vulnerability of a region. Deviating from this line of vulnerability assessment, Li, Li, Wu, and Hu (2013) estimated a spatiotemporal vulnerability for the Dongting Lake region of Hunan, China. In this study, flood disaster is considered as the product of disaster drivers, disaster environment, disaster bearer, and the severity. Flood disaster drivers were represented by coincidence of flood peaks among the considered rivers and highest precipitation over 3 days. Flood disaster environment was represented by ground elevation scale and ground cover runoff index, whereas population density, economic density represented the flood disaster bearer. These three groups of indicators were considered as the input indicators, whereas the severity of flood disaster, represented by direct economic loss, human casualties, and the area affected, were considered as the output for the DMUs representing various municipalities of the region. Apart from these literature, Hou et al. (2016) used superefficiency DEA and spatial autocorrelation method for a provincial-scale social vulnerability assessment in China. This study considered demography (represented by population density, age structure, and population literacy), economy (economic density, fixed assets density, per capita GDP), and society (road density, medical condition, prevention, and control density of the geological disaster) as input index. On the other hand, the extent of damage (represented by proportion of the injuries and deaths, the proportion of economic losses from the geological disaster) was incorporated as the output index.

Hence, we can infer from the aforementioned studies that environmental factors may play a crucial role in shaping the vulnerability of a place. However, the varying degree of socioeconomic attributes is equally important, if not more, to be factored for the assessment of social vulnerability to hazards. Besides, it also depends on the discretion of the research scientists that what attributes they want to explore while assessing vulnerability, given the data limitation one may encounter to represent the various spaces and scales. In this chapter, our selection of indicators is primarily guided by the socioeconomic and existing topographical aspects of vulnerability assessment as in how they shape the efficiency of DMUs on account of different losses and damages caused by floods in Bihar.

33.4 RESEARCH DESIGN: MATERIAL AND METHODS

In the context of disasters, SoVI construction using DEA requires defining the DMUs in alignment with the geographical boundary within which relative vulnerability can be assessed. This study considers only those districts of Bihar that are defined flood prone by the DMD, GoB. For social vulnerability assessment, we considered five output and five input indicators (Table 33.2). Values of indicators were estimated as average values for the recent 10 years period, that is, 2007−16. During this period, Bihar has experienced flooding of varying intensity, expansion, and attributes. The year 2007 is considered as one of the major flooding years in the known flood history of the state. In this year, the flooding was characterized by both the high discharge in various rivers in

Table 33.2 Input−Output Indicators for the Data Envelopment Analysis Evaluation of Social Vulnerability in Bihar.

Indicators	Measurement	Unit	Data Source
Output			
Economic damage	Real values of crop loss, public property damage, and house damage	INR	DMD, GoB
Human death	Total number of loss of lives	Absolute	DMD, GoB
Cattle death	Total number of cattle died	Absolute	DMD, GoB
Population affected	% of Total population affected	%	DMD, GoB
Cattle affected	Total number of cattle affected	Absolute	DMD, GoB
Input			
Mean elevation	As explained in Section 33.4.2.1	Meter from the sea level	As explained in Section 33.4.2.1
Population density	Population/area	Per km^2	Census of India, 2011
Per capita GDDP	GDDP/population	INR	Directorate of Economics and Statistics, GoB, Census of India, 2001 and 2011
Road density	Total road length/area	km/km^2	Economic Survey of Bihar, various issues
Medical	Total number of health institution/ 100,000 person	As explained in column 2	Economic Survey of Bihar, various issues

DMD, Disaster Management Department; *GDDP*, Gross District Domestic Product; *GoB*, Government of Bihar.

north Bihar along with high rainfall in the region. If the number of loss of lives can be an indicator of the scale of devastation, official estimates show that 1287 people died due to flood events across 22 districts.[1] 2008 flooding would not have gained worldwide attention, had there been no large-scale devastation brought by river Kosi due to the breach in the embankment in the upstream at Kusaha in Nepal. Five districts, Supaul, Saharsa, Madhepura, Purnea, and Araria, were severely affected by floods. As per the official estimates, just three districts Supaul, Saharsa, and Madhepura accounted for 537 deaths out of the total casualty figure of 626 in the state. A total of 2.5 million people were affected, and the damages on account of the economy and the number of affected were so severe that the region is still to recover from the loss. In 2016, flood expansion was of such a large scale that 31 out of 38 districts in the state reported flooding with varying intensity.[2] However, the damage was also on high scale, but comparatively, less than the flood year

---

[1]For detail, see: http://www.disastermgmt.bih.nic.in/Statitics/Copy%20of%20Daily%20reportIX%20format(Final%20Report%202007).pdf.

2. For detail, see: http://www.disastermgmt.bih.nic.in/Statitics/Copy%20of%20Final%20Flood%20Report%202008.pdf.

[2]This included the 6 districts out of 10 south Bihar districts which DMD does not consider flood prone. Nevertheless, flood damage intensity was moderate as per the official damage estimate. For detail, see http://www.disastermgmt.bih.nic.in/Statitics/FormIX2016.pdf.

of 2007. This period also includes the year 2015 when the state suffered least from the floods in the years after 2000.[3] Thus we can observe that the nature of flooding in the state has been stochastic. Therefore, the average value across the years for each indicator would provide a balanced view of the social vulnerability when the loss and damage across the years are factored in. However, it is to be noted that the selection of input indicators is sensitive to the availability of data. Therefore, we had to limit ourselves to only five indicators.

33.4.1 OUTPUT INDICATORS

Riverine flooding causes various damages to the residing population, their occupation, livestock, infrastructure, farmland, etc. Form IX of DMD of GoB provides interannual details on account of various losses and damages. We considered losses of lives on account of humans and cattle, affected population and cattle, and economic damage as our output indicators. For the economic damage, we considered price-adjusted real values of crop damage, house damage, and public property damage in Indian rupee terms. State GDP deflator was used for the price adjustment. Spatial variations across the loss and damage indicators are provided in Fig. 33.2A—E.

33.4.2 INPUT INDICATORS

Assessing current vulnerability to the floods requires a baseline to understand the impact of the future adaptation policies (Willner, Levermann, Zhao, & Frieler, 2018). However, widening the scope of the assessment brings various complications. Birkmann (2007) and also Birkmann et al. (2013) in this regard recognized that the literature on vulnerability has evolved to delineate various dimensions under various spatial and temporal circumstances and in the process brought some conceptual and empirical complications. Our assessment also suffers from this limitation. The considered study period does not allow us to incorporate various socioeconomic indicators available in the 2011 census; and the recent loss and damage are required to establish the symmetry among the input and output indicators simultaneously. Due to this, we restricted our analysis to only five indicators as shown in Table 33.2.

33.4.2.1 Mean elevation

Mean elevation is an important environmental factor that shapes the vulnerability of people in a given location. The expected risks triggered by climate change will have implications for the future human settlement, occupation, economic development, and urbanization, etc. The elevation of location makes inhabitants vulnerable to such risks. For example, low-lying coastal areas are susceptible to suffer more from the double whammy of weather extremes as well as the sea level rise to add substantial pressure on future adaptation and mitigation efforts (Liu et al., 2015). In the riverine locations, elevation plays an important role in determining the runoff as well as the land use pattern. Hence, elevation acts as a crucial intermediary interface in the event of hydrometeorological human—environment interaction. We estimated the mean elevation of flood-prone districts of Bihar by using GIS.

---

[3]For detail, see http://www.disastermgmt.bih.nic.in/Statitics/FormIX2015.pdf.

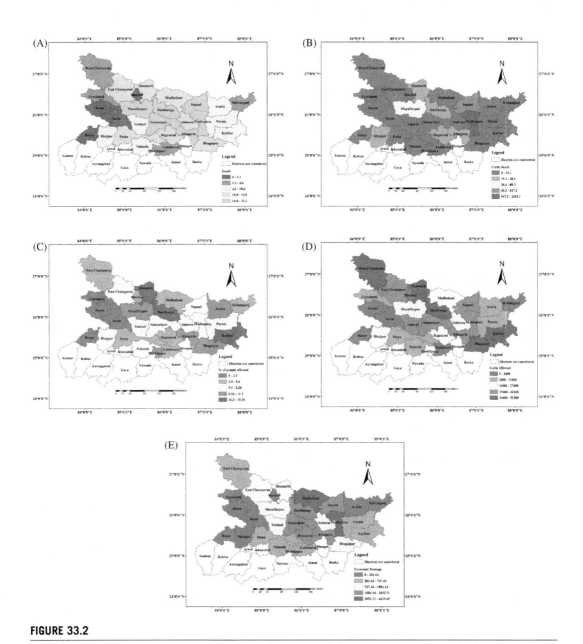

FIGURE 33.2

Spatial variation across the output indicators. (A) Loss of lives; (B) Cattle death; (C) % of people affected; (D) Number of cattle affected; and (E) Economic damage.

Using ArcGIS 10.1 software, we initially geo-referenced 1:1,000,000 Bihar toposheet and digitized various district boundaries in WGS_1984 coordinate system. Next, we overlaid the layer of ASTER Digital Elevation Model (DEM) by mosaicking 28 GDEM granules of 30 m resolution. ASTER GDEM data were retrieved through a formal e-mail request to the Earthdata, National Aeronautics and Space Administration. All the raster tiles were mosaicked following their respective latitude and longitude positions and series numbers. The assembled granules were then superimposed on the digitized boundary of districts following which district subsets of DEM were obtained for each digitized boundary. These subset images were reprojected under Universal Transverse Mercator zone 45N which is an apt category for projection for the location. We provide the spatial variation on account of district-wise mean elevation in Fig. 33.3A.

33.4.2.2 Population density

Population density has been considered a common indicator in the social vulnerability literature. Bihar has the highest population density among all the states in India. In the event of flooding, densely populated locations are more difficult for the evacuation (Sherly, Karmakar, Parthasarathy, Chan, & Rau, 2015). As we have considered 2007—16 as our study period, and as the census information in India is found for every 10 years, we had to estimate district-wise population density using two methods. For the years 2007—10, we used compound annual growth rate between 2001 and 2011 census population for every district and interpolated the data for the in-between period. For the period 2012—16, we employed the ratio method to compute the district population density from the projected state population density.[4] Spatial variation of the average value of population density across the districts is illustrated in Fig. 33.3B.

33.4.2.3 Per capita Gross District Domestic Product

Per capita income provides a buffer to individuals in the recovery process in the event of a disaster event. To measure the coping capacity, this indicator has been extensively used in the assessment of social vulnerability literature. Even in the place-based models, this indicator gains a prominent place (see, e.g., Chen et al., 2013; Cutter et al., 2003; Zhou et al., 2014a,b). However, Gross District Domestic Product (GDDP) data are only available up to the period 2011—12 for each district. For the remaining years in our study period, we interpolated the information from the gross state domestic product (GSDP) using 2011—12 GDDP proportion to the 2011—12 GSDP. Using population data, we estimated per capita GDDP across these years that were adjusted with the 2004—05 prices. District-wise spatial variation in per capita GDDP is shown in Fig. 33.3C.

33.4.2.4 Road density

Road density within a location reflects the reaction and recovery capacity in the event of a disaster (Hou et al., 2016). It has an intermediary function in two ways: first, it provides the evacuation medium for the individual as well as facilitates the state agencies to reach to the affected and second, it helps the disaster management relief and aid providing agencies to take stock of the situation for the immediate action and the reinforcement. The spatial variation in the district-wise road density is illustrated in Fig. 33.3D.

---

[4]https://www.census2011.co.in/census/state/bihar.html.

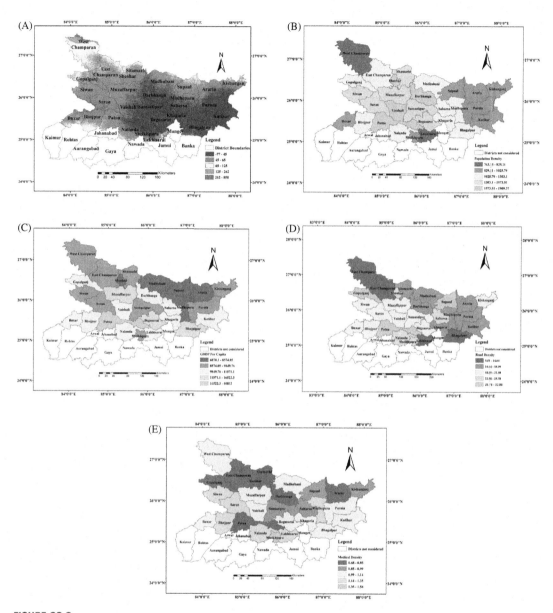

FIGURE 33.3

Spatial variation across the input indicators. (A) Elevation grade; (B) Population Density; (C) GDDP per capita; (D) Road density; (E) Medical density.

33.4.2.5 Medical density

State of the medical facility in the disaster affected location is considered as an additional reaction and recovery capacity in the social vulnerability model (Hou et al., 2016). It signifies disaster affected locations' recovery and coping potential in terms of providing immediate medical aid to people having suffered injuries of various degrees (Chen et al., 2013; Zhou et al., 2014a,b). Although these studies consider the unit as the number of beds available for defined population size, we have considered the number of health institutions per 100,000 population as the indicator due to lack of data on the number of beds at the district level. We have illustrated the district-wise spatial variation in available health institutions in Fig. 33.3E.

Except for the mean elevation for which we have provided the pixel gradation, the spatial variation has been illustrated through the choropleth[5] across all the indicators. We have adopted the Jenks natural breaks classification to cluster the variations into five classes. This is to ensure that variance within the classes is minimum whereas the variance between the classes is maximum (Jenks, 1967). We used ArcGIS 10.1 to do this exercise.

33.4.3 METHOD

DEA method can be used to understand the occurrence process of flood in the social vulnerability assessment. DEA has been employed under a given state of socioeconomic and environmental conditions, and the loss and damages suffered due to floods in the literature mentioned before. Popularly known as CCR, DEA was propagated by Charnes, Cooper, and Rhodes (1978) to evaluate the efficiency through a frontier analysis of inputs and outputs for a given set of DMUs. A basic CCR DEA model can be presented as

$$\max h_0 = \frac{\sum_{r=1}^{s} u_r y_{r0}}{\sum_{i=1}^{m} v_i x_{i0}} \tag{33.1}$$

Subject to:

$$\frac{\sum_{i=1}^{s} u_r y_{rj}}{\sum_{i=1}^{m} v_i x_{ij}} \leq 1; \quad j = 1, \ldots, n$$

$$u_r v_{ij} \geq 0; \quad r = 1, \ldots, s; \quad i = 1, \ldots, m$$

where the y_{rj} and x_{ij} (all positive) are the known outputs and inputs of jth DMU, and the $u_r v_i \geq 0$ are the variable weights to be determined by the solution of this problem. The efficiency of one member of this reference set of $j = 1, \ldots, n$ DMUs is to be rated relative to the others. Efficiency score ranges between 0 and 1, and the score reflects the radial distance from the estimated production frontier for the DMUs under assessment. Thus the DMUs with the score less than 1 are relatively inefficient compared to the DMUs on frontier. This means that a linear combination of other units from the sample could produce the same vector of outputs using a smaller combination of other unit inputs (Andersen & Petersen, 1993).

[5]*Choropleth* map is a thematic map depicting the variable under consideration.

The virtue of the DEA model lies in not requiring a priori weights on inputs and outputs. However, a drawback with this model is that a considerable number of DMUs are typically characterized as efficient, restricting the intercomparison. Superefficiency model in that sense can provide a solution as it facilitates comparison between the efficient DMUs. Following Seiford and Zhu (1999), a formal input-based constant return to scale superefficiency DEA model can be expressed as

$$\min \theta$$
$$s.t. \sum_{\substack{j=1 \\ j \neq 0}}^{n} \lambda_j x_j \leq \theta x_0$$
$$\sum_{\substack{j=1 \\ j \neq 0}}^{n} \lambda_j y_j \geq y_0;$$
$$\theta, \lambda_j \geq 0, \quad j \neq 0$$

$$(33.2)$$

For n DMUs, each DMU$_j$ ($j = 1, 2, \ldots, n$) consumes a vector x_j inputs to produce a vector of y_j outputs and where (x_0, y_0) represents DMU$_0$. In the superefficiency CCR model the problem of infeasibility occurs if there is a certain zero data pattern in the inputs and outputs.

33.4.4 SPATIAL ANALYSIS

To investigate whether there exists a spatial pattern and clustering among the districts of Bihar in relation to the social vulnerability, we used ESDA. We followed a standard two-step procedure, first by using global Moran's I and followed it with local Moran's I. Global Moran's I examines the extent to which similarity or dissimilarity pattern of a concern attribute—in our case the social vulnerability to floods—exists in all the neighboring areas within the defined space. On the other hand, local Moran's I assesses the clustering pattern at the local level. In simple words, Moran's I can be expressed as the cross product statistic between a location variable and its neighbor, while the variable is considered as a deviation from its mean. Mathematically, Moran's I can be expressed as

$$I = \frac{\sum_i \sum_j w_{ij} z_i . z_j / S_0}{\sum_i z_i^2 / n}$$

$$(33.3)$$

where I is global Moran's index, n is the number of observation; $z_i = x_i - \bar{x}$ for a location i; W_{ij} is the spatial weight matrix which reflects the relationship measure of units i and j; and $S_0 = \sum_i \sum_j w_{ij}$ as the sum of all the weights.

Moran's I typically follows the null hypothesis of spatial randomness. Opengeoda 1.4.1, instead of assuming the normality distribution or randomization of the statistic, calculates a reference distribution by randomly permuting the overserved values over the locations. A so-called pseudo P-value is calculated as

$$P = \frac{R + 1}{M + 1}$$

$$(33.4)$$

where R denotes the number of times the computed Moran's I from the permuted data set equals or go on to more extreme than the observed statistic, and M denotes the number of permutation that is usually considered as 99, 999, etc.

Moran's I scatter plot is centered around mean; in other words, Moran's I value lies between -1 and $+1$ around the four quadrants. A negative value denotes the negative spatial autocorrelation, and a positive value denotes the positive spatial autocorrelation. In addition to the P value, standard z score evaluates the significance test of spatial pattern.

Local Moran's I (Anselin, 1995), on the other hand, can be expressed as

$$I_i = z_i \sum_j w_{ij} z_j, \qquad (33.5)$$

where $z_i = x_i - \bar{x}$ for a location i, $z_j = x_j - \bar{x}$, and summation over j is such that only neighboring values $j \in J_i$ are considered.

In this way, local Moran's I detects the spatial clusters in terms of positive spatial autocorrelation [high–high (H–H) and low–low (L–L)] and the negative spatial autocorrelation [high–low (H–L) and low–high (L–H)].

33.4.5 STEPS FOR INDEX OPERATIONALIZATION PROCESS AND SPATIAL MAPPING

We followed the following procedure for the index generation and spatial mapping exercise.

Step 1: examine the linear relationship across the output and input indicators.

Step 2: normalize the indicators using z score if the indicators in any or both of the domains are found to be correlated.

Step 3: assess the suitability of PCA by employing Bartlett's test of sphericity and Kaiser–Mayer–Olkin (KMO) sampling adequacy test. KMO value, which varies between 0 and 1, is acceptable, when it is more than 0.5.

Step 4: perform PCA to extract PCs using the varimax rotation and limit the number of components as suggested by the Kaiser criterion (i.e., eigenvalues >1).

Step 5: transform the data across the indicators as the extracted PCs will have negative values that are prohibited in the DEA model. Further, superefficiency DEA model is infeasible if there is zero data pattern in the inputs and outputs. To avoid these problems a linear stretch of the data to a new data range (1–5) can be obtained using the following minimum–maximum standardization method.

$$V' = \frac{V - \min_A}{\max_A - \min_A} (5 - 1) + 1 \qquad (33.6)$$

Step 6: apply superefficiency DEA model if the number of efficient DMUs is found to be more than 1 in the classic CCR model.

Step 7: perform ESDA across the study units to examine the similarity or dissimilarity pattern using Opengeoda 1.4.1 software.

33.5 RESULTS AND DISCUSSION

Using SPSS 22 for the first four steps as mentioned in Section 33.4.5, we found the output domain indicators were correlated; however, in the input domain, we did not find indicators exhibiting a strong correlation. The significant correlation among the output indicators was intuitively expected as they reflect the damage and loss outcomes from the floods. Hence, to retain the maximum information from the original indicators and yet to have a set of uncorrelated indicators, we carried out a PCA. Two PCs were extracted with a cumulative explained variance of 81.43%. The KMO sample adequacy test (score = 0.71) and Bartlett's sphericity test ($P < .05$; chi square score = 61) showed that indicators were PCA suitable for the dimension reduction. The first PC had the dominant loading from the affected accounts, that is, population affected (0.90) and the cattle affected (0.89), whereas the second PC received the dominant loading from the loss accounts, that is, economic damage (0.76), human death (0.73), and cattle death (0.90). As the communalities scores were found to be more than 0.5, we could infer that the extracted factors were representing the original indicators well with the minimum loss of information. Due to the data of two extracted PCs having negative values and some of the input indicators across DMUs having zero values, we transformed the data in the linear stretch of 1−5 using the formula explained in step 5 in Section 33.4.5.

After data processing, we computed DEA-based SoVI using DEAP 2.1; the results varying efficiency scores across 28 DMUs (28 districts as the study units) showed that we had 7 districts (Bhagalpur, Sitamarhi, Supaul, Darbhanga, West Champaran, Katihar, and Madhepura) having efficiency scores 1 ($\theta = 1$). We used superefficiency DEA to rank these efficient DMUs using EMS 1.3 software. District-wise superefficiency scores and corresponding vulnerability ranks are shown in Table 33.3.

To illustrate the spatial variation across the districts, we imported these vulnerability scores in ArcGIS 10.1 for choroplething the geo-referenced district-wise shapefile. We adopted Jenks natural breaks classification method to classify the districts' vulnerability for the reason mentioned in Section 33.4.2. District-wise spatial variation in the different classes of social vulnerability is shown in Fig. 33.4. In Table 33.3, we find that Siwan is found to be the least vulnerable district (SE score 30.51%, rank 28), whereas the district Madhepura (SE score 249.48%, rank 1) receives

Table 33.3 District-Wise Superefficiency Vulnerability Scores and Corresponding Ranks.

Sr. No.	Districts	SE Score (Rank)	Sr. No.	Districts	SE Score (Rank)
1	Siwan	30.51 (28)	15	Madhubani	76.4 (14)
2	Saran	31.41 (27)	16	Sheikhpura	80.35 (13)
3	Buxar	42.58 (26)	17	Samastipur	85.71 (12)
4	Gopalganj	53.09 (25)	18	Araria	86.25 (11)
5	Nalanda	54.51 (24)	19	Sheohar	86.99 (10)
6	Kishanganj	58.48 (23)	20	Khagaria	97.2 (9)
7	Bhojpur	61.53 (22)	21	East Champaran	97.63 (8)
8	Vaishali	63.18 (21)	22	Bhagalpur	107.21 (7)
9	Muzaffarpur	65.32 (20)	23	Sitamarhi	121.53 (6)
10	Patna	67.49 (19)	24	Supaul	129.74 (5)
11	Lakhisarai	69.45 (18)	25	Darbhanga	141.45 (4)
12	Purnia	69.73 (17)	26	West Champaran	161.79 (3)
13	Begusarai	69.82 (16)	27	Katihar	171.66 (2)
14	Saharsa	74.49 (15)	28	Madhepura	249.48 (1)

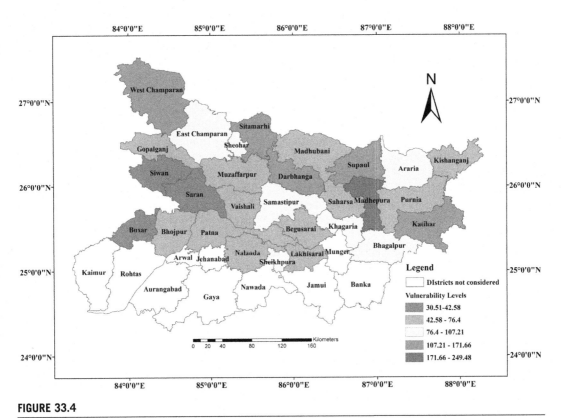

FIGURE 33.4

Spatial variation in social vulnerability to floods across the Bihar districts.

the highest vulnerability score. The mean of these district-wise social vulnerability scores stands at 89.46%, indicating the enormity of the task at hand in flood disaster reducing and coping capabilities enhancing exercise. In addition, the difference between the vulnerability scores of these two districts is quite large meaning intradistrict social vulnerability to floods difference is large within the state boundary. As per the Jenks natural breaks in Fig. 33.4, districts such as Katihar, Darbhanga, Supaul, Sitamarhi, and West Champaran come under the high category of vulnerability. Districts Sheikhpura, Samastipur, Araria, Sheohar, Khagaria, East Champaran, and Bhagalpur constitute the medium social vulnerability category. Whereas, Gopalganj, Nalanda, Kishanganj, Bhojpur, Vaishali, Muzaffarpur, Patna, Lakhisarai, Purnea, Begusarai, Saharsa, and Madhubani are the districts that exhibit low social vulnerability to floods. The districts Siwan, Saran, and Buxar have the least level of vulnerability. Districts in north Bihar are found to be more vulnerable than the districts in south.

To examine the spatial effect of the district-wise social vulnerability to floods, we performed the ESDA using Opengeoda 1.4.1. Initially, to generate a weight matrix, we opted the queen first-order contiguity and estimated global Moran's *I*. This was followed by estimating local Moran's *I* to analyze the similarity/dissimilarity pattern among the location polygons. Finally, we generated the spatial cluster maps to illustrate whether the districts are categorized under standard clusters of H−H, L−L, H−L, or L−H. We found that global Moran's *I* value is positive and statistically significant at *P* < .01. This explains that social vulnerability to floods in Bihar has significant global clustering effect. We further assessed local Moran's *I* to assess the local level spatial clustering pattern. The scatter plot shows three lead patterns: H−H, L−H, and L−L. The H−H concentration is confined to the district Katihar, whereas L−H concentration is located at the districts Saharsa, and Purnia. Lastly, Siwan, Saran, Bhojpur, and Patna are the districts that show the L−L spatial pattern (Fig. 33.5).

Starting with north Bihar, in the geo-cultural zone 1 formed by river Ghaghra−Gandak, districts West Champaran and East Champaran show medium to high vulnerability, whereas the districts Gopalganj, Siwan, and Saran reflect low to the lowest vulnerability pattern. Comparing to the spatial distribution pattern emerged across the output domain indicators and input domain indicators (Figs. 33.2 and 33.3), West Champaran presents an interesting case of a high level of social vulnerability. Even though the damages across the output indicators are in the low range (except the affected cattle) and the mean elevation of this district is highest, and population density is among the lowest, other socioeconomic indicators, namely per capita GDDP, road density, and medical density reflect the tattered state of this district. District East Champaran, on the other hand, shows a medium level of vulnerability mainly attributed to the varying degrees of outcome damages in the output and input indicators group. Remaining three districts, Siwan, Saran, and Buxar, are showing low to very low levels of vulnerability that is mainly attributed to low to very low levels of damages outcomes in the output domain.

In the geo-cultural zone 2, three districts, Muzaffarpur, Vaishali, and Begusarai, are reflecting the low level of social vulnerability, whereas Sheohar and Samastipur are showing the medium level of social vulnerability. Only Sitamarhi is placed in the group of districts with a high level of vulnerability. A medium vulnerability of Sheohar is attributed to low to very low levels of loss and damage account in the output domain, although medical density and per capita GDDP are in the lower side, and population density is on the higher side. In addition, the road density is on the higher side. District Muzaffarpur is found to be in the low vulnerability group as the socioeconomic input indicators perform well in comparison to the other districts, although the damage varies from medium to very high levels. In the same group, district Vaishali accompanies Muzaffarpur as the loss and damage across the output indicators range between very low and high, whereas road density, medical density, and per capita GDDP are in the medium range with very high population density. District Sitamarhi in this zone is found to be in the high vulnerable group due to two reasons: (1) except the cattle death, all other indicators are in the range of medium to very high group and (2) socioeconomic performance in the input groups are not performing well. With a high population density, Sitamarhi does not exhibit high per capita GDDP (in fact, it is placed in the low group), poor road network, and a mediocre level of medical facilities, this district shows all the elements of existing socioeconomic vulnerability.

Remaining in the geo-cultural zone 2 are the neighboring districts—Begusarai and Samastipur—which depict low-to-medium levels of social vulnerability, respectively. Begusarai exhibits very low to high levels of loss and damage outcomes across the output indicators, but the relative performance in the socioeconomic indicators in the input domain offsets the various losses and damages to place the district in the low vulnerable districts group. On the other hand, Samastipur does not perform better in offsetting socioeconomic indicators when compared to the

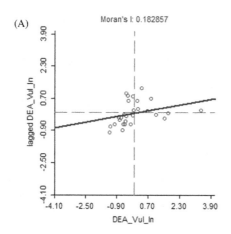

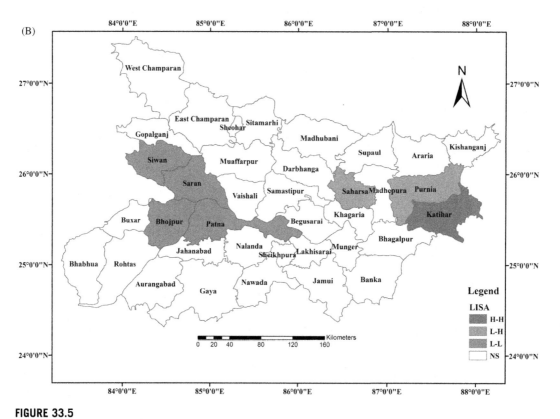

FIGURE 33.5

(A) Moran's *I* scatter plot for the social vulnerability to floods in Bihar and (B) LISA cluster map for the social vulnerability to floods in Bihar.

district Begusarai that makes this district placed in the medium vulnerability group. It is to be noted here that these northwest districts in Bihar have low to very high level of elevation depicting a varying degree of disaster environment in comparison the other districts of Bihar.

Geo-cultural zone 3, formed by the rivers Bagmati and Kosi, comprises the districts Darbhanga, Madhubani, Supaul, Saharsa, and Khagaria. These districts have suffered the most over the years. Adhwara group of rivers, Baghmati, and Kamla—Balan flood parts of districts Darbhanga and Madhubani, whereas river Kosi also brings floods in its western embankment area of these two districts. District Supaul, which was earlier a part of the district Saharsa, gets flooded every year. These two districts are situated in the eastern embankment of the river Kosi. The remaining district Khagaria is in unique location as it functions as the downstream of Bagmati (as a tributary of Kosi), Burhi Gandak, and the river Ganges. Geographically, most of the districts are low lying. Hence, in this geo-cultural zone, physical aspects play a crucial role in shaping the overall social vulnerability. However, our results show that two districts, Madhubani and Saharsa, are categorized in the group of low social vulnerability, whereas the two other districts, Darbhanga and Supaul, are placed in high social vulnerability group. District Khagaria is found to be in the medium social vulnerability zone. In the output domain, on account of the loss and damage outcomes, the loss of lives in the districts Madhubani and Saharsa is high but on account of the number of cattle deaths is low. In the group affected (both population and cattle), these two districts are in the medium to the low category. Madhubani may have experienced very high economic damage during the study period, but Saharsa is in the middle group of districts that have suffered the economic losses due to floods. On the socioeconomic front, these two districts are performing differently. Population density is in medium to low, per capita GDDP is in very low to medium, road density is in low to high, and medical density is in low to the medium category for both the districts, respectively. For the districts, Darbhanga and Supaul, high social vulnerability is attributed to low to very high loss and damages in the output domain. On the other hand, in the input domain, the relative very high to very low population density, medium to very low per capita GDDP, low to very high road density, and low to very low medical density are the main offsetting reasons for their respective positions in their group for both the districts. Finally, Khagaria's place in the medium social vulnerability group is attributed to the reasons cited earlier. All these districts are mostly in the low to very low-lying area. Some parts of districts Supaul and Madhubani are in medium elevation levels, but these parts function more as the upstream for the flood causing rivers.

In the geo-cultural zone 4, under the river Kosi and Mahananda, districts Madhepura, Araria, Purnia, Kishanganj, and Katihar show very differential grades of social vulnerability. On the one hand, results show that Madhepura is found as the most socially vulnerable districts of Bihar and Katihar are in the high group, on the other hand, Purnia and Katihar are in the low social vulnerability group. One of the reasons because of which Madhepura is found as an outlier in the Jenks natural classification scheme is the flood outcome of devastating scale in the district during 2008 Kosi floods. Notwithstanding, this district has been regularly flooded in other years except 2009, 2012, 2013, and 2015 during the considered study period. The other significant socially vulnerable district Katihar constitutes the most low-lying area in the state where the mean elevation is found to be only 25. 9 m from the sea level. Its relative position in the high vulnerability group is because of very low to very high loss and damage outcomes in the output domain as well as very low- to medium-socioeconomic conditions. The other three districts, Araria, Purnia, and Kishanganj, are placed in the low social vulnerability, even though the input indicators across these districts reflect the abysmal condition of people on account of per capita GDDP, road density, and medical density. As our model is contextual upon the loss and damage accounts due to floods, the low social vulnerability of these districts are mainly attributed to their overall low loss and damage outcomes during the assessment period.

From the geo-cultural zone 5, 6, and 7, districts Buxur and Bhojpur, districts Patna and Nalanda, and the districts Sheikhpura, Lakhisarai, and Bhagalpur are considered flood prone, respectively. These districts are in the south of the river Ganges, hence called as south Bihar districts. As per the Jenks classification, overall this area may not be considered socially vulnerable to floods as they are mostly classified in the low to very low vulnerability group. Sheikhpura is an exception that shows the medium level of vulnerability.

33.6 SUMMARY AND CONCLUSION

This chapter has attempted at assessing the district-wise social vulnerability to floods guided by generating the production efficiency scores of various losses and damages within the given socioeconomic conditions in the Indian state of Bihar. Following previous literature, we approached DEA and identified a set of output and input indicators while considering each district as DMU. Our findings show that districts in the north Bihar are more socially vulnerable to floods than the districts in the south. Most vulnerable districts are located in all the four geo-cultural zones. Further, the spatial analysis shows that there is a positive spatial clustering as well as the spatial outliers. We used the superefficiency DEA method for the twin purposes of (1) overcoming the a priori weight issues across the indicators; and (2) to reflect that why an area experiences differential damage in relation to the existing socioeconomic and given environmental conditions. This is not adequately addressed in the traditional additive methods.

In the line of DEA, we assumed that vulnerability is directly proportional to the losses and damages account and inversely proportional to the socioeconomic conditions. Efficiency scores from this ratio can provide important policy guidelines in reducing the disaster risk as well as enhancing the coping capacities. Proper disaster management efforts need to be in place and regularly reinforced by awareness, effective warning system at the local level with the use of current technology, updated flood hazard zones delineation, effective insurance tools, etc. Because the river floods in Bihar—and therefore the associated losses and damages—are transboundary in its characteristics, proper and timely coordination between the DMD and water resource department of GoB along with the coordination between responsible authorities in Nepal and Government of India would strengthen the warning system. On the other hand, enhancing the income opportunities, and access to various infrastructure, would increase people's capabilities to cope with the recurrent flood hazard in a location. When these efforts are targeted towards increasing then the policy tools would be more effective.

Although we have used (superefficiency) DEA as an assessment method with an assumption that it would overcome the shortcoming of traditional composite methods, the ranking of vulnerable districts is only relative to the efficient units. In addition, the data operationalization process in the DEA method limit the scope of the study in terms of examining the interannual variations in social vulnerability as zero values pattern across the indicators tend to make the method infeasible.

A recent 10-year average value consideration for all the indicators might have enhanced this study's value at the impact level, but we had to compromise on many other socioeconomically relevant indicators because the availability of data is a constraint for the considered period. Besides, vulnerability to floods at a location is driven by many environmental risk causing factors that can be represented by rainfall runoff, water level and discharge at the gauge stations, sedimentation, etc. However, data availability to these attributes across the considered flood-affected districts of Bihar is another constraint. This study also suffers from the problem of scales as the data across indicators are either added or they are averaged from the downscaled units such as census households, villages, and blocks. Adaptation efforts are more effective at the downscaled units. Hence, regular update in hazard zone delineation complemented by rich information at the downscaled units would make the assessment more precise. On these contours, this study provides important direction and scope for future research.

REFERENCES

Adger, W. N. (2006). Vulnerability. *Global Environmental Change*, *16*(3), 268–281. Available from https://doi.org/10.1016/j.gloenvcha.2006.02.006.

Adger, W. N., & Kelly, P. M. (1999). Social vulnerability to climate change and the architecture of entitlements. *Mitigation and Adaptation Strategies for Global Change*, *4*(3–4), 253–266. Available from https://doi.org/10.1023/A:1009601904210.

Andersen, P., & Petersen, N. C. (1993). A procedure for ranking efficient units in data envelopment analysis. *Management Science*, *39*(10), 1261–1264. Available from https://doi.org/10.1287/mnsc.39.10.1261.

Anselin, L. (1995). Local indicators of spatial association—LISA. *Geographical Analysis*, *27*(2), 93–115. Available from https://doi.org/10.1111/j.1538-4632.1995.tb00338.x.

Birkmann, J. (2007). Risk and vulnerability indicators at different scales: Applicability, usefulness and policy implications. *Environmental Hazards*, *7*(1), 20–31. Available from https://doi.org/10.1016/j.envhaz.2007.04.002.

Birkmann, J., Cardona, O. D., Carreño, M. L., Barbat, A. H., Pelling, M., Schneiderbauer, S., . . . Welle, T. (2013). Framing vulnerability, risk and societal responses: The MOVE framework. *Natural Hazards*, *67*(2), 193–211. Available from https://doi.org/10.1007/s11069-013-0558-5.

Blaikie, P., Cannon, T., Davis, I., & Wisner, B. (1994). *At risk: Natural hazards, people's vulnerability and disasters*. London: Routledge.

Brooks, N., & Adger, W. N. (2003). *Country level risk measures of climate-related natural disasters and implications for adaptation to climate change* (Vol. 26). Retrieved from <http://www.uea.ac.uk/env/people/adgerwn/wp26.pdf>.

Cannon, T. (2008). Vulnerability, "innocent" disasters and the imperative of cultural understanding. *Disaster Prevention and Management*, *17*(3), 350–357. Available from https://doi.org/10.1108/09653560810887275.

Chakraborty, A., & Joshi, P. K. (2016). Mapping disaster vulnerability in India using analytical hierarchy process. *Geomatics, Natural Hazards and Risk*, *7*(1), 308–325. Available from https://doi.org/10.1080/19475705.2014.897656.

Charnes, A., Cooper, W. W., & Rhodes, E. (1978). Measuring the efficiency of decision making units. *European Journal of Operational Research*, *2*(6), 429–444. Available from https://doi.org/10.1016/0377-2217(78)90138-8.

Chen, W., Cutter, S. L., Emrich, C. T., & Shi, P. (2013). Measuring social vulnerability to natural hazards in the Yangtze River Delta region, China. *International Journal of Disaster Risk Science*, *4*(4), 169−181. Available from https://doi.org/10.1007/s13753-013-0018-6.

Clark, G. E., Moser, S. C., Ratick, S. J., Dow, K., Meyer, W. B., Emani, S., ... Schwarz, H. E. (1998). Assessing the vulnerability of coastal communities to extreme storms: The case of Revere, MA, USA. *Mitigation and Adaptation Strategies for Global Change*, *3*(1), 59−82.

Cutter, S. L. (1996). Vulnerability to hazards. *Progress in Human Geography*, *20*(4), 529−539.

Cutter, S. L., Boruff, B. J., & Shirley, W. L. (2003). Social vulnerability to environmental hazards. *Social Science Quarterly*, *84*(2), 242−261.

Cutter, S. L., Emrich, C. T., Webb, J. J., & Morath, D. (2009). *Social vulnerability to climate variability hazards: A review of the literature*. Hazards & Vulnerability Research Institute, University of South Carolina. Available from http://adapt.oxfamamerica.org/resources/Literature_Review.pdf.

Cutter, S. L., & Finch, C. (2008). Temporal and spatial changes in social vulnerability to natural hazards. *Proceedings of the National Academy of Sciences of the United States of America*, *105*(7), 2301−2306.

Cutter, S. L., Mitchell, J., & Scott, M. S. (2000). Revealing the vulnerability of people and places: A case study of Georgetwon County, South Carolina. *Annals of the Association of American Geographers*, *90*(4), 713−737.

Eakin, H., & Luers, A. L. (2006). Assessing the vulnerability of social-environmental systems. *Annual Review of Environment and Resources*, *31*(1), 365−394. Available from https://doi.org/10.1146/annurev. energy.30.050504.144352.

Few, R. (2003). Flooding, vulnerability and coping strategies: Local responses to a global threat. *Progress in Development Studies*, *3*(1), 43−58. Available from https://doi.org/10.1191/1464993403ps049ra.

Gautam, D. (2017). Assessment of social vulnerability to natural hazards in Nepal. *Natural Hazards and Earth System Sciences*, *17*(12), 2313−2320. Available from https://doi.org/10.5194/nhess-17-2313-2017.

Goldberg, D. E. (1989). *Genetic algorithms in search, optimization, and machine learning*. New York: Addison-Wesley Publishing Company, INC.

Hou, J., Lv, J., Chen, X., & Yu, S. (2016). China's regional social vulnerability to geological disasters: Evaluation and spatial characteristics analysis. *Natural Hazards*, *84*(Suppl. 1), 97−111. Available from https://doi.org/10.1007/s11069-015-1931-3.

Huang, D., Zhang, R., Huo, Z., Mao, F., Youhao, E., & Zheng, W. (2012). An assessment of multidimensional flood vulnerability at the provincial scale in China based on the DEA method. *Natural Hazards*, *64*(2), 1575−1586. Available from https://doi.org/10.1007/s11069-012-0323-1.

Huang, J., Liu, Y., Ma, L., & Su, F. (2013). Methodology for the assessment and classification of regional vulnerability to natural hazards in China: The application of a DEA model. *Natural Hazards*, *65*(1), 115−134. Available from https://doi.org/10.1007/s11069-012-0348-5.

Jenks, G. F. (1967). The data model concept in statistical mapping. *International Yearbook of Cartography*, *7*, 186−190.

Kelly, P. M., & Adger, W. N. (2000). Theory and practice in assessing vulnerability to climate change and facilitating adaptation. *Climatic Change*, *47*(4), 325−352. Available from https://doi.org/10.1023/ A:1005627828199.

Li, C. H., Li, N., Wu, L. C., & Hu, A. J. (2013). A relative vulnerability estimation of flood disaster using data envelopment analysis in the Dongting Lake region of Hunan. *Natural Hazards and Earth System Sciences*, *13*(7), 1723−1734. Available from https://doi.org/10.5194/nhess-13-1723-2013.

Liu, J., Wen, J., Huang, Y., Shi, M., Meng, Q., Ding, J., & Xu, H. (2015). Human settlement and regional development in the context of climate change: A spatial analysis of low elevation coastal zones in China. *Mitigation and Adaptation Strategies for Global Change*, *20*(4), 527−546. Available from https://doi.org/ 10.1007/s11027-013-9506-7.

Mavhura, E., Manyena, B., & Collins, A. E. (2017). An approach for measuring social vulnerability in context: The case of flood hazards in Muzarabani district, Zimbabwe. *Geoforum, 86*, 103−117. Available from https://doi.org/10.1016/j.geoforum.2017.09.008.

Mazumdar, J., & Paul, S. K. (2016). Socioeconomic and infrastructural vulnerability indices for cyclones in the eastern coastal states of India. *Natural Hazards, 82*(3), 1621−1643. Available from https://doi.org/10.1007/s11069-016-2261-9.

Mazumdar, J., & Paul, S. K. (2018). A spatially explicit method for identification of vulnerable hotspots of Odisha, India from potential cyclones. *International Journal of Disaster Risk Reduction, 27*, 391−405. Available from https://doi.org/10.1016/j.ijdrr.2017.11.001.

Nicholls, R. J., Hoozemans, F. M. J., & Marchand, M. (1999). Increasing flood risk and wetland losses due to global sea-level rise: Regional and global analyses. *Global Environmental Change, 9*(Suppl. 1), S69−S87. Available from https://doi.org/10.1016/S0959-3780(99)00019-9.

Opiyo, F. E. O., Wasonga, O. V., & Nyangito, M. M. (2014). Measuring household vulnerability to climate-induced stresses in pastoral rangelands of Kenya: Implications for resilience programming. *Pastoralism, 4* (1), 1−15. Available from https://doi.org/10.1186/s13570-014-0010-9.

Ribot, J. C. (2010). Vulnerability does not just fall from the sky: Towards multi-scale pro-poor climate policy. In R. Mearns, & A. Norton (Eds.), *Social dimensions of climate change: Equity and vulnerability in warming world* (pp. 47−74). Washington, DC: The World Bank.

Rygel, L., O'Sullivan, D., & Yarnal, B. (2006). A method for constructing a social vulnerability index: An application to hurricane storm surges in a developed country. *Mitigation and Adaptation Strategies for Global Change, 11*(3), 741−764. Available from https://doi.org/10.1007/s11027-006-0265-6.

Seiford, L. M., & Zhu, J. (1999). Infeasibility of super-efficiency data envelopment analysis models. *INFOR: Information Systems and Operational Research, 37*(2), 174−187. Available from https://doi.org/10.1080/03155986.1999.11732379.

Sherly, M. A., Karmakar, S., Parthasarathy, D., Chan, T., & Rau, C. (2015). Disaster vulnerability mapping for a densely populated coastal urban area: An application to Mumbai, India. *Annals of the Association of American Geographers, 105*(6), 1198−1220. Available from https://doi.org/10.1080/00045608.2015.1072792.

Tapsell, S., McCarthy, S., Faulkner, H., & Alexander, M. (2010). *Social vulnerability to natural hazards in Brazil. CapHaz_Net WP4 report*. London: Flood Hazard Research Centre − FHRC, Middlesex University. Available from https://doi.org/10.1007/s13753-016-0090-9.

Wei, Y. M., Fan, Y., Lu, C., & Tsai, H. T. (2004). The assessment of vulnerability to natural disasters in China by using the DEA method. *Environmental Impact Assessment Review, 24*(4), 427−439. Available from https://doi.org/10.1016/j.eiar.2003.12.003.

Willner, S. N., Levermann, A., Zhao, F., & Frieler, K. (2018). Adaptation required to preserve future high-end river flood risk at present levels. *Science Advances, 4*(1), 1−9. Available from https://doi.org/10.1126/sciadv.aao1914.

Wisner, B., Blaikie, P., Cannon, T., & Davis, I. (2004). *At risk: Natural hazards, people's vulnerability and disasters* (2nd ed.). London: Routledge.

Zhang, N., & Huang, H. (2013). Social vulnerability for public safety: A case study of Beijing, China. *Chinese Science Bulletin, 58*(19), 2387−2394. Available from https://doi.org/10.1007/s11434-013-5835-x.

Zhou, Y., Li, N., Wu, W., & Wu, J. (2014a). Assessment of provincial social vulnerability to natural disasters in China. *Natural Hazards, 71*(3), 2165−2186. Available from https://doi.org/10.1007/s11069-013-1003-5.

Zhou, Y., Li, N., Wu, W., Wu, J., & Shi, P. (2014b). Local spatial and temporal factors influencing population and societal vulnerability to natural disasters. *Risk Analysis, 34*(4), 614−639. Available from https://doi.org/10.1111/risa.12193.

Zou, L. Le, & Wei, Y. M. (2009). Impact assessment using DEA of coastal hazards on social-economy in Southeast Asia. *Natural Hazards, 48*(2), 167−189. Available from https://doi.org/10.1007/s11069-008-9256-0.

SUBJECTIVE WELL-BEING IMPACTS OF NATURAL HAZARDS: A REVIEW

Olivia Jensen and Chitranjali Tiwari

LRF Institute for the Public Understanding of Risk, National University of Singapore, Singapore

34.1 INTRODUCTION

"Subjective well-being" (SWB) approaches, sometimes "life satisfaction" or "experienced utility" approaches are increasingly used by researchers and policy-makers to complement established methods for the valuation of public goods. This chapter reviews the small but growing body of literature which applies this method to evaluate the impacts of natural hazards.

The first section of the chapter focuses on the psychological and social impacts of natural hazards. The impacts on individual mental health are found to be severe and likely to be severely underestimated by evaluations based on cost of health care services. On the other hand, an interesting strand of literature shows mixed impact of natural hazards on social indicators. There is some evidence that disasters have a positive impact on social cohesion and trust which in turn positively correlated with well-being, while other studies find that disasters are associated with a decline in these social indicators.

The second section of the chapter looks at the application of SWB or "experienced utility" as a method in impact valuation. It presents the concept of experienced utility, how it is measured, and how it can be quantified for inclusion in benefit−cost analysis. It then presents the findings of studies applying this quantification method to natural hazards. The discussion in Section 34.3 considers the advantages and limitations of the use of SWB as a method to complement other approaches to valuing nonmarket economic impacts. The chapter argues that despite certain limitations, SWB is a useful complement to existing methods of disaster evaluation that may be particularly useful in low-income contexts, and for the evaluation of compensation and recovery interventions.

34.2 PSYCHOLOGICAL, SOCIAL, AND WELL-BEING IMPACTS OF NATURAL DISASTERS

Natural hazards affect survivors profoundly through loss of life and harm to others, loss of personal property, and widespread destruction and displacement. These lead to increases in specific psychological problems like acute stress disorder, posttraumatic stress disorder (PTSD), major depressive disorder and generalized anxiety; nonspecific psychological problems; and chronic problems in

Economic Effects of Natural Disasters. DOI: https://doi.org/10.1016/B978-0-12-817465-4.00034-0

583

living (Norris et al., 2002). Depression and stress may in turn lead to negative impacts on physical health (Krause, 1987). To take just one example, a study on the psychological impacts of Hurricane Katrina found the prevalence of PTSD stood at almost one-third of the population 30 days after the disaster, compared to 5% before (Davis, Grills-Taquechel, & Ollendick, 2010). Other studies confirmed the significant and long-lasting psychological impacts of Hurricane Katrina on residents (Mills, Edmondson, & Park, 2007; Paxson, Fussell, Rhodes, & Waters, 2012; Rhodes et al., 2010) and widespread short-term impacts even on those not directly affected by the disaster (Kimball, Levy, Ohtake, & Tsutsui, 2006).

This psychological distress falls more heavily on certain groups after disasters. Children are particularly vulnerable (Kar, 2009; Vigil & Geary, 2008); women (Dell'Osso et al., 2011; Priebe et al., 2011), the elderly (Okuyama & Inaba, 2017) and groups with lower income and lower education levels (Sekulova & Van den Bergh, 2016) are also found to experience more adverse psychological effects of natural disasters and experience slower and more incomplete recovery. Ethnicity is also found to be significant: in the United States, trauma rates are found to be higher among Latinos and African Americans than among whites (Calvo, Arcaya, Baum, Lowe, & Waters, 2015). Finally, low social support and poor preexisting health status are significantly correlated with psychopathology symptoms (Kamo, Henderson, & Roberto, 2011).

While the negative psychological impacts of disasters can be severe, they can be mediated by resilience or protective factors at the individual level. At the individual level, resilience is understood as the ability to avoid experiencing significant stress at any point in time. The characteristics of resilient individuals include positive emotion, optimism, and access to prompt intervention immediately following the disaster (Vázquez, Cervellón, Pérez-Sales, Vidales, & Gaborit, 2005).

A key concern with the psychological impacts of disasters is that these harms may not be easily identified, insured, or compensated. Furthermore, compensation for harms may be inequitably distributed, and in some circumstances, the process of seeking compensation itself may even add to stress and anxiety (Ritchie, 2012). The mitigating effect of compensation will be lower for marginalized individuals and those with few assets who may be excluded from the compensation process and for those who are uninsured or only partially insured (Botzen, Deschenes, & Sanders, 2019).

As physical recovery takes place over time, so too does psychological recovery, but the length of time and completeness of the recovery varies greatly. Some studies of specific disasters find enduring adverse effects. Ardalan et al. (2011) finds incomplete recovery 5 years after the Bam earthquake; Papanikolaou, Adamis, Mellon, Prodromitis, and Kyriopoulos (2012) find lower quality of life among those who experienced wildfires in Greece after 3 years; and Wang et al. (2000) find persistent negative effects 9 months after earthquakes in China. However, others find that well-being levels do return to ex ante levels after disasters when measured over a longer timeframe of 4 or more years (Calvo et al., 2015; Priebe et al., 2011), and some find that SWB recovers less than 2 years after a risk event (Krause, 1987; Luechinger & Raschky, 2009).

Disasters are also found to have an impact on the well-being of people outside affected areas but the direction of this effect differs across national contexts. In the United States, Kimball et al. (2006) found a significant but short-lived nationwide drop in SWB after Hurricane Katrina. In contrast, a study in Japan found a positive effect on well-being among young people in areas not directly affected by the Great East Japan Earthquake (Uchida, Takahashi, & Kawahara, 2014). In this study, the increase in well-being was associated with greater value which respondents placed on social connectedness and everyday life following the disaster and increases in prosocial behavior like making donations and volunteering. Related findings came from a nationwide study in Japan that found that six times as many people reported increases in SWB following the Great East Japan earthquake compound disaster as reported drops in SWB (Ishino, Kamesaka, Murai, & Ogaki, 2012). This result may be due to a comparative effect in which the disaster prompts people to reevaluate their own lives positively These findings suggest the possibility of cultural differences in response to disasters, for example, that Japanese may be more likely to actualize their happiness from perceived social support from others compared to Americans and thus to experience a net positive SWB impact.

Patterns in self-reported happiness among those directly affected by disasters are broadly consistent with the findings on prevalence of psychological disorders. Floods (Sekulova & Van den Bergh, 2016), storms (von Möllendorff & Hirschfeld, 2016), droughts (Carroll, Frijters, & Shields, 2009), earthquakes (Huang & Wong, 2014), and volcanic eruptions are all associated with a significant decline in self-reported well-being.

Going beyond single events, multiple experiences of natural hazards appears to have a profound impact on life satisfaction. Repeated direct and indirect experience of repeated hazards may induce continual anxiety among residents in exposed areas, reducing well-being and potentially leading to feelings of low self-efficacy when people feel that they are unable to move away from the location. Empirical studies have found large, negative, and statistically significant correlations between the frequency and severity of natural hazards events in a locality and average SWB (Carroll et al., 2009; Luechinger & Raschky, 2009; von Möllendorff & Hirschfeld, 2016). It is worth noting that these results were found in Germany (von Möllendorff & Hirschfeld, 2016) and Australia (Carroll et al., 2009) even though residents in these locations were covered by insurance and could expect compensation and support from government in the event of a disaster.

The negative relationship between exposure to hazards and SWB is even more pronounced in lower income contexts. For example, the negative correlation between SWB national-level hurricane frequency and severity measures found in a study of 80 countries was larger in lower and middle-income countries than in high-income countries (Berlemann, 2016). National income appears to act as a protective factor, perhaps by providing reassurance to people that they will receive compensation in the event of a disaster (Berlemann, 2016).

Although the majority of studies find a decline in SWB among those directly affected by a disaster, contrasting results have been found in Japan, where residents of areas impacted by the 1995 Hanshin earthquake in Kobe expressed higher levels of well-being than people outside those areas (Yamamura, Tsutsui, Yamane, Yamane, & Powdthavee, 2015). This surprising finding may be partly explained by the relationship between disasters and social connectedness and prosocial behavior.

It appears that social connections act as a buffer or mediating factor, weakening the negative impact of the shock on SWB (Lin et al., 2002). At the community level, social support is found to play a protective role quality of life postdisaster (Ardalan et al., 2011; Lin et al., 2002; Lowe, Chan, & Rhodes, 2010). Kaniasty (2012) highlights the role of perceptions of social support and aid in the 12 months after the disaster as protective factors that lessen the impact of disasters on SWB drawing on evidence from flood events in Poland. Conversely, the absence of social resources interacts with harm to amplify negative impacts on SWB (Hamama-Raz, Palgi, Leshem, Ben-Ezra, & Lavenda, 2017).

Studies also find large increases in prosocial behaviors following disasters, shown as cooperative behavior (Eckel, El-Gamal, & Wilson, 2009; Whitt & Wilson, 2007), altruism (Ishino et al., 2012), and social engagement (Okuyama & Inaba, 2017). One possible explanation is that resources of social trust and solidarity lie dormant within societies and only come to light in moments when the usual social order is suspended. These in turn have effects on SWB, with those who engage in pro-social behaviors postdisaster reporting higher SWB levels.

This relationship is also seen at the societal level in a positive correlation between disaster occurrence and trust, which appears to act as a protective factor (Toya & Skidmore, 2014). Several studies find significant correlations between adverse conditions, social networks, and social trust (Cassar, Healy, & Von Kessler, 2017; Phan & Airoldi, 2015; Veszteg, Funaki, & Tanaka, 2015). An interesting area of exploration is the possibility of reverse causality: those areas that are heavily impacted by disasters may have higher social cohesion. Adverse environmental conditions could strengthen social cohesion if people are obliged to work together to overcome harsh conditions. Conversely, repeated disasters could be expected to lead to a breakdown of social cohesion, trust, and prosocial behavior. Calo-Blanco, Kovářík, Mengel, & Romero (2017) finds support for the hypothesis that disasters have a positive impact on cohesion in Chile, using a careful study of areas with different levels of exposure to earthquakes.

However, the positive interaction between disasters and prosocial behavior appears to erode with time (Calo-Blanco et al., 2017; Solnit, 2010) and may even reverse in the medium-term as divergence in recovery paths become apparent (Sekulova & Van den Bergh, 2016).

Kaniasty (2012) finds that compassion is replaced by disillusionment and perceptions of the "harsh reality of grief, loss, and destruction" (Kaniasty, 2012, 22). Competition and polarization replace generosity and solidarity and survivors express disappointment that support from family and friends was not provided as readily as expected. Tensions may be exacerbated by the practical difficulties of distributing aid in the aftermath of the disaster. Less than 2 years after the disaster studied, victims were more likely to have withdrawn from social interactions. This resonates with the argument put forward by Hobfoll (2004) that the negative spiral of resource loss and declining well-being predominates.

34.3 SUBJECTIVE WELL-BEING: DEFINITION, MEASUREMENT, AND APPLICATIONS

This section provides an exposition of the concept of SWB, how it is measured, and the sources of data available for research. SWB refers to an individual's judgment of their own happiness or satisfaction with life. It is a self-reported measure, simply interpreted as thinking and feeling that life is going well (Diener, 2009). There are two dimensions of SWB: a short-term affective (emotional) dimension and a long-term (cognitive) dimension. The effect (positive or negative) captures the mood and this may change rapidly in response to stimuli, while the cognitive component is a more stable measure which is not expected to vary from day to day in the absence of a major shock in life circumstances. In the context of SWB and natural hazards, it is this latter dimension, cognitive life satisfaction, which is of primary interest and SWB data collection methods are designed to elicit assessments of longer term life satisfaction.

Critics have called into question whether self-reports provide a reliable way of measuring life satisfaction. In order to test for reliability, self-reports have therefore been compared with alternative measures of life satisfaction like assessments by experts and peers, experience sampling, recall of positive, and negative events and neurological activity (Diener, 1994). These comparisons find that on average self-reports converge with alternative measures, confirming that self-reports are a reasonably reliable measure of well-being.

Self-reports may nevertheless be a "noisy" measure of cognitive satisfaction as people's responses are influenced by short-term swings in their mood (Frey & Stutzer, 2010). This noisiness is less of a concern in large sample sizes when the measurement error would be unlikely to affect the results, but it may be a concern when SWB data are only available for a small sample. Under some conditions, however, emotions could lead to systematic bias in the data, if, for example, an external event such as a win for a local sports team leads to a short-term mood swing in a large number of respondents in a locality.

Life satisfaction self-reports are generally collected through large-scale surveys. SWB questions are now included as part of some large national or international social surveys or may be included in dedicated survey questionnaires for research projects in specific locations. The main SWB data sources are summarized in Table 34.1.

Reliance on survey data implies major challenges relating to the timing of surveys and the temporal nature of the impact of life events on happiness. Empirical work suggests that people have a baseline level of happiness to which they return relatively quickly after major life events such as marriage, accidents, and deaths or harm from hazards, a phenomenon known as the "hedonic treadmill" (Brickman & Campbell, 1971). While this phenomenon contributes to the stability of self-reported SWB as a measure, it also raises the issue that the impacts of shocks on well-being may be under- or overestimated depending on the time lapse between the shock and the self-report.

A further set of factors that may cause systematic bias is the structure of the survey itself and the inclusion of other topics in the survey. For example, survey questions on other topics such as living conditions can systematically influence life satisfaction reports (Stutzer & Frey, 2010). Given the survey context, people may also be influenced by social desirability/acceptability in their answers, creating systematic bias. Survey questionnaires should therefore be designed to address these issues as far as possible.

Table 34.1 Data Sources for Subjective Well-Being Data.

Survey	Period	Location
World Values Survey	6 waves 1981–2014	60 countries
Gallup World Poll	Biannual since 2005	140 countries
Latinobarometro	Annual	18 countries in Latin America
European Social Surveys	Biannual since 2002	38 countries in Europe
Eurobarometer	Semiannual since 1990	EU member + regional countries
European Values Study	5 waves 1981–2017	47 countries in Europe
General Social Survey	Biannual since 1994	United States
Japan Household Panel Survey	Annual	Japan
Australian Unity Wellbeing Index	Semiannual	Australia
Annual Population Survey	Annual	United Kingdom

There is no universally recognized question or set of survey questions to elicit SWB information from survey respondents. Typical survey questions include "Are you satisfied with your life [right now]?" or "If you were to consider your life in general these days, how happy or unhappy would you say you are?". Some surveys ask respondents to rate their agreement with negative statements such as "I lack many things that I would need in order to lead a happy life" or to compare themselves to others "Compared to others, I lead a happy life" (Hommerich, 2012).

Respondents are typically asked to rate their agreement with the statement on a scale. Among the measurement scales, two prominent scales are the five-item Satisfaction with Life Scale (Diener, Emmons, Larsen, & Griffin, 1985) and Cantril's Self-Anchoring Scale, which has the respondent rate his or her current life on a ladder scale in which 0 is "the worst possible life for you" and 10 is "the best possible life for you" (Kahneman & Deaton, 2010), although there are many variants employed.

The appropriate choice and structure of prompt questions and the rating scale used will depend on the context of the study and the planned uses of the data, but more consistency between survey organizations on SWB questions would allow for interesting comparative research across geographies.

The growing availability of SWB data has supported research in psychology investigating the personal and social factors correlated with SWB. This research shows a strong positive correlation between SWB and innate temperament and personality traits like optimism, extroversion and neuroticism, and physical health (Scheier & Carver, 1985). Many studies show a strong correlation between SWB and interpersonal factors in the form of rich and fulfilling relationships with a partner, family, and friends and measures of social participation, which are in turn connected to societal factors including trust, communal harmony, and the perception of being able to count on others (Diener & Seligman, 2002).

In economics, SWB data have been widely used to investigate the relationship between SWB and income. Numerous cross-sectional studies show a strong and robust positive relationship between SWB and income, a relationship that is found consistently within and across countries, cultures, and time periods (Veenhoven, Ehrhardt, Ho, & de Vries, 1993; Diener, 1994; Helliwell, Layard, & Sachs, 2012). On the other hand, Easterlin (1973) observed that increases in income over time in industrialized countries were not associated with equivalent increases in happiness. This paradox has given rise to a vigorous debate and different explanations have been put forward to explain it: the psychological process of adjustment (Diener, 2013), the selection and use of comparators (Duesenberry, 1949), and the effect of rising material norms in society (Easterlin, 1995). However, the existence of a positive relationship between SWB and income when measured at a point in time is robust and provides the foundation for the quantification of SWB, which we return to in the following section.

In the last two decades, the study of SWB has moved increasingly into the mainstream of economics (see reviews by Frey & Stutzer, 2010; Clark, Frijters, & Shields, 2008). The number and proportion of papers in peer-reviewed economics journals employing the concept have risen rapidly over this period (Levinson, 2013). While much of this work has looked at the relationship with income mentioned above, it has also been applied to other economic and social phenomena, including the overall level of unemployment (Blanchflower, Bell, Montagnoli, & Moro, 2014), the level of inequality (Schwarze & Härpfer, 2007) and political institutions (Frey & Stutzer, 2000). SWB is also widely accepted as an outcome measure in poverty and livelihood studies in developing countries (Narayan et al., 2000).

A relatively new but growing area of scholarship employs SWB to the valuation of environmental public goods (Welsch & Ferreira, 2014). It has been applied in the valuation of value environmental amenities (Ferreira et al., 2013); climate conditions (Maddison & Rehdanz, 2011; Van de Vliert, Huang, & Parker, 2004), environmental quality (Levinson, 2012; MacKerron & Mourato, 2009; Van Praag & Baarsma, 2005; Welsch, 2002), and urban regeneration (Dolan & Metcalf, 2008). Building on this, there has been increasing interest in applying SWB to the evaluation of natural disasters since the 2000s and these papers are reviewed in more depth in the following section, after a discussion of the method for quantification of SWB impacts.

34.4 QUANTIFICATION OF SUBJECTIVE WELL-BEING IMPACTS: METHOD AND FINDINGS

The quantification or monetization of SWB impacts relies on the robust relationship between income and SWB. Using survey data on income and SWB self-reports, regression analysis is employed to establish the correlation coefficient between income and SWB. The ratio of the marginal effect of the variable of interest on SWB to the marginal effect of income on SWB can then be calculated. This ratio, known as the compensating value, reflects the change (increase) in income that would compensate for the change (decrease) in SWB caused by the natural hazard, keeping the level of SWB constant.

Control variables with established correlations with SWB normally included in regressions are employment status, health, age, marital status, education level, race, gender, and, where available, extent and strength of personal and community relationships. Most studies use an Ordered Probit model, as the measure of SWB is a discrete noncontinuous number, although some studies use OLS (Sekulova & Van den Bergh, 2016) which may be equally robust for this type of analysis (Ferrer-i-Carbonell & Frijters, 2004).

Ideally, the compensating value between income and SWB would be generated from the study-specific dataset. In some cases, it is impossible to do this because the dataset does not include accurate and reliable data on respondent income. A further issue arises in datasets which include a financial SWB domain (e.g., responses to the question, "How happy are you with your financial situation?"). When a study-specific compensating factor cannot be established, estimates from other studies can be employed. Metaanalysis is a commonly used method for value transfer (Hudson, Botzen, Poussin, & Aerts, 2019a; Hudson, Pham, & Bubeck, 2019b). However, value transfer for SWB is subject to severe limitations, as is the case for other methods for nonmarket valuation (Rogers et al., 2019).

While SWB quantification methods are now being employed in nonmarket valuation, many challenges remain in their incorporation into impact evaluation. First, in studies that use general social survey data, it is difficult to identify the population affected by a hazard. Geographical location is usually used as a proxy but this may lead to measurement error, potentially leading to the underestimation of SWB effects if large numbers of unaffected respondents are resident in the hazard-prone area. Conversely, when a study-specific dataset is used, ex ante information on SWB levels is often missing. One possibility to address this is to match the affected communities with unaffected communities that share relevant characteristics using a "difference in differences" approach (Rehdanz, Welsch, Narita, & Okubo, 2015). A few studies are able to take advantage of coincidental timing when a survey with well-being information is conducted for other purposes in an area subsequently affected by a disaster (Vigil & Geary, 2008).

Table 34.2 summarizes the findings of published studies that use SWB for quantification of natural hazard impact.

Table 34.2 Quantification of Subjective Well-Being (SWB) Impacts of Natural Hazards.				
Author, Year of Publication	Hazard, Period	Location	SWB Cost/Value of Preventing a Sure Event	Note
Carroll et al. (2009)	Droughts, 2001–04	Australia	Drought in the spring period has a detrimental effect on life satisfaction on households in rural areas equivalent to an annual reduction in income of A $18,000 (38%) for a household with mean income.	Effects of spring drought on SWB large and significant only for rural households. SWB losses much higher in the region most severely affected by the drought.
Fernandez, Stoeckl, and Welters (2019)	Flood, 2008–13	Philippines	SWB reduction of US $2577–3221 due to flooding over a 5-year period. Equivalent to 3–4.4 times average annual income in the sample.	Intangible costs estimated to be equivalent to 40%–70% of tangible costs of flooding.
Hudson et al. (2019a)	Three flood-prone regions, time period indeterminate	France	At mean household income, flooding requires €150,000 in immediate compensation to attenuate SWB losses.	Intangible effects are twice as large as tangible effects. Self-protection in the form of elevating the house attenuates the negative impact on SWB worth about €39,000.
Hudson et al. (2019b)	Floods, June and September 2017	Vietnam	SWB impact of flood experience in the last 12 months equivalent to US $2100–4400 and US $480–1000 for respondents who had experienced a flood in the last 5 years. In the absence of a robust within-study value for the marginal effect of income on SWB, the study uses a range of three values for the marginal effect drawn from other studies (0.21–0.44).	Study finds significant gender differences in SWB impact of floods. The flood impact index (floods in the last 12 months, 5 years, and 10 years) is nearly 70% larger for women than for men.

Table 34.2 Quantification of Subjective Well-Being (SWB) Impacts of Natural Hazards. *Continued*

Author, Year of Publication	Hazard, Period	Location	SWB Cost/Value of Preventing a Sure Event	Note
Kountouris and Remoundou (2011)	Forest Fires, 1986–2003	Portugal, Spain, Italy, and Mediterranean France	Mean yearly welfare cost of forest fires per household estimated with SWB is €2900 in Spain, €317 in Mediterranean France, €1778 in Italy and €3165 in Portugal.	SWB effects of fires only significant in rural areas. Precipitation is used as an instrument for fire frequency and extent.
Luechinger and Raschky (2009)	Local flooding, 1973–98	16 European countries: regional data	Average WTP for the prevention of one sure event in the region of residence is $6505 or 23.7% of average household income. For a reduction of the probability of a flood by 2.6%, an individual would be willing to pay $195 or about 0.7% of annual household income.	The estimates of compensating surplus based on SWB are somewhat lower than estimates based on the hedonic pricing method. However, the difference is small and the estimates using SWB lie within the range of results for revealed preference method.
Ohtake, Yamada, and Yamane (2016)	Great East Japan Earthquake, 2011	Japan	Among those in directly affected regions, SWB impact equivalent to JPY 4.25 m per capita	Psychological distress in areas far from the quake epicenter was not significant.
Okuyama and Inaba (2017)	Great East Japan Earthquake, March 2011	Japan	Monetary value estimate for negative impact of reduced social interaction for respondents directly affected by the earthquake and in areas with medium seismic intensity equivalent to JPY 258,400 for interactions with neighbors and JPY 119,700 for interactions with friends and acquaintances.	Social engagement among disaster-affected individuals did not necessarily fall compared to nonaffected individuals. The direction and significance of the effect depend on the degree and level of damage.
Rehdanz et al. (2015)	Great East Japan Earthquake, 2011, tsunami, and Fukushima nuclear accident		The decline in "happiness with one's life in the previous year" in municipalities affected by the tsunami equivalent to JPY 1.7 million (72% of average annual income) and JPY 5.7 million (240% of average annual income) for those living in close	Employs a difference-in-differences approach. No effect was found on SWB measured as "happiness with one's whole life." No change in SWB found among people living close to nuclear facilities in general.

(Continued)

Table 34.2 Quantification of Subjective Well-Being (SWB) Impacts of Natural Hazards. *Continued*

Author, Year of Publication	Hazard, Period	Location	SWB Cost/Value of Preventing a Sure Event	Note
Sekulova and Van den Bergh (2016)	Flood events, 2005–10	Bulgaria	proximity (<150 km) to the Fukushima Daiichi power plant. The compensating surplus for a flood experience is 94% of average income level. For heavy damages, it is ~96% and for severe flood damages it is greater than the income level.	High estimates may be explained by the better quality of the data and by unobserved characteristics of the control group. Flood experience and worry about flood both negatively and significantly correlated with SWB.

Note: WTP, willingness to pay.

Our review of the scholarly literature on quantification of SWB impacts of natural hazards identified 10 relevant papers published from 2007 to 2019. The papers cover a range of natural hazards. Floods are the hazard type most commonly addressed by the papers (Fernandez et al., 2019; Hudson et al., 2019a, 2019b; Luechinger & Raschky, 2009; Sekulova & Van den Bergh, 2016), while one compound disaster event—the Great East Japan Earthquake, tsunami and Fukushima nuclear accident of 2011—is the focus of three papers (Ohtake et al., 2016; Okuyama & Inaba, 2017; Rehdanz et al., 2015). One paper was identified on the impact of droughts (Carroll et al., 2009), and one on the impact of forest fires (Kountouris & Remoundou, 2011).

Studies that are able to identify more accurately the population affected by a disaster find higher impacts than those studies which rely on residence in broad geographical regions as an indicator of whether a respondent is affected by a disaster. Consistent with expectations, estimates of SWB impacts are higher for those more exposed to the hazard and of those more vulnerable to its effects. For example, rural residents who are more likely to be engaged in agricultural activities experience a significant negative impact of drought on SWB while no significant effect is found in urban areas, where the impact of the drought is likely to be buffered by infrastructure and trade. In the cases of severe disasters triggered by natural hazards, like the compound disaster in Japan following the Great East Japan Earthquake of 2011, the SWB impacts can be very high, amounting to 240% of annual income for residents in proximity to the Fukushima power plant.

34.5 DISCUSSION

Given the relative novelty of the SWB approach, it is useful to compare the findings of these studies with standard cost assessments for natural hazards, which capture damage to property and costs of health services. Several of the SWB studies reviewed compare the SWB impact with standard direct cost estimates and find that the former yields much higher cost estimates (Hudson et al., 2019a; Sekulova and van den Bergh, 2016). These results are entirely consistent with the psychology literature which finds severe impacts from disasters on mental health. Hitherto, these costs have only been quantified at best through cost of mental health services provided, which is likely to have led to considerable underestimation. This gap between standard and SWB-based approaches therefore underlines the need to incorporate a measure of psychological costs in future natural hazard evaluations.

SWB results may also be compared to the other main methods for nonmarket valuation: revealed preference and stated preference. Revealed preference methods use information from markets associated with the nonmarket good being valued to infer values for the nonmarket good. The most common approaches are hedonic pricing, often applied to property valuations, and travel cost methods.

Luechinger and Raschky (2009) compare their SWB results with revealed preference studies that estimate the price discount for houses located in floodplains with an annual chance of a flood event of at least 1% in the United States. They find that their SWB estimates of the cost of flood hazards are lower than those found using hedonic pricing. The authors suggest that this may be because people adapt more quickly than they think of floods and therefore the discounts on house prices are larger than the measured loss in SWB.

A challenge in using revealed preference methods for natural hazards is that people's behavior reflects expected future risks. For example, Driscoll, Dietz, and Alwang (1994) and Bin and Polasky (2004) estimate larger price discounts in the aftermath of a flood than before the flood occurred. Furthermore, for revealed preference approaches to generate accurate estimates, markets must be complete and not distorted by regulation and other government interventions, and participants must have full information about the hazard risk. In many contexts, these conditions will not be met and SWB approaches may be a useful alternative approach.

The second main approach to valuing nonmarket goods is stated valuation. Stated preference methods employ survey data asking respondents directly how much they would be willing to pay for a good or service (contingent valuation), asking them whether they would accept to pay a specified price for a policy or outcome, or asking them to choose their preferred alternative between pairs of outcomes or features and prices or costs. While stated preference methods are increasingly used in cost estimation for natural hazards (Rogers et al., 2019), a concern is that people may give strategic answers seeking to influence a policy decision, or they may answer to express an attitude (e.g., of approval of a national park, or disapproval of a piece of infrastructure) rather than providing their true valuation. Respondents may also find it difficult to answer the questions because of their hypothetical nature and may ignore their budget constraints.

SWB approaches, on the other hand, may be less susceptible to strategic responses or to attitude expression because the link between life satisfaction questions and hazard experience is not made explicit. In studies that combine data on hazard rates with large-scale social survey data, the survey questionnaire may not include questions on hazards at all.

A further concern with the application of stated preference to natural hazards is that the answers will be biased if people tend to over- or underestimate the risk of an event. This may be particularly problematic when probability information is scarce (e.g., when no flood maps are publicly available), under conditions of nonstationarity, and in the direct aftermath of an event, which tends to inflate risk estimates as people are easily able to bring to mind a case of the event. In contrast,

SWB approaches estimate the effect when the risk materializes and so are not affected by biased risk perceptions.

While SWB approaches therefore prove to be a useful complement to other nonmarket valuation under certain conditions, it is necessary also to recognize their limitations. One important challenge with SWB methods is in the timing of data collection. In order to isolate the impact of the natural hazard on well-being, SWB data would ideally be collected prior to the disaster as well as at regular time intervals after the disaster. Of course, the nature of natural hazards is such that the location and timing of the event are not known beforehand and periodic national surveys may include only a small number of respondents from the hazard-stricken area. It is possible to address this through multievent studies, the approach taken by Carroll et al. (2009), Luechinger and Raschky (2009), and subsequent studies, but this is not applicable, of course, in studies of rare events or as a method for the assessment of costs of any particular event. In order to employ monetized values derived through SWB from realized disasters for policy evaluation, uncertainty about the likelihood and range of severity of future events must be taken into account. The reliability and validity of the estimates will therefore depend on the quality of the data on the hazard probability distribution.

One of the challenges with all nonmarket valuation techniques is in transferring benefits (or cost) valuations developed with data in one location and for one intervention (or risk) to other locations. Nonmarket valuation studies will of course be more accurate when conducted in the specific location of interest. However, this is not always possible given financial resource and other constraints. Under these conditions, "benefit transfer"—the use of value estimates from existing studies—may be possible, if the two study and policy contexts are sufficiently similar (Rogers et al., 2019). The growing number of stated preference studies for certain hazards and types of impacts support the validity of benefit transfer using this approach. At this stage, however, the limited number of studies employing SWB approaches means that it is unlikely to be possible to identify a sufficiently similar study for benefit transfer or to conduct a metaanalysis of study results. As the number of studies employing SWB methods grows, the scope for benefit transfer will increase.

34.6 CONCLUSIONS AND DIRECTIONS FOR FUTURE RESEARCH

In the context of rising natural hazard risks driven by climate change, there is a profound need for sound evaluation methods that can help to address the underestimation of costs and associated underinvestment in hazard protection. In this regard, SWB complements existing methods by capturing the important psychological costs of single and repeated experience of natural hazards. It adds further value by redressing some of the bias towards the protection of sites with high asset values over high human exposure caused by current property-based evaluation methods. In providing evidence of the psychological costs of worrying about natural hazards, SWB approaches also draw attention to the desirability of policy interventions to increase resilience of communities ex ante rather than providing ad hoc ex post compensation to those affected by disasters.

SWB applications to natural hazards are linked to a broad upsurge in well-being studies in economics and the social sciences (Stutzer & Frey, 2010). The development of this area of study has been enabled by growing data availability. Well-being questions are now included in global surveys, large-scale regional surveys in the Americas and Europe, and in national surveys in Japan,

Australia, and elsewhere. In the future the inclusion of well-being questions in more surveys in developing countries would help to extend the application of SWB methods in natural hazard cost assessment.

SWB estimates have important potential to inform policy design and are increasingly being used to this end. They provide evidence to ground debates on which policies contribute to the well-being of citizens; a method to capture unforeseen and unintended consequences of policy interventions, be they positive or negative; and a way to quantify or monetize these impacts to compare them with other costs and impacts. They also allow for high granularity to identify and can therefore support the effective targeting of policy interventions. For example, as noted previously, women, the elderly, and those with poor health and lower incomes may experience sharper declines in SWB as a result of natural hazards, suggesting that additional interventions or better targeting of interventions are needed to protect these groups. SWB data collected periodically postdisaster could be used to assess the effectiveness and distributional consequences of recovery interventions and to calibrate and improve the targeting of compensation.

Some governments have already recognized the value of SWB approaches in policy design. Scotland, Iceland, and New Zealand are among those which have adopted well-being targets, indicators and "national accounts" in recent years (Rehdanz & Maddison, 2005; Rehdanz et al., 2015). In the United Kingdom, official government guidelines on policy appraisal and evaluation, known as the "Green Book," recommend that well-being indicators be used directly to consider some policy options, when the intervention is targeting a social outcome, or be monetized as part of a social cost-effectiveness analysis (UK Government, 2018). However, government agencies may need a better understanding of how to use SWB indicators. Natural hazard impact assessments could be an appropriate domain in which to test and increase confidence with SWB indicators and methods in policy design.

Increasing data availability and familiarity with SWB quantification methods provide a promising basis for future research, which could address the analysis of the interaction between natural hazards, institutions, and governance; investigation of the relationship between risk tolerance and well-being in communities with high natural hazard exposure; and the assessment of the possible well-being benefits of investment in hazard protection infrastructure such as seawalls, levees, or seismic building codes in comparison to postdisaster compensation.

REFERENCES

Ardalan, A., Mazaheri, M., Vanrooyen, M., Mowafi, H., Nedjat, S., Holakouie Naieni, K., & Russel, M. (2011). Post-disaster quality of life among older survivors five years after the Bam Earthquake: Implications for recovery policy. *Ageing & Society, 31*(2), 179–196.

Berlemann, M. (2016). Does hurricane risk affect individual well-being? Empirical Evidence on the indirect effects of natural disasters. *Ecological Economics, 124*, 99–113.

Bin, O., & Polasky, S. (2004). Effects of flood hazards on property values: Evidence before and after Hurricane Floyd. *Land Economics, 80*(4), 490–500.

Blanchflower, D. G., Bell, D. N. F., Montagnoli, A., & Moro, M. (2014). The happiness trade-off between unemployment and inflation. *Journal of Money, Credit and Banking, 46*(S2), 117–141.

Botzen, W. J. W., Deschenes, O., & Sanders, M. (2019). The economic impacts of natural disasters: A review of models and empirical studies. *Review of Environmental Economics and Policy*, *13*(2), 167−188.

Brickman, P., & Campbell, D. T. (1971). Hedonic relativism and planning the good society. In M. H. Appley (Ed.), *Adaptation level theory: A symposium* (pp. 287−304). New York: Academic Press. Available from https://scholar.google.com/scholar_lookup?title = Hedonic%20relativism%20and%20planning%20the% 20good%20society&author = P.%20Brickman&author = D.T.%20Campbell&pages = 287- 302&publication_year = 1971.

Calo-Blanco, A., Kovářík, J., Mengel, F., & Romero, J. G. (2017). Natural disasters and indicators of social cohesion. *PLoS One*, *12*(6), e0176885.

Calvo, R., Arcaya, M., Baum, C. F., Lowe, S. R., & Waters, M. C. (2015). Happily ever after? Pre-and-post disaster determinants of happiness among survivors of Hurricane Katrina. *Journal of Happiness Studies*, *16* (2), 427−442.

Carroll, N., Frijters, P., & Shields, M. A. (2009). Quantifying the costs of drought: New evidence from life satisfaction data. *Journal of Population Economics*, *22*(2), 445−461.

Cassar, A., Healy, A., & Von Kessler, C. (2017). Trust, risk, and time preferences after a natural disaster: Experimental evidence from Thailand. *World Development*, *94*, 90−105.

Clark, A. E., Frijters, P., & Shields, M. A. (2008). Relative income, happiness, and utility: An explanation for the Easterlin Paradox and other puzzles. *Journal of Economic Literature*, *46*(1), 95−144.

Davis, T. E., Grills-Taquechel, A. E., & Ollendick, T. H. (2010). The psychological impact from Hurricane Katrina: Effects of displacement and trauma exposure on university students. *Behavior Therapy*, *41*(3), 340−349.

Dell'Osso, L., Carmassi, C., Massimetti, G., Daneluzzo, E., Di Tommaso, S., & Rossi, A. (2011). Full and partial PTSD among young adult survivors 10 months after the L'Aquila 2009 earthquake: Gender differences. *Journal of Affective Disorders*, *131*(1−3), 79−83.

Diener, E. (1994). Assessing subjective well-being: Progress and opportunities. *Social Indicators Research*, *31* (2), 103−157.

Diener, E. (2009). *The science of well-being: The collected works of Ed Diener* (37). Springer.

Diener, E. (2013). The remarkable changes in the science of subjective well-being. *Perspectives on Psychological Science*, *8*(6), 663−666.

Diener, E., Emmons, R. A., Larsen, R. J., & Griffin, S. (1985). The satisfaction with life scale. *Journal of Personality Assessment*, *49*(1), 71−75.

Diener, E., & Seligman, M. E. (2002). Very happy people. *Psychological science*, *13*(1), 81−84.

Dolan, P., & Metcalf, R. (2008). *Comparing willingness-to-pay and subjective well-being in the context of non-market goods* (Vol. 890). Centre for Economic Performance, London School of Economics and Political.

Driscoll, P., Dietz, B., & Alwang, J. (1994). Welfare analysis when budget constraints are nonlinear: The case of flood hazard reduction. *Journal of Environmental Economics and Management*, *26*(2), 181−199.

Duesenberry, J. S. (1949). *Income, saving, and the theory of consumer behavior*. Cambridge, Mass: Harvard University Press.

Easterlin, R. A. (1973). Does money buy happiness? *The Public Interest*, *30*, 3.

Easterlin, R. A. (1995). Will raising the incomes of all increase the happiness of all?. *Journal of Economic Behavior & Organization*, *27*(1), 35−47.

Eckel, C. C., El-Gamal, M. A., & Wilson, R. K. (2009). Risk loving after the storm: A Bayesian-network study of Hurricane Katrina evacuees. *Journal of Economic Behavior & Organization*, *69*(2), 110−124.

Fernandez, C. J., Stoeckl, N., & Welters, R. (2019). The cost of doing nothing in the face of climate change: A case study, using the life satisfaction approach to value the tangible and intangible costs of flooding in the Philippines. *Climate and Development*, *11*(9), 825−838.

Ferreira, S., Akay, A., Brereton, F., Cuñado, J., Martinsson, P., Moro, M., & Ningal, T. F. (2013). Life satisfaction and air quality in Europe. *Ecological Economics, 88,* 1−10.

Ferrer-i-Carbonell, A., & Frijters, P. (2004). How important is methodology for the estimates of the determinants of happiness? *The Economic Journal, 114*(497), 641−659.

Frey, B. S., & Stutzer, A. (2000). Happiness, economy and institutions. *The Economic Journal, 110*(466), 918−938.

Frey, B. S., & Stutzer, A. (2010). *Happiness and economics: How the economy and institutions affect human well-being.* Princeton University Press.

Hamama-Raz, Y., Palgi, Y., Leshem, E., Ben-Ezra, M., & Lavenda, O. (2017). Typhoon survivors' subjective wellbeing—A different view of responses to natural disaster. *PLoS One, 12*(9), e0184327.

Helliwell, J., Layard, R. & Sachs, J. (2012). *World happiness report.* The Earth Institute, New York, USA: Columbia University.

Hobfoll, S. E. (2004). *Stress, culture, and community: The psychology and philosophy of stress.* Springer Science & Business Media.

Hommerich, C. (2012). Trust and subjective well-being after the Great East Japan Earthquake, tsunami and nuclear meltdown: Preliminary results. *International Journal of Japanese Sociology, 21*(1), 46−64.

Huang, Y., & Wong, H. (2014). Impacts of sense of community and satisfaction with governmental recovery on psychological status of the Wenchuan Earthquake survivors. *Social Indicators Research, 117*(2), 421−436.

Hudson, P., Botzen, W. J. W., Poussin, J., & Aerts, J. C. J. H. (2019a). Impacts of flooding and flood preparedness on subjective well-being: A monetisation of the tangible and intangible impacts. *Journal of Happiness Studies, 20*(2), 665−682.

Hudson, P., Pham, M., & Bubeck, P. (2019b). An evaluation and monetary assessment of the impact of flooding on subjective well-being across genders in Vietnam. *Climate and Development, 11*(7), 623−637.

Ishino, T., Kamesaka, A., Murai, T., & Ogaki, M. (2012). Effects of the Great East Japan Earthquake on subjective well-being. *Journal of Behavioral Economics and Finance, 5,* 269−272.

Kahneman, D., & Deaton, A. (2010). High income improves evaluation of life but not emotional well-Being. *Proceedings of the National Academy of Sciences of the United States of America, 107*(38), 16489−16493.

Kamo, Y., Henderson, T. L., & Roberto, K. A. (2011). Displaced older adults' reactions to and coping with the aftermath of Hurricane Katrina. *Journal of Family Issues, 32*(10), 1346−1370.

Kaniasty, K. (2012). Predicting social psychological well-being following trauma: The role of postdisaster social support. *Psychological Trauma: Theory, Research, Practice, and Policy, 4*(1), 22.

Kar, N. (2009). Psychological impact of disasters on children: Review of assessment and interventions. *World Journal of Pediatrics, 5*(1), 5−11.

Kimball, M., Levy, H., Ohtake, F., & Tsutsui, Y. (2006). *Unhappiness after Hurricane Katrina.* National Bureau of Economic Research.

Kountouris, Y., & Remoundou, K. (2011). Valuing the welfare cost of forest fires: A Life satisfaction approach. *Kyklos, 64*(4), 556−578.

Krause, N. (1987). Exploring the impact of a natural disaster on the health and psychological well-being of older adults. *Journal of Human Stress, 13*(2), 61−69.

Levinson, A. (2012). Valuing public goods using happiness data: The case of air quality. *Journal of Public Economics, 96*(9−10), 869−880.

Levinson, A. (2013). *Happiness, behavioral economics, and public policy* (No. w19329). National Bureau of Economic Research.

Lin, M.-R., Huang, W., Huang, C., Hwang, H.-F., Tsai, L.-W., & Chiu, Y.-N. (2002). The impact of the Chi-Chi Earthquake on quality of life among elderly survivors in Taiwan—A before and after study. *Quality of Life Research, 11*(4), 379−388.

Lowe, S. R., Chan, C. S., & Rhodes, J. E. (2010). Pre-hurricane perceived social support protects against psychological distress: A longitudinal analysis of low-income mothers. *Journal of Consulting and Clinical Psychology*, *78*(4), 551.

Luechinger, S., & Raschky, P. A. (2009). Valuing flood disasters using the life satisfaction approach. *Journal of Public Economics*, *93*(3–4), 620–633.

MacKerron, G., & Mourato, S. (2009). Life satisfaction and air quality in London. *Ecological Economics*, *68*(5), 1441–1453.

Maddison, D., & Rehdanz, K. (2011). The impact of climate on life satisfaction. *Ecological Economics*, *70*(12), 2437–2445.

Mills, M. A., Edmondson, D., & Park, C. L. (2007). Trauma and stress response among Hurricane Katrina evacuees. *American Journal of Public Health*, *97*(Suppl. 1), S116–S123.

Narayan, D., Chambers, R., Shah, M. K., & Petesch, P. (2000). *Voices of the Poor: Crying out for Change*. New York: Oxford University Press for the World Bank.

Norris, F. H., Friedman, M. J., Watson, P. J., Byrne, C. M., Diaz, E., & Kaniasty, K. (2002). 60,000 disaster victims speak: Part I. An empirical review of the empirical literature, 1981–2001. *Psychiatry: Interpersonal and Biological Processes*, *65*(3), 207–239.

Ohtake, F., Yamada, K., & Yamane, S. (2016). Appraising unhappiness in the wake of the Great East Japan earthquake. *The Japanese Economic Review*, *67*(4), 403–417.

Okuyama, N., & Inaba, Y. (2017). Influence of natural disasters on social engagement and post-disaster wellbeing: The case of the Great East Japan Earthquake. *Japan and the World Economy*, *44*, 1–13.

Papanikolaou, V., Adamis, D., Mellon, R. C., Prodromitis, G., & Kyriopoulos, J. (2012). Trust, social and personal attitudes after wildfires in a rural region of Greece. *Sociology Mind*, *2*(1), 87.

Paxson, C., Fussell, E., Rhodes, J., & Waters, M. (2012). Five years later: Recovery from post traumatic stress and psychological distress among low-income mothers affected by Hurricane Katrina. *Social Science & Medicine*, *74*(2), 150–157.

Phan, T. Q., & Airoldi, E. M. (2015). A natural experiment of social network formation and dynamics. *Proceedings of the National Academy of Sciences of the United States of America*, *112*(21), 6595–6600.

Priebe, S., Marchi, F., Bini, L., Flego, M., Costa, A., & Galeazzi, G. (2011). Mental disorders, psychological symptoms and quality of life 8 years after an earthquake: Findings from a community sample in Italy. *Social Psychiatry and Psychiatric Epidemiology*, *46*(7), 615–621.

Rehdanz, K., & Maddison, D. (2005). Climate and happiness. *Ecological Economics*, *52*(1), 111–125.

Rehdanz, K., Welsch, H., Narita, D., & Okubo, T. (2015). Well-being effects of a major natural disaster: The case of Fukushima. *Journal of Economic Behavior & Organization*, *116*, 500–517.

Rhodes, J., Chan, C., Paxson, C., Rouse, C. E., Waters, M., & Fussell, E. (2010). The impact of Hurricane Katrina on the mental and physical health of low-income parents in New Orleans. *American Journal of Orthopsychiatry*, *80*(2), 237.

Ritchie, L. A. (2012). Individual stress, collective trauma, and social capital in the wake of the Exxon Valdez oil spill. *Sociological Inquiry*, *82*(2), 187–211.

Rogers, A. A., Dempster, F. L., Hawkins, J. I., Johnston, R. J., Boxall, P. C., Rolfe, J., ... Pannell, D. J. (2019). Valuing non-market economic impacts from natural hazards. *Natural Hazards*, 1–31.

Scheier, M. F., & Carver, C. S. (1985). Optimism, coping, and health: Assessment and implications of generalized outcome expectancies. *Health Psychology*, *4*(3), 219.

Schwarze, J., & Härpfer, M. (2007). Are People inequality averse, and do they prefer redistribution by the state?: Evidence from German longitudinal data on life satisfaction. *The Journal of Socio-Economics*, *36*(2), 233–249.

Sekulova, F., & Van den Bergh, J. C. J. M. (2016). Floods and happiness: Empirical evidence from Bulgaria. *Ecological Economics*, *126*(C), 51–57.

Solnit, R. (2010). *A paradise built in hell: The extraordinary communities that arise in disaster*. Penguin.

Stutzer, A., & Frey, B. S. (2010). Recent advances in the economics of individual subjective well-being. *Social Research: An International Quarterly, 77*(2), 679–714.

Toya, H., & Skidmore, M. (2014). Do natural disasters enhance societal trust? *Kyklos, 67*(2), 255–279.

Uchida, Y., Takahashi, Y., & Kawahara, K. (2014). Changes in hedonic and eudaimonic well-being after a severe nationwide disaster: The case of the Great East Japan Earthquake. *Journal of Happiness Studies, 15*(1), 207–221.

UK Government, 2018. HM Treasury. *The Green Book: Central Government Guidance on Appraisal and Evaluation*. Available at: https://assets.publishing.service.gov.uk/government/uploads/system/uploads/attachment_data/file/685903/The_Green_Book.pdf, Accessed 25 June 2020.

Van de Vliert, E., Huang, X., & Parker, P. M. (2004). Do colder and hotter climates make richer societies more, but poorer societies less, happy and altruistic? *Journal of Environmental Psychology, 24*(1), 17–30.

Van Praag, B. M. S., & Baarsma, B. E. (2005). Using happiness surveys to value intangibles: The case of airport noise. *The Economic Journal, 115*(500), 224–246.

Vázquez, C., Cervellón, P., Pérez-Sales, P., Vidales, D., & Gaborit, M. (2005). Positive emotions in earthquake survivors in El Salvador (2001). *Journal of Anxiety Disorders, 19*(3), 313–328.

Veenhoven, R., Ehrhardt, J., Ho, M. S. D., & de Vries, A. (1993). *Happiness in nations: Subjective appreciation of life in 56 nations 1946–1992*. Erasmus University Rotterdam.

Veszteg, R. F., Funaki, Y., & Tanaka, A. (2015). The impact of the Tohoku Earthquake and tsunami on social capital in Japan: Trust before and after the disaster. *International Political Science Review, 36*(2), 119–138.

Vigil, J. M., & Geary, D. C. (2008). A preliminary investigation of family coping styles and psychological well-being among adolescent survivors of Hurricane Katrina. *Journal of Family Psychology, 22*(1), 176.

von Möllendorff, C., & Hirschfeld, J. (2016). Measuring impacts of extreme weather events using the life satisfaction approach. *Ecological Economics, 121*, 108–116.

Wang, X., Gao, L., Zhang, H., Zhao, C., Shen, Y., & Shinfuku, N. (2000). Post-earthquake quality of life and psychological well-being: Longitudinal evaluation in a rural community sample in northern China. *Psychiatry and Clinical Neurosciences, 54*(4), 427–433.

Welsch, H. (2002). Preferences over prosperity and pollution: Environmental valuation based on happiness surveys. *Kyklos, 55*(4), 473–494.

Welsch, H., & Ferreira S. (2014). Environment, well-being, and experienced preference. *Oldenburg Discussion Papers in Economics*.

Whitt, S., & Wilson, R. K. (2007). Public goods in the field: Katrina evacuees in Houston. *Southern Economic Journal*, 377–387.

Yamamura, E., Tsutsui, Y., Yamane, C., Yamane, S., & Powdthavee, N. (2015). Trust and happiness: Comparative study before and after the Great East Japan Earthquake. *Social Indicators Research, 123*(3), 919–935.

ECONOMIC CONSEQUENCES OF SLOW- AND FAST-ONSET NATURAL DISASTERS: EMPIRICAL EVIDENCES FROM INDIA

35

Vikrant Panwar, Ashish Sharma and Subir Sen

Department of Humanities and Social Sciences, Indian Institute of Technology, Roorkee, India

35.1 INTRODUCTION

India remains highly vulnerable to natural disasters occurring at repeated intervals such as floods, cyclones, and droughts. According to the latest estimates of the United Nations Office for Disaster Risk Reduction (UNDRR), about 300 disaster events have been reported in the country over the past two decades which led to an estimated USD 79.5 billion of direct economic losses, approximately 76,000 deaths, and over 1 billion people affected including those who were left homeless following the disaster. Notably, natural hazard-induced disasters may be broadly classified into two categories based on the specific characteristics they demonstrate. First are the fast-onset disasters such as floods, cyclones, earthquakes that occur rapidly, after-effects stay for a relatively short period, impact a particular location or geography leading to property damages, loss of human lives and livelihood. The slow-onset disasters, on the other hand, are events such as droughts. The characteristics, causes, nature, and the ability of a drought event to harm individuals along with the impact on economic growth remain a complex issue in comparison to fast-onset disasters. Droughts often have no specific time of arrival or ending and may spread over a small or a large geographic area. Though droughts do not cause any direct property damage, they lead to land degradation, increases in the rate of desertification, and can adversely affect the individuals, community, and overall economic growth of the affected region. Therefore, most of the times, it becomes difficult to identify their severity and the loss and damages they cause. With more visible destruction, intensity, as well as heightened public and media attention, government and policy makers remain more concerned about fast-onset events than slow-onset disasters. Given the different modes and impacts of natural disasters on the economy, it would be prudent to analyze the fast and slow-onset disasters' impact separately.

With this motivation, the present study attempts to analyze the relationship between natural disasters and economic growth for a sample of selected 25 Indian states for the period 1990−2015. Using augmented panel vector autoregression (PVAR-X), the dynamic responses of the economic indicators [e.g., state gross domestic product (SGDP)] to disaster shocks are generated to identify the disaster impact not only in the year of the event but also in the following years. The fast-onset disasters are represented by flood events while droughts are used to approximate the slow-onset disasters. The results of this exercise may prove to be helpful for the policy makers at both central and the state levels of governance to formulate the appropriate disaster management policies for the respective disaster types.

The chapter is structured as follows. The second section describes the meaning and nature of slow- and fast-onset disasters along with an account of recent trends worldwide and specifically in India. The third section discusses the state of literature on the relationship between economic growth and natural disasters, with specific focus on droughts and floods. Data and methodologies employed in this study are then discussed in the fourth section of this chapter. The fifth section discusses the results followed by the sixth section which concludes the chapter.

35.2 SLOW AND FAST-ONSET NATURAL DISASTERS: RECENT TRENDS

Natural disasters are classified into two broad categories: slow- and fast-onset disasters, based on the specific characteristics they demonstrate. The fast-onset disasters, such as floods, cyclones, and earthquakes, occur rapidly at a particular point in time. Further, these extreme events continue for a short period with impacts mostly localized but causing significant property damage and human losses. The onset and the end time of such disasters are quick and measurable in general. The slow-onset disasters, on the other hand, are events such as droughts that manifest slowly and stay for a significantly longer period of time in comparison to a fast-onset disaster. Droughts are of several types such as agricultural, hydrological, ecological, and meteorological, among others. They neither occur at a particular time of a day nor end abruptly. These events are highly localized and the syndrome manifests over a prolonged period of time through various ways. Although droughts do not cause any direct property damages, they do adversely affect the individuals, community, and overall economic growth of the affected region.

Globally, around 9700 natural disaster events were recorded over the period 1990−2015 which caused an estimated economic loss of USD 2.54 trillion and directly or indirectly affected about 5.4 billion lives [Centre for Research on the Epidemiology of Disasters (CRED), n.d.]. Economic losses due to floods and droughts were approximately 25% and 5%, respectively. Further, of the 5.4 billion affected population, 52% and 30% were affected by floods and droughts, respectively. Fig. 35.1 shows that on an annual basis, droughts and floods (together) adversely impact a greater proportion of the population in comparison to any other disaster events. Notably, the direct economic losses are felt more in the developing economies. Among developing economies, India is considered a vulnerable country due to its climatic and topological conditions along with its large geographical area and population. Natural disasters such as floods, droughts, storms, and earthquakes frequently occur in India causing widespread human and property losses. The Indian states that are in the dry, arid, and semiarid regions frequently face acute to chronic droughts due to either deficient rainfall or extreme variations in the rainfall pattern. On the contrary, excess rainfall causes floods in many states including those states that are prone to droughts, almost on an annual basis (Fig. 35.2). Therefore many Indian states are vulnerable to both fast- and slow-onset disasters. For

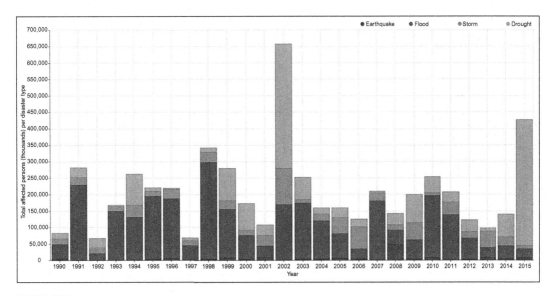

FIGURE 35.1

Total number of people affected (deaths, injured, homeless, or otherwise affected) caused by four major types of natural disasters worldwide during 1990−2015.

CRED-EMDAT.

example, Rajasthan, Maharashtra, Himachal Pradesh, and Madhya Pradesh witnessed a total of 16, 13, 12, 12 droughts, respectively, and 14, 14, 20, and 15 flood events, respectively, over the period 1990−2015. According to the disaster statistics from CRED-EMDAT, in India, droughts and floods on an average annual basis have affected 26 million and 22 million people, respectively. The recorded economic damages were substantially higher in the case of floods (average annual loss of USD 2 billion) than that for droughts (average annual loss of USD 0.2 billion) during the same period. The disaster statistics further confirm that floods are responsible for almost 70% of total average annual economic losses (about USD 3.1 billion per annum) making them the costliest natural hazard in India. Further, it is important to note that drought data (both the number of incidences and the associated loss data) may be highly underestimated within the EM-DAT since it incidences at the country level and not at the subnational (state) level are reported.

The Indian Meteorological Department (IMD) assesses and defines the meteorological drought based on the rainfall deficiency at a particular geography or location. According to IMD guidelines, drought occurs if the annual average rainfall in any state deviates negatively, equal, or less than 25% from the long term−average (generally, 30 years). Indian Meteorological Department (IMD) (2016) recently further modified its guidelines for drought declaration and the directive now identifies the "deficient" rainfall years and not "drought year." Similarly, if the annual average rainfall of any region/state deviates positively, equal, or above 20% from its long term−average then it is considered as "excess" rainfall that may lead to a flood-like situation. Following this, the drought and flood intensity variables are created for the purpose of this study and are discussed in detail in Section 35.4.1.

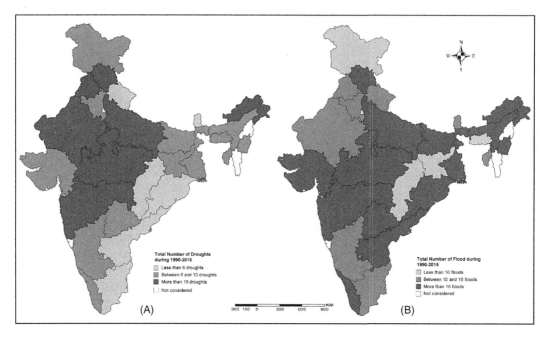

FIGURE 35.2

Total number of (A) drought and (B) flood events occurred across selected states in India during 1990−2015.

Note: *The number of drought events is decided based on the (negative) mean rainfall deviation criteria used in this study which is discussed in detail in Section 35.4.1. Data on the occurrence of flood event is taken from the Central Water Commission dataset on flood damages. Authors' own construction using IMD and CWC data.*

Still, the characteristics, causes, nature, and potential of drought and flood events that may potentially be harmful for the individuals and adversely affect the economy remain a complex issue. Therefore it would be timely and important to examine whether and how such disaster events may cause a decline in the aggregate output at a regional level. With this background and motivation, the present study examines the economic impact of both droughts (slow-onset disaster) and floods (fast-onset disaster) on the selected state economies. The two prominent climate-induced stressors are separately analyzed to examine the differences in the quantum (magnitude) and persistence of the impact in the successive years of the event.

35.3 REVIEW OF LITERATURE

In this section, we discuss the broader concerns and results emanating from the existing literature that explores the empirical relationship between natural disasters and economic growth. We limit the discussion to two types of natural hazards: droughts and floods. First, we present the review of selected literature on the economic impact of droughts followed by a discussion of selected literature focused on floods.

35.3.1 **ECONOMIC IMPACT OF DROUGHTS**

Droughts are considered to be slow-onset disasters that have a direct impact on the agricultural sector and those subsectors within the secondary sector which uses and employs heavily water-dependent production processes. There is a large volume of scientific studies confirming that scarcity of water or drought-like conditions negatively affect the agricultural sector as they directly affect production, productivity, and farm practices. Desai (2003) observed that in an agriculture dominant economy, droughts affect the supply of food and increase the production risk of cotton, affecting the textile industry. Therefore, due to droughts, the agro-based industries suffer that further reduces the agricultural value-addition. Droughts lead to not only difficulties in subsistence agriculture but may intensify health risks due to water stress; the agricultural wages are impacted which cause a reduction in demand for labor and therefore the economic output declines especially in labor-intensive economies (Hayati, Yazdanpanah, & Karbalaee, 2010; Mueller & Osgood, 2009; Singh, Feroze, & Ray, 2013; Udmale et al., 2015). Sen (1982) argues that due to inflation induced by crop losses, poor farmers become poorer by higher spending on procurement of food grains at inflated prices and are often forced to sell their productive assets (such as livestock, gold/silver, or land) to meet their livelihood challenges. This adversely affects the agrarian economy and aggregate rate of growth, especially in the absence of viable and alternative livelihood options (McPeak & Barrett, 2001). Droughts may also trigger migration of labor (Dallmann & Millock, 2017; Gray & Mueller, 2012; Murali & Afifi, 2014) which may further limit availability of labor and cause productivity losses due to the decreased supply of labor at the farm level in subsequent years. Further, severe droughts may reduce the profits and capacity of farmers to invest in advanced techniques which in turn causes a reduction in agricultural growth and thereby hampers aggregate growth (Sheng & Xu, 2019).

In the disaster literature the empirical relationship between droughts and economic growth (both agricultural and aggregate GDP growth) is mostly examined through cross-country analysis. For example, using panel autoregressive distributed lags model, Raddatz (2009) confirms that among climatic disasters, droughts hurt economic growth the most causing a reduction of about 1% in the GDP per capita. Loayza, Olaberria, Rigolini, and Christiaensen (2012) in a study considering a sample of 68 developing and 26 developed economies show that droughts lead to an adverse impact on agricultural growth, although the impact on GDP growth is observed to be negative, it is weak and statistically nonsignificant. However, for developing economies, droughts have had a statistically significant negative impact across all economic sectors and the growth rate of aggregate GDP. Fomby, Ikeda, and Loayza (2013) who analyzed a sample of 84 countries (including 60 developing countries) also reported similar findings with regard to droughts especially for the sample of developing countries. Felbermayr and Gröschl (2014) use a physical intensity variable from their own GeoMet data over the period 1979–2010 and observed that droughts negatively impact high-income economies rather than low-income economies. The results differ significantly from earlier studies as it also demonstrates that the level of development (determined by per capita income) is negatively correlated with the impact of disaster. In a recent attempt to examine the relationship between natural disasters and economic growth, Panwar and Sen (2019a) find that drought, along with other disasters, affect agricultural growth negatively especially in developing countries, while severe droughts have widespread economic consequences and may lower the aggregate GDP growth in developing countries by 3%.

There is limited systematic evidence of the growth effects of droughts at the subnational or regional level. Most of the studies are event-specific and there are differences in the estimated results for the same event. However, few of the important studies in this regard are discussed here. For instance, Diersen, Taylor, and May (2002) examine the direct economic impacts of the 2002 drought on crops and livestock and related "secondary effects" in South Dakota, United States. The study reports drought-induced losses of around $1.8 billion, while for the same event, Diersen and Taylor (2003) estimate USD 1.4 billion losses after adjusting the federal aid of USD 100 million to the state. Similarly, Horridge, Madden, and Wittwer (2005) observe 1.6% reduction in Australian GDP (1% reduction is due to losses in the agriculture sector and remaining 0.6% is owing to secondary effects on the economy) using computable general equilibrium model for 2002–03 drought. Kulshreshtha, Grant, Marleau, and Guenther (2003) estimate that the economic costs of the 2001 and 2002 droughts at the regional level in Canada were approximately USD 2.34 billion. In a recent study, Howitt, MacEwan, Medellín-Azuara, Lund, and Sumner (2015) find that the overall economic impact of the 2015 California drought may be as high as USD 2.74 billion which could also result in 21,000 job losses. Few studies examined the impact of rainfall variations as a proxy of droughts but the findings of these studies are also limited to an event or a region. For instance, Dercon (2004) examined the impact of reduced rainfall in Ethiopia and revealed that a 10% lower rainfall with 4–5 years' lag could have reduced the current economic growth of the country by at least 1%. Gadgil and Gadgil (2006) analyzed rainfall variations and their impact on the GDP growth rate to show that the negative 25%, 20%, and 15% variations cause a reduction in GDP by 7.04%, 5.13%, and 3.47%, respectively.

This review of the existing literature indicates that there are limited studies on the relationship between droughts and economic growth especially at the subnational level. As such this present study focuses on state-level data in India with an aim to fill the gap in the literature.

35.3.2 FLOODS AND ECONOMIC GROWTH

Floods by nature are fast-onset disasters and expected to cause human and monetary losses in a very short period of time. Floods may cause severe economic loss by destruction of human (deaths and injuries) and physical capital that result in the disruption of economic activities and out losses, for example, in agricultural or industrial output. Further, such losses often create an immediate fiscal burden for the government in order to finance recovery, rehabilitation, and reconstruction efforts (Miao, Hou, & Abrigo, 2018; Noy & Nualsri, 2011; Noy, 2009). Further, frequent flooding events may derail the long-term investment prospects and planning in an economy by creating an atmosphere of uncertainty (Chhibber & Laajaj, 2013). As pointed out by few studies (Chhibber & Laajaj, 2013; Cuaresma, Hlouskova, & Obersteiner, 2008; Fomby et al., 2013; Skidmore & Toya, 2002), floods may have positive effects in the long term because the reinvestment in physical capital followed by the destruction caused by floods may result in higher economic growth for a region. Hallegatte and Dumas (2009) called it the "productivity effect" where a reconstruction and upgradation of existing infrastructure following disaster leads to higher economic growth. Floods are also expected to have a positive impact on agricultural growth in the short to medium term especially the moderate flooding events since the abundance of water helps the current and subsequent cropping seasons to generate higher outputs (Loayza et al., 2012).

Like droughts, studies analyzing the economic impact of floods are conducted at the national-level using cross-country data. The extant research conducted at the national level reports a positive effect of floods that are assumed to be largely stimulated by the beneficial impact of flood for the agricultural growth (Klomp & Valckx, 2014; Noy & DuPont, 2016). Though, these effects are true for developing countries and flood events of moderate intensity, severe floods are often reported to have greater economic and fiscal consequences since they disrupt the industrial and services sector the most. Cunado and Ferreira (2014), Fomby et al. (2013), Loayza et al. (2012), and Panwar and Sen (2019a) are few of the studies conducted at the national level which report positive effects of floods on agricultural growth while negative effects on aggregate GDP growth in the case of severe floods. Further, the impact of floods is found to be stronger in the developing countries than in the developed countries (Noy, 2009).

At the regional level, just like droughts, there are limited systematic evidences on the economic impact of floods. Again, the impact may be both positive and negative. For instance, Panwar and Sen (2019b), who examine sector-specific impact of floods in 24 selected states in India, find that floods may reduce the aggregate state GDP growth by 0.3 percentage points in the medium term (over 5 years). Although floods of moderate intensity have a positive impact on the agricultural growth, they do not transmit to other economic sectors. Oosterhaven and Tobben (2017) find that the 2013 heavy flooding in eastern and southern Germany had a wider economic impact due to the interlinkages among different economic sectors based on the supply-chain relationships between and among them. Apart from this, there are studies that examined the impact of floods on local population which might affect the economic growth of the affected region. For example, Husby, de Groot, Hofkes, and Dröes (2014) found that outward migration of the population increases in the short term following the Great North Sea flood of 1953 in the Netherlands that may have affected the regional economy adversely. On similar lines, Boustan, Kahn, and Rhode (2012) find that people are attracted to the flood-affected region in the long term which may help to increase the growth prospects in the affected region. As before, the review of the existing literature indicates that there are limited studies on the relationship between droughts and economic growth, especially at the subnational level. Therefore the present study conducted using state-level data in India would fill this gap in the literature.

35.4 DATA AND METHODOLOGY

This section elaborates upon the data and variables used in this study followed by a discussion on the estimation methodology.

35.4.1 DATA AND VARIABLES

This study utilizes a panel data of 25 selected Indian states over the period 1990–2015. Table 35.1 presents the list of states included in this study along with their average annual GDP growth rates during 1990–2015 and an event count for droughts and floods, respectively. A subsample of 14 "highly vulnerable" states are also identified based on the frequency and severity of drought and

Table 35.1 List of States in the Sample.

States	SGDP Growth (%)	Number of Droughts	Number of Floods
Andhra Pradesh*	3.58	4	19
Arunachal Pradesh*	5.28	10	18
Assam*	3.31	6	21
Bihar*	1.68	8	24
Chhattisgarh	5.68	2	9
Gujarat*	5.52	4	18
Haryana	5.23	5	11
Himachal Pradesh*	6.09	12	20
Jammu & Kashmir	4.35	8	6
Jharkhand	12.5	5	3
Karnataka	6.21	7	11
Kerala	5.77	9	23
Madhya Pradesh*	1.87	12	17
Maharashtra*	3.76	13	16
Manipur*	3.63	9	22
Meghalaya	3.88	5	9
Orissa	6.19	1	17
Punjab*	3.56	13	11
Rajasthan*	3.88	16	14
Sikkim	7.28	0	12
Tamil Nadu	5.21	0	12
Tripura*	5.35	8	14
Uttar Pradesh*	2.84	13	25
Uttarakhand	7.03	2	10
West Bengal*	4.11	5	24

Note: Asterisk (*) represents the "high vulnerability" states in the sample. SGDP growth is in average annual per capita (real) terms during 1990–2015. Drought events are identified following the IMD drought declaration criterion (negative 25% deviation in annual rainfall). Information on flood events is extracted from the Central Water Commission dataset on state-wise flood damages. SGDP, State gross domestic product; IMD, Indian Meteorological Department. Authors' own preparation using RBI, CWC, and IMD data.

flood events during the selected sample period to highlight the impact of natural disasters in such states vis-à-vis other states which have witnessed lesser episodes of such events.

Data on state-level economic variables are extracted from the database of the Reserve Bank of India. The SGDP growth per capita is used as the dependent variable and expressed in real terms. The per capita output is obtained by dividing the annual SGDP by population figures of respective state-year. Following Levine et al. (2000), Dollar and Kraay (2004), and Mankiw et al. (1992), five state-level economic variables are used as the standard determinants of growth (endogenous variables) which include the investment rate, state government's final consumption expenditure, financial outreach or depth in terms of credit flows, gross rate of enrollment in higher education, and the state-level inflation rate. In addition, the disaster relief funds transferred from the union government to respective states are also considered endogenous to the SGDP growth since such funds are expected to reduce disaster impact (Becerra, Cavallo, & Noy, 2014). All the variables are deflated by the inflation rate considering 2010 as the base year. Table 35.2 summarizes the variables used in this analytical exercise, while Table 35.3 provides their basic descriptive statistics.

The data on the occurrences and intensity of droughts and floods is taken from the database of the IMD in the form of rainfall distribution across states in India. The Central Water Commission's data on floods is also used to gather any additional information on the occurrences of floods across Indian states. In the disaster literature, studies have mostly used human and economic loss data to approximate disaster intensity.

Table 35.2 Variables Description and Data Sources.

Variable	Description	Data Source
SGDP growth	Real state GDP growth rate per capita in log differences	*Handbook of Statistics on Indian States*, by RBI
Investments	Fixed capital formation as fraction of SGDP in logs	
Government's expenditure	Consumption expenditure as fraction of SGDP in logs	
Financial depth	Credit to private sector as fraction of SGDP in logs	
Inflation	State consumer price index growth rate in log differences	
Gross enrollment ratio	Rate of gross enrollment in higher education in logs	
Disaster relief	Relief from union government as fraction of SGDP in logs	
Drought Intensity		
Negative 20% deviation	If annual rainfall deviation is (−)20% or more than the long-term (30 years) average rainfall = 1 and 0, otherwise	IMD
Negative 25% deviation	If annual rainfall deviation is (−)25% or more than the long-term (30 years) average rainfall = 1 and 0, otherwise	
Negative 30% deviation	If annual rainfall deviation is (−)30% or more than the long-term (30 years) average rainfall = 1 and 0, otherwise	
Negative 50% deviation	If annual rainfall deviation is (−)50% or more than the long-term (30 years) average rainfall = 1 and 0, otherwise	
Flood Intensity		
Positive 20% deviation	If annual rainfall deviation is (+)20% or more than the long-term (30 years) average rainfall = 1 and 0, otherwise	IMD
Positive 25% deviation	If annual rainfall deviation is (+)25% or more than the long-term (30 years) average rainfall = 1 and 0, otherwise	
Positive 30% deviation	If annual rainfall deviation is (+)30% or more than the long-term (30 years) average rainfall = 1 and 0, otherwise	
Positive 50% deviation	If annual rainfall deviation is (+)50% or more than the long-term (30 years) average rainfall = 1 and 0, otherwise	

GDP, *Gross domestic product;* IMD, *Indian Meteorological Department;* RBI, *Reserve Bank of India;* SGDP, *state gross domestic product.*

Table 35.3 Descriptive Statistics.

Variables	Mean	Standard Deviation	Minimum	Maximum
Dependent and Endogenous Variables States (N) = 25, Obs. (n) = 650				
SGDP growth	0.011	0.007	− 0.013	0.076
Investments	0.835	1.748	− 6.714	6.246
Government's expenditure	− 1.334	0.584	− 2.908	1.967
Financial depth	2.711	0.878	− 2.405	4.590
Inflation	4.521	0.087	3.687	5.988
Gross enrollment ratio	2.144	0.747	− 0.163	3.842
Disaster relief	− 1.121	2.222	− 8.706	7.562
Disaster Intensity Variables ($N = 25$, $n = 650$)				
	Number of Observations (If Value = 1)		Number of Observations (If Value = 0)	
Drought intensity				
Negative 20% deviation	254		396	
Negative 25% deviation	173		477	
Negative 30% deviation	140		510	
Negative 50% deviation	54		596	
Flood intensity				
Positive 20% deviation	225		425	
Positive 25% deviation	191		459	
Positive 30% deviation	152		498	
Positive 50% deviation	76		574	

Note: *For description of variables, refer to Table 35.1. SGDP, State gross domestic product.*

As pointed out by Noy (2009) and Felbermayr and Gröschl (2014), human and economic loss data can pose several estimation difficulties. First, the human and economic loss data may be endogenous to GDP which, in the event of GDP being the dependent variable, may produce spurious results. Second, such data has many missing observations which may cause measurement errors in estimation. Therefore many earlier studies (e.g., Felbermayr & Gröschl, 2014; Klomp & Valckx, 2014; Panwar & Sen, 2019b) have encouraged adoption of physical measures of disaster intensity because they are considered to be inherently exogenous to GDP growth. Following this and considering the importance of the exogeneity of the disaster variable(s), this study adopts the rainfall distribution data as the principal measure to approximate droughts and floods.

The IMD provides monthly rainfall data collected by several weather stations set up across India. For the purpose of this study the monthly data is first converted to annual data and then the station-level data is brought to the state level. Next, the mean annual rainfall deviations from its long-term average are calculated based on the following formula:

$$\text{Mean rainfall deviation.}_{i,t} = \frac{\text{Long-term average rainfall}_i - \text{Average annual rainfall}_{i,t}}{\text{Long-term average rainfall.}_i} \times 100 \quad (35.1)$$

As discussed earlier, the IMD assesses and defines the meteorological drought based on the rainfall deficiency at a particular geography or location. According to IMD guidelines, the average annual rainfall of any state or region if deviates negatively by 25% or more from the long term−average rainfall (in general, for a period of 30 years), then the year is considered as a drought year. Following the aforementioned criteria, dummy variables depicting drought years are created which take a value of 1 if the average annual rainfall deviation is 25% or more from its long-term average and 0, otherwise. Notable here is that IMD has modified its guidelines regarding droughts (IMD, 2016), and instead of declaring a particular year as "drought year," it has now started declaring a "deficient year." To accommodate this change and to examine the state-wide economic impact and growth dynamics following droughts of varying intensities, different thresholds for rainfall deviation are used in this study (viz. 20%, 30%, and 50%) and the drought dummies are created accordingly. On similar lines the flood intensity variables are created depicting excess rainfall. The IMD guidelines consider the positive rainfall deviation of 20% or more from the long-term average as excess rainfall. To maintain consistency in the analysis, a benchmark flood intensity variable is created which takes a value one if there is a positive deviation of 25% or more in the average annual rainfall from its long-term average and 0, otherwise. As with droughts, different thresholds of rainfall deviations are used to examine the influence of the degree of flood intensity on SGDP growth.

35.4.2 ESTIMATION METHODOLOGY

This study uses augmented PVAR-X models to estimate the impact of slow- and fast-onset disasters on SGDP growth. This estimation technique has been widely used in the disaster literature to generate point estimates and mean growth responses of different macroeconomic and fiscal variables following a disaster shock (e.g., Cunado & Ferreira, 2014; Fomby et al., 2013; Miao et al., 2018; Noy & Nualsri, 2011; Panwar & Sen, 2019b; Raddatz, 2007, 2009). PVAR models allow the variables to be endogenous, which are regressed upon their own lag values and the lag values of other variables in a system of equations. A detailed note on the PVAR estimation methodology can be found in Holtz-Eakin, Newey, and Rosen (1988) and Love and Zicchino (2006).

PVAR-X models allow the inclusion of an exogenous variable into the system of equations which makes this estimation technique a natural choice for the present exercise since the disaster variables are widely considered to be exogenous in nature. As discussed earlier, the SGDP growth and other economic variables enter the estimation model as endogenous variables $(y_{i,t})$, while the drought (or flood) variables enter as exogenous $(x_{i,t})$ variables along with the year dummies representing time-variant shocks common for all the states. Exogeneity, being a vital prerequisite for the disaster variables, has been statistically checked using Granger causality test (Granger, 1969, 1980, 1988), and the results are presented in Table 35.4. In this test the failure to reject the null hypothesis indicates that the endogenous variable(s) do(es) not granger cause droughts/floods both individually and in combination, and therefore droughts and floods can be treated as weakly exogenous. Considering the endogeneity and exogeneity conditions, the following model specification is used in this study:

$$A_0 z_{i,t} = \alpha_i + \sum_{j=0}^{q} A_j z_{i,t-j} + \eta_i + \mu_t + \varepsilon_{i,t} \tag{35.2}$$

where $i \in \{1, 2, 3, 4, \ldots, 25\}$, $t \in \{1, 2, 3, 4, \ldots, 26\}$.

In Eq. (35.2), subscripts i and t denote states and years, respectively. The contemporaneous coefficients of the endogenous $(y_{i,t})$ and exogenous $(x_{i,t})$ variables, that is, $z'_{i,t} = (y'_{i,t}, x'_{i,t})$ are

Table 35.4 Granger Causality Test for Exogeneity of the Disaster Variables.

	Droughts		Floods	
	χ^2 Value	*P* Value	χ^2 Value	*P* Value
Variables				
SGDP growth	0.047	.678	0.052	.822
Investment	0.852	.433	0.064	.796
Government's financial burden	0.584	.155	0.198	.841
Financial depth	1.004	.189	0.853	.269
Inflation	0.412	.754	1.562	.375
Gross Enrollment Ratio				
Disaster relief	1.229	.122	0.745	.817
Combined (all variable)	4.534	.542	5.621	.852

Note: *The exogeneity of the drought and flood intensity variables (using ± 25% mean deviation criterion) was checked using 1-year lag with null hypothesis that the respective endogenous variables do not granger cause droughts and floods, both individually and in combination.* SGDP, *State gross domestic product.* Authors' own calculation.

represented by matrix A_0. Matrix A_j represents the estimated coefficients. η_i and μ_t denote the state fixed effects and the year fixed effects, respectively. $\varepsilon_{i,t}$ represents the error term in the model.

Prior to the estimation of the model portrayed in Eq. (35.2), the stationarity conditions of the variable series are checked using Im, Pesaran, and Shin (2003) test for the presence of unit root. The rejection of the null hypothesis in this test would mean that the series is stationary. As in Table 35.5, the series pertaining to SGDP per capita, investment rate, government expenditure, and inflation rate contain unit root in levels, while after the log-transformation and inclusion of time trend, they seem to be stationary. Another important prerequisite for PVAR estimation is the lag selection criteria which is decided based on Akaike's information criterion (Akaike, 1973) and Bayesian information criterion (Akaike, 1979). A lag-length of one is prescribed by both these criteria and the same has been incorporated in the estimation. Finally, to address the problem of endogeneity which may arise in dynamic panel settings while using fixed effects (refer Wooldridge, 2013), the generalized method of moments (GMM) technique is used to estimate the system of equations following Holtz-Eakin et al. (1988). Further, forward mean differencing is used to avoid the possibility of a negative bias (popularly known as the "Nickel Bias") in the estimates which may arise due to the demeaning process used while removing fixed effects in GMM estimation (refer to Nickell, 1981).

Finally, after estimating PVAR-X model, the dynamic multiplier functions (DMFs) are generated which depict the average impact of a disaster shock in the year of the event (year t) and also in the following 2 years, that is, year $t+1$ and year $t+2$. Beyond 3 years the isolated average effects of disaster shock are found to be largely very low in magnitude and also statistically nonsignificant. The point estimates are also generated portraying the average cumulative impact over 3, 5, and 10 years following the disaster event to show the short-, medium-, and long-term impact of droughts and floods, respectively, along with the rate of decay of the estimated impact on SGDP

Table 35.5 Unit Root Test.

	Im–Pesaran–Shin Test (In Levels)		Im–Pesaran–Shin Test (In Logs and with Time Trend)	
	w-Statistics	Probability	*w*-Statistics	Probability
SGDP per capita	2.522	**1.255**	− 4.856	**.000**
Investments	0.962	**.524**	− 0.012	**.004**
Government's expenditure	− 0.743	**.129**	− 4.226	**.000**
Financial depth	− 1.298	.042	− 7.514	.000
Inflation	− 0.284	**.321**	− 5.819	**.000**
Gross enrollment ratio	− 4.482	.008	− 6.552	.000
Disaster relief	− 2.172	.011	− 4.015	.000
Drought intensity (negative 25% mean deviation)	− 5.983	.000	− 7.217	.000
Flood intensity (positive 25% mean deviation)	− 4.854	.000	− 6.751	.000

Note: *The variable series that are not stationary in levels are indicated in bold figures that seem to be stationary after log-transformation and inclusion of time trend. The SGDP and inflation series are transformed into growth rates and used in log differences.* SGDP, *State gross domestic product.*

growth. The DMFs depict the average impact of one exogenous shock (floods or droughts) at a time, ceteris paribus. Monte Carlo simulations (500 repetitions) are used to calculate the confidence intervals and standard errors of the point estimates. With the help of confidence intervals, the statistical significance of the point estimates is determined at 5% and 10% levels of significance. In the next section, the results of this study are discussed.

35.5 RESULTS AND DISCUSSION

The point estimates generated following the estimation of PVAR-X models as in Eq. (35.2) are reported in Tables 35.6 and 35.7. The corresponding dynamic response functions are graphically presented in Figs. 35.3 and 35.4. As mentioned in the foregoing sections, the mean growth responses following a drought or flood shock are reported both individually (for the first 3 years) and cumulatively (over 3, 5, and 10 years). The point estimates are to be interpreted as $\pm x$ percentage point (or pp) change in the SGDP growth following a drought or flood year in comparison to a nondrought or nonflood year.

35.5.1 BENCHMARK RESULTS

The results presented in Part A of Table 35.6 pertain to the full sample of states using the standard mean rainfall deviation criteria of $\pm 25\%$ for both droughts and floods, respectively. The results for the subsample of "high vulnerability" states are presented in Part B of Table 35.6. First, the impact of droughts on SGDP growth is discussed followed by that of floods.

Table 35.6 Mean Growth Responses Following Droughts and Floods: Benchmark Results.

Dependent Variable: Real State GDP Growth Per Capita

	Degree of Rainfall Deviation ($\pm 25\%$ From Long-Term Average Annual Rainfall)	
	Slow-Onset Disasters: Droughts (Negative 25% Deviation)	Fast-Onset Disasters: Floods (Positive 25% Deviation)
(A) Full Sample of States (25)		
Year t	-0.311**	-0.205**
Year $t+1$	-0.221**	-0.342*
Year $t+2$	-0.001	0.012
Short term: cumulative (3 years)	-0.542**	0.523*
Medium term: cumulative (5 years)	-0.601**	-0.498
Long term: cumulative (10 years)	-0.622*	-0.385
(B) High Vulnerability States (14)		
Year t	-0.681**	-0.66**
Year $t+1$	-0.368**	-0.261**
Year $t+2$	-0.021	-0.07
Short term: cumulative (3 years)	-1.02**	-0.987**
Medium term: cumulative (5 years)	-1.06**	-1.01**
Long term: cumulative (10 years)	-1.12**	-0.822

Note: Asterisks ** and * indicate statistical significance at 5% level and 10% level, respectively. GDP, Gross domestic product.

35.5.1.1 Slow-onset disasters: droughts

Droughts are observed to have a negative relationship with SGDP growth. In the year of the event (t), the SGDP growth on average declined by 0.31pp which may reduce further by 0.22pp in the following year ($t+1$) while remaining statistically significant at the 5% level. After the first 2 years of the event, the average annual impact of droughts starts to decay rapidly and becomes marginal and statistically nonsignificant. However, the cumulative impact of droughts remains negative and statistically significant. It persists from short (over 3 years, -0.54pp) to medium term (over 5 years, -0.6pp) and even extends to the long term (over 10 years, -0.62) suggesting the adverse and compounding effects of droughts on SGDP growth. The impact of droughts is more pronounced in terms of magnitude of the estimated coefficients for a subsample of 14 "high vulnerability" states. While the overall relationship of droughts with SGDP growth remains the same, the severity of the impact increases significantly. In "high vulnerability" states, droughts are expected to reduce the SGDP growth by 0.68pp in the year of the event, while they may cause a cumulative decline in the SGDP growth of about 1pp and 1.12pp in the medium and long term, respectively.

The results obtained through this analysis would imply that droughts have a significantly negative effect on aggregate SGDP growth which persists in the medium to long term as well. There are wseveral empirical studies (e.g., Berlemann & Wenzel, 2016; Christiaensen & Subbarao, 2005; Dercon, 2004; Fomby et al. 2013; Hlavinka, Trnka, Semeradova, Dubrovsky, Zalud, & Monzy,

Table 35.7 Mean Growth Responses Following Droughts and Floods: Varying Degree of Rainfall Deviations.

Dependent Variable: Real State GDP Growth Per Capita

	Degree of Rainfall Deviation from Long-Term Average Annual Rainfall			
	± 20%	± 25%	± 30%	± 50%
Slow-Onset Disasters: Droughts (Negative Deviation)				
Year t	− 0.214**	− 0.311**	− 0.352**	− 0.62**
Year $t + 1$	− 0.222	− 0.221**	− 0.301**	− 0.322*
Year $t + 2$	0.000	− 0.001	− 0.001	− 0.374**
Short term: cumulative (3 years)	− 0.51**	− 0.532**	− 0.641**	− 1.25**
Medium term: cumulative (5 years)	− 0.542*	− 0.601**	− 0.684**	− 1.34**
Long term: cumulative (10 years)	− 0.544	− 0.622*	− 0.72**	− 1.56**
Fast-Onset Disasters: Floods (Positive Deviation)				
Year t	− 0.322**	− 0.205**	− 0.413**	− 0.521**
Year $t + 1$	− 0.035	− 0.342*	− 0.292**	− 0.234*
Year $t + 2$	0.004	0.012	0.047	0.001
Short term: cumulative (3 years)	− 0.354	− 0.523*	− 0.633**	− 0.681**
Medium term: cumulative (5 years)	− 0.262	− 0.498	− 0.598**	− 0.672**
Long term: cumulative (10 years)	− 0.211	− 0.385	− 0.462	− 0.533

Note: Asterisks ** and * indicate statistical significance at 5% level and 10% level, respectively. GDP, Gross domestic product.

2009; Kilimani, Van Heerden, Bohlmann, & Roos, 2018; Panwar & Sen, 2019a) in the existing literature which report a strong, negative impact of droughts on agricultural growth. Considering that about 60% of India's total population is dependent on agriculture and allied services, the negative impact of droughts on the aggregate SGDP growth does not come as a surprise. However, what is interesting is the fact that the impact of droughts is observed to be persistent and could have long-term economic consequences. This may be true since droughts, due to their slow-onset characteristics, eat up the economic assets, and livelihoods of the affected population especially of the poorest. In the absence of affordable credit and insurance access and effective drought response mechanisms in India, the drought-hit population is often forced to dispose of their remaining productive assets which can further trap them into poverty (Loayza et al., 2012; Sen, 1982). Further, the persistence of the economic impact may hamper economic development in the long term (Panwar & Sen, 2019a; Pelling, Özerdem, & Barakat, 2002). The results obtained in this exercise are reminiscent of a few earlier studies (e.g., Berlemann & Wenzel, 2016; Fomby et al., 2013; Loayza et al., 2012; Panwar & Sen, 2019a), which estimated the impact of droughts on national-level GDP growth.

35.5.1.2 Fast-onset disasters: floods

The point estimates for flood impact are reported in Column 2 of Table 35.6. Floods exhibit a negative relationship with SGDP growth. Floods on an average may reduce the SGDP growth by 0.21pp in the year of the event (t), while it may cause a decline of 0.34pp in the following year ($t + 1$) of

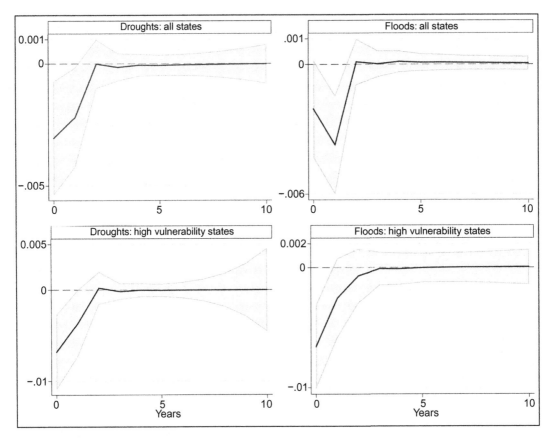

FIGURE 35.3

Mean growth responses to droughts and floods shocks: Benchmark results.

Note: Dynamic multiplier functions are generated with 95% confidence intervals. The mean growth responses are generated following one standard deviation shock of droughts/floods.

the event. The growth effects of floods remain negative in the successive years as well, but they fail to be statistically significant. The cumulative impact of floods on SGDP growth is statistically significant over 3 years only, while it fails to persist in the medium to long term. Contrary to the full sample of states, in the case of the "high vulnerability" states, the flood impact is observed to be stronger in magnitude and found to be present in the following years of the flood event. The point estimate in the year of the flood event is more than double in magnitude. The cumulative impact of floods over 3 (−0.99pp) and 5 years (−1.01pp) remains statistically significant which is not the case in the long term over 10 years. Similar results are reported by Panwar and Sen (2019b) where floods were showing a statistically significant negative impact on SGDP growth of Indian states.

In general, the studies conducted at the country level have reported positive effects of floods on agricultural growth especially in developing countries which are further expected to be transmitted to the aggregate GDP growth. Most of these studies (e.g., Cunado & Ferreira, 2014; Fomby et al.,

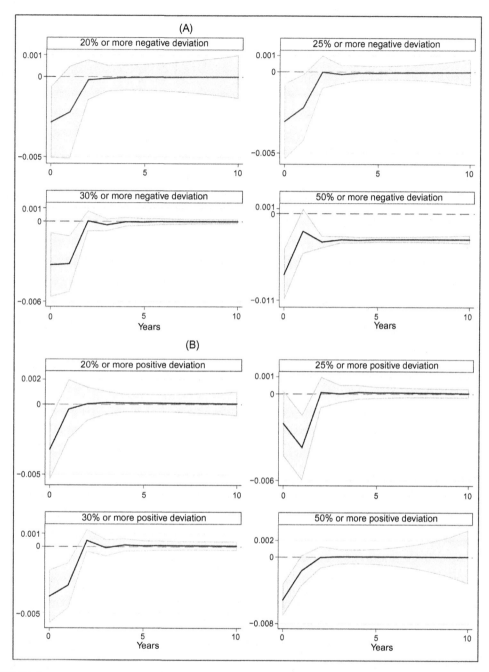

FIGURE 35.4

Mean growth responses to droughts and floods shocks of varying intensities: (A) droughts and (B) floods of varying intensities.

Note: Dynamic multiplier functions are generated with 95% confidence intervals. The mean growth responses are generated following one standard deviation shock of droughts/floods.

2013; Loayza et al., 2012) have argued that since developing countries are predominately agrarian economies, therefore the positive effects of floods on agriculture sector are transmitted to the other economic sector and also to the aggregate GDP due to the interlinkages among different sectors. However, as explained by Panwar and Sen (2019b), this may not be true for India since the agricultural sector's contribution to the GDP has been going down significantly in recent decades (from 32% in 1990 to 15.11% in 2015), and the positive effects of floods on the agricultural growth may not be strong enough to stimulate the growth of other economic sectors. Instead, the damages and destruction caused by floods may cause output losses by disrupting businesses and affecting potential labor (human deaths and injuries) and capital (assets and property losses). The next subsection presents a comparative picture of the growth effects of slow- and fast-onset disasters.

35.5.2 COMPARING THE GROWTH EFFECTS OF SLOW- AND FAST-ONSET DISASTERS

From the benchmark results obtained using standard rainfall deviation criteria of $\pm 25\%$, it can be observed that both droughts and floods exhibit a negative relationship with SGDP growth. However, the extent of the negative impact in terms of the magnitude of estimates is different for different disaster types. Further, there are differences in terms of the statistical significance of the estimated coefficients. While the impact of droughts on SGDP growth is statistically significant in both the medium and long term, the same for floods is only significant in the short term, that is, cumulatively over 3 years (-0.52pp, in Table 35.6). The DMFs in graphical form can be very useful in comparing the difference in mean growth responses following droughts and floods. As shown in Fig. 35.3, the SGDP growth is observed to be significantly deviated from its normal or original trajectory following a drought shock and it appears to be taking at least 5 years in attaining the same. On the contrary, the SGDP growth attains its normal trajectory in the third year following a flood shock ($t + 2$). In fact, it appears to be settling in a higher growth trajectory in the medium to long term. A similar scenario can be observed in the case of the "high vulnerability" states where growth seems to be settling down on its normal trajectory relatively quickly following a flood shock in comparison to a drought shock. Though, the impact of both droughts and floods is significantly greater in magnitude for high vulnerability states. The persistence of the negative impact of droughts on SGDP growth in the medium to long term also underlines the fact that the rate of decay of the drought impact is far less than that of floods.

To further compare the mean growth responses following droughts and floods of varying intensities, the degree of mean rainfall deviation is adjusted with $\pm 5\%$ from the benchmark criteria. Thus drought and flood intensity variables having underlying mean rainfall deviation of $\pm 20\%$ and $\pm 30\%$ are created. Further, variables based on the $\pm 50\%$ mean rainfall deviation criteria are also used to check the growth effects of extreme droughts and floods. These results would also work as a robustness check for the benchmark results discussed in the previous subsection. As shown in Table 35.7, the magnitude of the adverse impact of droughts and floods increases with an increase in their underlying intensities. In other words, the SGDP growth experiences a relatively greater decline with increasing intensity of the excess and deficient rainfall. However, droughts on average have a stronger economic impact than floods across different levels of disaster intensities. The impact of severe droughts (with $\pm 50\%$ deviation) remains strongly negative and may reduce the SGDP growth by 1.3pp and 1.6pp in the medium and long term, respectively. This implies that severe droughts may have a larger and long-lasting negative economic impact. On the contrary, the magnitude of the estimated growth effects of floods declines in the medium (−0.67pp) to long term (0.53pp) since floods largely start to have a positive impact on SGDP growth from the third year ($t + 2$) onward. Therefore the extent of a drought impact on average may be twice in magnitude than that of floods cumulatively in the short and medium term. Further, the impact of floods is not found to be persisting in the long term since it remains statistically nonsignificant across different intensity variables. Although, the magnitude of the estimated coefficients increases with the increase in the positive mean rainfall deviation which confirms the fact that severe floods have stronger and lasting economic consequences.

Another important distinction between the economic impact of droughts and floods can be observed by using the graphical representation of the estimated mean growth responses (Fig. 35.4). The SGDP growth takes a relatively longer recovery period to attain its normal trajectory following a drought shock in comparison to a flood shock where it appears to be recovering faster. In the case of extremely severe droughts the growth is observed to be settled in a lower trajectory than its original or normal trajectory. This confirms the observations made by the World Bank (2007) and Chhibber and Laajaj (2013) that a disaster shock of severe nature may force the economies into lower growth paths. On the contrary, in most cases, the SGDP growth appears to be settling in a slightly higher trajectory than normal after recovering from initial shocks of floods suggesting a beneficial economic impact in the long term. As before, the rate of decay of the impact is relatively higher in the case of floods than in the case of droughts which implies that droughts have long-term adverse economic effects.

35.6 SUMMARY

This chapter analyses the impact of slow- and fast-onset disasters on the state-level GDP growth using a panel dataset of 25 selected Indian states compiled over the period 1990−2015. Using PVAR-X models, the dynamic responses of GDP growth following droughts and floods are generated separately which depict the mean growth effects not only in the year of the event but also in successive years following the events. Further, the cumulative impact of both droughts and floods over 3, 5, and 10 years is estimated to examine the persistence of their impact on the state economies. The results of the study indicate that while both droughts and floods adversely affect GDP growth, the impact of droughts is relatively stronger in magnitude. The economic impact of droughts remains statistically significant and stronger in magnitude in the medium to long term as well compared to that in the case of floods. In other words, GDP growth is found to be recovering rather quickly following floods, while it is taking significantly longer time to recover and achieve its normal growth trajectory following droughts. Further, the economic impact of both droughts and floods become stronger in magnitude as they move along with the increases in the physical intensity of disaster variables. This further confirms the fact that more severe disasters have significantly stronger economic consequences. Notable here is that the economic impact of droughts remains relatively stronger than floods.

The results of this study may be very useful for policy makers both at national and state levels of governance. Interestingly, the existing disaster management policies in India treat both these types alike raising a serious concern toward such existing approaches and practices. Although different ministries manage different disasters, the policy (in form of Disaster Management Act, 2005 and National Disaster Management Plan, 2016) guiding them remains the same. The results of this exercise certainly point into the direction where different disaster types should be treated differently and proactive risk management practices should be encouraged. Dedicated and proactive investments in risk reduction activities would therefore be prioritized. For example, building disaster-resilient infrastructure, constructing multipurpose dams and river channel would be beneficial in safeguarding the people and assets against fast-onset disasters like floods. Similarly, sensitizing and educating people regarding alternate irrigation facilities and developing the same at the state level would contribute toward drought mitigation and management. In general, such risk mitigation strategies should be part of the mainstream developmental agenda for both national and state government.

REFERENCES

Akaike, H. (1973). Information theory and an extension of the maximum likelihood principle. *International Symposium on Information Theory*, 267–281. <https://doi.org/10.1007/978-1-4612-1694-0>.

Akaike, H. (1979). A Bayesian extension of the minimum AIC procedure of autoregressive model fitting. *Biometrika, 66*(2), 237–242. <https://doi.org/10.1093/biomet/66.2.237>.

Becerra, O., Cavallo, E., & Noy, I. (2014). Foreign aid in the aftermath of large natural disasters. *Review of Development Economics, 18*(3), 445–460. <https://doi.org/10.1111/rode.12095>.

Berlemann, M., & Wenzel, D. (2016). Long-term growth effects of natural disasters—Empirical evidence for droughts. *Economics Bulletin, 36*(1), 464–476.

Boustan, L. P., Kahn, M. E., & Rhode, P. W. (2012). Moving to higher ground: Migration response to natural disasters in the early twentieth century. *American Economic Review, 102*, 238–244. <https://doi.org/10.1257/aer.102.3.238>.

Centre for Research on the Epidemiology of Disasters (CRED). (n.d.) *Emergency events database (EM-DAT)*. Available from <www.emdat.be/database>.

Central Water Commission of India. (n.d.). *State wise flood damage statistics*. New Delhi: Flood Forecast Monitoring Directorate, Central Water Commission of India, Government of India.

Chhibber, A., & Laajaj, R. (2013). The interlinkages between natural disasters and economic development. In D. Guha-Sapir, & I. Santos (Eds.), *The economic impacts of natural disasters* (pp. 28–56). New York: Oxford University Press.

Christiaensen, L., & Subbarao, K. (2005). Towards an understanding of household vulnerability in rural Kenya. *Journal of African Economies, 14*(4), 520–558. <https://doi.org/10.1093/jae/eji008>.

Cuaresma, C., Hlouskova, J., & Obersteiner, M. (2008). Natural disasters as creative destruction? Evidence from developing countries. *Economic Inquiry, 46*, 214–226.

Cunado, J., & Ferreira, S. (2014). The macroeconomic impacts of natural disasters: The case of floods. *Land Economics, 90*(1), 149–168.

Dallmann, I., & Millock, K. (2017). Climate variability and inter-state migration in India. *CESifo Economic Studies, 63*(4), 560–594. <https://doi.org/10.1093/cesifo/ifx014>.

Dercon, S. (2004). Growth and shocks: Evidence from rural Ethiopia. *Journal of Development Economics, 74* (2), 309–329. <https://doi.org/10.1016/j.jdeveco.2004.01.001>.

Desai, B. M. (2003). Drought impact and vision for proofing. *Economic and Political Weekly, 38*, 2023–2024.

Diersen, M. A., Taylor, G. (2003). Examining economic impact and recovery in South Dakota from the 2002 drought (no. 32028). *Economics Staff Paper*. South Dakota State University, Department of Economics.

Diersen, M. A., Taylor, G., May, A. (2002). Direct and indirect effects of drought on South Dakota's economy. *Economics Commentator, Paper 423*.

Dollar, D., & Kraay, A. (2004). Trade, growth, and poverty. *Economic Journal, 114*(493), F22−F49. <https://doi.org/10.1111/j.0013-0133.2004.00186>.

Felbermayr, G., & Gröschl, J. (2014). Naturally negative: The growth effects of natural disasters. *Journal of Development Economics, 111*, 92−106. <https://doi.org/10.1016/j.jdeveco.2014.07.004>.

Fomby, T., Ikeda, Y., & Loayza, N. V. (2013). The growth aftermath of natural disasters. *Journal of Applied Econometrics, 28*(3), 412−434.

Gadgil, S., & Gadgil, S. (2006). The Indian monsoon, GDP and agriculture. *Economic and Political Weekly, 41*, 4887−4895.

Granger, C. W. (1969). Investigating causal relations by econometric models and cross-spectral methods. *Econometrica: Journal of the Econometric Society, 37*, 424−438.

Granger, C. W. J. (1980). Testing for causality. A personal viewpoint. *Journal of Economic Dynamics and Control, 2*(C), 329−352. <https://doi.org/10.1016/0165-1889(80)90069-X>.

Granger, C. W. (1988). Some recent development in a concept of causality. *Journal of Econometrics, 39* (1−2), 199−211. <https://doi.org/10.1016/0304-4076(88)90045-0>.

Gray, C., & Mueller, V. (2012). Drought and population mobility in rural Ethiopia. *World Development, 40*(1), 134−145. <https://doi.org/10.1016/j.worlddev.2011.05.023>.

Hallegatte, S., & Dumas, P. (2009). Can natural disasters have positive consequences? Investigating the role of embodied technical change. *Ecological Economics, 68*(3), 777−786. <https://doi.org/10.1016/j.ecolecon.2008.06.011>.

Hayati, D., Yazdanpanah, M., & Karbalaee, F. (2010). Coping with drought: The case of poor farmers of south Iran. *Psychology and Developing Societies, 22*(2), 361−383. <https://doi.org/10.1177/097133361002200206>.

Hlavinka, P., Trnka, M., Semeradova, D., Dubrovsky, M., Zalud, Z., & Monzy, M. (2009). Effect of drought on yield variability of key crops in Czech Republic. *Agricultural and Forest Meteorology, 149*(3-4), 431−442.

Holtz-Eakin, D., Newey, W., & Rosen, H. S. (1988). Estimating vector autoregressions with panel data. *Econometrica, 56*(6), 1371. <https://doi.org/10.2307/1913103>.

Horridge, M., Madden, J., & Wittwer, G. (2005). The impact of the 2002−2003 drought on Australia. *Journal of Policy Modeling, 27*(3), 285−308. <https://doi.org/10.1016/j.jpolmod.2005.01.008>.

Husby, T. G., de Groot, H. L. F., Hofkes, M. W., & Dröes, M. I. (2014). Do floods have permanent effects?: Evidence from the Netherlands. *Journal of Regional Science, 54*(3), 355−377. <https://doi.org/10.1111/jors.12112>.

Howitt, R., MacEwan, D., Medellín-Azuara, J., Lund, J., & Sumner, D. (2015). *Economic analysis of the 2015 drought for California agriculture*. University of California.

Im, K. S., Pesaran, M. H., & Shin, Y. (2003). Testing for unit roots in heterogeneous panels. *Journal of Econometrics, 115*(1), 53−74. <https://doi.org/10.1016/S0304-4076(03)00092-7>.

Indian Meteorological Department. (n.d.). *District-wise rainfall information*. New Delhi: Hydromet Division, Indian Meteorological Department, Ministry of Earth Science, Government of India.

Indian Meteorological Department (IMD). (2016). *Terminologies and glossary*. Indian Meteorological Department, Ministry of Earth Science, Government of India. <http://imd.gov.in/section/nhac/termglossary.pdf>.

Kilimani, N., Van Heerden, J., Bohlmann, H., & Roos, L. (2018). Economy-wide impact of drought induced productivity losses. *Disaster Prevention and Management: An International Journal, 27*(5), 636−648.

Klomp, J., & Valckx, K. (2014). Natural disasters and economic growth: A meta-analysis. *Global Environmental Change.* <https://doi.org/10.1016/j.gloenvcha.2014.02.006>.

Kulshreshtha, S. N., Grant, C. W., Marleau, R., & Guenther, E. (2003). *Technical report: Canadian droughts of 2001 and 2002.* Saskatoon, SK: Saskatchewan Research Council.

Levine, R., Norman, L., & Beck, T. (2000). Financial intermediation and growth. *Journal of Monetary Economics, 46,* 31−77. <https://doi.org/10.1016/S0304-3932(00)00017-9>.

Loayza, N. V., Olaberria, E., Rigolini, J., & Christiaensen, L. (2012). Natural disasters and growth: Going beyond the averages. *World Development, 40*(7), 1317−1336. <https://doi.org/10.1016/j.worlddev.2012.03.002>.

Love, I., & Zicchino, L. (2006). Financial development and dynamic investment behavior: Evidence from panel VAR. *The Quarterly Review of Economics and Finance, 46*(2), 190−210. <https://doi.org/10.1016/j.qref.2005.11.007>.

Mankiw, N. G., Romer, D., & Weil, D. N. (1992). A contribution to the empirics of economic-growth. *Quarterly Journal of Economics, 107*(2), 407−437. <https://doi.org/10.2307/2118477>.

McPeak, J. G., & Barrett, C. B. (2001). Differential risk exposure and stochastic poverty traps among East African pastoralists. *American Journal of Agricultural Economics, 83*(3), 674−679.

Miao, Q., Hou, Y., & Abrigo, M. (2018). Measuring the financial shocks of natural disasters: A panel study of U.S. states. *National Tax Journal, 71*(1), 11−44. <https://doi.org/10.17310/ntj.2018.1.01>.

Mueller, V. A., & Osgood, D. E. (2009). Long-term impacts of droughts on labour markets in developing countries: Evidence from Brazil. *The Journal of Development Studies, 45*(10), 1651−1662. <https://doi.org/10.1080/00220380902935865>.

Murali, J., & Afifi, T. (2014). Rainfall variability, food security and human mobility in the Janjgir-Champa district of Chhattisgarh state, India. *Climate and Development, 6*(1), 28−37. <https://doi.org/10.1080/17565529.2013.867248>.

Nickell, S. (1981). Biases in dynamic panel models with fixed effects. *Econometrica.* <https://doi.org/10.2307/1911408>.

Noy, I. (2009). The macroeconomic consequences of disasters. *Journal of Development Economics, 88*(2), 221−231. <https://doi.org/10.1016/j.jdeveco.2008.02.005>.

Noy, I., & DuPont, W. (2016). The long-term consequences of natural disasters-A summary of the literature. Working Papers in Economics and Finance, 02/2016, School of Economics and Finance, Victoria Business School. Available at: https://researcharchive.vuw.ac.nz/xmlui/bitstream/handle/10063/4981/Working%20Paper.pdf?sequence = 1.

Noy, I., & Nualsri, A. (2011). Fiscal storms: Public spending and revenues in the aftermath of natural disasters. *Environment and Development Economics, 16*(01), 113−128. <https://doi.org/10.1017/S1355770X1000046X>.

Oosterhaven, J., & Tobben, J. (2017). Wider economic impacts of heavy flooding in Germany: A non-linear programming approach. *Spatial Economic Analysis, 12*(4), 404−428. <https://doi.org/10.1080/17421772.2017.1300680>.

Panwar, V., & Sen, S. (2019a). Economic impact of natural disasters: An empirical re-examination. *Margin: The Journal of Applied Economic Research, 13*(1), 109−139. <https://doi.org/10.1177/0973801018800087>.

Panwar, V., & Sen, S. (2019b). Examining the economic impact of floods in selected Indian states. *Climate and Development.* Available from https://doi.org/10.1080/17565529.2019.1614897.

Pelling, M., Özerdem, A., & Barakat, S. (2002). The macro-economic impact of disasters. *Progress in Development Studies, 2*(4), 283−305. <https://doi.org/10.1191/1464993402ps042ra>.

Reserve Bank of India. (n.d.). *Database on Indian economy.* Reserve Bank of India. Retrieved from <https://dbie.rbi.org.in/DBIE/dbie.rbi?site = statistics>.

Raddatz, C. (2007). Are external shocks responsible for the instability of output in low-income countries? *Journal of Development Economics, 84*(1), 155−187. <https://doi.org/10.1016/j.jdeveco.2006.11.001>.

Raddatz, C. (2009). The wrath of God: Macroeconomic costs of natural disasters. *World Bank Policy Research Working Paper Series No. 5039*. Washington, DC: The World Bank.

Sen, A. (1982). *Poverty and famines: An essay on entitlement and deprivation*. Oxford University Press.

Sheng, Y., & Xu, X. (2019). The productivity impact of climate change: Evidence from Australia's Millennium drought. *Economic Modelling, 76*, 82–191. <https://doi.org/10.1016/j.econmod.2018.07.031>.

Singh, R., Feroze, S. M., & Ray, L. I. (2013). Effects of drought on livelihoods and gender roles: A case study of Meghalaya. *Indian Journal of Gender Studies, 20*(3), 453–467. <https://doi.org/10.1177/0971521513495293>.

Skidmore, M., & Toya, H. (2002). Do natural disasters promote long-run growth? *Economic Enquiry, 40*, 664–687.

UN-General Assembly. (2015). Sendai framework for disaster risk reduction. In *Third United Nations world conference on disaster risk reduction* (pp. 1–25). https://doi.org/10.1016/a/conf.224/crp.1.

Udmale, P. D., Ichikawa, Y., Manandhar, S., Ishidaira, H., Kiem, A. S., Shaowei, N., & Panda, S. N. (2015). How did the 2012 drought affect rural livelihoods in vulnerable areas? Empirical evidence from India. *International Journal of Disaster Risk Reduction, 13*, 454–469. <https://doi.org/10.1016/j.ijdrr.2015.08.002>.

Wooldridge, J.M. (2013). Introductory econometrics: A modern approach (5th international ed.), CENGAGE Learning Custom Publishing, pp. 534–535.

World Bank. (2007). *Disasters, climate change, and economic development in sub-Saharan Africa - lessons and future directions*. Washington, DC: The World Bank.

Index

Note: Page numbers followed by "*f*" and "*t*" refer to figures and tables, respectively.

Printed in the United States
By Bookmasters